THE PRIMATE NERVOUS SYSTEM
PART II

HANDBOOK OF CHEMICAL NEUROANATOMY

Series Editors: A. Björklund and T. Hökfelt

Volume 14

THE PRIMATE NERVOUS SYSTEM, PART II

Editors:

F.E. BLOOM
Department of Neuropharmacology, The Scripps Research Institute, San Diego, CA, USA

A. BJÖRKLUND
Department of Medical Cell Research, University of Lund, Lund, Sweden

T. HÖKFELT
Department of Histology and Neurobiology, Karolinska Institute, Stockholm, Sweden

1998

ELSEVIER

Amsterdam – Lausanne – New York – Oxford – Shannon – Singapore – Tokyo

ISBN 0-444-82912-1 (volume)
ISBN 0-444-90340-2 (series)

This book is printed on acid-free paper.

Published by:
Elsevier Science B.V.
P.O. Box 211
1000 AE Amsterdam
The Netherlands

Library of Congress Cataloging-in-Publication Data

The primate nervous system / editors, F.E. Bloom, A. Björklund, T. Hökfelt.
p. cm.-- (Handbook of chemical neuroanatomy ; v. 13, etc.)
Includes bibliographical references and index.
ISBN 0-444-82558-4 (pt. 1 : alk. paper). -- ISBN 0-444-90340-2 (series : alk. paper)
1. Neurochemistry. 2. Neuroanatomy. 3. Primates -- Physiology. I. Bloom, Floyd E. II. Björklund, Anders. 1945– .III Hökfelt, Tomas. IV. Series.
[DNLM: 1. Central Nervous System--chemistry. 2. Central Nervous System--physiology. 3. Central Nervous System--anatomy & histology. 4. Primates. W1 HA51J v. 13 etc./ WL 300 P9525 1997]
QM451.H24 1983
[QP356.3]
573.8'331 s--dc21
[573.8198]
DNLM/DLC
for Library of Congress

97-3750
CIP

Printed in the Netherlands

List of Contributors

P.R. HOF
Neurobiology of Aging Laboratories
Fishberg Research Center for Neurobiology
Department of Geriatrics and Adult Development
Department of Ophthalmology
Mount Sinai School of Medicine
New York, NY 10029
U.S.A.

G.W. HUNTLEY
Fishberg Research Center for Neurobiology
Mount Sinai School of Medicine
New York, NY 10029
U.S.A.

E.G. JONES
Department of Anatomy and Neurobiology
University of California, Irvine
Irvine, CA 92697
U.S.A.

J.H. MORRISON
Neurobiology of Aging Laboratories
Fishberg Research Center for Neurobiology
Department of Geriatrics and Adult Development
Mount Sinai School of Medicine
New York, NY 10029
U.S.A.

Preface

This volume is the second in the planned coverage of the neurochemical circuitry of the primate central nervous system. While this volume contains only two chapters, their topics and the extra-ordinarily comprehensive coverage with which their authors have dealt with their topics will nevertheless likely contribute equal amounts of knowledge, wisdom, and opportunities for future research extensions as have every volume in this unique series. As such, these chapters extend the goals of this primate series to develop a broadly based coverage of human and non-human primate chemical neuroanatomic details in a volume which makes clear the known and desirable appreciation for differences between and among subsets of primate brains.

Jones covers the Primate Thalamus with equal emphases on New World, Old World, pro-simian and human anatomic details and their differences. After establishing the major histochemically definable nuclei of the primate thalamus, he establishes the morphological properties of the major cell types and their chemical attributes and circuitry, and the details of the innervation of these neurons from ascending and descending non-thalamic locations. Lastly he provides an integrated overview of the thalamic circuitry in terms of the functionally defined forms of neuronal information that pass through it.

Morrison, Hof and Huntley undertake a comparably comprehensive examination of one of the most intensively studied regions of the primate brain, namely the primate visual cortex. Their coverage includes the primary, secondary and tertiary cortical areas known to participate in visual functions, and likewise covers the characteristic laminations, cells types, intrinsic and extrinsic connectivity patterns and importantly, the degree to which specific neurotransmitters, their receptors, and other neurochemical markers, with or without functional attributions have been mapped throughout this complex system, and offer their interpretations as to the functional insights these localizations may provide.

While much has been studied, both chapters also reveal how much remains for future efforts in these enormously important regions which are the archetypes of primate sub-cortical and cortical function.

FLOYD E. BLOOM   ANDERS BJÖRKLUND   TOMAS HÖKFELT

Contents

CHAPTER I

The thalamus of primates

E.G. JONES

1. INTRODUCTION

The primate thalami that current investigators are most likely to encounter are those of selected prosimians, of a range of New World and Old World monkeys, and of humans. Consequently the following description will focus on these three. The prosimian thalamus, in many aspects of its nuclear parcellation, differs considerably from that of both monkeys and humans, reflecting the early divergence of the evolutionary lines leading to the prosimian and anthropoid primates. Many seeming discrepancies between the two, however, may be determined as much by different traditions in terminology as by organizational differences. Apart from the dorsal lateral geniculate nuclei which can show wide variations in cytoarchitecture, the thalami of New World and Old World monkeys, despite evolutionary divergence of their ancestors some 50 million years ago, are remarkably similar, the major differences being those of size. In this case, a common nomenclature has historically facilitated the recognition of similarities in nuclear parcellation in the two families of monkeys. The principal differences between the human and simian thalamus are also those of size, the well known overgrowth of the pulvinar in humans not being accompanied by any nuclear divisions dramatically different from those of the monkey pulvinar. The human thalamus, in being so large, offers a higher level of resolution in defining histological borders between nuclei, a task that is further helped by the fact that size differences among cells forming the populations characteristic of adjacent nuclei, in many cases, tend to be more pronounced in the human thalamus. To anyone with an eye for the monkey thalamus, that of the human can readily be described using the customary nomenclature of monkeys. Unfortunately there is a time-honored precedent for describing the human thalamus in a nomenclature which, although deriving from that originally applied to monkeys in the early years of this century, has diverged to such an extent that an ill-informed reader could be forgiven for believing that the thalami of monkeys and humans are fundamentally different. One of the aims of the following account will be to emphasize that a common nomenclature is not only feasible but desirable. The desirability of a common nomenclature is especially compelling in describing patterns of histochemical and immunocytochemical localization and gene expression which is the primary aim of this review.

Handbook of Chemical Neuroanatomy, Vol. 14: The Primate Nervous System, Part II
F.E. Bloom, A. Björklund and T. Hökfelt, editors

2. BASIC SUBDIVISIONS OF THE THALAMUS

In all mammals, the dorsal part of the diencephalon can be divided into the epithalamus, dorsal thalamus and ventral thalamus, three multicellular conglomerates that are derivatives of three of the four major zones of neuroepithelial proliferation found in the wall of the embryonic third ventricle on each side and illustrated in the human embryo by Müller and O'Rahilly (1988). (The fourth is the hypothalamus). There is no reason to discard the useful tripartite division of the thalamus for descriptive purposes, because the three divisions represent thalamic components that make different connectional relationships with other parts of the forebrain and possess different functional connotations. There are very few examples, indeed, that cause us to modify the basic defining principles which, in addition to their embryonic origins, distinguish the three divisions of the thalamus: the dorsal thalamus has two-way connections with the cerebral cortex; the ventral thalamus receives fibers from, but does not project to the cerebral cortex; the epithalamus is not connected with the cerebral cortex (Rose 1942 a,b; Jones, 1985a).

Dominating the thalamus of primates by its sheer size is the dorsal thalamus – 'the thalamus' of many accounts. It is divided up into numerous nuclei, descriptions of which form a large part of this account and whose number reflects in large part the high degree of structural and functional parcellation of the primate cerebral cortex. The dorsal lateral geniculate nucleus, although commonly split off by fiber laminae from the rest of the thalamus in primates, is developmentally and in all other respects part of the dorsal thalamus. Covering the anterior, lateral and much of the ventral surfaces of the dorsal thalamus is the reticular nucleus of the ventral thalamus. The other components of the ventral thalamus are the ventral lateral geniculate nucleus, zona incerta and fields of Forel. The subthalamic nucleus is not developmentally part of the thalamus and is excluded from it. The epithalamus consists of the stria medullaris thalami, the associated paraventricular nuclei, which form a thin strip along the dorsomedial margin of the dorsal thalamus, and the habenular nuclei. Developmentally, the pineal body and parts of the pretectum may be derived from the same precursor region as the epithalamus but, by tradition, they are excluded from the adult thalamus.

Sheets and bundles of myelinated fibers, forming the white matter of the diencephalon, serve to delimit the dorsal thalamus from the components of the ventral thalamus and divide the dorsal thalamus itself up into its principal nuclear groups. These sheets and bundles can be visualized by staining for myelinated fibers (Fig. 1) and commonly by histochemical stains for enzymes such as acetyl cholinesterase (see below). They are composed in large part of afferent fibers entering the thalamus from the brainstem and from other parts of the forebrain, and of efferent fibers leaving the dorsal thalamus for the cerebral cortex and basal ganglia.

The *external medullary lamina* covers the anterior, lateral, ventral and posterior surfaces of the dorsal thalamus, separating it from the reticular nucleus, zona incerta and pretectum. Medially, it is thickened by fibers of the ansa lenticularis and brachium conjunctivum and, posteriorly, by those of the medial and trigeminal lemnisci. In the primates, the dorsal lateral geniculate and medial geniculate nuclei have broken through the lamina, as it were, to reach the surface of the diencephalon. Posteriorly, the external medullary lamina separates the dorsal thalamus from the pretectum along a very discrete boundary. Where the corticotectal tract and brachium of the superior colliculus traverse it, it can be considerably thickened.

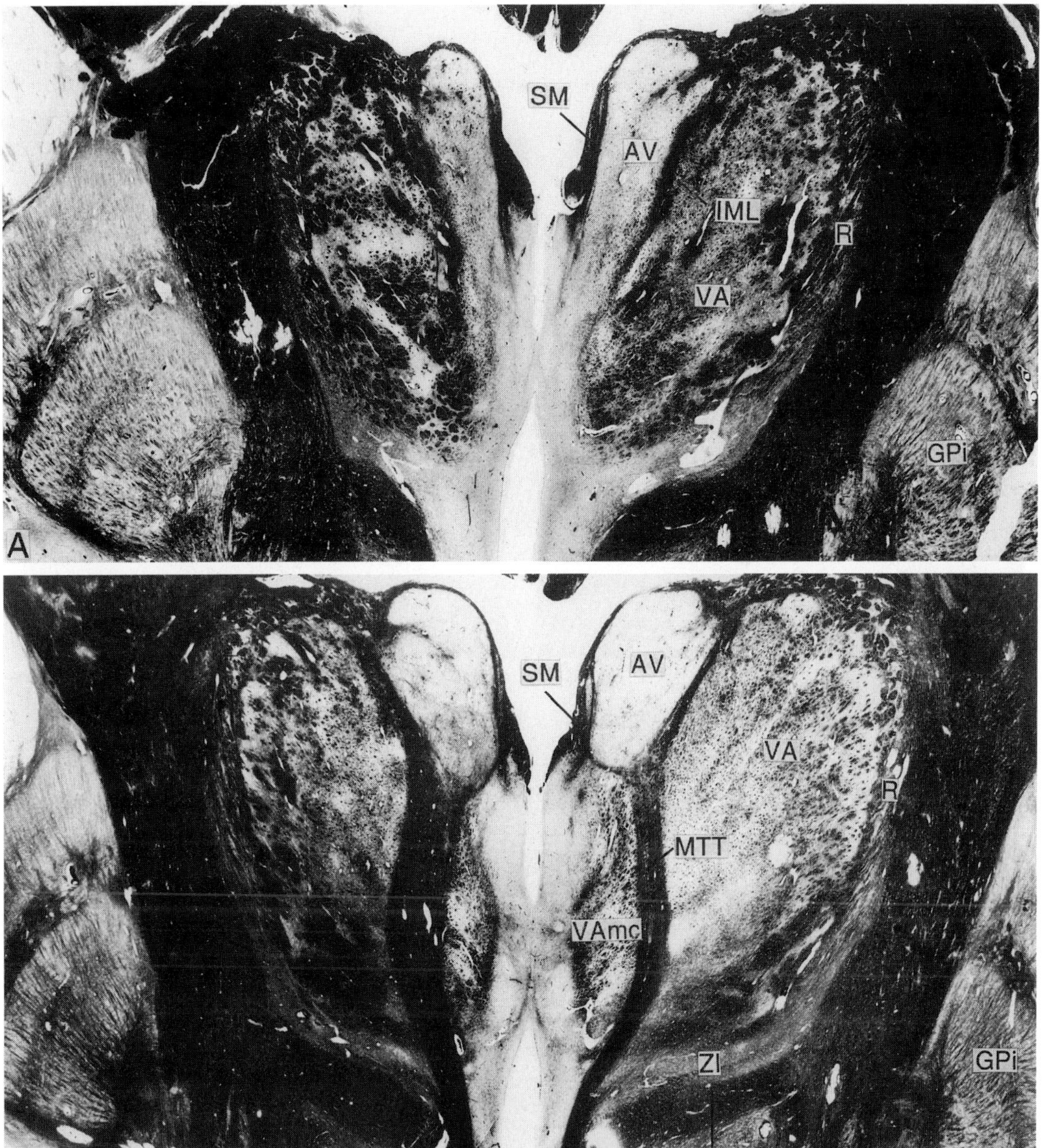

Fig. 1. Frontal sections from a human thalamus, in anterior (A) to posterior (E) order and stained with hematoxylin for myelinated fibers. These show the position of the internal medullary lamina (IML) at different anteroposterior levels, its invasion by the mamillothalamic tract (MTT) anteriorly, its splitting around the centre médian nucleus (CM) at middle levels, and its breakup over the posterior surface of the mediodorsal nucleus (MD) posteriorly. For other abbreviations, see page 244. Bar 2 mm.

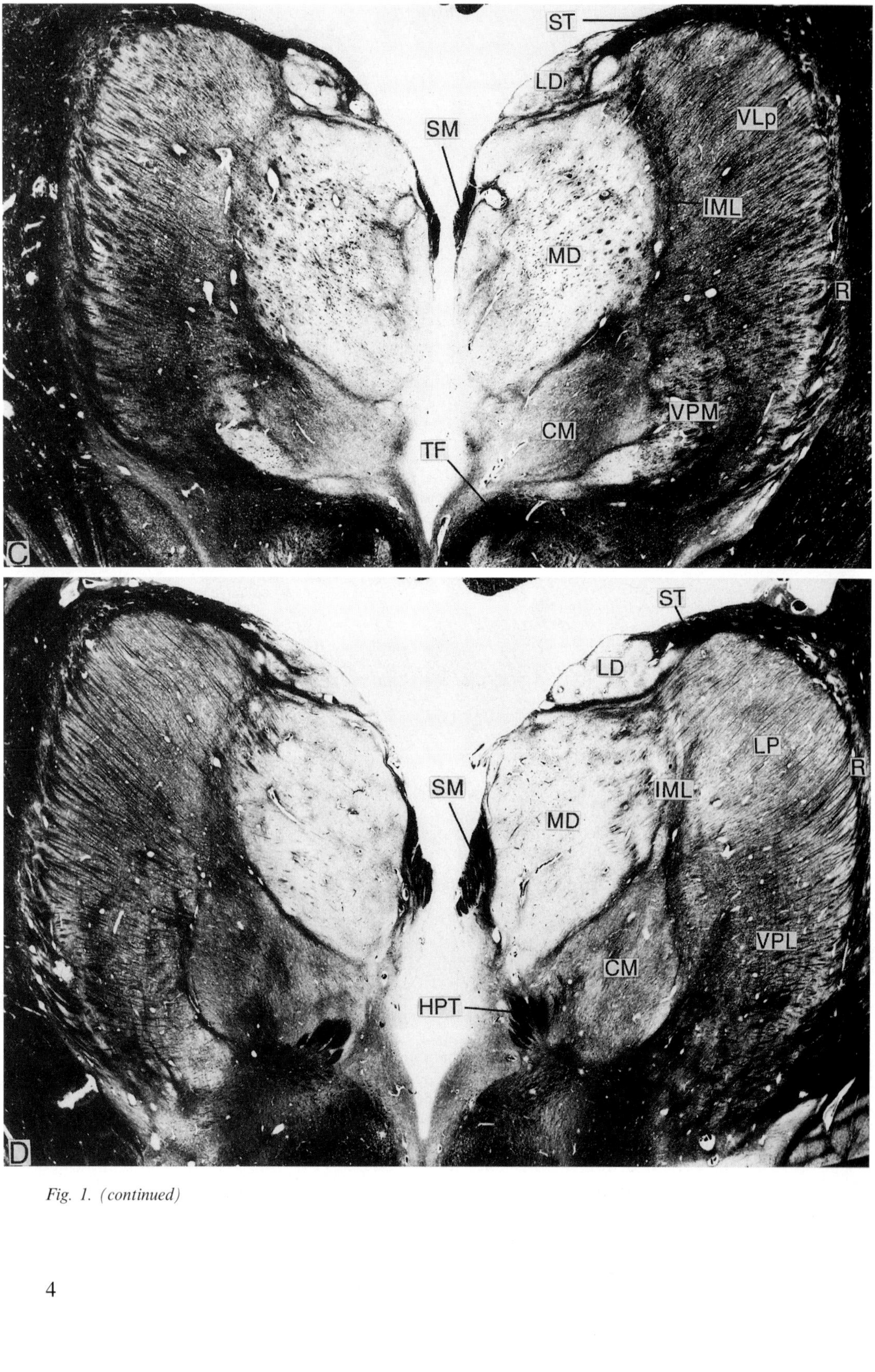

Fig. 1. (continued)

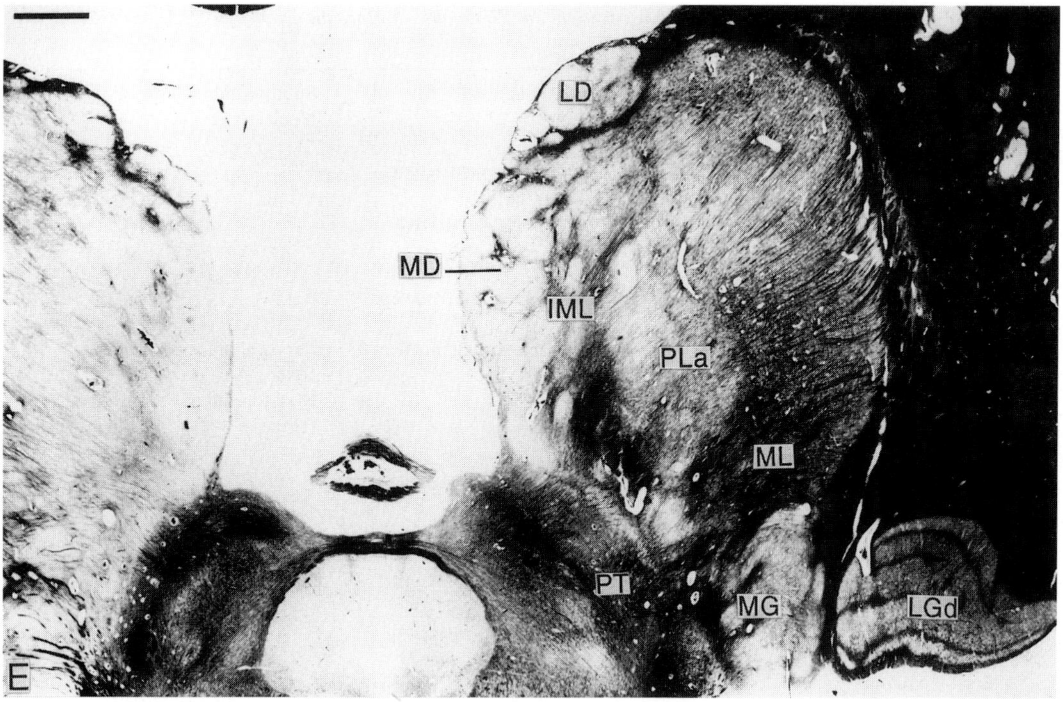

Fig. 1. (continued)

Within the dorsal thalamus, the *internal medullary lamina*, is the great defining landmark of the dorsal thalamus. The lamina is Y-shaped in horizontal sections, and serves to delimit the territories that form the anterior, medial and lateral groups of nuclei. Members of these three groups are often called the principal or relay nuclei. Table 1 lists the nuclei customarily designated within each of the major nuclear groups of the monkey thalamus.

The internal medullary lamina itself contains large numbers of nerve cells which constitute the intralaminar nuclei (Table 1). An anterior group of intralaminar nuclei fills the arms and stem of the Y-shaped lamina, meeting at the midline where the laminae of the two thalami join. Bulging posteriorly and considerably distorting the internal medullary laminae, is the posterior group of intralaminar nuclei, made up of the parafascicular and centre médian nuclei, which are relatively much larger in primates than in any other species.

The two thalami meet in the midline at a relatively narrow region termed the massa intermedia in humans. Sometimes in humans, the union across the third ventricle is absent. In non-human primates, the fusion in the midline is so extensive, as in other animals, that the use of the term massa intermedia is inappropriate. At the midline, the cell populations of the anteromedial, anterior intralaminar, mediodorsal and ventral nuclei merge. Fiber offshoots of the internal medullary lamina can often partially separate the merging nuclei, carrying with them cells of the anterior intralaminar group that have sometimes been given the name 'central' or 'midline' nuclei. There are few or no fiber connections across the midline fusion of the left and right thalami.

TABLE 1. *Major Nuclei and their Subdivisions in the Thalamus of Macaque Monkeys (Olszewski, 1952; Jones, 1985)*

Division of Diencephalon	Nuclear Group	Major Nuclei	Subdivisions of Major Nuclei
Epithalamus	Paraventricular Nuclei (Para)	Anterior paraventricular nucleus Posterior paraventricular nucleus	
	Habenular Nuclei	Medial Habenular nucleus (Hm)	
		Lateral habenular nucleus (Hl)	Parvocellular subnucleus (Hlpc) Magnocellular subnucleus (Hlmc)
Ventral Thalamus		Reticular nucleus (R) Zona incerta (ZI) Nucleus of the fields of Forel (FF) Ventral lateral geniculate nucleus (Pre-geniculate nucleus, Prg)	
Dorsal Thalamus	Anterior Group	Anterodorsal nucleus (AD) Anteroventral nucleus (AV) Anteromedial nucleus (AM) Lateral dorsal nucleus (LD)	
	Medial Group	Parataenial nucleus (Pt) Medioventral nucleus (MV) (Reuniens nucleus)	
		Mediodorsal nucleus (MD)	Magnocellular (medial) division (MDmc) Parvocellular (central) division (MDpc) Multiform (lateral) division (MDmf) Densocellular division (part of the central lateral nucleus, CL)
	Anterior Intra-laminar Group	Rhomboid nucleus (Rh) Central medial nucleus (CeM) Paracentral nucleus (Pc) Midline extensions of CL and CeM ('central' or 'midline' nuclei)	
	Posterior Intra-laminar Group	Parafascicular nucleus (Pf) Centre médian nucleus (CM)	

(continued)

TABLE 1. *(continued)*

Ventral Nuclei	Ventral anterior nucleus (VA)	Principal division (VMp)
		Magnocellular division (VAmc) (part of CeM)
	Ventral lateral anterior nucleus (VLa)	
	Ventral lateral posterior nucleus (VLp)	Anterior division
		Dorsal division
		Ventral division
	Ventral posterior lateral nucleus (VPL)	Anterodorsal shell
		Central core
	Ventral posterior medial nucleus (VPM)	Core
		Matrix
	Ventral posterior inferior nucleus (VPI)	
	Basal ventral medial nucleus (VMb)	
Lateral Posterior/ Pulvinar Group	Lateral posterior nucleus (LP)	
	Anterior pulvinar nucleus (Pla)	
	Lateral pulvinar nucleus (Pll)	
	Inferior pulvinar nucleus (Pli)	
	Medial pulvinar nucleus (Plm)	
Posterior Group	Posterior nucleus (Po)	
	Limitans/suprageniculate nucleus (L/SG)	Limitans division (L)
		Suprageniculate division (SG)
Medial Geniculate Complex	Ventral nucleus (MGv)	
	Anterodorsal nucleus (MGad)	
	Posterodorsal nucleus (MGpd)	
	Magnocellular nucleus (MGmc)	
Lateral Geniculate Body	Dorsal lateral geniculate nucleus (LG)	Parvocellular layers (3–6)
		Magnocellular layers (1,2)
		S layers (s) and interlaminar plexuses

3. NUCLEAR PARCELLATION OF THE THALAMUS IN REPRESENTATIVE PRIMATES

3.1. PROSIMIAN THALAMI

The six families of prosimian primates encompass a large number of species with a wide range of habitats and behavioral adaptations. Few have served as experimental animals so there have been comparatively few descriptions of prosimian thalami, although species variations in the dorsal lateral geniculate nuclei are known. Perhaps the only prosimian that has received any degree of attention in recent years is the (lesser) bush baby, *Galago senegalensis*. *Galago crassicaudatus* is also in relatively common use. The *Galago* will serve as an example for comparison of the prosimian thalamus with the thalami of monkeys and humans.

The *Galago* thalamus (Fig. 2) presents a number of features, notably the large pulvinar and posterior intralaminar nuclei, that are unmistakably characteristic of primates. Other features, for example, the relatively large size and clear-cut nuclear parcellation of the anterior group of nuclei more closely resemble those of non-primates than they do those of monkeys or humans.

The anterior nuclei are prominent, protuberant and the group as a whole resembles more closely that of rodents or diprotodent marsupials than it does that of simians, pongids or humans (Figs. 2 B, C). The anteromedial and anteroventral nuclei are clearly distinguishable by differences in cell size, staining and packing density and show much less of the tendency to merge found in monkeys and nothing resembling the virtually complete merging found in humans. The anterodorsal nucleus is relatively large and dorsally placed, not attenuated and medially displaced as in monkeys and especially humans. There is a clearly defined band of cells linking the anterodorsal nuclei to the rhomboid nucleus at the midline. This is the equivalent of the strip of cells sometimes called the interanterodorsal nucleus in rodents and absent or highly discontinuous in monkeys. The lateral dorsal nucleus is relatively large, and clearly a continuation of the anterior group, as in non-primates (Fig. 2 D-F). There is a large parataenial nucleus (Fig. 2 A-C), also a feature not found in anthropoid primates. The anterior intralaminar nuclei are relatively thick, although not unusually large; the paracentral nucleus is not reduced to the thin, often interrupted strip seen in monkeys and humans (Fig. 2 C-F). The overall appearance of the anterior intralaminar nuclei is more reminiscent of that seen in the thalamus of the cat than in that of monkeys. By contrast, the two nuclei of the posterior intralaminar complex, the parafascicular and centre médian, are relatively much larger than in any non-primate thalamus and there are clear indications of the presence of at least two cell populations of different sizes in the centre médian nucleus, also a primate feature (Fig. 2 G-I).

The mediodorsal nucleus is clearly divided into four distinct subnuclei, based on differences in cell size and packing density (Fig. 2 D-G). Divisions of the mediodorsal nuclei of this type are not unique to primates but the relative proportions and distributions of the four populations of cells make the nucleus virtually a facsimile of that of the monkey and human, albeit in miniature. There is a prominent medial or magnocellular division, an encompassing parvocellular division which shows signs of dorsolateral and posteroventral subdivisions, and laterally placed populations of cells that merge with the central lateral and paracentral nuclei.

All of the major divisions of the ventral nuclei found in anthropoid primates are evident in the *Galago* (Fig. 2 B-I). Borders between them are perhaps less clear-cut than in monkeys, especially that between the ventral lateral anterior (VLa) and ventral lateral posterior (VLp) nuclei, but the same characteristics of relative cell sizes, packing densities and intensities of cell staining serve to define the cell populations characteristic of each nucleus. The VLa nucleus, for example, is characterized by islands and clouds of closely packed cells, separated by finger-like extensions of the less densely

Fig. 2. Nissl-stained frontal sections in anterior (A) to posterior (L) order through the thalamus of a bush baby (*Galago crassicaudatus*). Bar 1 mm.

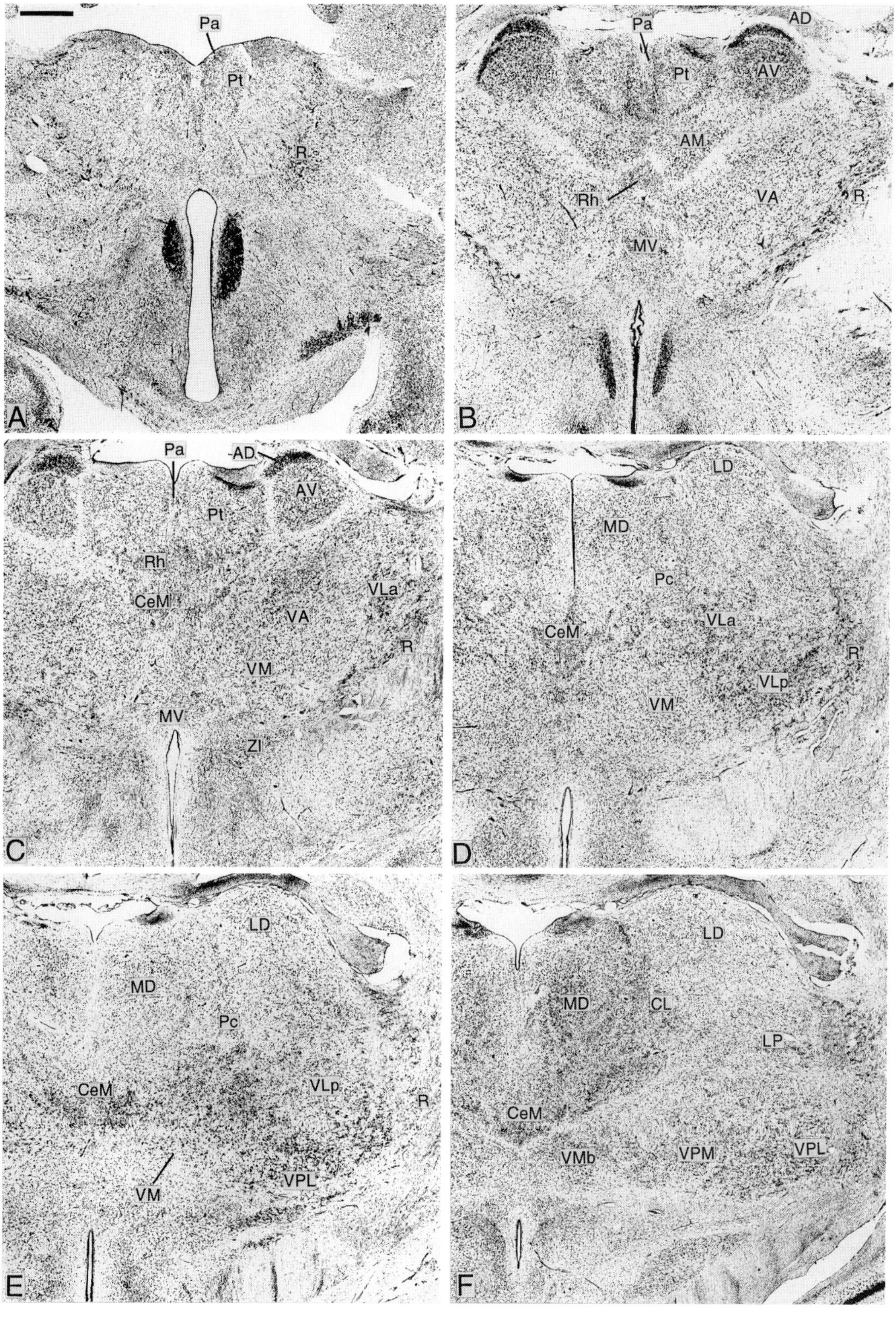
Pa
Pt
R
A
Pa
AD
Pt
AV
AM
Rh
VA
R
MV
B
Pa
AD
AV
Pt
Rh
CeM
VLa
VA
R
VM
MV
ZI
C
LD
MD
Pc
CeM
VLa
R
VLp
VM
D
LD
MD
Pc
CeM
VLp
R
VM
VPL
E
LD
MD
CL
LP
CeM
VMb
VPM
VPL
F

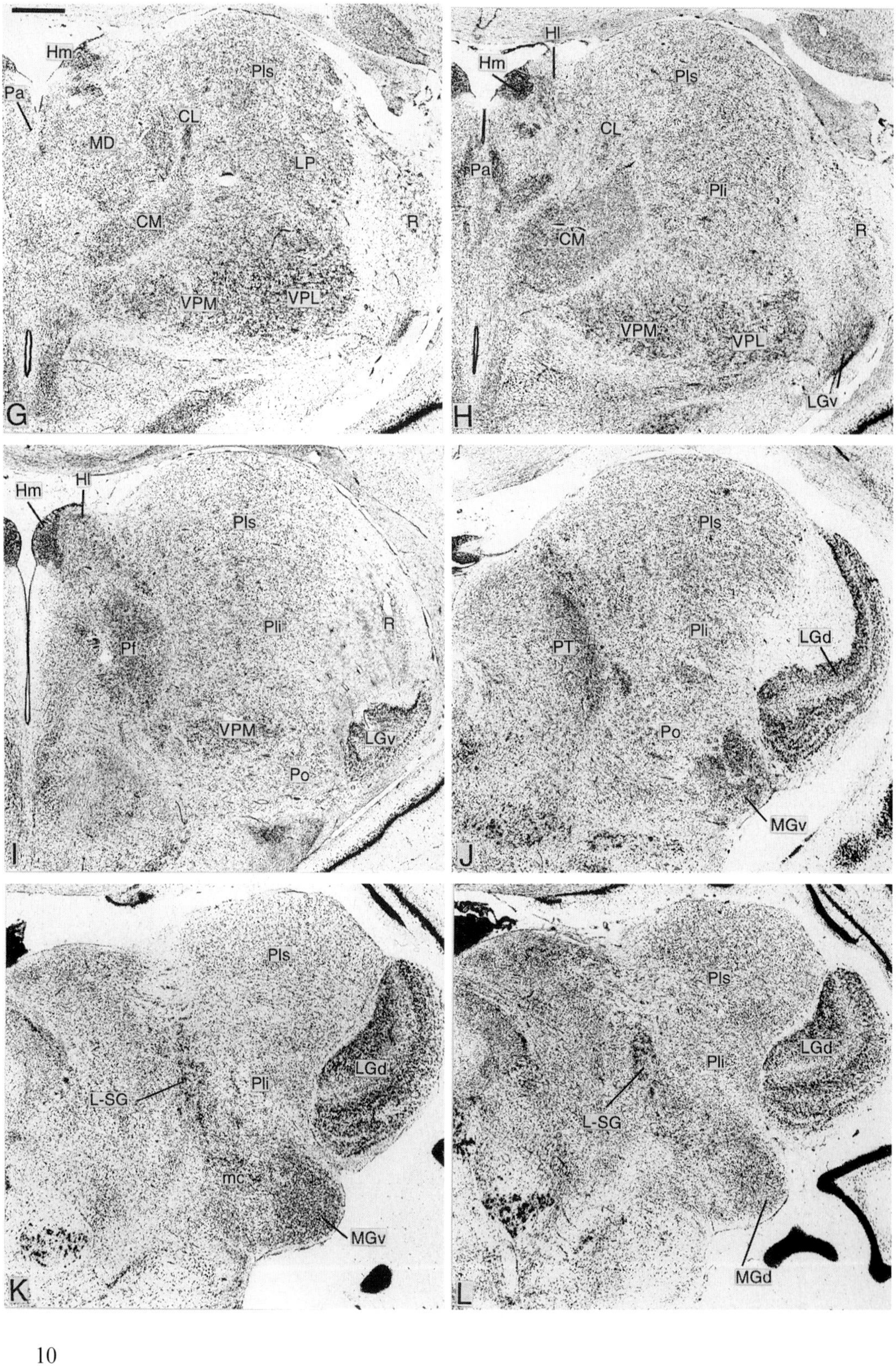
Hm
Pa
Pls
CL
MD
LP
CM
R
VPM
VPL
G
Hl
Hm
Pls
CL
Pa
Pli
R
CM
VPM
VPL
LGv
H
Hm
Hl
Pls
Pli
R
Pf
VPM
LGv
Po
I
Pls
PT
Pli
LGd
Po
MGv
J
Pls
L-SG
Pli
LGd
mc
MGv
K
Pls
LGd
Pli
L-SG
MGd
L

Fig. 2. (continued)
←

packed cells of the VLp nucleus. One notable feature of the ventral nuclei is their ventral compression by the lateral posterior (LP)/pulvinar complex and the commensurately small size of the VLp nucleus (Fig. 2 D, E). In comparison with monkeys, all the ventral nuclei appear as though flattened ventrally and, unlike in monkeys, the VLp nucleus at no point reaches the dorsal surface of the thalamus.

The posterior half of the *Galago* thalamus is dominated by an LP/pulvinar complex that is unusually large in comparison with the other nuclei (Fig. 2 G-L). Although protruding posteriorly, it does not overhang the lateral and medial geniculate nuclei as in monkeys. The posterior pole of the pulvinar and the two geniculate bodies form three prominent bumps of approximately equal size on the posterior surface of the thalamus. Within the LP/pulvinar complex, the small, relatively densely packed cells typical of LP in monkeys and humans occupy the same relative position. By convention in prosimians, the pulvinar is divided into a dorsal nucleus, composed of relatively pale stained, loosely-packed cells, and a ventral nucleus of more heterogeneous cell groups. The dorsal nucleus has cytoarchitectural similarities to the medial pulvinar nucleus of monkeys and humans, while the ventral nucleus could with some justification be divided into nuclei equivalent to the lateral, inferior and anterior pulvinar nuclei of monkeys and humans. The relative positions of these nuclei, especially with respect to the dorsal lateral geniculate nucleus are, however, quite different.

The posterior complex is quite distinct and its limitans/suprageniculate and posterior nucleus components are as clearly defined as in monkeys (Fig. 2 J-L).

The medial geniculate complex shows evidence of parcellation into magnocellular, dorsal and ventral nuclei but borders between the latter two in particular are not very distinct (Fig. 2 J-L). The ventral nucleus, typically composed of closely-packed, darkly staining cells, as in most mammals, appears to be relatively small and displaced into a dorsal position within the complex in comparison with that of monkeys.

The dorsal lateral geniculate nucleus of the *Galago* is typical of most lemuroid and lorisoid primates but quite unlike that of monkeys or humans. It occupies a dorsolateral rather than ventrolateral position and is flattened around the lateral curvature of the thalamus (Fig. 2 I-L). There are six layers but they are not identical to the six of monkeys or humans. Two external magnocellular layers (layers 1 and 2) are found and two parvocellular layers (3 and 6). The latter pair are probably comparable to layers 3 through 6 of the anthropoids. The two parvocellular layers are separated by layers 4 and 5 which are composed of very small cells probably comparable to the cells of the interlaminar plexuses between the layers of the monkey dorsal lateral geniculate nucleus. They have sometimes been called 'koniocellular layers' (Casagrande, 1994).

Of the ventral thalamic components, the reticular nucleus is much more like that of rodents than monkeys since it does not encircle the dorsal lateral geniculate nucleus. The other components are similar to those found in non-primate species.

In the epithalamus, the nuclei are similar in relative position and cellular composition to those of monkeys. The lateral habenular nucleus, however, although containing populations of large and smaller cells, is not clearly divided into magno- and parvocellular subnuclei as in monkeys and humans.

3.2. THALAMI OF MONKEYS

This description will focus on the thalamus of Old World monkeys, as exemplified by that of the various macaque species (Fig. 3). Comments will be made in passing about features in which the thalami of some of the better known New World monkeys differ. In general, however, apart from size and the degree of laminar differentiation in the dorsal lateral geniculate nuclei, the thalami of the Old World and New World simians are not substantially different.

Features in which the monkey thalamus differs from that of the prosimians include a relatively less prominent anterior group (Fig. 3 A-G), considerable relative enlargement of the ventral lateral nuclei, especially of the VLp nucleus (Fig. 3 F-K), enormous overgrowth and increased nuclear differentiation in the LP/pulvinar complex (Fig. 3 K-P), increased nuclear differentiation in the medial geniculate complex (Fig. 3 J-M), and a change in relative position, shape and lamination of the dorsal lateral geniculate nucleus. The mediodorsal nucleus is enlarged, as also are the posterior group of intralaminar nuclei. Contrary to popular perception, the anterior group of intralaminar nuclei is also enlarged.

The anterior nuclei do not dominate cross sections of the anterior part of the thalamus, as in the *Galago* and on each side are drawn into an obliquely lying band. In this band the anteroventral and anteromedial nuclei are merged and only dorsolateral and ventromedial differences in cell packing density permit the two to be separately named.

Medially, the fusion of the anteromedial nuclei across the midline is invaded by an anterior extension of the anterior intralaminar group forming the rhomboid (or 'central') nucleus (Fig. 3 C, D). The anterodorsal nucleus is distinct, made up of large, densely staining cells and separated from the other nuclei by a fiber capsule. It is relatively large in comparison with that of the human but is dorso-medially rather than dorsally located in comparison with that of the *Galago*. Old World monkeys show only slight evidence of the medial extensions of the anterodorsal nucleus which form the incipient interanterodorsal nucleus described in the *Galago,* but these can be identified in some New World monkeys.

The lateral dorsal nucleus remains clearly an extension of the anterior nuclear group (Fig. 3H) but is relatively much narrower in cross section than in the *Galago*, although still extending as far posteriorly as the medial pulvinar nucleus with which it merges. The parataenial nucleus is relatively much smaller than in the *Galago* (Fig. 3 A,B).

The anterior intralaminar nuclei, in terms of cell population, are quite large, especially the central medial nucleus and the anterior midline components that collectively form the rhomboid or central nucleus (Fig. 3 C-H). The central lateral nucleus is also large, especially when the so-called densocellular division of the mediodorsal nucleus is included in it (see below). The perception that the anterior intralaminar nuclei are small in anthropoid primates probably stems from the thinning and interruption by fiber bundles of the paracentral nucleus and from failure to recognize the extent of the midline and central lateral nuclei. The inclusion of part of the central lateral nucleus as a densocellular component of the mediodorsal nucleus in monkeys, stems from Olszewski's (1952) decision to describe it (as he said, 'temporarily') with the mediodorsal nucleus, although he recognized that it more correctly belonged to the intralaminar

Fig. 3. Nissl-stained frontal sections in anterior (A) to posterior (P) order through the thalamus of a rhesus monkey (*Macaca mulatta*). Bar 1 mm.

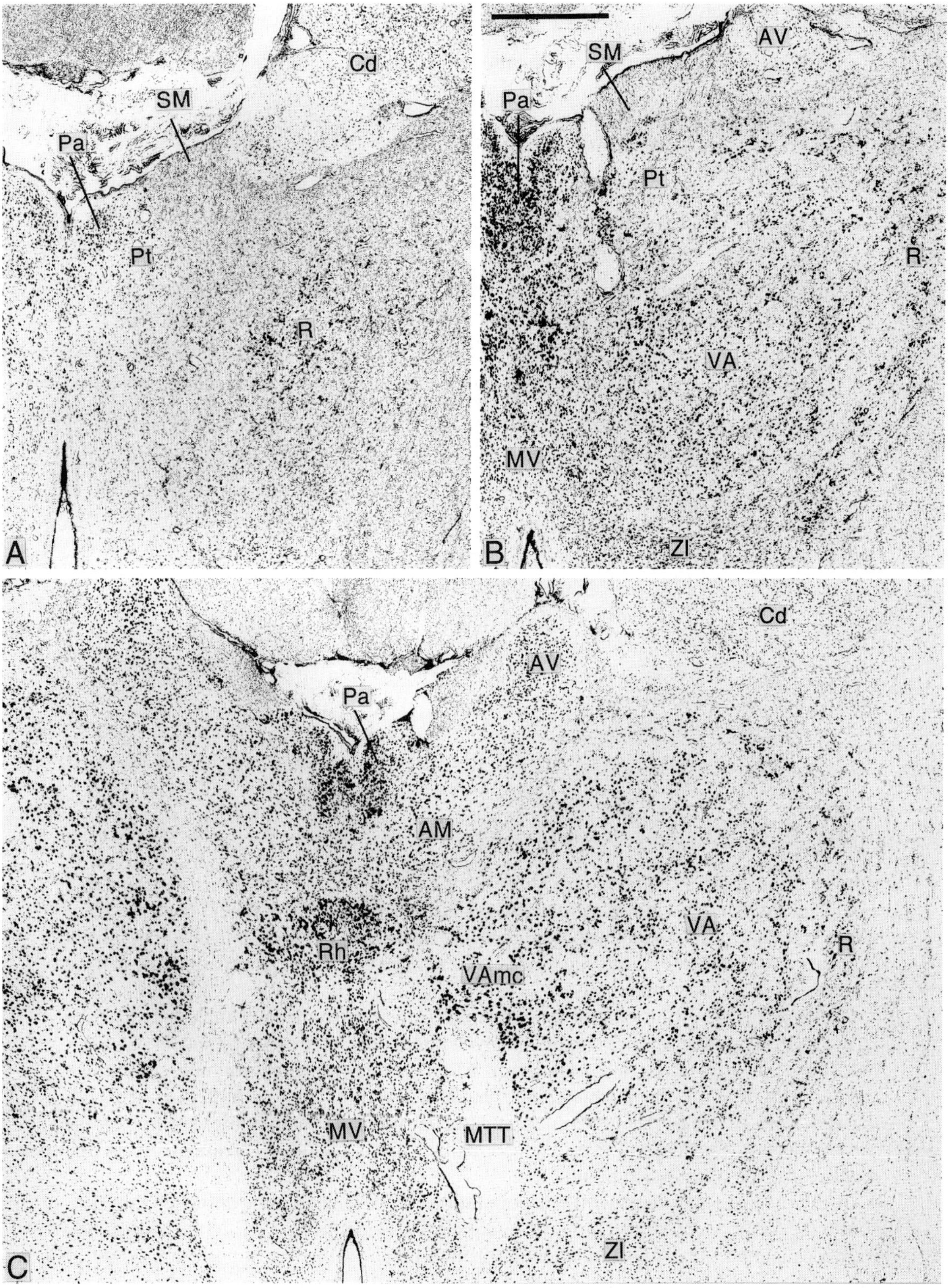
Cd
SM
Pa
Pt
R
A
AV
SM
Pa
Pt
R
VA
MV
ZI
B
Cd
AV
Pa
AM
VA
R
Rh
VAmc
MV
MTT
ZI
C

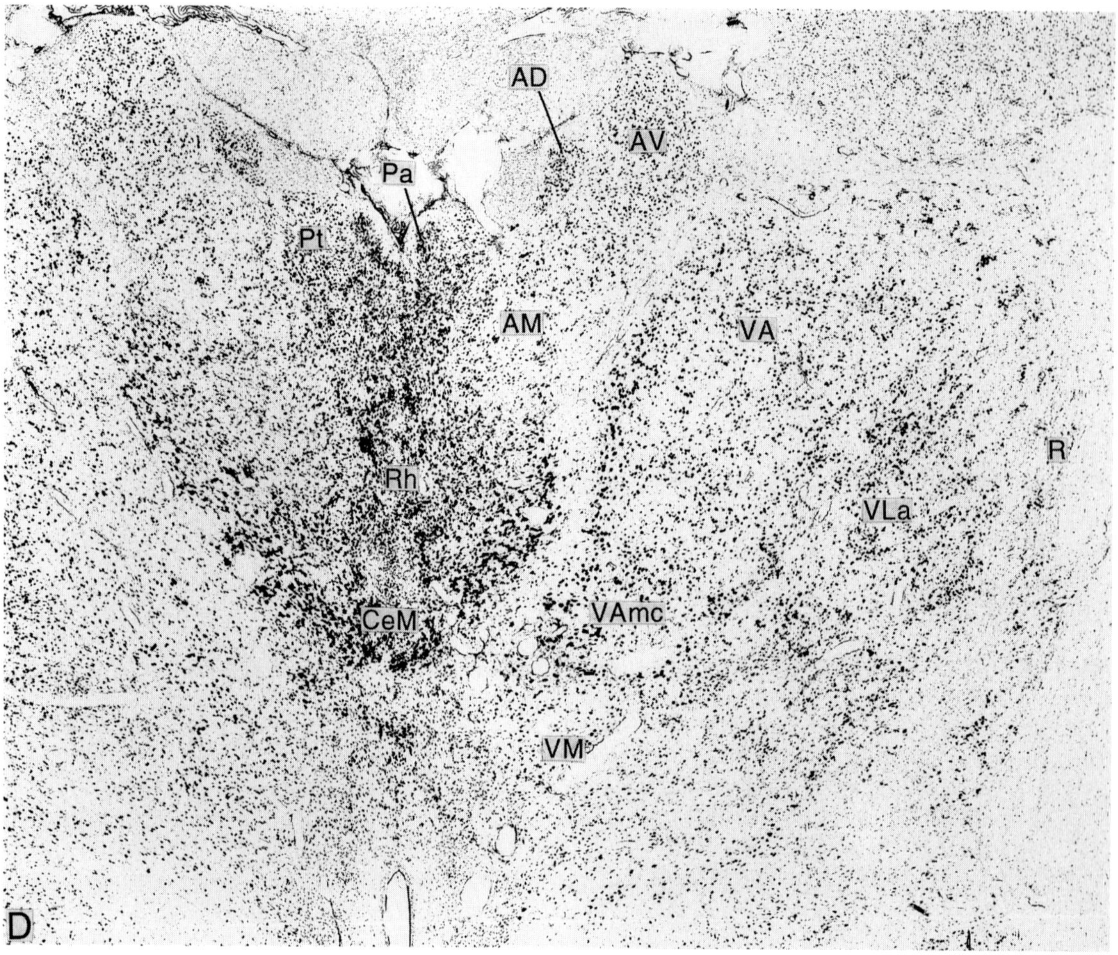

Fig. 3. (continued)

system. These cells are larger than those in the rest of the central lateral nucleus and extend in interrupted clumps across the posterior surface of the mediodorsal nucleus where they are often erroneously included in the medial pulvinar nucleus. The posterior part of the central lateral nucleus is thus composed of two populations of cells, a feature found in most mammals.

Of the posterior group of intralaminar nuclei, the centre médian nucleus is particularly large in monkeys, with small, palely-stained cells predominating (Fig. 3 G-L). Medially, however, there is a population of larger, more deeply staining cells which merges with that of the parafascicular nucleus. A thin line of darkly-stained cells derived from the central lateral nucleus extends over the dorsomedial surface of the centre médian nucleus and has also sometimes been included, erroneously, in the mediodorsal or parafascicular nucleus.

In the mediodorsal nucleus, three populations of cells, excluding those of the 'densocellular' component alluded to above, are clearly evident (Fig. 3 E-M). The medially located, magnocellular division extends through much of the anteroposterior extent of the nucleus, embraced by the larger, laterally located, parvocellular division which

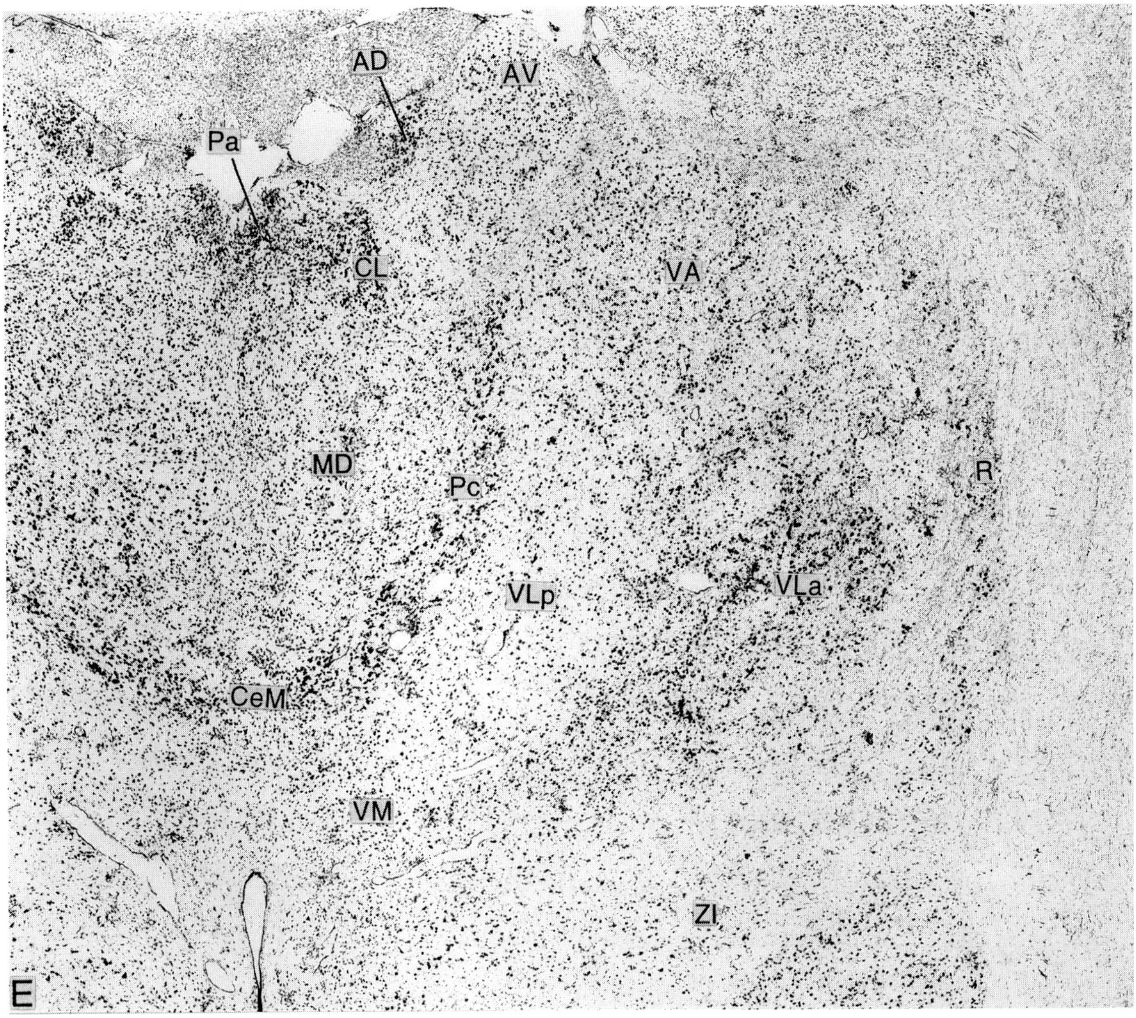

Fig. 3. (continued)

displaces it from the anterior and posterior poles and ventral surface. In myelin preparations, the magnocellular division is characteristically more densely stained than the other subnuclei, a feature of most mammals. It does not reach to the midline. Here there is a thin, paramedian strip of small cells. Along the lateral margin of the parvocellular division is the multiform division, lying adjacent to the internal medullary lamina until replaced posteriorly by the large celled component of the central lateral nucleus which some authors label the densocellular division of the mediodorsal nucleus. These large cells cover the posterior pole of the mediodorsal nucleus and merge with the limitans nucleus.

The lateral mass of the monkey dorsal thalamus, over most of its anteroposterior extent, is dominated by the ventral nuclei (Fig. 3 B-M), the large pulvinar being expanded posteriorly in comparison with that of the *Galago* (Fig. 3 L-P). The ventral lateral complex is especially large, a feature that is also remarkable in the human thalamus. The ventral anterior nucleus dominates the anterior pole, merging with

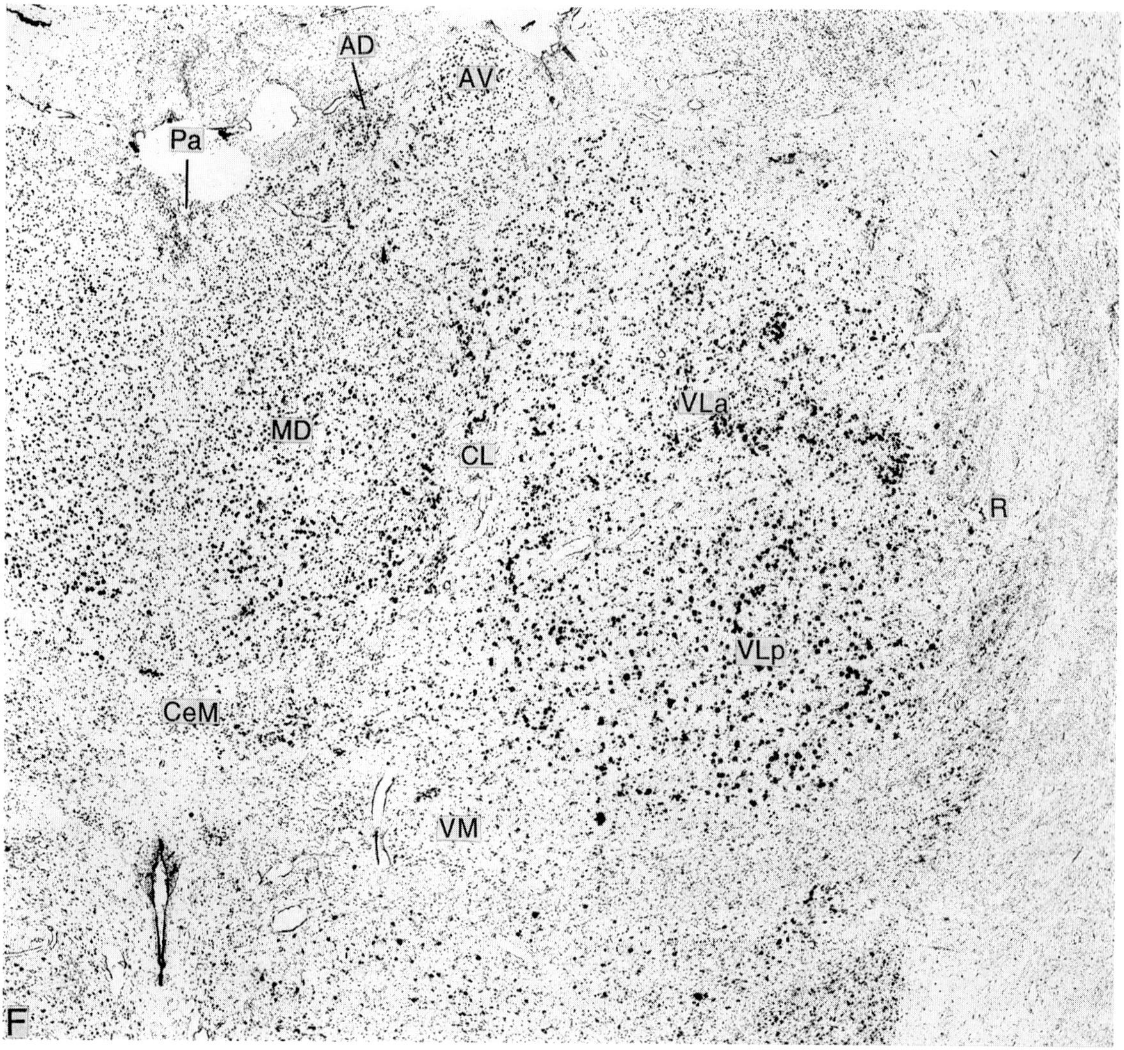

Fig. 3. (continued)

the medioventral (reuniens) nucleus at the midline and posteroventrally with the principal ventral medial nucleus. The cytoarchitectonic distinctions between these nuclei are not particularly clear and there is a good deal of disagreement about their borders. Around the mamillothalamic tract, there is a well defined focus of large, deeply staining neurons, customarily referred to as the magnocellular ventral anterior nucleus (Fig. 3 C, D). These cells are probably to be regarded more correctly as part of the anterior intralaminar system (Jones, 1989b). Posterior to the principal ventral anterior nucleus, nuclear parcellation becomes more distinct, with the appearance of the clouds and clusters of closely packed, densely-stained neurons of the ventral lateral anterior nucleus (VLa) which for a short distance occupy the full cross-sectional extent of the lateral mass. These are displaced laterally by the anteromedial pole of the ventral lateral posterior nucleus (VLp) which is interposed between the VLa nucleus and the internal medullary lamina. The VLp nucleus then expands to occupy the full

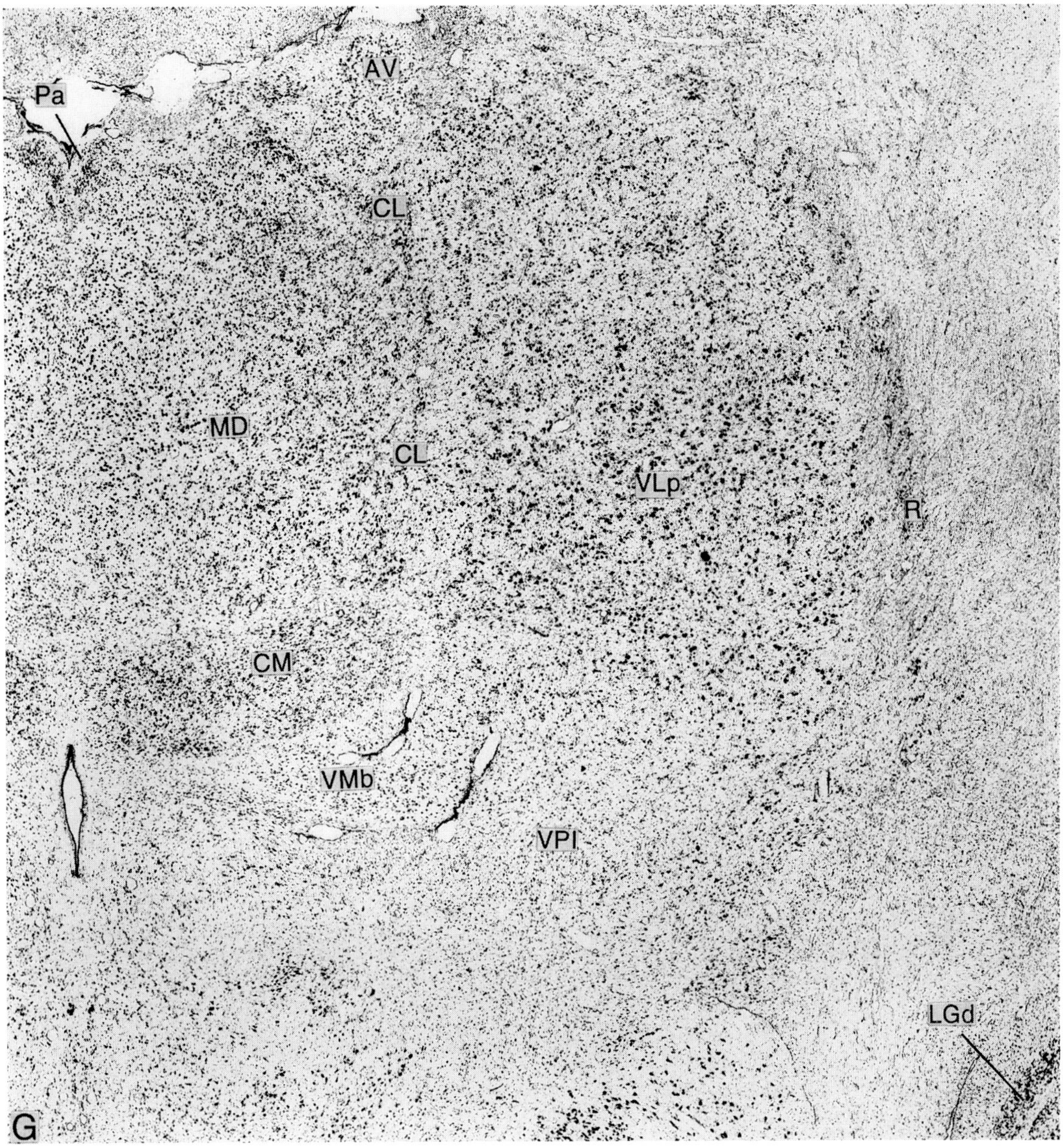

Fig. 3. (continued)

dorsoventral and mediolateral extent of the lateral mass. It is characterized by large, deeply staining cells – among the largest in the thalamus – which are quite widely separated from one another. In sagittal sections the nucleus is sickle-shaped, with a narrow ventral aspect and a curved dorsal aspect which expands posteriorly over the ventral posterior nuclei. The ventrally located cells are larger than those of the dorsal part and the VLp nucleus has sometimes been divided into subnuclei as a consequence (see below). Posteriorly, the ventral posterior nucleus occupies the ventral half of the

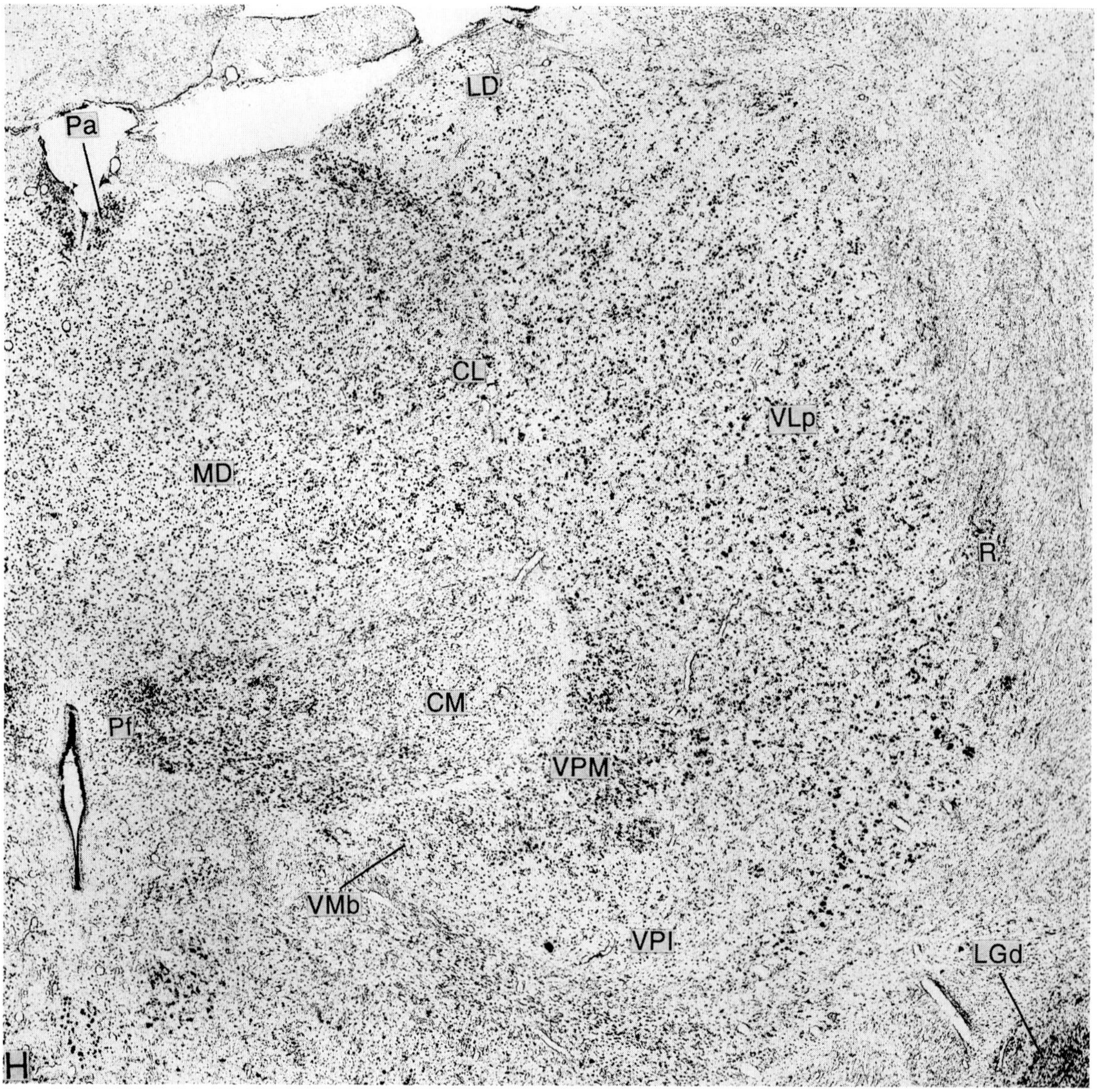

Fig. 3. (continued)

lateral nuclear mass, narrowing to a medial tongue ventral to the centre médian nucleus and to a posterior pole dorsal to the medial geniculate complex. Many of its cells are as large as those of the ventral lateral posterior nucleus, but there are many interspersed smaller cells as well, giving the nucleus a more compact appearance. The nucleus is a complex of two nuclei, the ventral posterior lateral (VPL) and ventral posterior medial (VPM), separated by a fiber lamina (Fig. 3 H-K). The VPM nucleus is clearly lobulated in frontal section, the lobules actually forming elongated rods that extend anteroposteriorly through the nucleus. The ventromedial aspect of VPM is separated off as a separate nucleus of closely packed smaller cells, that have very different immunocytochemical properties from the VPM cells proper (see below). Tra-

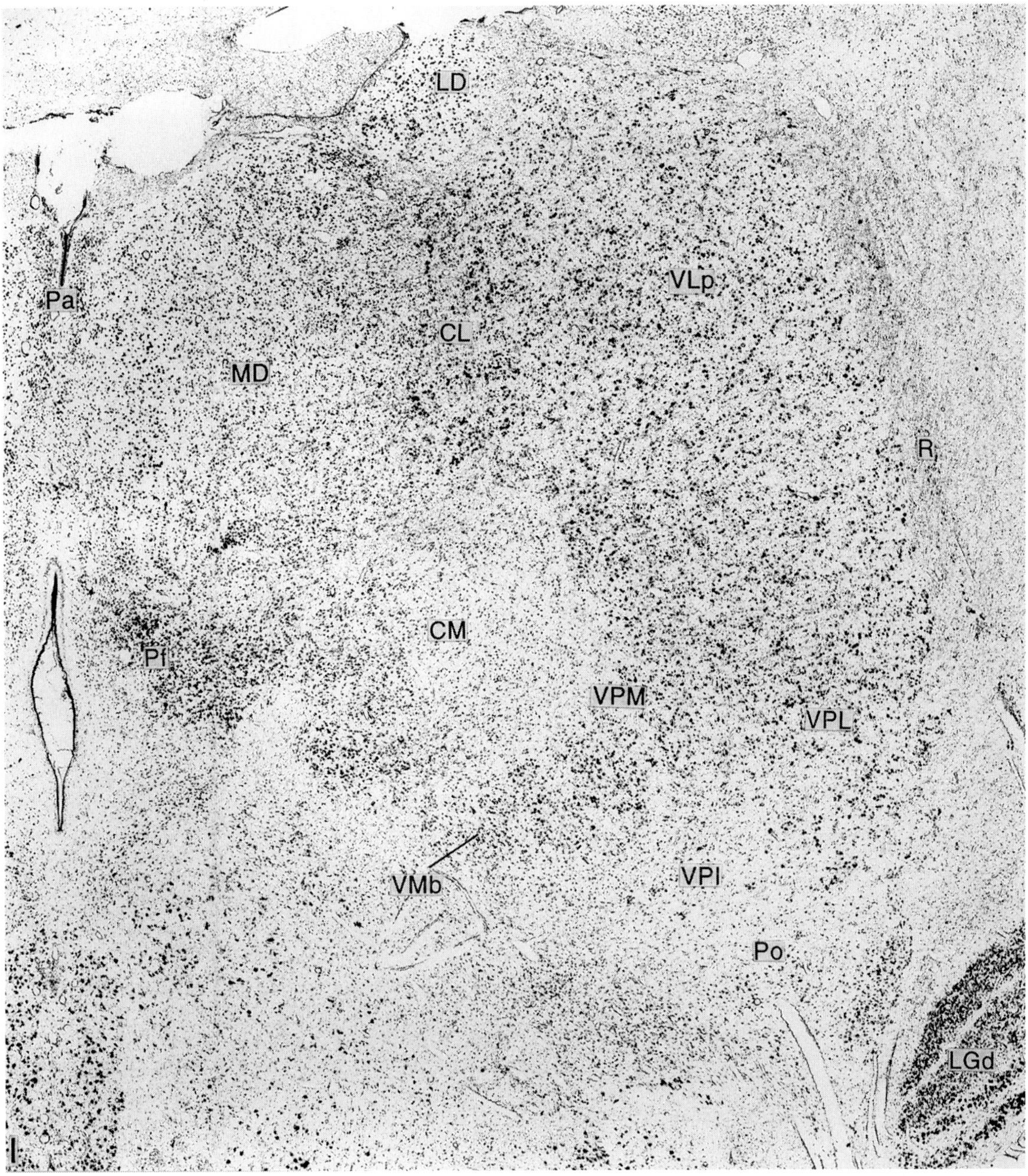

Fig. 3. (continued)

ditionally, this is referred to as the parvocellular division of VPM but in recent years it has come to be called the basal ventral medial nucleus (VMb). The ventral posterior lateral nucleus (VPL) has a tendency to be divided into curved, dorso-ventral lamellae, but these are less prominent than in the *Galago*. Cytologically, it consists of a thin anterodorsal shell dominated by large cells and a larger core in which the cell popula-

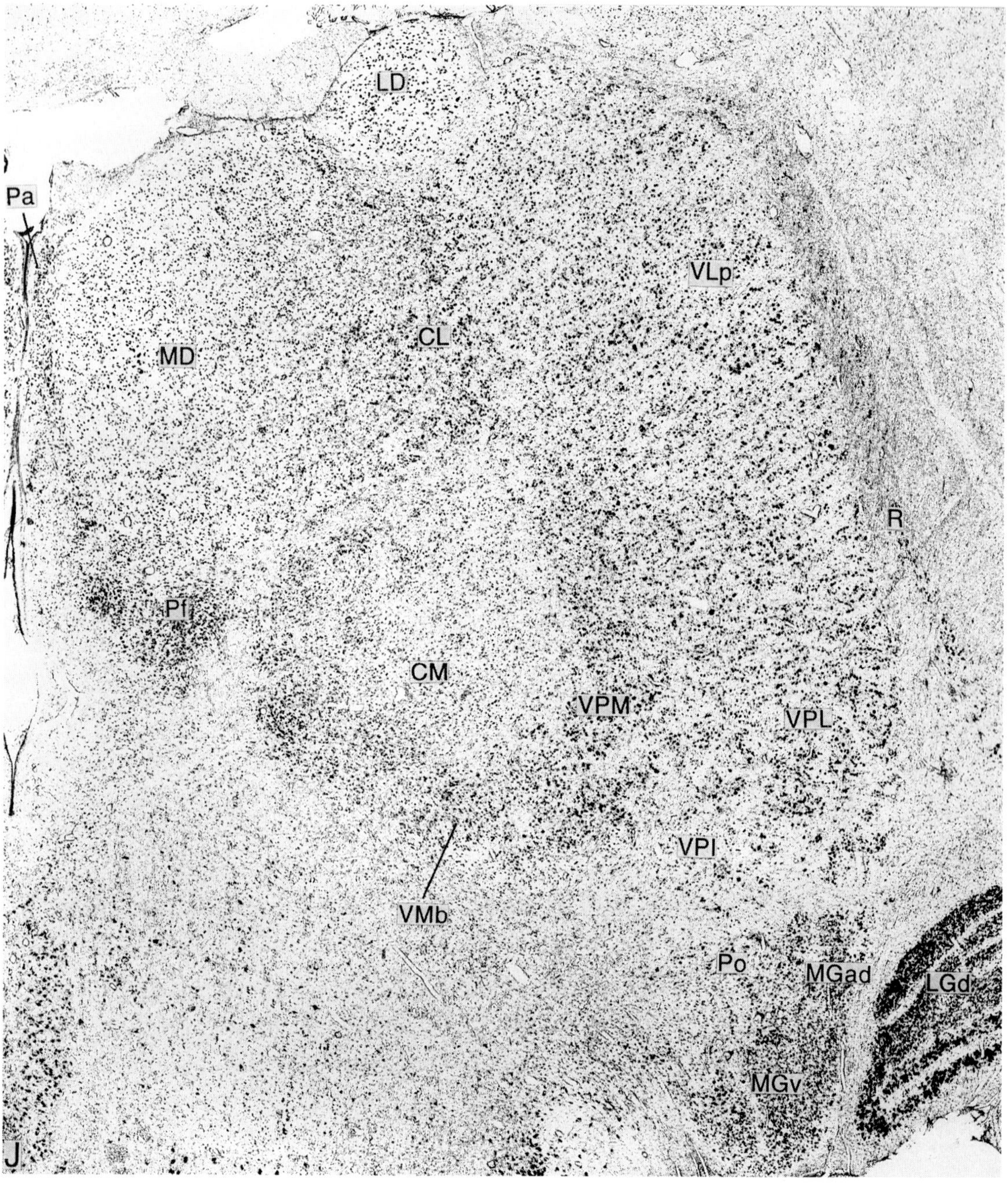

Fig. 3. (continued)

tion is more evenly mixed (Friedman and Jones, 1981; Jones and Friedman, 1982). These divisions may be clearer in some species of monkeys than others, perhaps warranting their designation as separate subnuclei (Kaas et al., 1984a, b). They are very much clearer in the human thalamus (Hirai and Jones, 1989a) (Section 3.3).

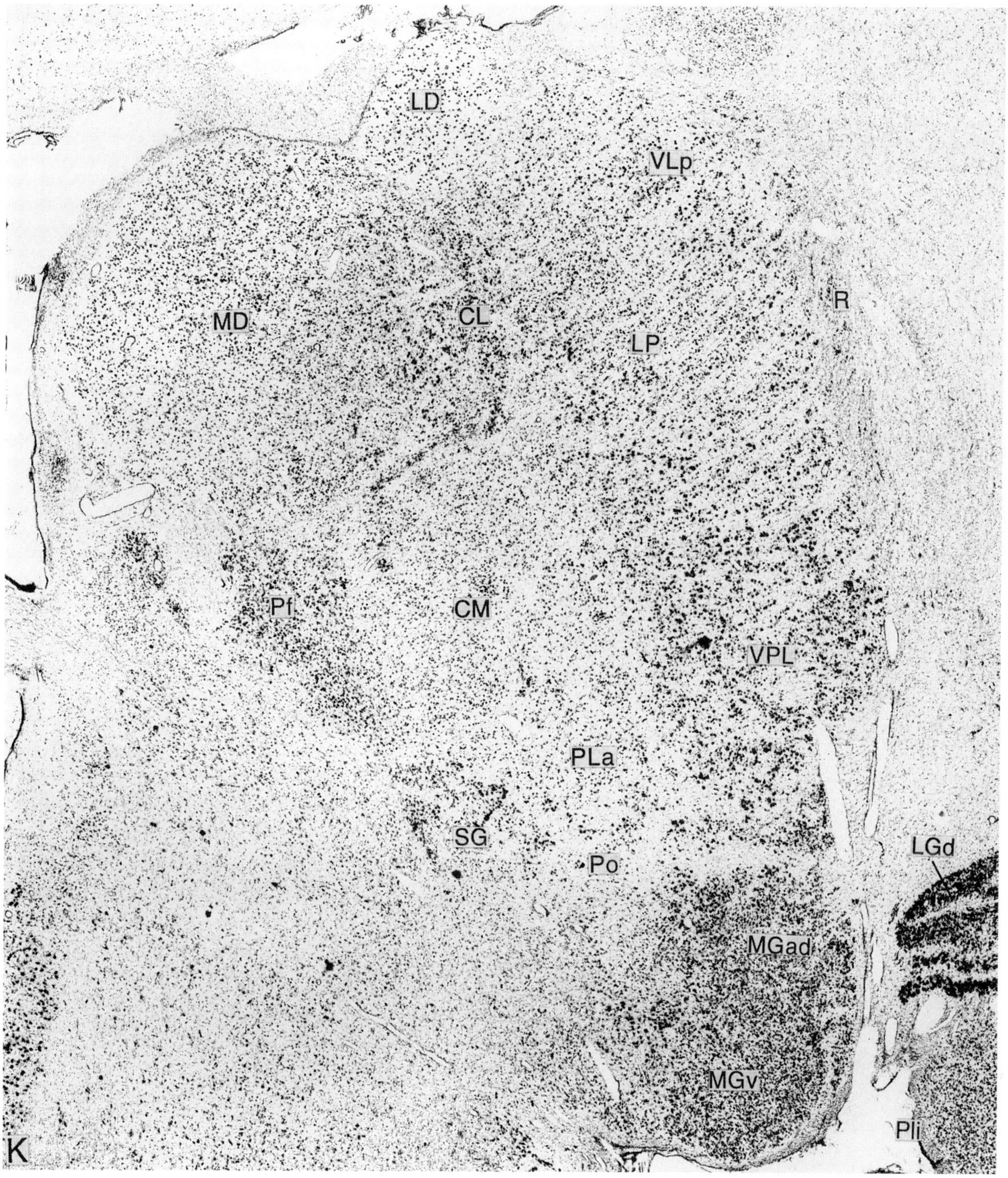

Fig. 3. (continued)

Intercalated between the ventral aspects of the VPM and VPL nuclei, along the ventral part of the external medullary lamina, is the ventral posterior inferior nucleus (VPI), the region through which most of the fibers of the medial lemniscus enter the thalamus and, thus, dominated by neuroglial cells, but containing numerous small

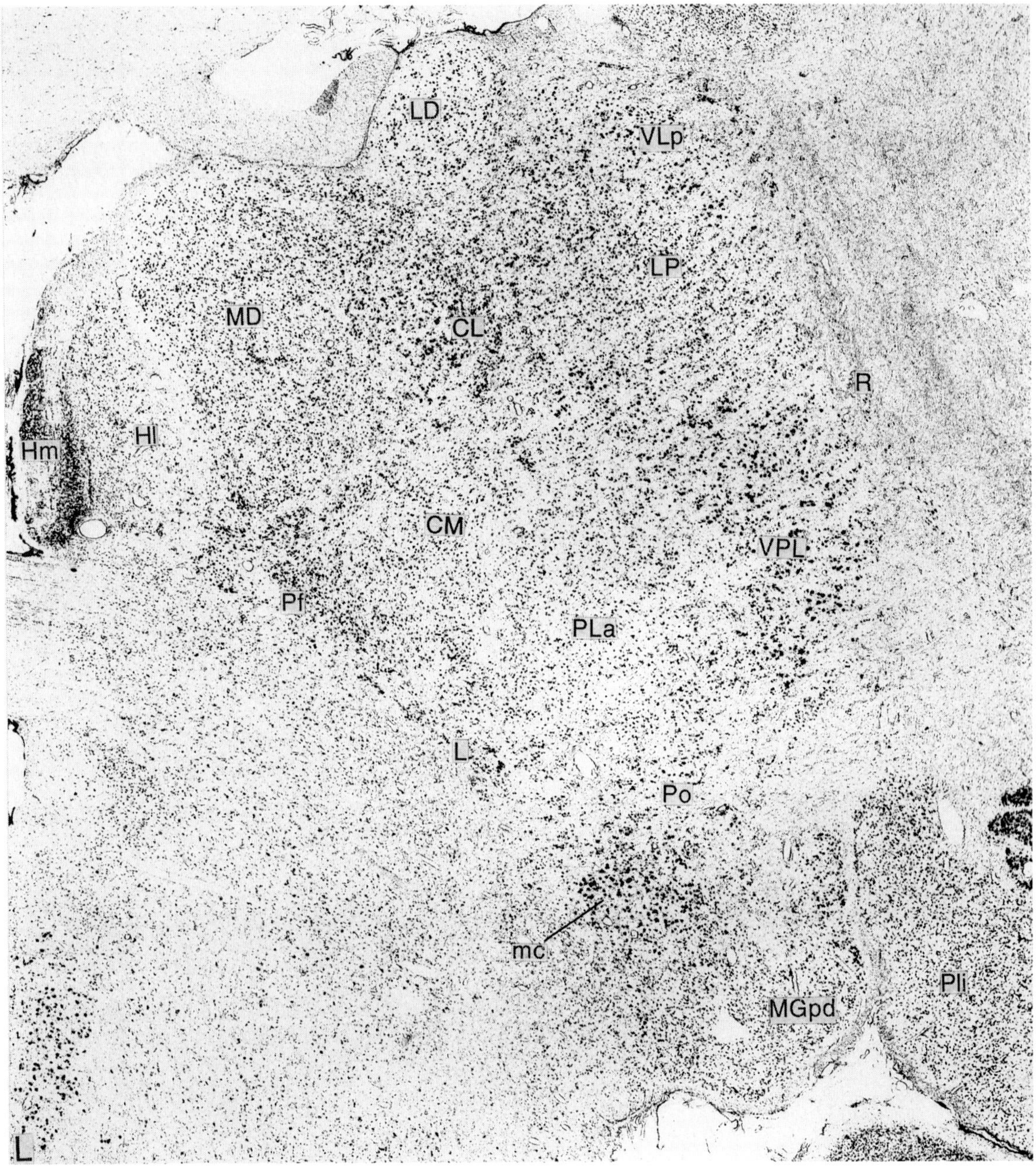

Fig. 3. (continued)

neurons as well (Fig. 3 F-J). These have immunocytochemical properties similar to those of the VMb nucleus and to a matrix of small cells that infiltrates the VPM nucleus and covers its medial and posterior surfaces (see below).

The anterior aspect of the lateral posterior/pulvinar complex is made up of the small cells of the lateral posterior nucleus (LP) which are insinuated between the dorsal

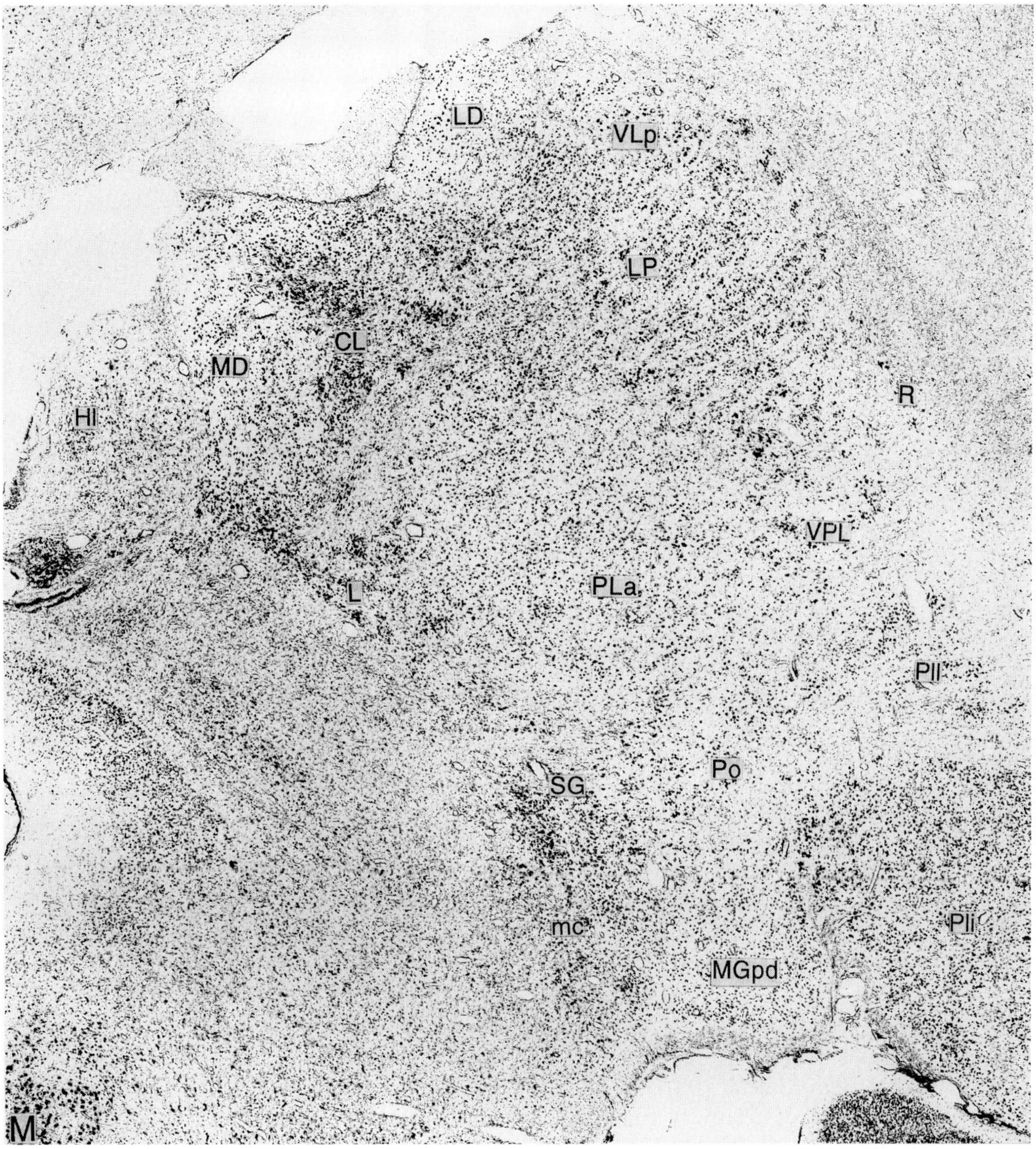

Fig. 3. (continued)

aspect of the VPL nucleus and the posterior pole of the VLp nucleus (Fig. 3K). The LP nucleus expands posteriorly before being itself replaced by the nuclei of the pulvinar. The anterior pulvinar nucleus (Pla), also a small-celled nucleus but with less densely-stained cells than LP, is first interposed between the VPM and centre médian nuclei, expanding over the posterior pole of the latter. Where the pulvinar forms the expanded posterior pole of the whole thalamus, it is composed of three large nuclei;

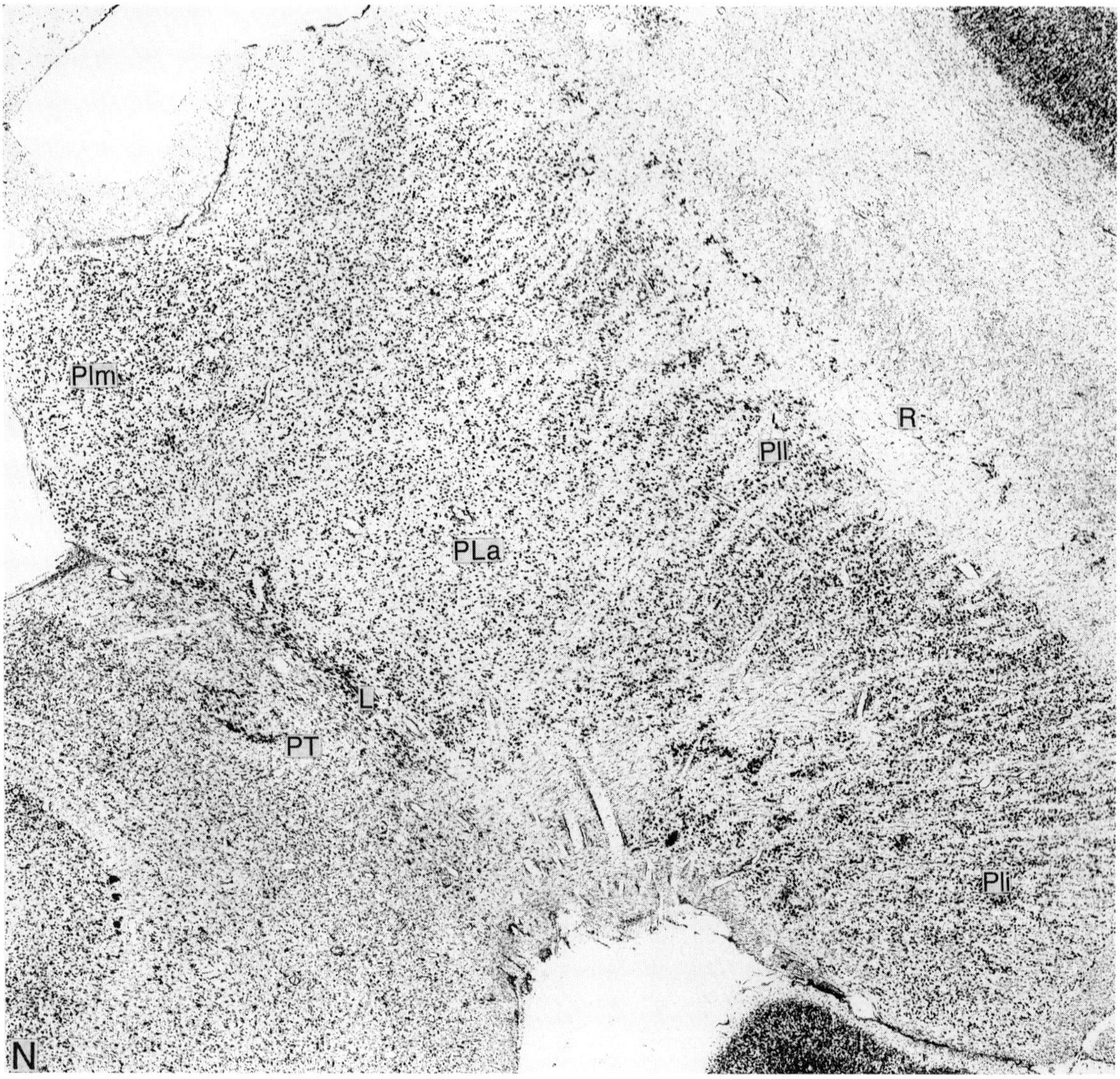

Fig. 3. (continued)

the medial (Plm), lateral (Pll) and inferior (Pli) pulvinar nuclei (Fig. 3 O). The Plm nucleus consists of relatively small, relatively pale-staining cells. Those of the Pll nucleus are larger and more deeply stained but relatively widely separated. Those of the Pli nucleus are mostly small but close packed and densely stained; among them are some widely dispersed, large cells.

The posterior complex consists of a clearly-defined linear strip of dark cells forming the nucleus limitans, extending from the posterior pole of the central lateral nucleus along the border between the medial pulvinar nucleus and the pretectum (Fig. 3 K-N). Laterally, the nucleus limitans expands into the dorsal lobe of the medial geniculate complex as the suprageniculate nucleus (SG). Anterolateral to the SG nucleus is an ill-defined zoned of smaller, loosely-packed cells, continuous with the VPl nucleus and referred to as the posterior nucleus (Po).

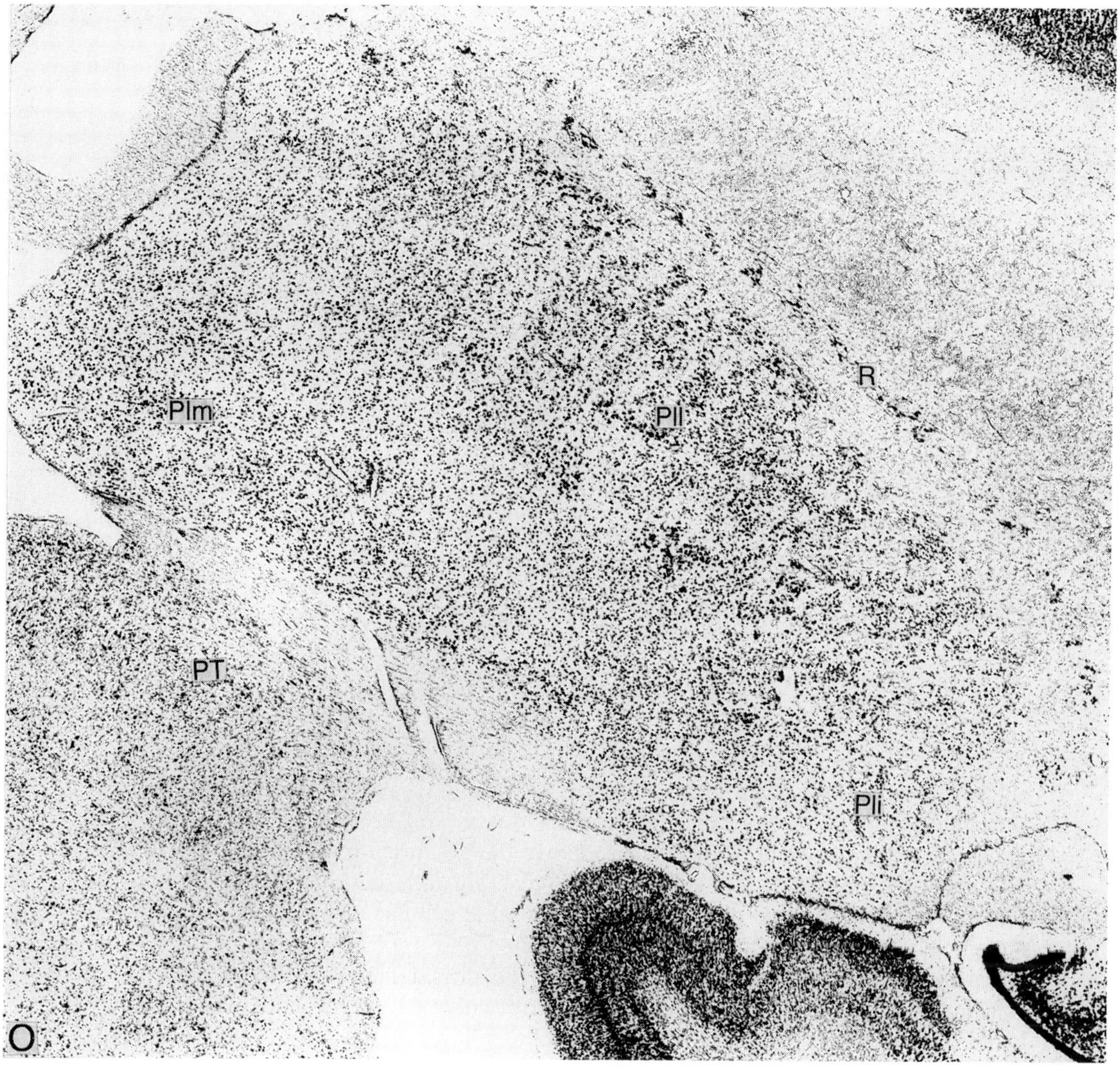

Fig. 3. (continued)

The medial geniculate complex (Fig. 3 J-M) consists of a medially located magnocellular nucleus (MGmc), actually containing small and medium cells as well as the dominating large cells, a ventral (actually ventral-lateral) nucleus (MGv) of close-packed, darkly staining cells and a dorsal nucleus. The dorsal nucleus arches over the ventral nucleus and expands posterior to the ventral nucleus as the posterior pole of the complex. Its anterior division, or anterodorsal nucleus (MGad), is composed of relatively closely-packed cells not unlike those of the ventral nucleus. The posterior division, or posterodorsal nucleus (MGpd), is composed of palely-staining, less densely-packed cells. The different cell classes of the medial geniculate complex have unique immunocytochemical properties and connections (see below). In some of the smaller New World monkeys, such as the marmosets, the divisions of the dorsal

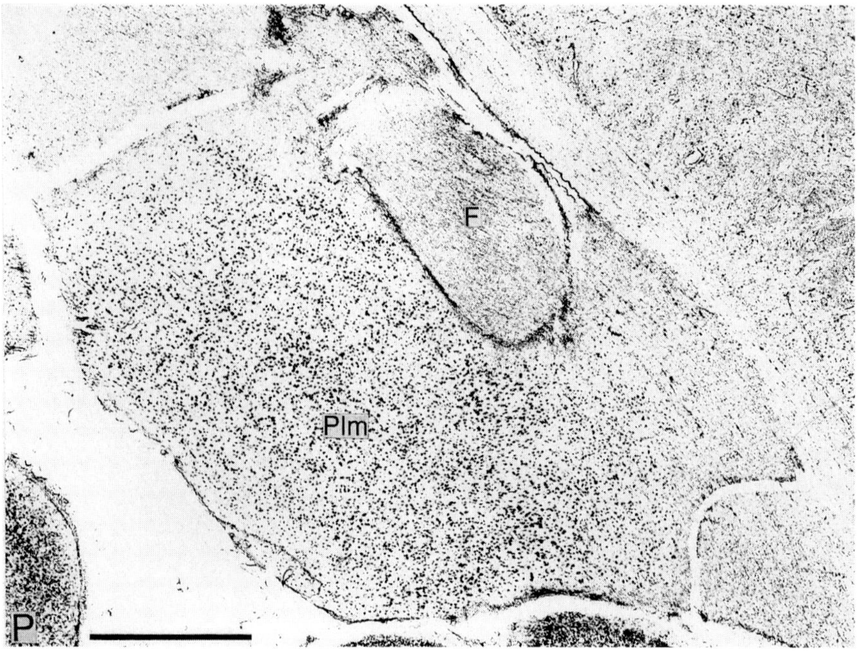

Fig. 3. (continued)

nucleus may be less clear than in Old World monkeys and the ventral nucleus, although distinct, can be drawn into a more dorsal and horizontal position.

The dorsal lateral geniculate nucleus shows the greatest range of variation in monkeys (Fig. 4), probably to a large extent reflecting differentiation in relation to the demands of nocturnal or diurnal vision. The same cellular elements are, nonetheless, present in all species that have been investigated and appear to be common to other primates as well. The well-known form of the dorsal lateral geniculate nucleus in macaques is the typical dome-shaped structure with two magnocellular layers (layers 1 and 2) located ventrally and four parvocellular layers (layers 3–6) located dorsally, the layers being curved around a hilus through which blood vessels and many fibers of the optic tract enter the nucleus (Fig. 4). Ventral to layer 1, among the fibers of the optic tract is a strip (often bilaminated) of small cells, called the S layer(s). Similar small cells with identical immunocytochemical properties intrude between all the principal layers and around the perimeter of layer 6. These are referred to as the intercalated or interlaminar cells and are clearly homologous to the small cells of layers 4 and 5 in prosimians.

The parvocellular layers are incompletely split into four, the four being only clearly evident in the center of the nucleus i.e. in the representation of the central visual field. This is also evident in humans (Fig. 4E). Around the perimeter of the nucleus, layers 2 and 4, and 3 and 5 merge. This indicates the more fundamental division of the parvocellular layers into two layers, one (layers 2 and 4) innervated by the contralateral eye and a second (layers 3 and 5) innervated by the ipsilateral eye (Kaas et al., 1972). In many of the guenons (*Cercopithecus* species), the splitting of the parvocellular laminae is least complete.

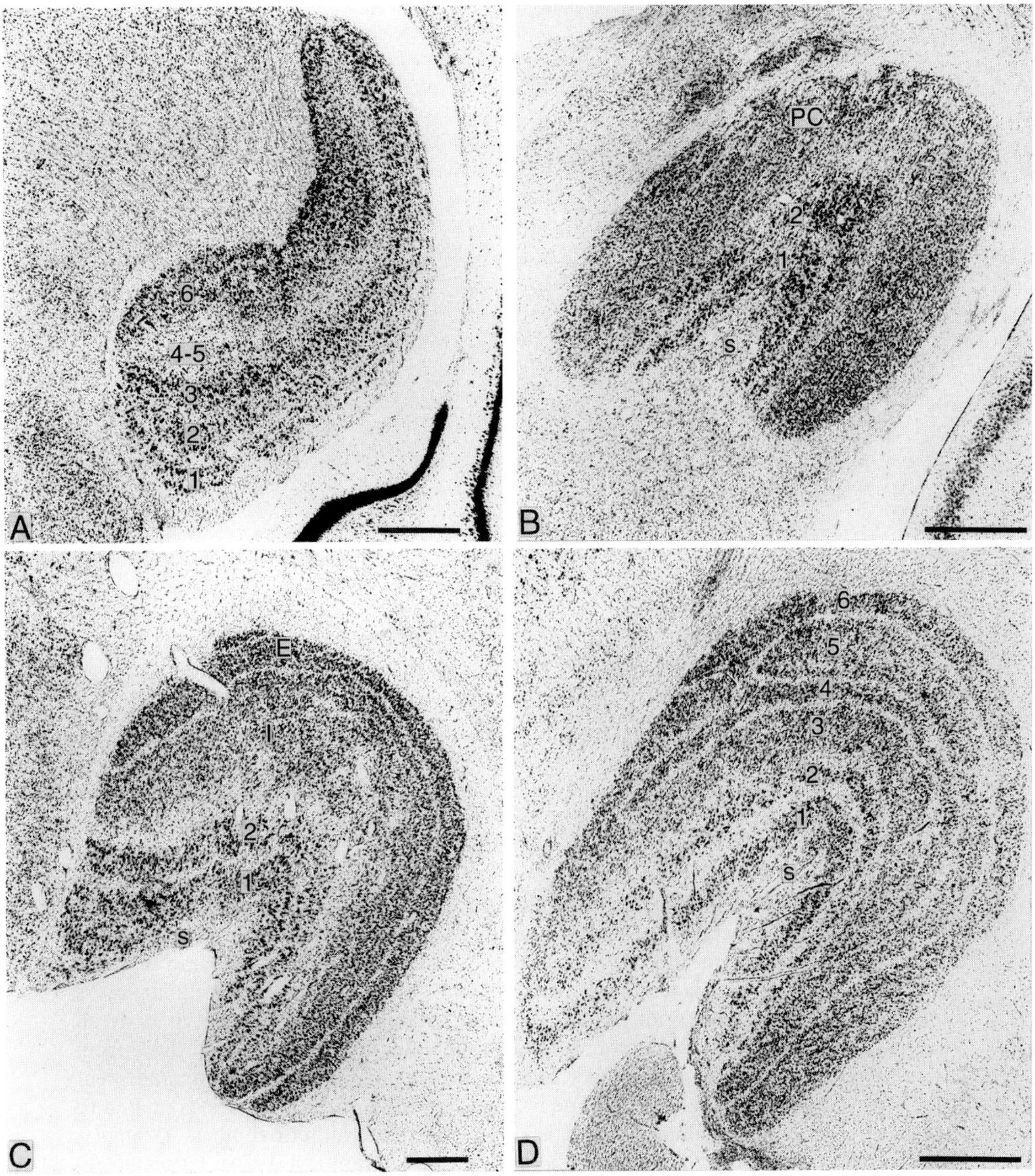

Fig. 4. Nissl-stained sections in the frontal (A, C-E) or parasagittal (B) planes through the dorsal lateral geniculate nuclei of a *Galago* (A), a squirrel monkey (*Saimiri sciureus*) (B), a capuchin monkey (*Cebus albifrons*) (C), a rhesus monkey (*M. mulatta*) (D) and a human (E). 1, 2 indicate magnocellular layers in all cases; parvocellular layers (3 and 6 in A) are fused into a common mass (PC) in B and divided into external (E) and internal (I) layers in C which represent the partially split, layers 3,5 and 4,6 respectively in D and E. Layers 4–5 in A are small cells represented by the S layer(s) and cells in the interlaminar plexuses between the main layers in B-E. Bars 0.5 mm.

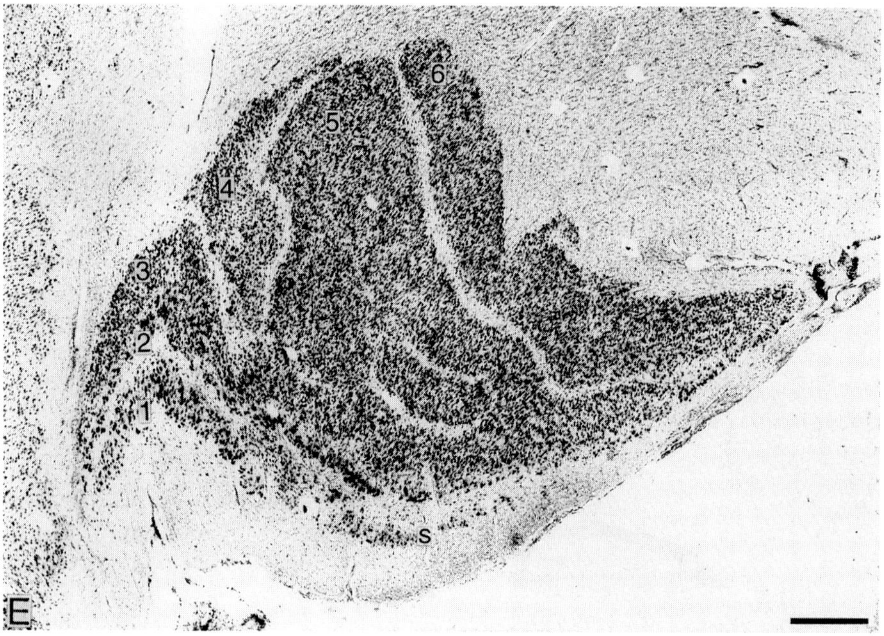

Fig. 4. (continued)

The presence of only two, or of two only slightly split parvocellular laminae, is found in many New World monkeys, including the capuchin, owl monkey, and many of the marmosets. Splitting into four is most evident in some of the cebus species (Fig. 4C). Lack of splitting is found among the apes in the gibbon and orangutan; splitting is present but not particularly strong in the gorilla and only well-marked in the chimpanzee. In a number of species of New World monkey, including, for example, the squirrel monkey and spider monkey, the parvocellular region is not laminated at all, simply appearing as a solid mass of cells (Fig. 4B). A lack of lamination, or an incomplete lamination in the parvocellular region is not accompanied however, by corresponding differences in the distribution of fibers from the two eyes. These invariably sort themselves into alternating laminae of terminations corresponding to those seen in the parvocellular layers of the macaque and human. The magnocellular layers show little variation among all primates, two invariably being found. In virtually all species, fibers from the contralateral eye end in the outer layer (layer 1) and those from the ipsilateral eye in the inner layer (layer 2). In *Tarsius*, however, this arrangement may be inverted (Rosa et al., 1996). The incipient splitting of the S layer(s) is related to alternating zones of termination of ipsi- and contralateral retinal fibers (Kanagasumtheram et al., 1969a; Kanagasuntheram and Krisharamurti, 1970; Tigges et al., 1977).

In the ventral thalamus of monkeys, the same elements can be discerned as in other species. The reticular nucleus is thin but very extensive, covering the anterior and lateral surfaces of the dorsal thalamus and, unlike in the *Galago* but like in humans, continuing over the lateral surface of the dorsal lateral geniculate nucleus. The ventral lateral geniculate nucleus, which tends to be bilaminar, has come, with the downward rotation of the dorsal lateral geniculate nucleus during brain development, to occupy a position dorsomedial to the dorsal lateral geniculate nucleus. There, it is customarily referred to as the pregeniculate nucleus (Prg). The zona incerta, made up of mixed small and medium sized cells, dominates the ventral surface of the thalamus, grading medially into the dispersed larger cells that form the nucleus of the field of Forel. Dorsomedial to the nucleus of the field of Forel and slung beneath the thinned external medullary lamina on the ventral surface of the centre médian and parafascicular nuclei, is an indistinct zone of small cells called the subfascicular or subparafascicular nucleus. It is likely that this is an extension of the midbrain tegmentum or periaqueductal gray matter but there is a tendency on the part of some workers to include it in the thalamus. In being outside the external medullary lamina, it is probably not part of the dorsal thalamus but this could be contested.

The components of the epithalamus are distinct, although the paraventricular nuclei are reduced to narrow strips. The lateral habenular nucleus is very clearly divided into lateral, magnocellular, and medial, parvocellular, divisions. In this, it has a closer resemblance to the human than to the prosimian habenula. Unlike in the human, however, the habenular nuclei are still relatively dorsally located.

3.3. THE HUMAN THALAMUS

The dominating feature of the human thalamus is its size (Figs. 5, 6). The nuclear parcellation is, however, demonstrably similar to that of monkeys, the larger size of the nuclei and the commonly greater size differences between the cells of adjacent nuclei usually making nuclear parcellation a much easier task. As is well-known, the pulvinar is especially large, extending not only dorsally over the remainder of the dorsal thalamus but also down its posterior surface and over the midbrain. Less frequently noted is the marked enlargement of the ventral lateral nuclei, especially the VLp nucleus, relative to other nuclei. The relative proportions of the VLp nucleus are much greater in the human than in monkeys.

The anterior nuclei of humans are unusual in comparison with other mammals, not only on account of their small size relative to the rest of the thalamus, but also on account of the lack of a clear distinction between anteroventral and anteromedial nuclei (Fig. 6 D-H). These are merged into a common nucleus whose overall position is comparable to that occupied by the two nuclei in monkeys but in which there is little differentiation of two cell populations. The general appearances of cell size, packing density and staining density resemble those of the anteroventral nucleus of monkeys. The cells closer to the midline are somewhat less densely packed and may be regarded as an incipient anteromedial nucleus. The anterodorsal nucleus forms a very thin strip of large, dark cells along the dorsomedial edge of the anteroventral nucleus (Fig. 6 H, I). The laterodorsal nucleus is relatively large and flattened over the surface of the mediodorsal nucleus (Fig. 6 J-N). Although quite extensive anteroposteriorly, the large posterior growth of the pulvinar makes its extent appear relatively less than in the monkey thalamus. The parataenial nucleus is not readily distinguishable from the mediodorsal nucleus.

The anterior intralaminar nuclei are large overall (Fig. 6 E-M) but the attenuation of the internal medullary lamina and the concomitant thinning of the paracentral and of large parts of the central lateral nucleus convey a false impression of a reduced size and a sheet-like character. A very large number of cells, however, reside in the central medial nucleus in particular, and this is very prominent. The posterior intralaminar nuclei are very large, much larger relative to the rest of the thalamus than in other primates, especially the centre médian (Fig. 6 I-M). The centre médian nucleus, more

Number	Species[1]	Family
1	*Cheirogaleus major*	Prosimians
2	*Microcebus murinus*	
3	*Lepilemur ruficaudatus*	
4	*Lemur fulvus*	
5	*Avahi l. occidentalis*	
6	*Propithecus verreauxi*	
7	*Loris tardigradus*	
8	*Periodicticus potto*	
9	*Galago crassicaudatus*	
10	*Galago demidovii*	
11	*Tarsius sp.*	
12	*Cebuella pygmaea*	New World monkeys
13	*Saguinus oedipus*	
14	*Ateles geoffroyi*	
15	*Lagothrix lagotricha*	
16	*Cebus sp.*	
17	*Saimiri sciureus*	
18	*Cercocebus albigena*	Old World monkeys
19	*Papio anubis*	
20	*Cercopithecus ascanius*	
21	*Nasalis larvatus*	
22	*Colobus badius*	
23	*Hylobates lar*	Apes
24	*Pan troglodytes*	
25	*Gorilla gorilla*	
26	*Homo sapiens sapiens*	Humans

[1] The species names used by Stephan et al. have been retained. Some of them were subsequently modified.

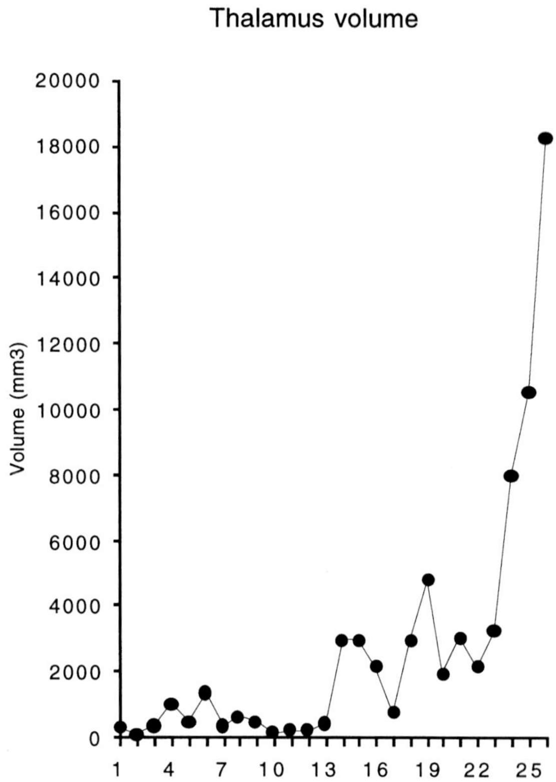

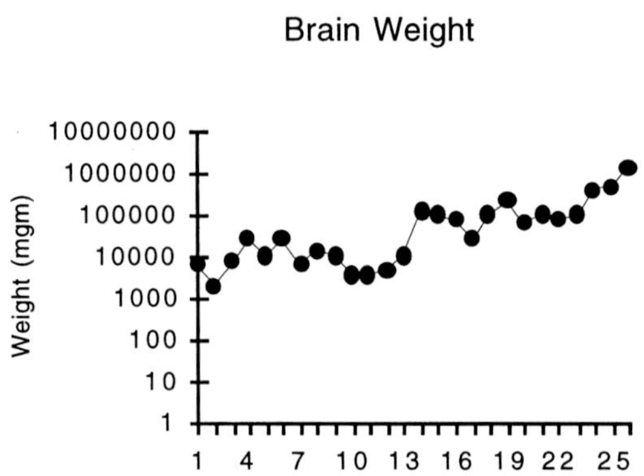

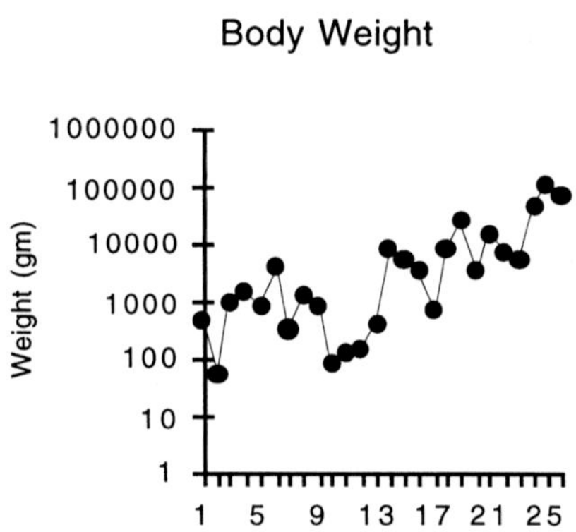

Fig. 5. Graphs of thalamic volume (top right) relative to brain weight (lower left) and body weight (lower right) for 26 species of primate. The graphs were made using figures derived from sections of paraffin-embedded material and published in Stephan et al. (1981). The species names used by Stephan et al. are given at top left. The species numbers correspond to those on the abscissae of the graphs.

clearly than in any other species, displays a dimorphism with thze typical small, pale, densely packed cells interspersed with larger, darker cells, especially in its medial half.

The mediodorsal nucleus, although large, is not disproportionately so (Fig. 6 G-N). It displays the same three subdivisions (magnocellular, parvocellular and multiform) as

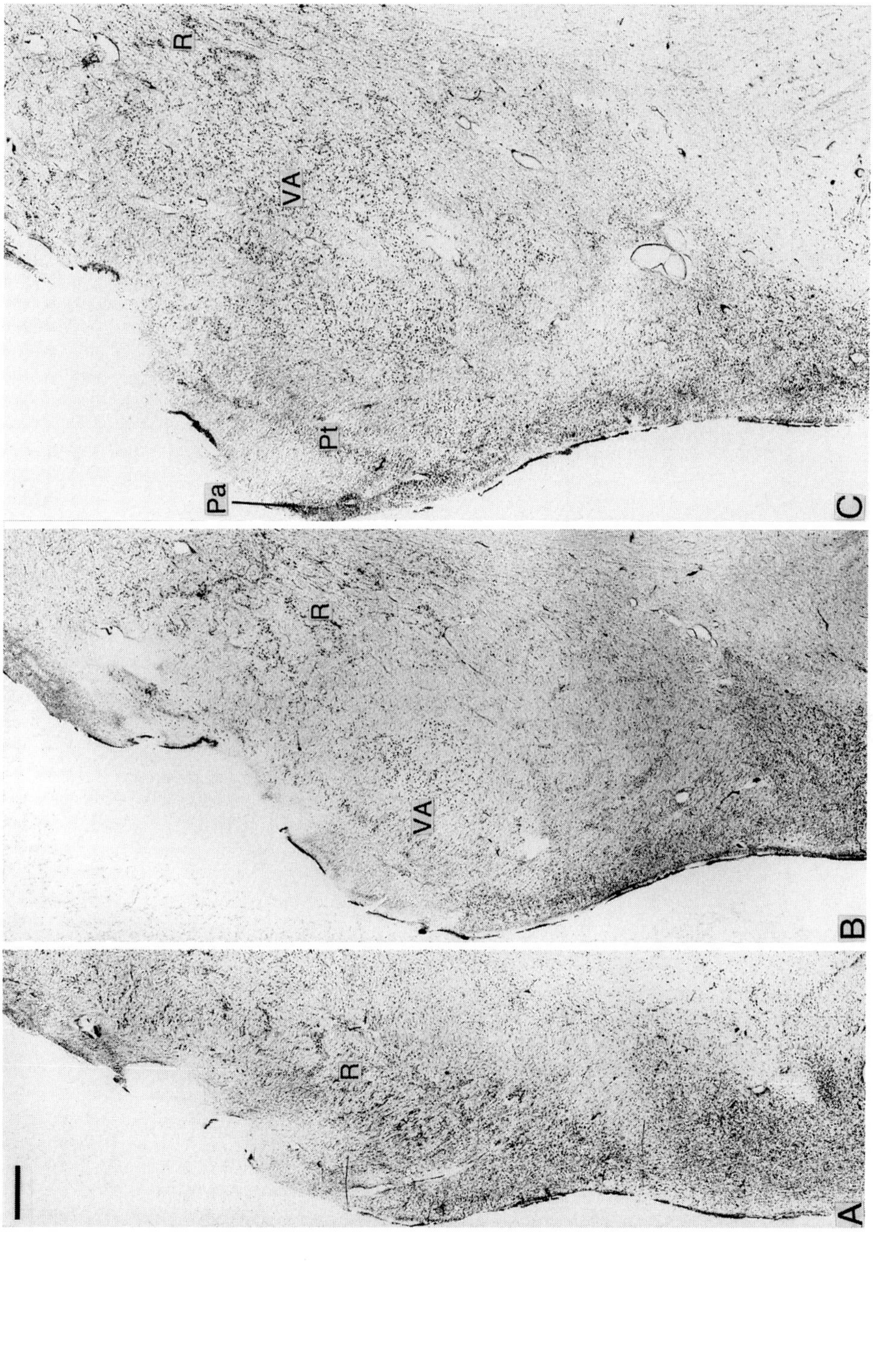

Fig. 6. Nissl-stained frontal sections in anterior (A) to posterior (T) order from the thalamus of a human. Bar 1 mm

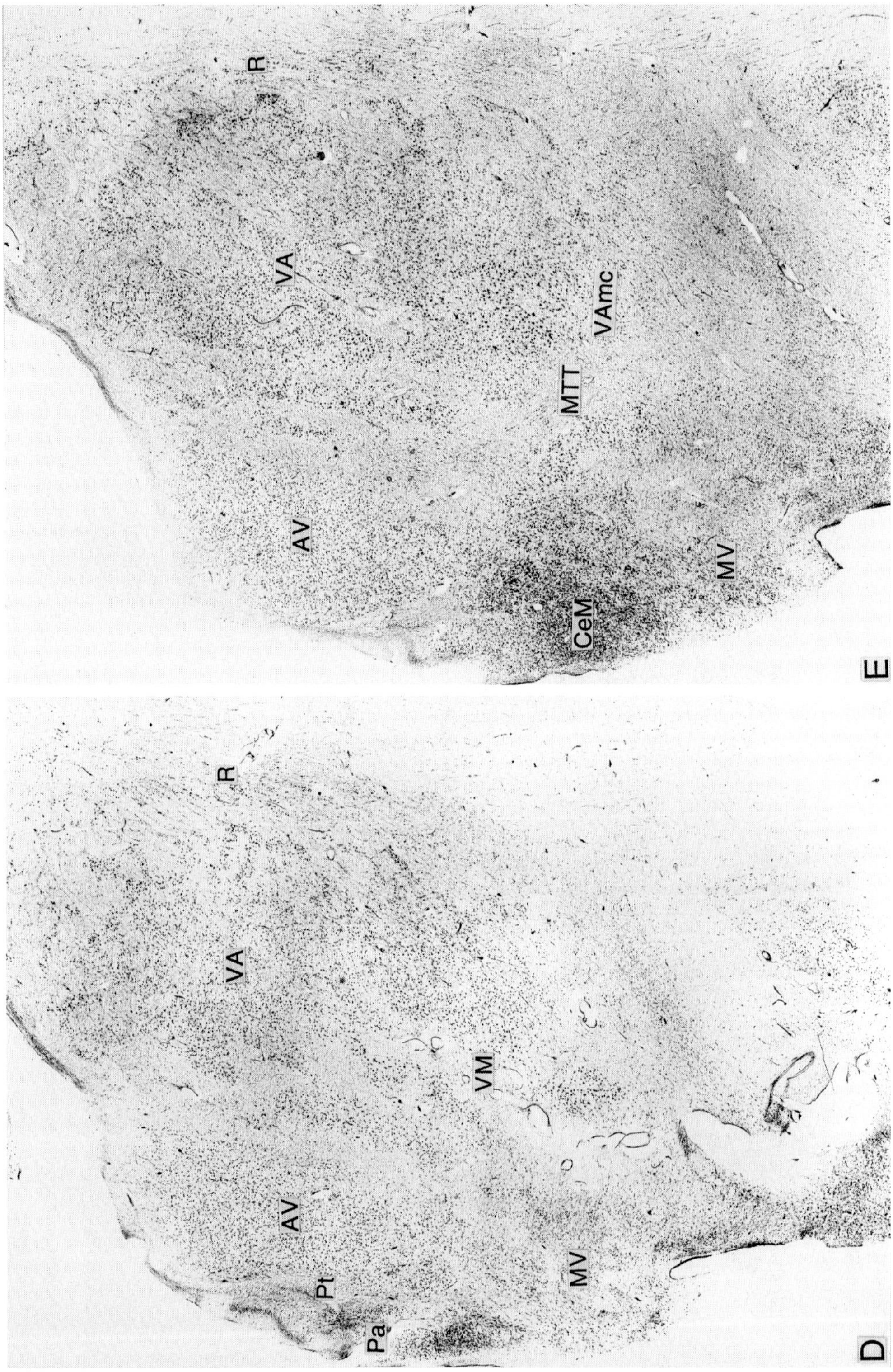

Fig. 6. (continued)

in monkeys, the three having the same relative proportions as in monkeys. The part of the central lateral nucleus which forms the so-called fourth or densocellular division of the mediodorsal nucleus in monkeys, is present but less evident on account of the more prominent break-up of the central lateral nucleus into islands of cells on the posterior surface of the mediodorsal nucleus.

The nuclear divisions of the lateral mass of the human thalamus so closely resemble those of the monkey in relative disposition and cytoarchitecture that it would be redundant to describe them anew. The differences are those of overall size and distinctness of borders between the nuclei, especially among those of the ventral group. The ventral group occupies virtually the whole anteroposterior extent of that part of the lateral mass contained within the heart of the diencephalon, the LP/pulvinar

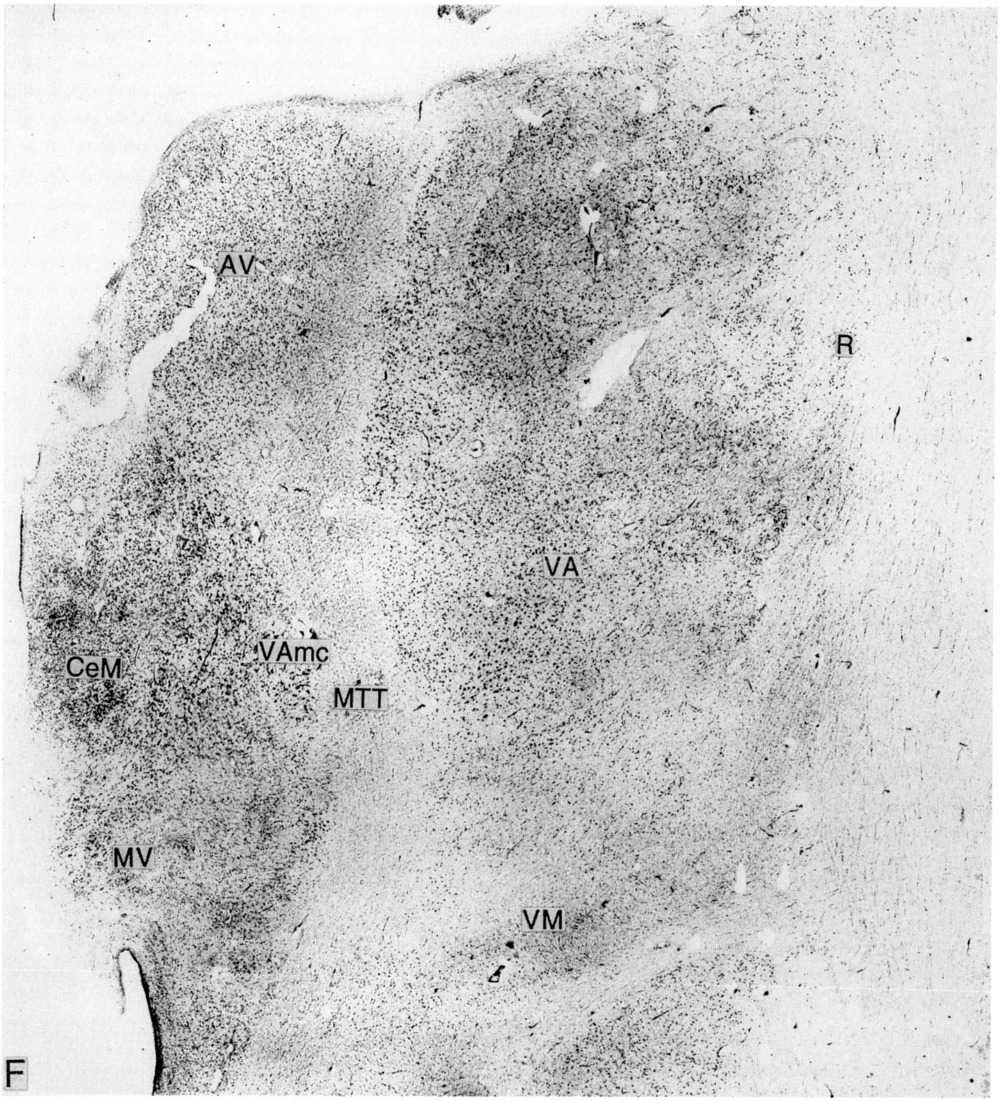

Fig. 6. (continued)

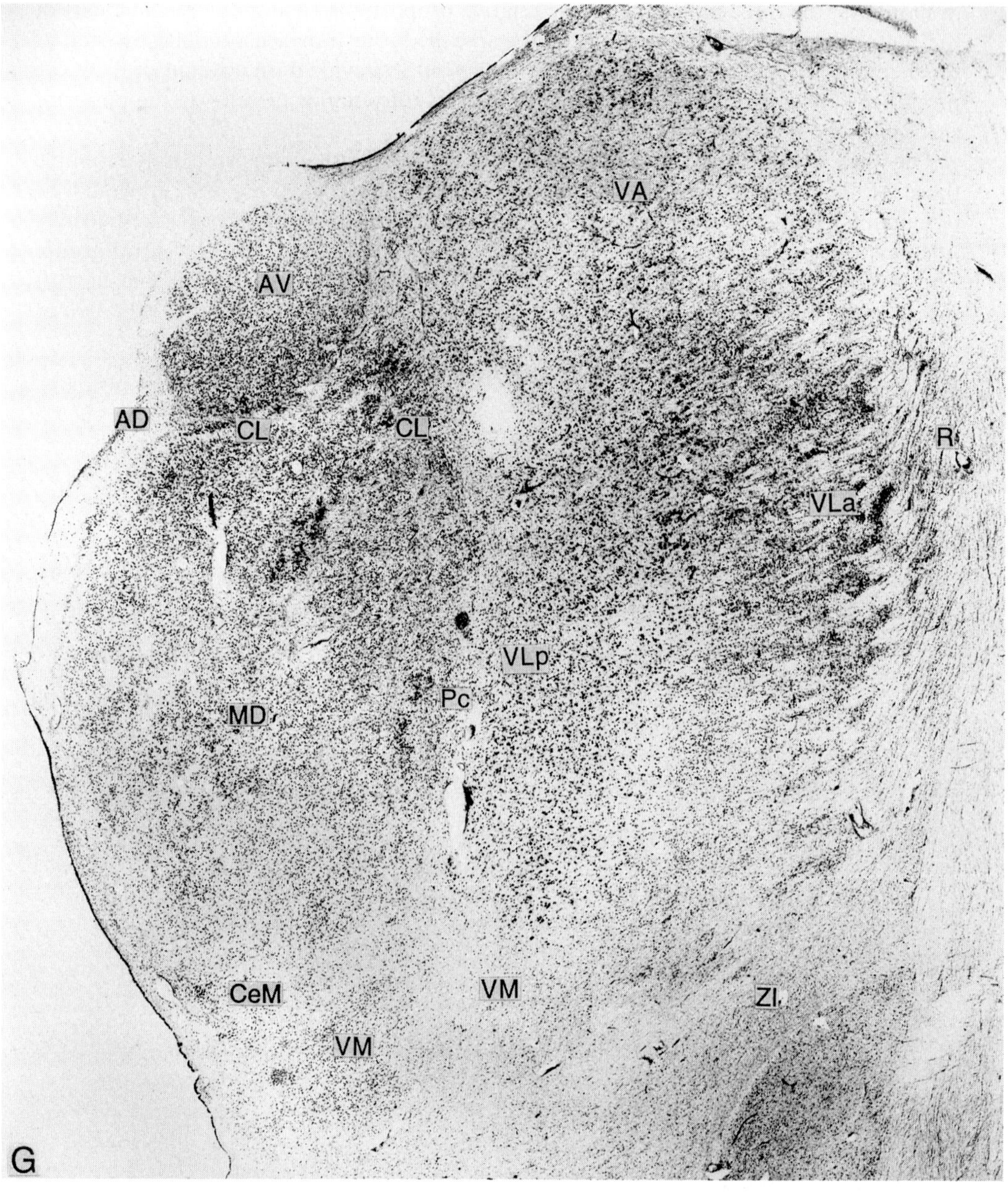

Fig. 6. (continued)

complex occupying virtually all of that part of the thalamus which protrudes posteriorly over the midbrain.

In the ventral group, clearly defined ventral anterior, ventral medial, ventral lateral anterior, ventral lateral posterior and ventral posterior nuclei are visible, all with cytological characteristics similar to those of monkeys (Fig. 6 D-Q). The division of the VLp nucleus into dorsal and ventral subnuclei, the ventral containing especially

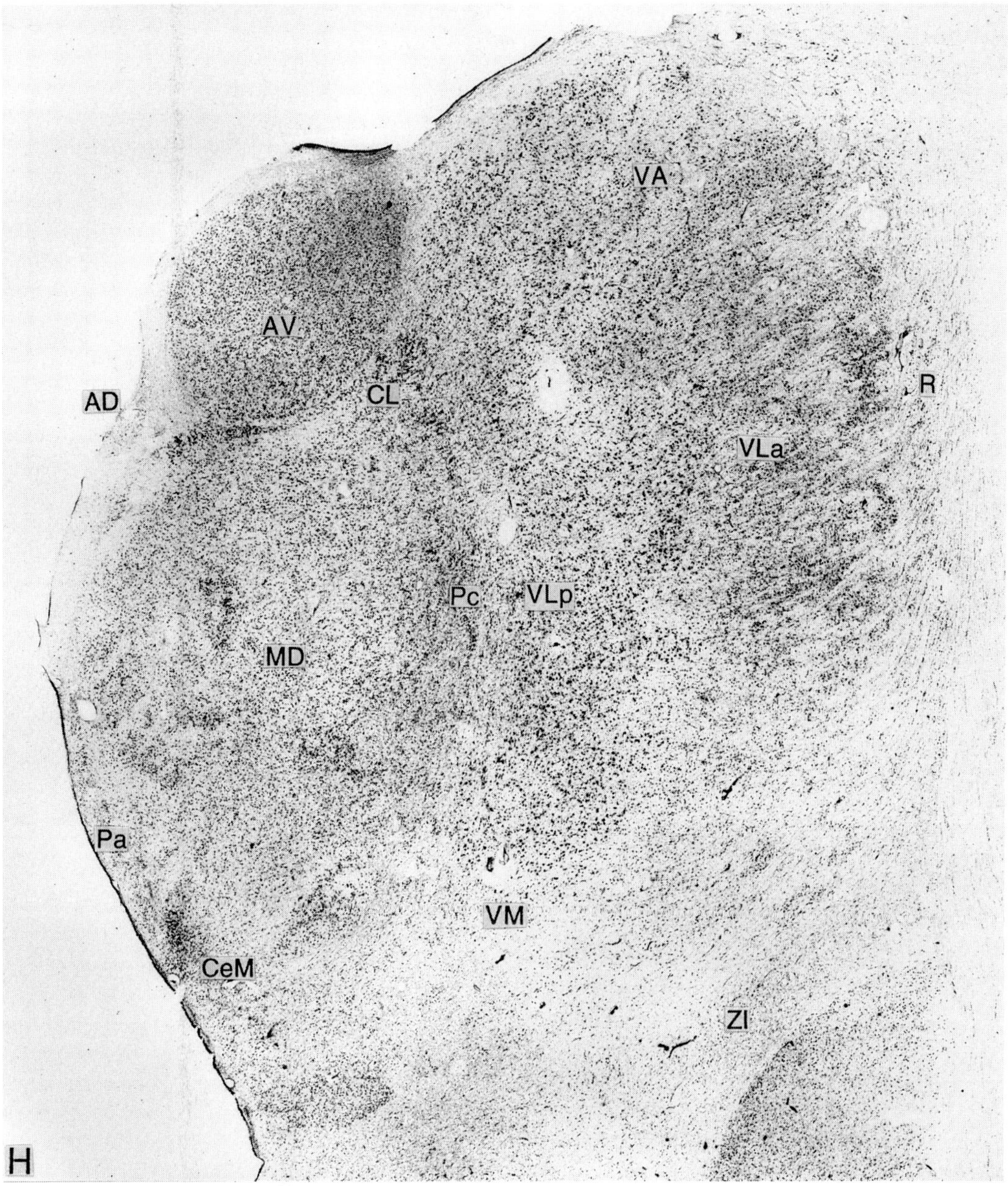

Fig. 6. (continued)

large cells, is clearly evident (Fig. 6J). Of special note is the clear division of the VPL nucleus into an anterodorsal subnucleus (VPLa) dominated by large cells and a posteroventral subnucleus (VPLp) with a more mixed cell population (Fig. 6 K-N). These represent the same divisions that are only incipiently visible in monkeys.

The VMb and VPI nuclei and the accompanying matrix of small cells in the VPM nucleus are as distinct as in monkeys (Fig. 6 I-M).

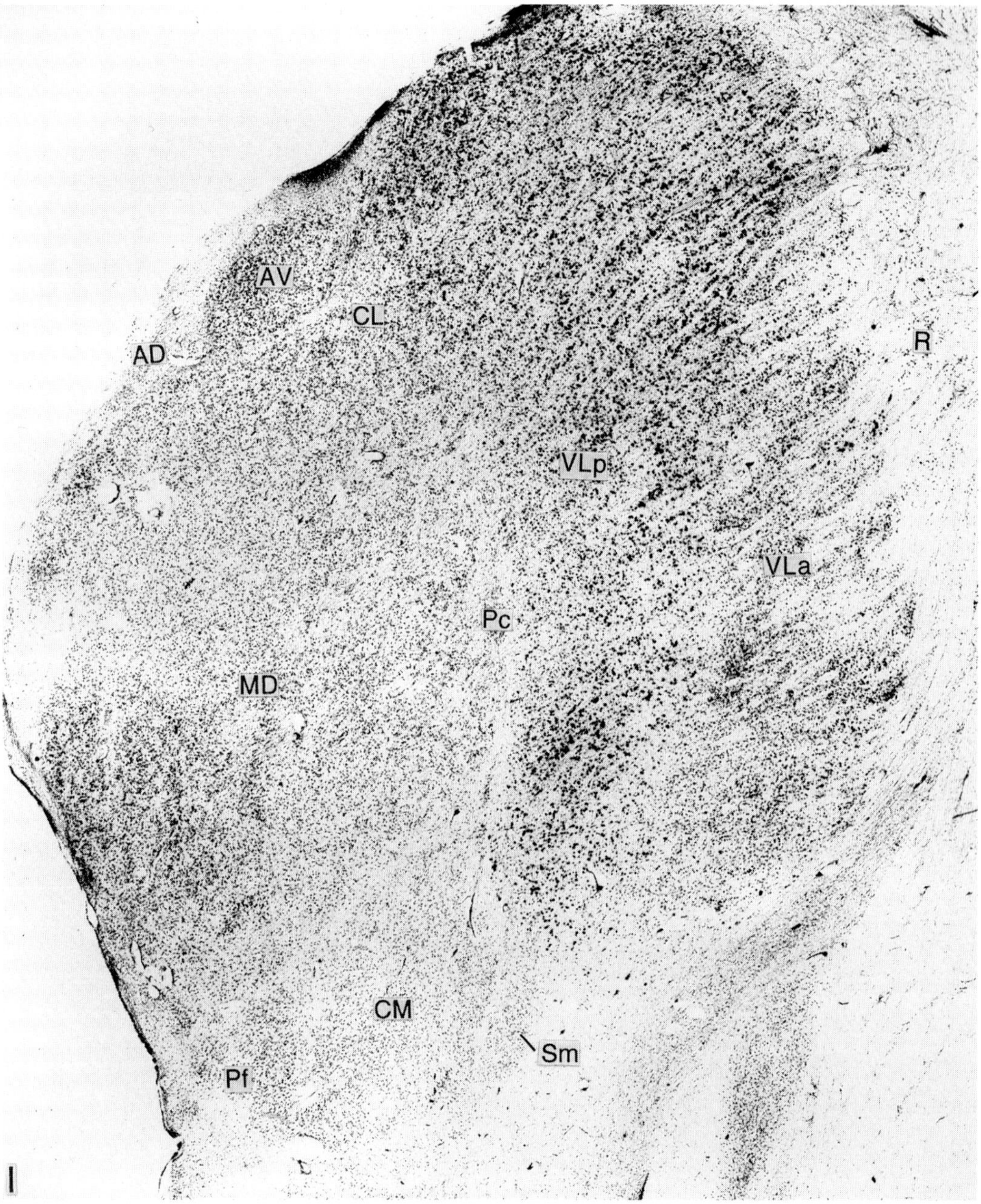

Fig. 6. (continued)

The nuclei of the human LP/pulvinar complex are also remarkably similar in their cytology, relative sizes and dispositions to those of the monkey (Fig. 6 L-T). LP, Pla, Plm, Pll and Pli nuclei can be identified. The only major difference is in overall size. Close examination reveals large islands of cells with somewhat different packing densities, particularly in the Pll and the Pli nuclei, which may reflect incipient nuclear divisions, but the cytoarchitecture of the Plm nucleus is monotonous.

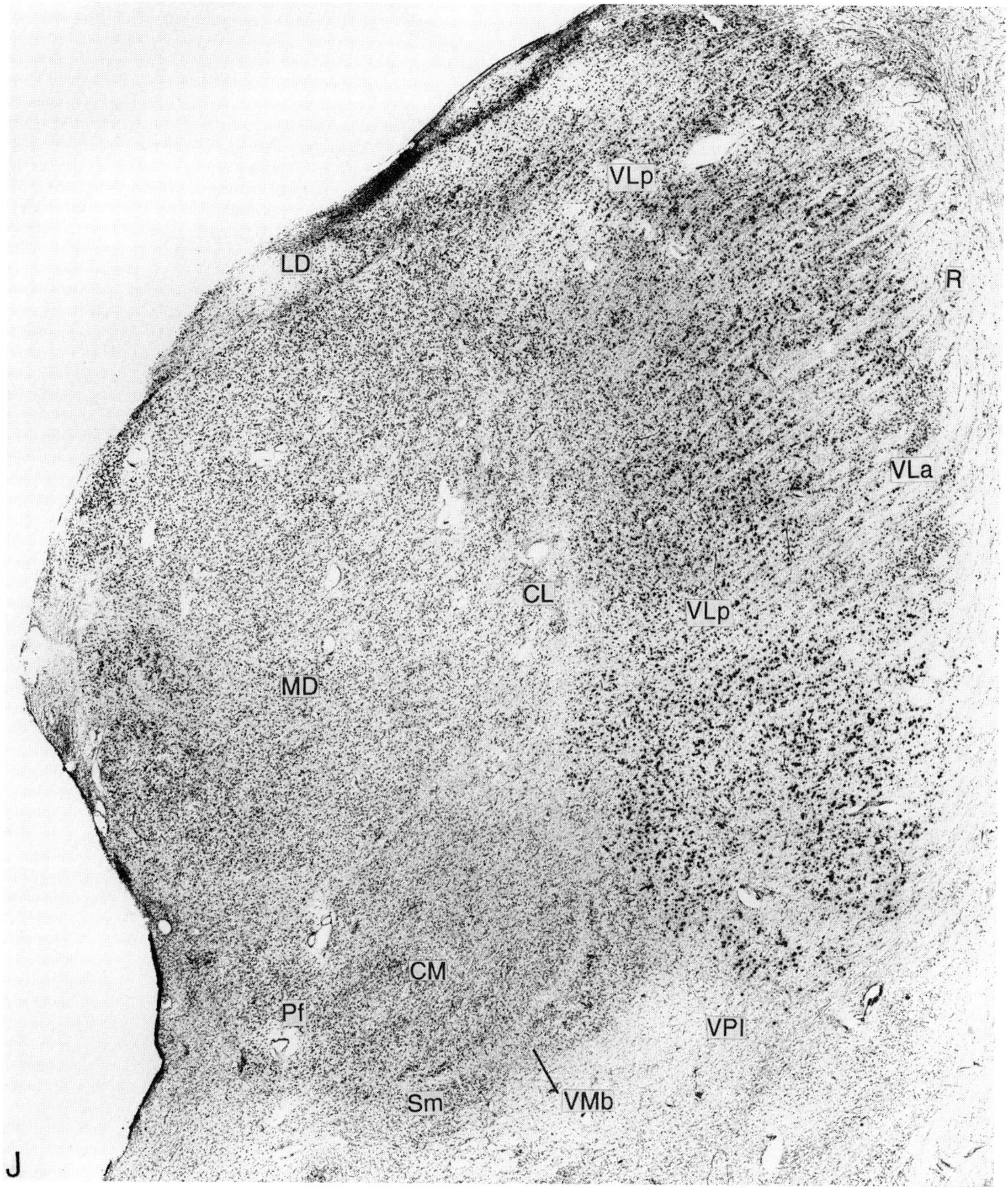

Fig. 6. (continued)

In the posterior complex, the limitans-suprageniculate and posterior nuclei have the same cell types and locations relative to other nuclei, as in monkeys (Fig. 6 N-Q). The SG and posterior nuclei form a prominent zone intercalated between the midbrain, medial geniculate complex and ventral posterior nuclei, which some have likened to a hilus or portal or entry to the thalamus from the midbrain.

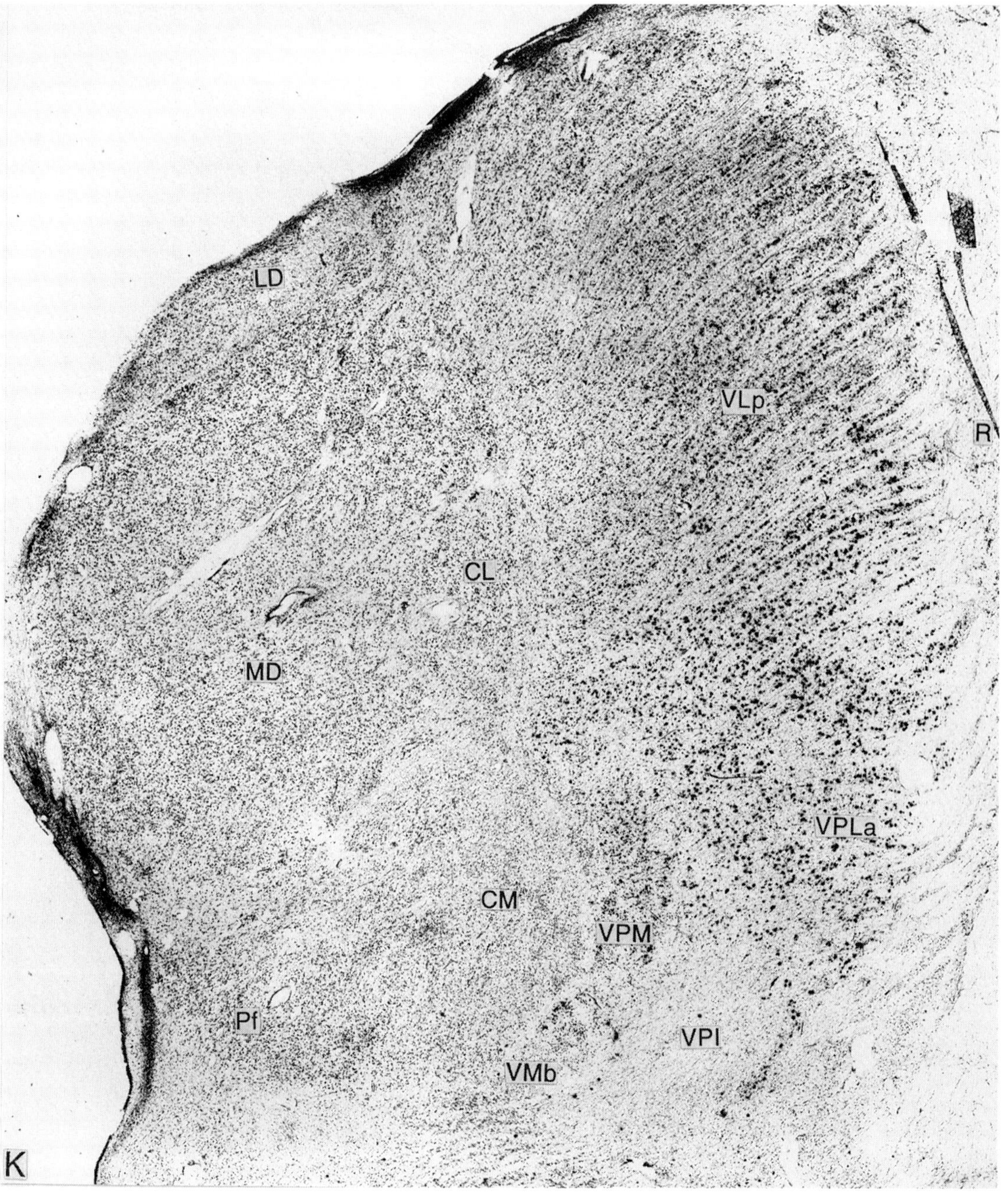

Fig. 6. (continued)

The medial geniculate complex is divided into the same nuclei as in Old World monkeys, with the same cytoarchitecture and relative sizes (Fig. 6 N-P).

The dorsal lateral geniculate nucleus, although very large, is smaller relative to the remainder of the thalamus than in monkeys (Fig. 4E). It has an everted shape in

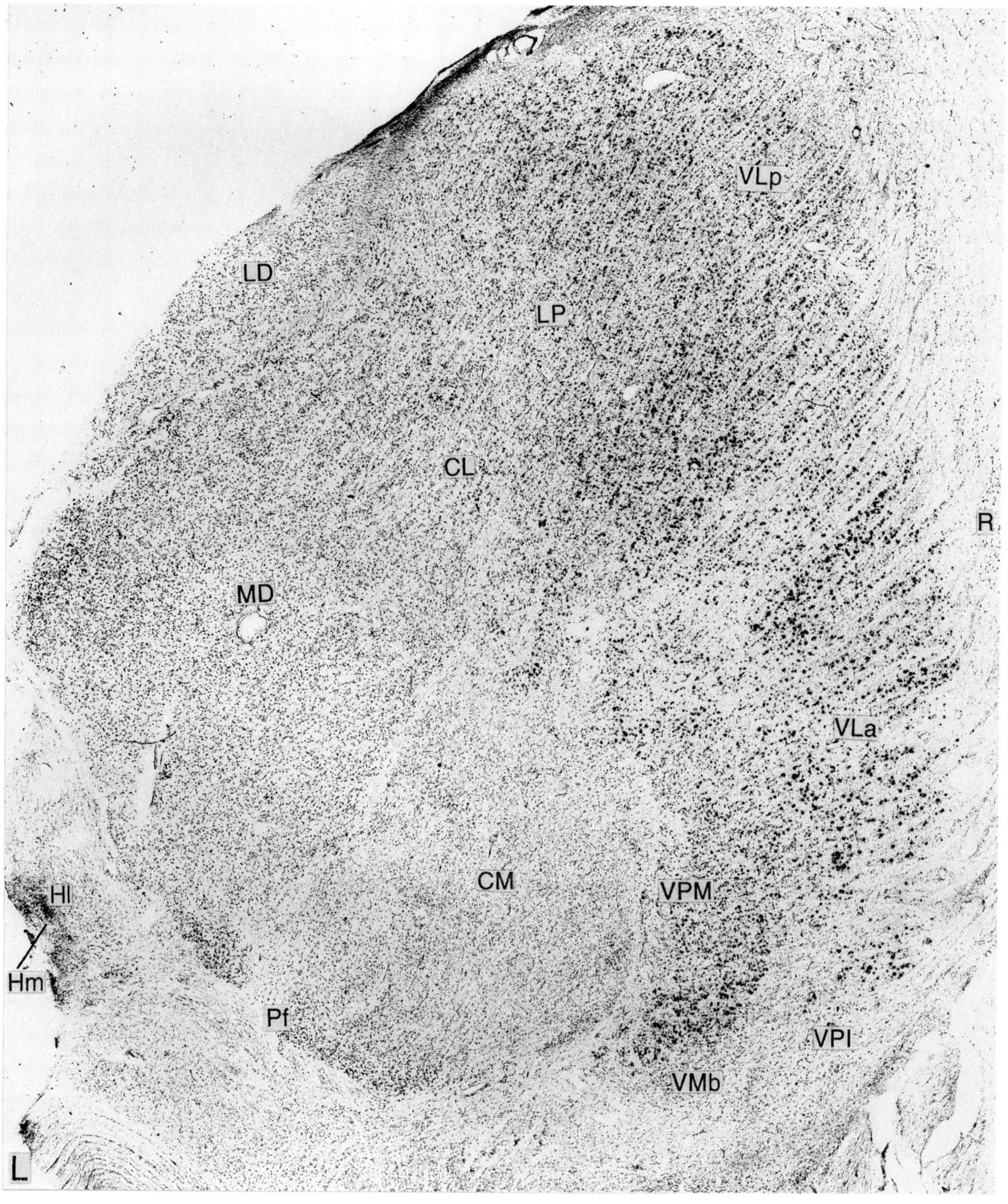

Fig. 6. (continued)

comparison with the dome-like configuration of the nucleus in macaques and many other species of monkeys. It is clearly divided into two magnocellular (1 and 2) and four parvocellular layers (3–6), with layers 3 and 5, and 4 and 6 showing approximately the same degree of peripheral fusion as in monkeys. An S layer (often paired as

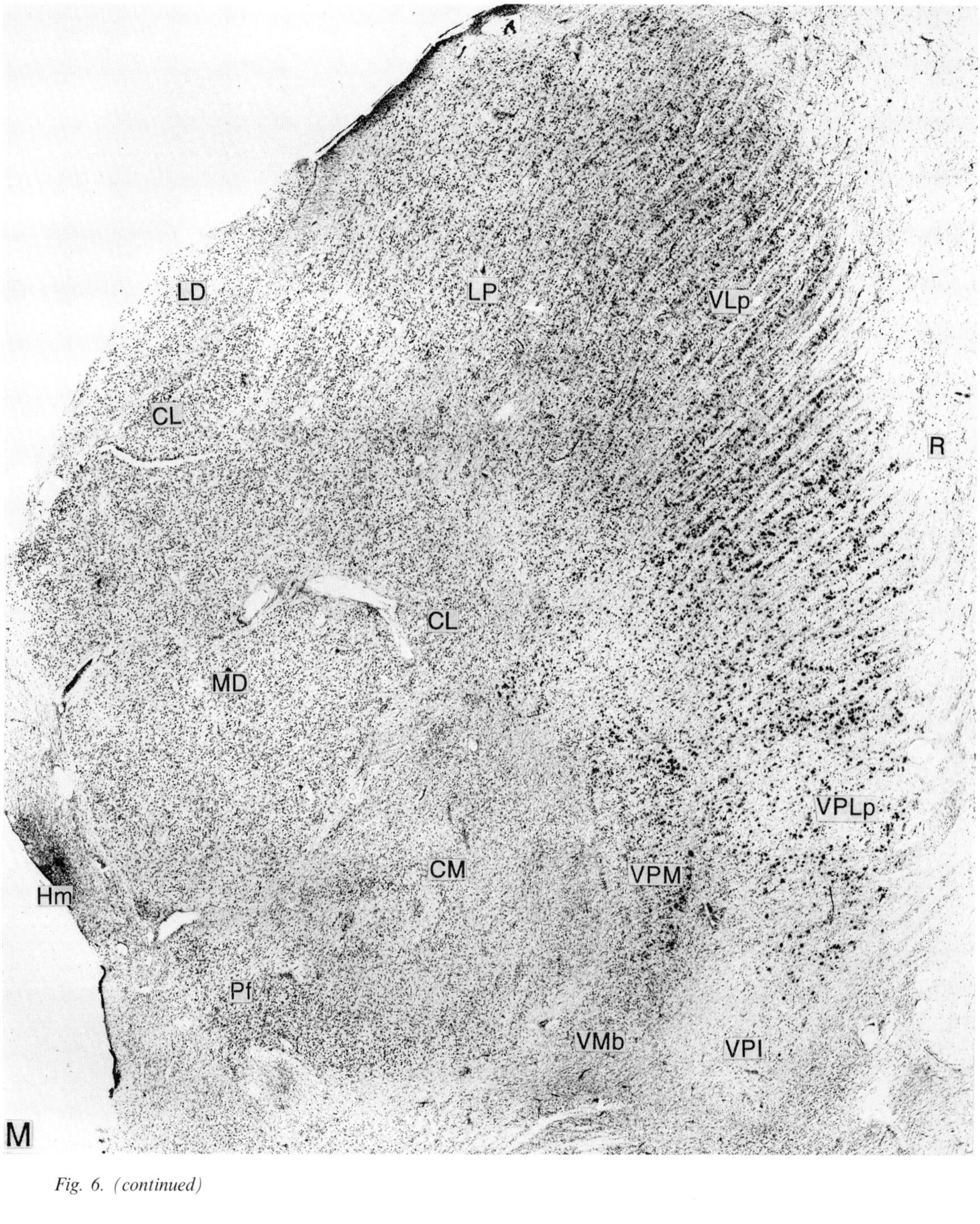

Fig. 6. (continued)

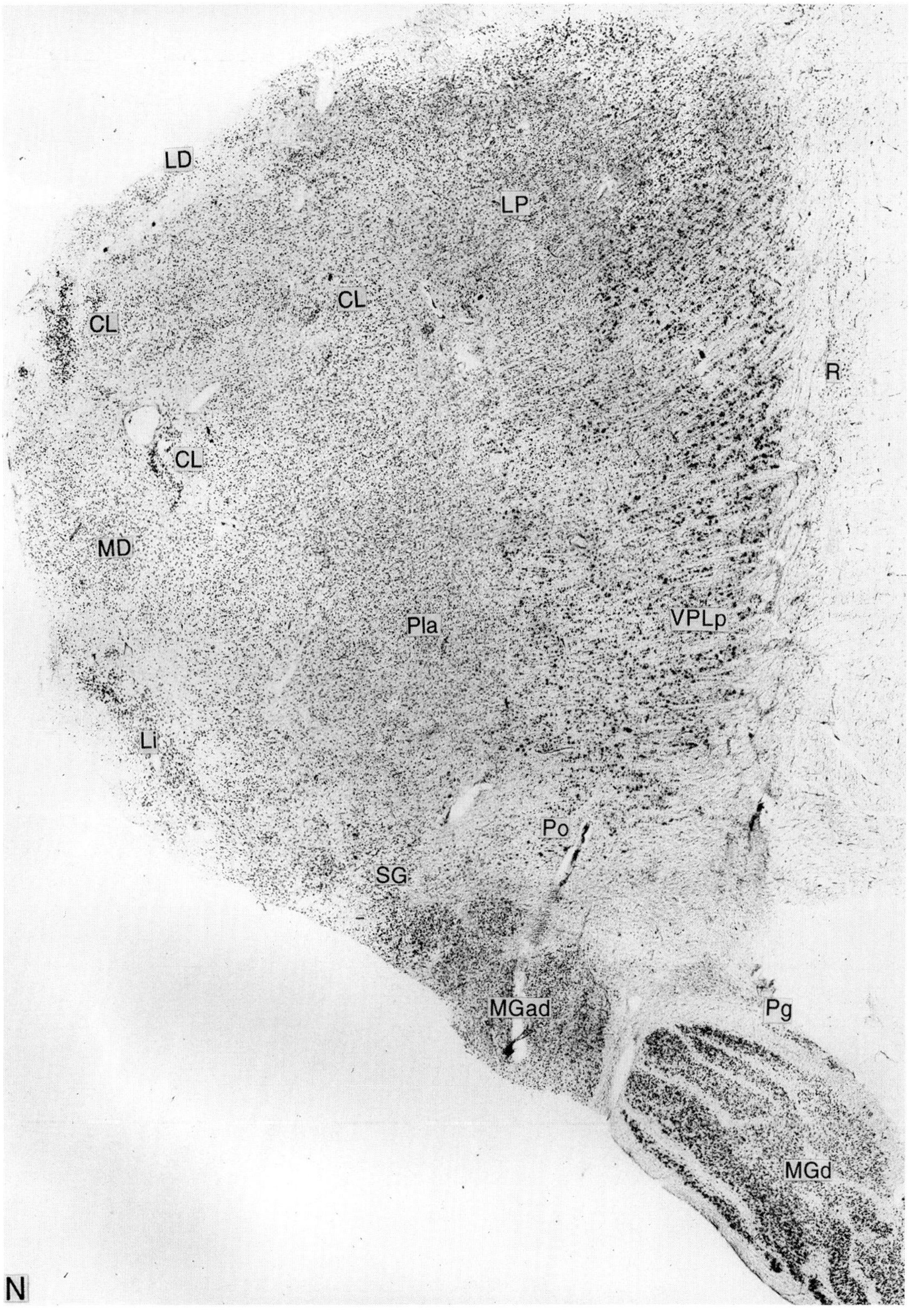

Fig. 6. (continued)

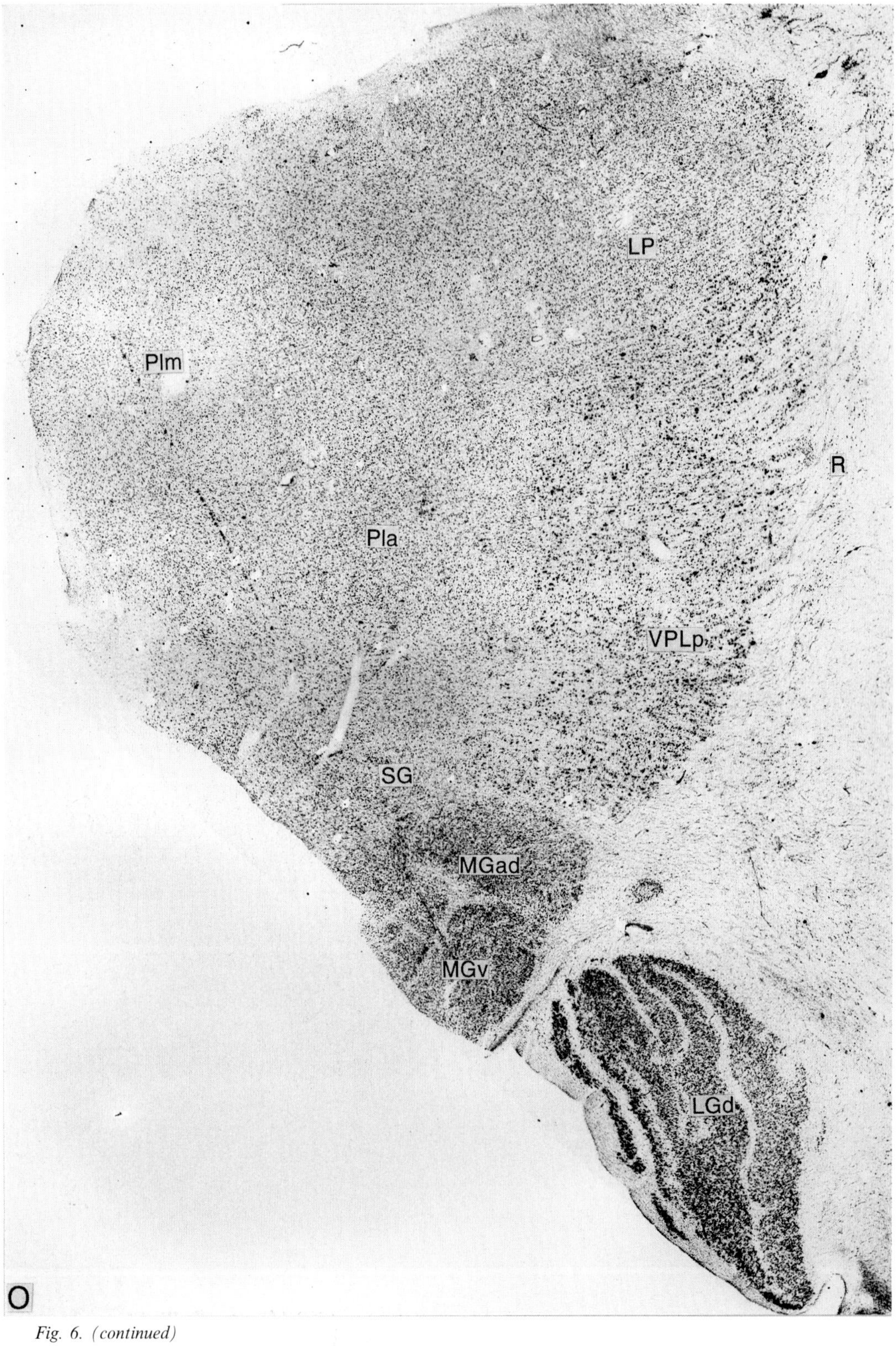

Fig. 6. (continued)

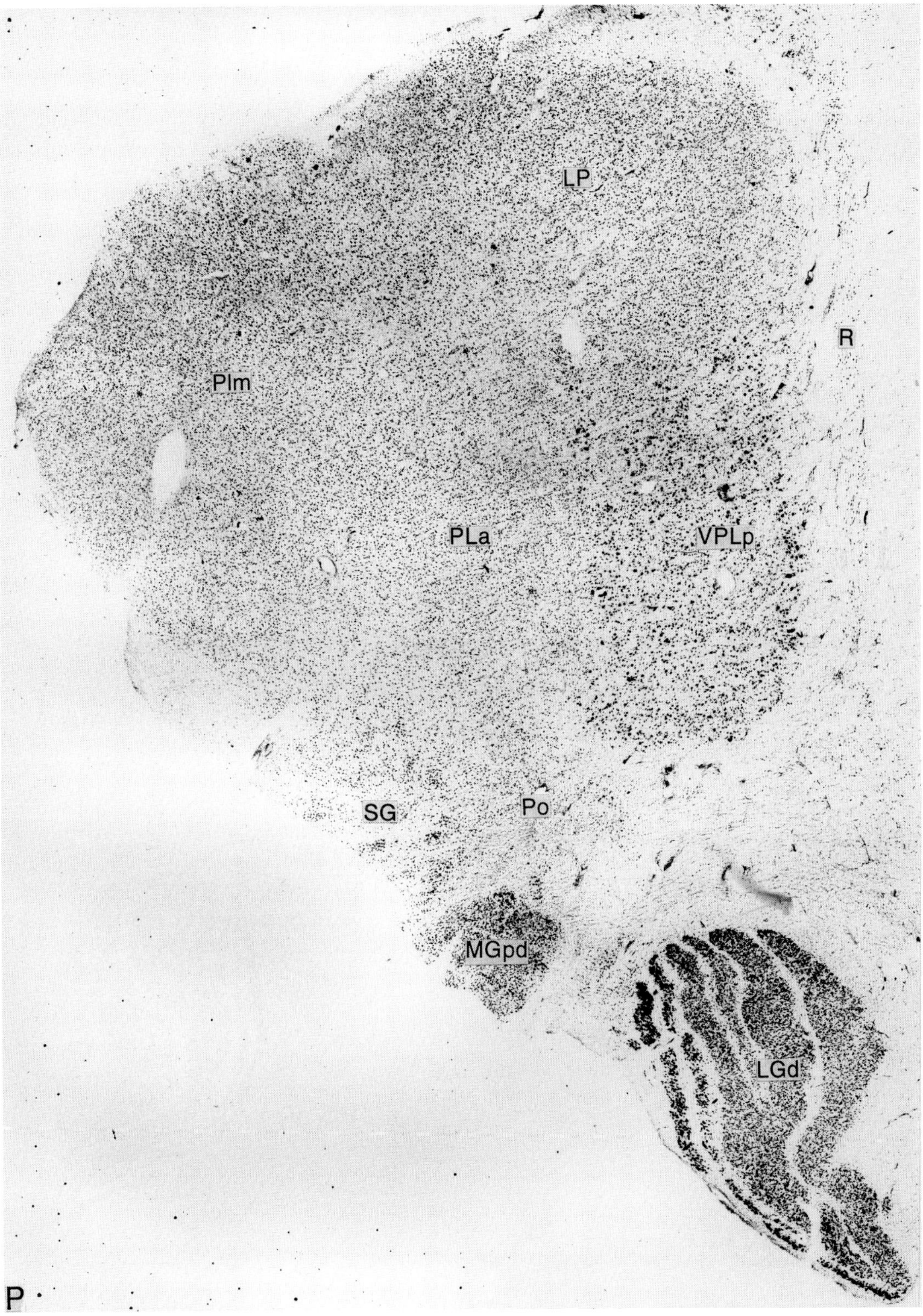

Fig. 6. (continued)

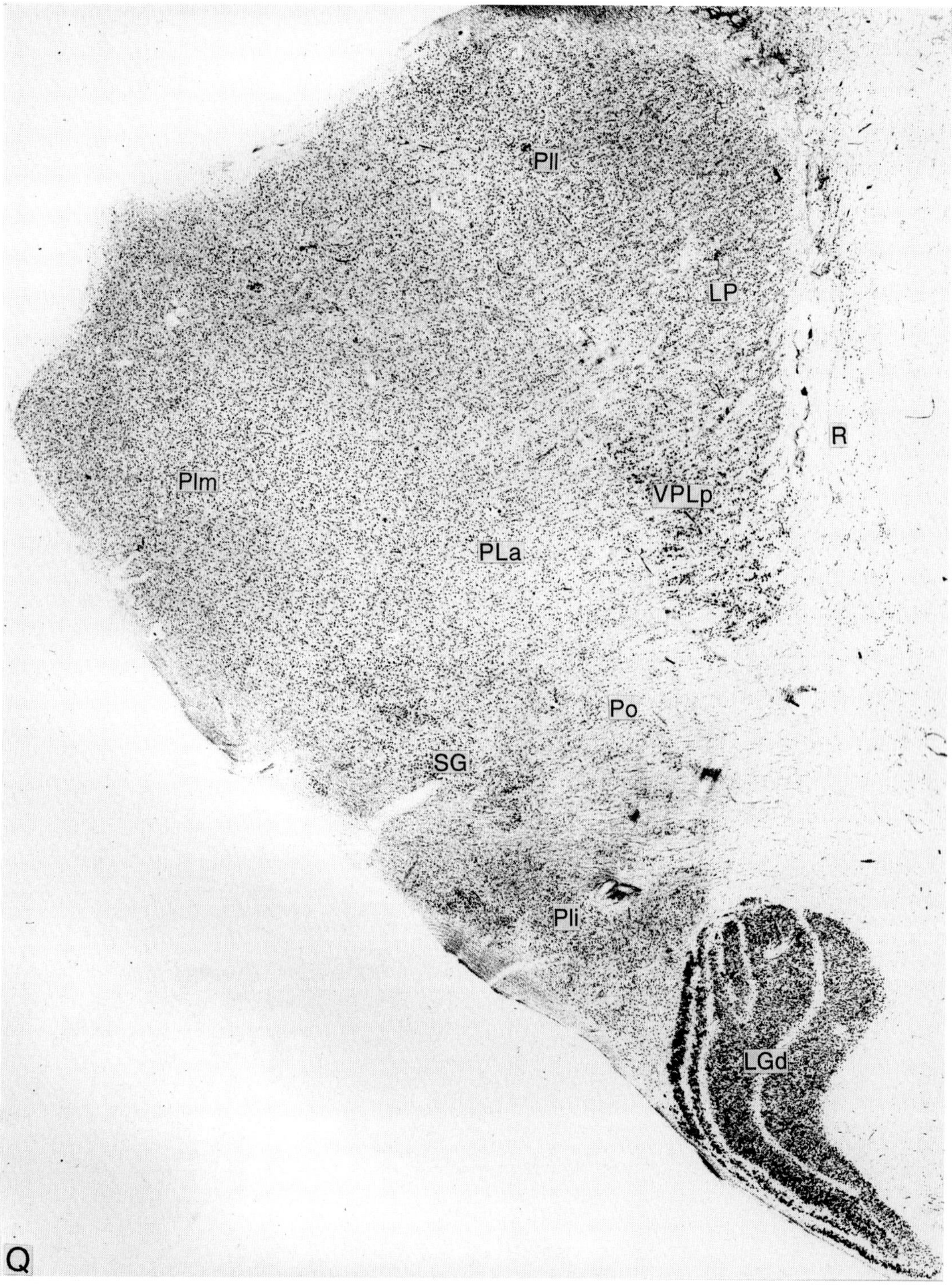

Fig. 6. (continued)

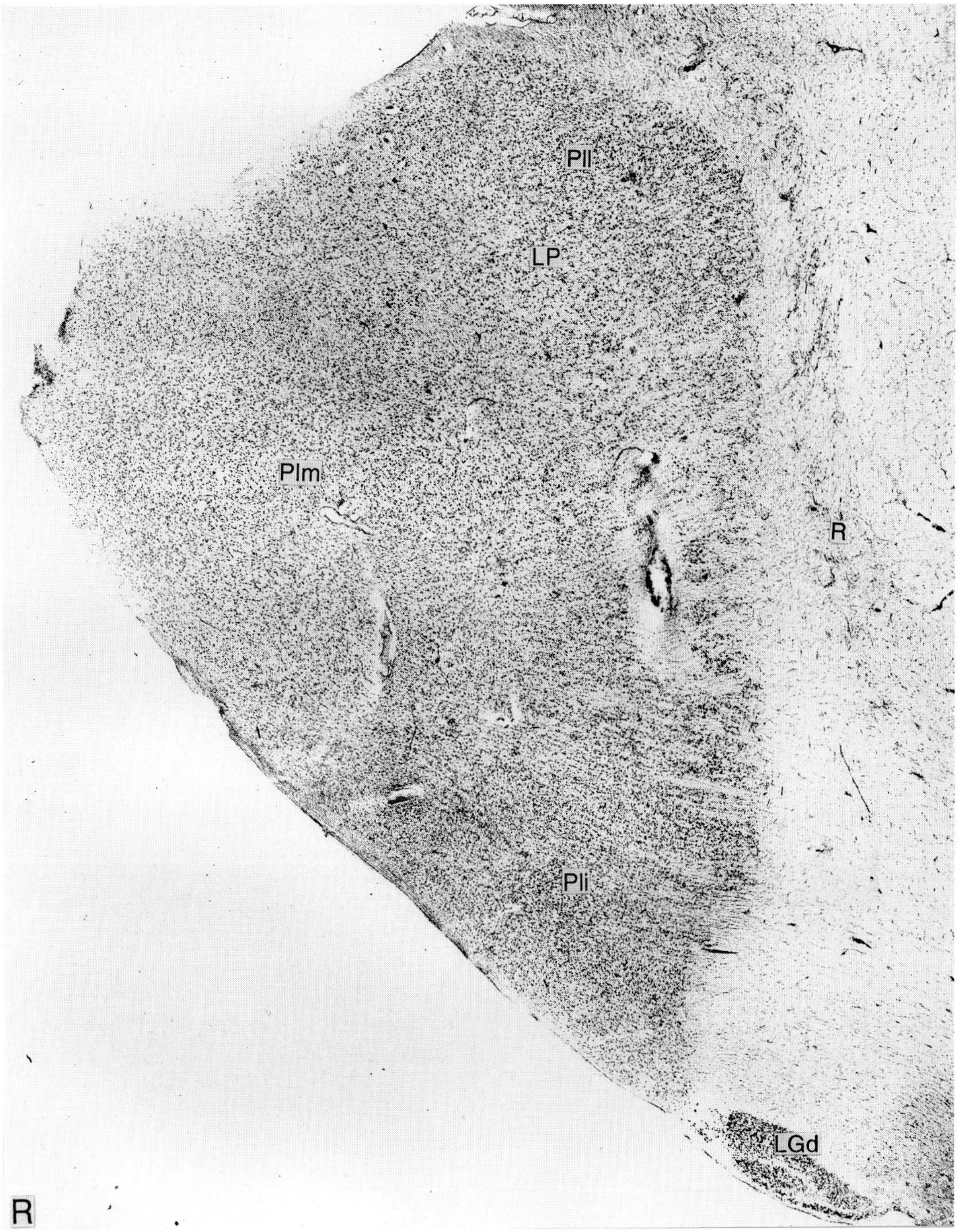

Fig. 6. (continued)

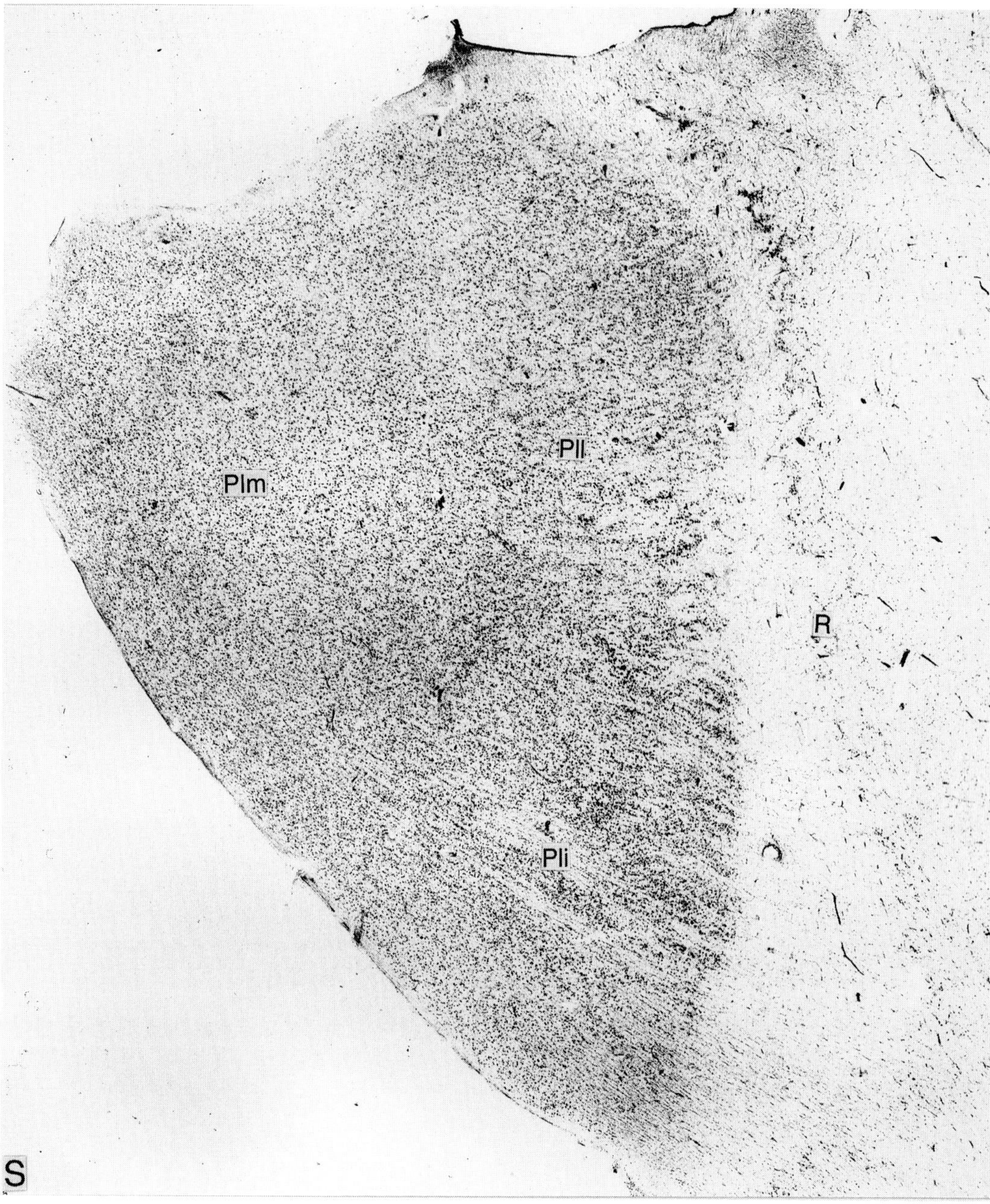

Fig. 6. (continued)

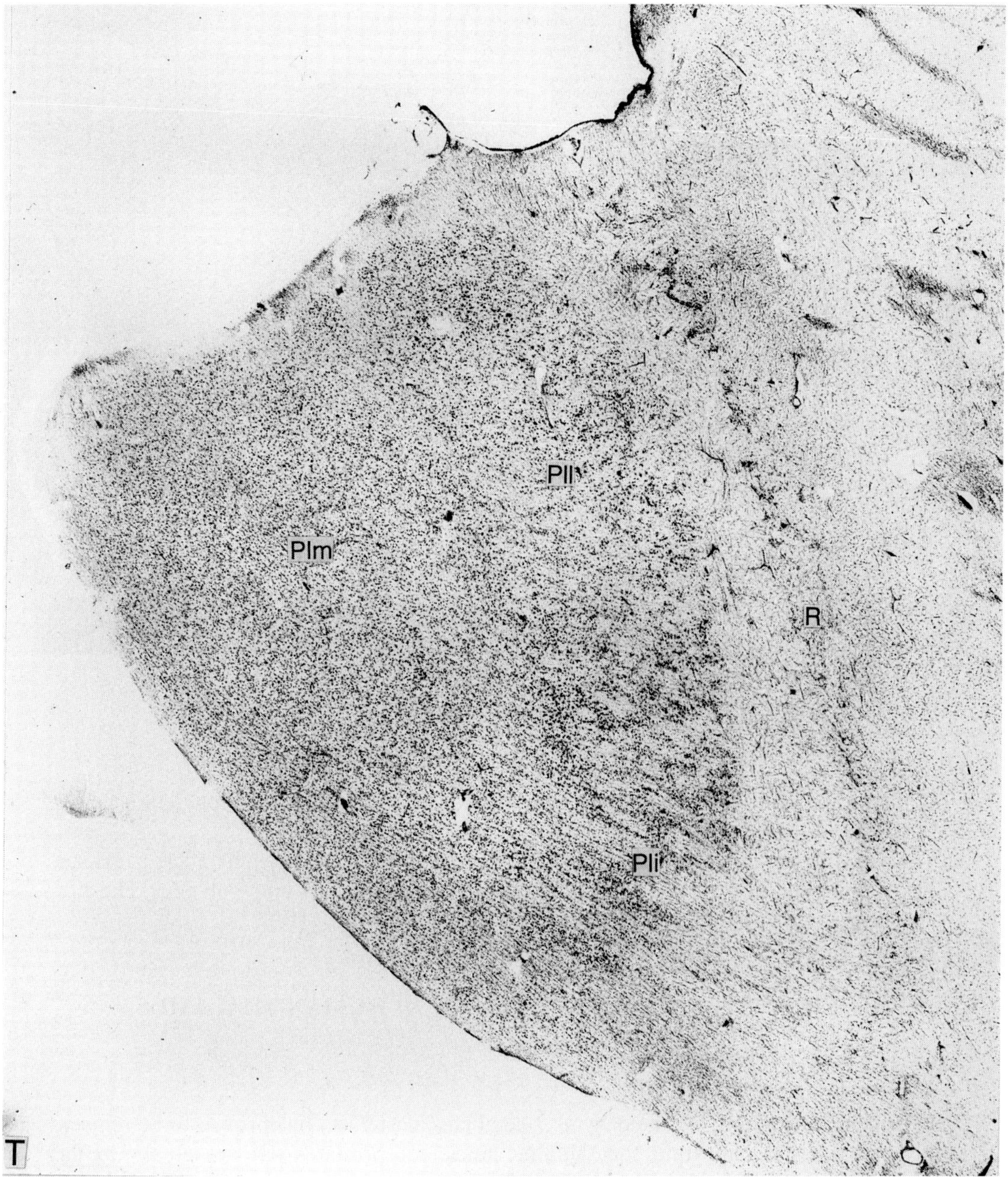

Fig. 6. (continued)

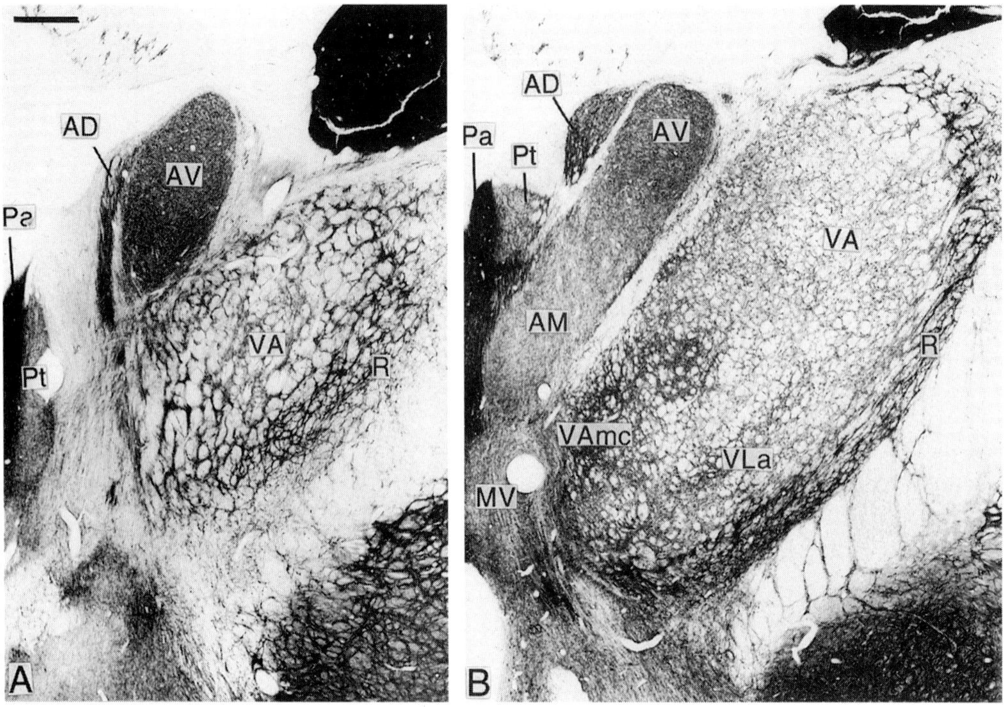

Fig. 7. Frontal sections in anterior (A) to posterior (N) order through the thalamus of a cynomolgus monkey (*Macaca fascicularis*), stained histochemically for acetyl choline-esterase activity. Bar 1 mm.

in monkeys) and other small cells intercalated between the principal laminae are clearly seen.

The ventral thalamus is also like that of monkeys with the reticular nucleus extending over the surfaces of the dorsal lateral geniculate nucleus and pulvinar. The pregeniculate nucleus is relatively small.

4. HISTOCHEMISTRY OF THE MONKEY AND HUMAN THALAMUS

4.1. INTRODUCTION

Histochemical staining for the enzyme, acetyl cholinesterase, has proven to be remarkably effective in delineating the thalamic nuclei of primates (Olivier et al., 1970a; Jones, 1985a; Hirai and Jones, 1989a). Each nucleus or functionally related group of nuclei has its own particular histochemical signature, based mainly on the intensity of staining. Differences in staining density can be extremely sharply defined at nuclear and subnuclear borders, often providing a better indication of the location of a border than Nissl- or myelin-stained sections, and in some instances providing confirmation of subnuclear parcellations suggested by connectional or physiological investigations but only indistinctly visible with conventional stains. In what follows, I shall provide a common description of the histochemical parcellation of the human and monkey thalamus, alluding to any differences in the course of the description but concentrating

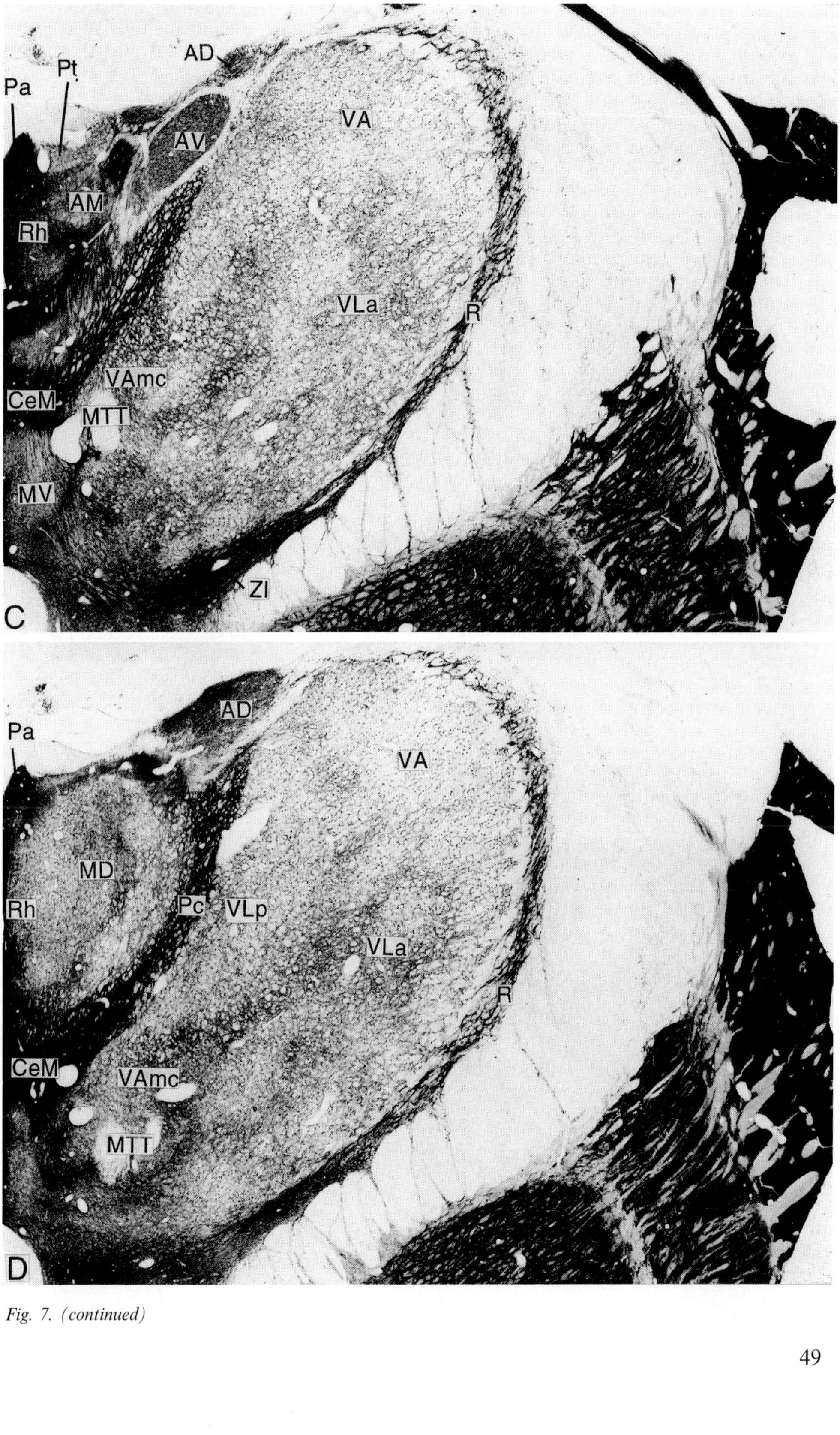

Fig. 7. (continued)

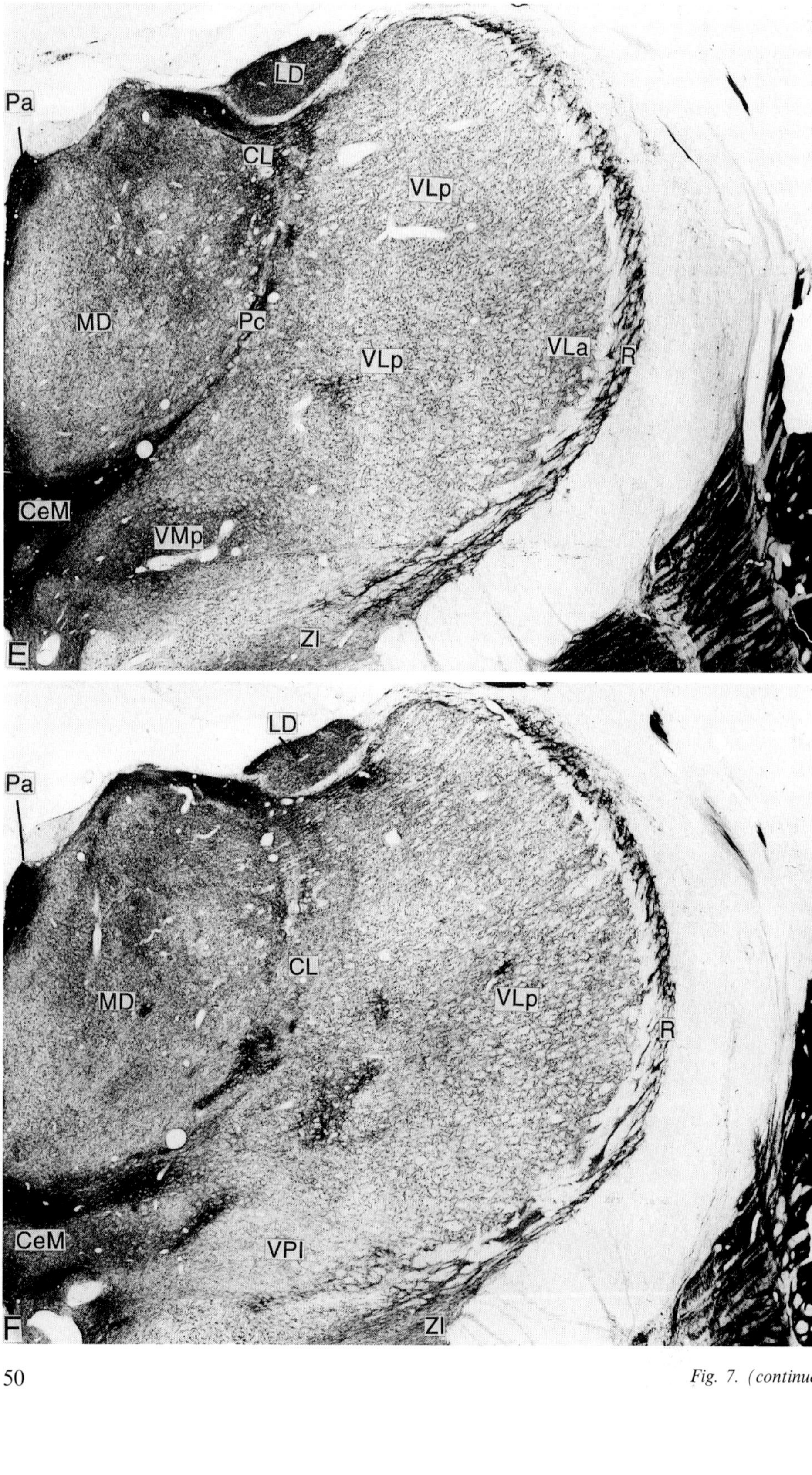

Fig. 7. (continued)

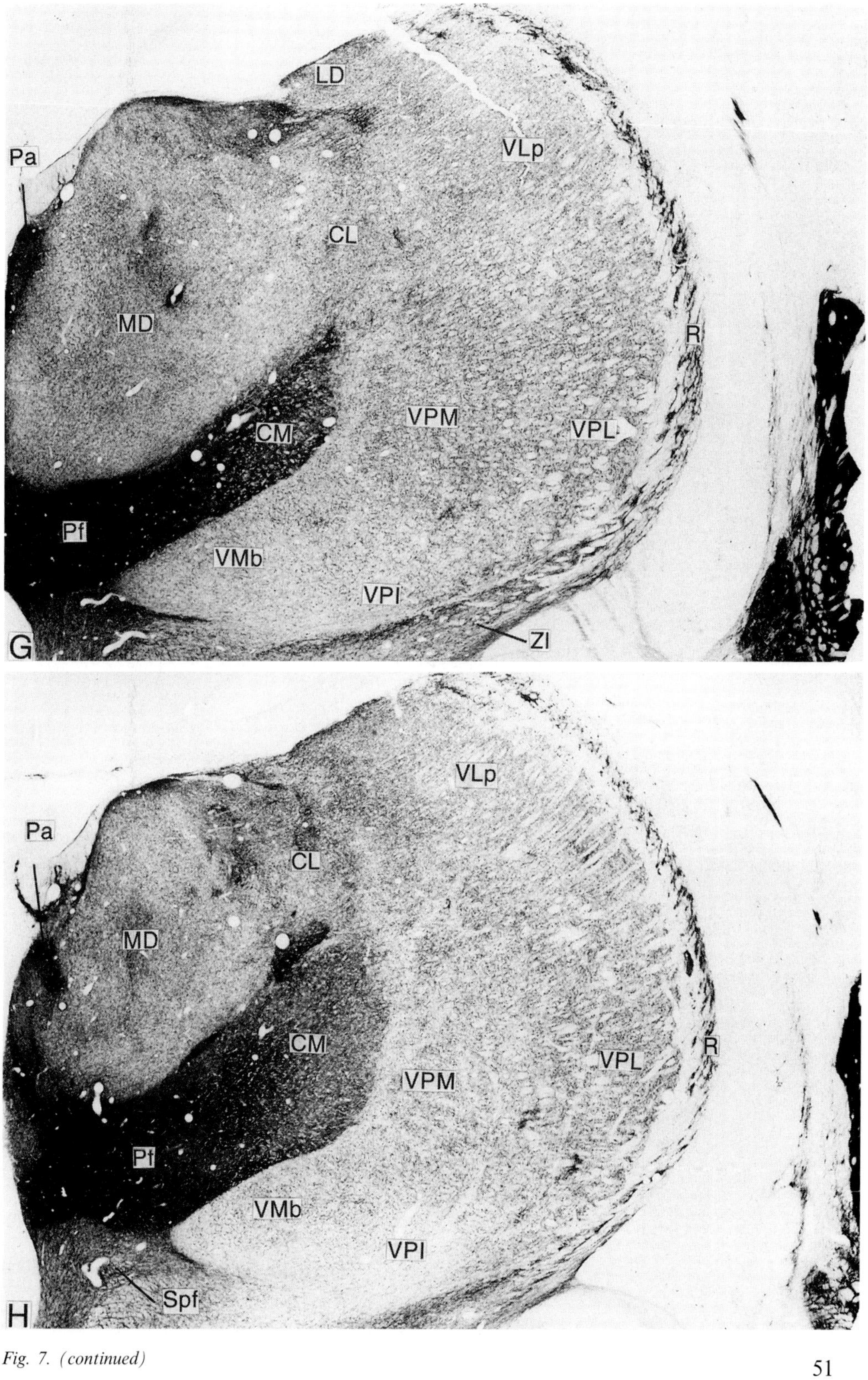

Fig. 7. (continued)

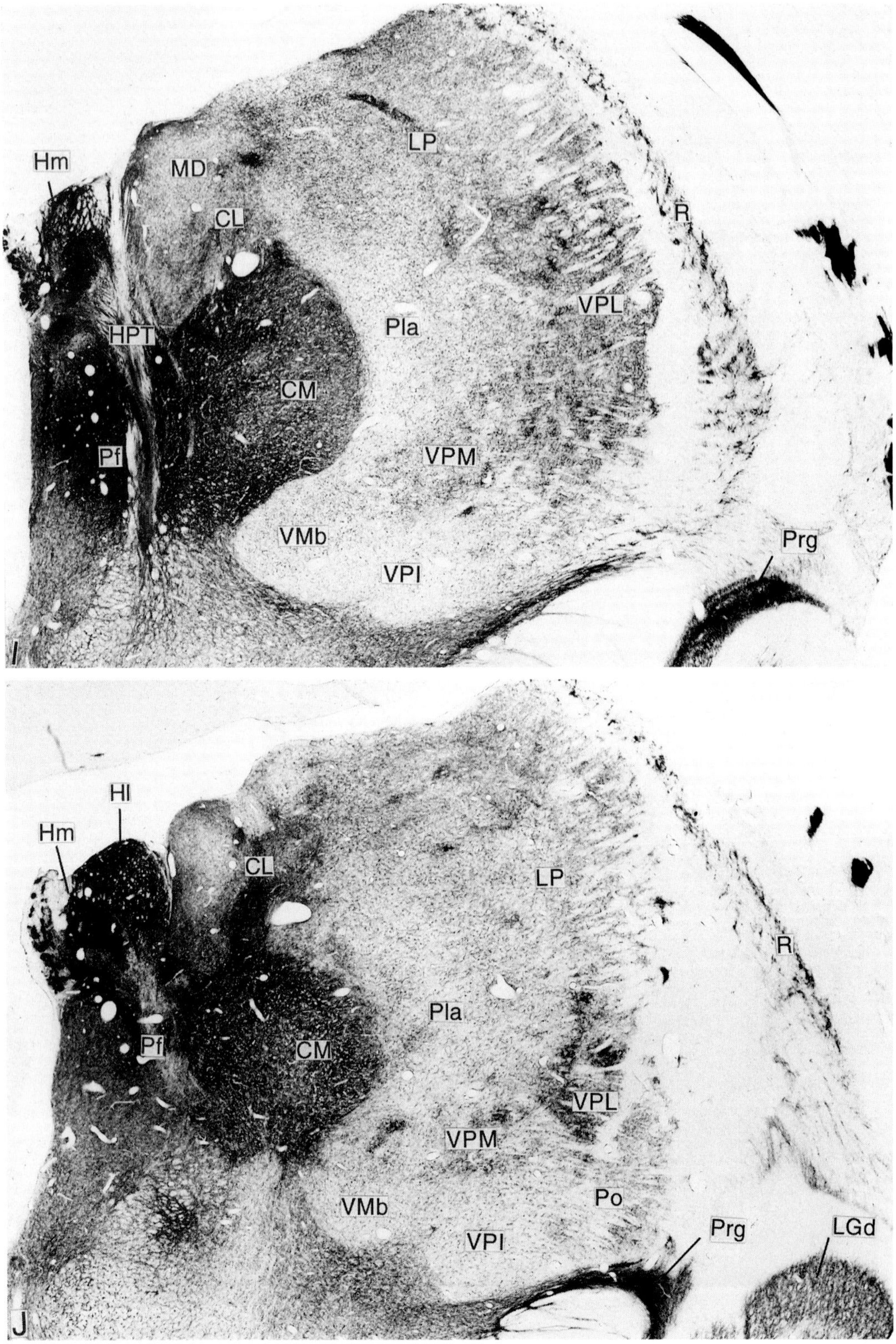

Fig. 7. (continued)

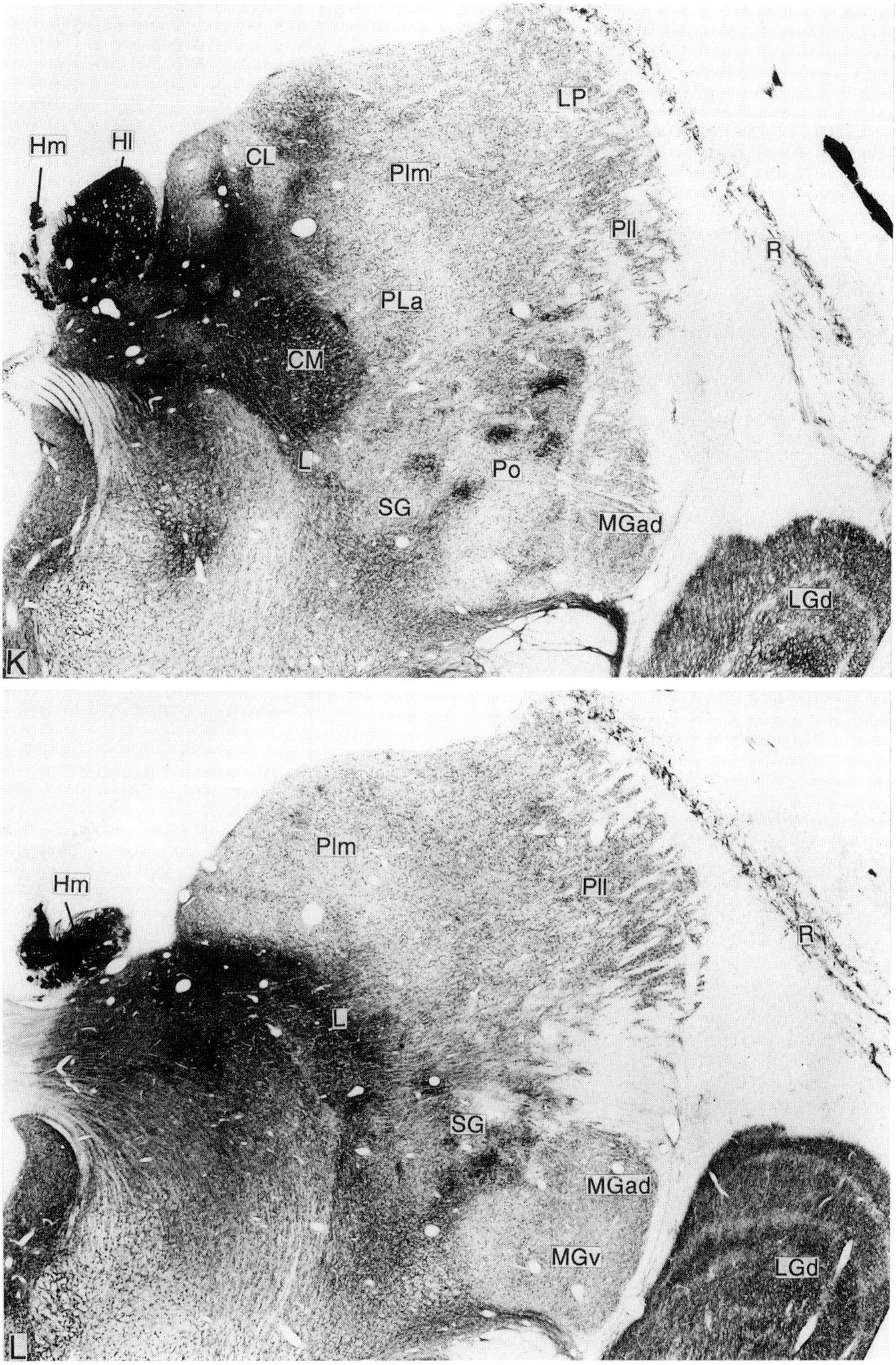

Fig. 7. (continued)

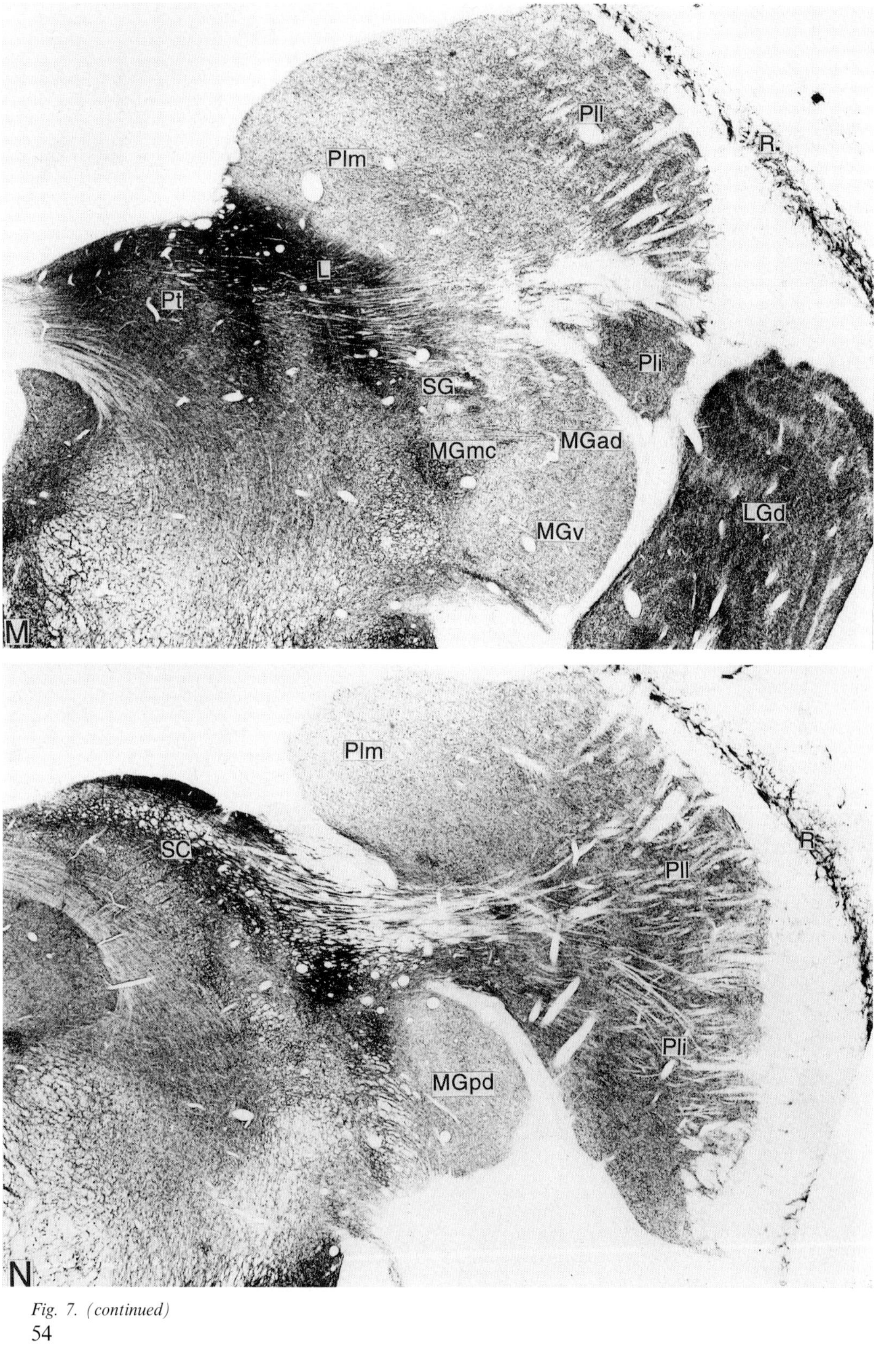

Fig. 7. (continued)

on the relationship of histochemical to cytoarchitectonic delineations of the thalamic nuclei.

4.2. INTERNAL MEDULLARY LAMINA AS AN ACETYL CHOLINESTERASE-STAINED LANDMARK

The internal medullary lamina has a dense and consistent pattern of acetylcholinesterase staining and forms a conspicuous landmark from which one can begin in dividing up the dorsal thalamus into its component nuclei. Commencing from the most anterior of a series of frontal sections through the monkey or human thalamus (Figs. 7, 8), the acetylcholinesterase-stained lamina can be seen at the midline, on the anterior surfaces of the central (rhomboid) and central medial nuclei. From there, it expands with its contained paracentral nucleus, over the anterior surface of the mediodorsal nucleus and, as the mediodorsal nucleus expands, this part of the lamina is reduced to a thin sheet on its lateral surface. Ventral to the mediodorsal nucleus, the lamina remains thick and coextensive with the central medial nucleus. The dorsal aspect of the thinned lateral part of the lamina divides into the characteristic Y-like configuration, embracing the ventral aspects of the anterior and lateral dorsal nuclei and expanding across the dorsal surface of the mediodorsal nucleus (Figs. 1, 7, 8).

From the region ventral to the mediodorsal nucleus, the lamina continues posteriorly on either side of the mamillothalamic tract. This region is relatively wide in

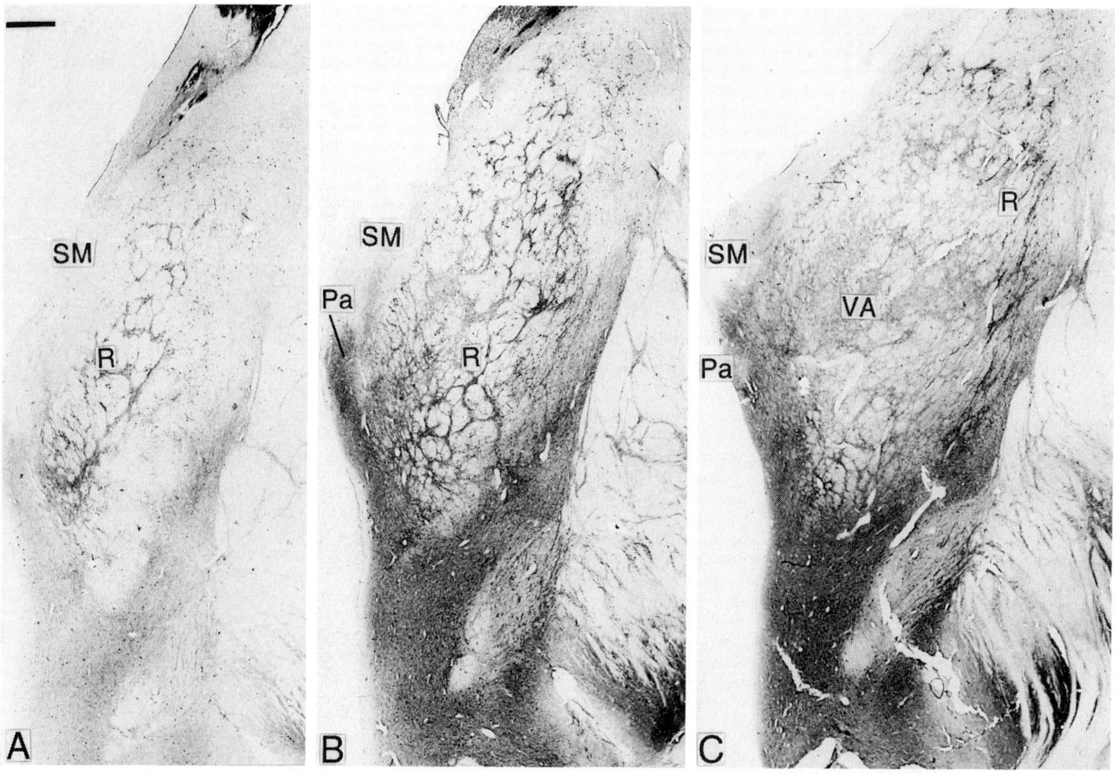

Fig. 8. Frontal sections in anterior (A) to posterior (Q) order through the thalamus of a human, stained for acetyl cholinesterase. Bar 1 mm.

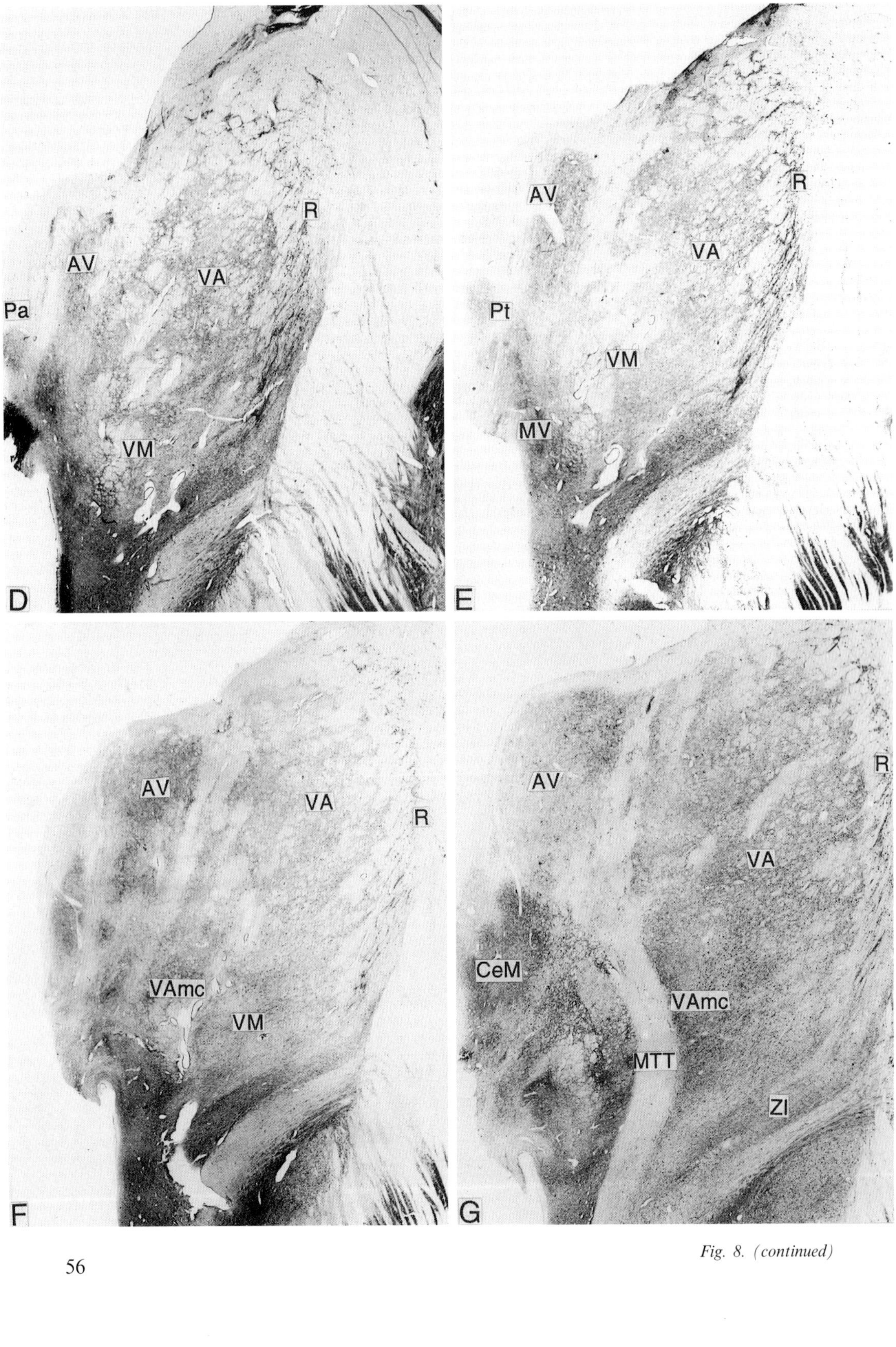

Fig. 8. (continued)

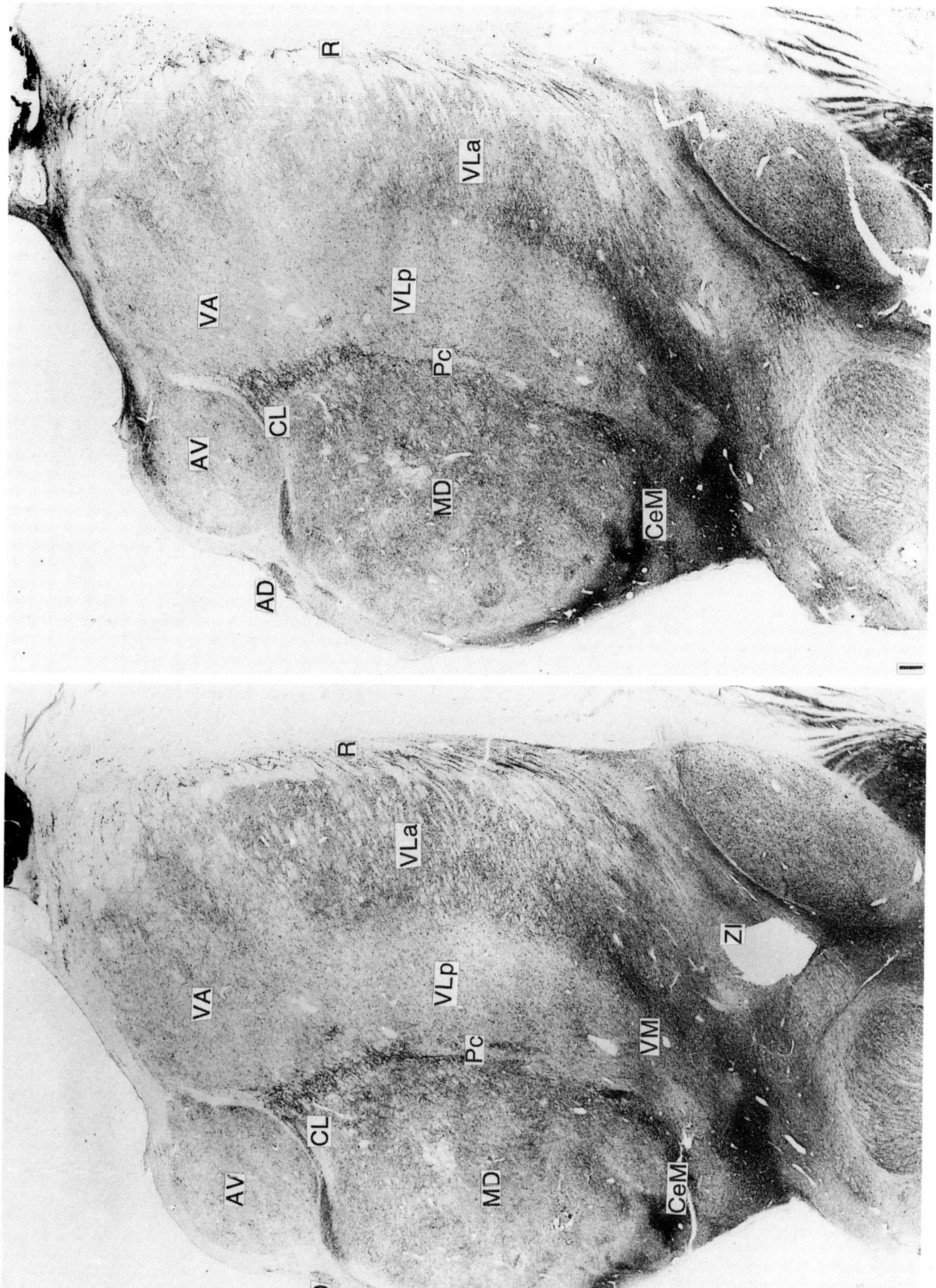

Fig. 8. (continued)

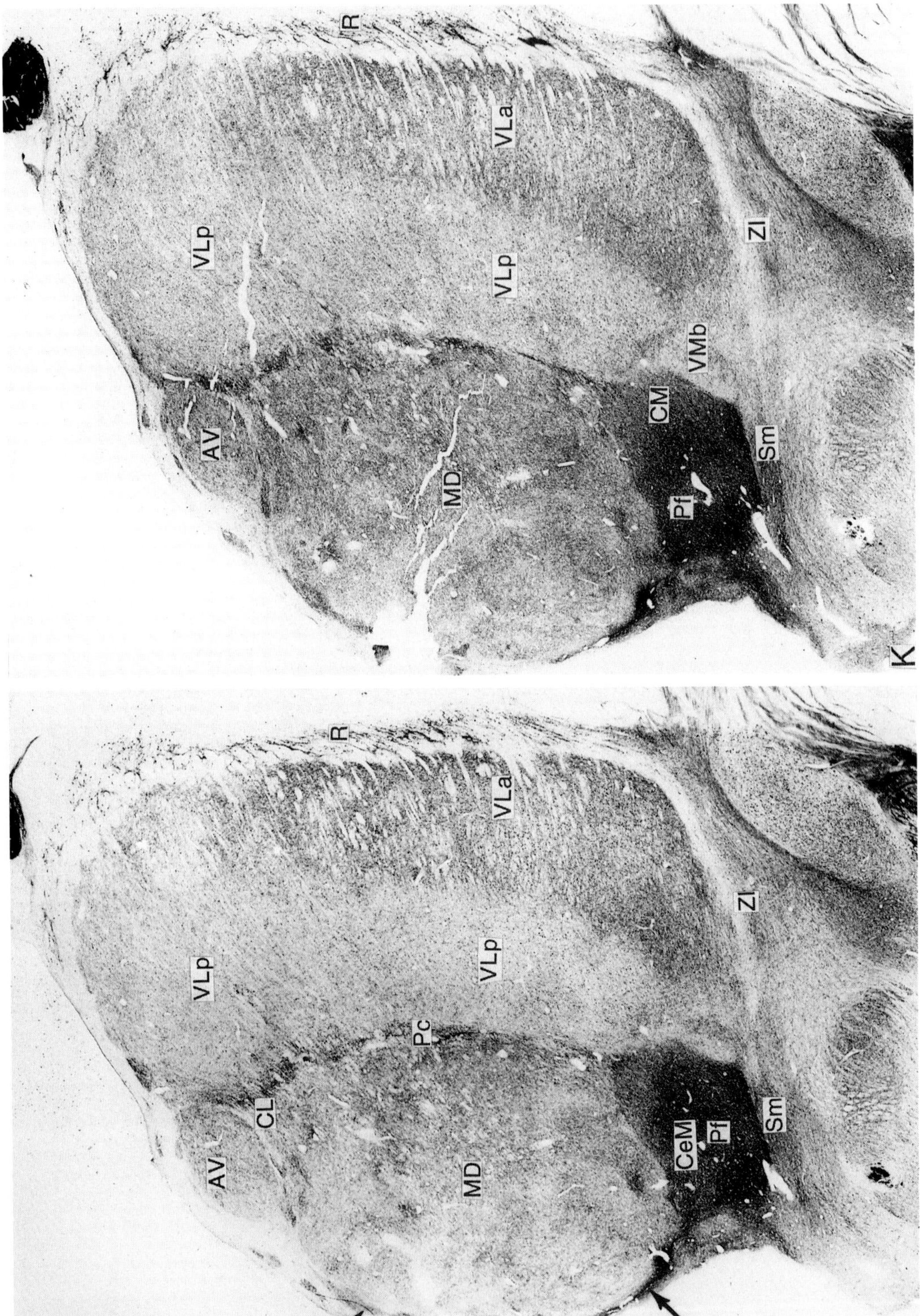

Fig. 8. (continued)

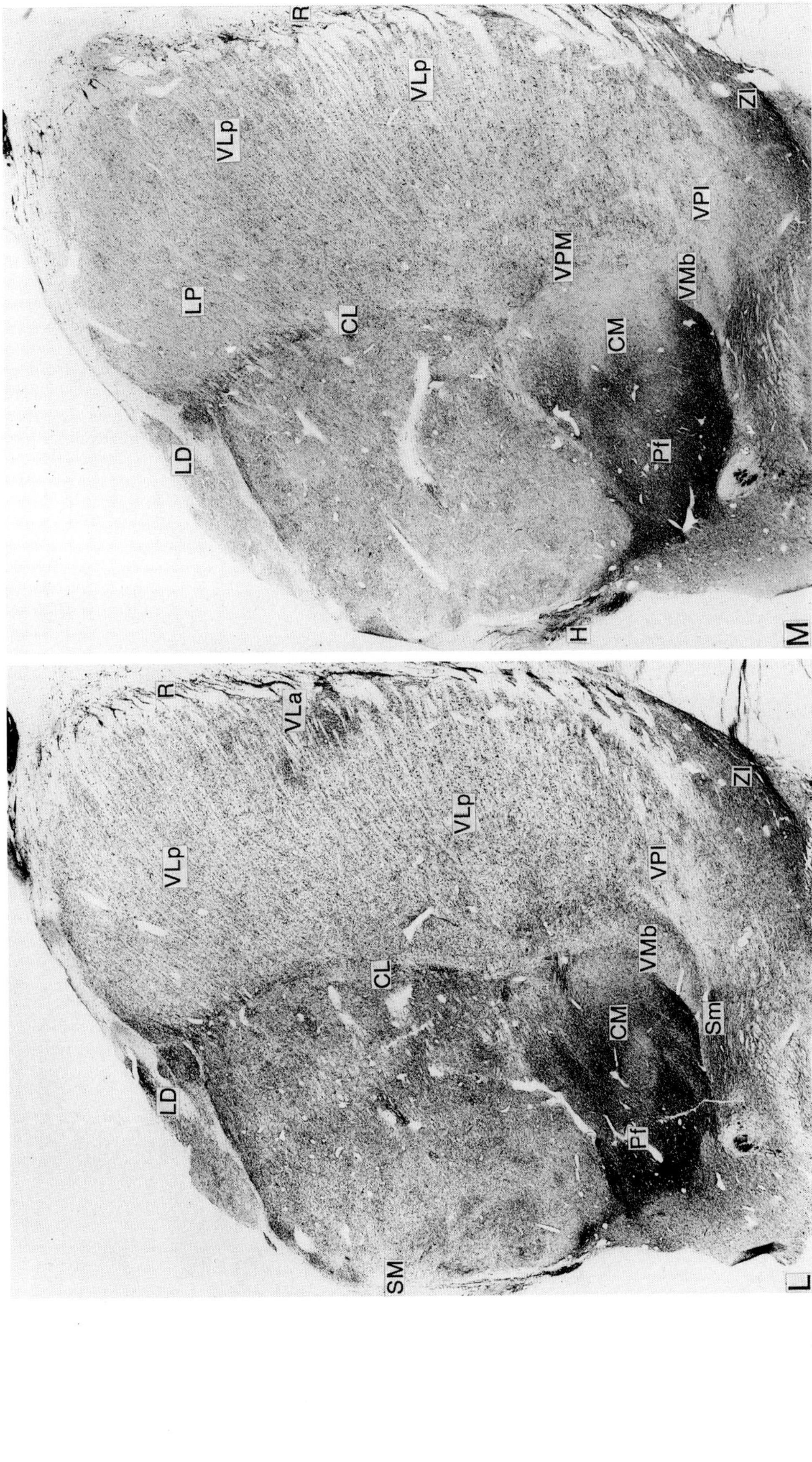

Fig. 8. (continued)

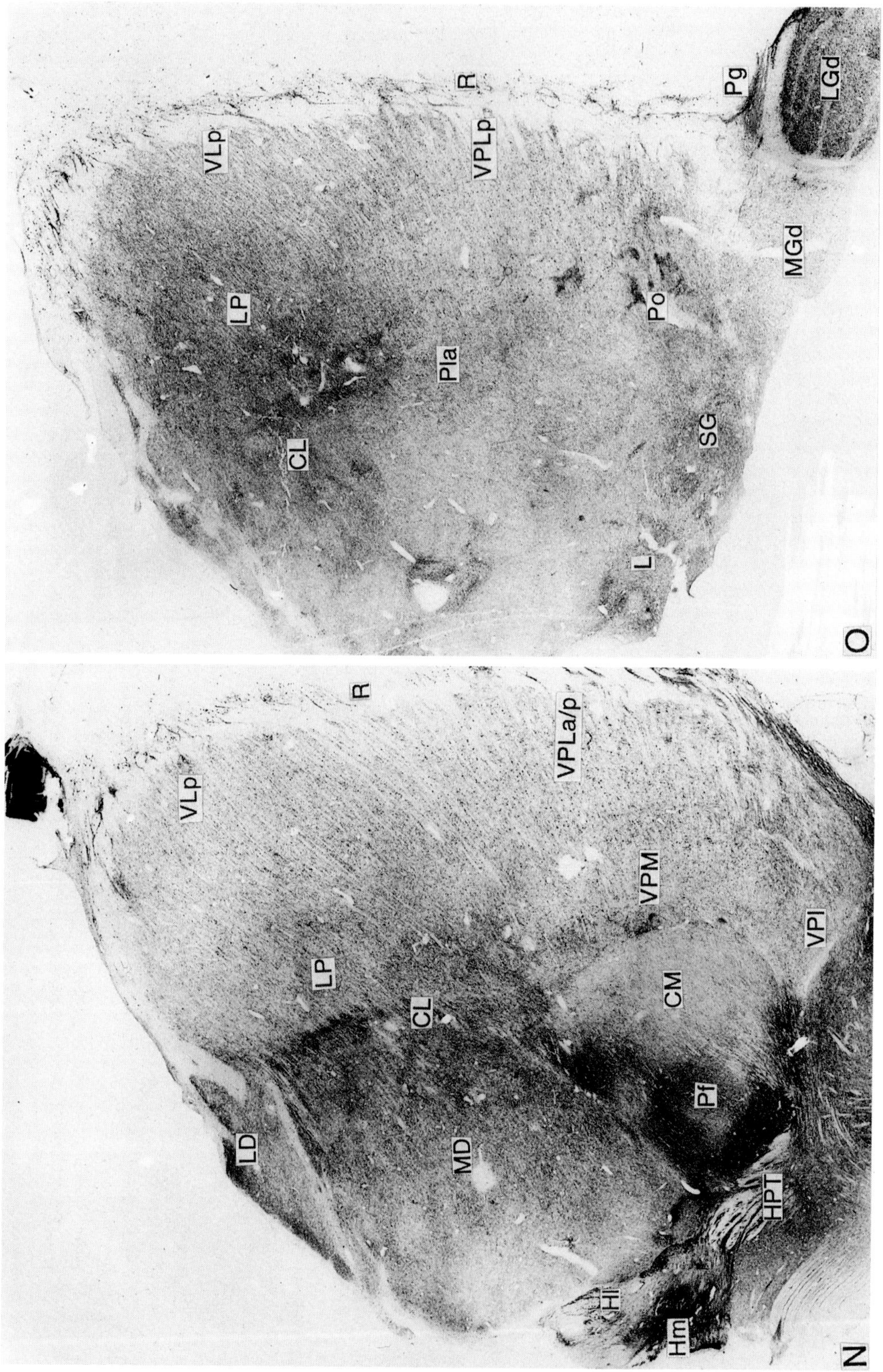

Fig. 8. (continued)

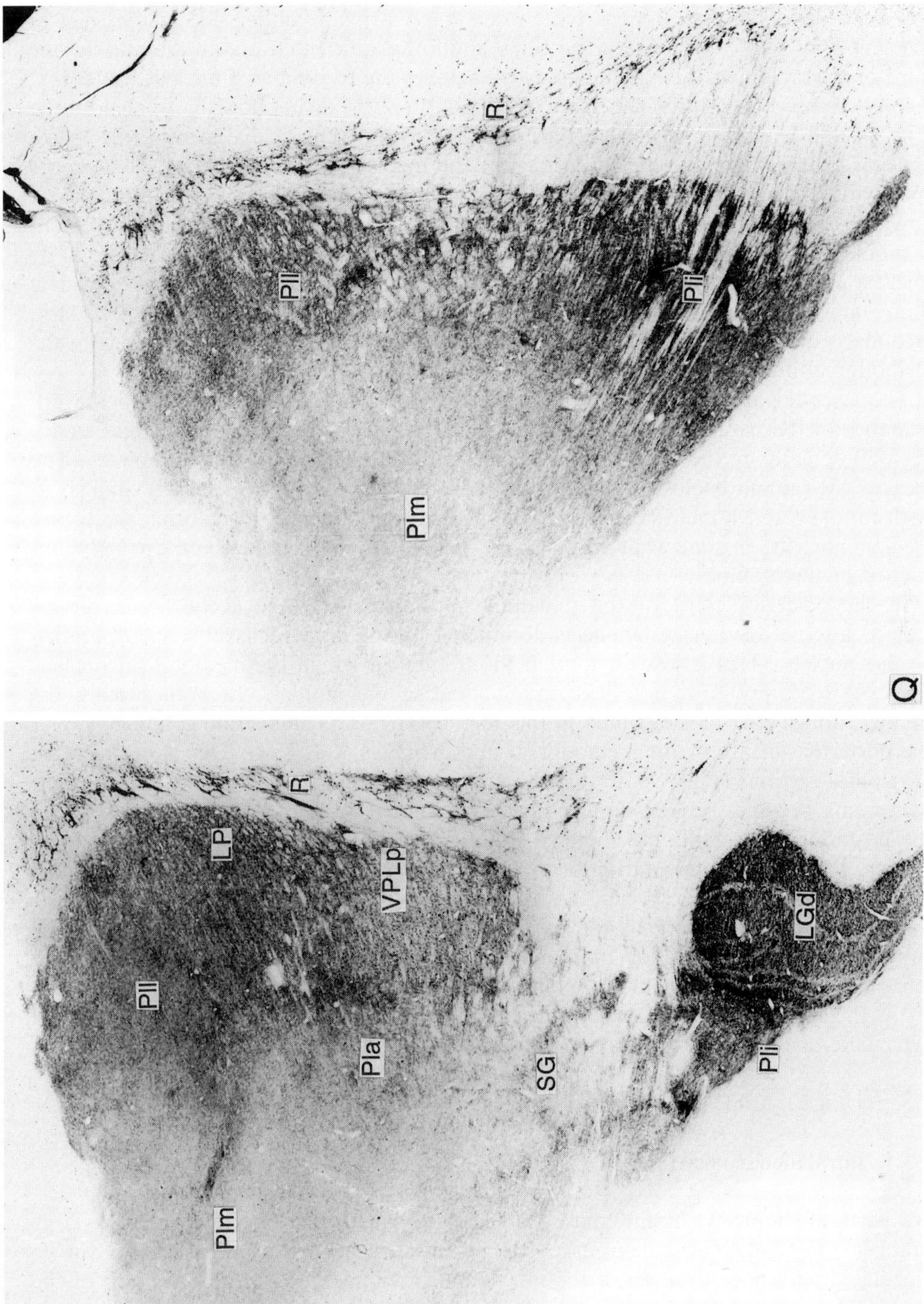

Fig. 8. (continued)

monkeys but reduced to thin partially interrupted strips on either side of the tract in humans (Figs. 7C, 8G). Posterior to the mamillothalamic tract, and in the middle one third of the thalamus, the lamina virtually encircles the mediodorsal nucleus. Laterally, it forms a thin but well-defined sheet which continues to divide dorsally to enclose the anterior and lateral dorsal nuclei (Figs. 7 D-F, 8 H-M). The medial arm of this Y-like division extends over the dorsal surface of the mediodorsal nucleus just beneath the ependyma of the third ventricle, reaching to the stria medullaris in the human and down between the fused mediodorsal nuclei of the two sides for a considerable extent in monkeys. Ventral, to the mediodorsal nucleus, the internal medullary lamina is at its thickest and fills this part of the thalamus (Figs. 7 E-H, 8 I-M). Medially, it extends a thin, stained strip up the medial surface of the mediodorsal nucleus, deep to the stria medullaris in the human and almost reaching the similar strip descending between the mediodorsal nuclei in the monkey.

The ventral part of the densely stained lamina expands even further with the appearance of the parafascicular and centre médian nuclei, the expanded region extending well dorsally up the lateral surface of the mediodorsal nucleus in the monkey. The part of the lamina enclosing the parafascicular nucleus is equally densely stained in both monkey and human (Figs. 7 G,H, 8 K,L). The lateral part enclosing the centre médian nucleus, is somewhat less densely stained in the monkey and virtually unstained in the human.

In the posterior third of the thalamus in the monkey, the acetylcholinesterase-stained internal medullary lamina is dominated by the large, densely stained parafascicular nucleus. This is less obvious in the human in which the mediodorsal nucleus narrows to its posterior pole less rapidly than in the monkey. In both species, the thinner, dorsal part of the lamina in this region extends as a covering completely over the posterior surface of the mediodorsal nucleus (Figs. 7 J, 8 O). This is best seen in horizontal sections (Figs. 9, 10). On the posterior surface of the mediodorsal nucleus, the lamina becomes broken up into clustered patches of stain that intervene between the posterior pole of the mediodorsal nucleus and the medial pulvinar nucleus. The clusters reflect the fragmentation of the internal medullary lamina that can be seen in myelin-stained sections, especially in humans (Fig. 1 E). The expanded ventral part of the lamina, corresponding to the parafascicular nucleus, extends as a line of interrupted clumps of stain along the junction of thalamus and pretectum, following the line of the limitans-suprageniculate nucleus and expanding laterally (Figs. 7 L, M, 8 O, P) in the region of the posterior nucleus and magnocellular nucleus of the medial geniculate complex.

4.3. NUCLEI ASSOCIATED WITH THE INTERNAL MEDULLARY LAMINA

4.3.1. Intralaminar nuclei

All parts of the acetylcholinesterase-stained, internal medullary lamina contain large numbers of Nissl-stained neurons. Aggregations of neurons of similar type can be correlated with named nuclei of the intralaminar complex. With the exception of the part enclosing the centre médian nucleus, all parts of the lamina contain relatively large, deeply staining, close-packed cells (Figs. 3, 6, 9, 10) typical of the intralaminar complex. Contained within the anterior, midline part of the lamina are the rhomboid (central) and central medial nuclei and cells of similar type are found within the midline extensions of the lamina between the mediodorsal nuclei. The cells of the central

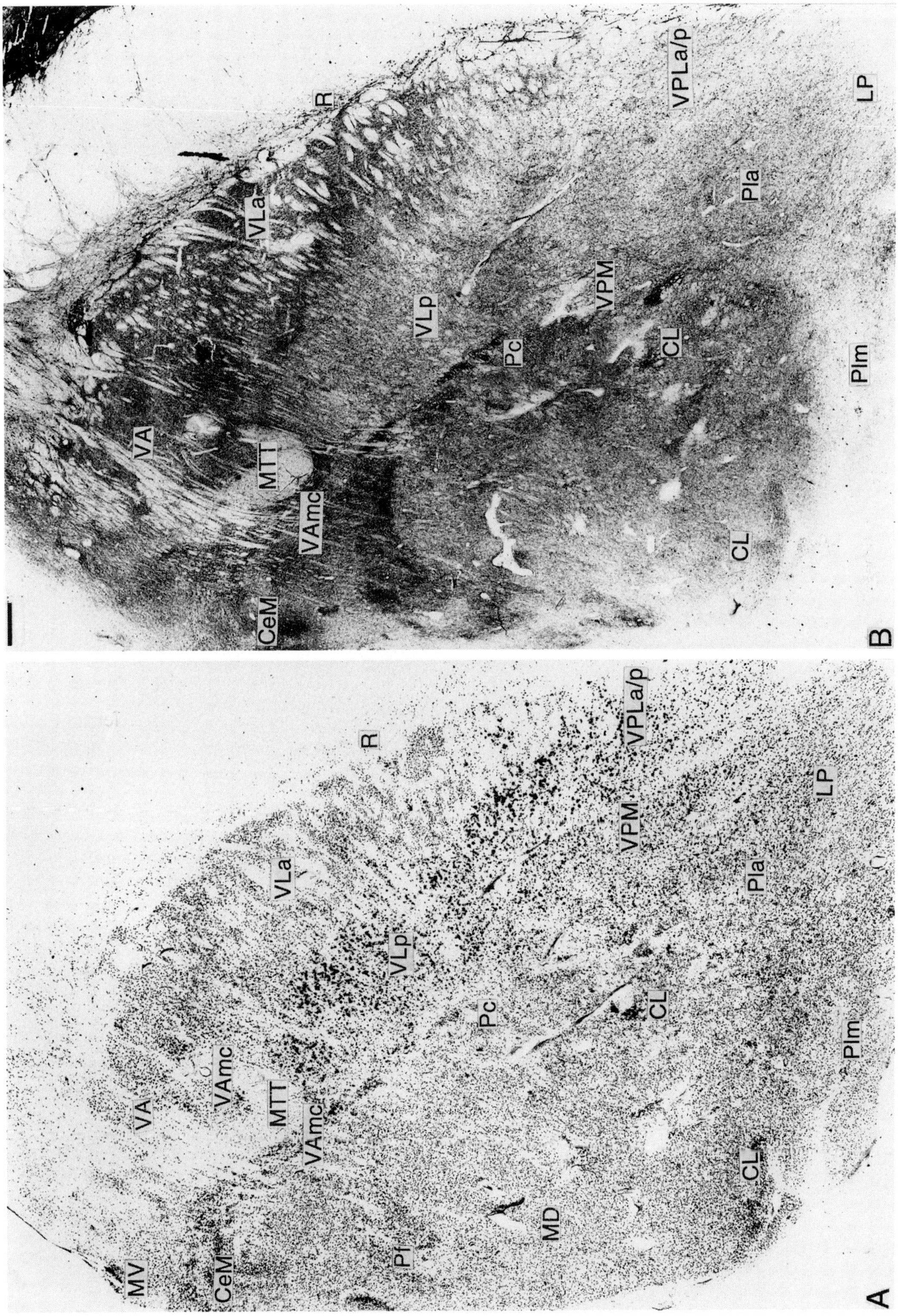

Fig. 9. Adjacent horizontal sections through a human thalamus, stained with thionin (A) or for acetyl cholinesterase (B). Note how the acetyl cholinesterase-rich internal medullary lamina breaks up posteriorly, carrying islands of cells of the central lateral nucleus (CL) over the posterior surface of the mediodorsal nucleus where they are commonly misinterpreted as part of the medial pulvinar nucleus. Bar 1 mm. C.f. Fig. 10.

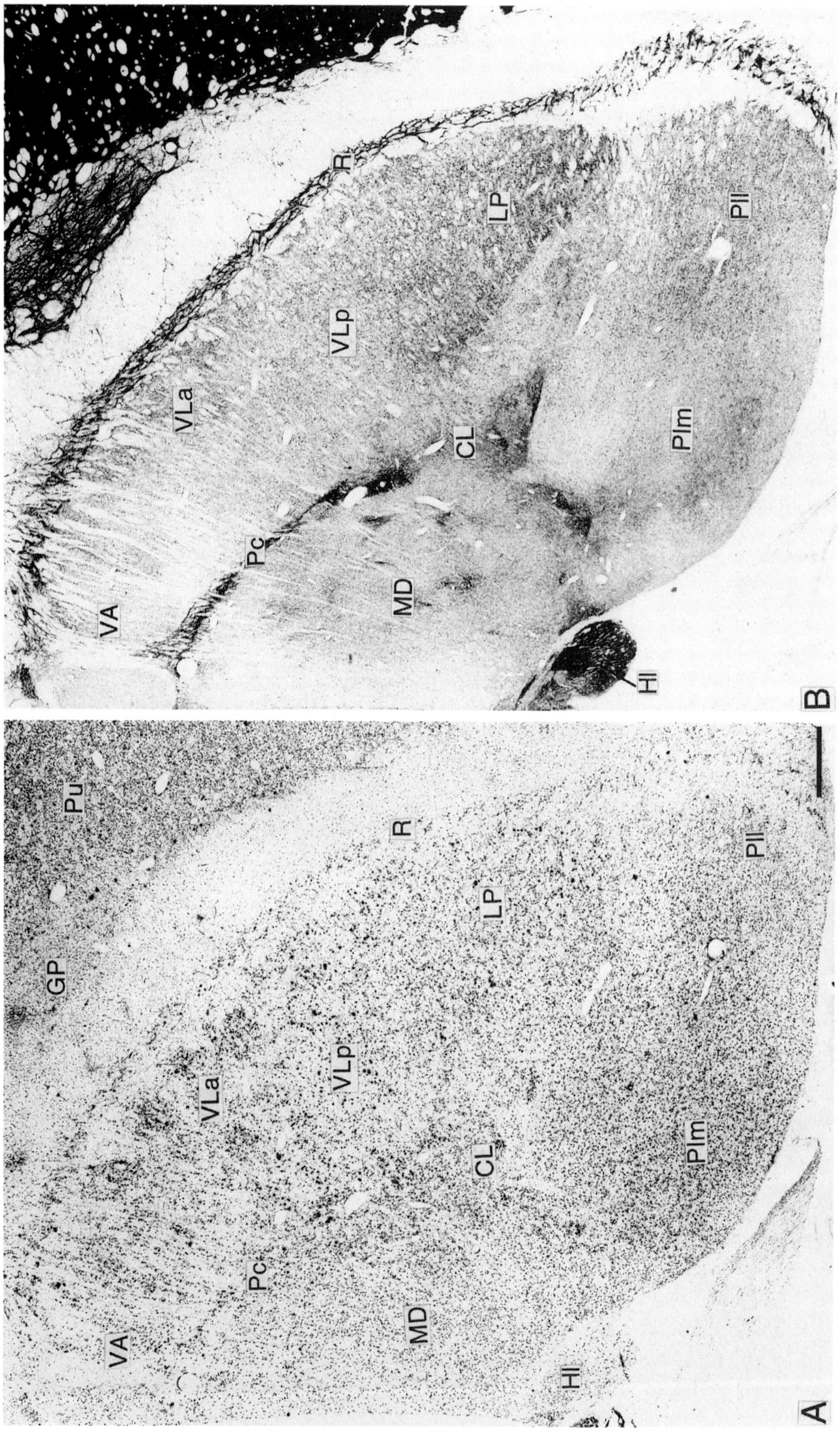

Fig. 10. Adjacent horizontal sections through the thalamus of a cynomolgus monkey (*M. fascicularis*), showing the encircling of the mediodorsal nucleus by the acetyl cholinesterase-rich internal medullary lamina and the central lateral nucleus. Bar 1 mm. C.f. Fig. 9.

medial nucleus are continuous posterolaterally with accumulations of large, deeply staining but more dispersed cells, coextensive with the patches of acetylcholinesterase staining on either side of the mamillothalamic tract (Figs. 3 C, 7 C). These patches, therefore, correspond to the magnocellular ventral anterior nucleus (VAmc).

The thinner parts of the acetylcholinesterase-stained internal medullary lamina covering the lateral surface of the mediodorsal nucleus and the ventral surfaces of the anterior and lateral dorsal nuclei, contain large, deeply staining cells corresponding to the paracentral and central lateral nuclei (Figs. 3 E-L, 7 C-H). The dorsal expansions of the lamina also contain the large, darkly stained cells typical of the central lateral nucleus. The intralaminar cells on the lateral surface of the mediodorsal nucleus (MD) are especially large posteriorly where they invade the lateral aspect of MD and extend around its posterior surface as densely stained clumps coextensive with the clumps of acetylcholinesterase staining present on that surface (Figs. 3 M, N, K, 6, 7). Posteriorly, the CeM nucleus expands within the extensive, deeply stained, ventral part of the internal medullary lamina (Figs. 3, 6–8). Further posteriorly, however, the cells associated with this part of the lamina become somewhat smaller and more densely packed, although still densely stained. They occupy the part of the acetyl cholinesterase-stained lamina surrounding the habenulo-peduncular tract and correspond to the parafascicular (Pf) and centre médian nuclei of the monkey and to the parafascicular nucleus and medial part of the centre médian nucleus in the human. The greater part of the centre médian nucleus in the human corresponds to the region of much reduced acetylcholinesterase staining, encapsulated by thin sheets of the internal medullary lamina. The patchy, posterolateral extension of the acetylcholinesterase-stained internal medullary lamina along the junction of thalamus and pretectum, is coextensive with islands of deeply stained, large and medium sized cells corresponding to the limitans (L) – suprageniculate (SG) nucleus. The magnocellular medial geniculate nucleus also seems to be in continuity with this part of the lamina.

4.3.2. Ventral nuclei

Most of these nuclei can be distinguished by different densities of acetylcholinesterase staining. The ventral anterior nucleus (VA) is filled by light-to-moderate staining and is perforated by numerous bundles of unstained axons. The region ventromedial to this nucleus is more homogeneously stained for acetylcholinesterase and extends, tongue-like, towards the mamillo-thalamic tract and under the internal medullary lamina as the principal ventral medial nucleus (VM). The so-called magnocellular ventral anterior nucleus (VAmc) is the most densely stained in this region (Figs. 7 B, C, 8 F, G).

The clumped acetyl cholinesterase staining corresponding to the VA nucleus is replaced posteriorly by two, radically different zones of staining. A medial zone adjacent to the internal medullary lamina, and extending to the dorsal surface of the thalamus shows very weak acetylcholinesterase staining. It corresponds to the ventral lateral posterior nucleus (VLp). A lateral, more heavily acetylcholinesterase-positive region corresponds to the ventral lateral anterior nucleus (VLa) (Figs 7 D, 8 H-K). As the VLp nucleus expands posteriorly, it retains its weak acetyl cholinesterase staining. There is no differentiation in terms of histochemical staining that follows the division of the nucleus into dorsal and ventral subnuclei on the basis of cell size.

Staining becomes somewhat enhanced with the appearance of the ventral posterior medial (VPM) and basal ventral medial (VMb) nuclei (Figs. 7G, 8L). Immediately lateral to these nuclei and lying adjacent to the external medullary lamina, the ventral posterior inferior nucleus (VPI) is distinguished as a zone of very weak acetylcholinesterase staining.

Lateral to VPM, the ventral posterior lateral nucleus (VPL) is also characterized by moderate to dense acetyl cholinesterase staining (Figs. 7 H, 8 N). In the human, there is no evidence of variations in the density of histochemical staining that would correspond to the division of the nucleus into anterior (VPLa) and posterior (VPLp) subnuclei.

4.3.3. Lateral posterior and pulvinar nuclei

The pale acetyl cholinesterase staining of the posterior extension of the dorsal division of VLp is replaced medially by a zone of medium density staining corresponding to the lateral posterior nucleus (LP) (Figs. 7 I, 8 M). Medial to LP and expanding across the posterior surface of the thalamus posterior to the internal medullary lamina where it encloses the posterior surface of the mediodorsal nucleus, is a large region of extremely weak acetylcholinesterase staining that corresponds to the medial pulvinar nucleus (Plm). Anterolateral to this, the anterior pulvinar nucleus (Pla) is distinguished as a region of slightly enhanced acetylcholinesterase staining, approaching in density that of LP. Posterior to LP is a region of slightly enhanced acetylcholinesterase staining that fills most of the lateral half of the posterior surface of the thalamus and is coextensive with the lateral pulvinar nucleus (Pll) (Figs. 7 K, 8 P). Ventrally, and extending on to the posterior surface of the thalamus, the inferior pulvinar nucleus (Pli) shows acetylcholinesterase staining denser than in any of the nuclei of the lateral mass. Variations in density of staining within it and adjacent parts of the lateral pulvinar nucleus appear to correlate with terminal zones of pretectal and tectal afferents (Lysakowski et al., 1986).

4.3.4. Anterior nuclei and lateral dorsal nucleus

Three major regions of moderate acetylcholinesterase staining, each separated from the others by offsets of the internal medullary lamina, can be distinguished (Figs. 7 A-D, 8 D-K). The largest, corresponding to the anteroventral nucleus in monkeys and to the merged anteroventral (AV) and anteromedial (AM) nuclei in humans, is a rounded zone, tapering to a narrow tail on the dorsal surface of the mediodorsal nucleus. In humans, the acetylcholinesterase staining is somewhat heavier in its dorsolateral aspect, suggesting a division of the nuclei comparable to that in the monkey that is not evident in Nissl-stained preparations. At the medial edge of the AV nucleus in humans and at its dorsomedial edge in monkeys, the anterodorsal nucleus is characterized as a small focus of moderate acetylcholinesterase staining.

As the AV nucleus narrows to its posterior pole, a new zone of moderately dense acetylcholinesterase staining corresponding to the lateral dorsal nucleus appears in its place and continues posteriorly towards the medial pulvinar nucleus in both species (Figs. 7 E, 8 L).

4.3.5. Medial nuclei

The mediodorsal nucleus is not homogeneous in its acetylcholinesterase staining (Figs. 7 D-I, 8 H-N). Throughout, there are irregular patches of lighter and denser staining and anterior parts are more lightly stained than posterior parts. These inhomogeneities do not obviously correlate with the major cellular subdivisions of the nucleus, except that the most lateral ('densocellular') part, here regarded as part of the central lateral nucleus, is consistently more densely stained than any part of the mediodorsal nucleus proper.

The medioventral (reuniens) nucleus (MV) is typically a region of relatively dense but patchy acetylcholinesterase staining. The small parataenial nucleus (Pt) beneath the stria medullaris is weakly stained in monkeys (Fig. 7 C) and not separately identifiable in humans.

4.3.6. Medial and lateral geniculate nuclei

The medial geniculate complex (MG) shows weak acetylcholinesterase staining overall (figs. 7 K-N, 8 O). The magnocellular and dorsal medial geniculate nuclei show enhanced staining in comparison with the ventral medial geniculate nucleus which has one of the lowest densities of acetylcholinesterase staining seen in the thalamus.

The dorsal lateral geniculate nucleus (GLd) has very dense acetylcholinesterase staining overall with no marked differences among its laminae (Figs. 7 J-M, 8 O, P).

4.3.7. Posterior complex

The limitans (L) – suprageniculate (SG) and posterior (Po) nuclei, which are densely stained for acetyl cholinesterase, are described with the internal medullary lamina (Figs. 7 K-M, 8 O, P).

4.3.8. Ventral thalamus

Reticular (R) and pregeniculate (Prg) nuclei and the zona incerta (ZI) and fields of Forel (FF), all display heavy staining for the enzyme.

4.3.9. Epithalamus

The paraventricular (Pv) and habenular nuclei (H) are densely stained, the first as a compact linear streak lying ventromedial to the stria medullaris. This can be traced, tenuously, into a larger zone of dense acetylcholinesterase staining that is coextensive with the medial habenular nucleus (Hm). The lateral habenular nucleus (Hl) is only lightly stained.

4.4. INTEGRATION OF NOMENCLATURES APPLIED TO HUMAN AND MONKEY THALAMIC NUCLEI

The nomenclatures customarily used for describing the human and monkey thalami, while having the same historical roots, have diverged during the course of their evolution to the extent that they are barely recognizable as referring to what cannot be other than homologous structures (Table 2). Many of the names used in either species were introduced in the first detailed myeloarchitectonic (Vogt, 1909) and cytoarchitectonic (Friedemann, 1911) studies, carried out probably on sections from the same cercopithecine monkeys in the Berlin Brain Research Institute. Many of Friedemann's names for the Nissl-stained nuclei have survived in the nomenclature of the monkey thalamus which became codified by Aronson and Papez (1934) and Crouch (1934),

TABLE 2. *A concordance of the nomenclature used to designate the human thalamic nuclei by Hassler (1959) and by Hirai and Jones (1989a). Equivalent nuclei in the macaque monkey, as delineated by Olszewski (1952), are shown in the right column and the nomenclature of the middle column is applicable to both species.*

Human (Hassler, 1959)	Hirai and Jones (1989a)	Monkey (Olszewski, 1952)
N. Anterior principalis (A.Pr.)	AV	AV
N. Anteromedialis (A.m.)	AM	AM
N. Anteroinferior (A.if.)	Part of AM	Alaris
N. Anteroreuniens (A.r.)	Part of AM	Part of AM
N. Anterodorsalis (A.d.)	AD	AD
N. Medialis dorsalis (M)	MD	MD
fibrosus (M.ft)	mc	mc
fasciculosis (M.fa.)	lateral	pc
caudalis (M.c.i. and M.c.e.)	ventral	mf
paralamellaris (M.pL.)	part of CL	dc
N. Endymalis (Edy)	MV (Reuniens)	Reuniens
N. Parataenialis (Pt)	Pt	Part of Cdc
parvocellularis (Pt.p.c.)		
interstitialis (Pt.ist.)		
Envelope (Involucrum medial)	Internal medullary lamina	Internal medullary lamina
N. Fasciculosus (Fa)	Part of MV	?
N. Intralamellaris (iLa)	CL and Pc	CL and Pc
oralis (iLa.o)	ant. CL & Pc	Part of Clc & Pc
ventralis (iLa.v.)	mid CL	CL
interpolaris (iLa.p.)	mid Cl	CL •
caudalis (iLa.c.)	post. CL	(MD dc)
N. Commissuralis (Co)	CeM	CLc, Cdc, Cif, Cim
N. Cucullaris (Cu)	Dorsomedial CL	Csl & Cs
N. Centralis thalami (Ce)	CM	CM
parvocellularis (Ce.pc)	(CM proper)	
magnocellularis (Ce.mc)	(part merging with Pf)	
N. Parafascicularis (Pf)	Pf	Pf
N. Limitans (Li)	Li-Sg	
medialis (Li.m.)	CL post. to MD	
opticus (Li.opt)	Li	Li
portae (Li.por)	Sg (and Po)	Sg
N. Ventro-caudalis posterior externus (V.c.p.e.)	VPLp	VPLc
N. Ventro-caudalis anterior externus (V.c.a.e.)	VPLa	

(continued)

TABLE 2. *(continued)*

Human (Hassler, 1959)	Hirai and Jones (1989a)	Monkey (Olszewski, 1952)
N. Ventro-caudalis anterior internus (V.c.a.i.)	VPM	VPM
N. Ventro-caudalis portae (V.c.por)	Pla	Plo
N. Ventro-caudalis parvocellularis internus (V.c.pc.i.)	VMb (and Sm)	VPMpc
N. Ventro-caudalis parvocellularis externus (V.c.pc.e.)	VPI	VPI
N. Zentrolateralis caudalis (Z.c.) internus (Z.c.i.) externus (Z.c.e.)	Posterodorsal parts of VPLa	Part of VPL
N. Ventro-intermedius externus (V.im.e.) N. Ventro-intermedius internus (V.im.i.)	} Ventral parts of VLp	} VPLo
N. Dorso-intermedius externus (D.im.e.) N. Dorso-intermedius internus (D.im.i) N. Dorso-intermedius superior (D.im.s.)	} Dorsal parts of VLp	} VLc
N. Dorso-lntermedius externus magno-cellularis (D.im.e.mc)		VLps
N. Zentrolateralis intermedius (Z.im) internus (Z.im.i) externus (Z.im.e)	Parts of VLp/VPLa	Parts of VLc/VPLc
N. Ventro-oralis posterior (V.o.p.)	VLa/ part of VLp	VLo
N. Ventro-oralis anterior (V.o. a.)	VLa	
N. Ventro-oralis medialis (V.o.m)	VM	VLm
N. Ventro-oralis internus (V.o.i)	Anteromedial part of VLp	X
N. Zentrolateralis oralis (Z.o)	Part of VLa	Part of VLo
N. Dorso-oralis (D.o) internus (D.o.i) externus (D.o.e.)	} Parts of VA	} ?
N. Lateropolaris (L.po) externus (L.po.e) basalis (L.po.b) internus (L.po.i) superior(L.po.s)	} VA	} VA
magnocellularis (L.po.mc)	VAmc	VAmc
N. Dorsocaudalis (D.c)	LP	LP

(continued)

TABLE 2. *(continued)*

Human (Hassler, 1959)	Hirai and Jones (1989a)	Monkey (Olszewski, 1952)
N. Dorsalis superficialis (D.s)	LD	LD
N. Pulvinaris superficialis (Pu.sf)	Parts of LD/LP/Plm	?
N. Pulvinaris lateralis (Pu.l)	Pll	Pll
N. Pulvinaris medialis (Pu.m) internus (Pu.m.i) dorsalis (Pu.m.d,) zentralis (Pu.m.z) ventralis (Pu.m.v)	Plm	Plm
N. Pulvinaris ventralis (Pu.v) N. Pulvinaris intergeniculatus (Pu.ig) N. Pulvinaris suprabrachialis (Pu.sb)	Pli	Pli
N. Pulvinaris oralis (Pu.o) orolateralis (Pu.o.l) oromedialis (Pu.o.m) oroventralis (Pu.o.v.)	Mostly LP and dorsal parts of Pla Dorsal Pla	Mostly LP and dorsal parts of Plo Plo
Corpus Geniculatum laterale (G.l)	GLD	GLD
N. Praegeniculatus griseus (pG.) griseus (pG.gr) fibrosus (pG.fi)	Pg GLvo GLmc	
Corpus Geniculatum mediale (G.m)		
N. Geniculatus medialis (G.m) fasciculosus (G.m.fa) fibrosus (G.m.fi) magnocellularis (G.m.mc) limitans (G.m.li)	 MGv MGd MGmc	 GMpc GMmc
Reticulatum thalami (Rt)	R	R
Zona incerta (Zi)	ZI	ZI
N. Paramedianum oralis (Pm.o) principalis (Pm) caudalis (Pm.c)	Pv Pa Pac	
N. Subhabenularis (S.Hb)		
N. Habenularis medialis (Hb.m)	Hm	Hm
N. Habenularis lateralis (Hb.l) magnocellularis (Hb.l.mc) interstitialis (Hb.l.ist) mixtocellularis (Hb.l.mix)	Hl Hlmc Hlpc	

working under the influence of the Michigan school of neuroanatomy, and popularized in the experimental studies of Walker (1938). Many of the names of Vogt for the myelin-stained nuclei, on the other hand, formed the basis of the expansion of the human nomenclature commenced by Vogt and Vogt (1941) and completed by their student, Hassler (1959), possibly because myelin staining remained the prevalent method for histologically examining the postmortem human brain. Although attempts have been made at intervals to arrive at a concordance that would permit the use of a common nomenclature (Dewulf, 1971; Hirai and Jones, 1989a; Percheron et al., 1993, 1996; Macchi and Jones, 1997), these have had little impact and the two nomenclatures remain ingrained, the one in the experimental literature and the other mainly in that of functional neurosurgery.

Anyone coming to the human thalamus with the experience of the monkey thalamus cannot help but be impressed by the similarity between the two, whether visualized by Nissl or myelin staining. The recent application of acetylcholinesterase staining permitted a parcellation of the human thalamus that was independent of these traditional methods (Hirai and Jones, 1989a). It, too, shows a remarkable similarity to descriptions of the Old World and New World monkey thalamus based on this method (Olivier et al., 1970; Jones, 1985; Stepniewska et al., 1994a). Hence a common nomenclature for the human and monkey thalamus is a realistic posibility. Apart from facilitating dialogue between experimentalists and neurosurgeons, a common nomenclature can only help in the transfer of knowledge about functional connections derived from experimental work in non-human primates to the human.

Table 2 compares the nomenclature used in the classical account of the human thalamus by Hassler (1959) with that used in the present account, which is the same as that introduced by Hirai and Jones (1989a). The different names applied to the same nuclei appear on the camera lucida drawings of the human thalamus in Figs. 11, 12 and 13. Fig. 11 is drawn from sections of the brain illustrated in Fig. 6. Most of the nuclear divisions of the human thalamus made by Hirai and Jones did not greatly differ from those originally made by Hassler (1959). The only significant disagreement was in the region in which the lateral posterior (LP), anterior pulvinar (Pla) and ventral posterior lateral (VPL) nuclei come together at the anterior end of the pulvinar. Here, Hassler's oral pulvinar nucleus (Pu.o) which had three subdivisions, represents the LP nucleus and the dorsal part of the anterior pulvinar (Pla) nucleus of the present account. His nucleus ventralis caudalis portae (V.c.por) seems to be equivalent to the ventral part of Pla. It is difficult to draw sharp boundaries between the nuclei in this region, especially in myelin stained preparations which were mainly used by Hassler, so the differences are explicable. Nissl staining, which shows the differences in cell density makes the borders between the nuclei somewhat clearer, so that a parcellation similar to that of Olszewski (1952) in the macaque monkey could be made in the human. The medial pulvinar nucleus (Plm) of the human (Pu.m of Hassler) is identical in general cytoarchitecture and relative position to that of the monkey. It is disproportionately large in the human and in it there are regions of cytoarchitectonic variation that are not so readily identifiable in monkeys. Each of the regions of cytoarchitectonic variation received separate names from Hassler (Table 2). The large size probably reflects the enlargement of the parietotemporal association cortex with which Plm is connected and the incipient divisions of Plm may reflect the differentiation of this cortex into a larger number of functional areas than found in the monkey. Hassler's lateral pulvinar nucleus (Pu.l) is identical to the present nucleus of the same name (Pll) and he made no subdivisions of it, indicating its homogeneous cytoarchi-

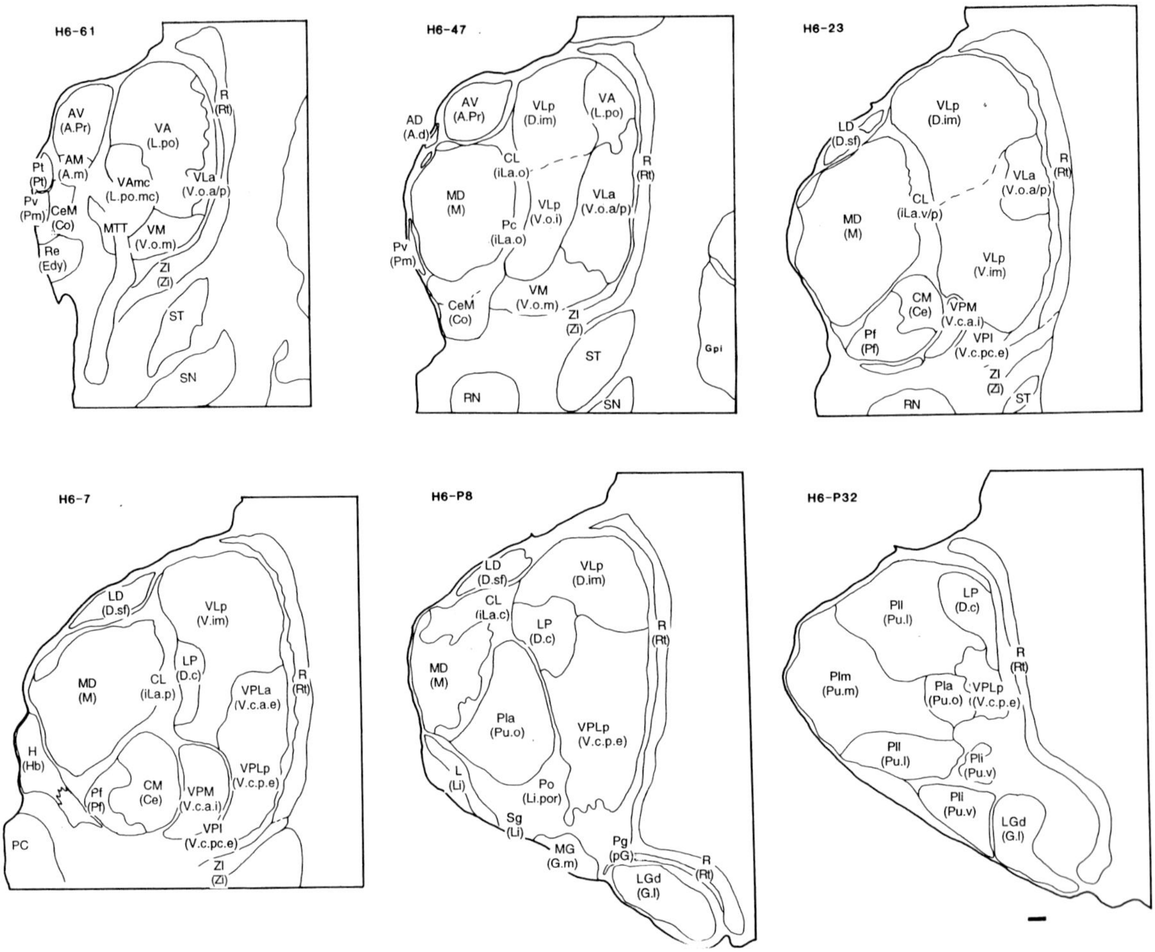

Fig. 11. Camera lucida drawings from frontal sections of a human thalamus, in anterior (top left) to posterior (lower right) order, showing the nuclear outlines as presented by Hirai and Jones (1989a), with the nomenclature of those authors and that of Hassler (1959) (in parenthesis). See Table 2.

tecture. Hassler has no inferior pulvinar nucleus. The region included in it in the present account, as in Olszewski's acount of the monkey, includes the ventral (Pu.v), intergeniculate (Pu.ig) and suprabrachial (Pu.sb) pulvinar nuclei of Hassler. The names reflect Hassler's reliance on myeloarchitecture which is affected by the passage of many fiber tracts through Pli, rather than upon cytoarchitecture which, apart from the evenly scattered large cells, is relatively homogeneous. Hassler's superficial pulvinar nucleus is the region where the LP and LD nuclei merge with the Plm nucleus, a region of transition, elevated to the status of a separate nucleus.

Among the ventral and lateral nuclei, Hassler recognized a large number of subnuclei and gave them lengthy and complex names. His basic subdivisions, however, are essentially the same as those of Hirai and Jones. What appear to be an excessive number of finer nuclear subdivisions of Hassler, can with some justification be attributed to the presence of regions of different cell size and/or packing density within a nucleus and/or to transitions between these regions. Some of these subnuclei are

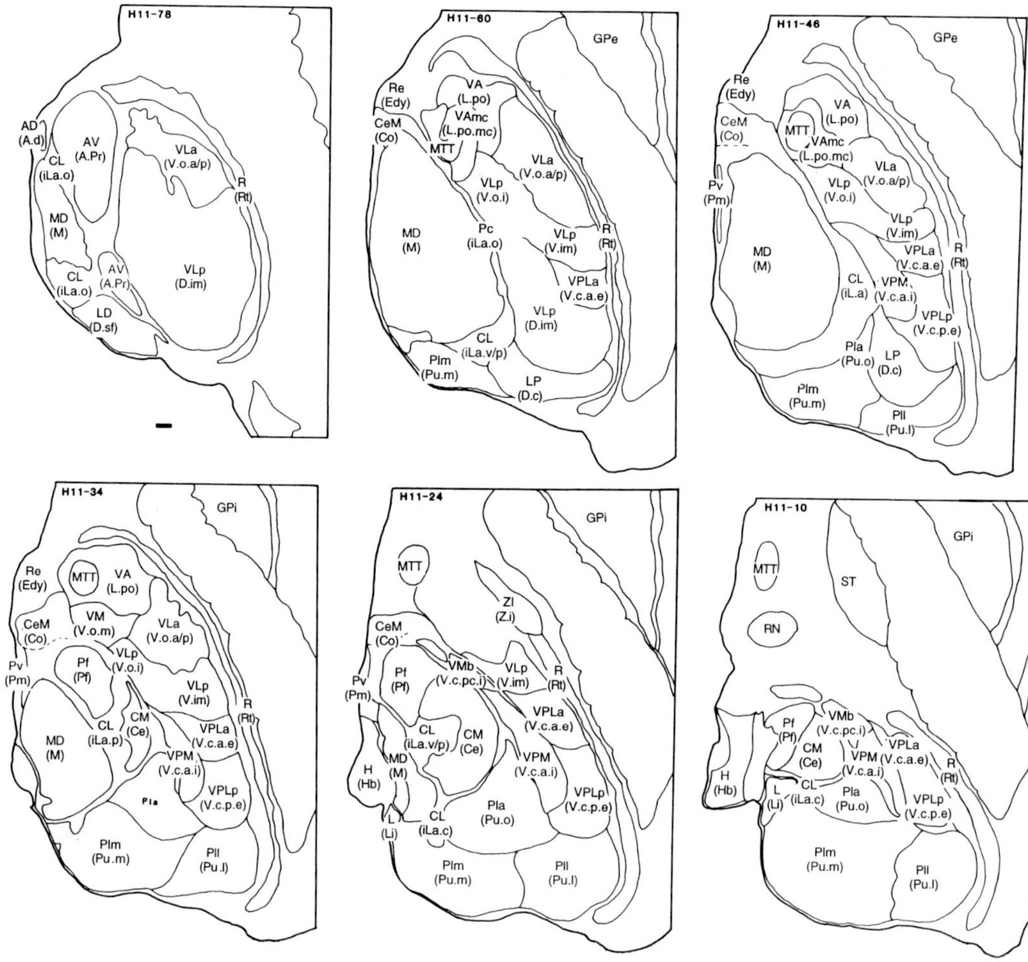

Fig. 12. Camera lucida drawings of horizontal sections from a human thalamus, in dorsal (top left) to ventral (lower right) order, with the nuclear outlines as drawn by Hirai and Jones (1989a) and the nomenclature of those authors and that of Hassler (1959) (in parenthesis). See Table 2.

recognizable even in the monkey. For example, divisions of the VLp nucleus based on cell size and called D.im and V.im by Hassler, are the same as those called VLc and VPLo in the monkey by Olszewski (1952). The rationale for bringing them together in a common nucleus depends upon the knowledge that in the monkey the major concentration of cerebellar fiber terminals spreads without interruption from one into the other (Asanuma et al., 1983a,b). Similarly, Hassler's nucleus ventralis oralis internus (V.o.i) is also subsumed into VLp on the grounds that, in the monkey, cerebellar terminals extend into it as well.

Transition zones between nuclei, which Hassler recognized as separate nuclei, can also be seen, for example, in the interdigitation of cells of the VLa and VLp nuclei. This region, Hassler called nucleus ventro-oralis posterior (V.o.p). Other regions which he clearly identified as regions of cytoarchitectutral transition, Hassler labeled 'zentro' nuclei. One of them, the nucleus zentrolateralis intermedius, is obviously the region of

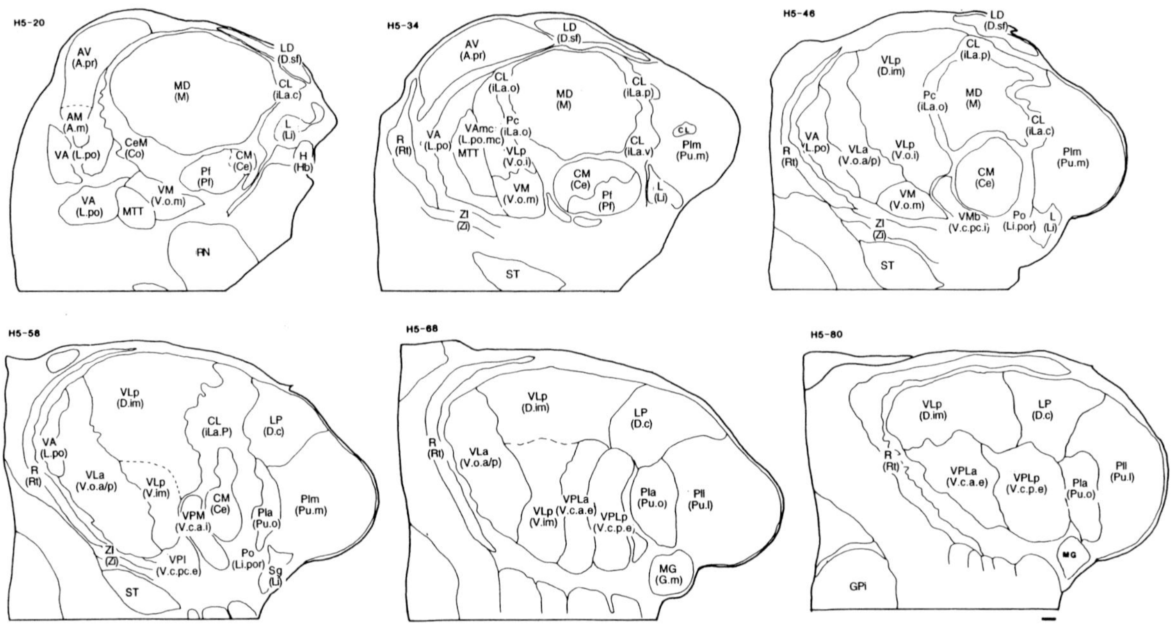

Fig. 13. Camera lucida drawings of parasagittal sections from a human thalamus, in medial (top left) to lateral (lower right) order, showing the nuclear outlines as drawn by Hirai and Jones (1989a) and the corresponding names of those authors and of Hassler (1959) (in parenthesis).

transition between the two differently sized cell populations of VLp mentioned in the preceding paragraph. These transitions, unless justified by fiber connections were excluded when the nuclei were renamed by Hirai and Jones.

One example of a region which appears as a zone of cytoarchitectonic transition in the monkey and becomes clearly differentiated as a separate nucleus in the human, is the VPLa nucleus (Hassler's V.c.a.e). Its elevation to the status of a separate nucleus can be justifed on connectional and physiological grounds (Section 10.7). Some workers consider that even in monkeys it may be sufficiently distinct to warrant a separate name (Kaas et al., 1984a). That has not been done here. Nuclear borders in general are more clear-cut in the human thalamus than in that of monkeys. This reflects not only the larger size of the human thalamus (approximately 5:1; Fig. 5) but also the fact that size differences between cells in many nuclei are commensuarately greater in the human, making the drawing of borders easier. This is facilitated further by the acetylcholinesterase stained preparations.

Perhaps surprizingly, in view of the enlargement of the frontal association cortex in the human, the mediodorsal nucleus, unlike the medial pulvinar nucleus, does not show any incipient or actual parcellation beyond the three (or four) divisions seen in monkeys; the subnuclei do not seem to be any more well-defined in humans than in monkeys and their cytology is indistinguishable. Even Hassler, for all his efforts in other nuclei, did not divide the human mediodorsal nucleus into more than the recognizable three, although giving them names that reflect their fiber architecture, rather than their cytoarchitecture (Table 2).

The reader will be able to recognize in Table 2 other nuclei identified by different names in the two nomenclatures and will be able to see the further divisions of most of them made by Hassler, the names of these divisions generally reflecting myeloarchitectonic differences but sometimes cytorchitectonic differences as well. Many of these variations can be found mentioned in the above description of the human thalamus. One cannot help but admire the clear eye that Hassler had for anatomical variation in the thalamus even while regarding his parcellation as excessive.

5. THE CELL TYPES OF THE PRIMATE THALAMUS

5.1. RELAY NEURONS AND GABAERGIC NEURONS

All mammalian thalami contain three fundamental types of neurons: relay neurons, interneurons intrinsic to the dorsal thalamus, and neurons of the reticular nucleus. The primate thalamus is no exception in this regard. Relay neurons are so called because their axon projects to the cerebral cortex or striatum; intrinsic interneurons, which may not be present in all nuclei of rodents and some other non-primate species, have their synaptic connections confined to the dorsal thalamic nucleus in which they lie; reticular nucleus neurons have their somata contained within the confines of the reticular nucleus or one of its satellites and possess an axon that projects into the dorsal thalamus. The morphology of each cell type is characteristically different. The relay cells are also distinguished from the other two types by being glutamatergic. The intrinsic interneurons and the reticular nucleus neurons are all GABAergic.

The earliest Golgi studies of the thalamus were conducted on that of humans and date to Kölliker (1896) who recognized the chief characteristic of relay cells, namely the presence of dendrites that radiate out in all directions from the soma, branching as they do so. This creates a symmetrical, bush-shaped dendritic field. (Hence the cells were called Buschzellen by Kölliker). Interneurons which are much smaller and were recognized later in the thalamus of the cat by Cajal (1904), have fewer, angular and less branched dendrites which are studded with lengthy appendages. Reticular nucleus cells, also probably recognized for the first time by Kölliker in the human, are large cells with ovoid somata, from the poles of which dendrites radiate out within the confines of the nucleus to form elongated or discoidal dendritic fields. Some examples of cells from a human thalamus are given in Figs. 14 and 18.

5.2. MORPHOLOGY OF RELAY NEURONS

The morphologies of thalamic relay neurons have received relatively little attention in primates. Much work, however, has been devoted to the cat in which, in the dorsal lateral geniculate nucleus and to a lesser extent in the ventral posterior nucleus, variations which occur in the fundamental bushy morphology of relay cells are associated with differences in receptive field properties of the neurons and with differences in the extent and laminar terminations of their axons in the cerebral cortex (Guillery, 1966; Friedlander et al., 1981; Stanford et al., 1981 Yen et al., 1985a). Asymmetry of the dendritic fields of relay neurons can be associated with a cell's conforming to the constraints imposed by the borders of a nucleus and/or adjacent fiber laminae. The relay cells of the intralaminar nuclei which in every other respect resemble typical bushy relay cells, are particularly affected in this manner (Hazlett et al., 1976; Pearson et al., 1984; Tseng and Royce, 1986; Tömböl et al., 1990).

Golgi studies or other forms of investigation that reveal the full morphology of

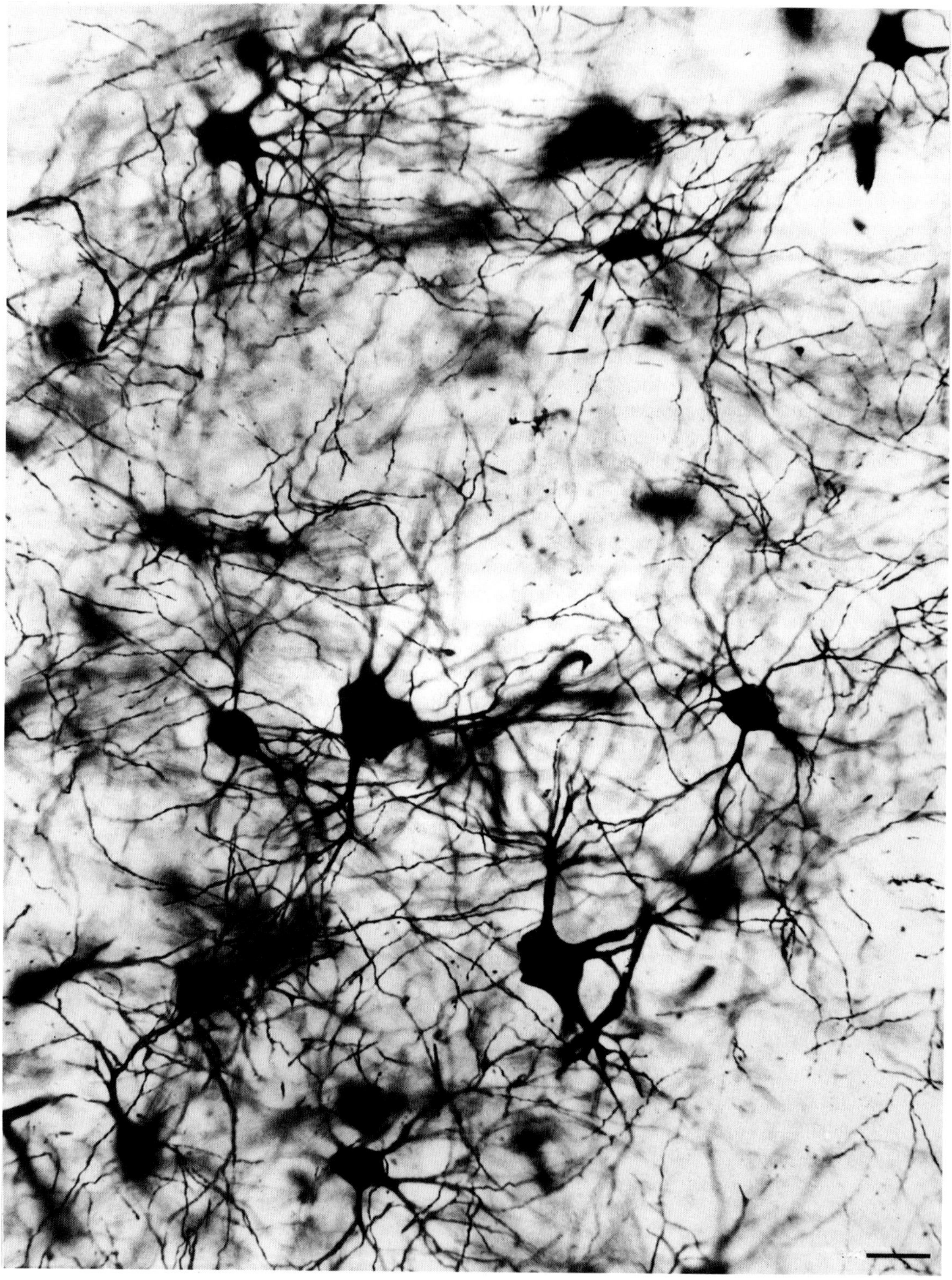

Fig. 14. Photomicrograph of a Golgi-stained section from the human VLp nucleus, showing the large relay cells with their characteristically bushy dendrites and a smaller cell (arrow) tentatively identified as an interneuron. From a preparation by Dr. T. Hirai. Bar 50 μm.

neurons in the thalamus of primates have been few and not very comprehensive. As described, however, in the ventral posterior nucleus of the *Galago* (Pearson and Haines, 1980), or squirrel monkey (Havton and Ohara, 1994), or in the dorsal lateral geniculate nucleus of the *Galago* (Conley et al., 1985) and of macaque monkeys (Wilson, 1989), the relay cells appear in the traditional bushy form and show few, if any, of the morphological variations that accompany the possession of defined physiological properties in the cat.

Golgi studies of the dorsal lateral geniculate nucleus in macaque monkeys and in the *Galago* show relay cells in the parvocellular laminae that have smaller somata and a bitufted dendritic organization confined to the lamina in which they lie, while those in the magnocellular laminae have larger somata and more radial dendritic fields which cross laminar boundaries (Saini and Garey, 1981; Wilson and Hendrickson, 1981; Conley et al., 1985). The anatomical distinctions between the cells are, however, far less dramatic than those of relay cells in the dorsal lateral geniculate nucleus of the cat. In a study involving intracellular injection of horseradish peroxidase into physiologically-characterized cells of the macaque dorsal lateral geniculate nucleus, Wilson (1989) confirmed the differences in cell size and dendritic field organization and showed that the cells in the parvocellular layers were wavelength-sensitive and that most of those in the magnocellular layers had linear responses to visual stimuli composed of different grating patterns, a means of testing for spatial frequency detection and linearity of spatial summation in the responses of a visual cell. Neither type of relay cell possessed significant populations of dendritic appendages and thus did not possess another feature that is a distinguishing mark of one of the physiological cell types in the cat lateral geniculate nucleus.

Even in the cat, the division of relay neurons into morphological classes with distinct physiological properties is mainly a feature of the dorsal lateral geniculate nucleus, although modest differences between rapidly and slowly adapting relay neurons have been observed in the ventral posterior nucleus (Yen et al., 1985a). Most intracellularly injected neurons had somata of similar sizes (ca. 400 μm^2), and radially symmetrical dendritic fields without the appendages typical of one form of cat geniculate cell. Cells injected in the ventral posterior nucleus of squirrel monkeys also had radially symmetrical dendritic fields with few or no appendages and appeared to conform to a single morphological type (Havton and Ohara, 1994).

The thalamic relay cells of primates (Fig. 14), so far as they have been studied, thus, appear to conform to the typical thalamic bushy cell, with a relatively symmetrical dendritic field and occasional dendritic appendages or protrusions but without major morphological variations which might follow the possession of particular physiological properties.

5.3. SIZES AND PROJECTIONS OF RELAY NEURONS

Most thalamic nuclei in monkeys show a range of cell sizes. Relay cells are on the whole larger than GABAergic interneurons but the relay cells themselves can show a broad range of sizes. The sizes of non-GABAergic cells in the ventral posterior and ventral lateral nuclei of monkeys range from less than 70 μm^2 to more than 400 μm^2 (Hunt et al., 1991). In the dorsal lateral geniculate nucleus, cells in the magnocellular laminae are 25 to 40 μm in diameter and in the parvocellular laminae 16 to 25 μm in diameter (Saini and Garey, 1981), while those in the S and interlaminar zones are 8 to 10 μm in diameter (Hendry and Yoshioka, 1994). The smaller relay cells represent a particular cell class that projects, in the main, to superficial layers of the cerebral cortex and are immunocytochemically different from the larger cells which project to middle layers of the cortex (section 6.2).

In the prosimian, *Galago*, applications of anatomical tracers to the surface of the visual cortex result in retrograde labeling specifically of cells in the small-celled layers (4 and 5) of the dorsal lateral geniculate nucleus (Carey et al., 1979a,b; Lachica and Casagrande, 1992; Johnson and Casagrande, 1995). In monkeys, similar experiments result in labeling specifically of the small cells in the S laminae and intercalated zones (Fitzpatrick et al., 1983; Weber et al., 1983b; Hendry and Yoshioka, 1994). Applications of tracers to the surface of the somatic sensory cortex labeled cells that were consistently less than half the size (mean 100–250 μm^2) of those labeled after injections into the middle layers (200–400 μm^2) (Rausell et al., 1992a; Fig. 15).

The small-celled, superficially-projecting population of relay neurons in the ventral lateral, ventral posterior and dorsal lateral geniculate nuclei of the monkey, and in the dorsal lateral geniculate and ventral posterior nuclei of the *Galago* is invariably calbindin-immunoreactive (Rausell and Jones, 1991a; Rausell et al., 1992a; Diamond et al., 1993; Johnson and Casagrande, 1995). These calbindin positive neurons form a matrix of superficially-projecting thalamic relay cells that is not constrained by traditional, architectonically-defined nuclear borders (Rausell et al., 1992). The cells of the matrix appear to be preferentially innervated by small fibers of spinal cord or retinal and perhaps other origins (see section 6.2).

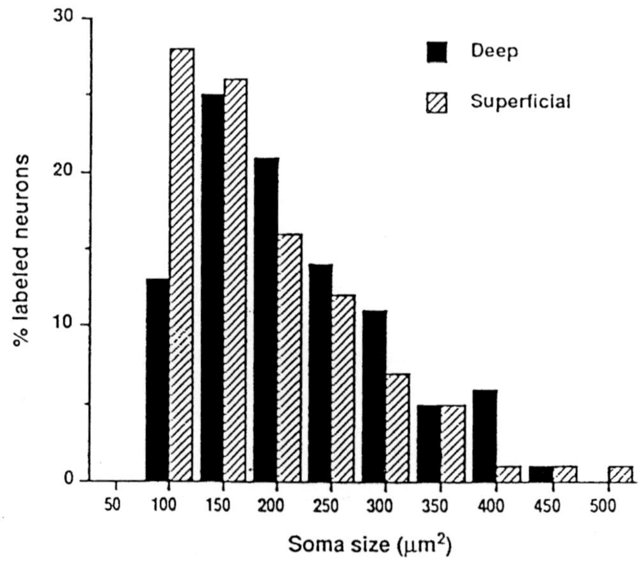

Fig. 15. Sizes of relay neurons retrogradely labeled in the ventral posterior nucleus of a monkey following injection of tracer into the middle layers (deep), or its application to the surface of the somatosensory cortex, affecting layers I-II or IIIA (superficial). Cells projecting to superficial layers are consistently smaller than those projecting to middle layers. From Rausell et al. (1992).

5.4. INTERNEURONS

The GABAergic intrinsic interneurons of the thalamus are small cells that are found in all nuclei. Early reports of their absence from the intralaminar nuclei have proven incorrect. Typically, as revealed in morphological studies in non-primates, the cells are small, with somata less than 10 μm in diameter. The dendrites are few, short and angular but give off numerous, lengthy, branched appendages which end in boutons terminaux and form dendro-dendritic synapses. Whether all the cells or only those in the dorsal lateral geniculate nucleus of the cat possess an axon is still uncertain. Two, injected in the dorsal lateral geniculate nucleus of macaque monkeys, did not (Wilson, 1989).

In situ hybridization histochemistry for glutamic acid decarboxylase (GAD) mRNA (Benson et al., 1991b, 1992b; Huntsman et al., 1996), or immunocytochemical staining for GAD or its product, gamma aminobutyric acid (GABA) (Montero and Zempel, 1986; Smith et al., 1987; Clark et al., 1989; Hunt et al., 1991), show that in monkeys GABA cells form approximately 30% of the neurons in every thalamic nucleus (Figs. 16, 17). Reports of much lower numbers in the intralaminar nuclei (Smith et al., 1987), have proven unfounded (Hunt et al., 1991).

In monkeys, in nuclei other than the lateral geniculate nucleus, GABA immunoreactive cells have somal areas ranging from 65–100 μm^2, in comparison with somal areas of 150–230 μm^2 for relay neurons (Hunt et al., 1991) (Figs. 16, 20). The GABA cells tend to be smaller in nuclei in which the sizes of the relay cells are small but, invariably, two distinct populations are revealed. In the dorsal lateral geniculate nucleus (Montero and Zempel, 1986), GABA cells in the parvocellular layers have mean areas of 97 μm^2 in comparison with 200 μm^2 for relay cells, and in the magnocellular layers they have areas of 137 μm^2 in comparison with 316 μm^2 for the relay cells (Fig. 17). In the mediodorsal nucleus (Clark et al., 1989), GABA immunoreactive neurons in the parvocellular division are consistently slightly smaller than those in the magnocellular division.

The density of GABA cells is relatively high in all nuclei, accounting for not less than 30% of the volume of the nucleus and in some cases (e.g. the ventral lateral posterior nucleus and the magnocellular layers of the dorsal lateral geniculate nucleus) as much as 45% (Montero and Zempel, 1986; Hunt et al., 1991; Figs. 16, 17). The density of GABA cells in the relay nuclei is approximately 10–15% greater than in the intralaminar nuclei. Most of these differences are reflected in the different levels of gene expression for GAD seen in the different nuclei and in different laminae of a structure such as the lateral geniculate nucleus (Benson et al., 1991b) (see below).

5.5. RETICULAR NUCLEUS NEURONS

5.5.1. Introduction

The cells of the reticular nucleus in monkeys and humans are large, with somal diameters of 25 to 50 μm. The cells tend to be flattened over the surface of the dorsal thalamus so that the dendritic field is commonly discoidal. The thickness of the nucleus at any level is filled by overlapping dendritic fields (Fig. 18). Cells on the anterior surface of the dorsal thalamus are flat in the frontal plane. Those on the lateral and ventral surfaces are flattened in the parasagittal or horizontal plane. The degree of flattening depends upon the relative thickness of the nucleus in each region of the

Table 2. Density of GABA-immunoreactive and non-immunoreactive neurons in selected thalamic nuclei (1-μm plastic sections)

Nucleus	Number of cells		P_A		P_V		Mean diameters of cells (μm)		Sample area (mm^2)
	G(+)*	G(−)†	G(+)	G(−)	G(+)	G(−)	G(+)	G(−)	
CeM	12‡	30	29	71	38	62	23	36	1.15
	15§	44	25	75	36	64	22	36	1.59
CL	26	43	37	63	44	56	25	33	1.83
	18	44	29	71	36	64	29	39	1.54
CM	17	30	36	64	42	58	23	30	1.34
	28	51	35	65	40	60	28	34	2.29
Pf	9	18	33	67	43	57	26	40	0.67
	20	32	38	62	48	52	26	39	2.33
VPM	23	49	33	67	42	58	24	36	1.74
	29	54	35	65	44	56	29	41	1.75
VPL	19	32	37	63	47	53	26	39	2.43
	24	46	34	66	46	54	30	49	2.03
VLP	32	40	44	56	56	44	29	46	2.04
	20	37	35	65	48	52	29	49	2.09
AM	24	53	31	69	36	64	25	32	3.74
	25	58	30	70	35	65	25	32	2.37
MDl	11	20	35	65	41	59	23	30	0.83
	16	37	30	70	36	64	29	38	1.09
MDm	21	54	28	72	39	61	24	40	1.82
	26	55	32	68	42	58	27	41	2.02
LP	41	54	43	57	54	46	25	39	2.17
	39	55	41	59	48	52	27	38	2.33

*G(+), GABA immunoreactive.
†G(−), Non-immunoreactive.
‡§Two sections sampled from each nucleus.

Table 3. Mean somal areas of GABAergic and non-GABAergic neurons in selected thalamic nuclei of two monkeys (CM216 and CM202)

Nucleus	No. of GABA neurons	Area of GABA neurons ($\mu m^2 \pm$ S.D.)	No. of non-GABA neurons	Area of non-GABA neurons ($\mu m^2 \pm$ S.D.)
CM216				
CeM	235	61.1 ± 17.0	547	176.1 ± 53.7
CL	191	75.9 ± 21.1	429	193.1 ± 57.5
CM	213	67.6 ± 20.2	315	152.0 ± 41.6
Pf	244	65.7 ± 16.5	432	186.0 ± 48.5
VPM	198	84.4 ± 17.0	252	206.0 ± 67.5
VLp	239	80.9 ± 15.6	286	231.1 ± 65.7
AM	280	87.4 ± 20.4	346	210.4 ± 54.2
MDl	276	73.0 ± 14.0	374	166.2 ± 39.8
MDm	196	78.3 ± 19.5	433	221.9 ± 65.5
CM202				
CeM	85	73.1 ± 17.5	165	174.0 ± 47.0
CL	255	93.0 ± 28.0	221	216.5 ± 62.8
CM	158	74.1 ± 18.3	268	147.5 ± 39.0
Pf	112	83.6 ± 16.4	227	198.7 ± 48.7
VPM	166	91.9 ± 22.7	202	218.8 ± 67.2
VLp	155	85.3 ± 15.7	235	234.6 ± 81.5
AM	112	107.3 ± 23.6	117	232.5 ± 48.2
MDl	276	83.2 ± 14.7	232	188.2 ± 55.8
MDm	256	81.3 ± 17.5	333	213.0 ± 48.8

Fig. 16. Tables from Hunt et al. (1991), showing the relative densities per area (P_A) and volume (P_V) and the sizes of GABAergic interneurons and of non-GABAergic relay cells in certain intralaminar and principal nuclei of the thalamus in cynomolgus monkeys (*M. fascicularis*). Abbreviations as on page 244. The proportions of GABAergic neurons do not vary greatly between the two types of nuclei.

←

Table 1. Parameters of GABA(+) and GABA(-) cells in monkey LGN (see text for explanation)

	N	%xA	Area sampled (mm²)	$\bar{d}$ (μm)	$\bar{D}$ (μm)	%xV	Circ. coef.
Magnocel. layers							
GABA(-) cells	656	74	2.59	19.9	25.3	65	0.73
GABA(+) cells	228	26	2.59	13.1	16.7	35	0.80
Parvocel. layers							
GABA(-) cells	2371	81	5.18	15.9	20.2	75	0.72
GABA(+) cells	552	19	5.18	11.0	14.0	25	0.79
Total cells analyzed 3807							

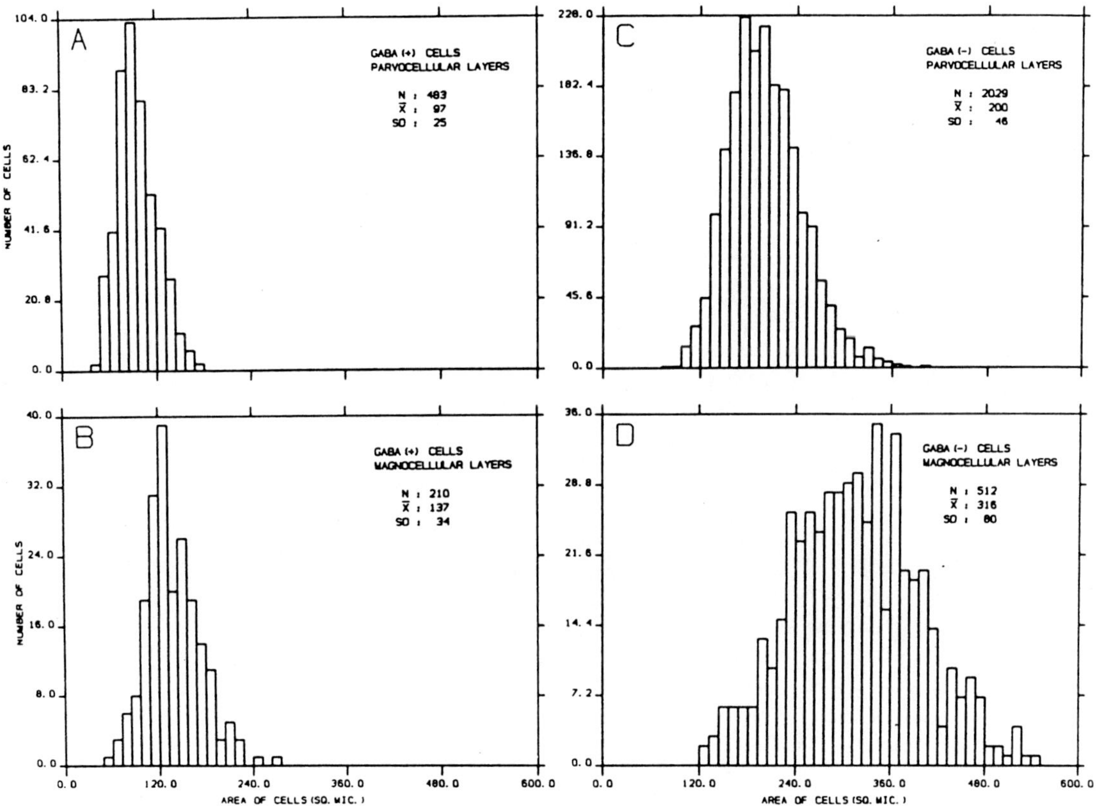

Fig. 17. Tables from Montero and Zempel (1986) showing the relative percentages by area (%xA) and volume (%xV) and the sizes of GABAergic interneurons (GABA (+)) and relay neurons (GABA(-)) in the magno- and parvocellular layers of the monkey dorsal lateral geniculate nucleus.

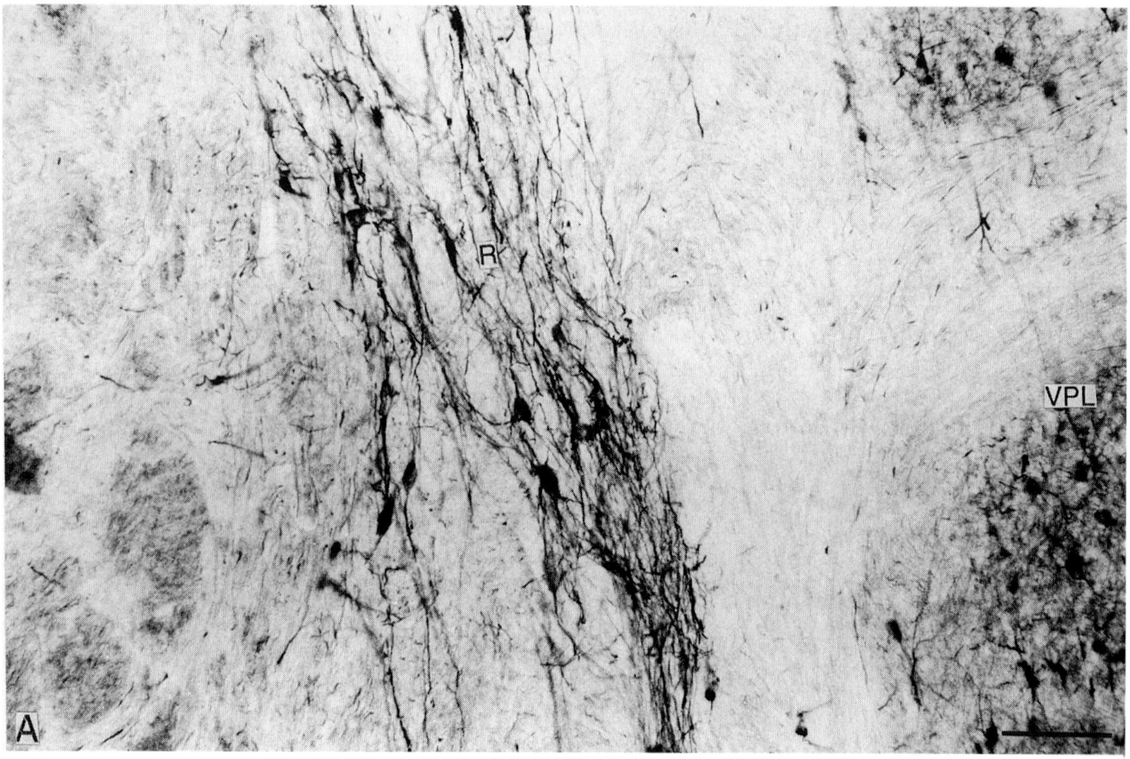
R
VPL
A

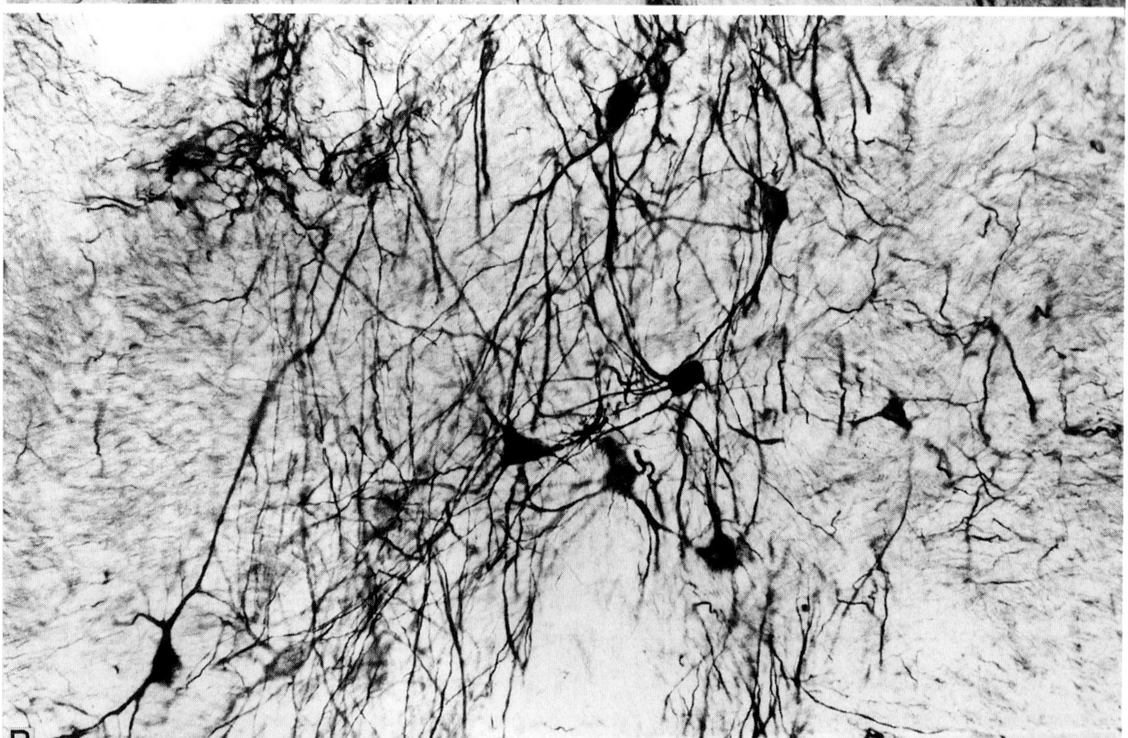
B

Fig. 18. Low (A) and high (B) magnification photomicrographs of cells in the reticular nucleus of a human thalamus, stained immunocytochemically for non-phosphorylated neurofilament protein by the antibody SMI32. The dendrites of the cells are extended in the plane of the reticular nucleus but many branches extend across its full thickness. Bars 100 μm (A), 40 μm (B).

reticular nucleus, so that the flattest reticular cells with the most elongated dendritic fields are found laterally where the nucleus is thinnest and more bushy cells with symmetrical dendritic fields are found dorsally and anteriorly where the nucleus is thickest. These morphological variations may reflect little more than adaptations to fit the dendritic field into available space. The axons of the cells possess intranuclear collaterals of relatively short extent and in some species the terminal dendritic branches form dendro-dendritic synapses with neighboring cells (Yen et al., 1995b; Deschênes et al., 1985; Rinvik et al., 1987). These are reported as common in monkeys in some studies (Asanuma, 1994) but as rare in others (Williamson et al., 1994).

All reticular nucleus cells in monkeys and humans are GABAergic (Benson et al., 1991b) (Figs. 19, 21). This, and their projection into the dorsal thalamus (Section 8.3), makes them a second population of thalamic inhibitory interneurons.

Among other substances expressed in reticular nucleus neurons of primates are parvalbumin (Jones and Hendry, 1989), the CAT301 antigen (Hendry et al., 1988d) (Section 6.3), certain neuropeptides such as somatostatin (Molinari et al., 1987; Conley et al., 1991b) (Section 8.5); growth factor receptors such as the low affinity NGF receptor are found in other species (Fusco et al., 1991). When present, these molecules are normally found in all of the reticular nucleus cells, although in the *Galago* there appears to be a population that does not stain for somatostatin (Conley et al., 1991b). Of special note is the absence of gene expression for most of the known polypeptide subunits of the $GABA_A$ receptor (Huntsman and Jones, 1993; Huntsman et al., 1995, 1996), despite their expression in cells of the dorsal thalamus and despite the presence of bicuculline-sensitive GABA receptors in the reticular nucleus (Section 8.3).

5.5.2. Circuitry of the reticular nucleus

Thalamocortical and corticothalamic fibers passing through the reticular nucleus are the major sources of afferent inputs to the nucleus (Jones, 1975b). Thalamostriatal and pallidothalamic fibers also contribute collaterals to the nucleus as they pass through it (Jones, 1975b; Hazrati and Parent, 1991; Sadikot et al., 1992a; Cornwall et al., 1990; Deschênes et al., 1996). The collateral terminations of the thalamocortical and corticothalamic inputs are arranged topographically in the reticular nucleus so that a sector of the reticular nucleus is dedicated to each major dorsal thalamic nucleus (Jones, 1975b; Montero et al., 1977; Hale et al., 1982; Shosaku et al., 1984; Ohara and Lieberman, 1985; Sumitomo et al., 1988a; Crabtree and Killackey, 1989; Shosaku et al., 1989; Conley and Diamond, 1990; Ray et al., 1992; Crabtree, 1992a,b, 1996; FitzGibbon et al., 1995; Tai et al., 1995) and the cells of a sector take on many receptive field characteristics of the relay cells in a sensory nucleus which provides the input to that sector (Cleland et al., 1971; Dubin and Cleland, 1977; So and Shapley, 1981; Pollin and Rokyta, 1982; Shosaku and Sumitomo, 1983; Yen et al., 1985b; Shosaku, 1985; Jones and Sillito, 1994c). The topographically ordered distribution of the terminations of thalamocortical and corticothalamic fiber collaterals (Yen et al., 1985b; Agmon et al., 1995; Murphy and Sillito, 1996; Bourassa et al., 1995 a,b) creates a crude topographic representation of the appropriate system in each

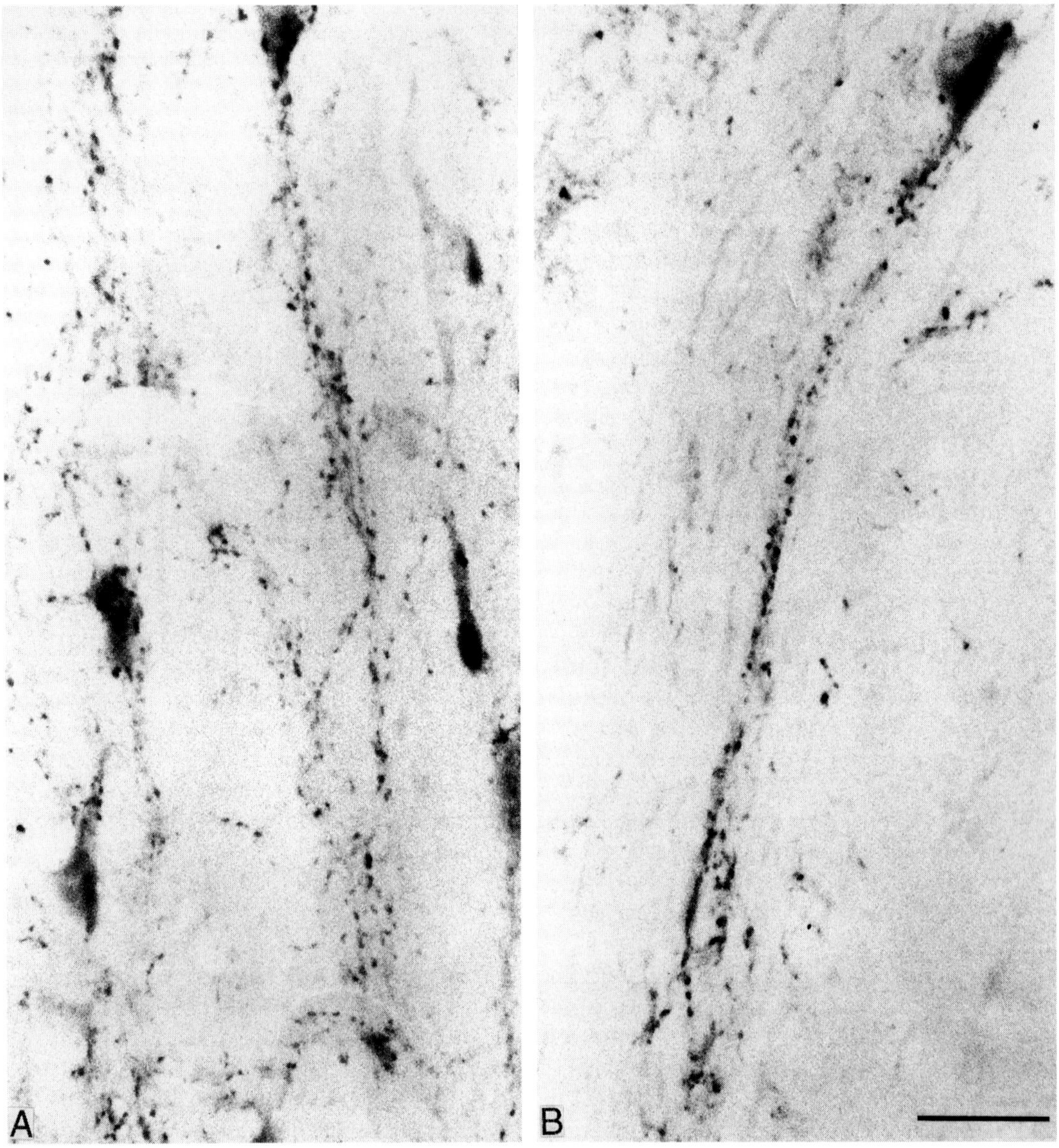

Fig. 19. immunocytochemical staining for glutamic acid decarboxylase (GAD) in the reticular nucleus of a monkey (*M. fascicularis*). The long dendrites of the GAD positive reticular nucleus cells are studded with GAD positive puncta which represent terminals of axon collaterals and dendrodendritic contacts between the cells Bar 20 μm.

sector of the reticular nucleus (Montero et al., 1977; Symonds and Kaas, 1978; Agmon et al., 1995; Jones, 1975b; Crabtree and Killackey, 1989; Bourassa et al., 1995a,b; Crabtree, 1992; Crabtree and Killackey, 1989; Harting et al., 1991b; Cicirata et al., 1990; Conley et al., 1991a; Cornwall et al., 1990; Lozsádi, 1995) and topographic maps have been reported from physiological studies (Hale et al., 1982; Shosaku and Sumitomo, 1983; Shosaku et al., 1984). The extensive overlap of the dendrites of

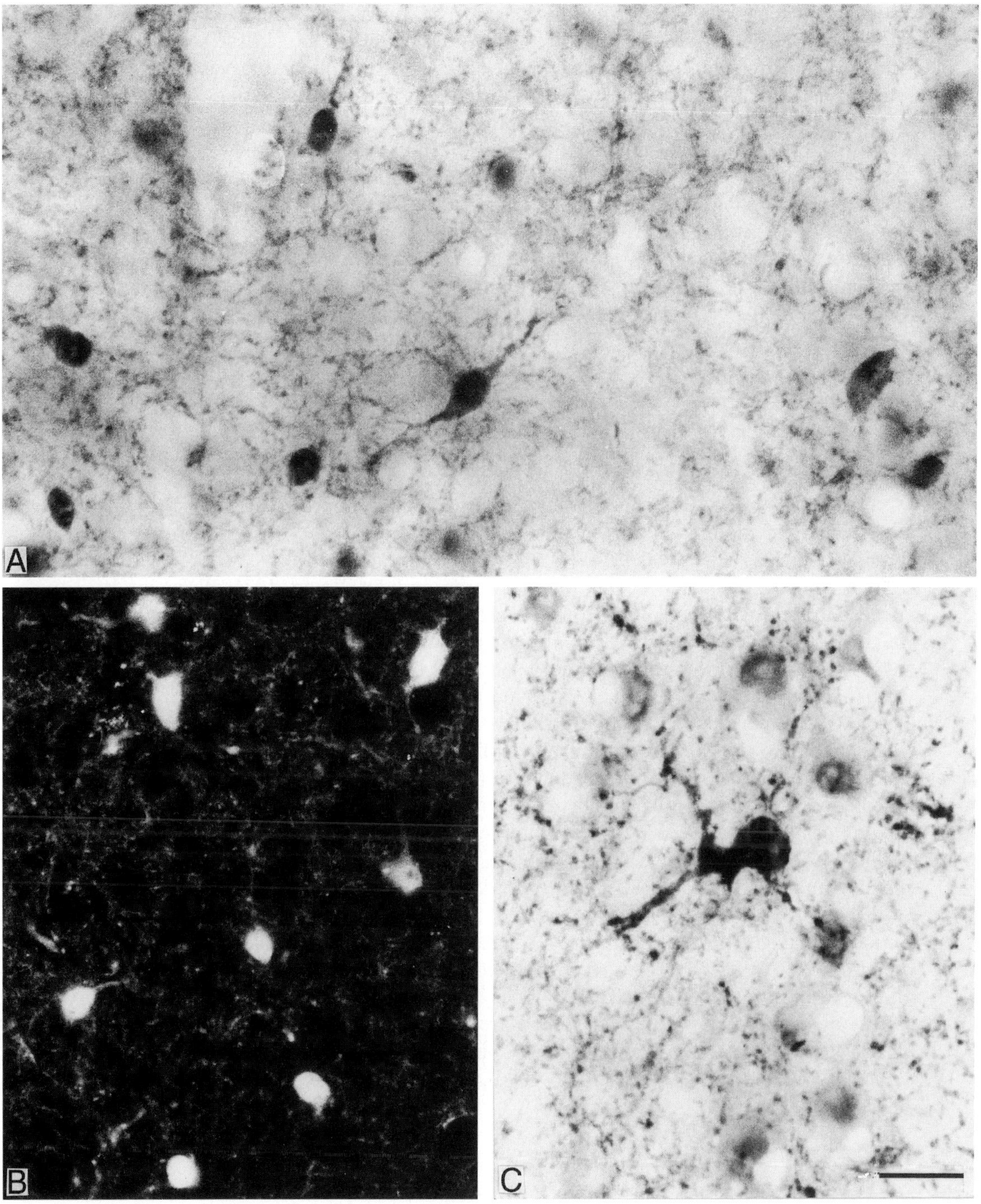

Fig. 20. Typical immunocytochemical staining of intrinsic thalamic interneurons for GABA (A, B) and GAD (C) in the macaque monkey thalamus. A and C are from an immunoperoxidase and B from an immunofluorescent preparation. Bar 50 μm (AB), 20 μm (C).

reticular nucleus cells makes for a high degree of convergence and the receptive fields of neurons in sensory sectors of the reticular nucleus reflect this. In the somatosensory sector, neurons have large receptive fields and, unlike cells in the ventral posterior nucleus and somatosensory cortex, are not modality specific (Pollin and Rokyta, 1982; Yen et al., 1985b) or directionally relective (Sugitani 1979; Shosaku, 1985). In the visual sector, cells with large, binocular, on-off center receptive fields, which are rare in the dorsal lateral geniculate nucleus, are common (Dubin and Cleland, 1977; Hale et al., 1982; Uhlrich et al., 1991) and indicative of convergence of collateral inputs from two classes of geniculate cell.

The reticular nucleus also contains some of the major thalamic concentrations of terminations of fibers arising in the brainstem and basal forebrain. The brainstem fibers arise in the cholinergic, noradrenergic and serotoninergic nuclei (Swanson and Hartman, 1975; Hallanger et al., 1987; Edley and Graybiel, 1983; Cropper et al., 1984; de Lima et al., 1985; Woolf and Butcher, 1986; Rye et al., 1987; Paré et al., 1988; Hallanger et al., 1987; Raczkowski and Fitzpatrick, 1989), while basal forebrain afferents arise mainly in GABAergic cells associated with the globus pallidus and the basal nucleus of Meynert (Nauta, 1979a,b; Levey et al., 1987; Steriade et al., 1987; Hallanger et al., 1987b; Smith et al., 1988; Parent et al., 1988; Jourdain et al., 1989; Asanuma, 1989, 1994; Cornwall et al., 1990; Asanuma and Porter, 1990; Bickford et al., 1994). The cholinergic cells in the basal forebrain reportedly do not project to the reticular nucleus (Bickford et al., 1994). Afferents also arise in the pretectum and rostral midbrain reticular formation (Edwards and DeOlmos, 1976; Berman, 1977; Benevento et al., 1977). Those from the pretectum, which mainly arise in the nucleus of the optic tract are GABAergic, at least in cats (Cucchiaro et al., 1993).

Axons of cells in the reticular nucleus give off one or two collaterals in the nucleus (Cajal, 1904; Scheibel and Scheibel, 1966; Yen et al., 1985b; Lübke, 1993; Liu et al., 1995b) then enter the underlying dorsal thalamus where they terminate. The projection is organized so that a sector defined by collateral terminations of the thalamocortical and corticothalamic fibers projects back to the nucleus or nuclei from which it receives it collateral thalamocortical input. The topographic representation in the dorsal thalamic nucleus is respected by the terminations of the returning reticular nucleus fibers (Jones, 1975b; Yen et al., 1985b; Crabtree and Killackey, 1989; Uhlrich et al., 1991; Conley and Diamond, 1990; Harting et al., 1991b; Liu et al., 1995b; Pinault et al., 1995a,b; Cox et al., 1996). The focused nature of the return projection from the reticular nucleus to the relevant dorsal thalamic nucleus belies a functional divergence conditioned by the convergent receptive fields of the reticular nucleus cells (Eysel et al., 1986). Projections to the intralaminar nuclei and to relay nuclei other than the sensory nuclei may arise from more widespread areas of the reticular nucleus (Steriade et al., 1984; Velayos et al., 1989) and they may be less topographically organized (Pinault et al., 1995; Crabtree, 1996; Cox et al., 1996), although this has only been demonstrated in non-primates.

Fig. 21. Dark field photomicrographs of pairs (AB, CD, EF, GH) of adjacent frontal sections from the same thalamus of a Japanese monkey (*M. fuscata*) in anterior to posterior order, labeled by in situ hybridization histochemistry for the mRNAs for 67 kDa GAD (A, C, E, G) and for the α subunit of type II calcium/calmodulin dependent protein kinase (B, D, F, H). Sections have been inverted to appear as though from opposite sides of the brain. GAD mRNA is expressed in reticular nucleus neurons and in the intrinsic interneurons of the dorsal thalamus, CAMIIK-α is expressed only in the relay neurons of the dorsal thalamus. For abbreviations, see page 244. Bar 2 mm. From material illustrated in Benson et al. (1991b).

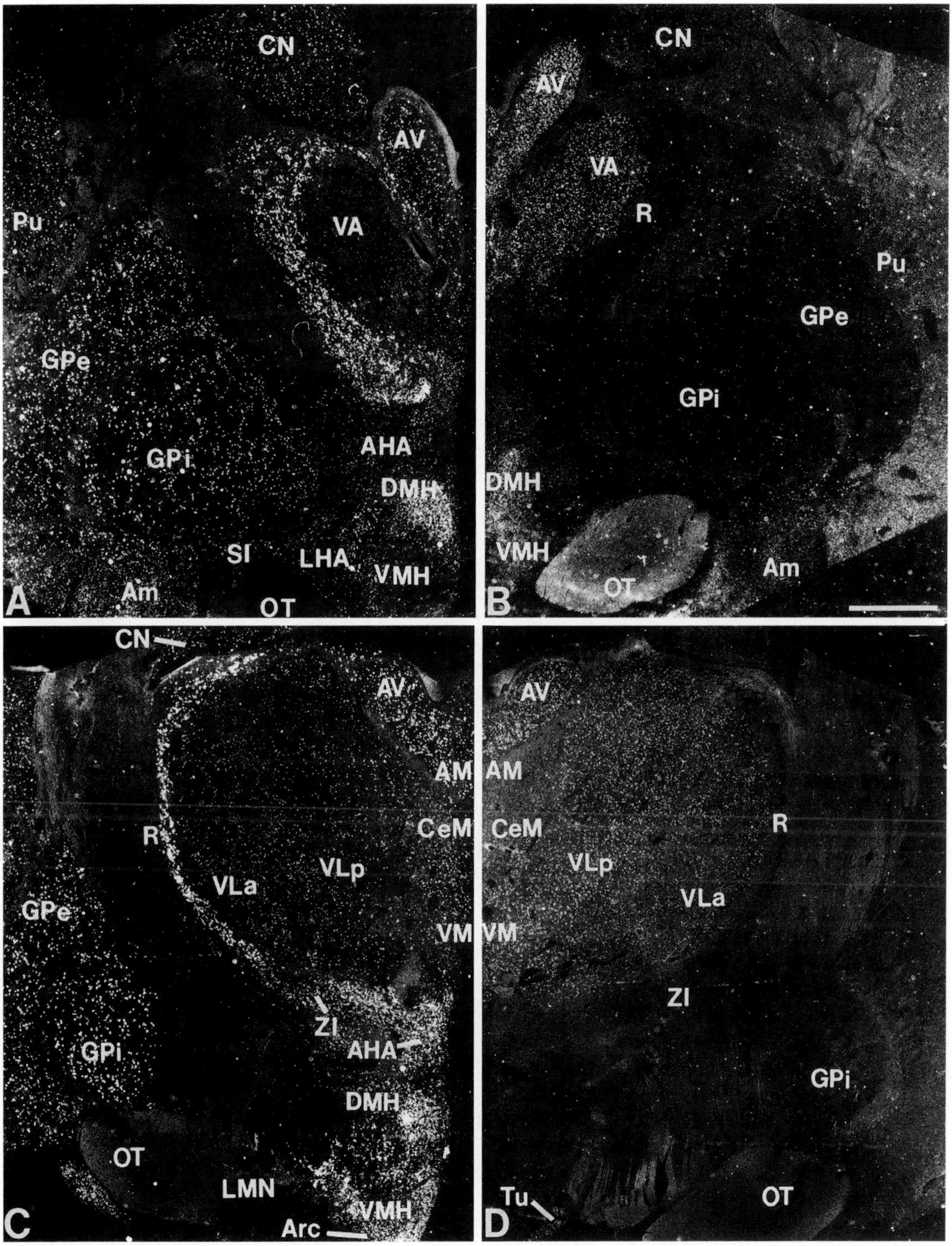

Fig. 21. (continued)

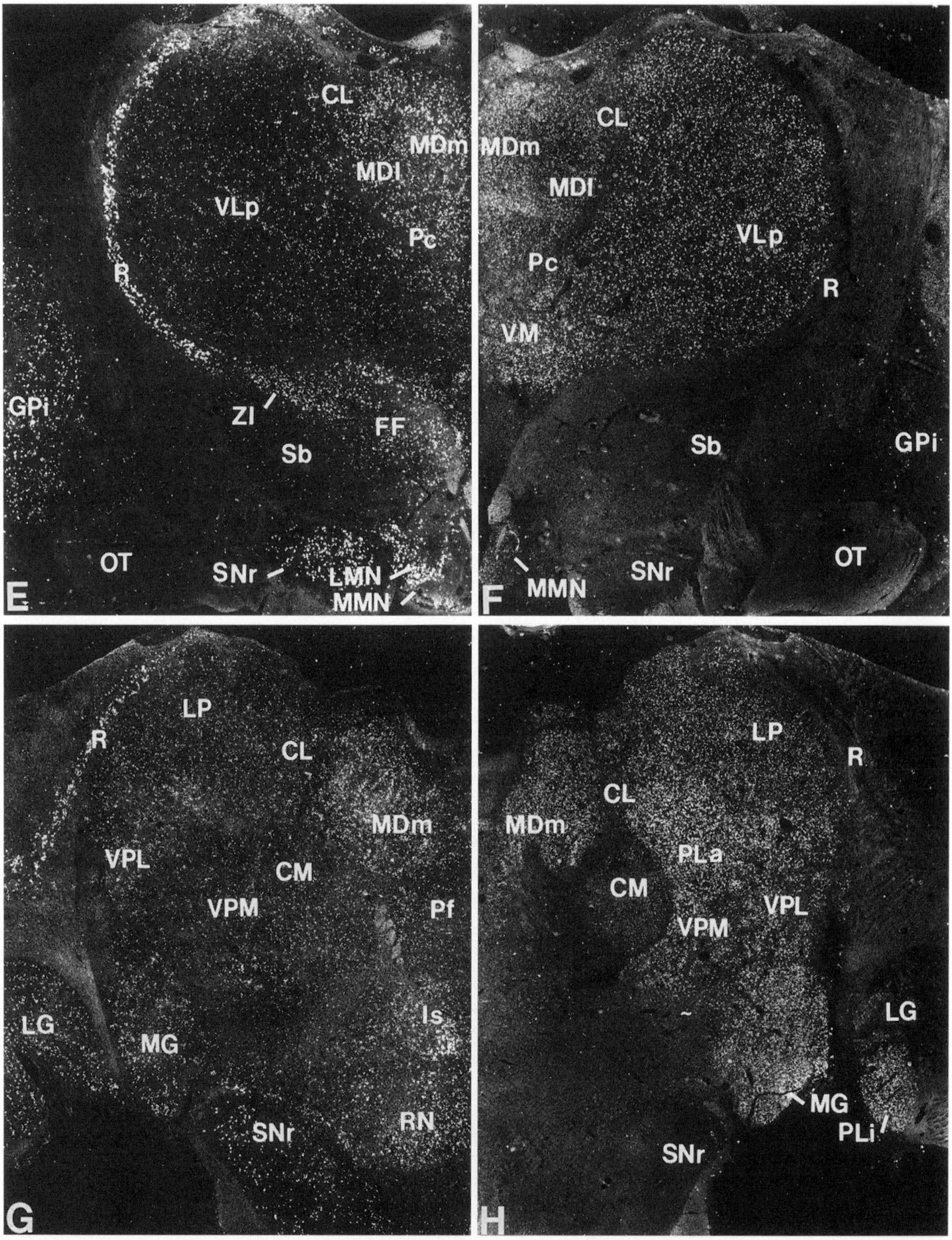

Fig. 21. (continued)

All dorsal thalamic nuclei except the anterior nuclei of cats are reported to receive a return projection from the reticular nucleus (Jones, 1975b; Steriade et al., 1984; Rouiller et al., 1985; Velayos et al., 1989; Harting et al., 1991; Ray et al., 1992; Lozsádi, 1995; Crabtree, 1996). The anterior nuclei of primates do receive a reticular nucleus projection (Kultas-Ilinsky et al., 1995).

6. CHEMICAL IDENTITIES OF RELAY NEURONS

6.1. SPECIFIC EXPRESSION OF ALPHA TYPE II CALCIUM/CALMODULIN-DEPENDENT PROTEIN KINASE

Relay cells can be distinguished from interneurons by lack of gene expression for glutamic acid decarboxylase (Benson et al., 1991b, 1992) and by lack of immunostaining for this enzyme or its product, GABA, in monkeys and other species (Ohara et al., 1983; Penny et al., 1983; Fitzpatrick et al., 1984; Penny et al., 1984; Montero and Singer, 1985; Montero and Zempel, 1986; Smith et al., 1987; Hunt et al., 1991; Fig. 21). Expression of the alpha subunit of type II calcium calmodulin dependent protein kinase (CAM II K-α), an enzyme that is specifically associated with neurons utilizing excitatory amino acid transmitters in the forebrain, represents a positive marker for relay neurons in the thalamus of primates and other species (Benson et al., 1991a,b, 1992; Jones et al., 1994; Figs. 21, 22). Immunoreactivity for the excitatory amino acid transmitter, glutamate, is found in relay cells of the lateral geniculate nucleus and in the terminals of their axon collaterals in the reticular nucleus (Montero, 1990). In monkeys, all thalamic relay neurons, except a large population in the dorsal lateral geniculate nucleus, express CAM II K-α. In monkeys, for nuclei other than the dorsal lateral geniculate, the sum of CAM II K-α and GAD expressing cells accounts for the total neuronal population. In the dorsal lateral geniculate nucleus, only the small cells of the S laminae and interlaminar zones express CAM II K-α (Benson et al., 1991b; Fig. 21, 22). CAM II K-α is not expressed in the reticular nucleus, other components of the ventral thalamus, or in the epithalamus. The other three subunits of the kinase are expressed in the magno- and parvocellular layers of the dorsal lateral geniculate nucleus, and two of them (β and δ), selectively, in the reticular nucleus (Fig. 23) (E.G. Jones, B. Tighilet and M.M. Huntsman, in preparation). CAM II K-α is particularly associated with the postsynaptic membrane densities of corticothalamic terminals; it is absent from both the pre- and postsynaptic sides of GABA synapses (Liu and Jones, 1996). Its association with NMDA receptors and the induction of long-term potentiation in the hippocampus, makes it a candidate for mediating synaptic plasticity in the thalamus as well.

6.2. CALCIUM BINDING PROTEINS

Thalamic cells show differential expression for the calcium binding proteins, parvalbumin, 28KD calbindin and 29KD calretinin (Fig. 24), although there are significant species differences. In the monkey thalamus, with few exceptions, relay neurons are stained either for parvalbumin or calbindin and virtually never for both (Jones and Hendry, 1989). Calretinin cells, which have not been comprehensively studied in primates, appear to represent a subpopulation of the calbindin cells (Campbell and Pandya, 1994). Virtually no GABAergic interneurons stain for any of the calcium

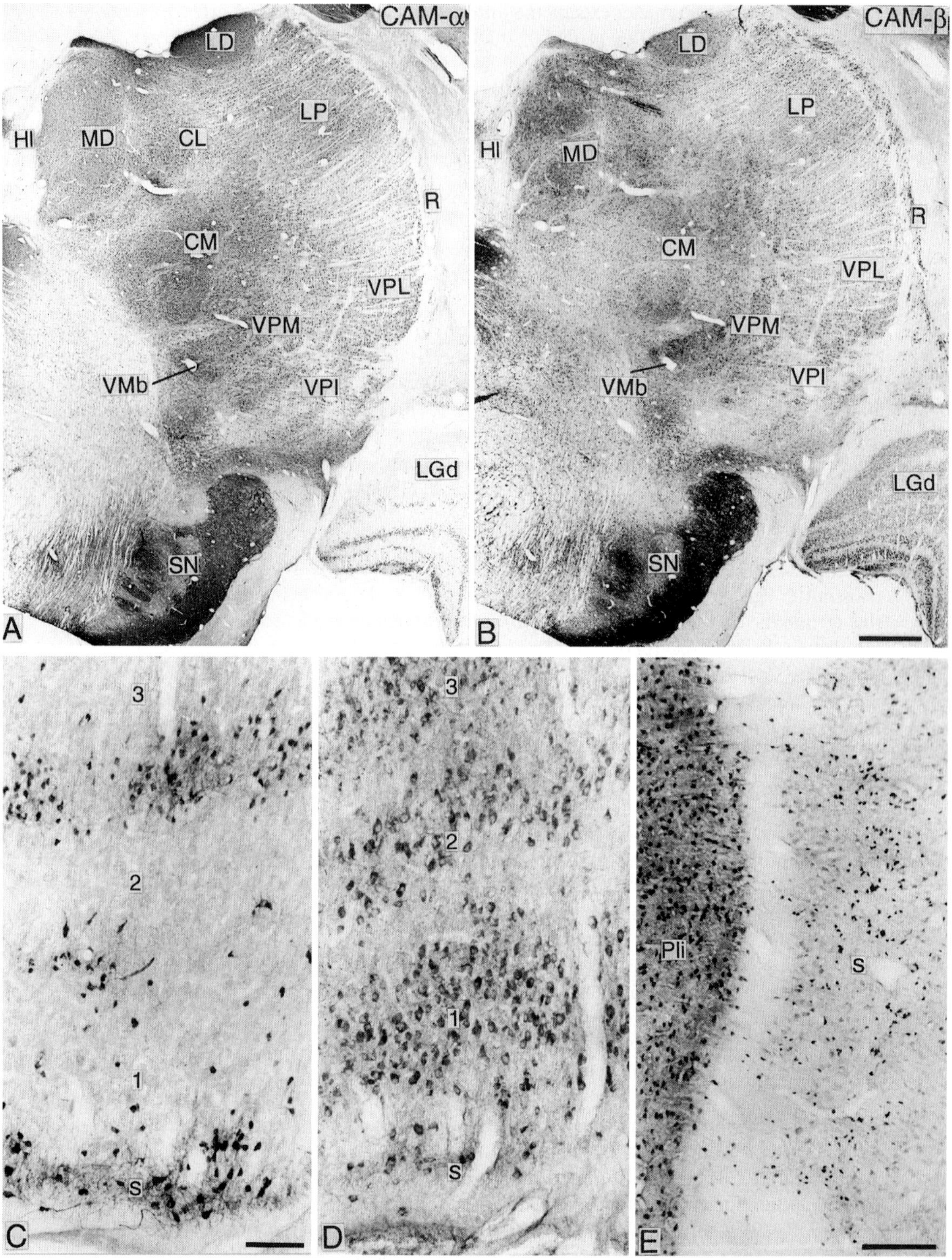
CAM-α
LD
LP
Hl
MD
CL
R
CM
VPL
VPM
VMb
VPl
LGd
SN
A
CAM-β
LD
LP
Hl
MD
R
CM
VPL
VPM
VMb
VPl
LGd
SN
B
3
2
1
s
C
3
2
1
s
D
Pli
s
E

Fig. 22. A, B: Adjacent frontal sections from a macaque thalamus, stained immunocytochemically for alpha (CAM-α) or beta (CAM-β) isoforms of type II calcium/calmodulin dependent protein kinase. Both isoforms label thalamic relay cells but CAM-α is expressed only in the cells of the S laminae and interlaminar plexuses of the lateral geniculate nucleus. In the reticular nucleus, only CAM-β is expressed (c.f. Fig. 21). C,D: adjacent sections through the outer layers of the dorsal lateral geniculate nucleus stained immunocytochemically for CAM-α (C) or CAM-β (D), showing restriction of the former to the S layers and interlaminar zones. E: Continuation of the CAM-α matrix of small cells across the lamina separating the lateral geniculate nucleus from the inferior pulvinar nucleus (Pli). Bars 1 mm (A, B), 100 μm (C,D), 250 μm (E). From unpublished material of B. Tighilet and the author.

binding proteins in the monkey dorsal thalamus. This contrasts with other species in which the GABA cells stain selectively for parvalbumin (Stichel et al., 1988; Demeulemeester et al., 1989). In the monkey, only the GABA cells of the reticular nucleus stain for parvalbumin.

In the monkey thalamus (Figs. 24–27), the calbindin and parvalbumin populations of relay cells show a marked reciprocity of distribution so that some nuclei, for example the ventral posterior, ventral medial geniculate, dorsal lateral geniculate, centre médian and parafascicular, possess a preponderance of, or exclusively parvalbumin cells, while others such as the anterior intralaminar, lateral dorsal, anterior pulvinar and dorsal medial geniculate, show a preponderance of calbindin cells (Jones and Hendry, 1989). The enhanced cellular staining of a particular nucleus for one of the calcium binding proteins is usually accompanied by staining of afferent fibers ending in that nucleus for the same calcium binding protein. Thus, the ventral posterior, ventral medial geniculate and dorsal lateral geniculate nuclei, which contain a majority of parvalbumin cells, are innervated by large numbers of parvalbumin fibers in the medial lemniscus, brachium of the inferior collliculus and optic tract. By contrast, heavier calbindin staining of nuclei such as the posterodorsal medial geniculate nucleus is associated with innervation by large numbers of calbindin stained fibers ascending in the lateral tegmental tracts of the brainstem (Molinari et al., 1995).

The intralaminar nuclei in the monkey display a remarkable reciprocity of immunostaining for parvalbumin and calbindin. All the anterior intralaminar nuclei, including those at the midline, the magnocellular ventral anterior nucleus and the part of the central lateral nucleus commonly included in the mediodorsal nucleus, are heavily stained for calbindin, as is the limitans-suprageniculate nucleus. By contrast, the posterior intralaminar nuclei, that is the parafascicular and centre médian, are heavily stained for parvalbumin. Within some of the anterior intralaminar nuclei, there are patches of parvalbumin stained cells among the majority, calbindin population but in the parafascicular and centre médian nuclei the populations are almost purely parvalbumin positive.

Among the principal relay nuclei, some, such as the anterior pulvinar and lateral dorsal nuclei, contain only calbindin cells while others, such as the principal laminae of the dorsal lateral geniculate nucleus, contain only parvalbumin cells. Most of the relay nuclei, however, although dominated by one cell type, contain mixed populations of cells. Only rarely are the two calcium binding proteins colocalized in the same cell.

Calbindin and parvalbumin immunostaining of particular populations of relay neurons has been investigated in the ventral posterior nuclei of monkeys and prosimian primates and in the medial and lateral geniculate and inferior pulvinar nuclei of monkeys (Jones and Hendry, 1989; Rausell and Jones 1991a,b; Rausell et al., 1992a; Hashikawa et al., 1991; Diamond et al., 1993; Cusick et al., 1993; Molinari

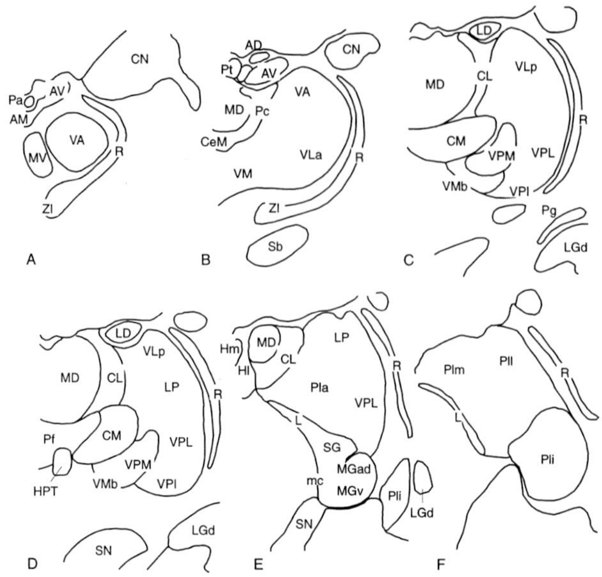

Fig. 23. Part 1. Camera lucida drawings of Nissl-stained frontal sections from the same macaque brain and adjacent to those illustrated in the succeeding four parts of this figure. In all parts, sections are in anterior (A) to posterior (F) order. Parts 2 through 5 are from autoradiograms showing in situ hybridization of RNA probes complementary to mRNAs of alpha, beta, gamma and delta subunits of type II calcium/calmodulin dependent protein kinase. Gamma subunit is expressed very weakly. Beta and delta subunits, but not the alpha subunit are expressed in the reticular nucleus and in principal layers of the dorsal lateral geniculate nucleus. The subunits also show differential expression in the dorsal thalamus, especially in the intralaminar nuclei. Bars 1mm. From unpublished material of B. Tighilet and the author.

et al., 1995). In all of these nuclei, the calbindin cells are relatively small and diffusely distributed, forming a background matrix to the nucleus. In the ventral posterior and the geniculate nuclei, they project mainly or exclusively to superficial layers of the cerebral cortex. Parvalbumin cells, on the other hand, are large and form the highly topographically organized set of relay cells that projects to middle layers of the cortex.

In the ventral posterior nuclei (Rausell and Jones, 1991a; Rausell et al., 1992), calbindin cells of small size (ca. 200 μm^2) form a matrix to all the nuclei that make up this complex and are especially concentrated in nuclei and subnuclei in which histochemical staining for the mitochondrial enzyme, cytochrome oxidase (CO), is low (Fig. 27; Section 10–7). This includes the small celled matrix of the ventral posterior medial nucleus (VPM), irregular islands of the ventral posterior lateral nucleus (VPL), the ventral posterior inferior nucleus and the basal ventral medial nucleus. The calbindin positive matrix extends without interruption at nuclear borders into adjacent nuclei such as the ventral lateral posterior, and forms the predominant population of cells in other adjacent nuclei such as the anterior pulvinar and posterior in which cytochrome oxidase (CO) staining is weak. By contrast, parvalbumin positive relay cells are all of larger size (250 μm^2) and are concentrated in nuclei and subnuclei showing higher densities of larger neurons and dense histochemical staining for CO. Thus, the elongated, rod-like aggregations of cells in VPM, most parts of VPL, and the ventral lateral posterior nucleus have high densities of parvalbumin cells and relatively few calbindin cells. The fibers of the medial lemniscus, spinothalamic tract and ascending trigeminal pathways mainly terminate in relation to one or other of the two domains

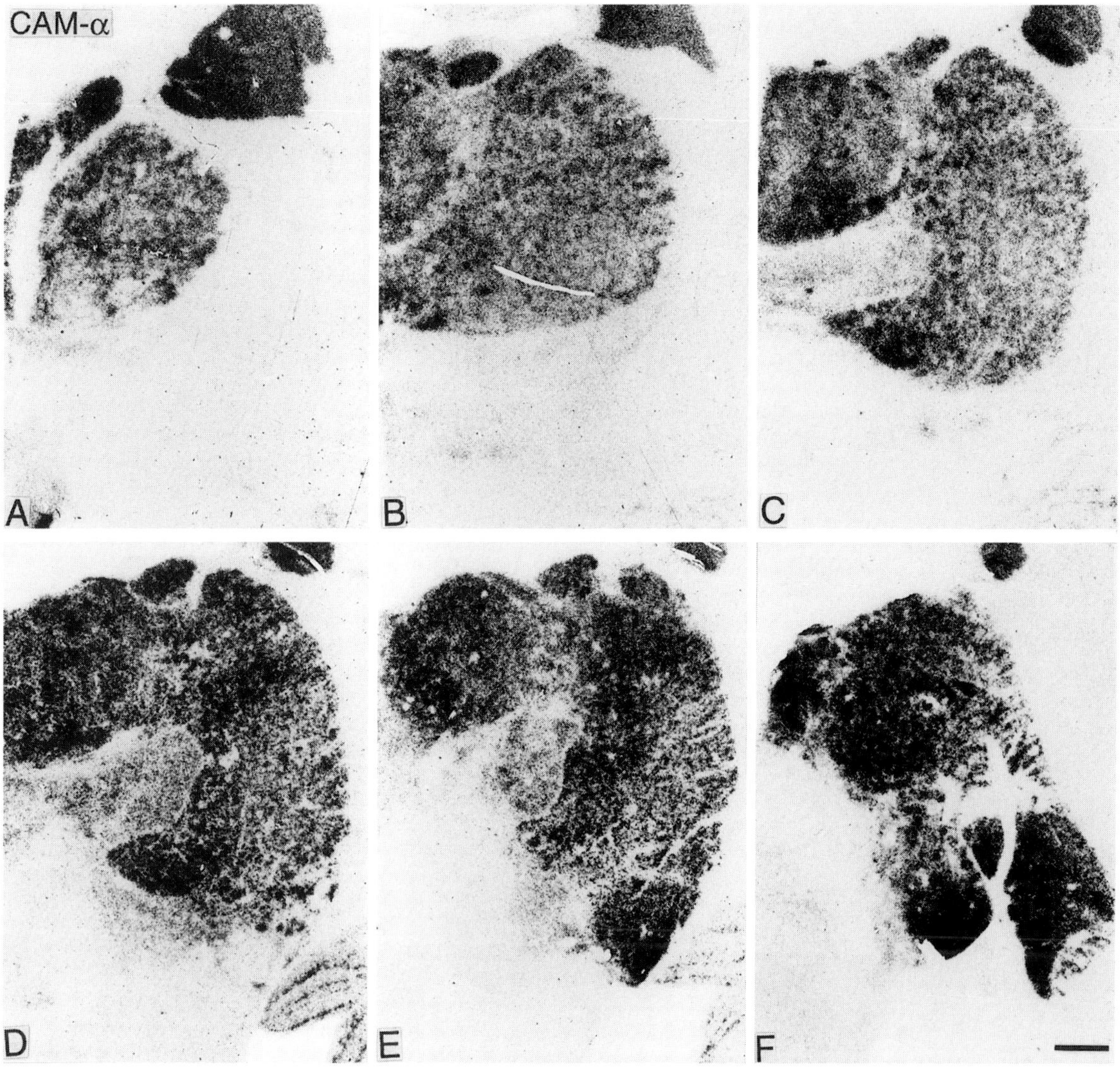

Fig. 23. (continued)

(Rausell and Jones, 1991b; Rausell et al., 1992a). Spinothalamic and caudal trigeminothalamic fibers terminate predominantly in the CO-weak, smaller-celled, calbindin positive matrix of VPM or VPL, extending their terminals into this matrix where it forms parts of the posterior, anterior pulvinar and ventral lateral nuclei. By contrast, medial lemniscal and principal trigeminothalamic fibers terminate in relation to the CO-dense, larger celled, parvalbumin positive zones, exclusively in VPL or VPM.

The calbindin positive cells of the matrix appear to project predominantly to superficial layers of the somatosensory cortex since they can be selectively labeled by retrograde transport when fluorescent dyes are applied to the surface of the cerebral cortex, affecting layers I and II (Rausell and Jones, 1991b; Rausell et al., 1992a). Those in the ventral posterior nuclei can also be labeled from the motor cortex, implying a relatively difffuse projection. The parvalbumin cells, by contrast, project only to middle

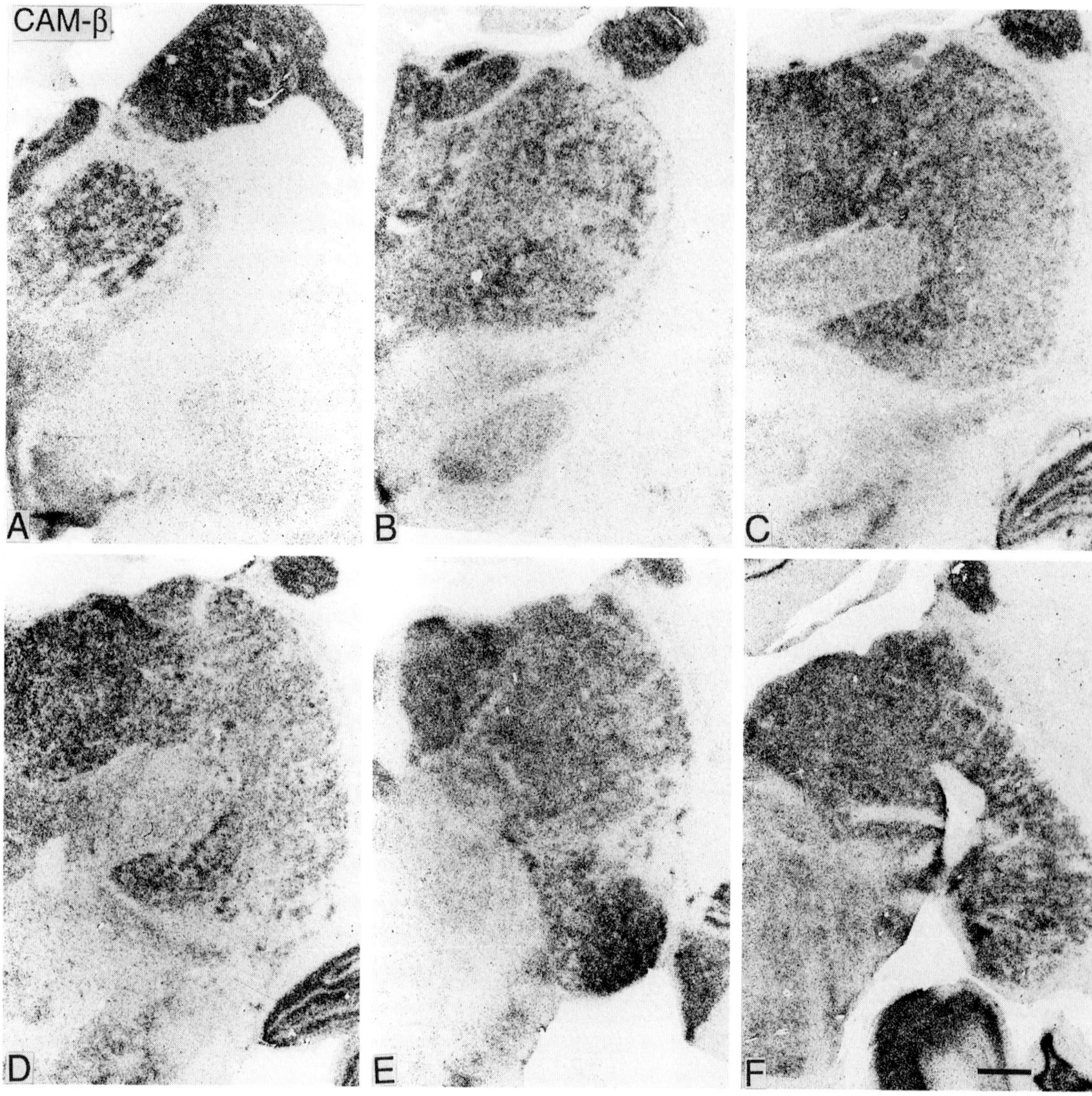

Fig. 23. (continued)

layers of the somatosensory cortex, being retrogradely labeled by dyes injected into the middle layers of the cortex. The focal nature of this labeling indicates a topographically ordered projection.

The connectional relationships of calbindin and parvalbumin cells in the ventral posterior nucleus still remain fully to be explored by electron microscopy. As defined by light microscopy, however, there are two parallel streams running through the ventral posterior nuclei of primates: one, formed by a diffuse matrix of small, calbindin cells, transfers spinothalamic and caudal trigeminothalamic influences to superficial layers of the sensory-motor and adjacent areas of the cerebral cortex; a second, formed by a core of larger, topographically-organized parvalbumin cells, transfers leminiscal influences specifically to middle layers of the somatosensory cortex.

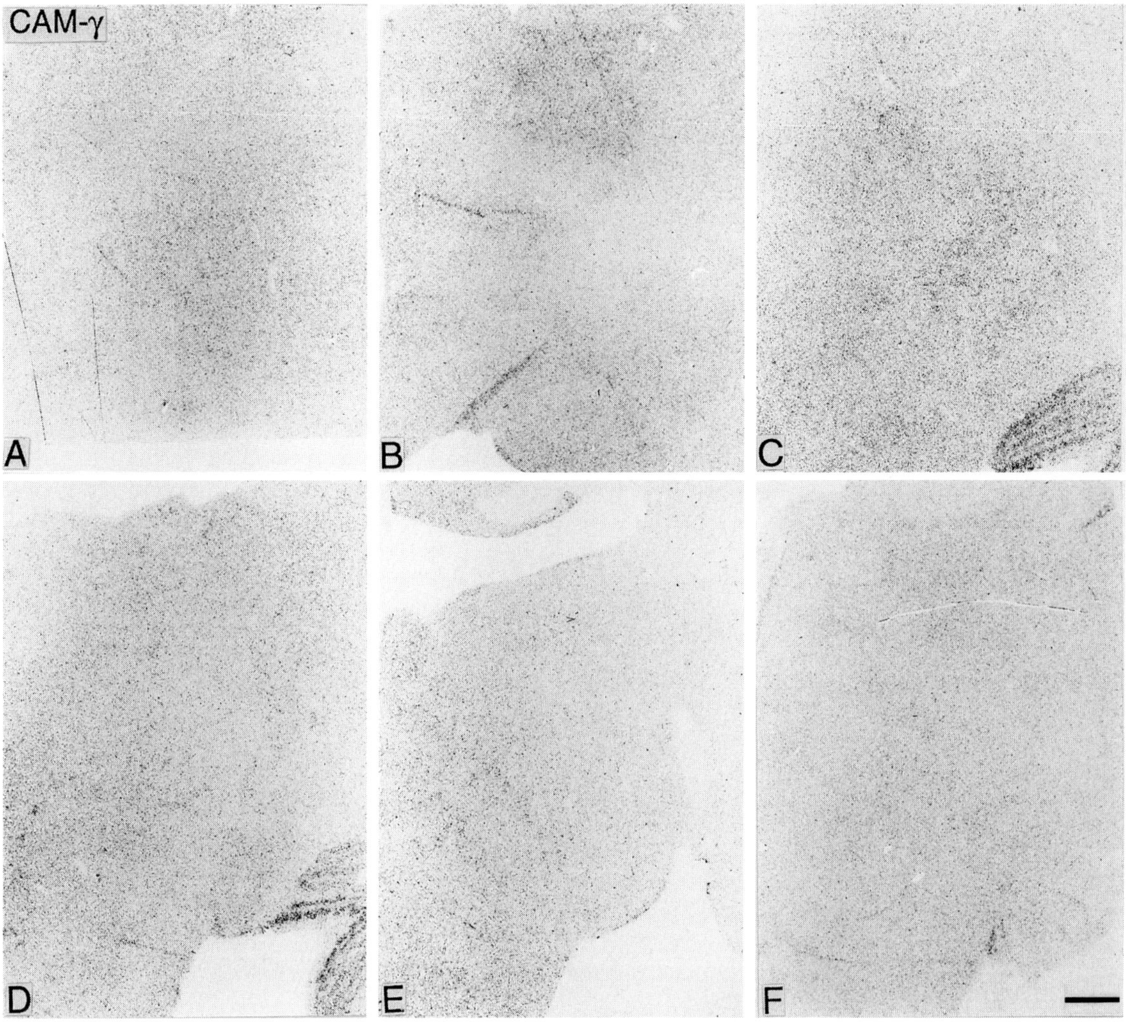

Fig. 23. (continued)

Similar parallel streams of connections based on the differential distributions and area- and layer-specific projections of calbindin and parvalbumin positive relay cells can be found in the medial and lateral geniculate nuclei of monkeys. The four nuclei of the medial geniculate complex in Old World monkeys are distinguished by the presence of different proportions of parvalbumin and calbindin relay cells (see Section 10.2). The ventral nucleus contains almost exclusively parvalbumin cells and only very rare calbindin cells. The anterodorsal nucleus contains a predominance of parvalbumin cells but moderate numbers of calbindin cells as well. The posterodorsal nucleus contains approximately equal proportions of parvalbumin and calbindin cells but calbindin cells dominate the posterior pole of the nucleus. The magnocellular nucleus contains both kinds of cell arranged in clusters. The calbindin cells of each nucleus project relatively widely over two or more fields of the auditory cortex and their axons terminate in superficial layers, including layer I. The parvalbumin cells, on

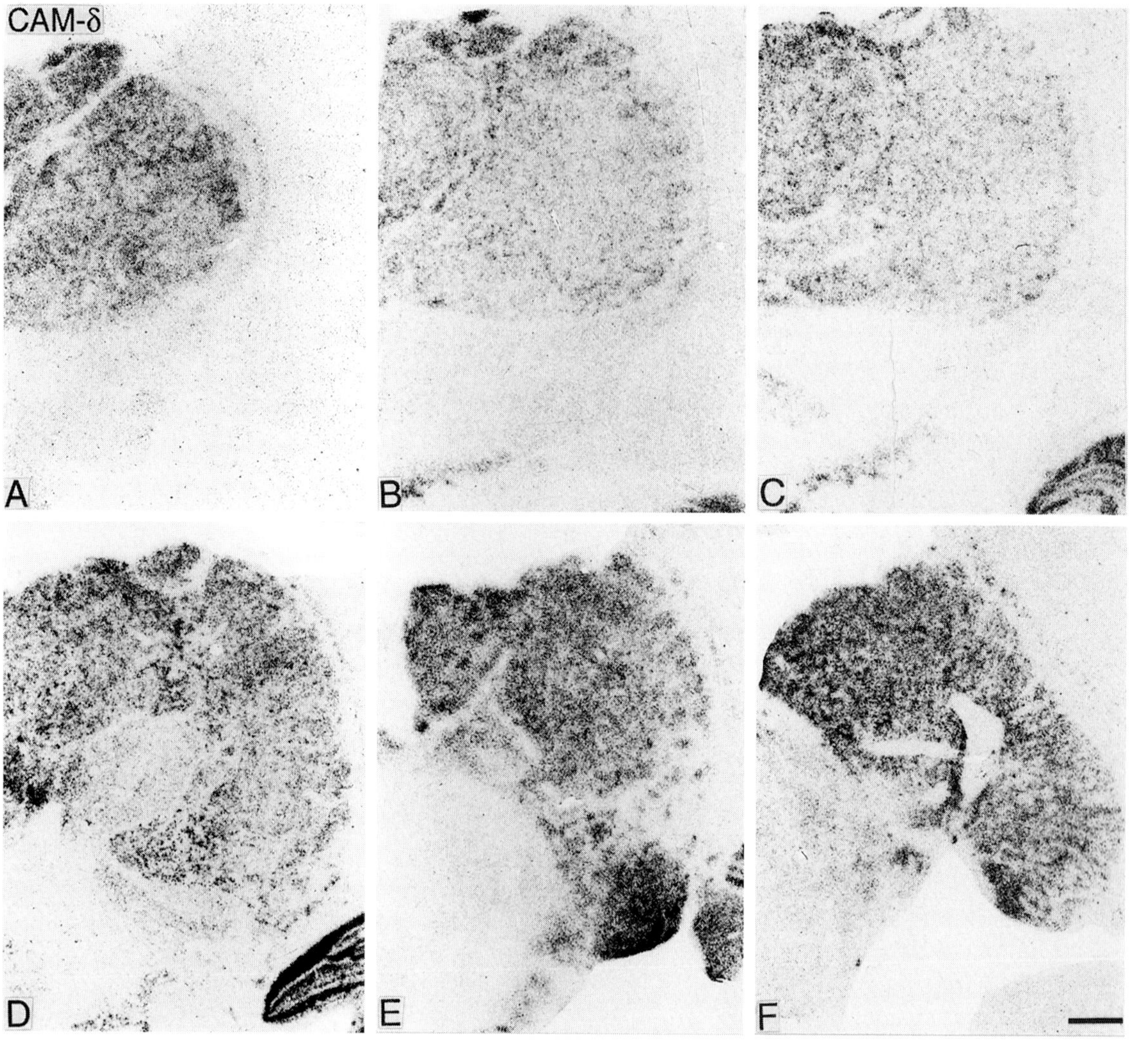

Fig. 23. (continued)

the other hand, project in an area-specific and topographically-organized manner to specific fields of the auditory cortex (Molinari et al., 1995), their axons ending in middle layers (Hashikawa et al., 1991, 1995). Differences in the density of parvalbumin staining in each of the auditory cortical fields reflect differences in the numbers of parvalbumin positive medial geniculate cells projecting to each field (Jones et al., 1995). When a retrogradely transported tracer is injected in an auditory cortical field and affects all layers, a topographically-related focus of parvalbumin cells is labeled in a single nucleus and a relatively diffuse population of labeled calbindin cells is scattered across two or more nuclei. The calbindin cells, thus, seem to form a diffusely projecting small-celled matrix to the medial geniculate complex, as in the ventral and adjacent nuclei. The calbindin matrix and the parvalbumin stained cells may be specifically innervated by different ascending auditory pathways, as in the somatosensory thalamus. The ventral and anterodorsal medial geniculate nuclei are innervated by large numbers of parvalbumin-immunoreactive fibers ascending in the brachium of the inferior colliculus from the central nucleus of the inferior colliculus; the posterodorsal nucleus is innervated by calbindin immunoreactive fibers ascending in the midbrain tegmentum. The magnocellular nucleus appears to be innervated by both sets of fibers and by fibers from other sources.

The differential projections of parvalbumin and calbindin relay cells in the medial geniculate complex also indicate the presence of two parallel pathways through this part of the thalamus, one characterized by parvalbumin immunoreactivity and focused on middle layers of specific cortical areas, the other characterized by calbindin immunoreactivity and distributed to superficial layers over wide areas of cortex. The parvalbumin pathway predominates in those nuclei of the medial geniculate complex characterized by frequency specificity and reliability of neuronal responses to auditory stimuli, while the calbindin pathway predominates in those nuclei in which neurons are not frequency specific, less responsive, labile, or respond to convergent stimuli (Aitkin 1973; Calford and Webster, 1981; Calford 1983; Imig and Morel, 1985a,b; Allon and Yeshurun, 1985).

In the dorsal lateral geniculate nucleus of monkeys, parvalbumin positive relay cells are found in large numbers exclusively in the six principal laminae of the nucleus, while calbindin positive, smaller relay cells occupy only the S laminae and interlaminar zones (Jones and Hendry, 1989) in a distribution identical to that of CAM II K–α positive cells (Fig. 24 and see Section 6.1). Both parvalbumin and calbindin positive fibers are found in the optic tract but it has not yet been determined if the two sets of relay cells are innervated selectively by parvalbumin or calbindin fibers of retinal origin. The S laminae and interlaminar zones are reportedly innervated mainly by the superior colliculus (see Jones, 1985 and Harting et al., 1991a). The two sets of cells appear to preserve the layer-specific types of projections found in the somatosensory and auditory thalamic nuclei. The principal laminae project to layer IV of area 17, while the S laminae and interlaminar zones project to superficial layers of areas 17 and 18 (Fitzpatrick et al., 1983; Weber et al., 1983; Hendry and Yoshioka, 1994). The small calbindin cells of the S laminae and interlaminar zones can thus be seen as resembling the matrix of the ventral posterior nucleus and medial geniculate complex. In the prosimian, *Galago*, the two small-celled layers (4 and 5) which can be likened to the interlaminar zones of the monkey lateral geniculate nucleus (see above), also stain specifically for calbindin and project to superficial layers of the visual cortex (Diamond et al., 1993; Johnson and Casagrande, 1995). The cells of the two parvocellular layers (3 and 6) stain only for parvalbumin and project to middle layers, as in the monkey. The layers comparable to the magnocellular layers of the monkey (1 and 2) contain both calbindin and parvalbumin cells in the *Galago*.

In the inferior pulvinar nucleus, large patches of weak CO activity coincide with subdivisions of the nucleus relatively lacking in parvalbumin immunoreactive cells and dominated by calbindin immunoreactive cells (Cusick et al., 1993). Other CO-rich subdivisions contain both parvalbumin and calbindin cells while others show very few calbindin cells. The calbindin cells which are also strongly immunoreactive for the alpha isoform of type II calcium/calmodulin-dependent protein kinase, appear to form a matrix of small cells continuous with that in the S layers and interlaminar zones of the dorsal lateral geniculate nucleus (Tighilet et al., 1998).

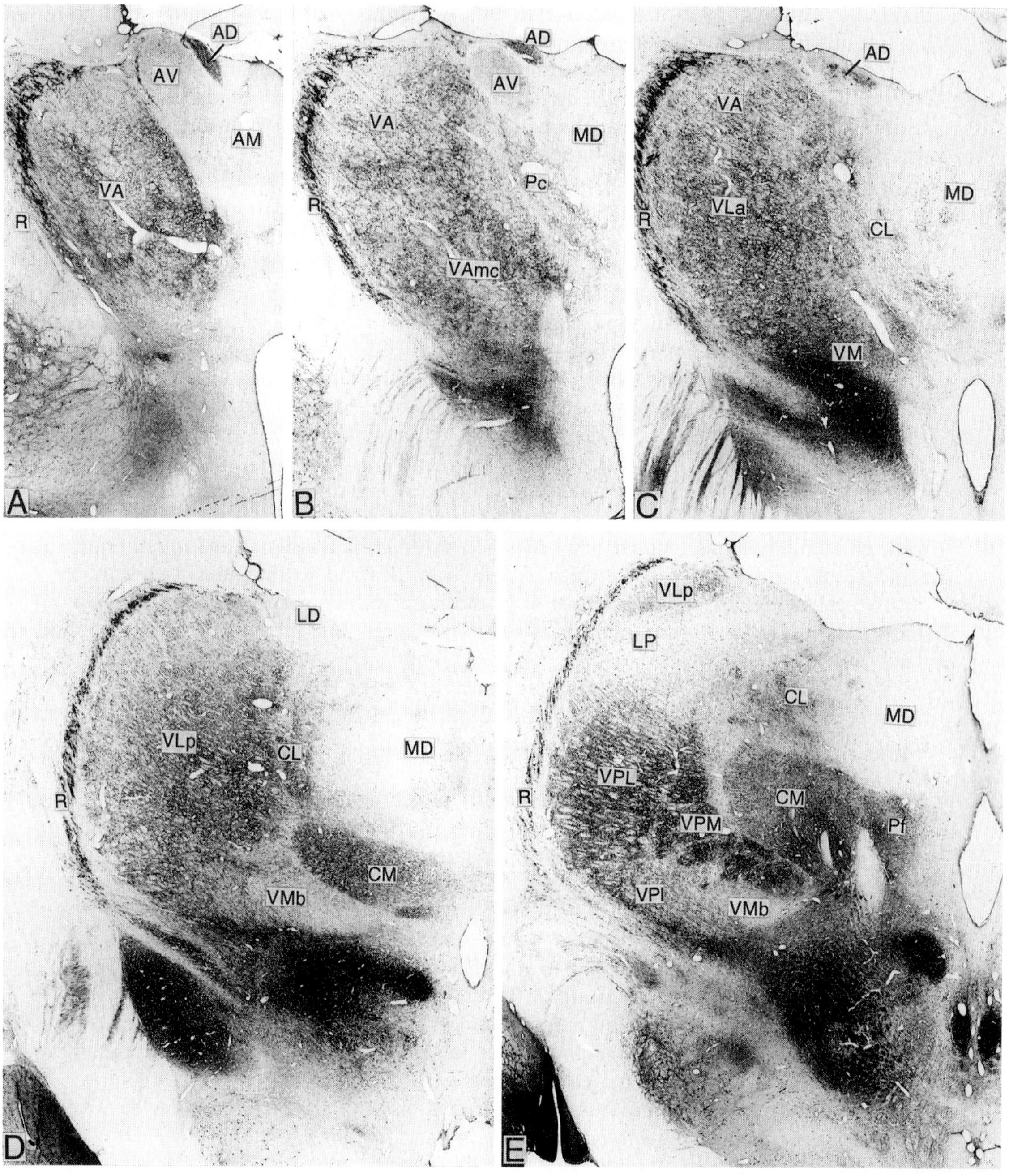

Fig. 24. A-E and K-N are frontal sections in anterior (A) to posterior (N) order through the thalamus of a macaque, stained immunocytochemically for parvalbumin. F-J and O-R are matched sections at approximately the same levels stained for 28kDa calbindin. Parvalbumin is heavily expressed in cells of certain nuclei only, while calbindin cells form a matrix throughout all nuclei, becoming more heavily concentrated in nuclei and parts of nuclei in which parvalbumin staining is weak or absent. Bars 1 mm. From material illustrated in Jones and Hendry (1989).

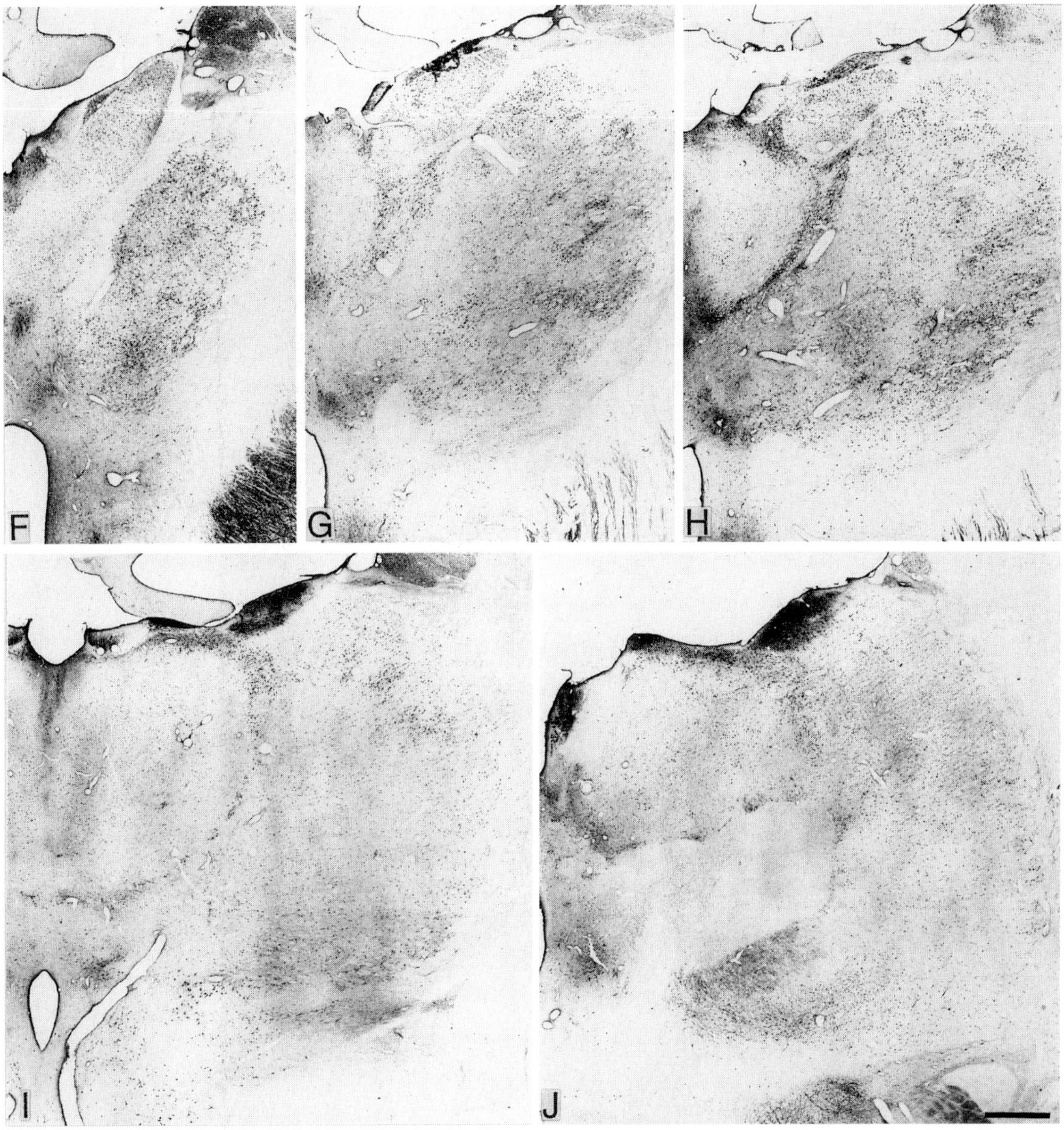

Fig. 24. (continued)

6.3. CAT301

CAT301, a monoclonal antibody that recognizes a polysaccharide-rich proteoglycan on the surfaces of specific populations of neurons (Hockfield and McKay, 1983, 1985; McKay and Hockfield, 1983; Hockfield et al., 1983; Hendry et al., 1988d), stains relay cells only in certain nuclei of the monkey and shows an unusual affinity for certain cell types (Hendry et al., 1988d) (Fig. 28). In the monkey thalamus, the dorsal and ventral lateral geniculate nuclei, the ventral nuclei, the anterior intralaminar nuclei, the ante-

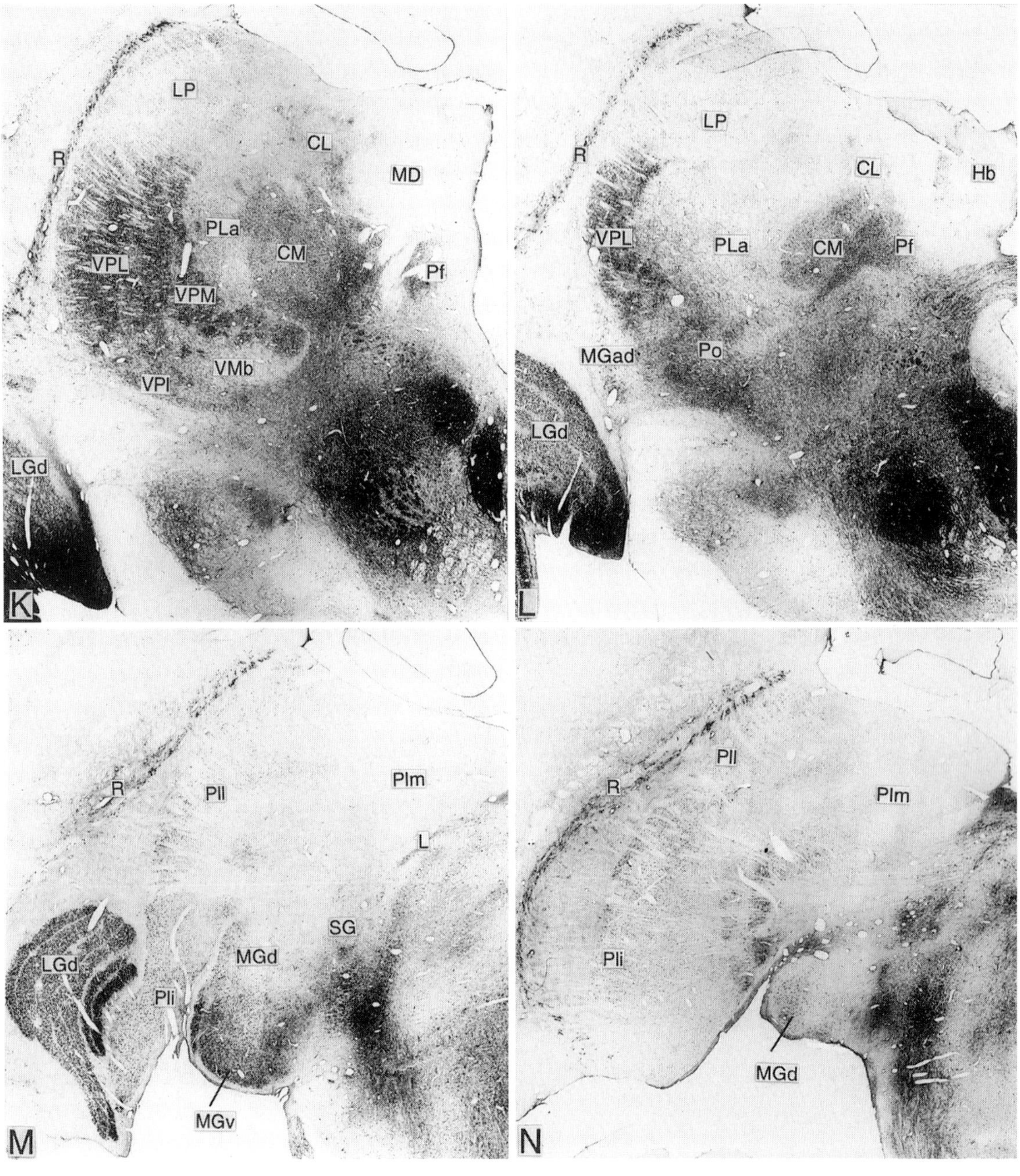

Fig. 24. (continued)

rodorsal nucleus, magnocellular medial geniculate nucleus, posterior complex, zona incerta and reticular nucleus are stained by CAT301. In the dorsal lateral geniculate nucleus, there is a differential staining pattern that may represent a particular association with one class of relay cell; in the magnocellular laminae, 70% of the cells are CAT301 positive and densely stained; in the parvocellular laminae, although 40% of

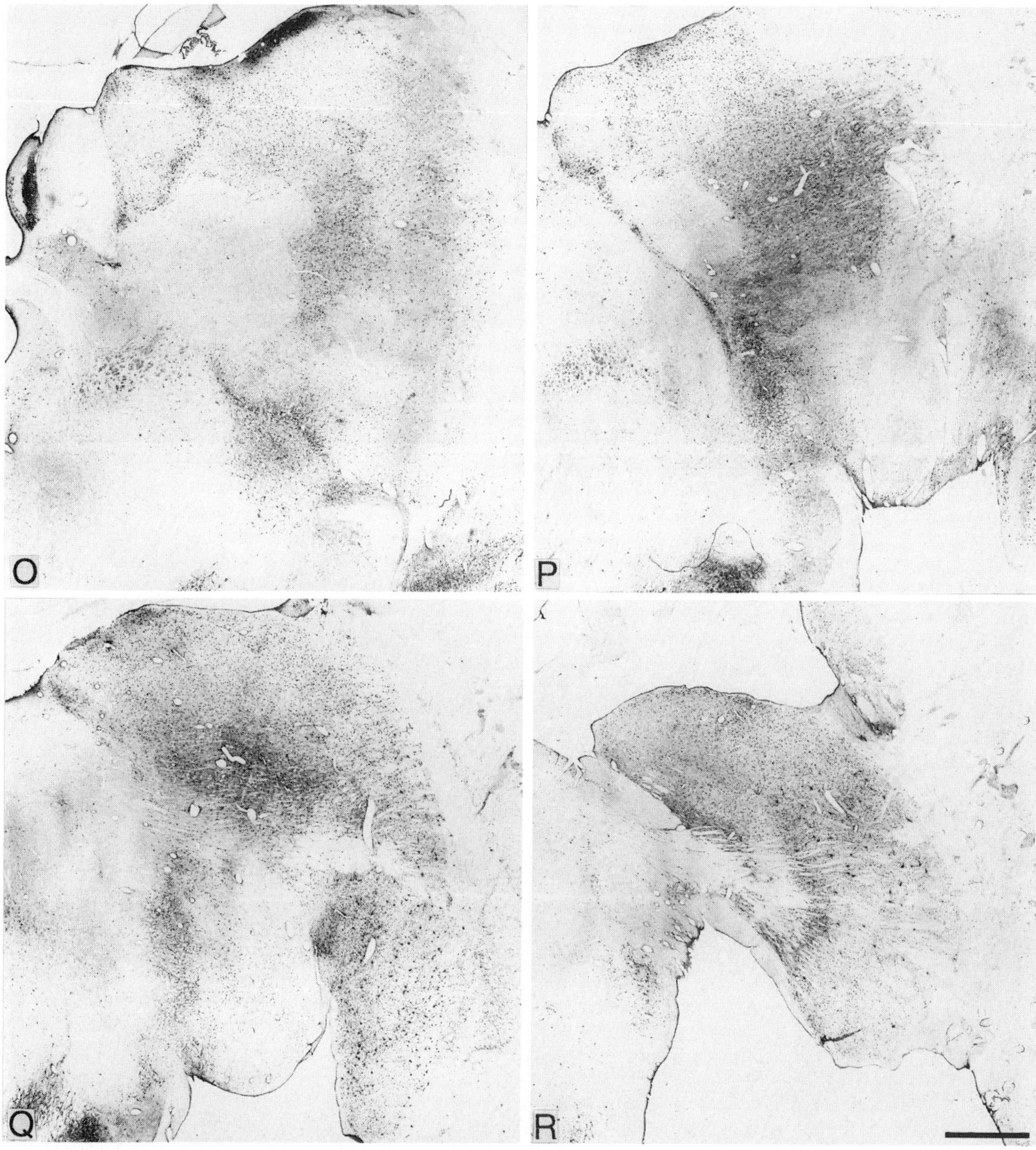

Fig. 24. (continued)

the cells are stained, the staining is weaker. The higher degree of staining of large cells has led to the suggestion that CAT301 is especially associated with the broad band/ motion sensitive relay cells of the dorsal lateral geniculate nucleus (Hendry et al, 1988d). In the ventral posterior nucleus of the monkey, CAT301 staining is associated only with the larger, parvalbumin positive relay cells, and not with the smaller, calbindin positive relay cells typical of the matrix zones (Rausell and Jones, 1991a).

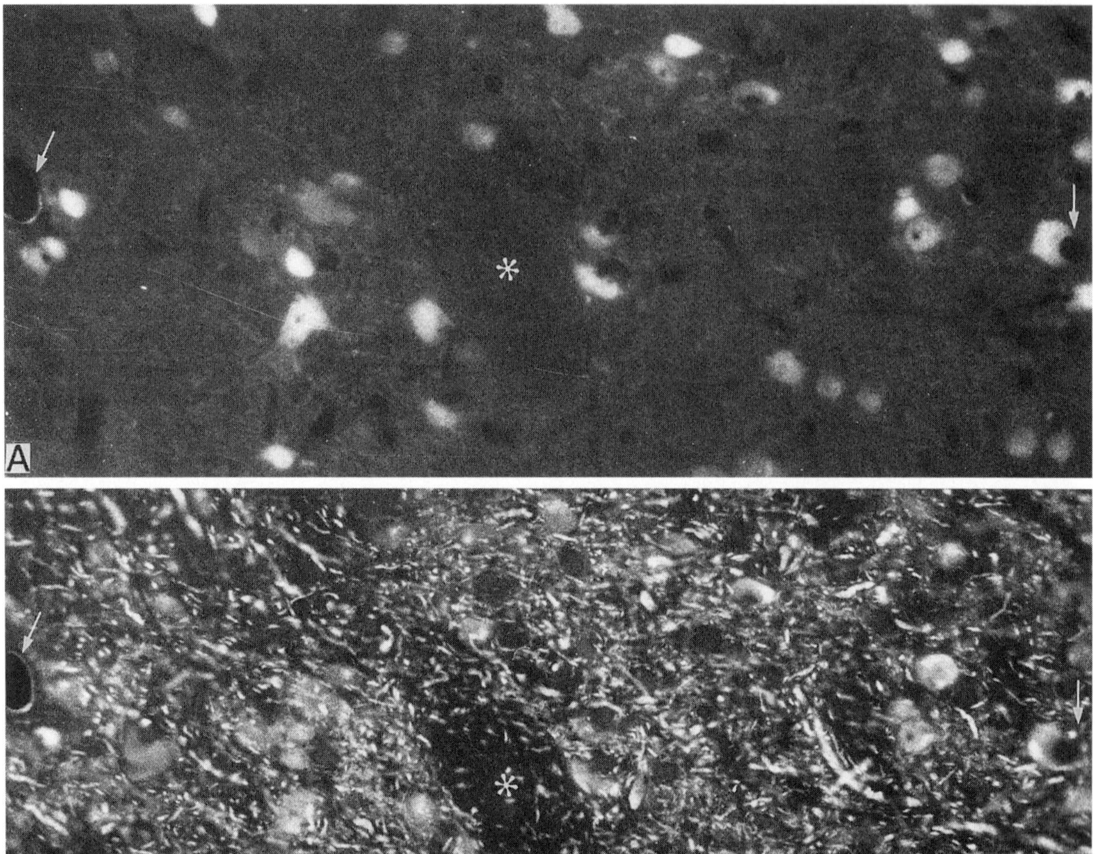
A
B
C

Fig. 25. Fluorescent photomicrographs from the same section through a part of the VPM nucleus of a macaque thalamus, showing cells retrogradely labeled by Fast Blue transported from an injection site in the ipsilateral somatosensory cortex (A), the same cells stained immunocytochemically for parvalbumin by fluorescein immunofluorescence (B), and the different population of intrinsic interneurons stained for GABA by rhodamine immunofluorescence. Arrows indicate same blood vessels, asterisk the same fiber bundle. Bar 100 μm. From material illustrated in Rausell and Jones (1991a).

7. BASIC CIRCUITRY OF THE PRIMATE THALAMUS

7.1. SYNAPTIC STRUCTURE

Although there has been relatively less work devoted to the fine structural and functional organization of the primate thalamus, there have been sufficient anatomical studies, at least, to indicate that the elements of the circuitry are the same as those that have been unraveled by more extensive studies in the cat and other experimental animals. Basic electron microscopical studies that indicate this are those of Colonnier and Guillery (1964); Pecci-Saavedra et al. (1968); Guillery and Colonnier, (1970); Wong-Riley, (1972a); Harding and Powell, (1977); Ralston et al., (1988). Inevitably, however, much of the following discussion depends upon data derived from studies of non-primates.

Most of the principal dorsal thalamic nuclei, including the intralaminar nuclei, have been sampled electron microscopically (Jones, 1985a) but the most extensive studies have been conducted on the dorsal lateral geniculate and ventral posterior nuclei of cats. In the neuropil of both nuclei in cats four types of morphologically distinguishable synaptic terminals can be identified. These have their equivalents in monkeys. They are termed RL, RS, F and PSD terminals (Fig. 29). These terms depend upon the presence in the terminals of round (R) or flattened (F) synaptic vesicles, upon the size of the terminals (L or S) or their identification as presynaptic dendrites (PSD). RL terminals are relatively electron dense, contain large mitochondria and round synaptic vesicles, and terminate in asymmetrical synaptic thickenings. They are present in moderate numbers and represent the terminals of the principal subcortical afferents (medial lemniscus, optic tract, brachium of inferior colliculus, mamillothalamic tract, cerebellothalamic pathway) to the nucleus (Pecci-Saacedra et al., 1968; Guillery and Colonnier, 1970; Harding, 1972; Harding and Powell, 1973, 1977; Wilson and Hendrickson, 1981; Ralston et al., 1988; Kultas-Ilinsky and Ilinsky, 1991; Ralston and Ralston, 1994; Mason et al., 1996). RS terminals, which outnumber RL terminals by at least 3 to 1 (Wong-Riley 1972a), are small, electron dense, contain smaller mitochondria and round synaptic vesicles, and also terminate in asymmetrical synaptic thickenings. They are the terminals of corticothalamic fibers arising in layer VI of the cerebral cortex (Jones and Powell, 1969b; Wong-Riley, 1972b; Harding and Powell, 1973b, 1977; Winfield et al., 1974). F terminals are relatively electron lucent, contain flattened synaptic vesicles in aldehyde-fixed material and terminate in symmetrical synaptic thickenings (Ralston et al., 1988; Mason et al., 1996). They are GABAergic (Ohara et al., 1983, 1989) and have been shown in non-primates to derive from the axons of reticular nucleus cells (Cucchiaro et al., 1991b; Liu et al., 1995b) and probably from axons of intrinsic interneurons when these are present (Montero, 1987). PSD terminals are electron lucent, usually contain only small aggregations of flattened synaptic vesicles and terminate in symmetrical synaptic thickenings. They commonly also contain significant quantities of rough endoplasmic reticulum and clusters of free ribosomes,

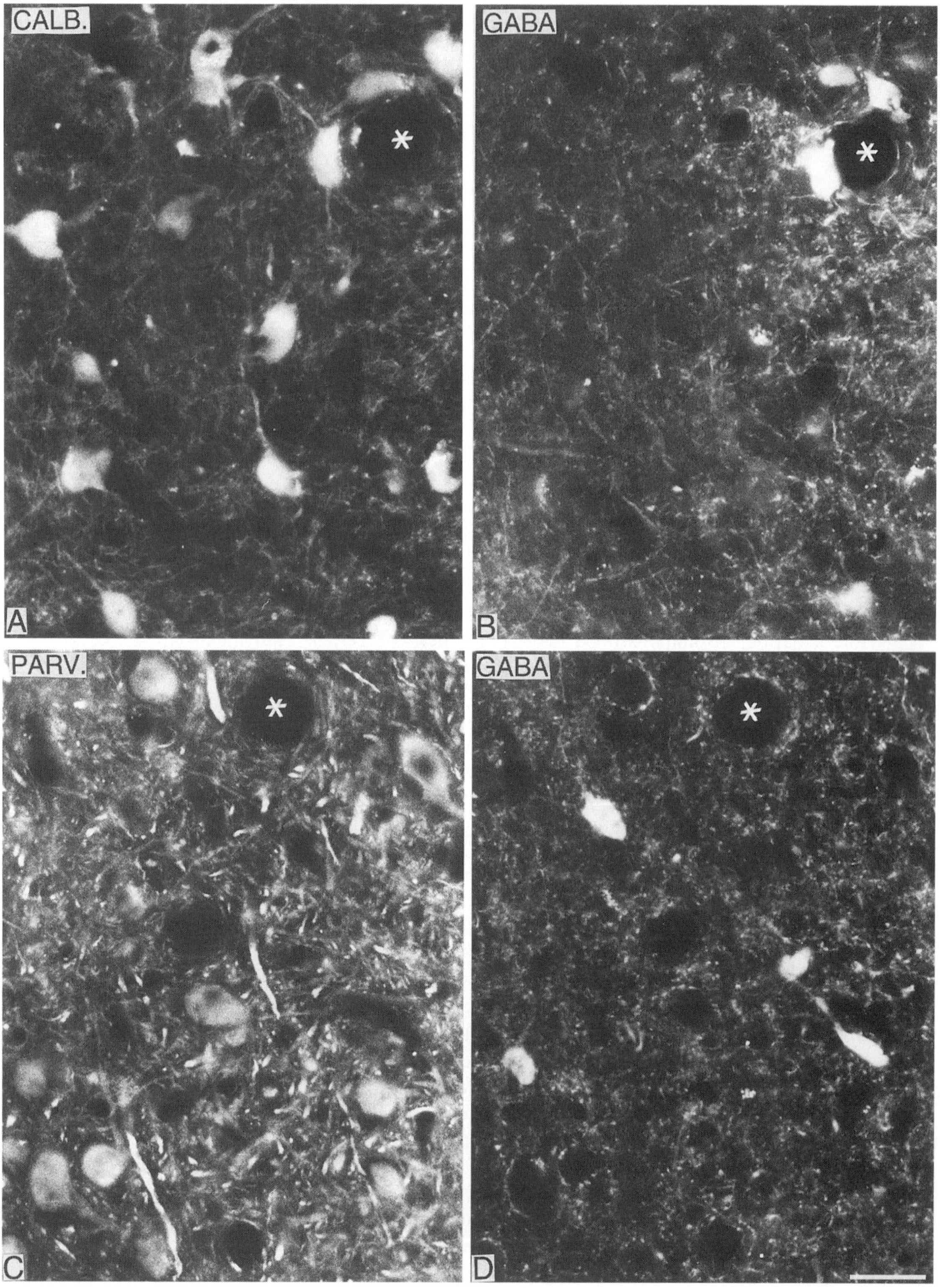
CALB.
GABA
PARV.
GABA
A
B
C
D

Fig. 26. Pairs of fluorescence photomicrographs from the same microscopic fields through the VPM nucleus of macaques, stained immunocytochemically for calbindin (CALB.) and GABA or for parvalbumin (PARV.) and GABA. Asterisks indicate same blood vessels. The cells containing the calcium binding proteins are separate from the population of GABAergic intrinsic neurons. Bars 20 μm. From material illustrated in Rausell and Jones (1991a) and Jones and Hendry (1989).

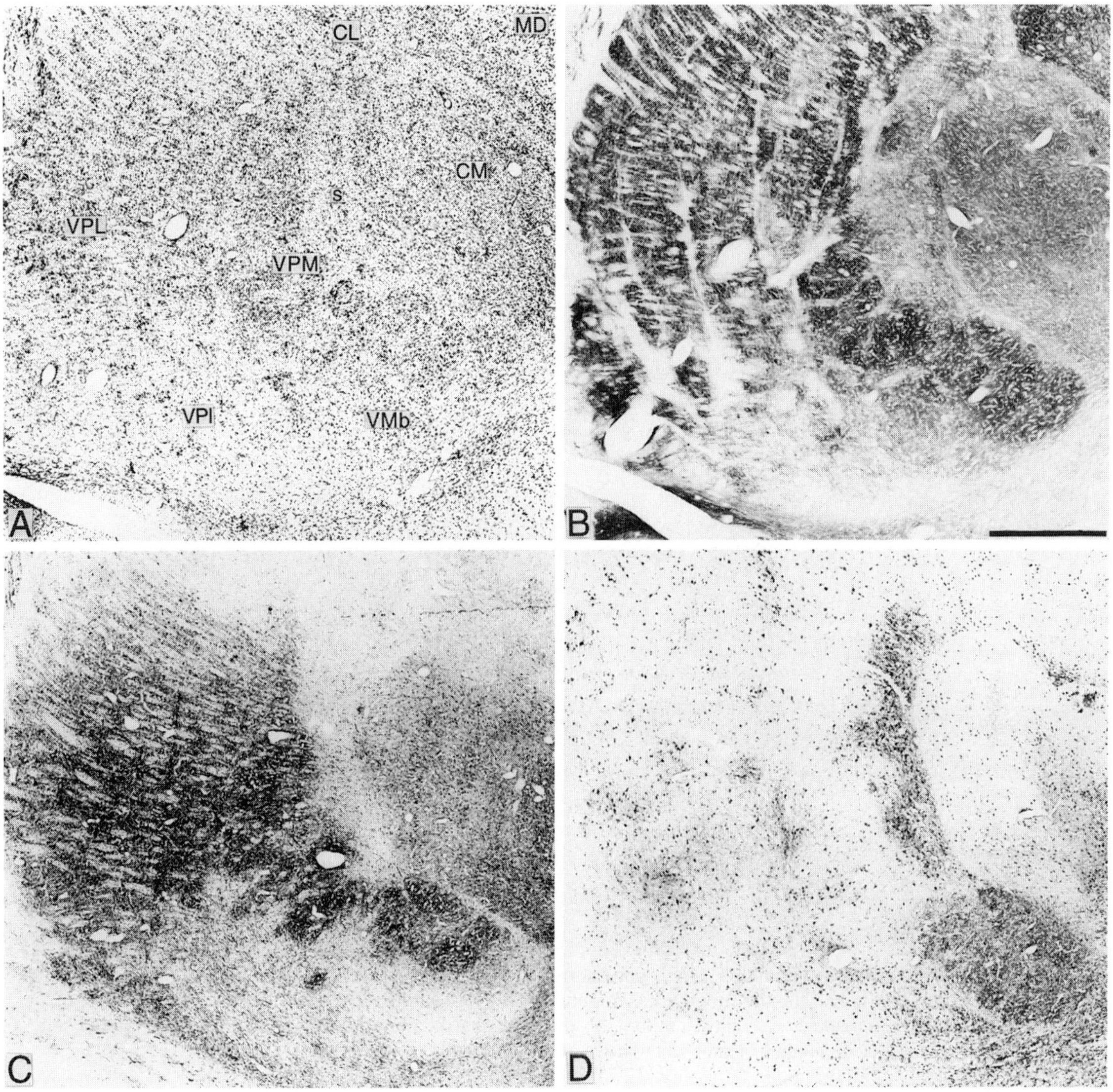

Fig. 27. Frontal sections at the same anteroposterior level through the ventral posterior and adjacent nuclei of a macaque monkey, stained with thionin (A), for cytochrome oxidase (B), or immunocytochemically for parvalbumin (C) or calbindin (D). These show the complementarity in parvalbumin and calbindin expression and the association of high cytochrome oxidase activity with the former. S identifies the small-celled, cytochrome oxidase-weak, calbindin-rich matrix of the VPM nucleus which expands posteriorly (see Fig. 86). Bar 1 mm. From material illustrated in Rausell and Jones (1991b) and Rausell et al. (1992).

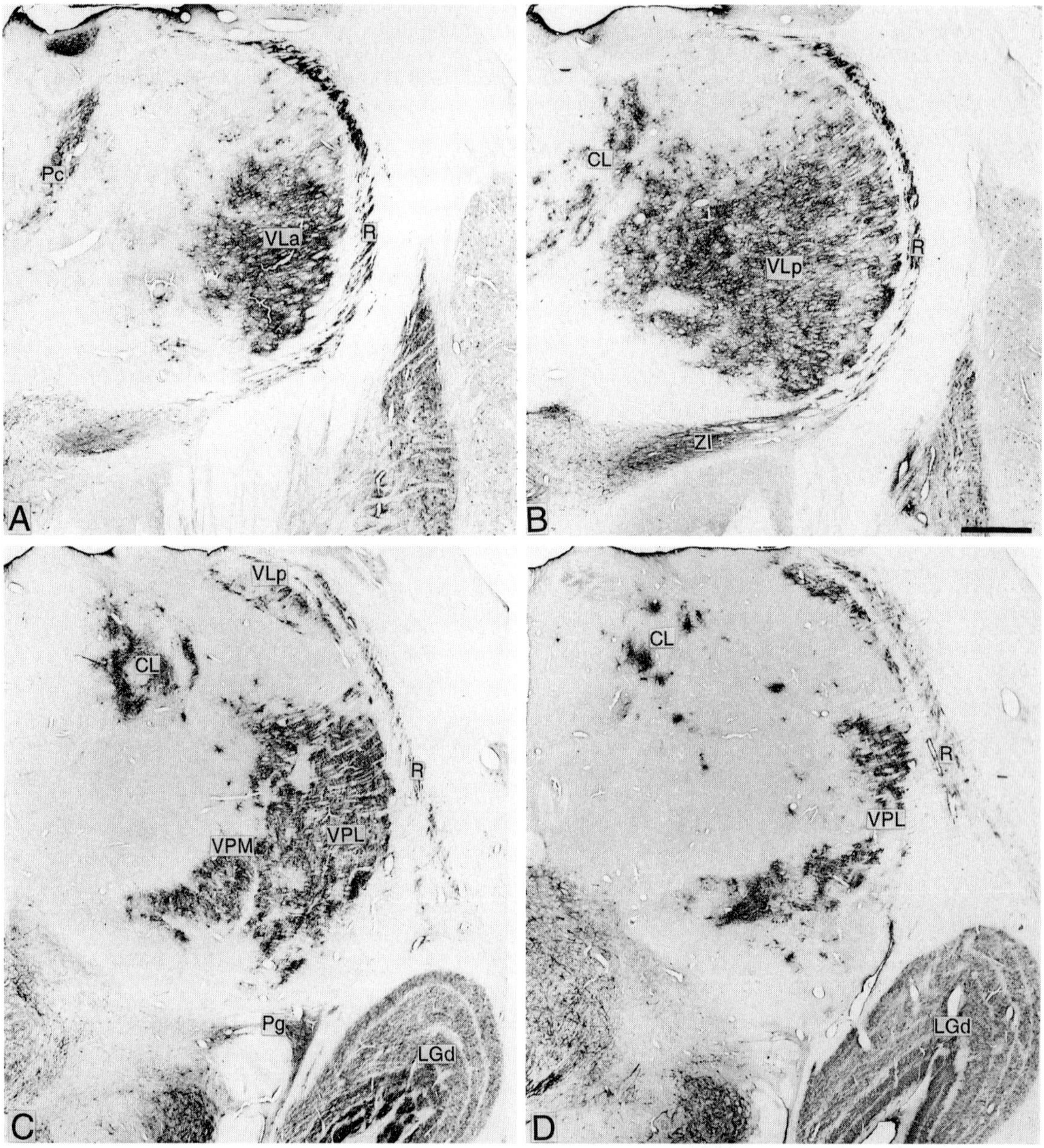

Fig. 28. Frontal sections in anterior (A) to posterior (D) order through the thalamus of a macaque, stained immunocytochemically with the monoclonal antibody, CAT301, which recognizes a proteoglycan on the surfaces of certain thalamic cells only, notably those of the ventral, lateral geniculate, reticular and anterior intralaminar nuclei. Bar 1 mm. From material illustrated in Hendry et al. (1988d).

Figs 29. Parts 1–3. Electron micrographs from the ventral posterior (29A) and dorsal lateral geniculate nuclei (all others) of macaques, showing the various forms of axon terminals and their synaptic arrangements. In B of each part, the afferent terminals of optic tract fibers have been labeled immunocytochemically for parvalbumin. Arrows indicate postsynaptic thickenings. A: axon; D: dendrite; F: terminal of reticular nucleus axon or of intrinsic interneuron; I: interneuron; PSD: presynaptic dendrite of interneuron; RL: lemniscal or optic tract terminal; RS: terminal of corticothalamic fiber. Bars 0.5 μm. From unpublished material of C.S. Bae and the author.

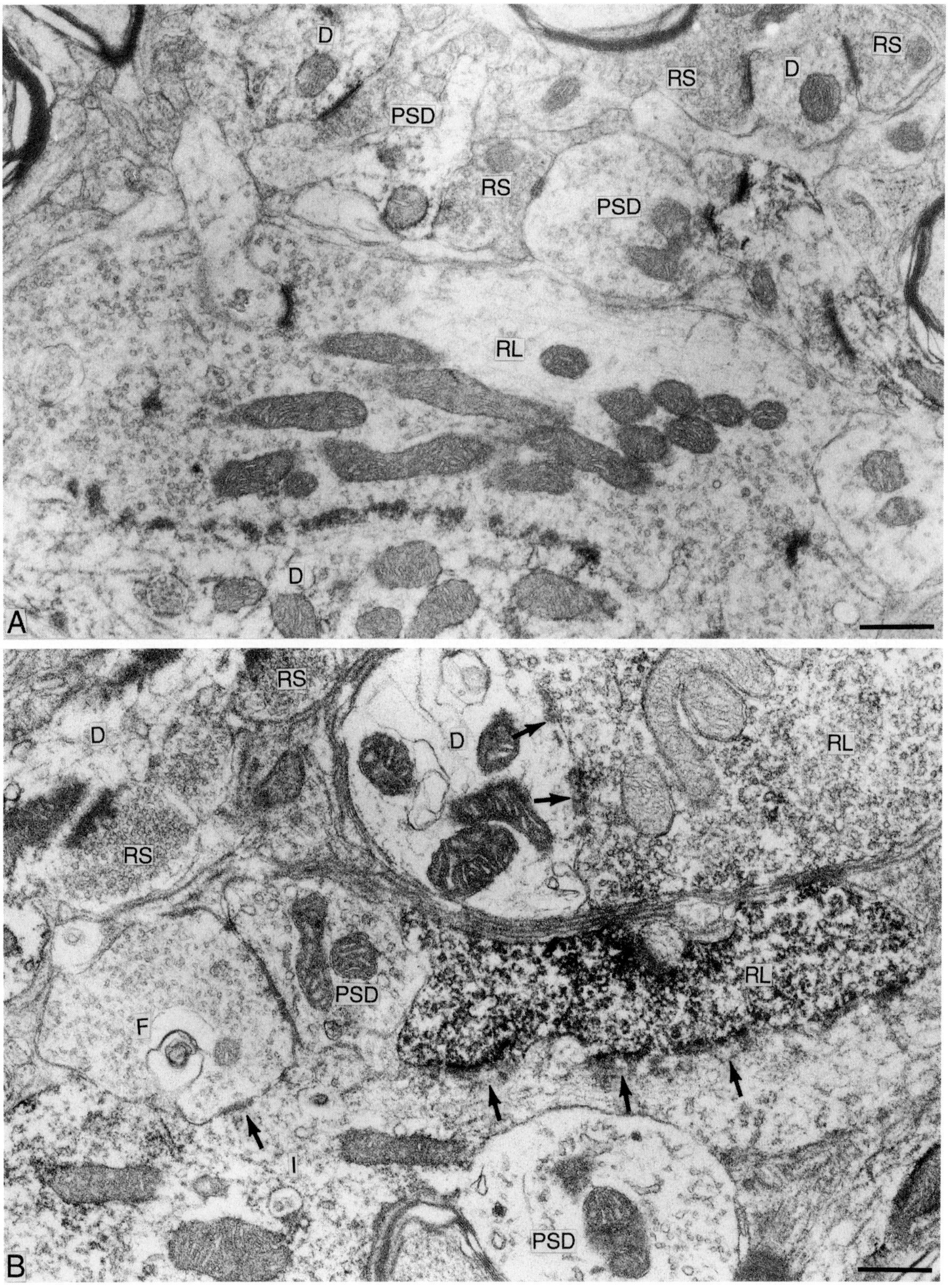
D
RS
D
RS
PSD
RS
PSD
RL
D
A
RS
D
D
RL
RS
PSD
RL
F
I
PSD
B

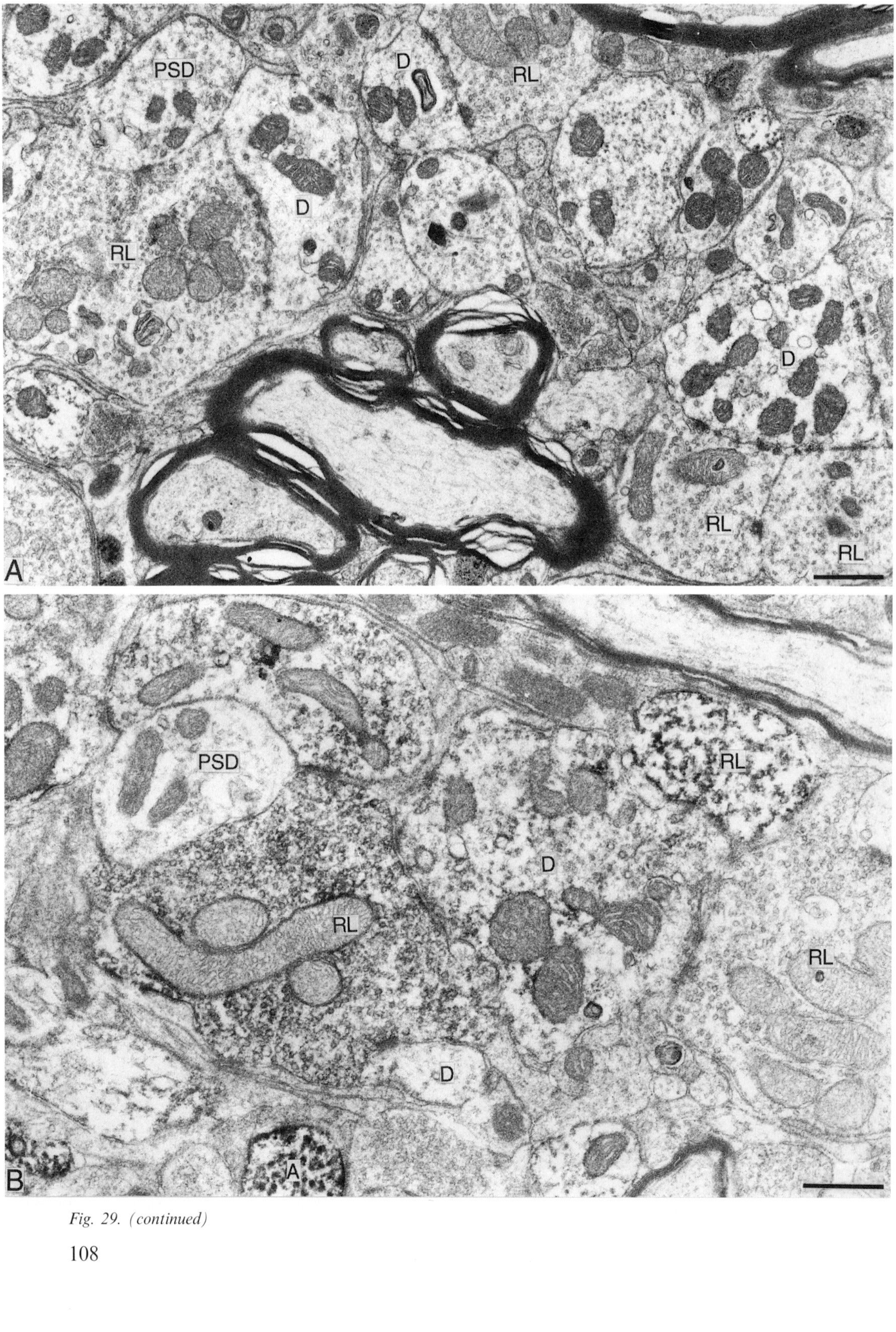

Fig. 29. (continued)

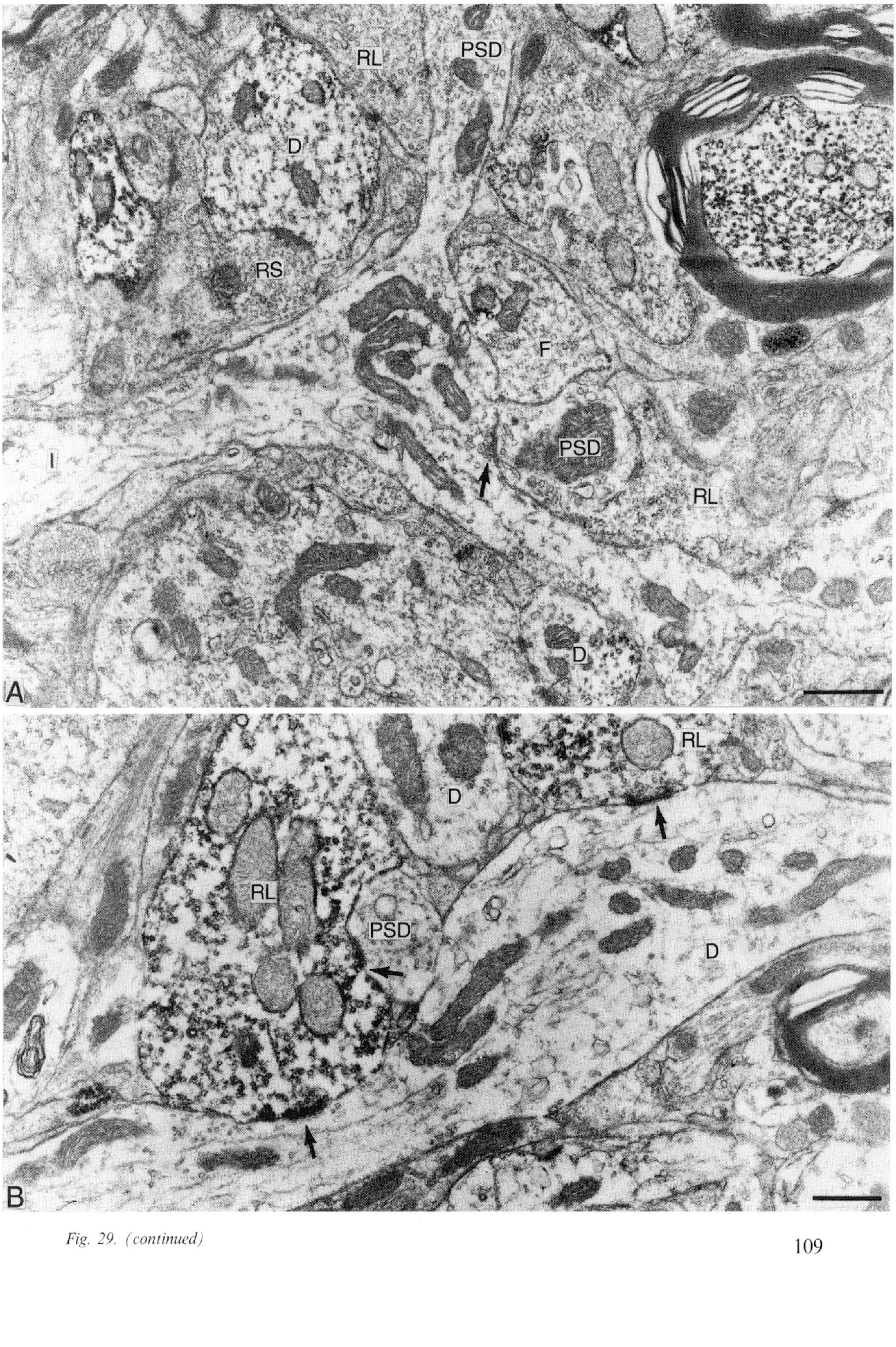

Fig. 29. (continued)

indicative of their status as dendrites (Harding, 1971; Wong-Riley, 1970, 1972a; Pasik et al., 1973, 1976; Ralston, 1971). They are GABAergic (Spreafico et al., 1983a; Montero, 1986a; Ohara et al., 1989). Other forms of terminal, arising from non-specific afferents, are described in Section 8.4.

Again taking information mainly from the cat, but with supportive evidence from monkeys (Jones, 1985a; Ralston et al., 1988; Ohara et al., 1989; Sherman, 1988; Ralston and Ralston, 1994; Mason et al., 1996), a general pattern of synaptic circuitry can be built up in which RL terminals and PSDs terminate in large clusters proximally on the dendrites and dendritic protrusions of relay neurons, RS terminals terminate in very large numbers on distal dendrites of relay neurons and to a lesser extent on dendrites of interneurons, while F terminals end on somata and proximal dendrites of relay neurons and interneurons and on the presynaptic dendrites of the latter. In the proximal synaptic aggregations, RL terminals end on the relay cell dendrites and on the PSDs. PSDs end on the dendrites and on one another, and F terminals end in small numbers on the PSDs and on the relay cell dendrites.

The distally located RS terminals on the dendrites of relay neurons are the endings of corticothalamic fibers, as originally reported in cats (Jones and Powell, 1969b) and confirmed in monkeys (Wong-Riley, 1972b; Harding, 1972; Harding and Powell, 1977; Feig and Harting, 1994).

Quantification of the distribution of the above four types of synapse on physiologically characterized cells has been carried out on a number of relay neurons and on one interneuron in the dorsal lateral geniculate and ventral posterior nuclei of cats (Wilson et al., 1984; Hamos et al., 1985; Raczkowski et al., 1988; Liu et al., 1995a). Depending on physiological class, the relay cells receive between 4000 and 8000 synapses, with a density of 0.4–1 synapse per μm of dendritic length. RL terminals account for 2–29%, F terminals for 20–44%, RS terminals for 40–80% and PSD terminals for 2–7%. In a study with different techniques, Montero (1991) found that 12% of the terminals on relay cells of the dorsal lateral geniculate nucleus were RL, 58% RS and 24% GABAergic (F and PSD).

In comparison with the cat figures, neurons in the dorsal lateral geniculate nucleus of the monkey have a lower synaptic density. Wilson (1989), from electron microscopic observations on physiologically identified cells, calculated that neurons in the parvocellular layers received approximately 500 synapses, with a synaptic density of approximately 0.2 per μm of dendrite, while neurons in the magnocellular layers received approximately 3000 synapses with a density of approximately 0.5 per μm. RL terminals formed 30–34% of the total synapses, F terminals 40–45% and RS terminals 25–30%.

On proximal dendrites of thalamic relay neurons in cats, RL terminals form 43–62% of the total synaptic contacts, F terminals form 38–54%, there are few PSD contacts and none are formed by RS terminals (Liu et al., 1995a). RL terminals making synaptic contacts on intermediate dendrites, fall to 16–35% of the total, F terminals remain relatively constant at 38–50%, PSDs form 8–9% of the contacts and RS terminals 9–42%. Few RL terminals end on distal dendrites, and account for only 5–13% of the total synaptic contacts there; F terminals continue to form 25–40%, PSDs 8–9%, while contacts formed by RS terminals increase to 46–66%. The concentration of subcortical afferent terminals on the proximal dendrites of relay neurons, the concentration of corticothalamic ternminals on distal dendrites and the more even distribution of inhibitory tertminals, has been generally confirmed in all species, including monkeys. In the monkey ventral posterior nucleus, Ralston and Ralston (1994) estimated that 54% of the synaptic contacts of RL terminals deriving from medial lemniscal fibers were on relay cell dendrites, 42% on GABAergic PSDs and 4% on GABAergic dendritic shafts. 85% of RL terminals derived from spinothalamic fibers, however, synapsed on relay cell dendrites and 15% on PSDs.

In the dorsal lateral geniculate nucleus of cats, there is relatively little evidence for significant convergence of different types of optic tract axons on relay cells, each cell class being specifically innervated by one type of retinal ganglion cell axon (Hoffman et al., 1972; Bullier and Norton, 1979). In the ventral posterior nucleus of primates, convergence of terminals of lemniscal and spinothalamic fibers may occur. Ralston and Ralston (1994), estimated that 75% of the RL terminals ending on relay cells were formed by axons arising in the dorsal column nuclei and 25 % by axons arising in the spinal cord. Direct convergence of lemniscal and spinothalamic terminals on the same dendrite was not directly observed, however. Most lemniscal terminals ended on relay cell dendrites and on PSDs, whereas most spinothalamic terminals ended on relay cell dendrites only. Convergence may underlie the capacity of the majority of ventral posterior cells which respond to high intensity stimuli to respond also to innocuous stimuli. Hence, they are called wide dynamic range cells (Willis, 1988a,b). This may also, however, reflect their input from wide dynamic range cells in the spinal dorsal horn.

The large number of RS terminals on the distal dendrites of a relay cell suggests that they arise from many corticothalamic fibers. The fact that EPSPs evoked in relay cells by cortical stimulation rise in a series of steps (Deschênes et al., 1984) is consonant with this view. Cortical stimulation elicits initial, fast EPSPs in relay cells, based on activation of ionotropic excitatory amino acid receptors; these are succeeded by a slow, long lasting depolarization, dependent on activation of metabotropic, excitatory amino acid receptors (McCormick and Von Krosigk, 1992; Golshani et al., 1998).

In addition to terminations on relay neurons, corticothalamic (RS) terminals also end on interneurons (Jones and Powell, 1969a,b; Guillery 1969a,b; Wong-Riley, 1972a; Harding and Powell, 1973, 1977) but the relative proportion has been the subject of debate. Weber et al. (1989) reported that three times as many labeled RS terminals in the cat dorsal lateral geniculate nucleus ended on GABA immunoreactive dendrites as on non-immunoreactive relay cell dendrites. Montero (1991), however, reported that twice as many RS terminals ended on relay neurons as on the GABA immunoreactive dendrites of interneurons. In agreement with Montero, Vidnyánsky and Hámori (1994) reported that identified corticothalamic terminals on dendrites not immunoreactive for GABA, formed 76–87% of the total observed and those on dendrites positively immunostained for GABA formed only 7–17.5%.

The small RS terminals may be derived only from corticothalamic cells located in layer VI of the cerebral cortex. There is mounting evidence from non-primates that terminals derived from layer V corticothalamic cells, whose axons may have a more widespread distribution in the thalamus than those arising from layer VI cells, may end in large RL-like terminals on proximal dendrites of relay cells. Corticothalamic terminals in the pulvinar of squirrel monkeys resemble RL terminals and end on relay cell dendrites in a form resembling the terminals of retinal fibers in the dorsal lateral geniculate nucleus (Mathers, 1972b). Two types of corticothalamic terminals have also been identified in the mediodorsal nucleus of monkeys, one resembling typical RS terminals and terminating abundantly in the neuropil and a second larger, rarer, resembling typical RL terminals and terminating on dendrites and PSDs at the centers of proximal synaptic aggregations (Schwartz et al., 1991). In the ventral posterior nucleus of rodents and in the medial geniculate nuclei of rodents and cats, thicker axons arising from layer V cells give rise to boutons that are approximately 2 μm in diameter while the smaller axons arising from layer VI cells give rise to boutons that are approximately 1 μm in diameter (Hoogland et al., 1987; Welker et al., 1988; Rouiller and DeRibaupierre, 1990; Rouiller et al., 1991; Rouiller and Welker, 1991; Ojima 1994; Bourassa et al., 1995; Bajo et al., 1995). In the mouse and rat, corticothalamic fibers ending in small boutons are typically distributed in the ventral posterior and adjacent part of the reticular nucleus while fibers ending in large boutons are found only in the posterior nucleus (Heegelund et al., 1987) or central lateral nucleus (Bourassa et al., 1995). In the auditory thalamus and visual thalamus of the cat, the thicker fibers are distributed widely through the dorsal and medial nuclei of the medial geniculate complex and in the lateral division of the posterior group of nuclei or lateral posterior nucleus; branches continue on to the inferior colliculus (Ojima, 1994; Ojima et al., 1996). In all regions they terminate in grape-like clusters of large boutons. The thinner corticothalamic fibers give collaterals to the reticular nucleus, and then terminate only in the ventral nucleus of the medial geniculate complex or dorsal lateral geniculate nucleus. There, they end in small boutons, in the ventral medial geniculate nucleus forming plexuses that follow the contours of the isofrequency lamellae in the ventral nucleus. The thin fibers, therefore, form a tonotopically organized, feedback projection to the nucleus which provides the main thalamic input to the primary auditory cortex. The thicker fibers, by contrast, form a dispersed projection to thalamic nuclei in which neurons are more broadly tuned, have more labile responses (Aitkin, 1973; Calford and Webster, 1981; Calford, 1983; Imig and Morel, 1985) and tend to have more diffuse projections to the cerebral cortex (Hashikawa et al., 1991; Molinari et al., 1995). The thin-fibered projection to the ventral nucleus may be involved in gain control of the sensory relay to the primary auditory cortex while the thicker fibered projection could be more involved in selective attention.

F terminals arising from axons of the reticular nucleus end intermittently along the full length of relay cell dendrites (Cucchiaro et al., 1991a; Liu et al., 1995b) and appear to account for the majority of the F terminals present, at least in the ventral posterior nucleus of the cat.

GABAergic fibers arising in the globus pallidus or substantia nigra, pars reticulata, which terminate in the ventral anterior, ventral lateral, centre médian or ventral medial nuclei, were thought to end in typical F terminals (Kultas-Ilinsky et al., 1983, 1985) on proximal dendrites, but the origins of these terminals was not experimentally verified. In the centre médian nucleus of the monkey, the pallidal fibers were said to end in typical RS terminals (Harding and Powell, 1977), which was surprising, given their presumed GABAergic character. Recent experimental evidence shows that the pallidal fibers do, indeed, end in typical F terminals (Sidibé et al., 1997), the earlier contrary results, which were based on lesion-induced degeneration, perhaps stemming from the interruption of corticothalamic fibers.

The distinguishing feature of intrinsic interneurons at the synaptic level is the presence of presynaptic dendrites which form synapses with symmetrical membrane thickenings on one another and on relay cell dendrites (Fig. 29; Harding, 1971; Harding and Powell, 1973, 1977; Pasik et al., 1973, 1976; Ralston, 1971).

Interneurons probably receive a less dense innervation by RL terminals than relay neurons. In the magnocellular layers of the monkey lateral geniculate nucleus, Winfield (1980) found that 68% of RL synapses were on relay cell dendrites and 32% on PSDs, while in the parvocellular layers 85% were on relay cell dendrites and 14% on PSDs.

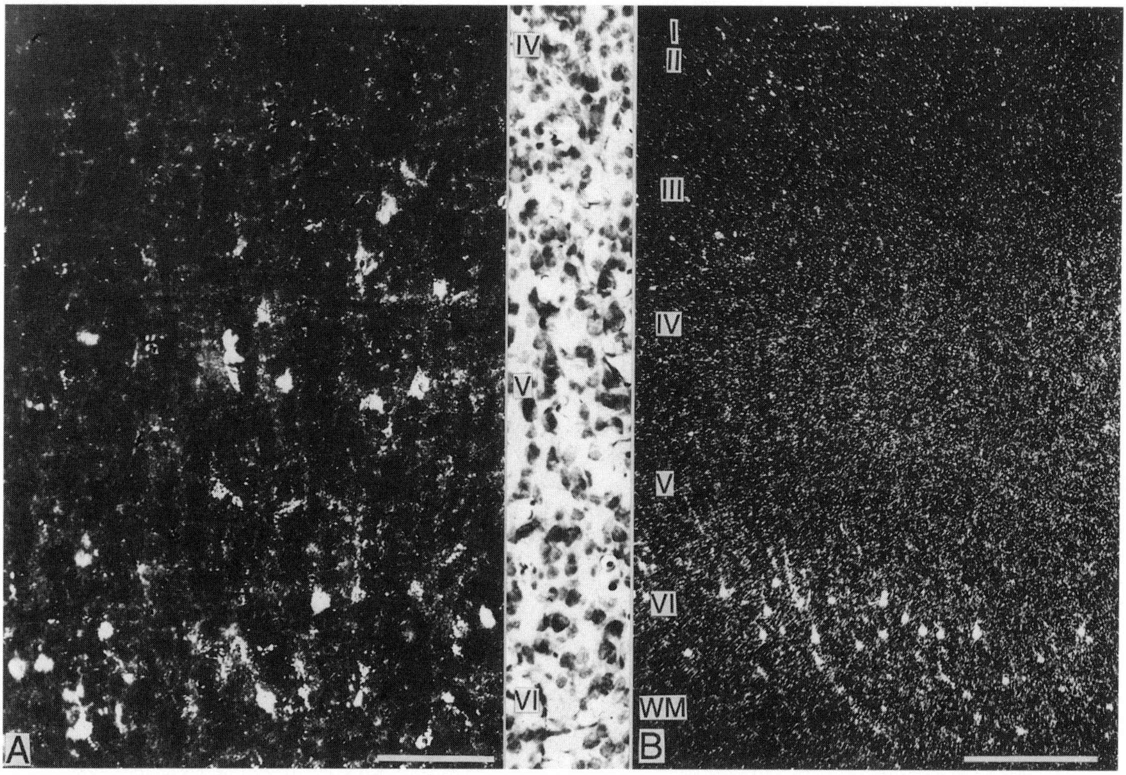

Fig. 30. Darkfield photomicrograph showing retrogradely labeled corticothalamic cells in layers V and VI of the somatosensory cortex of a squirrel monkey following injections of horseradish peroxidase (A) or [3H] D-asparatate (B) in the ventral posterior nucleus. In B Label in layer IV represents material transported anterogradely in thalamocortical fibers and recurrently in axons of labeled cells in layer VI.

This is similar to the observations made on a single identified interneuron in the cat lateral geniculate nucleus (Hamos et al., 1985). Kultas-Ilinsky and Ilinsky (1991) claim that about 10% of RL terminals formed by cerebellar afferents in the ventral lateral nucleus end on dendrites of interneurons. The synapses formed by single cerebellar terminals on relay neurons outnumber those on interneurons by 3 or 4 to 1 (Mason et al., 1996)

There is probably also a much less dense corticothalamic (RS) innervation of interneurons than of relay neurons (see above). However, stimulation of corticothalamic fibers results in rapid, disynaptic IPSPs in relay neurons (Soltesz et al., 1989b; Deschênes and Hu, 1990), indicating effective activation of intrinsic interneurons.

The majority of F terminals, including those arising from the reticular nucleus end on relay cell dendrites, and relatively few end on dendrites or PSDs of interneurons (Ohara et al., 1980, 1989; Montero and Scott, 1981; Norita and Katoh, 1987; Cucchiaro et al., 1991b; Harting et al., 1991b; Liu et al., 1995b). Montero (1991) in a study of the types of terminal profiles making synaptic contact with GABA immunoreactive processes in the cat dorsal lateral geniculate nucleus, calculated that 25% of the terminals ending on interneurons were retinal (RL), 37% cortical (RS) and 26% GABAergic (F and PSD).

The other major set of terminals ending on interneurons is that made up of PSDs which commonly make multiple and sometimes reciprocal contacts with one another, usually in proximity to the RL terminals. The degree to which PSD to PSD contacts occur between appendages of the same or different interneurons is not known.

The relative isolation of PSD terminals from the parent cell by lengthy, thin processes suggests that synaptic potentials generated, e.g., by RL terminals, will be attenuated and have little or no effect upon the membrane potential of the soma and therefore upon generation of action potentials. Hamos et al. (1985; see also Bloomfield and Sherman, 1989), calculated that for their injected cell, the amplitudes at the soma of postsynaptic potentials generated in a PSD would be 10 to 100 times less than those generated by a synapse on a dendritic shaft. Hence, interactions in the aggregations of RL and PSD terminals, on the proximal dendrites of relay cells, may be quite local and serve mainly to set the gain of RL synaptic inputs to relay cells. Possible examples of the effects of PSD inputs to relay cells are seen in experiments in which stimulation of the mamillothalamic tract (Paré et al., 1991) or optic tract (Soltesz and Crunelli, 1992b) led to an initial excitation of relay cells in the anteroventral or lateral geniculate nucleus, followed by a series of three IPSPs. This series consisted of a brief initial, miniature IPSP, thought to be due to local excitation of PSDs and not found after stimulation of corticothalamic input, followed by a $GABA_A$-mediated IPSP lasting 10–60 msec, and a long (300 msec) $GABA_B$-mediated IPSP, both thought to be due to the axonal inputs, via F terminals, of interneurons. (Reticular nucleus inputs which may influence this sequence in other nuclei are thought to be absent from the anterior nuclei of the cat; Steriade et al., 1984; Velayos et al., 1989).

7.2. CIRCUITRY INVOLVING THE RETICULAR NUCLEUS

Within the reticular nucleus, collaterals of thalamocortical fibers (Jones, 1975b; Updyke, 1977; Friedlander et al., 1981; Yen et al., 1985a; Ahlsén et al., 1978; Ahlsén and Lindström, 1982a; Stanford et al., 1983; Harris, 1987; Montero, 1988, 1989b), and collaterals of corticothalamic fibers (Jones, 1975b; Ohara and Lieberman, 1981, 1985; Montero, 1989b; Bourassa et al., 1995a,b; Murphy and Sillito, 1996) probably form the majority of afferent terminations in the nucleus. Certain sectors of the nucleus also receive GABAergic terminals of pallidothalamic (Asanuma, 1994; Hazrati and Parent, 1991) and of thalamostriatal fibers (Jones, 1975b; Sadikot et al., 1992a; Deschênes et al., 1996). Throughout the nucleus, the terminals of monoaminergic and cholinergic fiber systems arising in the brainstem (Swanson and Hartman, 1975; Mesulam et al., 1983; Cropper et al., 1984) are also found, along with the terminals of GABAergic fibers that arise in the basal forebrain (Nauta, 1979b; Levey et al., 1987; Steriade et al., 1987b; Parent et al., 1988; Jourdain et al., 1989; Asanuma and Porter, 1990; Cornwall et al., 1990; Chen and Bentivoglio, 1993; Bickford et al., 1994). Each of these sets of terminals has been identified, as discussed above. Other synapses in the reticular nucleus are formed by terminals of the collateral branches of the axons of reticular nucleus cells (Mulle et al., 1986; Yen et al., 1985b; Uhlrich et al., 1991; Liu et al., 1995b) and in some species, at least, by presynaptic dendrites of the cells (Yen et al., 1985b; Deschênes et al., 1985; Mulle et al., 1986). In monkeys, the presynaptic dendrites are reported in one study as relatively common (Asanuma, 1994) but in another as rare (Williamson et al., 1994).

The collateral innervation of reticular nucleus cells by thalamocortical and corticothalamic fibers passing through it links the reticular nucleus tightly to the underlying dorsal thalamus. EPSPs can be evoked at di-synaptic latencies in reticular nucleus cells by electrical stimulation of the medial lemniscus or optic tract (Ahlsén and Lindström, 1982a,b; Hale et al., 1982), and reticular nucleus cells in monkeys (Pollin and Rokyta, 1982) and other species (Sugitani, 1979; Yen et al., 1985b; Hale et al., 1982) have receptive fields indicative of input from particular sensory nuclei of the dorsal thalamus (Sugitani, 1979; Yen et al., 1985b). Prolonged spike barrages occur at short latency in reticular nucleus cells after stimulation of corticothalamic fibers (Frigyesi, 1972; Steriade and Wyzinski, 1972; Waszak, 1974).

When the cortex is electrically stimulated in the cat, reticular nucleus cells are excited monosynaptically, mainly by thalamocortical collaterals (Ahlsén and Lindström, 1982a; Ahlsén et al., 1982b). This excitation is followed at disynaptic latency by a hyperpolarization which is likely to be due to inhibitory connections between the reticular cells (Ahlsén and Lindström, 1982b). Stimulation within the reticular nucleus itself is also accompanied by inhibition of reticular nucleus cells (Ulrich and Huguenard, 1996). It is possible that the role of intra-reticular nucleus inhibitory connections is to bring the resting membrane potential of the cells to a level of hyperpolarization sufficient to deinactivate the low threshold calcium conductance which should predispose the neurons to discharge in bursts of action potentials when returning to a less hyperpolarized resting membrane potential. Cortical stimulation is a potent influence in inducing burst discharges in reticular nucleus neurons and in setting up, as a consequence, oscillations in the interconnected thalamic network (Steriade and Wyzinzki, 1972; Steriade et al., 1990d; Contreras and Steriade, 1995). IPSPs induced in reticular nucleus cells are typical, fast $GABA_A$ mediated responses (Huguenard and Prince, 1994b; Bal et al., 1995c) but $GABA_B$ mediated IPSPs have also been demonstrated (Ulrich and Huguenard, 1996). The former is interesting in view of the fact that reticular nucleus cells do not express genes for the majority of the known subunit polypeptides that combine to form receptors with classical $GABA_A$ properties (Huntsman and Jones, 1993; Huntsman et al., 1996; Section 8.3).

The projection of the reticular nucleus to the dorsal thalamus is topographically organized so that nuclei, the collaterals of whose thalamocortical fibers innervate a sector of the reticular nucleus, receive fibers back from that sector (Jones, 1975b) (Fig. 32). The F type terminals made in the dorsal thalamus of the cat by axons arising in the reticular nucleus end on the somata, dendrites and dendritic appendages of relay cells (Section 7.1) and to a lesser extent on presynaptic dendrites of interneurons (Cucchiaro et al., 1991a ; Liu et al. 1995b). Bal et al. (1995c) could not demonstrate an input from the reticular nucleus to interneurons in the ferret dorsal lateral geniculate nucleus. In the ventral posterior nucleus of the monkey, Ohara et al. (1989) reported that most F terminals end on non-GABA immunoreactive dendrites which also implies a relative lack of reticular nucleus terminals on interneurons.

7.3. CIRCUITRY INVOLVING THE CORTICOTHALAMIC PROJECTION

Every dorsal thalamic nucleus receives a corticothalamic projection (Jones, 1985a) (Figs. 30, 31). The components of the ventral thalamus (reticular nucleus, zona incerta and ventral lateral geniculate nucleus) receive corticothalamic fibers but not those of the epithalamus. Corticothalamic fibers returning from a cortical area to the dorsal thalamic nucleus that provides the dominant thalamic input to that area are distributed in a manner that matches the topographic organization of the thalamocortical projection (Wise and Jones, 1977; Saporta and Kruger, 1977; Jones et al., 1979; Jones, 1985a; Winer and Larue, 1987; Chmielowska et al., 1989), but only in a general

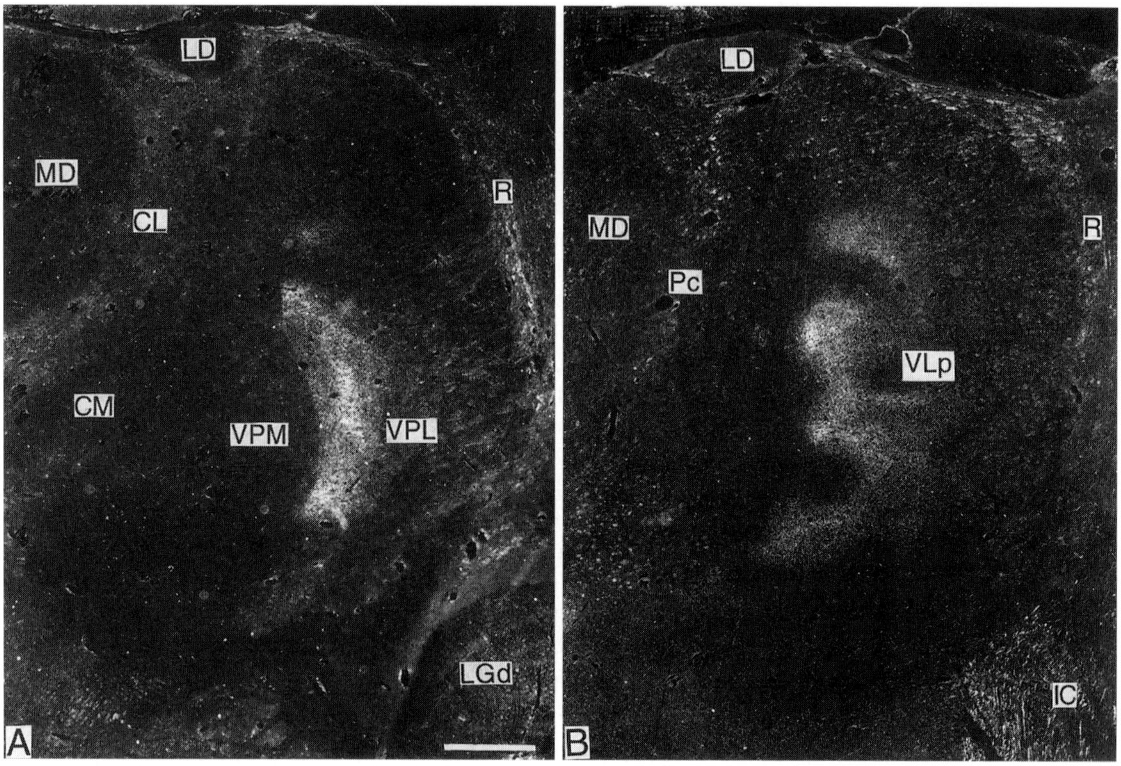

Fig. 31. Autoradiographic labeling of corticothalamic fiber terminations in the ventral lateral posterior (A) and in the ventral posterior lateral nucleus (B) of macaque monkeys following injection of [3H] amino acids in the somatosensory cortex (A) or the primary motor (B) Bar 1 mm. From material illustrated in Jones et al. (1979).

way. When examined in detail, the corticothalamic fibers arising in a part of a cortical representation spread beyond the confines of the zone of cells in the thalamic relay nucleus providing input to that part of the representation, at least in non-primates (Hoogland et al., 1988; Murphy and Sillito, 1996). The zone of terminals can exceed by 2–5 times the region of thalamic cells containing the representation of the area of the periphery providing input to the cortical locus from which they arise.

All the corticothalamic fibers that return to the thalamic nucleus providing the principal thalamic input to a cortical area arise from relatively small, modified pyramidal neurons in layer VI of the cerebral cortex (Jacobson and Trojanowski, 1975; Lund et al., 1975; Jones and Wise, 1977; Arikuni et al., 1983; Jones, 1983, 1984; Giguére and Goldman-Rakic, 1988; Chmielowska et al., 1989; Yeterian and Pandya, 1994). In the visual cortex of monkeys, cells at different levels of layer VI project independently to the magno- and parvocellular layers of the dorsal lateral geniculate nucleus (Lund et al., 1975). In most areas of the cerebral cortex, a divergent corticothalamic projection also arises from larger pyramidal cells located in layer V. These terminate in other thalamic nuclei (see Section 7.2), often in the intralaminar nuclei or in other nuclei that send diffuse projections ending in layer I of the cortex (Catsman-Berrevoets and Kuypers, 1978; Royce, 1982; Jones, 1985a). The primary visual cortex, for example, sends a reciprocal projection from layer VI to the dorsal lateral geniculate

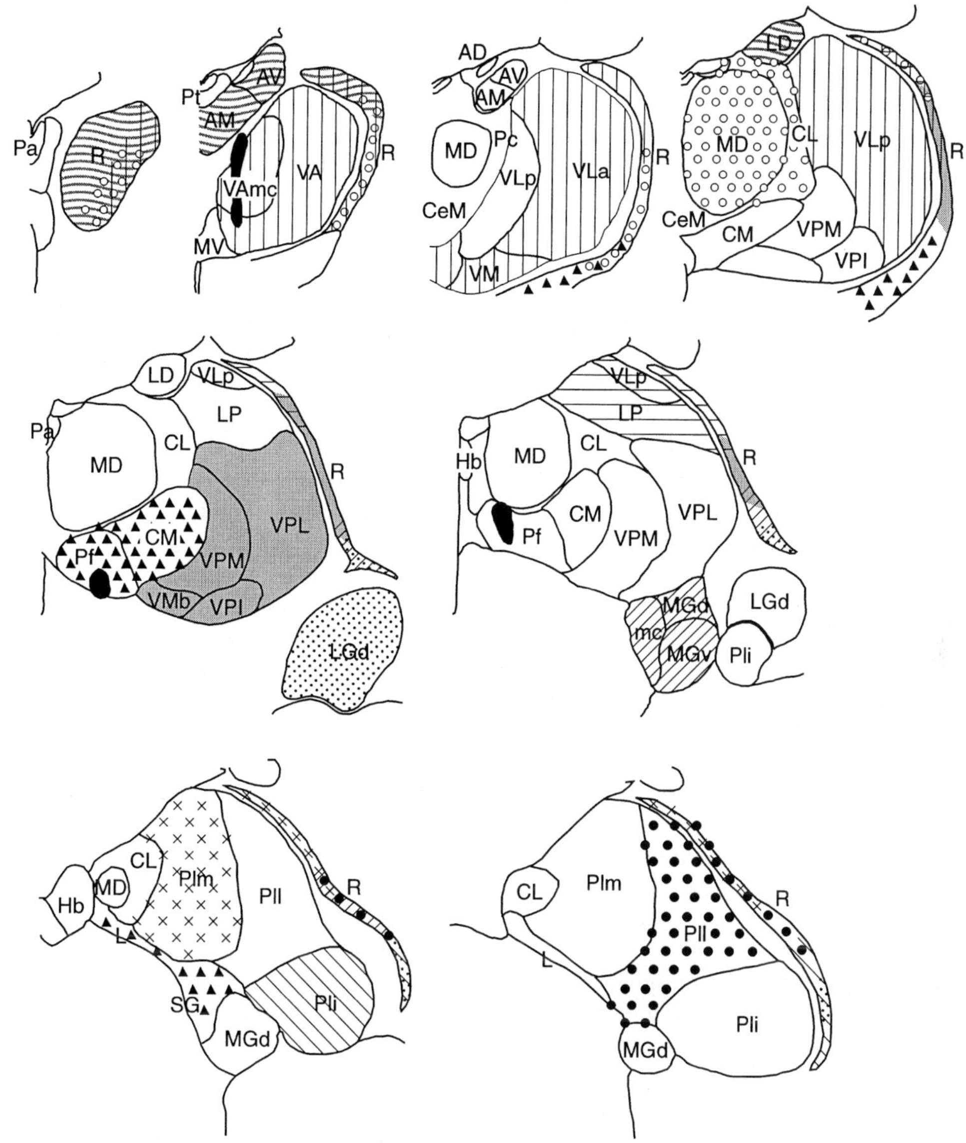

Fig. 32. A series of frontal sections in anterior to posterior order through the thalamus of a macaque, showing the sectors of the reticular nucleus (R) that are related to each of the nuclei or groups of nuclei of the underlying dorsal thalamus. A symbol appears only once on a dorsal thalamic nucleus but appears throughout the related sector of the reticular nucleus. From unpublished material of the author. For abbreviations see page 244.

nucleus and a divergent corticothalamic projection from layer V to parts of the lateral posterior/pulvinar complex (Gilbert and Kelly, 1975; Lund et al., 1975; Harvey, 1980a, b. These nuclei themselves receive a typical, layer VI-orginating corticothalamic projection from the cortical areas to which they send their principal thalamocortical projection.

The non-reciprocal projection, in arising from relatively large cells in layer V of the cerebral cortex, is composed of larger, more rapidly conducting fibers than the reciprocal projection. The non-reciprocal corticothalamic projection may therefore serve as a kind of feed-forward projection system in a chain of cortico-cortical associations beginning in a primary sensory area such as the visual cortex and extending into the association cortex, or it may be involved in some kind of alerting function, directing attention towards relevant stimuli by priming thalamic nuclei outside the principal sensory relays.

Stimulation or deactivation of the corticothalamic projection results in facilitation and/or inhibition of thalamic relay neurons (Ogden, 1960; Shimazu et al., 1965; Hull, 1968; Kalil and Chase, 1970; Burchfiel and Duffy 1974; Tsumoto et al., 1975, 1978; Richard et al., 1975; Baker and Malpeli, 1977; Geisert et al., 1981; Marrocco et al., 1982; Albe Fessard et al., 1983; Kayama et al., 1984; McClurkin and Marrocco, 1984; Marrocco and McClurkin, 1985; Yuan et al., 1985, 1986; Giuffrida et al., 1988; Molotchnikoff et al., 1980, 1984; Shin and Chapin, 1990; Rapisarda et al., 1992; Marrocco et al., 1996; He et al., 1997). Inhibitory effects may extend away from the part of the thalamic nucleus topographically corresponding to the part of the cortical representation stimulated. This is on account of the divergence of corticothalamic fibers in the nucleus and because of the similar divergence of axons of inhibitory reticular nucleus cells which are simultaneously stimulated; these are the major sources of the inhibition (Rapisarda et al., 1992; Murphy and Sillito, 1996).

The temporal structure of thalamocortical spike trains generated by sensory stimulation can be modulated by cortical stimulation (McClurkin et al., 1994; Sillito et al., 1994) or deactivation (Marrocco et al., 1996), and in the dorsal lateral geniculate nucleus non-dominant eye inhibition (Varela and Singer, 1987), surround-mediated attenuation of center responses in the concentrically organized receptive fields (Murphy and Sillito, 1989), spatial frequency tuning (McClurkin and Marrocco, 1984), and length tuning are also influenced (Murphy and Sillito, 1987; Sillito et al., 1993). Corticothalamic feedback also affects orientation-dependent modulations of lateral interactions between relay cells in the dorsal lateral geniculate nucleus (Marrocco et al., 1982; Murphy and Sillito, 1987; Sillito et al., 1993, 1994). This has the effect of enhancing the surround antagonism of a center response in the concentrically arranged receptive fields of the geniculate cells (Cudeiro and Sillito, 1995).

One of the most powerful effects of the corticothalamic projection is the facilitation and augmentation of synchronized behavior in the interconnected thalamo-cortico-thalamic network. The corticothalamic projection exerts a powerful effect in the initiation and maintenance of thalamic rhythms that underlie changes in the electroencephalogram that are accompaniments of changes in conscious state (Steriade et al., 1989; Steriade and Contreras, 1995; Contreras and Steriade, 1996). Volleys of corticothalamic impulses elicit bursts of spikes in neurons of the thalamic reticular nucleus which are followed by repetitive bursts in the network of thalamocortical cells, occurring at 7–14 Hz (spindles). These are associated with slow wave sleep (Steriade et al., 1971, 1972). During wakefulness, desynchronized sleep and directed attention, higher frequency (ca. 40Hz) oscillations induced by corticothalamic activation of the same thalamocortical network may be important in 'binding' different features of a stimulus and its context into a single frame of reference for the perceptive centers of the cerebral cortex (Llinás and Ribary, 1993).

8. NEUROTRANSMITTERS AND THEIR RECEPTORS IN THE THALAMUS

8.1. INTRODUCTION

Experimental studies indicate that inputs and outputs of all dorsal thalamic nuclei are based upon excitatory amino acid transmission. Acompanying the transmission of a sensory message through the thalamus are inhibitory effects that are mediated by GABAergic interneurons intrinsic to each nucleus and by the GABAergic neurons of the reticular nucleus. Extrinsic inputs that are less specific than the principal afferent pathways in terms of their intrathalamic distribution, include noradrenergic, serotoninergic, cholinergic and peptidergic fiber systems arising mainly in the brainstem. Within each of the transmitter-defined systems, many receptor types and subtypes are found and differential expression of these confers a high degree of complexity on the thalamus. Because knowledge of the functional attributes of each system in the thalamus depends upon studies in non-primates, it will be necessary to review many of them here. However, there have been a number of investigations devoted to the distributions of transmitter-defined fiber systems and the nuclear specific localization of neurotransmitter-related molecules in monkeys and a few in humans, and these will be illustrated.

8.2. EXCITATORY AMINO ACID RECEPTORS

Neuropharmacological investigations show that afferent fibers ending in the principal sensory relay nuclei of the thalamus release excitatory amino acid transmitters (Crunelli et al., 1987a; Salt, 1987; Salt and Eaton, 1989, 1990a,b, 1996; Sillito et al., 1990a,b; Scharfman et al., 1990; Deschênes et al., 1992; Ramoa and McCormick, 1992). The terminals of afferent fibers in the ventral posterior and dorsal lateral geniculate nucleus or the parent cell bodies of these fibers have been found to contain immunoreactive glutamate (Rustioni et al., 1988; Montero and Wenthold, 1989; Broman and Ottersen, 1992; De Biasi et al., 1994; Broman, 1994; Montero, 1994; Ericson et al., 1995; Kechagias and Broman, 1995; Liu and Jones, 1996). The terminals of afferent fibers in the thalamus possess a high affinity uptake system for excitatory amino acids so that injection of glutamate or aspartate into the thalamus results in retrograde labeling of the parent cell bodies of the afferent fibers (Baughman and Gilbert, 1981; Rustioni et al., 1983; Jones, 1983; Matute and Martinez-Millan, 1985; Magnusson et al., 1987; Ray et al., 1992; Broman, 1994). A similar range of experimental techniques has demonstrated that the thalamocortical projection is also glutamatergic (reviewed in Tsumoto, 1990; Hicks et al., 1991; Johnson and Burkhalter, 1992; Kharazia and Weinberg, 1994).

The excitatory response of a thalamic relay cell to afferent stimulation is determined by the action of the released excitatory amino acid transmitter upon NMDA and non-NMDA receptors (Kemp and Sillito, 1981; Salt, 1986a, 1987; Crunelli et al., 1987; Moody and Sillito, 1987; Crunelli et al., 1987; Salt and Eaton, 1989, 1990b, 1991a; Hartveit and Heggelund, 1990; Jones and Sillito, 1992; Heggelund and Hartveit, 1990; Sillito et al., 1990a,b; Funke et al., 1991; Kwon et al., 1991; Scharfman et al., 1990; Hu et al., 1994a,b; Ramoa and McCormick, 1994). The amplitude and fast rise of EPSPs engendered by retinal or lemniscal stmulation is strongly dependent on non-NMDA receptors but the sustained elements of the response depend on NMDA receptors. Metabotropic glutamate receptors may also play a part in the generation of slower, longer-lasting EPSPs (Eaton et al., 1993). NMDA and non-NMDA receptors are also involved in the response of intrinsic thalamic interneurons to afferent stimulation (Williams et al., 1994; Pape and McCormick, 1995).

The corticothalamic input returning to each thalamic nucleus from the cerebral cortex is also glutamatergic (Fig. 30 B) (Baughman and Gilbert, 1980; Fonnum et al., 1981; Young et al., 1981; Kvale and Fonnum, 1983; Jones, 1983; Fosse et al., 1986; Giuffrida and Rustioni, 1988; Montero and Wenthold, 1989; McCormick and Von Krosigk, 1992; De Biasi et al., 1994; Eaton and Salt, 1996) and activates thalamic relay cells via NMDA, non-NMDA and metabotropic glutamate receptors (Deschênes and Hu, 1990a; Scharfman et al., 1990; McCormick and Von Krosigk, 1992; Eaton and Salt, 1996). Typically following electrical stimulation of corticothalamic fibers, an early, short-lasting EPSP, mediated by non-NMDA receptors, is succeeded by longer lasting EPSPs, mediated by NMDA and metabotropic glutamate receptors. Corticothalamic stimulation also exerts a strong excitatory effect on reticular nucleus cells via both NMDA and non-NMDA receptors (Deschênes and Hu, 1990a; Warren et al. 1994). Although intrinsic interneurons, which receive a small but significant corticothalamic input (Jones and Powell, 1969b; Montero, 1989a) have not been directly examined, facilitation of the excitatory response of relay neurons to repetitive stimulation of corticothalamic fibers is brought about in part by a reduction in GABA-based inhibition (McCormick and Von Krosigk, 1992). There is little evidence for metabotropic glutamate receptor-based effects on intrinsic thalamic interneurons (Pape and McCormick, 1995).

The distributions of thalamic cells expressing some of the better known glutamate receptors have been reported in a general way in non-primate thalami (reviewed in Steriade et al., 1997) and Figs. 33–37 show the distributions of selected examples in the monkey thalamus (E.G. Jones, B. Tighilet and M.M. Huntsman, unpublished).

Of the nine identified genes that code for mammalian polypeptides that can form ionotropic receptors with different affinities for excitatory amino acid receptor agonists, (GluR1–4 for AMPA receptors; GluR5–7 for high affinity kainate receptors), most are expressed at low levels throughout the dorsal thalamus of rodents, with the exception of, GluR4 and GluR7 which are expressed at high levels, especially in sensory relay nuclei (Keinänen et al., 1990; Sommer et al., 1990; Petralia and Wenthold, 1992; Sato et al., 1993; Martin et al., 1993; Molnar et al., 1993; Spreafico et al., 1994a; Page and Everitt, 1995). GluR4 is the most highly expressed and there are differences among nuclei. In the thalamus of monkeys, recent studies from the author's laboratory reveal very low levels of gene expression for GluR1, GluR2, GluR5 and GluR7, moderate levels for GluR3 and GluR6 and relatively high levels for GluR4 (Fig. 33). Where mRNA labeling or immunostaining with subunit specific antibodies can be detected by in situ hybridization histochemistry, nuclei of the dorsal thalamus and the reticular

Fig. 33. Part 1 Camera lucida drawings of Nissl-stained frontal sections in anterior (A) to posterior (F) order through the thalamus of a macaque and adjacent to the sections from which the autoradiograms in the following parts 2–10 were taken, showing the locations of nuclei in the autoradiograms. Parts 2–10 are from autoradiograms showing in situ hybridization of radioactive RNA probes complementary to the mRNAs for AMPA-kainate glutamate receptor subunits, GluR1-GluR7, and NMDA receptor subunits, NR1-NR2D. For the weakest expressed subunits (GluR2, GluR5, GluR7 and NR2C), only a single representative section is shown (Part 3). Other subunits are represented by sections at the same levels as those shown in Part 1. Note the different patterns of expression in the anterior (CeM, CL, Pc with L-SG) and posterior (CM, Pf) intralaminar nuclei, in the reticular nucleus, habenular nuclei and principal sensory relay nuclei (MG, VP, LGd). Bars 1 mm. From unpublished material of B. Tighilet and the author.

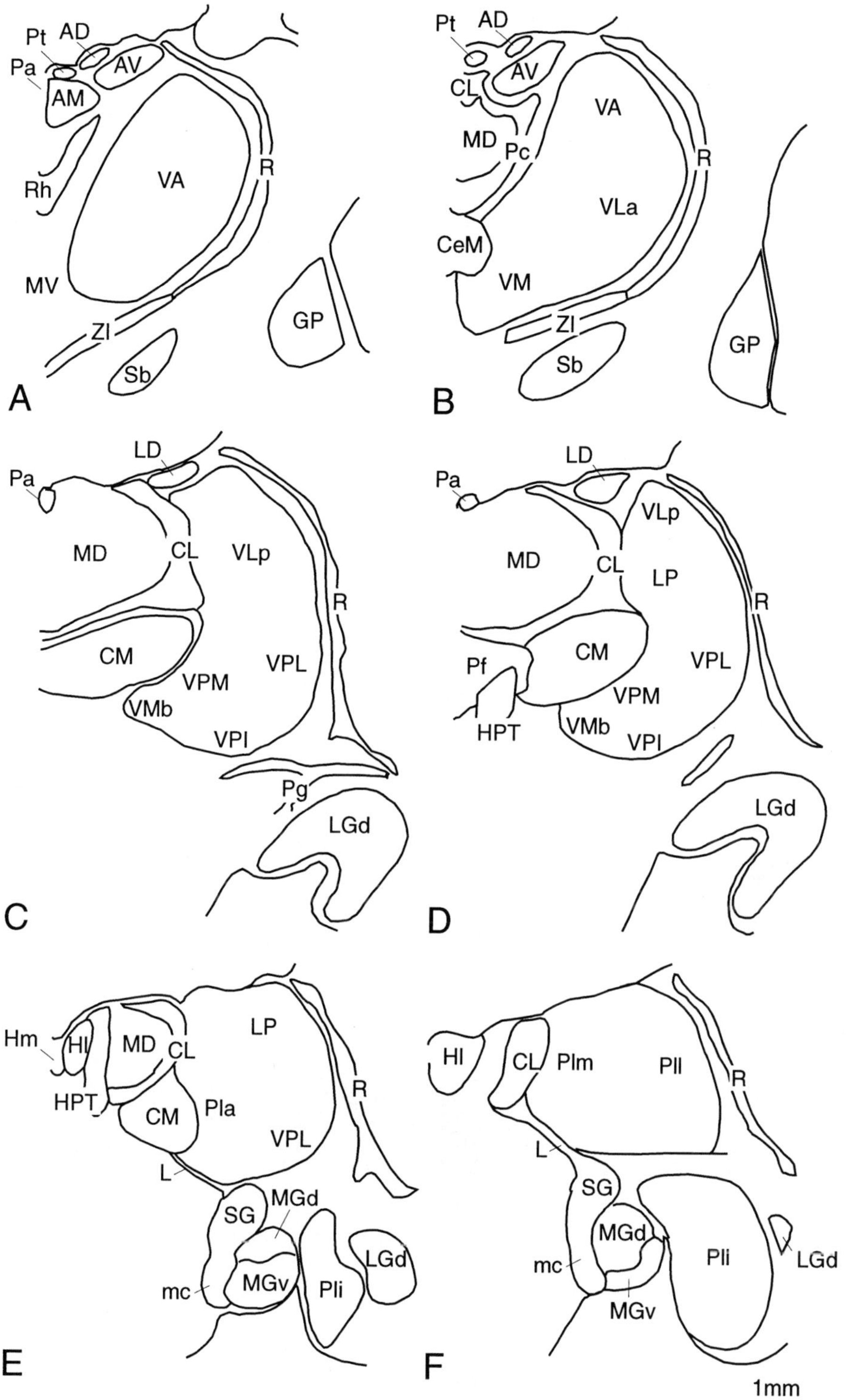

Pt
AD
Pa
AV
AM
Rh
VA
R
MV
ZI
Sb
GP
A
Pt
AD
AV
CL
VA
MD
Pc
R
VLa
CeM
VM
ZI
Sb
GP
B
LD
Pa
MD
CL
VLp
R
CM
VPM
VPL
VMb
VPl
Pg
LGd
C
LD
Pa
VLp
MD
CL
LP
R
CM
VPL
Pf
VPM
HPT
VMb
VPl
LGd
D
LP
Hm
Hl
MD
CL
R
HPT
CM
Pla
VPL
L
SG
MGd
mc
MGv
Pli
LGd
E
Hl
CL
Plm
Pll
R
L
SG
MGd
mc
Pli
LGd
MGv
F
1mm

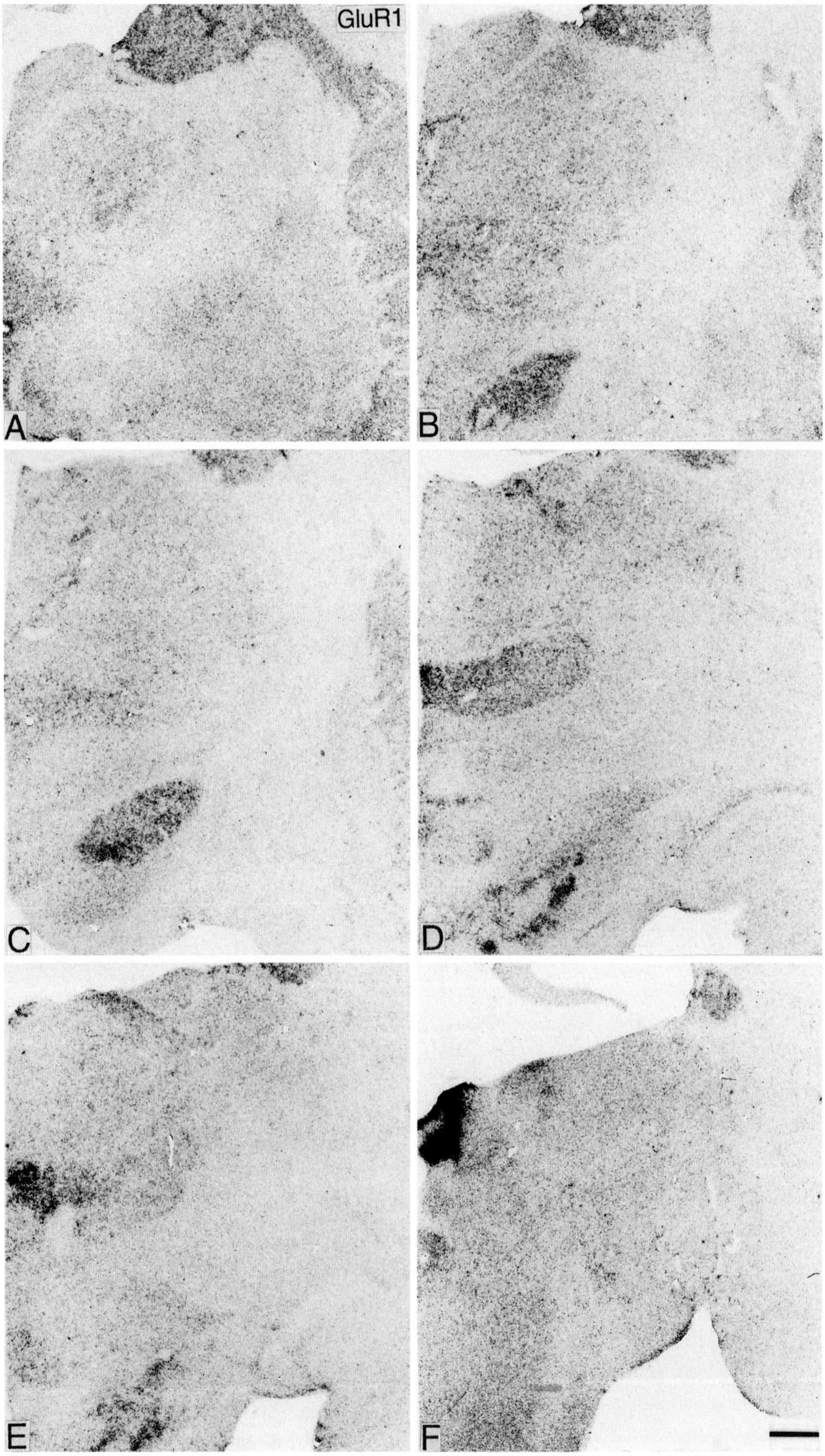

Fig. 33. (continued)

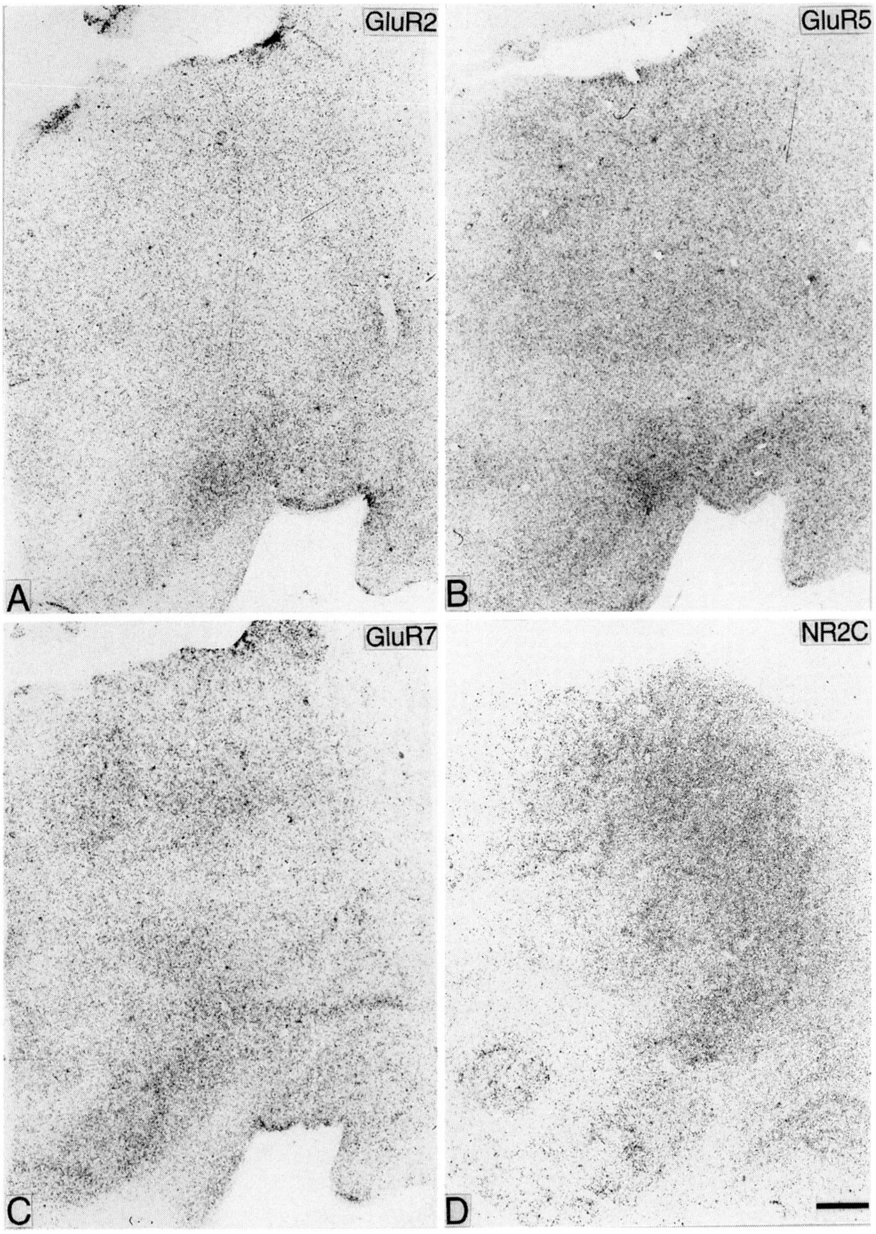

Fig. 33. (continued)

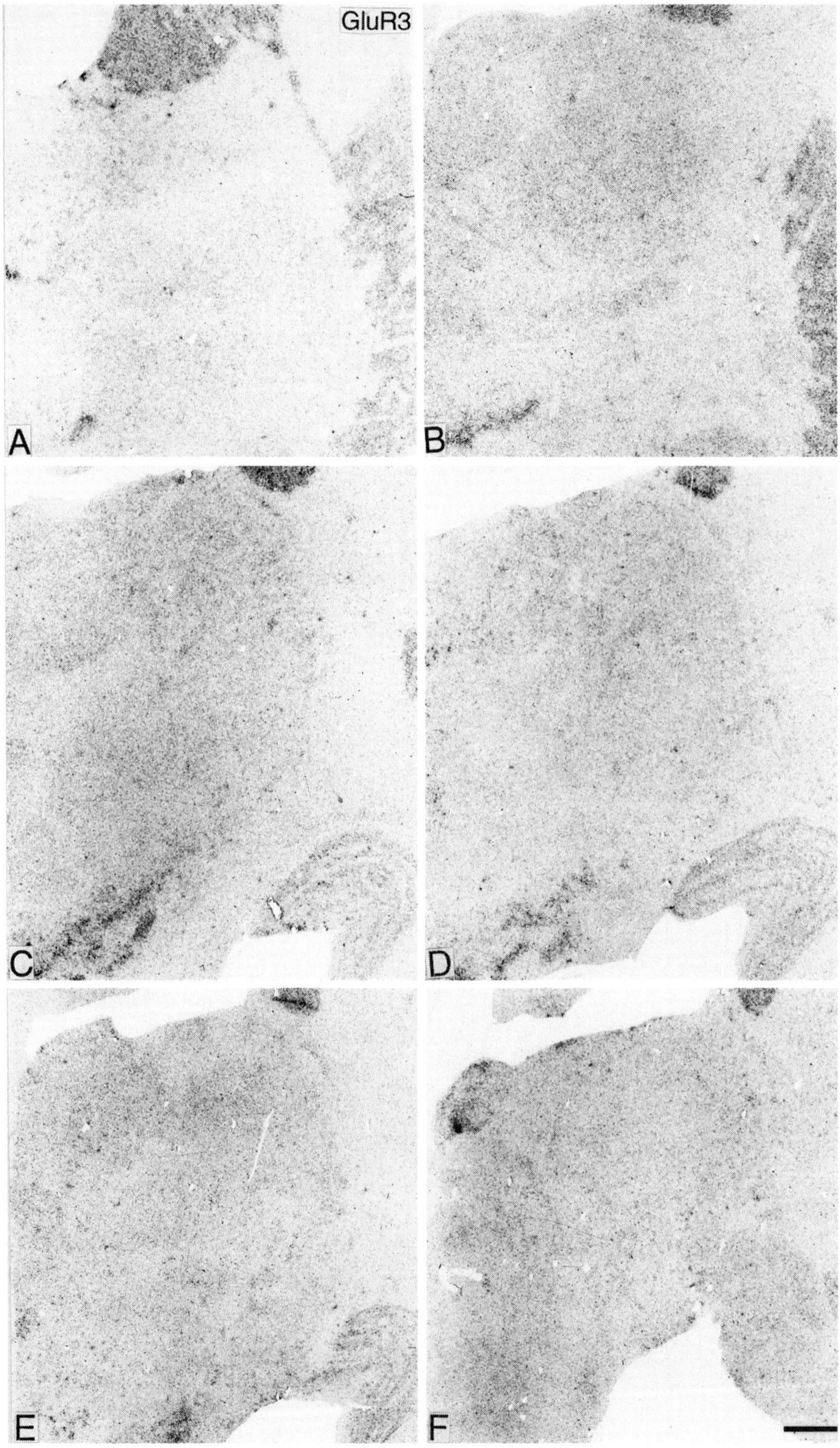

Fig. 33. (continued)

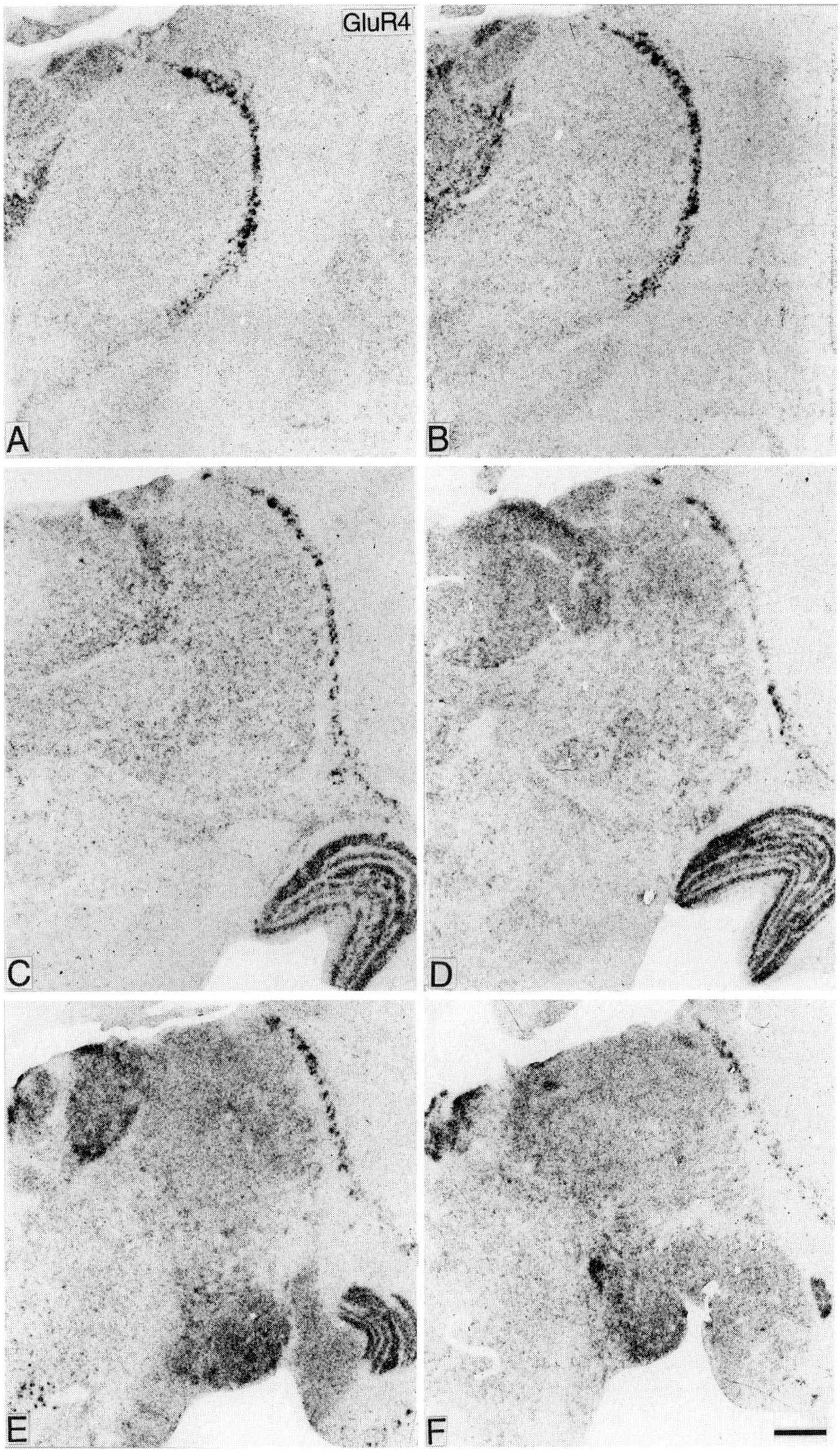

Fig. 33. (continued)

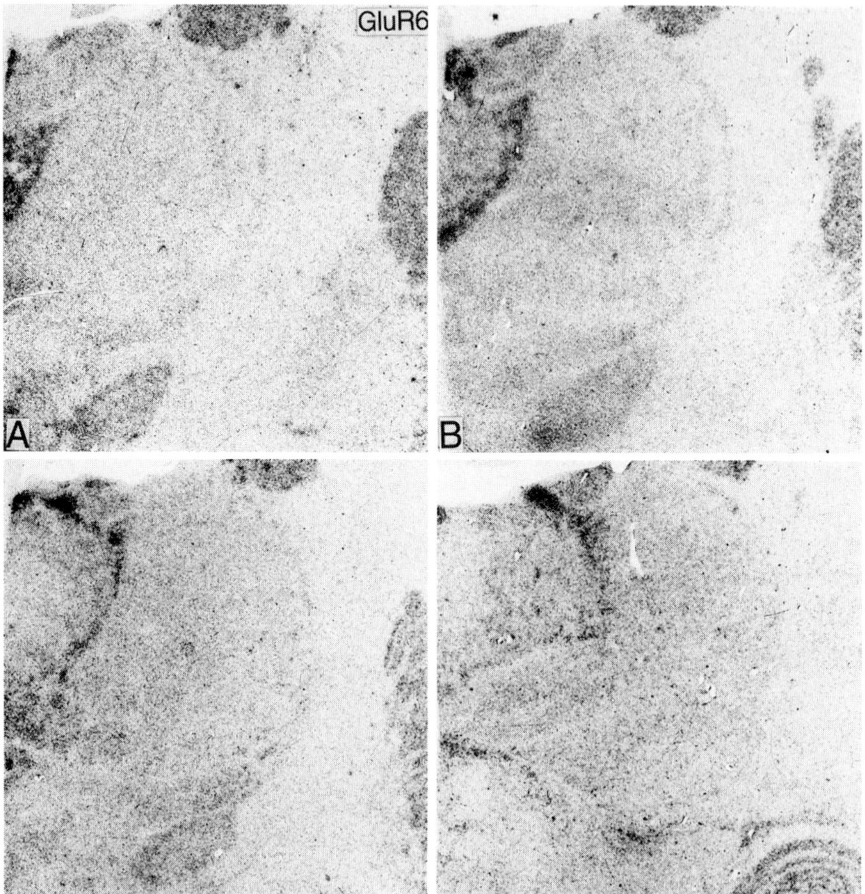

Fig. 33. (continued)

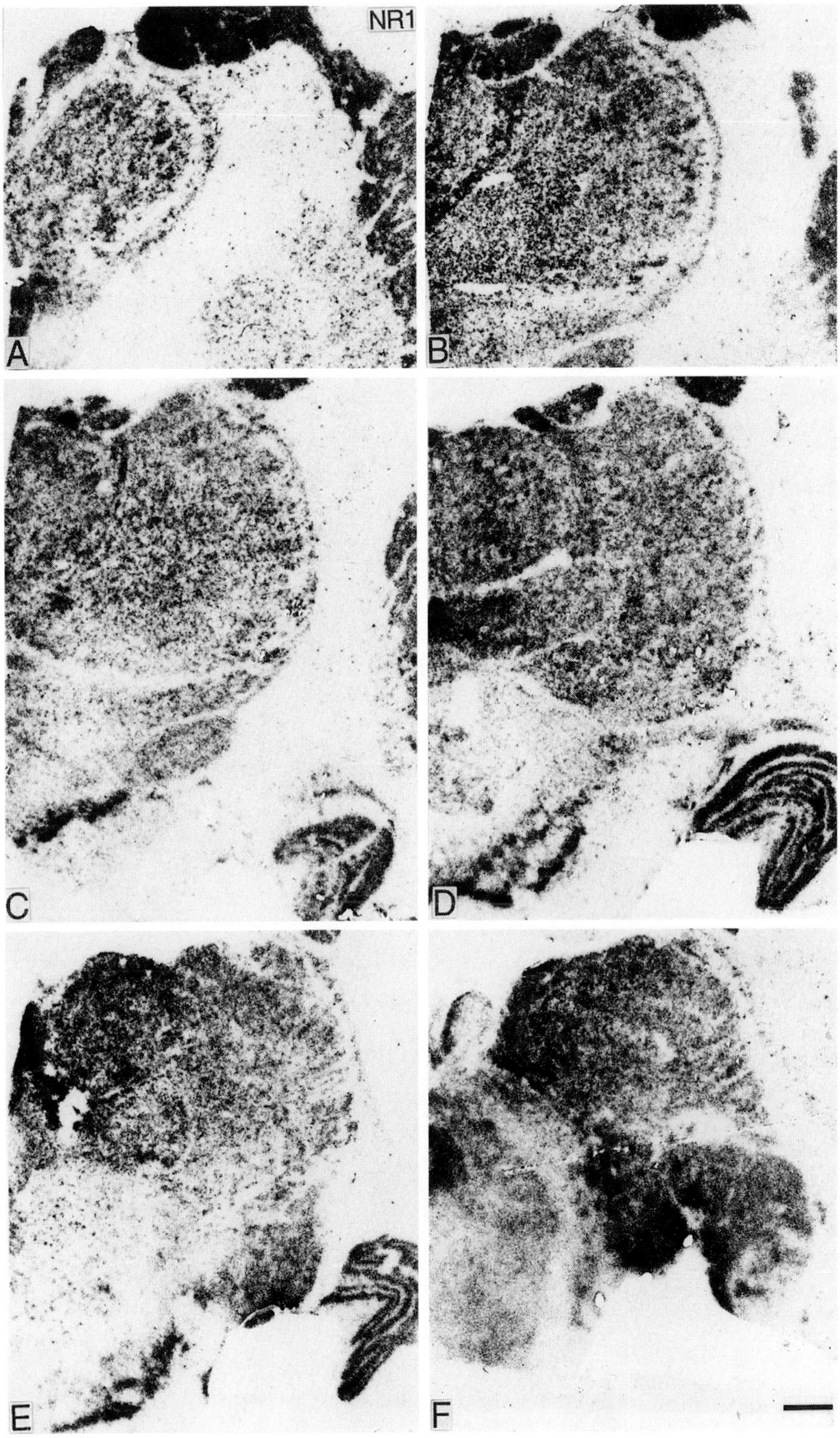

Fig. 33. (continued)

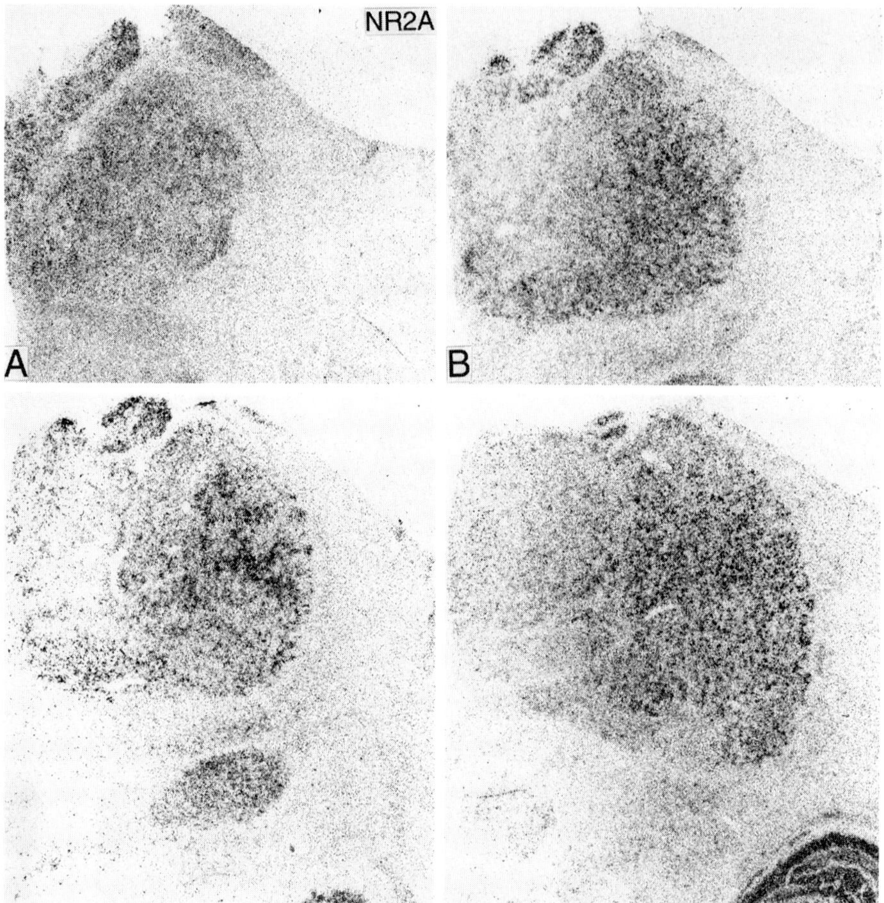

Fig. 33. (continued)

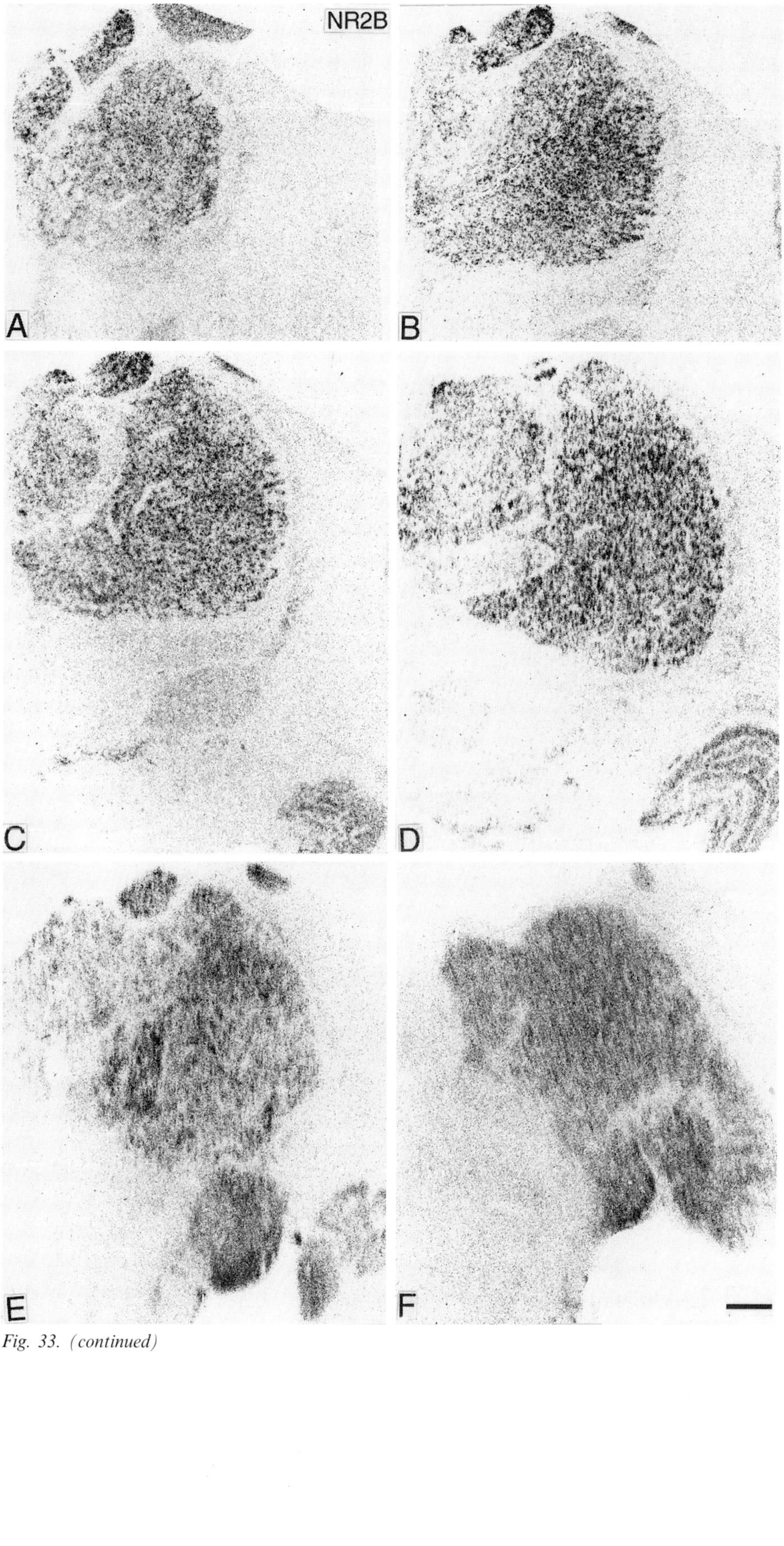

Fig. 33. (continued)

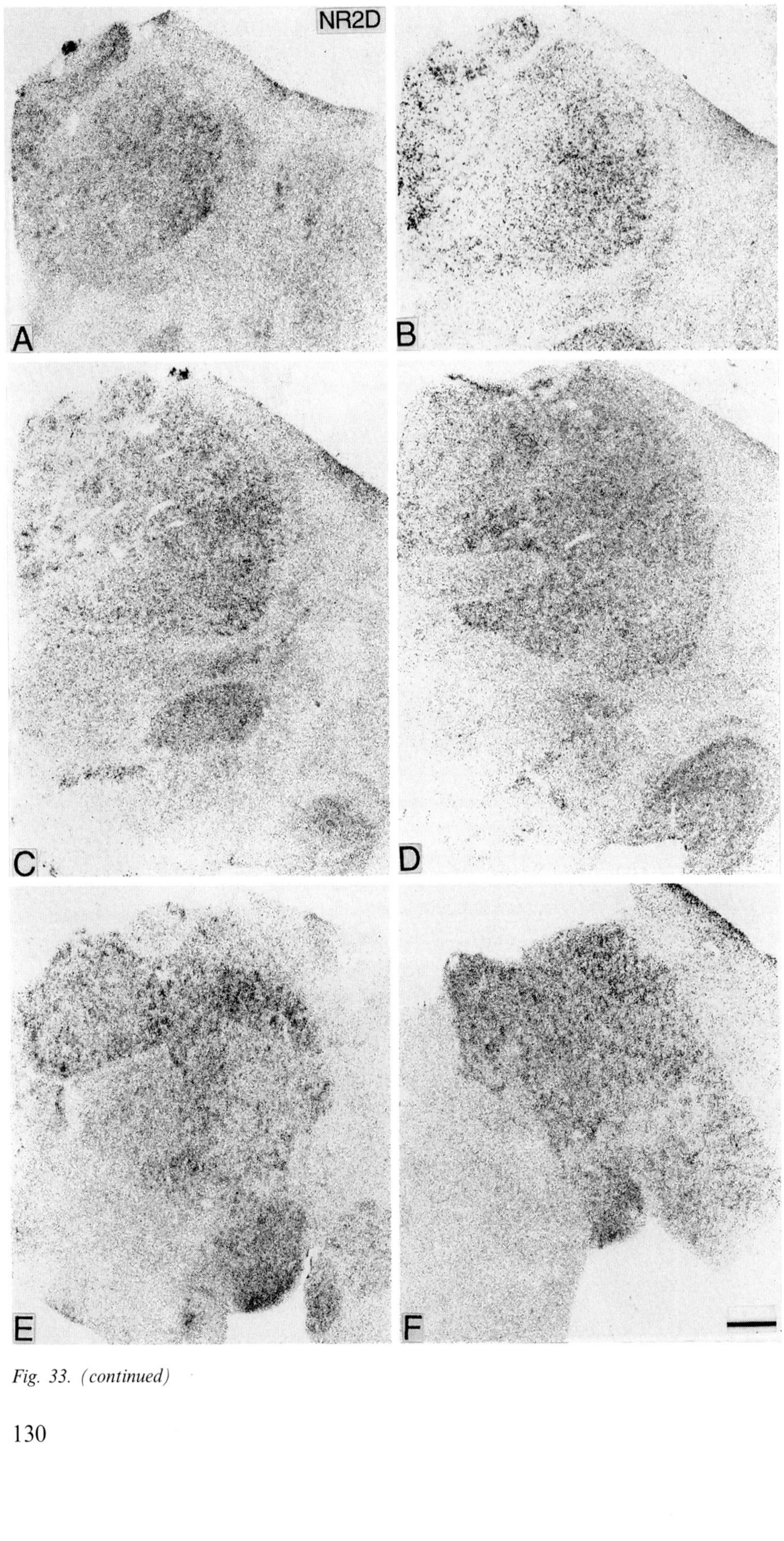

Fig. 33. (continued)

nucleus are labeled (Figs. 34, 35). There are, however, differences in levels, primarily affecting the reticular nucleus and anterior group of intralaminar nuclei. GluR3 and GluR6 expression, for example, is denser in the anterior intralaminar nuclei, extending into the VAmc, limitans/suprageniculate and magnocellular medial geniculate nuclei in a manner that supports the notion that these are part of a common nuclear system (Sections 4.3, 10.3). GluR4 is strongest in the reticular nucleus, followed by the anterior intralaminar nuclei. GluR7 is only at significant levels in the central medial and rhomboid nuclei and in the epithalamus. GluR1 staining appears to have a special affinity for interneurons (Tighilet et al., 1998). The habenular nuclei also show variations in the specificity with which different subunits are expressed (Figs. 33–37).

Of the five known genes that encode subunits involved in the formation of NMDA receptors (NR1 and NR2A-D) (reviewed in Hollmann and Heinemann, 1994), all are expressed to some extent in the thalamus. Throughout the dorsal and ventral thalamus of primates and non-primates, expression is moderately high for the NR1 and NR2B subunits and low for the NR2A and NR2D subunits (NR2A is higher in the ventral posterior nucleus and NR2D in the paraventricular nuclei) (Monyer et al., 1992, 1994; Keinänen et al., 1992; Watanabe et al., 1993; Buller et al., 1994a; Petralia et al., 1994) (Fig. 33, Parts 7–9). NR2C is not expressed or only at extremely low levels (Fig. 33, Part 3 D). A gene encoding a further NMDA receptor-like sequence (NMDAR-L) is expressed at high levels in the thalamus during development (Sucher et al., 1995; Ciabarra et al., 1995).

Of eight known genes that encode polypeptides that can form metabotropic glutamate receptors (mGluR1–8; Nakanishi, 1994), most are expressed in the thalamus, at least of non-primates. mRNAs for mGluR1 (Meguro et al., 1992; Monyer et al., 1992), mGluR4 (Tanabe et al., 1993), mGluR5 (Ohishi et al., 1993a), and mGluR7 (Okamoto et al., 1994; Saugstad et al., 1994) are high in the dorsal thalamus and mGluR3 mRNAs are high in the reticular nucleus (Shigemoto et al., 1992; Martin et al., 1992; Ohishi et al., 1993b; Tanabe et al., 1993; Catania et al., 1994; Neki et al., 1996). mGluR2 expression is low throughout the thalamus (Ohishi et al., 1993b) and unlike the other mGluR mRNAs, may be mainly associated with neuroglial cells. In monkeys, mGluR1α immunostaining gives strong labeling of the neuropil and also labels scattered cells throughout most nuclei, while mGluR2 and 3 and mGluR5 immunostaining is confined to cells (Fig. 34–36). At the electron microscopic level, mGluR1α immunostaining is particularly associated with the terminations of corticothalamic fibers and mGluR5 immunostaining with the terminals of interneurons, while mGluR2/3 immunostaining is mainly in neuroglial processes (Liu and Jones, 1996b; Xiong et al., 1996). Putting together data from published descriptions of glutamate receptor mRNA localization in non-primates leads to the suggestion that individual dorsal thalamic nuclei can be distinguished from one another by differences in the relative densities of, or in the range of expressed mRNAs and, therefore, potentially of the functional type(s) of glutamate receptor.

Autoradiographic studies of glutamate receptor localization in non-primates (Monaghan and Cotman, 1982, 1985; Rainbow et al., 1984; Olsen et al., 1987; Monaghan et al., 1988; Sakurai et al., 1993; Buller et al., 1994) show patterns of thalamic localization that in general coincide with the nuclear-specific patterns of subunit gene expression. For example, the distribution of NMDA receptors localized by autoradiography is very similar to the distribution of NR2A-D mRNAs and the pharmacological profiles of native receptors in the ventral thalamic nuclei (Sakurai et al., 1993) and paraventricular nuclei (Beaton et al., 1992) show differences that may be

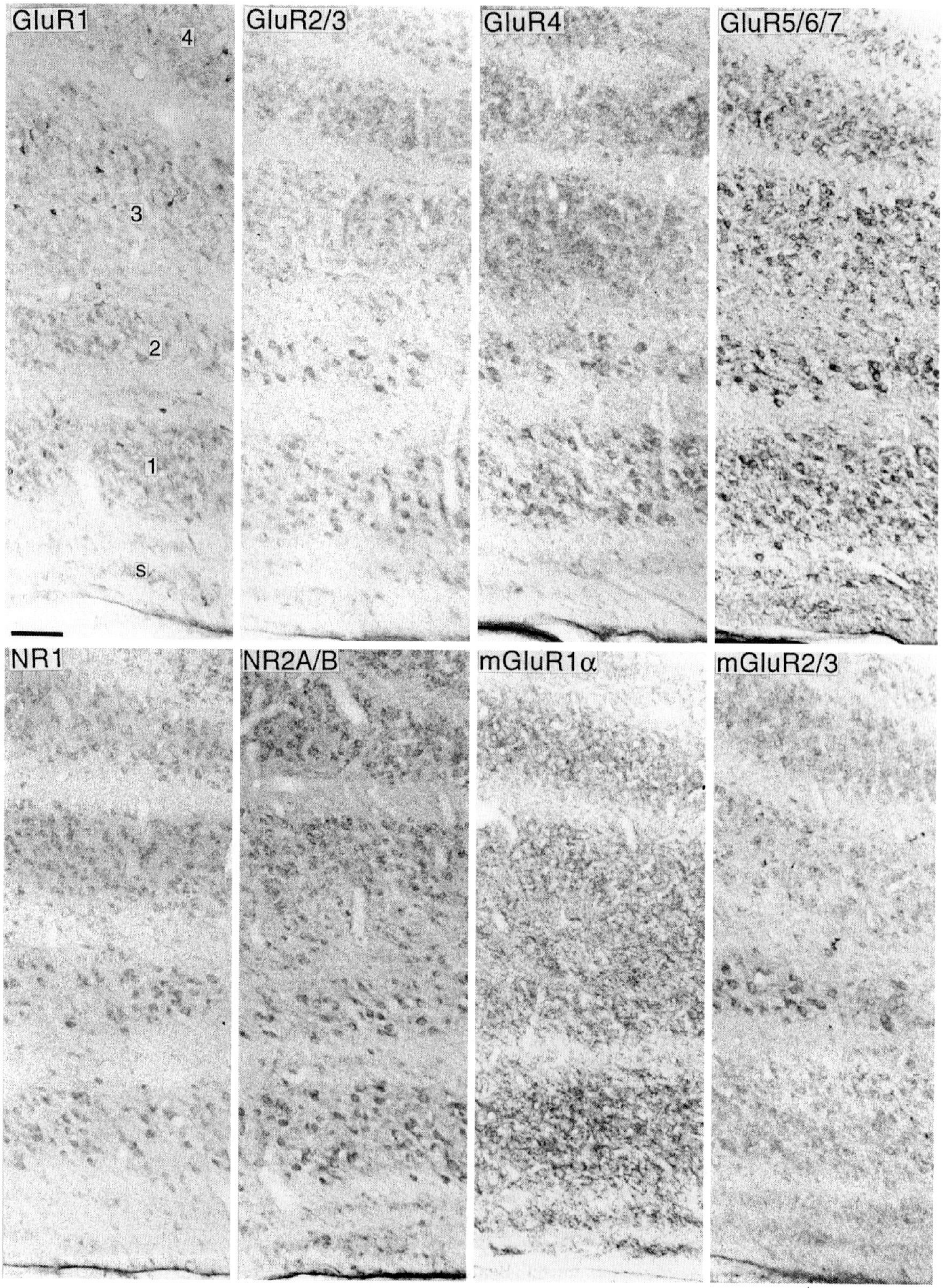
GluR1
4
3
2
1
s
GluR2/3
GluR4
GluR5/6/7
NR1
NR2A/B
mGluR1α
mGluR2/3

Fig. 34. Photomicrographs from adjacent sections through layers 1–4 and the S layers of the dorsal lateral geniculate nucleus of a macaque, showing the different densities and patterns of immunostaining for the major ionotropic glutamate receptor subunits (GluR1-GluR5/6/7, NR1, NR2A/B) and for representatives of group I and Group II metabotropic glutamate receptors (mGluR1α, mGluR2/3). Note that all stain the S layers as well as the principal layers; note also the dense staining of putative interneurons for GluR1. Bar 100 μm. From material illustrated in Tighilet et al. (1998).

correlated with the differential expression of NR2A and NR2D subunits in those nuclei, and with different preferences for the binding of antagonists or agonists.

8.3. GABA RECEPTORS

8.3.1. Introduction

The GABAergic system is the second major neurotransmitter system of the thalamus, the transmitter, GABA, being released from the dendritic terminals of intrinsic interneurons, from the axon terminals of neurons whose cell bodies lie in the reticular nucleus and from certain afferent fiber systems. Thalamic GABAergic cells have been identified in monkeys and other species by immunocytochemistry for GABA or glutamic acid decarboxylase (GAD) or by in situ hybridization for GAD mRNAs (Figs. 19–21, 38). In non-primates, at least, all GABA neurons in the thalamus express the genes encoding both the 67kDa and 65kDa isoforms of GAD (Benson et al., 1991b, 1992; Esclapez et al., 1994). In the reticular nucleus of all species that have been studied, all cells are GABAergic (Houser et al., 1980; Oertel et al., 1983; Yen et al., 1985b; Benson et al., 1991b, 1992). GABAergic neurons form approximately 25% of the neurons in all dorsal thalamic nuclei of species other than rodents (Montero, 1989; Hunt et al., 1991; Bentivoglio et al., 1991b; Benson et al., 1991b; Spreafico et al., 1993c, 1994b). In rodents, intrinsic GABAergic neurons are confined to the dorsal lateral geniculate nucleus. Earlier reports of lower numbers in the intralaminar and certain other dorsal thalamic nuclei of cats and monkeys (Rinvik et al., 1987; Penny et al., 1983; Smith et al., 1987) can now be discounted (Hunt et al., 1991; Bentivoglio et al., 1991b) (Section 5.3). Figs. 16 and 17 show data from counts of GABA immunoreactive neurons in the intralaminar and adjacent nuclei and in the layers of the dorsal lateral geniculate nucleus of macaque monkeys.

Most of the GABA synapses in the thalamus are formed by the axons of cells in the reticular nucleus and where present (e.g. in monkeys and cats) from the intrinsic interneurons. Most of the synapses made by the intrinsic neurons are dendro-dendritic. GABAergic afferent fibers from extrinsic sources reach certain thalamic nuclei: these include the anterior ventral lateral and ventral medial nuclear regions which are innervated by axons of GABA cells in the internal segment of the globus pallidus and the so-called ventral pallidum and in the pars reticulata of the substantia nigra (Uno and Yoshida, 1975; Uno et al., 1978; Ueki et al., 1977; Fonnum et al., 1978; Nagy et al., 1978; Deniau et al., 1978; Kilpatrick et al., 1980; MacLeod et al., 1980; Starr and Kilpatrick, 1981; Yoshida, 1981; Penney and Young, 1983; Young et al., 1984; Araki et al., 1985; Zahm and Heimer, 1987; Pearson et al., 1987; Yamamoto et al., 1988). GABAergic terminals have been described in these nuclei in monkeys, although experimental verification of their extrinsic origins has not been made (Kultas-Ilinsky et al., 1978, 1983, 1985; Ilinsky and Kultas-Ilinsky, 1990). The input from the globus pallidus to the centre médian nucleus may also be GABAergic, although electron

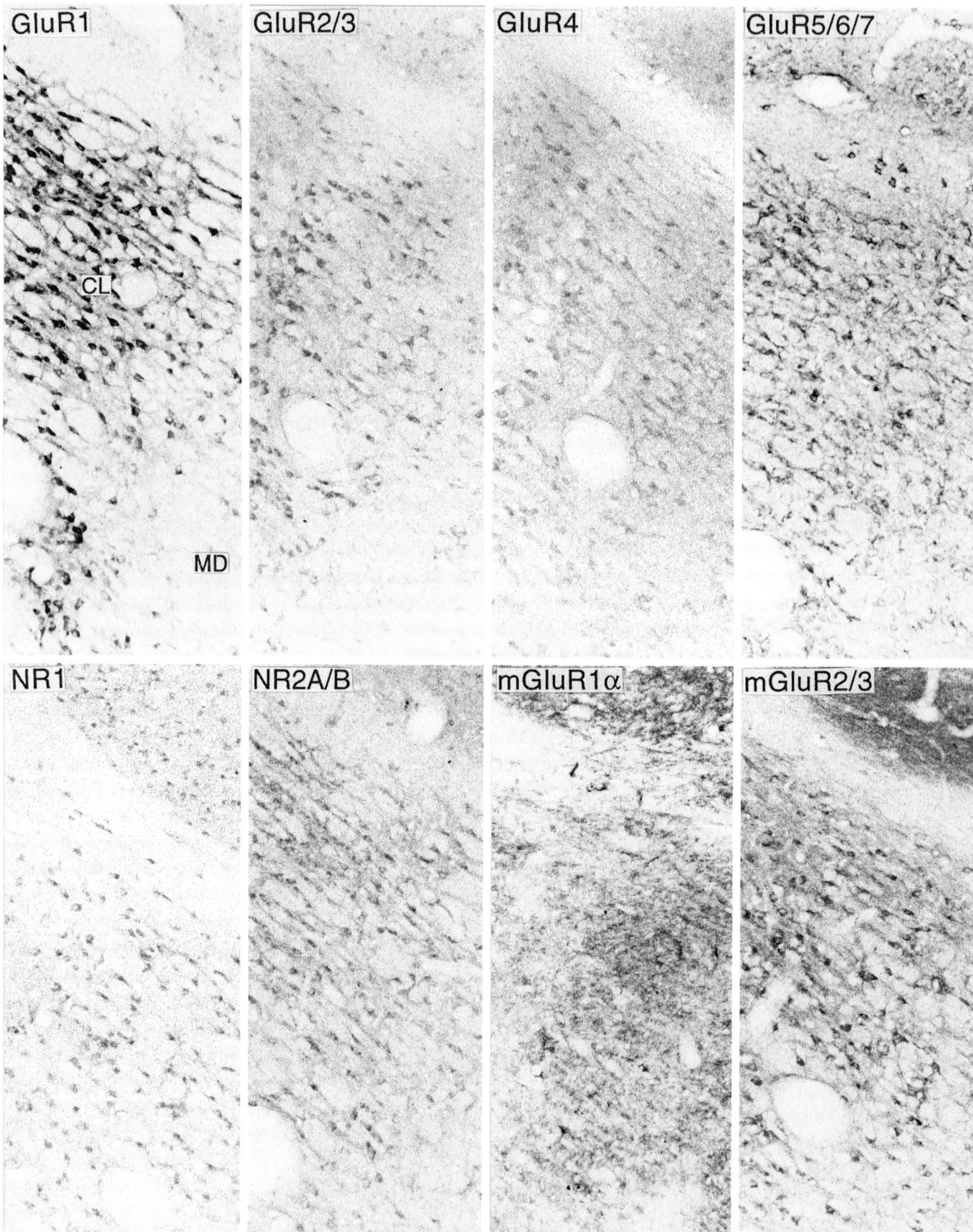
GluR1
CL
MD
GluR2/3
GluR4
GluR5/6/7
NR1
NR2A/B
mGluR1α
mGluR2/3

Fig. 35. Adjacent sections through the central lateral (CL) and adjacent nuclei of the same monkey as in Fig. 34, showing different patterns of labeling for the same set of glutamate receptor subunits. Bar 100 μm. From unpublished material of B. Tighilet and the author.

microscopic evidence is equivocal (Harding and Powell, 1977). GABAergic projections from the basal forebrain terminate in parts of the reticular nucleus, and a small GABAergic projection from the nucleus of the optic tract in the pretectum of the cat reaches the A and C laminae of the dorsal lateral geniculate nucleus and the perigeniculate part of the reticular nucleus (Cucchiaro et al., 1991a). A similar GABA projection has not been described in primates.

GABA-mediated effects which accompany afferent stimulation influence receptive field organization (Sillito and Kemp, 1983b) and receptive field size (Hicks et al., 1986a) of thalamic relay neurons; the effects include various forms of lateral inhibition (Eysel et al., 1986; Pape and Eysel, 1986) and modulations of stimulus-response coupling (Gottschaldt et al., 1983a). GABA-mediated inhibition is also a critical element in the genesis and maintenance of oscillatory phenomena in the thalamus (Warren et al., 1994; Huguenard and Prince, 1994a; Bal et al., 1995a).

Both bicuculline-sensitive, ionotropic, $GABA_A$ receptors and bicuculline-insensitive, metabotropic, $GABA_B$ receptors (Bowery et al., 1987, 1990; Olsen and Tobin, 1990) have been described in the thalamus. $GABA_C$ receptors (Johnstone, 1994a,b; Bormann and Feigenspan, 1995; Enz et al., 1995; Feigenspan et al., 1993) have not yet been identified there. In many experimental paradigms, fast IPSPs mediated by $GABA_A$ receptors, which are localized postsynaptically, have been described in the thalamus of many species. Some recent studies include: Crunelli et al. (1988), Paré et al. (1991), Soltesz and Crunelli (1992d) and Salt (1989). There have been fewer studies of thalamic $GABA_B$ receptors, which are localized pre- and postsynaptically, but the long, slow IPSPs typical of activation of these receptors are starting to be observed (Crunelli et al., 1988; Thomson, 1988a,b; Huguenard and Prince, 1994b; Bal et al., 1995a; Ulrich and Huguenard, 1996; Warren et al., 1997).

The ionotropic $GABA_A$ receptor is made up of hetero-oligomeric combinations of polypeptide subunits which form the chloride channel. The range of subunits potentially involved in the formation of native $GABA_A$ receptors is potentially large since the subunit polypeptides are the products of at least 15 different genes, 12 of which are expressed in the forebrain (Wisden et al., 1992; Persohn et al., 1992; Fritschy and Möhler, 1995). Ten of these have been described in the monkey and human thalamus (Huntsman et al., 1995, 1996; Figs. 39–45). $GABA_A$ receptor subunit genes are classified by sequence identity into five classes (α, β, γ, δ and ρ), within each of which there are variants (α_{1-6}, β_{1-3}, γ_{1-3}, δ and $\rho_{1,2}$) (Wisden and Seeburg, 1992; Wisden et al., 1992; MacDonald and Olsen, 1994). In *in vitro* systems, the combination of subunits determines the kinetics of the channel and the pharmacological properties of the receptor, especially in relation to the binding of benzodiazepines (DeLorey and Olsen, 1992; MacDonald and Olsen, 1994). Although the structure of native $GABA_A$ receptors is not known, it is generally accepted that a combination of $\alpha1$, $\alpha3$ or $\alpha5$ subunits with $\beta2$ and $\gamma2$ subunits is essential for the formation of $GABA_A$ receptors with the full range of 'classical' physiological and pharmacological properties (Olsen et al., 1991; Lüddens and Wisden, 1991; Sieghart, 1991; Wieland et al., 1992; Wisden and Seeberg, 1992; Möhler et al., 1995). One of the strongest pieces of evidence in favor of this view is the fact that $\alpha1$, $\beta2$ and $\gamma2$ subunit mRNAs are the most highly expressed of all the subunits in the adult brain. The thalamus, as indicated in Figs. 39–45, is no

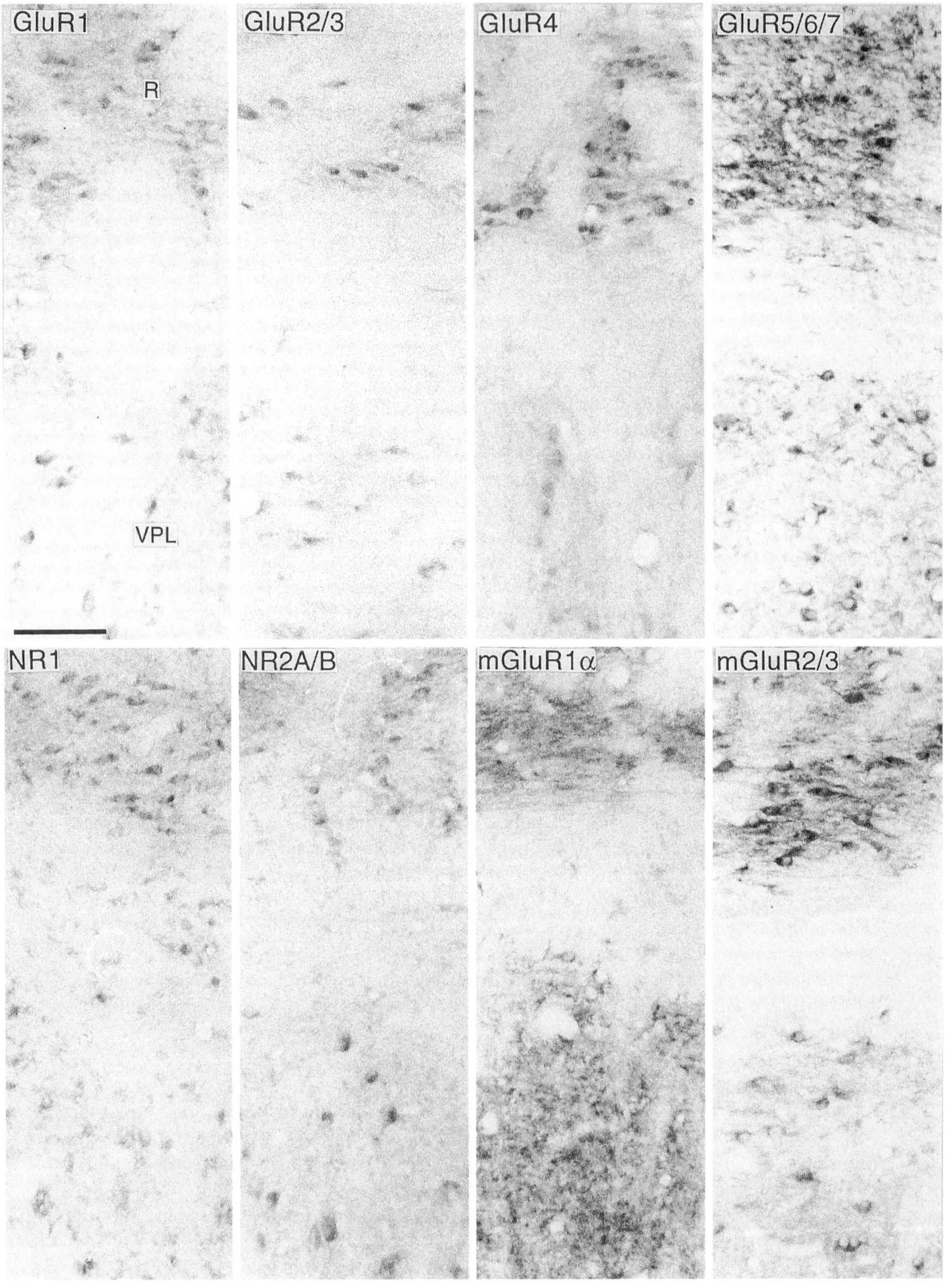
GluR1
R
VPL
GluR2/3
GluR4
GluR5/6/7
NR1
NR2A/B
mGluR1α
mGluR2/3

Fig. 36. Adjacent sections through the reticular nucleus (R) and an adjacent part of the VPL nucleus from the same monkey as in Figs 34 and 35, stained for the same set of glutamate receptor subunits. Bar 100 μm. From unpublished material of B. Tighilet and the author.

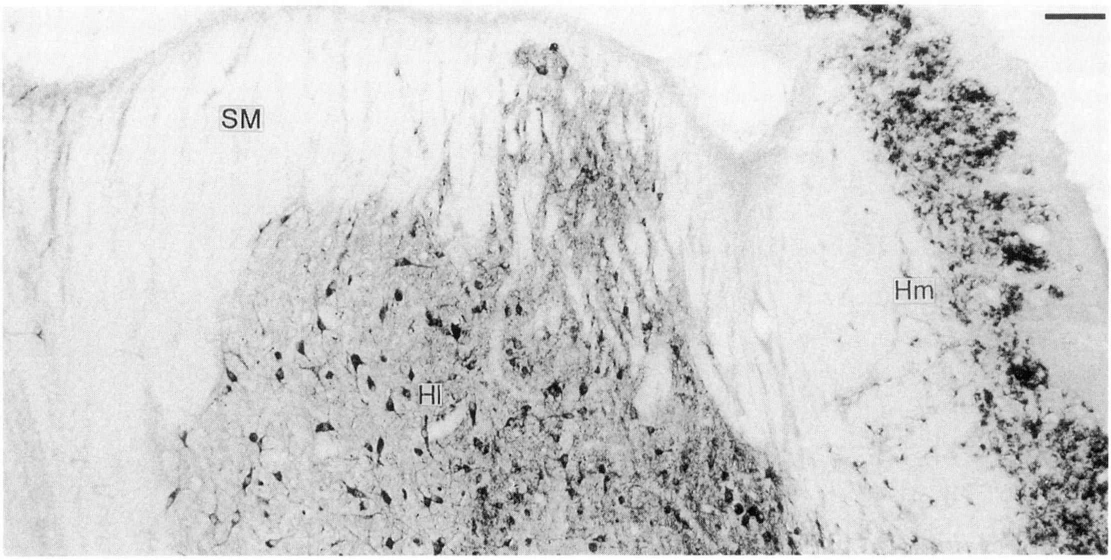

Fig 37. Dense immunostainjng for glutamate receptor subunit, GluR1, in the habenular nuclei. C.f. Figs 34–36. Bar 100 μm. From unpublished material of B. Tighilet and the author.

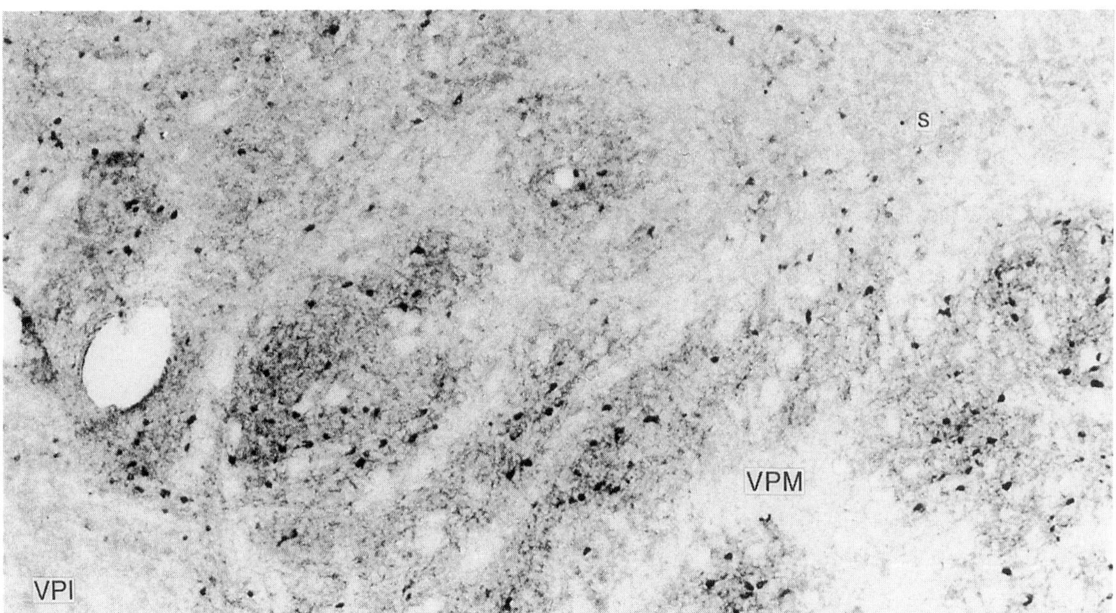

Fig. 38. Frontal section through part of the VPM nucleus of a macaque, showing distribution of GABA immunoreactive intrinsic interneurons and concentrations of their immunoreactive processes in the matrix zone of the nucleus (see text and Figs 24, 83, 86). Bar 100 μm. From material illustrated in Rausell and Jones (1991a).

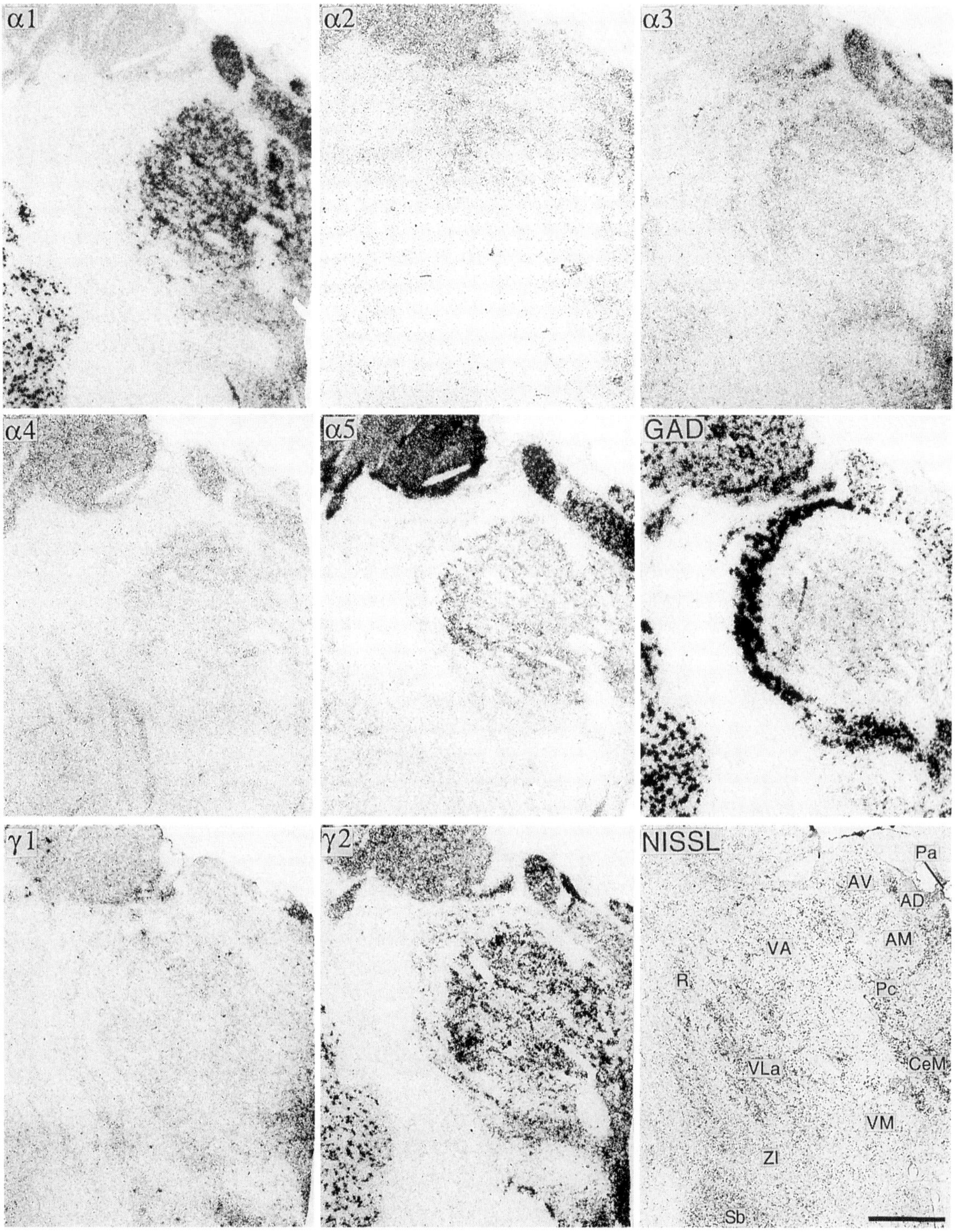
α1
α2
α3
α4
α5
GAD
γ1
γ2
NISSL
Pa
AV
AD
AM
VA
R
Pc
VLa
CeM
VM
ZI
Sb

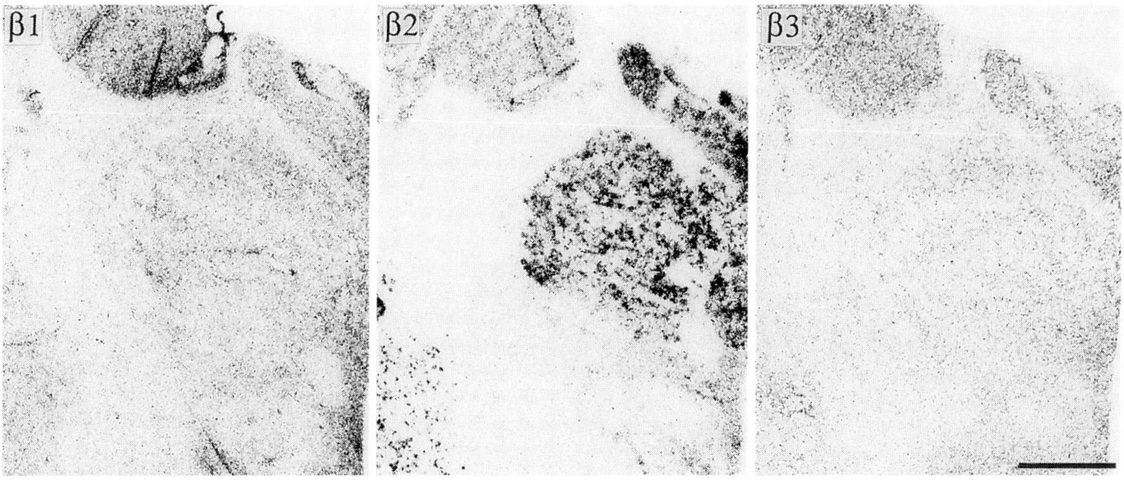

Fig. 39. (continued)

exception. $GABA_A$ receptor density, as localized by radioligand binding and autoradiography, is high in all thalamic nuclei except the reticular nucleus and different densities among dorsal thalamic nuclei do not necessarily correlate with the distribution of GABAergic afferents. In monkeys, for example, benzodiazepine binding sites are rather evenly distributed throughout the dorsal thalamus and zona incerta while muscimol binding sites are concentrated in the anterior and medial nuclei (Kultas-Ilinsky et al., 1988). Modest species differences may also be present, especially in benzodiazepine binding (Palacios et al., 1981; Doble et al., 1987; Benavides et al., 1984; Kultas-Ilinsky et al., 1988; Olsen et al., 1990). These differences might reflect the differential expression of subunit genes referred to below.

Autoradiographically localized $GABA_B$ receptors are reported to be particularly dense in the thalamus of rodents (Bowery et al., 1987; Liu et al., 1993; Turgeon and Albin, 1994) although this has not been confirmed in studies of primates (Kultas-Ilinksy and Ilinsky, 1988). Physiologically, they may form two or more classes (Emri et al., 1996; Guyon and Leresche, 1995). At the time of writing, the $GABA_B$ receptor gene or genes had not been cloned and information about the distribution of $GABA_B$ receptors in the thalamus comes from receptor binding studies. See Muñoz et al. (1998).

All of the known $GABA_A$ receptor subunit genes, with the exception of α6, expression of which is restricted to the cerebellum, are expressed in the thalamus. There are, however, large nucleus-specific differences in the levels of gene expression for particular subunits in adults and other subunits are preferentially expressed in the developing thalamus.

In the dorsal thalamus of monkeys and humans (Huntsman et al., 1995, 1996) (Figs. 39–45), α1, β2 and γ2 subunit mRNAs are expressed at the highest overall levels,

←

Fig. 39. Autoradiograms from adjacent frontal sections and a matched thionin-stained section (NISSL) through the anterior part of the thalamus of a macaque monkey, showing in situ hybridization of radioactive RNA probes complementary to $GABA_A$ receptor subunit mRNAs α1-α5, β1-β3, γ1, γ2 and to GAD67 mRNA. Note relative difference in levels of expression of the subunits and absence of all except γ2 from the reticular nucleus (R). Bar 2 mm. From material illustrated in Huntsman et al. (1996).

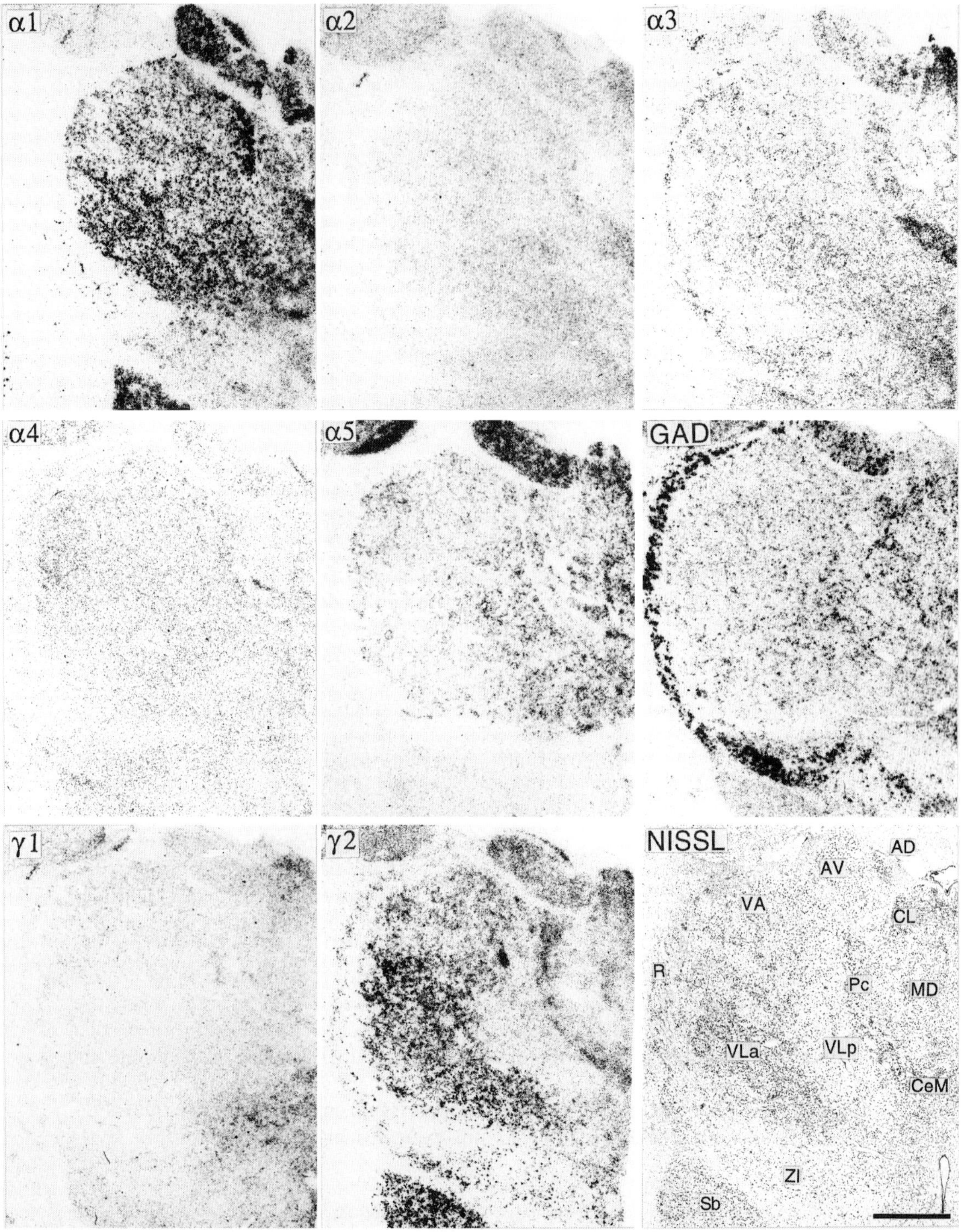

Fig. 40. Autoradiograms and Nissl-staining from sections through levels posterior to those illustrated in Fig. 39. Note differences in the labeling of the intralaminar nuclei (CeM, Pc). Bar 2 mm. From material illustrated in Huntsman et al. (1996).

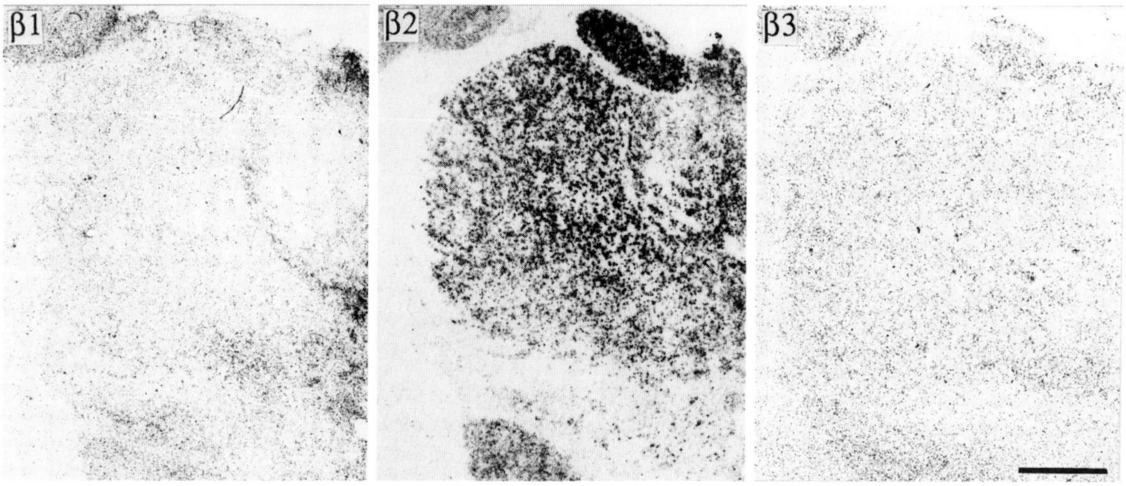

Fig. 40. (continued)

followed in order by α5, β1, α3, γ1, β3, α4 and α2 subunit mRNAs. In the thalamus of rodents, α5 transcripts are rare and α4 and β2 transcripts are the most abundant (Khrestchatisky et al., 1989; Wisden et al., 1992). Other rare mRNAs such as those for δ subunits, not studied in monkeys, are also expressed. The three sensory relay nuclei of the monkey and human (ventral posterior, ventral medial geniculate and dorsal lateral geniculate) show the highest levels and the widest range of subunits expressed. The dorsal lateral geniculate nucleus is highest of all, and in it expression for α1, β2 and γ2 subunits mRNAs in the magnocellular layers exceeds that in the parvocellular layers (Fig. 44). ^{3}H benzodiazepine binding demonstrates a similar difference between the magnocellular and parvocellular layers (Shaw and Cynader, 1986). Expression of other $GABA_A$ receptor subunit mRNAs is equal across the geniculate layers (Fig. 44), a pattern which is similar to that found with ^{3}H muscimol binding. Some dorsal thalamic nuclei display complementary patterns of expression. For example, α4 and α5 subunit mRNAs are expressed in sensory relay nuclei but not in the intralaminar nuclei (Fig. 44) while α2, α3, β1, β3 and γ1 subunit mRNAs are expressed at moderate levels in the intralaminar nuclei but at very low levels in the sensory relay nuclei.

In the two primates that have been studied, the cells of the reticular nucleus express only γ2 receptor subunit mRNAs at significant levels and α3 subunit mRNAs at levels bordering on undetectability. In rodents only α3 is expressed (Wisden et al., 1992). The lack of expression of β2 and β3 subunits, at least, has been confirmed by immunocytochemistry (Fig. 46) (unpublished observations and see Bentivoglio et al., 1991a; Spreafico et al., 1993b; Fritschy and Möhler, 1995). Functional $GABA_A$ receptors are unlikely to be assembled from homomeric combinations of γ2 or α3 subunits. Moreover, reticular nucleus cells are joined by GABAergic synapses (Deschênes et al., 1985; Yen et al., 1985b) and many examples of bicuculline-sensitive IPSPs with reversal potentials typical of those mediated by $GABA_A$ receptors have been reported in the reticular nucleus (McCormick and Prince, 1986; Hugenard and Prince, 1994a,b; Bal et al., 1995). It is likely, therefore, that one or more novel $GABA_A$ receptor subunit polypeptides that are joined heteromerically with the γ2 or α3 receptor subunits remain to be discovered in the reticular nucleus.

Based on the sizes of cells labeled by in situ hybridization histochemistry, it appears

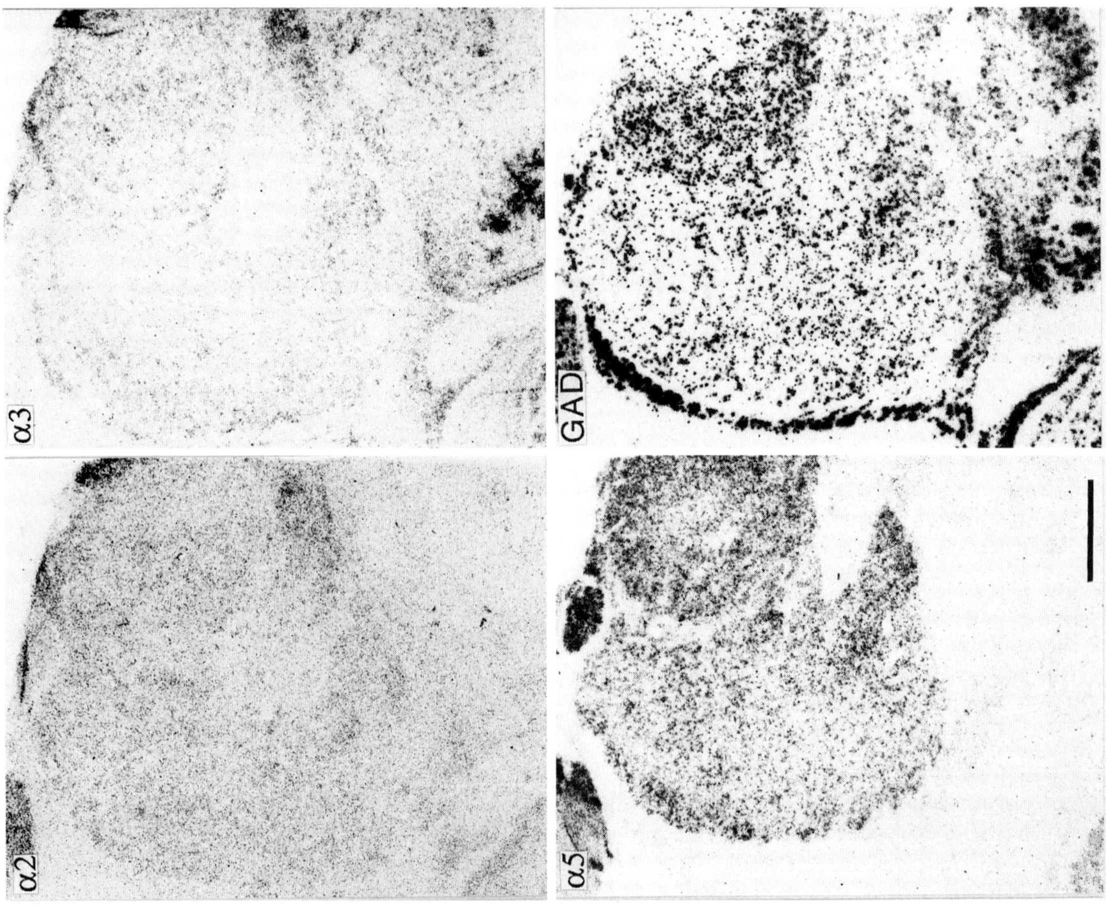

Fig. 41. Autoradiograms and Nissl-staining from adjacent sections through middle levels of the same thalamus shown in Figs 39 and 40. Note differences in the labeling of the posterior intralaminar nuclei (CM, Pf). Bar 2 mm. From material illustrated in Huntsman et al. (1996).

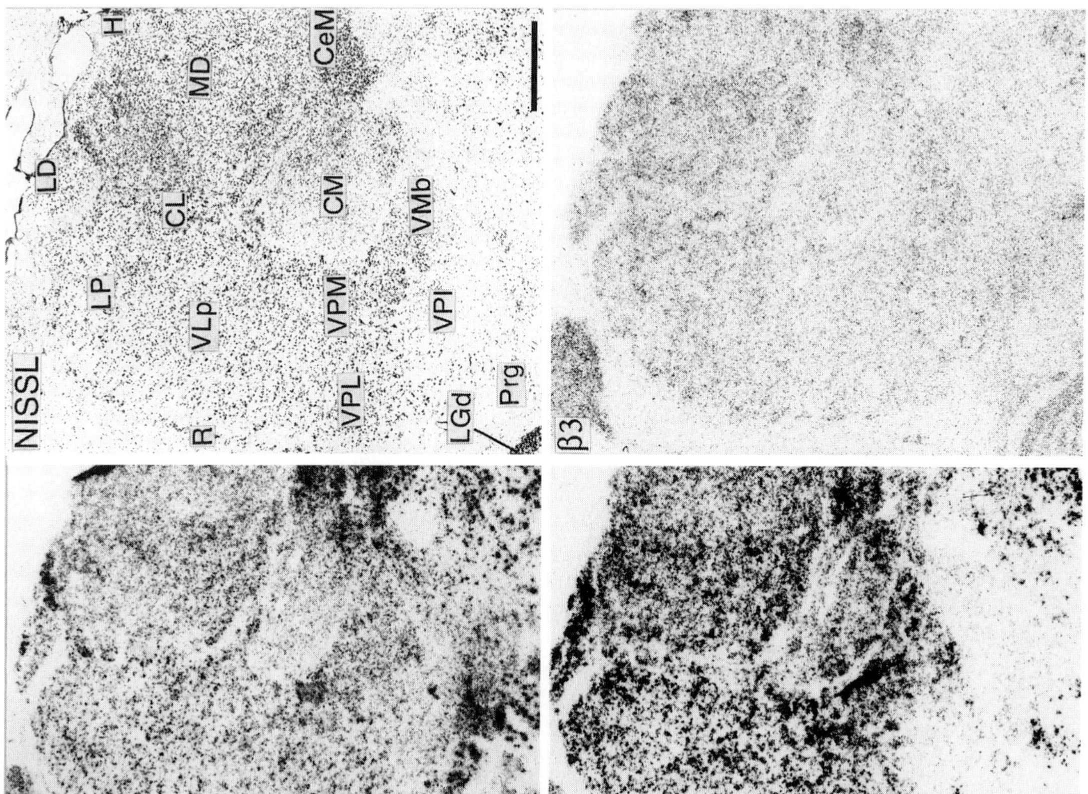

Fig. 41. (continued)

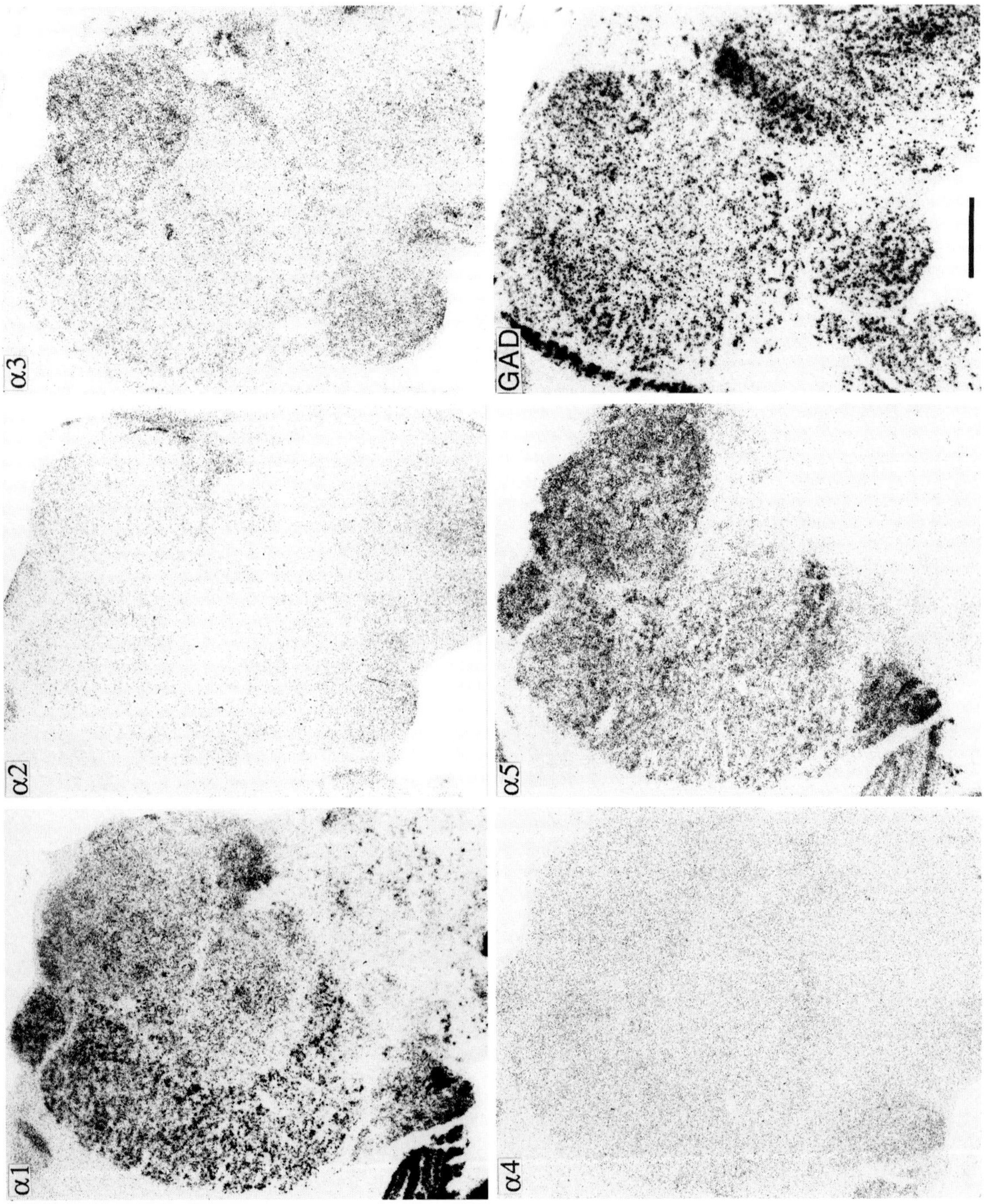

Fig. 42. Autoradiograms and Nissl-staining from adjacent sections at more posterior levels of the same thalamus illustrated in Figs 39–41. Bar 2 mm. From material illustrated in Huntsman et al. (1996).

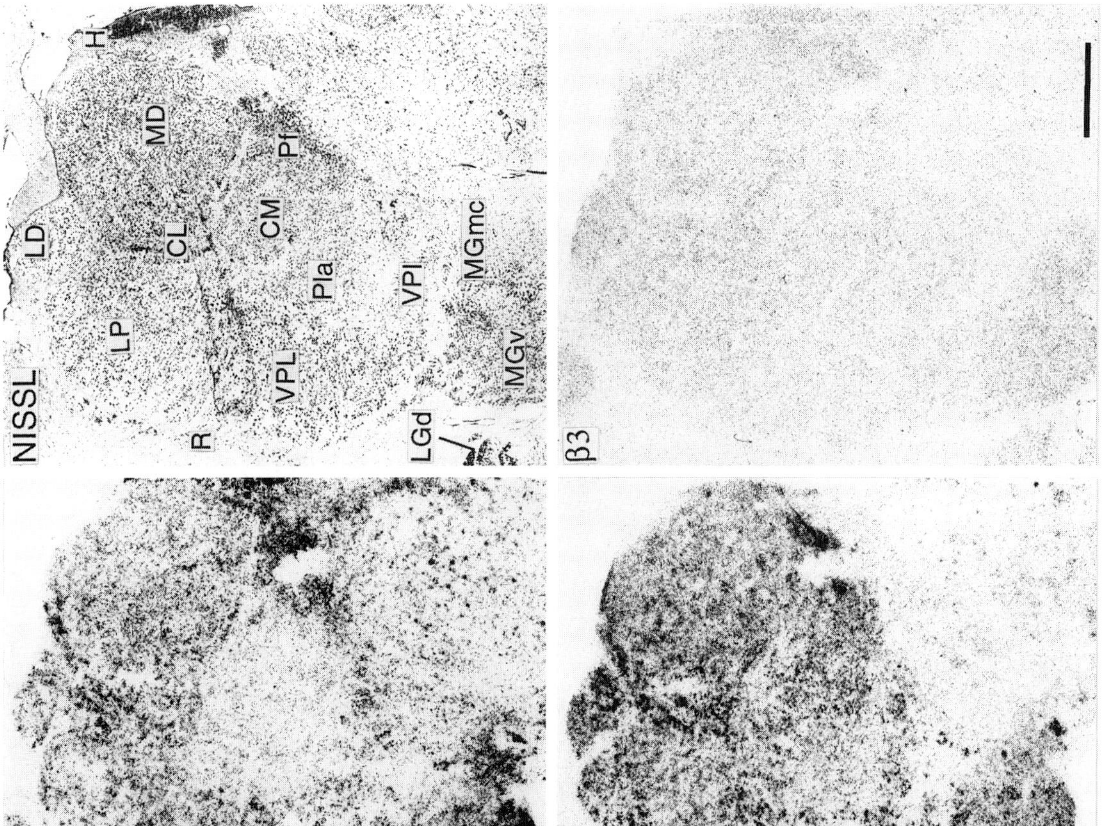

Fig. 42. (continued)

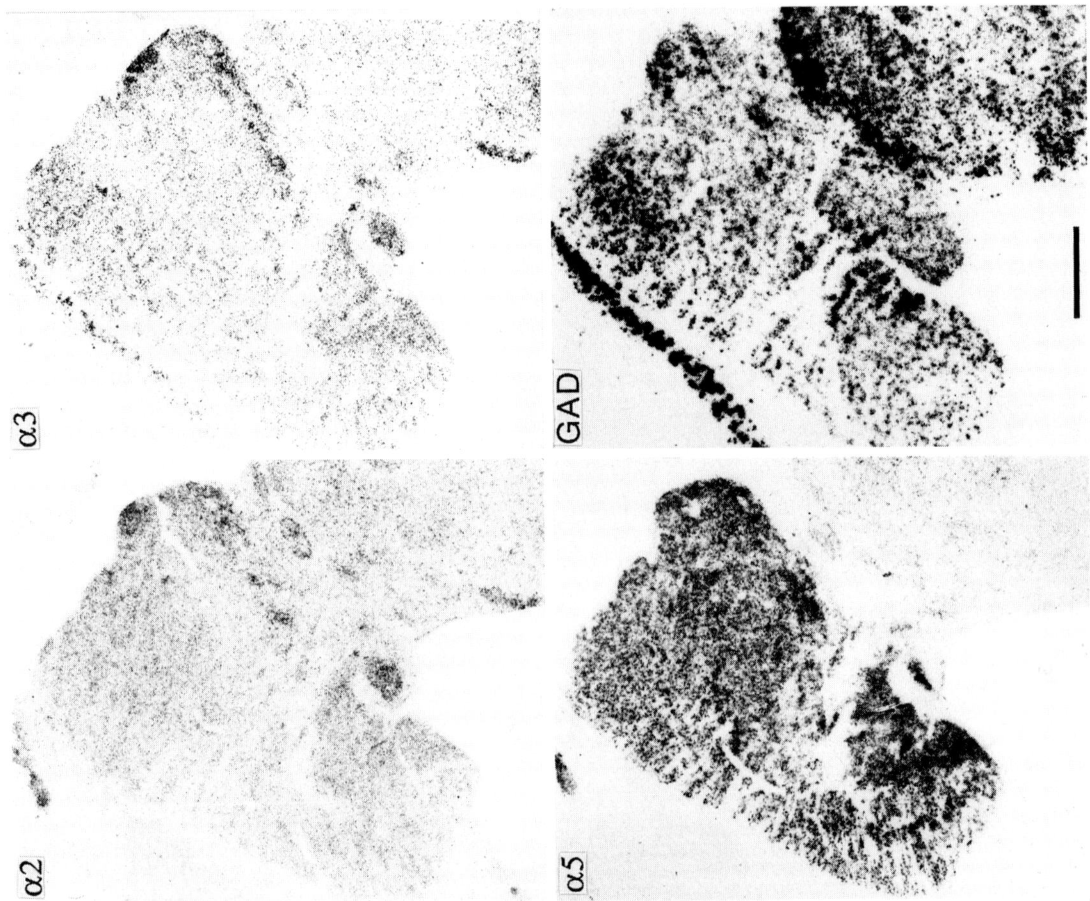

Fig. 43. Autoradiograms and Nissl-staining section from adjacent sections at the posterior pole of the same thalamus illustrated in Figs 39–42. Note that labeling patterns in the limitans (L) and suprageniculate (SG) nuclei tend to match those in the anterior intralaminar nuclei. Bar 2 mm. From material illustrated in Huntsman et al. (1996).

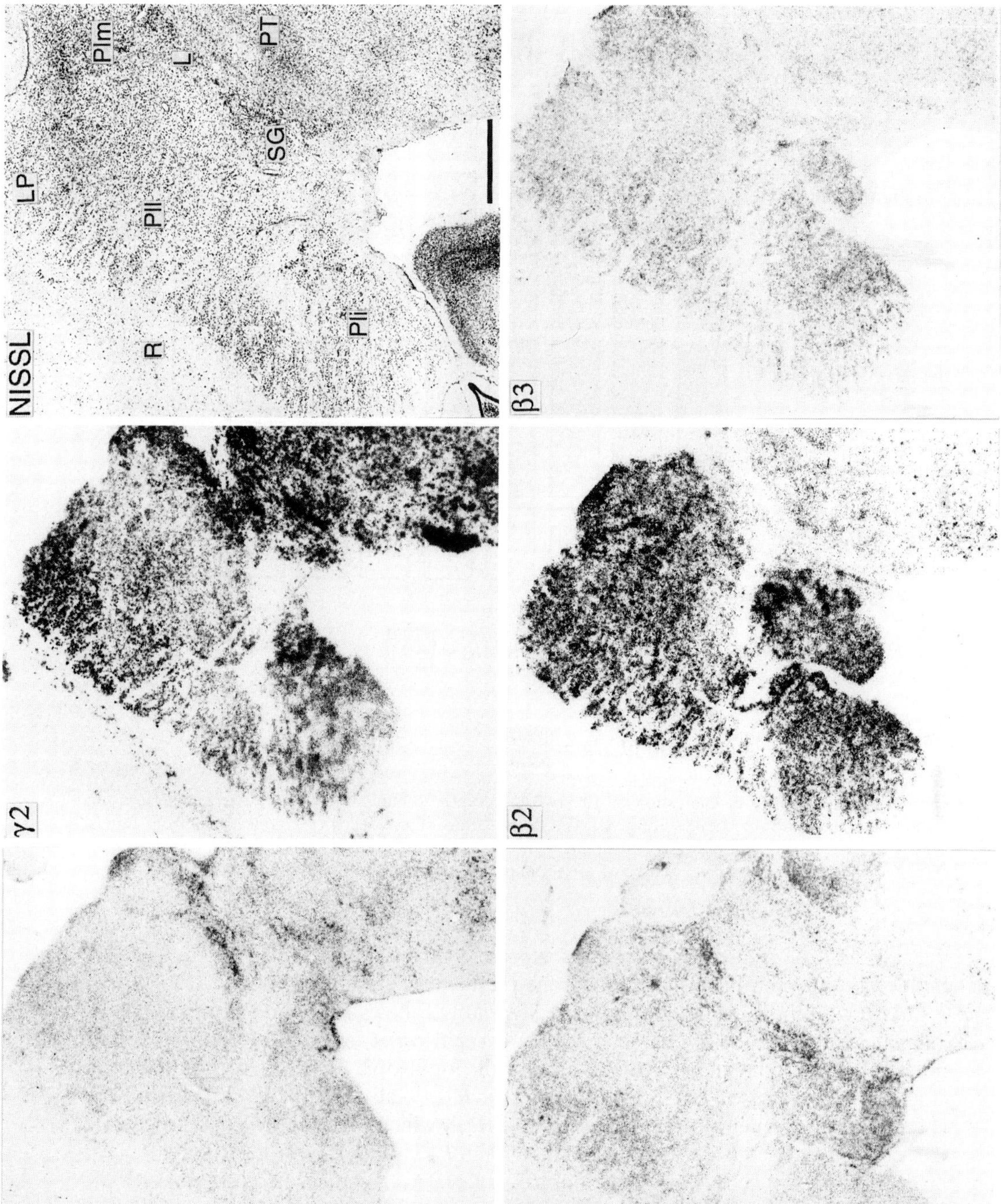

Fig. 43. (*continued*)

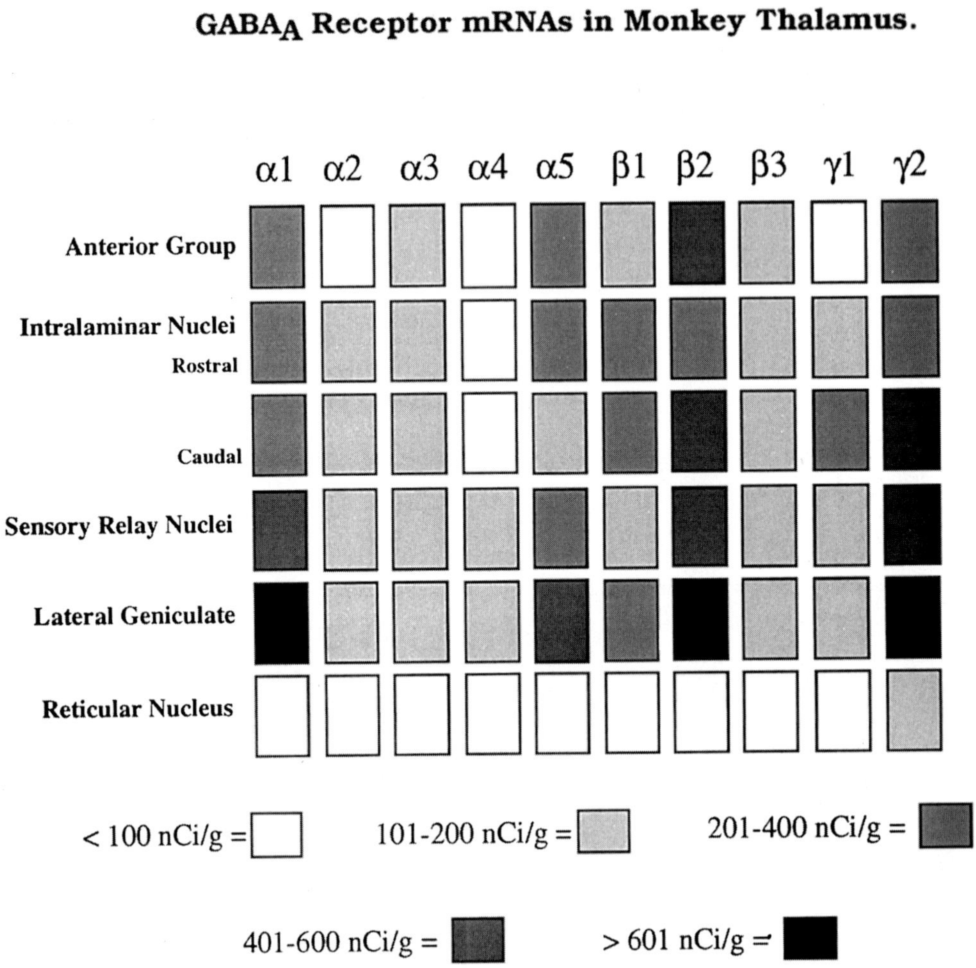

Fig. 44. Relative density of expression of 10 $GABA_A$ receptor subunit mRNAs in selected nuclei of the thalamus of the macaque. Levels are given as nCi per gram of radioactivity, as measured from optical density readings of film autoradiograms. Alpha1, beta2 and gamma2 subunit mRNAs are those most highly expressed in all nuclei but alpha5 and beta1 are prominent in the anterior intralaminar nuclei. From Huntsman et al. (1996).

that all the $GABA_A$ receptor subunits expressed in the dorsal thalamus are found in both relay cells and intrinsic interneurons (Huntsman et al., 1995, 1996) (Fig. 47).

$GABA_A$ receptor-mediated inhibition invariably follows activation of relay cells by afferent stimulation. It is characterized by a rapid onset, steep IPSP with a reversal potential of approximately -75 mV typical of the ionotropic $GABA_A$ receptor (Crunelli et al., 1988; Paré et al., 1991; Soltesz and Crunelli, 1992b). Commonly, the early IPSP is followed by a much slower, longer lasting IPSP which is dependent upon increased K+ conductance, and reverses at about -100mV. This is, therefore, thought to be brought about by activation of metabotropic $GABA_B$ receptors (Soltesz et al., 1988, 1989b; Bloomfield and Sherman, 1988; Paré et al., 1991). The two forms of GABA-mediated IPSPs can be generated by stimulation of virtually any extrinsic afferent pathway to the thalamus. In the anterior nuclei of the cat, which are thought to be

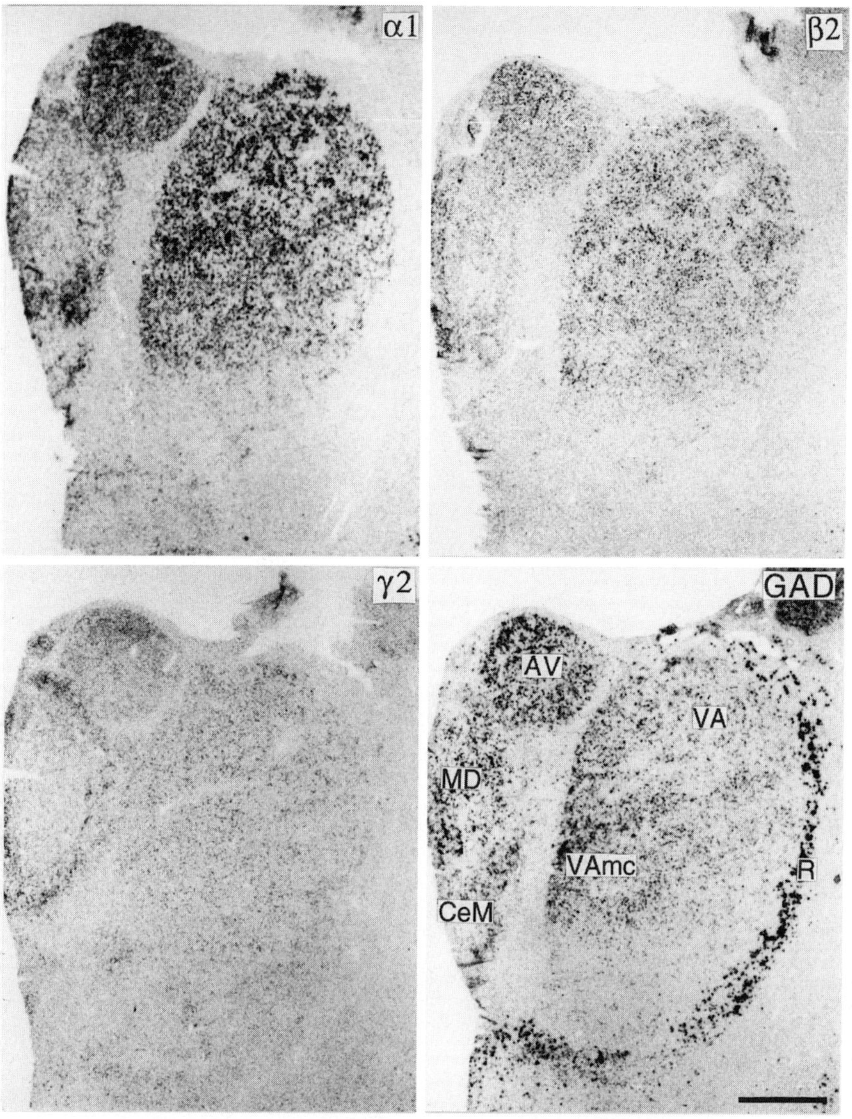

Fig. 45. Photomicrographs of film autoradiograms, showing in situ hybridization of radioactive cRNA probes specific for $\alpha1$, $\beta2$ and $\gamma2$ $GABA_A$ receptor mRNAs and for GAD67 mRNA in frontal sections through the anterior aspect of a human thalamus. Patterns of expression are identical to those found in the monkey. Bar 2 mm. From unpublished material of M.G. Leggio and the author.

devoid of afferents from the reticular nucleus (Steriade et al., 1984; Velayos et al., 1989), stimulation of the mamillo-thalamic tract still produces the two IPSPs (Paré et al., 1991), suggesting that the synapses made by intrinsic interneurons on relay cells are associated with both $GABA_A$ and $GABA_B$ receptors. The typical early $GABA_A$-mediated IPSP may be preceded by a very early, short, Cl^--dependent IPSP. This may have its origins in the local release of GABA from dendritic terminals of interneurons as the result of local depolarization by afferent synapses (Paré et al., 1991; Soltesz and Crunelli, 1992b; Pape and McCormick, 1995).

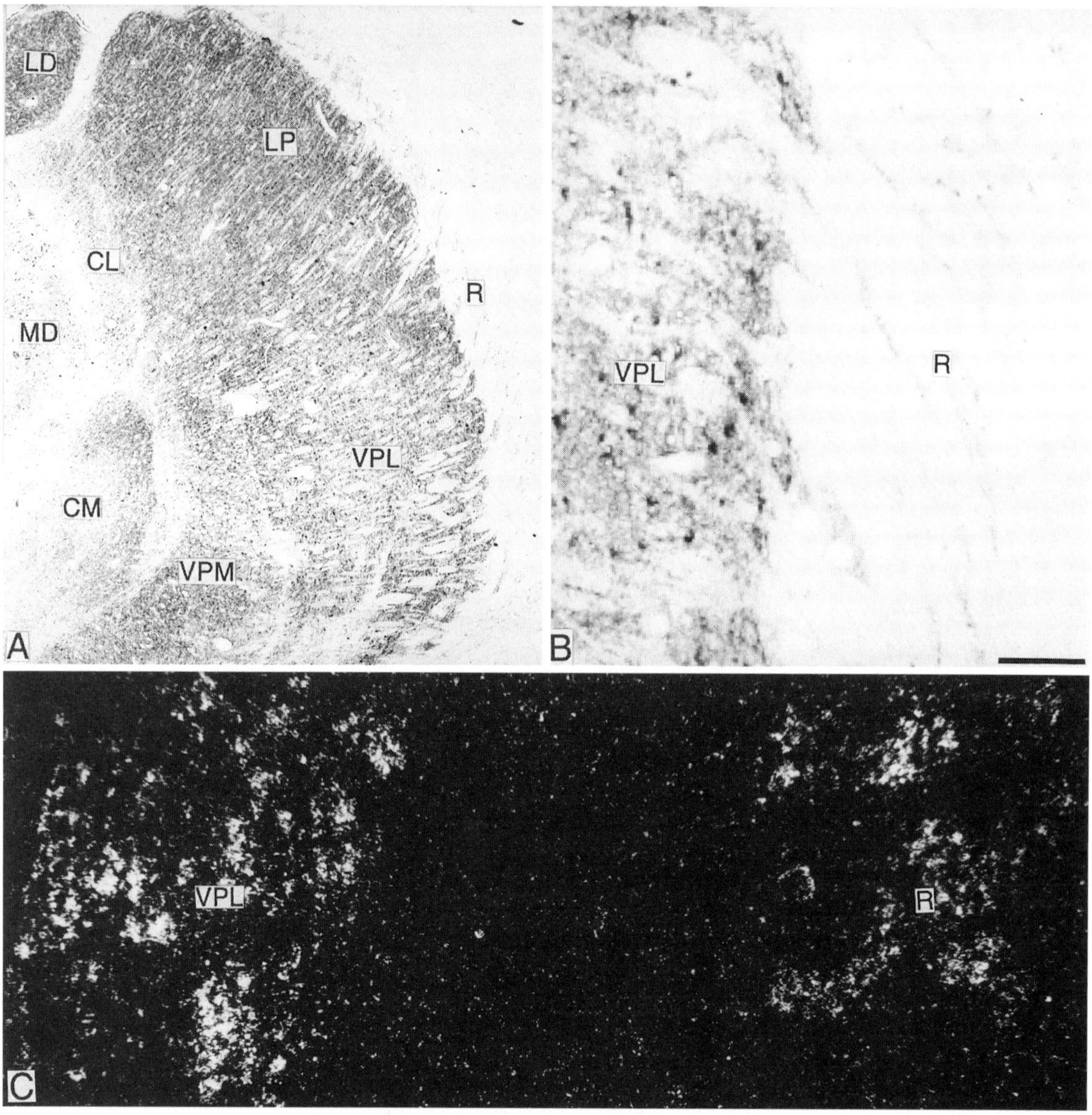

Fig. 46. Brightfield (A, B) and darkfield (C) photomicrographs from frontal sections of macaque monkey thalami, showing lack of immunocytochemical staining for $\beta2$ and $\beta3$ $GABA_A$ receptor subunits in the reticular nucleus (A, B) and expression of $\gamma2$ receptor subunit mRNA in the reticular nucleus (R) and adjacent VPL nucleus (C). Bar 1 mm (A), 250 μm (B), 100 μm (C). C, from material illustrated in Huntsman et al. (1996).

→

Fig. 47. Photomicrographs of autoradiographs, showing in situ hybridization of cRNA probes specific for $\gamma2$ $GABA_A$ receptor subunit mRNA in the mediodorsal (A) and dorsal lateral geniculate nuclei (B) of a macaque. Note lack of hybridization signal over smaller nuclei of neuroglial cells. Bar 100 μm (A), 50 μm (B). From material illustrated in Huntsman et al. (1996).

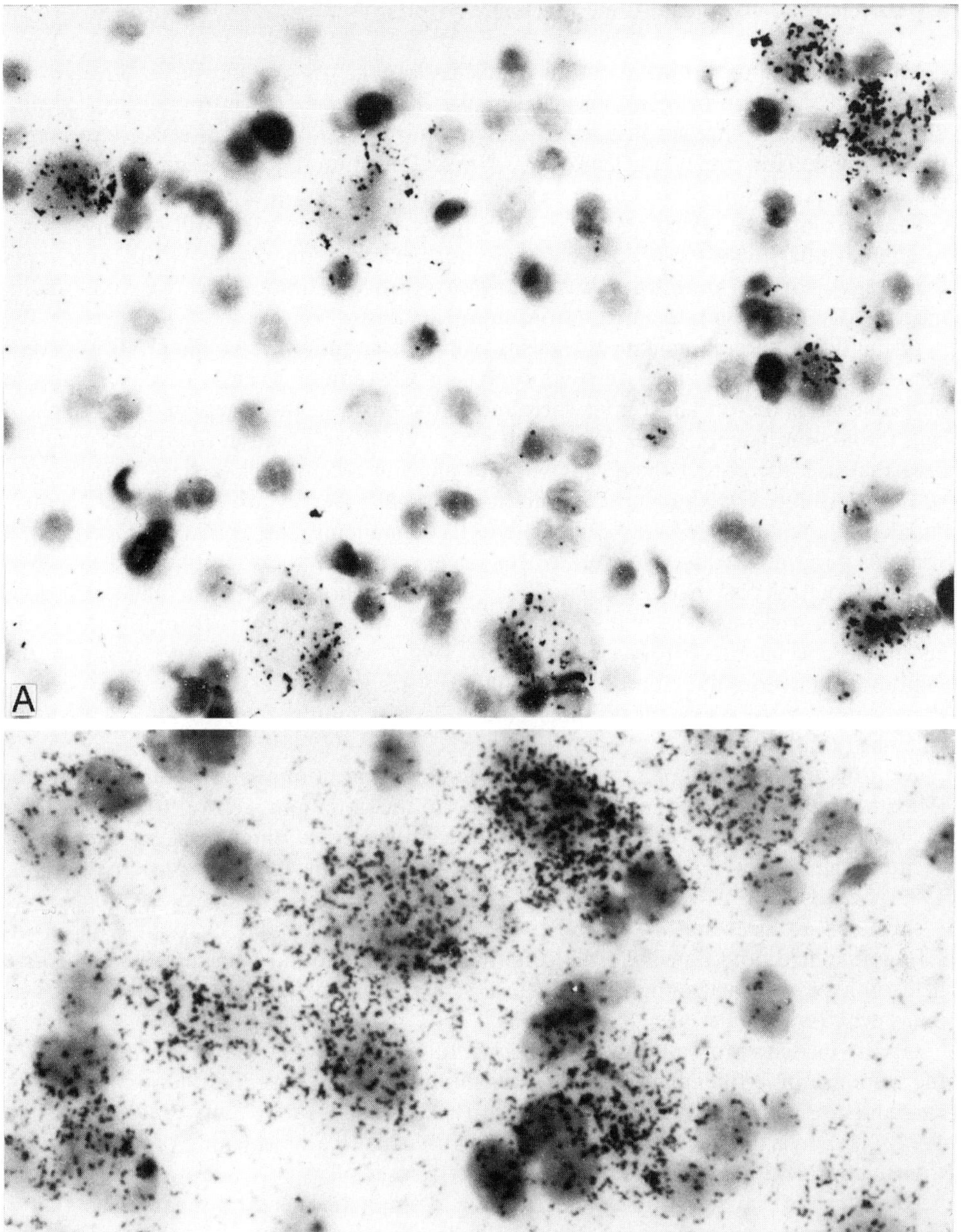
A
B

8.3.2. Role of GABA in thalamic function

GABA plays important roles in the transmission of sensory messages through the thalamus and also in bringing large ensembles of thalamic cells into collective oscillations that underlie changes in conscious state. During the passage of sensory messages through a thalamic relay nucleus, GABA-mediated inhibition does not cause a profound reorganization of receptive fields or cause changes in stimulus specificity but enhances resolution, offsetting the loss that tends to occur as the result of convergence of afferent fibers whose receptive fields are not perfectly congruent. At the same time, the strength of the excitatory drive that convergence promotes is preserved. Again, most work on this subject comes from non-primates.

In the dorsal lateral geniculate nucleus of cats, iontophoresis of the $GABA_A$ receptor antagonist, bicuculline, onto on-center or off-center, X and Y cells results in degradation of the normal center-surround antagonism in the receptive fields of these cells (Sillito and Kemp, 1983), bringing it back to the levels seen in the retinal ganglion cells that are afferent to them, but without altering receptive field size. This demonstrates that the $GABA_A$ based inhibition normally serves to sharpen the center-surround antagonism of the geniculate cells, enhancing the spatial contrasts in a stimulus without modifying the extent or organization of the receptive fields.

In the ventral posterior nucleus of cats, the iontophoresis of bicuculline also does not alter the sizes of the receptive fields of neurons receiving inputs from low threshold cutaneous mechanoreceptors (Hicks et al., 1986; Salt, 1986) but the depth of the inhibition that normally succeeds the excitatory response to a stimulus is reduced so that responses to a second stimulus coming a few hundred msecs after a first are enhanced (Salt, 1986).

Neurons in the VPM of rats, which respond to deflections of whiskers, when subjected to bicuculline show enhanced responses to the dominant, central whisker in their multi-whisker receptive fields, but the number of surrounding whiskers capable of exciting the cell is not altered and there is no loss of direction selectivity (Salt, 1989; Lee et al., 1994b). The $GABA_B$ receptor antagonist, 2-hydroxy saclofen, however, increases the excitatory responses to both the dominant and surround whiskers in a receptive field, and especially the responses to the surround whiskers. The receptive field, unlike with bicuculline, also increases in size to include two or three times as many whiskers (Lee et al., 1994b).

GABA-mediated inhibition exerts a powerful influence over the temporal aspects of the response of a thalamic neuron to a stimulus. In the ventral posterior nucleus of cats and rats, iontophoresis of bicuculline (Hicks et al., 1986; Salt, 1989; Lee et al., 1994b) or 2 hydroxy saclofen (Lee et al., 1994b) leads to a broadening of the responses to mechanical stimuli and the transient, on-responses of rapidly adapting neurons can become lost in the resulting enhanced train of high frequency discharges so that a rapidly adapting neuron can resemble a slowly adapting neuron. The receptor-specific character of the response is nevertheless maintained.

The relative contributions in GABAergic intrinsic neurons and reticular nucleus neurons to inhibitory effects on thalamic relay cells have still not been completely separated but the reticular nucleus seems to predominate. Feedback inhibition from the reticular nucleus to relay cells in the underlying dorsal thalamus can be demonstrated following stimulation of the optic tract or medial lemniscus (Sumitomo et al., 1979, 1988; Dubin and Cleland, 1977; Ahlsén et al., 1978, 1985; Ahlsén and Lindström, 1982a; Hale et al., 1982; French et al., 1985; Shosaku et al., 1989). If reticular nucleus neurons are made to discharge tonically by iontophoresis of glutamate, the background discharges of relay neurons are reduced (Mushiake et al., 1984; Warren and Jones 1994). This serves to enhance the signal-to-noise ratio when relay cells are activated by peripheral stimuli (Warren and Jones, 1994). Tonically discharging reticular nucleus neurons are characteristic of the awake state (Steriade et al., 1986, 1990, 1997) during which, this effect of the reticular nucleus may serve to focus sensory inputs to the cerebral cortex.

In the rat VPM nucleus from which intrinsic GABAergic neurons are absent (Benson et al., 1991b; Sperafico et al., 1993b), the responses of cells to whisker deflections are characterized by an initial excitation, succeeded by an inhibition lasting up to 200 msecs (Salt and Eaton, 1990). This must be a manifestation of the GABAergic input from the reticular nucleus. Cytotoxic lesions of the reticular nucleus in rats correspondingly lead to higher levels of spontaneous activity in VPM neurons (Lee et al., 1994a).

The importance of the reticular nucleus in setting up low frequency oscillatory behavior in large ensembles of thalamic neurons has been demonstrated in *in vitro* preparations of the ferret lateral geniculate nucleus (Von Krosigk et al., 1993; Bal et al., 1995a) and rodent ventral posterior nucleus (Huguenard and Prince, 1994a; Warren et al., 1994). Stimulation of the internal capsule or of the reticular nucleus leads to short latency EPSPs in reticular nucleus neurons and in the relay neurons of the underlying dorsal thalamus. A reticular nucleus neuron typically responds with a short burst of high frequency action potentials while the relay cells usually responds with a single, small EPSP which is rapidly succeeded by a large IPSP, generated by the burst of discharges in the reticular nucleus cell. The IPSP is blocked by bicuculline and is therefore a $GABA_A$-mediated response. On recovering from the reticular nucleus-imposed hyperpolarization, the low threshold calcium conductance (Jahnsen and Llinás, 1984) is activated in the relay cell and the resulting depolarization causes the cell to fire a high frequency burst of Na^+/K^+ action potentials. Transmitted by the collaterals of the thalamocortical axon of the relay cell to the reticular nucleus, this burst leads, in turn, to a barrage of recurrent EPSPs in the reticular nucleus cell that commonly reaches burst threshold. The new burst of action potentials fired by the reticular nucleus cell leads to further recurrent, bicuculline-sensitive IPSPs in the relay cells and to new bursts of discharges which again excite the reticular nucleus cell. Thus, the cycle repeats itself as part of a low frequency (4–10Hz) oscillation, very similar to that seen during slow wave sleep in the intact animal (Steriade et al., 1990, 1993d). Blockade of the $GABA_A$-mediated hyperpolarizations of the relay cells will prevent them from bursting and, thus, blocks the recurrent EPSPs in the reticular nucleus cells, preventing the cycle from occurring. Blockade of the recurrent EPSPs in the reticular nucleus cells by glutamate receptor antagonists also prevents the reticular nucleus cells from bursting; the recurrent IPSPs then do not occur in the relay cells and the cycle is again prevented. The close synaptic coupling of reticular nucleus cells and relay cells upon which these oscillatory phenomena depend, can also be demonstrated in vivo (Deschênes et al., 1984; Steriade et al., 1985, 1987a). The rhythmic hyperpolarizations of relay cells and the oscillations of the whole thalamic network that follow from them, fail when the input from the reticular nucleus to the dorsal thalamus is interrupted (Steriade et al., 1985; Von Krosigk et al., 1993). However spontaneous or evoked rhythmic bursting continues in the isolated reticular nucleus, implying that the reticular nucleus serves as a 'conditional' pacemaker for the rest of the thalamus in the low frequency oscillations typically called spindling (Steriade et al., 1987a; Bal and McCormick, 1993).

Stimulation of the reticular nucleus elicits fast IPSPs typical of $GABA_A$ receptor-based responses in relay cells of the dorsal thalamus (Thomson, 1988a,b; Huguenard and Prince, 1994a; Warren et al., 1994; Bal et al., 1995a, b; Ulrich and Huguenard, 1996). Slower, $GABA_B$-mediated IPSPs can also be revealed (Ulrich and Huguenard, 1996; Warren et al., 1997), especially under conditions in which the discharges of the reticular nucleus cells are repetitive and highly synchronized (Bal et al., 1995a,b). $GABA_A$-mediated, but not $GABA_B$-mediated IPSPs have been elicited in identified interneurons of the dorsal thalamus (Williams et al., 1994; Pape and McCormick, 1995). In the reticular nucleus itself, connections between the cells can be demonstrated to elicit effects via both $GABA_A$ and $GABA_B$ receptors (Ulrich and Huguenard, 1996).

The relative roles of $GABA_A$ and $GABA_B$ receptors in maintaining the oscillations of large ensembles of interconnected thalamic relay cells, reticular nucleus cells and cortical cells that accompany the EEG manifestations of changes in conscious state (see Contreras and Steriade, 1996) have not been fully worked out. Undoubtedly, from the work reviewed above, both are involved. One particular effect of the $GABA_B$ receptors is suggested by the capacity of $GABA_B$ receptor antagonists to block the low frequency spike and wave discharges in animal models of petit mal epilepsy in which the spike and wave discharges are thought to depend upon the interconnected network of thalamic cells (reviewed in Gloor and Fariello, 1988; Coulter et al., 1990; Juhász et al., 1994). Spike and wave discharges resemble normal 7–10Hz spindle oscillations and are thought to depend upon deinactivation of the low threshold calcium conductance with generation of bursts of action potentials as thalamic relay cells recover from the reticular nucleus-imposed hyperpolarization described above.

8.4. MONOAMINERGIC AND CHOLINERGIC PATHWAYS AND THEIR RECEPTORS

8.4.1. Introduction

The glutamatergic and GABAergic systems of the thalamus clearly interact to ensure the rapid and secure transfer of information through the thalamus to the cerebral cortex during attentive wakefulness. However, the two systems cannot be divorced from the activity of the thalamus during drowsy inattentiveness and slow wave sleep. During this time the monoaminergic and cholinergic inputs to the thalamus from the brainstem engage the glutamatergic and GABAergic circuits and interact with them.

The serotoninergic, noradrenergic and cholinergic fiber systems converging on the thalamus from the brainstem (Fig. 48) are probably distributed to all thalamic nuclei, but the fibers concentrate their terminations in certain nuclei. The ventral lateral geniculate nucleus is heavily innervated by all three systems; noradrenergic termina-

Fig. 48. Camera lucida drawings from transverse sections of the brainstem of a macaque monkey on which are plotted the locations of cells retrogradely labeled with Fast Blue from an injection in the ipsilateral ventral posterior nucleus, and the locations of cells immunocytochemically stained for choline acetyl transferase, tyrosine hydroxylase (TH) and serotonin (5HT). The labeled cells occupy regions containing the cholinergic, noradrenergic and serotoninergic cells. Abbreviations: BC: brachium conjunctivum; CG: central gray; DR: dorsal raphe nucleus; DTN: dorsal tegmental nucleus; IC: inferior colliculus; IP: interpeduncular nucleus; LC: locus coeruleus; LL: lateral lemniscus; LM: medial lemniscus; MLF: medial longitudinal fasciculus; MR: median raphe nucleus; Pm: medial parabrachial nucleus; PPN: peripeduncular nucleus; RTN: reticulotegmental nucleus; SCN: superior central nucleus; SNc: pars compacta of substantia nigra (dopaminergic cells); VTA: ventral tegmental area; IV: trochlear nucleus. From Jones in Steriade et al. (1997).

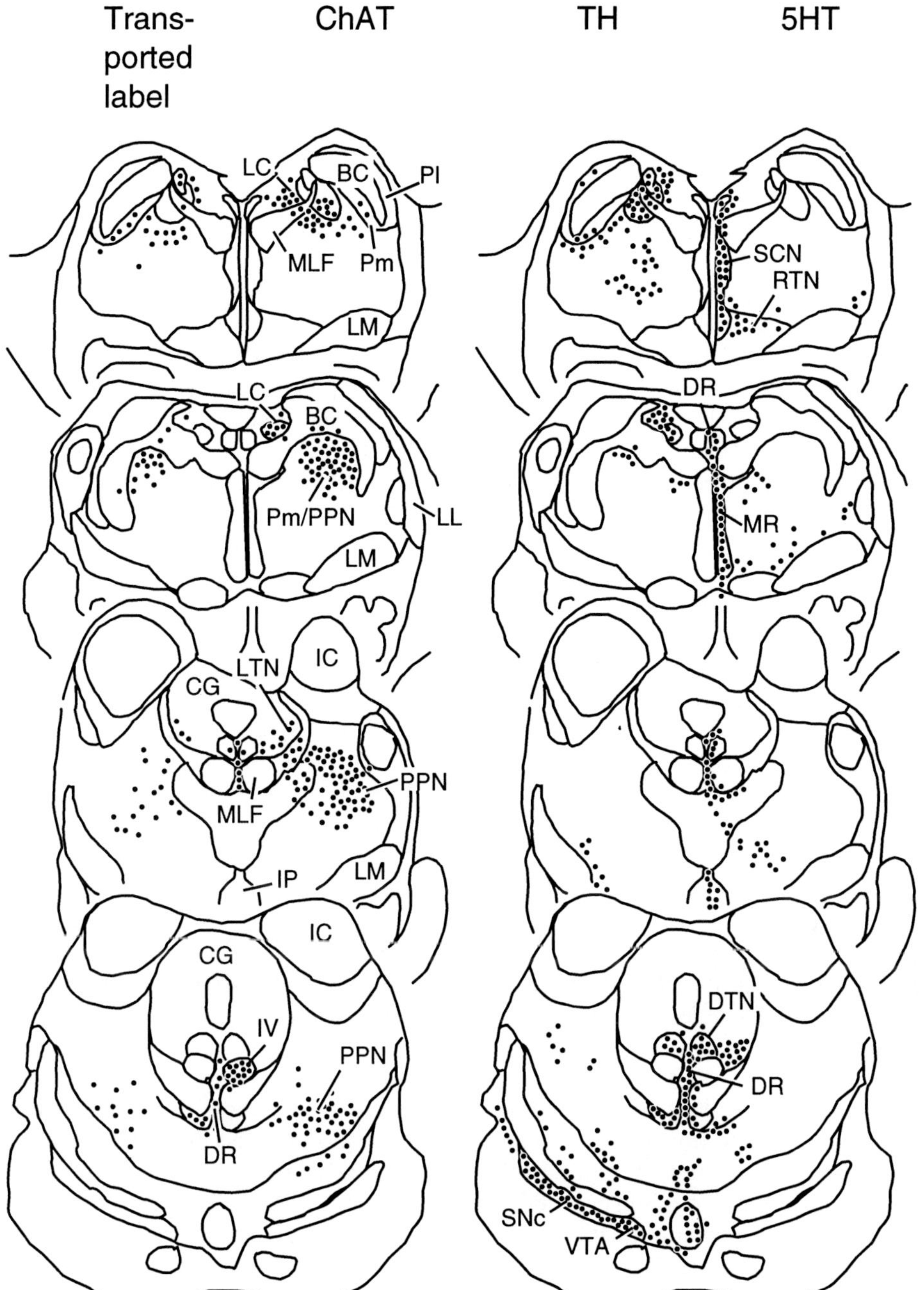
Trans-
ported
label
ChAT
TH
5HT
LC
BC
PI
MLF
Pm
LM
SCN
RTN
DR
Pm/PPN
LL
MR
LTN
IC
CG
PPN
IP
DTN
IV
SNc
VTA

tions are heavily concentrated in the anterior nuclei, at least of rats (Swanson and Hartman, 1975) and noradrenergic and serotonin fibers are densely concentrated in the nuclei of the monkey LP/pulvinar complex and in the reticular nucleus (Morrison and Foote, 1986). Cholinergic fibers are more densely distributed in the dorsal lateral geniculate nucleus than in other nuclei in many species (Fitzpatrick and Raczkowski, 1991), including humans (Heckers et al., 1992).

Dopaminergic inputs to the thalamus are usually reported to be absent or extremely sparse (Moore and Bloom, 1978; Lindvall et al., 1974). One recent report, however, describes a significant dopamine innervation of the rat lateral geniculate nucleus (Papadopoulos and Parnavelas, 1990a) and effects of iontophoresed dopamine on neurons in this nucleus have been described (Albrecht et al., 1996). Autoradiography generally demonstrates low levels or absence of dopamine receptors (Dawson et al., 1986; Boyson et al., 1986; Dubois et al., 1986; Savasta et al., 1986; Mansour et al., 1990; Young and Wilcox, 1991; Levesque et al., 1992), although nuclei such as the lateral dorsal and medial geniculate and the habenula and the zona incerta may show densities of D_2 binding sites, especially in development (Sales et al., 1989). mRNAs for D_1 receptors have been reported to be expressed in several nuclei of the rat thalamus (Freeman et al., 1991) and the D_{1B} (sometimes called D_5) subtype is reportedly found at high levels but confined to the parafascicular nucleus (Meador-Woodruff et al., 1992). D_2 receptor mRNAs have been reported only in the habenular nuclei and zona incerta (Mansour et al., 1990). Immunocytochemistry for D_1 and D_2 receptors in rats, monkeys and humans reveals no staining in the diencephalon other than the reticular nucleus (Huang et al., 1992). Therefore, the receptor mRNAs although present, may not be translated (Levey et al., 1993). mRNAs for other dopamine receptor subtypes are absent or expressed at extremely low levels in the thalamus (Meador-Woodruff et al., 1989; Najleramin et al., 1989; Mengod et al., 1989; Mansour et al., 1990, 1991; Tiberi et al., 1991; Weiner et al., 1991).

8.4.2. The noradrenergic pathway and adrenergic receptors

The noradrenergic innervation of the thalamus arises from cells of the locus coeruleus (Swanson and Hartman, 1975; Leger et al., 1975; Kromer and Moore, 1980; Hughes and Mulliken, 1984; de Lima and Singer, 1987a). Fibers entering the thalamus may all be collaterals of fibers passing by, en route to more rostral forebrain sites (Swanson and Hartman, 1975). All thalamic nuclei appear to receive noradrenergic fibers (Fig. 49) including the reticular nucleus. The densest innervation in many species is found in the anterior nuclei and dorsal lateral geniculate nuclei and the least in the intralaminar and reticular nuclei (Lindvall et al., 1974; Swanson and Hartman, 1975; Moore and Bloom, 1979; de Lima and Singer, 1987a; Liu and Jones, 1991, 1992) but in monkeys, the dorsal lateral geniculate nucleus is reported to have a very low density (Morrison and Foote, 1986; Fitzpatrick et al., 1989). Electron microscopy reveals that noradrenergic terminals in the thalamus are small and end in relation to dendrites of both relay neurons and interneurons, commonly without morphologically-identifiable synaptic thickenings (de Lima and Singer, 1987a; Liu and Jones, 1991) (Fig. 49 B). Iontophoretic application of noradrenaline or stimulation of the locus coeruleus elicits tonic firing in thalamic relay cells and in cells of the reticular nucleus (Nakai and Takaori, 1974; Rogawski and Aghajanian, 1980a,b, 1982; Kayama et al., 1982; Kayama, 1985). Similar excitatory effects can be induced by acetylcholine and to some extent by serotonin, glutamate and a number of neuropeptides, implying involvement

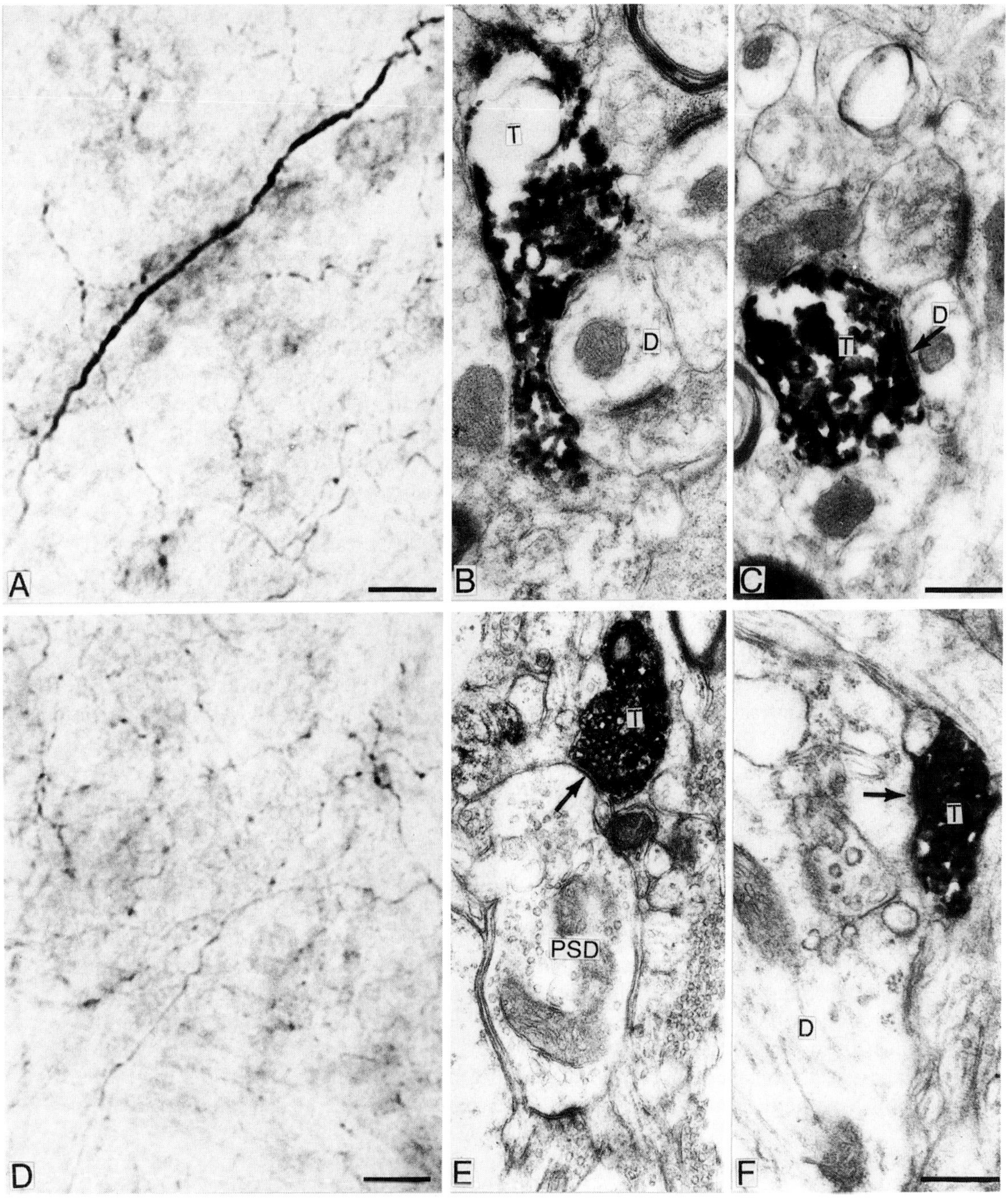

Fig. 49. Light (A,D) and electron micrographs (B, C, E, F), showing immunoreactivity for tyrosine hydroxylase (A-C) indicative of the presence of noradrenaline and for serotonin (D-F) in fibers and terminals in the ventral posterior nucleus of macaque monkeys. Terminals of both sets of fibers commonly make close contacts with dendrites (D) of relay neurons and presynaptic dendrites (PSD) of interneurons, but membrane specializations indicative of conventional synapses at these points (arrows) are rare. Bars 25 μm (A, D), 0.25 μm (B, C, E, F). From material illustrated in Liu and Jones (1991).

of a common set of membrane mechanisms (McCormick, 1992). The excitatory effect of noradrenaline on thalamic relay cells and on cells of the reticular nucleus is a slow depolarization (McCormick and Prince, 1988; McCormick and Pape, 1990b; McCormick, 1991b), which depends on suppression of a potassium current, I_{KL}, mediated by α_1 adrenergic receptors (McCormick and Wang, 1991). This inhibits burst firing and promotes single spike activity (McCormick and Prince, 1988), thus reducing low frequency oscillations in the thalamic network. α_2 receptor agonists, by contrast, tend to promote oscillation (Buzsáki et al., 1991). Blockade of α_1 receptors in the reticular nucleus, in vivo, reduces discharge rates in reticular nucleus neurons (Pinault and Deschênes, 1992b) which should also help to reduce oscillations.

The same slow depolarization generated in relay cells and reticular nucleus cells by noradrenaline can be induced in the same cells by stimulation of muscarinic cholinergic, H_1 histaminergic and metabotropic glutamate receptors. Its presence in intrinsic interneurons is uncertain (Pape and McCormick, 1995). Effects of activating β adrenoreceptors on thalamic relay cells consist of enhancement of a hyperpolarization-induced cation current, I_h, which leads to depolarizations distinguished by their small size and shape from the slow, α_1 -mediated depolarization (McCormick and Pape, 1990b). This effect can also be induced by activation of serotoninergic, and H_2 histaminergic receptors.

Autoradiographic studies of adrenergic receptor localization in the thalamus have been few and have not involved primates. α_{1B} receptor sites are reported to be more prominent than α_{1A} sites, except in the reticular nucleus (Morrow et al., 1985; Morrow and Creese, 1986; Wilson and Minneman, 1989) and α_{1B} mRNA is more highly expressed than other α_1 adreno-receptor mRNAs. It is expressed at relatively high levels in all nuclei except the reticular and habenular (Pieribone et al., 1994). $\alpha_{1A/D}$, mRNA, by contrast, is highly expressed in the reticular nucleus and absent from the remainder of the thalamus (Pieribone et al., 1994).

8.4.3. The serotonin pathway and serotonin receptors

Serotonin fibers (Fig. 49 D) entering the thalamus have their origins in cells of the dorsal nucleus of the midbrain raphe (Conrad et al., 1974; Leger et al., 1975; Azmitia and Segal, 1978; Parent et al., 1981; Steinbusch, 1981; Hughes and Mulliken, 1984; Cropper et al., 1984; Consolazione et al., 1984; Peschanski and Besson, 1984; Morrison and Foote, 1986; de Lima and Singer, 1987b; Nothias et al., 1988; Smith et al., 1988a; Westlund et al., 1990; Lavoie and Parent, 1991). Serotonin fibers innervate at relatively high density all thalamic nuclei, including the ventral thalamus and epithalamus. In monkeys, the density of innervation is higher in the magnocellular than in the parvocellular layers of the dorsal lateral geniculate nucleus (Morrison and Foote, 1986). Serotonin terminals in the thalamus make close membrane appositions with small dendrites of relay neurons, of intrinsic GABA cells and of cells in the reticular nucleus, commonly without obvious junctional specializations (de Lima and Singer, 1987b; Pasik et al., 1988; Liu and Jones, 1991, 1992) (Fig. 49 E).

$5HT_1$ and $5HT_2$ receptor sites, as demonstrated by autoradiography in the rat brain, are densely distributed in the dorsal thalamus and reticular nucleus. Higher $5HT_1$ binding tends to be higher in the anterior and intralaminar nuclei than in the principal relay nuclei or reticular nucleus and $5HT_2$ binding is generally weak except in the intralaminar nuclei (Young and Kuhar, 1980; Meibach et al., 1980; Biegon et al., 1982; Deshmukh et al., 1983; Marcinkiewicz et al., 1984; Pazos and Palacios, 1985;

Pompeiano et al., 1992; Bruinvels et al., 1993). Immunocytochemical staining for the 5-HT_{1A} receptor reveals modest densities of fiber staining in the intralaminar and medial habenular nuclei and higher densities in the reticular nucleus and zona incerta (Kia et al., 1996). A few cell bodies are stained but only in the intralaminar nuclei and zona incerta. This follows the lack of $5HT_{1A}$ mRNA expression in the thalamus, except for the reticular nucleus, geniculate nuclei and habenula (Chalmers and Watson, 1991; Pompeiano et al., 1992). There are high levels of $5HT_{1C}$ mRNA in the thalamus of the rat, especially in the lateral dorsal, lateral posterior, medial geniculate, posterior, habenular and paraventricular nuclei (Julius et al., 1988; Molineaux et al., 1989) and expression, although not described, can also be seen in the reticular nucleus.

Serotonin is reported to facilitate responses of relay neurons in the thalamus of the rat to excitatory amino acids and to inhibit excitatory responses to cholinomimetic agents (Eaton and Salt, 1989). At the membrane level, the effects of serotonin on thalamic neurons have certain similarities to those of noradrenaline (see above). Application of serotonin to relay cells elicits a prolonged, slow depolarization dependent on suppression of the hyperpolarization-activated cation current, I_h; this is virtually identical to that elicited by β adrenergic and H_2 histaminergic receptors (McCormick and Wang, 1991). This current cannot be elicited by application of serotonin to neurons in the reticular nucleus. In reticular nucleus neurons, the slow depolarization dependent on the K^+ leak current, I_{KL}, and mediated by $5HT_2$ receptors is found instead. This cannot be elicited in relay neurons.

8.4.4. The cholinergic pathway and cholinergic receptors

Most cholinergic fibers entering the thalamus arise from cells in the lateral dorsal tegmental and pedunculo-pontine nuclei of the brainstem (MacKay-Sim, 1983; Sofroniew, 1985; Woolf and Butcher, 1986; Hallanger et al., 1987; Levey et al., 1987; Jones, 1987; Paré et al., 1988; Smith et al., 1988; Steriade et al., 1988; Fitzpatrick et al., 1989) (Fig. 50). Parts of the pulvinar and dorsal lateral geniculate nucleus are also innervated by cholinergic fibers arising in the parabigeminal nucleus of the midbrain (Diamond et al., 1992). There are few or no cholinergic inputs from the basal forebrain. All thalamic nuclei are innervated by cholinergic fibers but the anterior, intralaminar, medial habenular, reticular, and dorsal and ventral lateral geniculate nuclei receive by far the densest innervation (Hoover et al., 1978; Houser et al., 1988; Levey et al., 1987; Fitzpatrick and Raczkowski, 1991; Ichikawa and Hirata, 1986; Vincent and Reiner, 1987; Fitzpatrick et al., 1988, 1989; Heckers et al., 1992). The density of innervation of the dorsal lateral geniculate nucleus tends to vary with species. In humans, the highest densities of cholinergic fibers are in the anterior group of intralaminar nuclei, the anterodorsal nucleus, the magnocellular division of the mediodorsal nucleus, the reticular nucleus and the dorsal lateral geniculate nucleus (Heckers et al., 1992).

The terminals of cholinergic fibers in the dorsal thalamus and in the reticular nucleus (de Lima and Singer, 1987a; Houser et al., 1988; Raczkowski and Fitzpatrick, 1989) are reported to make either symmetrical or asymmetrical synaptic contacts but most look symmetrical in published electron micrographs. In the dorsal lateral geniculate nucleus, the terminals end on dendrites of relay cells and on presynaptic dendrites of interneurons, while in the reticular nucleus the terminations are on both large and small dendrites.

Muscarinic and nicotinic cholinergic receptors have been recognized in the thalamus

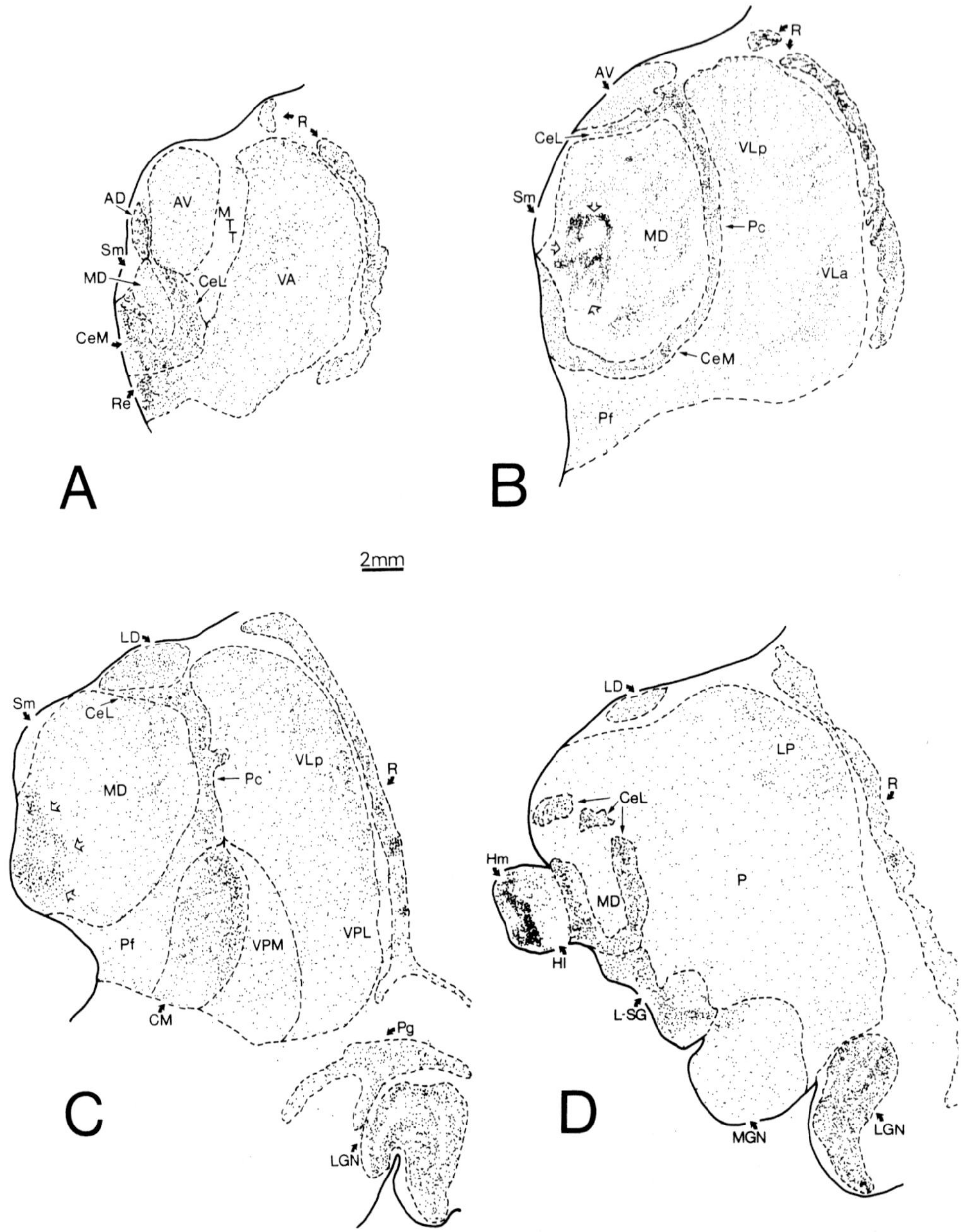

Fig 50. Drawings, showing the distribution of choline acetyl transferase immunoreactive fiber ramifications in frontal sections of the human thalamus. From Heckers et al. (1992).

by neuropharmacological studies (Phillis, 1970; Tēbēcis, 1972; Krnjević, 1974; Rotter et al., 1979; Wamsley et al., 1980b, 1984; Sillito et al., 1983; Hu et al., 1989a-c) and by autoradiography (Rainbow et al., 1984; London et al., 1985; Fuchs and Schwark, 1993). When applied by iontophoresis or released from cholinergic terminals following stimulation of the brainstem, acetylcholine enhances spontaneous and stimulus-evoked discharges in relay neurons and suppresses them in intrinsic interneurons and in reticular nucleus neurons (Phillis et al., 1967; Godfraind, 1975, 1978; Ben-Ari et al., 1976; Dingledine and Kelly, 1977; Singer, 1977b; Godfraind, 1978; Sillito et al., 1983; Berardi and Morrone, 1984b; Eysel et al., 1986; McCormick and Prince, 1987; McCormick and Pape, 1988; Francesconi et al., 1988; Hu et al., 1989a,b; Deschênes and Hu, 1990b; Curró Dossi et al., 1991). The excitation of relay neurons by acetyl choline is characterized by a rapid depolarization dependent upon nicotinic receptor activation (Curró Dossi et al., 1991), followed by a slower, longer lasting depolarization dependent upon muscarinic receptor activation (Steriade and Deschênes, 1984; Deschênes and Hu, 1990b; Curró Dossi et al., 1991). The slow depolarization results from the suppression of the potassium leak current, I_{KL}, and is identical to that found in relay cells following iontophoretic application of noradrenaline, histamine or metabotropic glutamate receptor agonists, as described above. Small hyperpolarizing effects on relay neurons, dependent on an increase in the postsynaptic potassium conductance, I_{KG}, and mediated by muscarinic M_2 receptors have also been described (McCormick and Prince, 1987).

The inhibition of intrinsic thalamic interneurons and of reticular nucleus neurons by acetyl choline (McCormick and Pape, 1988; McCormick and Prince, 1986; Francesconi, 1988; Hu et al., 1989c; Funke and Eysel, 1993; Kayama et al., 1986a; Murphy et al., 1994; Ahlsén et al., 1984; Curró Dossi et al., 1992b) is dependent on a muscarinic, M_2 receptor effect, leading to hyperpolarization of the GABAergic neurons (McCormick and Pape, 1988; Pape and McCormick, 1995) and *in vivo* to reduction of afferent-induced IPSPs in relay cells (Singer, 1977b; Hu et al., 1989c). The muscarinic inhibitory response in reticular nucleus neurons is preceded by a short, nicotinic receptor-based depolarization (Lee and McCormick, 1995). Stimulation of cholinergic nuclei in the basal forebrain will inhibit reticular nucleus cells (Pinault and Deschênes, 1992a) but this is thought to depend upon a GABAergic projection (Asanuma and Porter, 1990; Asanuma, 1994), since few cholinergic fibers actually project to the thalamus from the basal forebrain.

Muscarinic and nicotinic cholinergic receptor binding, as demonstrated by autoradiography, is reportedly low in all dorsal and ventral thalamic nuclei (Hunt and Schmidt, 1978; Segal et al., 1978; Rotter et al., 1979; Wamsley et al., 1980; Rainbow et al., 1984; London et al., 1985). In rats, muscarinic receptors are reported to be most dense in the anterior or ventral nuclei (Rotter et al., 1979; Cortes and Palacios, 1986; Mash and Potter, 1986) and nicotinic receptors in the ventral lateral geniculate nucleus (Hunt and Schmidt, 1978; Segal et al., 1978; Schulz et al., 1991). In monkeys, [^{3}H] nicotine binding is relatively high throughout the dorsal thalamus (Cimino et al., 1992) but is absent from the reticular nucleus. This nucleus is, however, heavily labeled by [^{3}H] – bungarotoxin binding (Jones, 1983, 1985a; Cimino et al., 1992).

Of the several types of acetylcholine binding (α_{2-7}) and structural subunits (β_{2-4}) of neuronal nicotinic receptors, α_3, α_4 and β_2 are widely expressed in the thalamus of rats, with some subunit mRNAs more richly expressed in certain nuclei (Swanson et al., 1987; Goldman et al., 1987; Wada et al., 1989, 1990). In monkeys, α_3 is most densely expressed in the medial habenular and dorsal lateral geniculate nuclei (Cimino et al., 1992). Others are expressed in single nuclei, β_4 for example, only in the medial habenular nucleus (Duvoisin et al., 1989; Dineley-Miller and Patrick, 1992) and others such as α_2 are not expressed at all (Wada et al., 1989). The lateral habenular nucleus lacks any of the subunits. Immunocytochemistry using subunit specific antibodies in rats, reveals cells and fibers immunoreactive for α and $\beta 2$ subunit polypeptides concentrated in all thalamic nuclei except the lateral habenular nucleus, especially in the anteroventral, interanteroventral, anterior paraventricular and medial habenular nuclei (Deutch et al., 1987; Swanson et al., 1987).

Immunocytochemistry for the α_7 subunit, an α bungarotoxin binding polypeptide, revealed high densities of labeled cells in most thalamic nuclei (Del Toro et al., 1994) (see also Broide et al., 1995).

Of the five types of neuronal muscarinic receptor (M_{1-5}), M_2 and M_3 receptor subunit mRNAs and immunoreactivity for their translated polypeptides are found in all dorsal thalamic nuclei and in the reticular nucleus (Buckley et al., 1988; Kubo et al., 1988; Vilaró et al., 1992; Amadeo et al., 1995; Hersch et al., 1994). M_1 and M_4 mRNAs are reported to be absent (Buckley et al., 1988).

Fibers immunoreactive for nerve growth factor receptor permeate the thalamus in small numbers and are concentrated in the reticular nucleus in rats (Pioro and Cuello, 1990; Bentivoglio et al., 1993). In humans, they are also found in the anterior intralaminar and reticular nuclei and in the magnocellular division of the mediodorsal nucleus (Heckers et al., 1992). They are probably associated with the cholinergic innervation of the thalamus.

8.5. PEPTIDES AND OTHER NEUROACTIVE MOLECULES IN THE THALAMUS AND ITS AFFERENT FIBERS.

8.5.1. Peptides

Many of the better-known neuropeptides have been localized by immunocytochemistry in fibers entering the thalamus from the brainstem and hypothalamus. Some of the peptides are found in fibers of afferent pathways such as the medial lemniscus and optic tract, where they are presumed to be colocalized with the predominant excitatory amino acid transmitter. In monkeys, for example, tachykinin imunoreactivity is found in fibers of the trigeminal lemniscus, although only in those ending in the representation of the contralateral side of the face in the ventral posterior nucleus (Liu et al., 1989). Here, the fibers terminate in synapses typical of the medial lemnsicus (Fig. 51). Major species differences can be present. Cholecystokinin immunorectivity, for exmple, is found in the dorsal column-lemniscal fibers ending in the ventral posterior nucleus of the rat (Hunt et al., 1987) but not in those of monkeys (Molinari et al., 1987). Immunoreactivity for tachykinins and enkephalin is found in spinothalamic fibers ending in basal and medial aspects of the rat and human thalamus (Coffield and Miletic, 1987; Battaglia et al., 1988; Gall et al., 1987; Hirai and Jones, 1989b; Nishiyama et al., 1995). Fibers showing immunoreactivity for certain neuropeptides have uncertain origins and very restricted terminations in the thalamus, e.g. calcitonin gene related peptide positive fibers in the rat are confined to the basal ventral medial nucleus and adjacent regions (Kruger et al., 1988a,b; Williamson and Ralston, 1993).

Large numbers of tachykinin, somatostatin and neuropeptide Y immunoreactive fibers in monkeys arise in hypothalamic regions (Molinari et al., 1987; Fig. 52–55), ascend the anterior surface of the thalamus and terminate in the paraventricular and habenular nuclei. The majority of the other peptide immunoreactive fibers entering the thalamus in primates, arise from ill-defined brainstem sources and ascend in the lateral midbrain tegmentum, entering the thalamus from the pretectum and reticular nucleus. Many of the peptide-containing fibers arise from cells in the lateral dorsal tegmental nucleus, peribrachial region, locus coeruleus and periaqueductal gray matter (Sim and

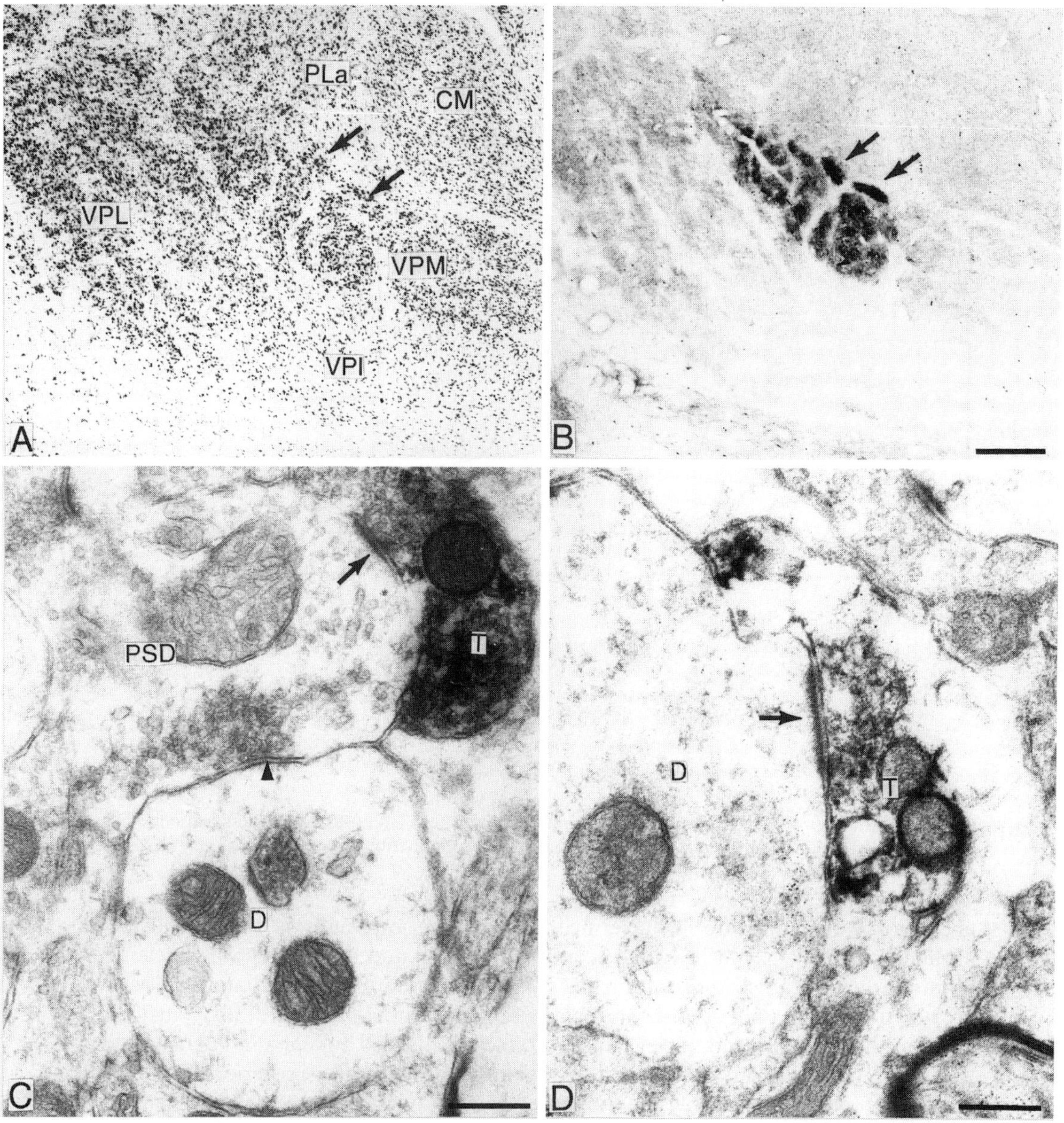

Fig. 51. Light (A, B) and electron (C, D) micrographs from the ventral posterior nucleus of a late fetal (A, B) and an adult (C, D) macaque monkey, showing the high concentration of tachykinin immunoreactivity in the VPM nucleus and the presence of tachykinin immunoreactivity in the terminals (T) of lemniscal fibers ending on presynaptic dendrites (PSD) of interneurons and on dendrites (D) of relay neurons. Arrows in A, B indicate same aggregations of cells and tachykinin terminals; arrows and arrowhead in C, D indicate sites of synaptic contact. Bars 500 μm (A, B), 0.25 μm (C, D). From material illustrated in Liu et al. (1989).

Joseph, 1992; Lechner et al., 1993). Cholecystokinin, tachykinin, neuropeptide Y and somatostatin immunoreactive fibers have been described in these pathways in monkeys (Molinari et al., 1987) and tachykinin and metenkephalin immunoreactive fibers in humans (Hirai and Jones, 1989b) (Figs. 55–57). The major concentrations of all these

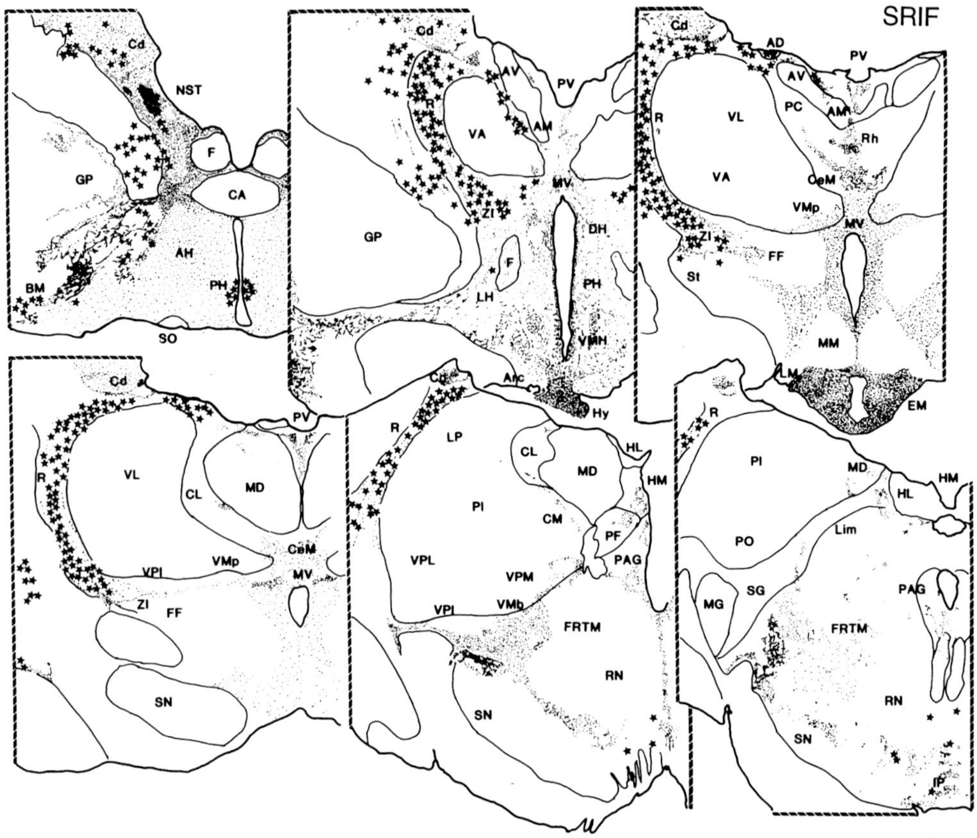

Fig. 52. Distribution of somatostatin (SRIF) immunoreactive cells (stars) and fiber ramifications (dots) in frontal sections of a macaque monkey thalamus. Sections are in anterior (top left) to posterior (lower right) order. From Molinari et al. (1987).

fibers in the thalamus are in nuclei near the midline, especially in the central medial intralaminar nucleus and in the adjacent mediodorsal and paraventricular nuclei. Retrograde tracing studies in non-primates confirm that more peptide containing neurons project to medial than to lateral thalamus. Tachykinin fibers are also particularly dense in the anteroventral nucleus and both tachykinin and enkephalin fibers dominate the small posterobasal region of the thalamus beneath the posterior intralaminar nuclei (Fig. 58 A-C), possibly the human equivalent of the submedial nucleus (Section 10.8.5). Less dense concentrations of fibers are found in the reticular nucleus and in the pregeniculate nucleus. Apart from these concentrations, a few peptide-immunoreactive fibers permeate every nucleus of the thalamus. The fewest are found in the ventral lateral and dorsal lateral geniculate nuclei.

The four peptides mentioned in the preceding paragraph, although subject to certain species differences, are generally found in the same nuclei of non-primate species. Other peptides not yet described in primates but demonstrable by immunocytochemistry in fibers innervating the rat thalamus, include neurotensin (Jennes et al., 1982; Inagaki et al., 1983; Ray and Price, 1990; Kuroda et al., 1991), galanin (Skofitsch and Jacobowitz, 1985; Melander et al., 1986; Blasco et al., 1989), bombesin (Panula et al., 1982), and angiotensin II (Quinlan and Phillips, 1981).

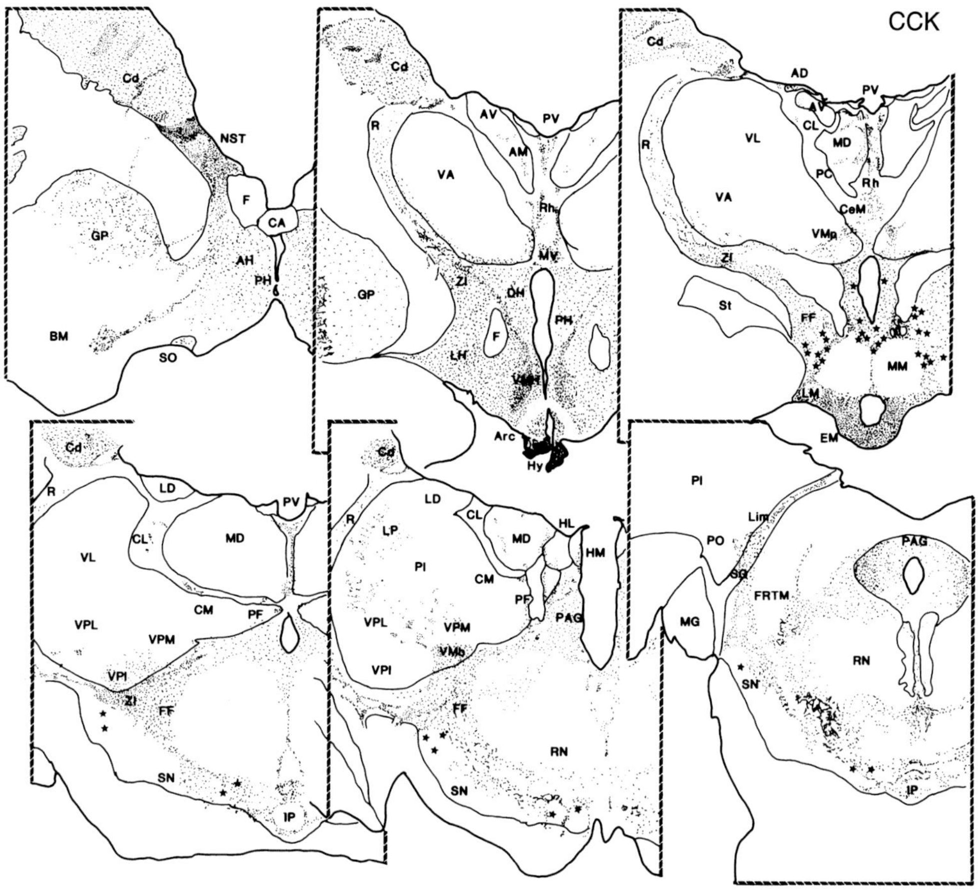

Fig. 53. Distribution of cholecystokinin (CCK) immunoreactive fibers in frontal sections of the same macaque thalamus as in Fig. 52. From Molinari et al. (1987).

Reports vary, especially across species, but most have reported few cells immunoreactive for any of the commoner neuropeptides in the thalamus, with the exception of somatostatin, which is expressed in every neuron of the reticular nucleus in cats and monkeys (Graybiel and Elde, 1983; Bouras et al., 1986, 1987; Molinari et al., 1987) (Figs. 52, 59) but may be less exclusively expressed in those of certain other species (Conley et al., 1991b). Neuropeptide Y immunoreactive cells are found in small numbers in the ventral lateral geniculate (pregeniculate) nucleus (Card and Moore, 1982; Mantyh and Kemp, 1983; Jones, 1985a; Molinari et al., 1987). A few cells immunoreactive for calcitonin gene related peptide are found at the thalamus-midbrain border and in the thalamic reticular nucleus of rats (Kruger et al., 1988a,b). Angiotensin II immunoreactive somata are widely distributed in the rat thalamus (Coveñas et al., 1989). Immunoreactivity for cholecystokinin has also been reported in certain thalamic cells (Ogawa et al., 1989; Miceli et al., 1987), especially in intralaminar and anterior nuclei of the cat (Sugimoto et al., 1985). Vasoactive intestinal polypeptide immunoreactivity is reported by some to be absent from the thalamus (Lorén et al., 1979b) but others describe its presence in intralaminar and reticular nucleus neurons of the cat (Kaneko et al., 1985).

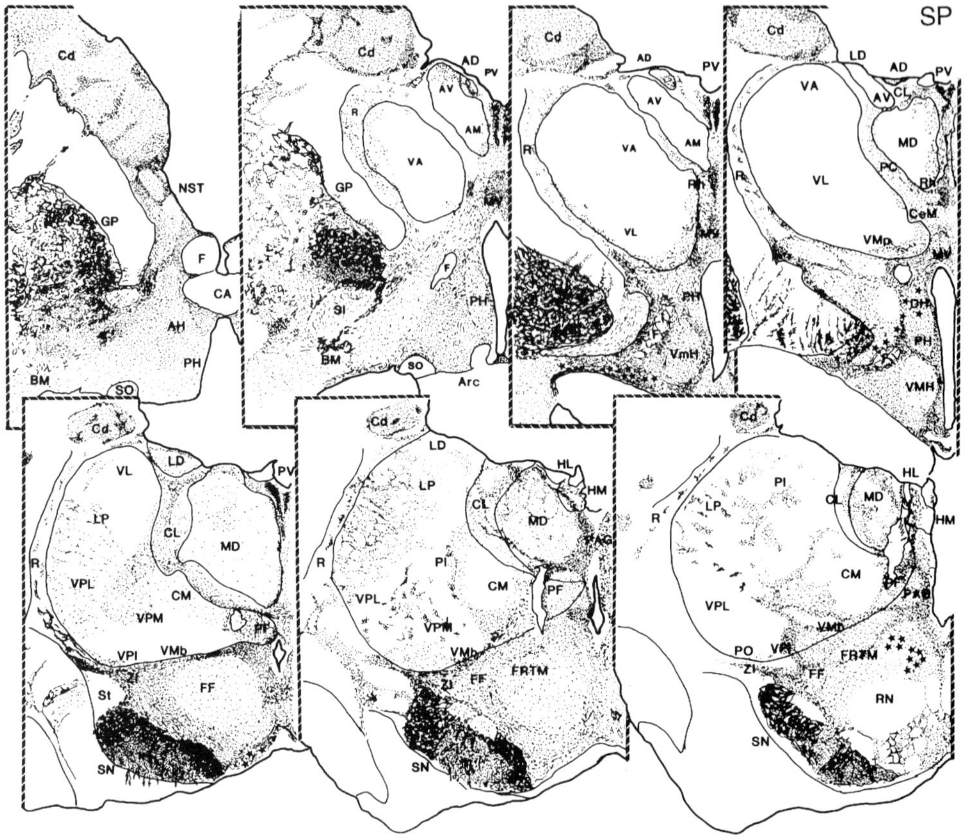

Fig. 54. Distribution of tachykinin (SP) fibers in the same macaque thalamus as in Figs. 52 and 53. From Molinari et al. (1987).

Localization of prepropeptide mRNAs by situ hybridization reveals that gene expression for neuropeptides may be more widespread in thalamic cells than evident in immunocytochemical preparations. mRNAs not only for somatostatin, but also for neuropeptide Y (Morris, 1989) and cholecystokinin have now been demonstrated in the thalamus of rats, especially in the reticular nucleus. The transcript for preprocholecystokinin can also be demonstrated in relay neurons (Sugimoto et al., 1985; Wahle and Albus, 1985; Voigt and Uhl, 1988; Burgunder and Young, 1988, 1990, 1992; Savasta et al., 1988; Ingram et al., 1989; Lanaud et al., 1989). Cholecystokinin may also be found in corticothalamic fibers. In monkeys, immunoreactivity for the peptide in the cerebral cortex has only been demonstrated in nonpyramidal cells (Hendry et al., 1983c) and not in pyramidal cells of layers V and VI from which the corticothalamic projection arises (Jones, 1984). In rats, however, cortical cells immunoreactive for cholecystokinin or containing preprocholecystokinin mRNA, can be retrogradely labeled from injections of anatomical tracers in the vicinity of the thalamus (Burgunder and Young, 1990; Senatorov et al., 1995).

The localization of neuropeptide receptors in the thalamus of rats does not match the distribution of the immunoreative fibers. ^{135}I CCK binding reveals receptors for cholecystokinin only in the reticular nucleus (Pelaprat et al., 1987) and ventral lateral

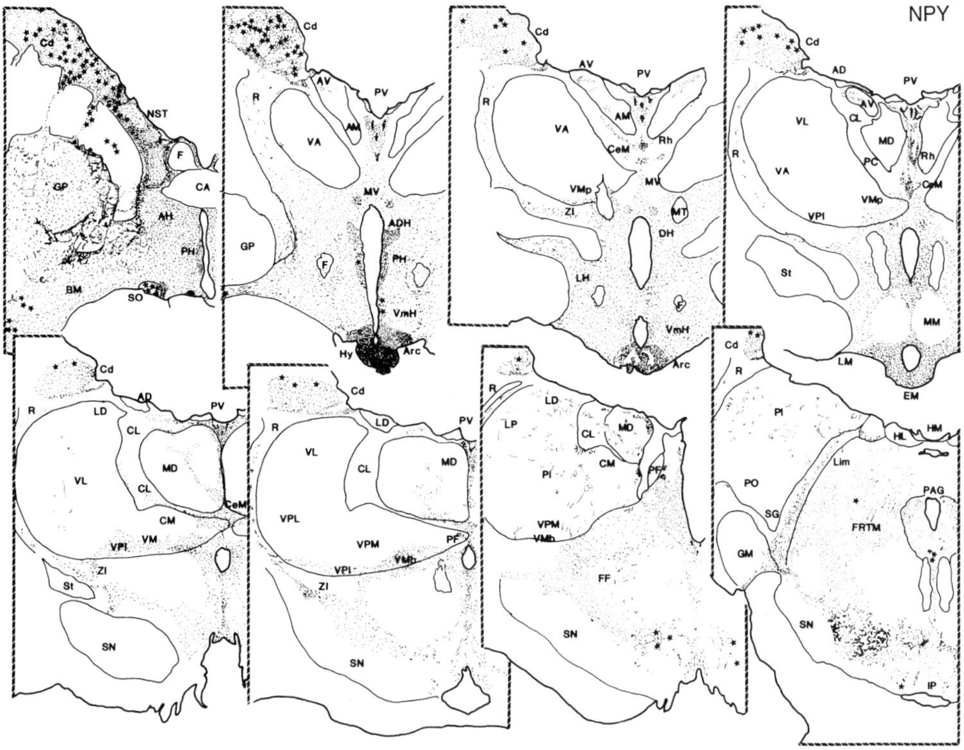

Fig. 55. Distribution of neuropeptide Y (NPY) immunoreactive fibers in the same macaque thalamus as in Figs. 52–54. From Molinari et al. (1987).

geniculate nucleus (Zarbin et al., 1983) and not in the dorsal thalamus. Tachykinin receptor gene expression is reportedly restricted to the habenular nuclei and zona incerta (Warden and Young, 1988). Both of these observations are surprising in view of the moderately dense innervation of the thalamus by tachykinin and cholecystokinin fibers. Cholecystokinin weakly excites thalamic cells (Albrecht et al., 1994), and when applied to reticular nucleus neurons *in vitro*, results in a prolonged depolarization, dependent on suppression of the potassium leak current, I_{KL}, similar to that mediated by $\alpha 1$ adrenoreceptors, histamine H_1, muscarinic, or metabotropic glutamate receptors (Cox et al., 1995). The effects of other peptides have not been described.

Autoradiography for neuropeptide Y receptor binding shows high densities of receptors in the intralaminar nuclei (Widdowson, 1993; Dumont et al., 1993) which follows the distribution of neuropeptide Y fibers, and also in the medial geniculate nucleus which does not.

Opiate receptors, as demonstrated by autoradiography, have their highest concentrations in the anterior intralaminar nuclei and in the paraventricular and habenular nuclei and are low in the sensory relay nuclei (Pert et al., 1976; Atweh and Kuhar, 1977; LaMotte et al., 1978; Wamsley et al., 1982; Khachaturian et al., 1983; Morris and Herz, 1986; Herkenham, 1987; Mansour et al., 1987; Herkenham et al., 1988). This is much more in keeping with the distribution of enkephalin imunoreactive fibers (Hirai and Jones, 1989b). Opiate receptors are reported to be low in the reticular and ventral lateral geniculate nuclei by some authors but relatively high by others. mRNAs

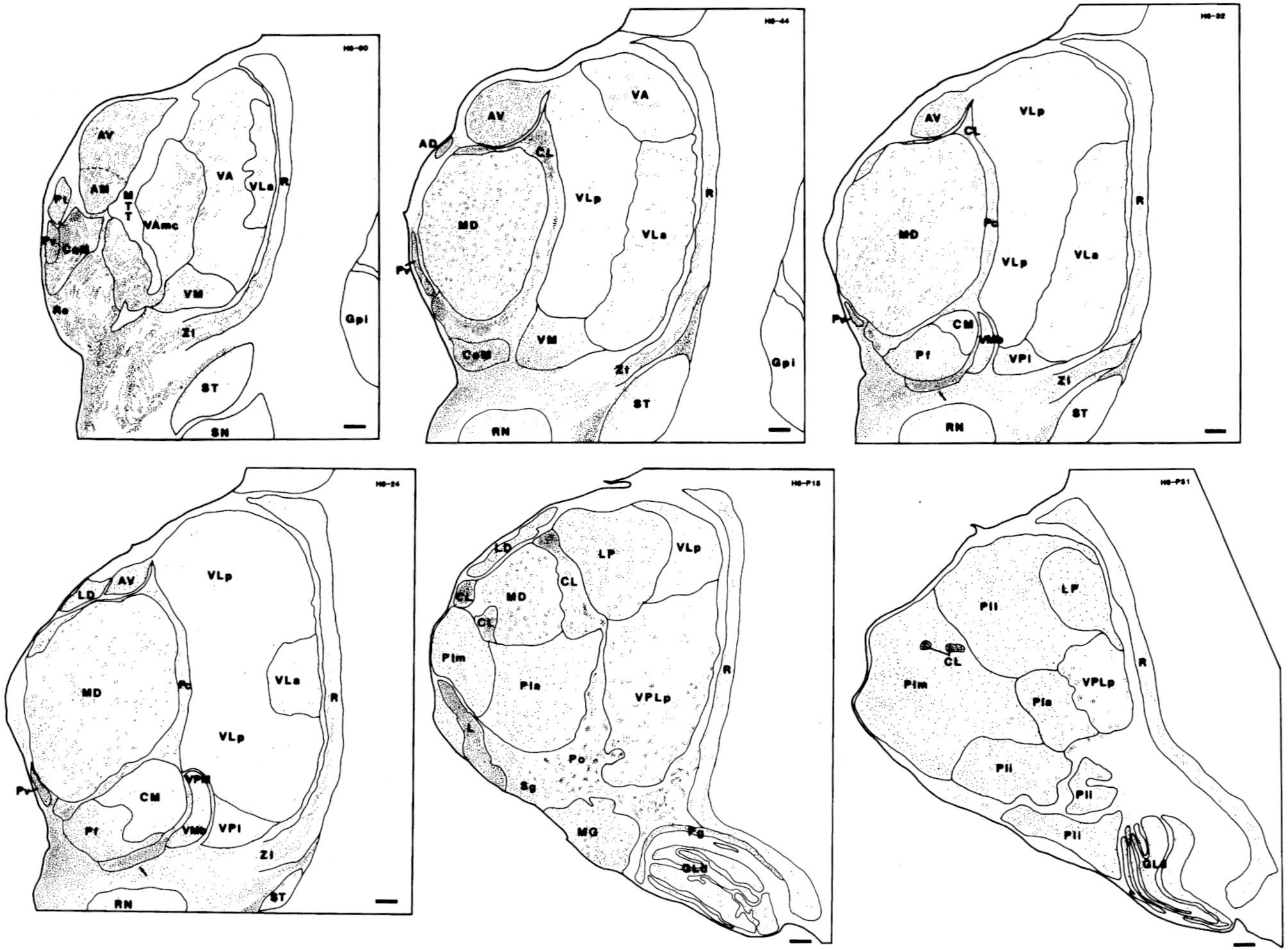

Fig. 56. Distribution of tachykinin immunoreactive fibers in frontal sections of a human thalamus. Arrows indicate putative nucleus submedius. Sections are in anterior (top left) to posterior (lower right) order. From Hirai and Jones (1989b).

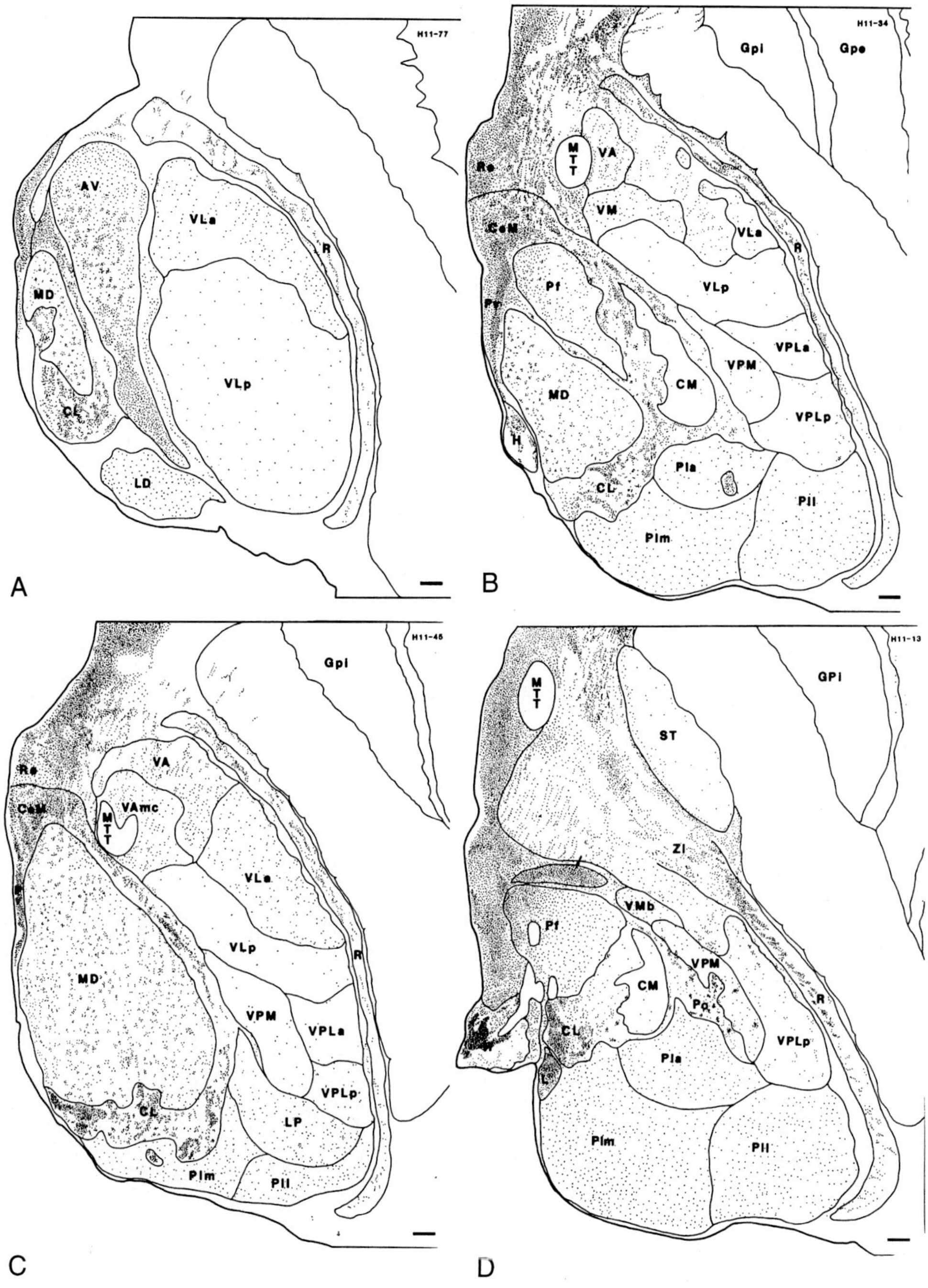

Fig. 57. Distribution of tachykinin immunoreactive fibers in horizontal sections from a human thalamus. Sections are in dorsal (A) to ventral (D) order. From Hirai and Jones (1989b).

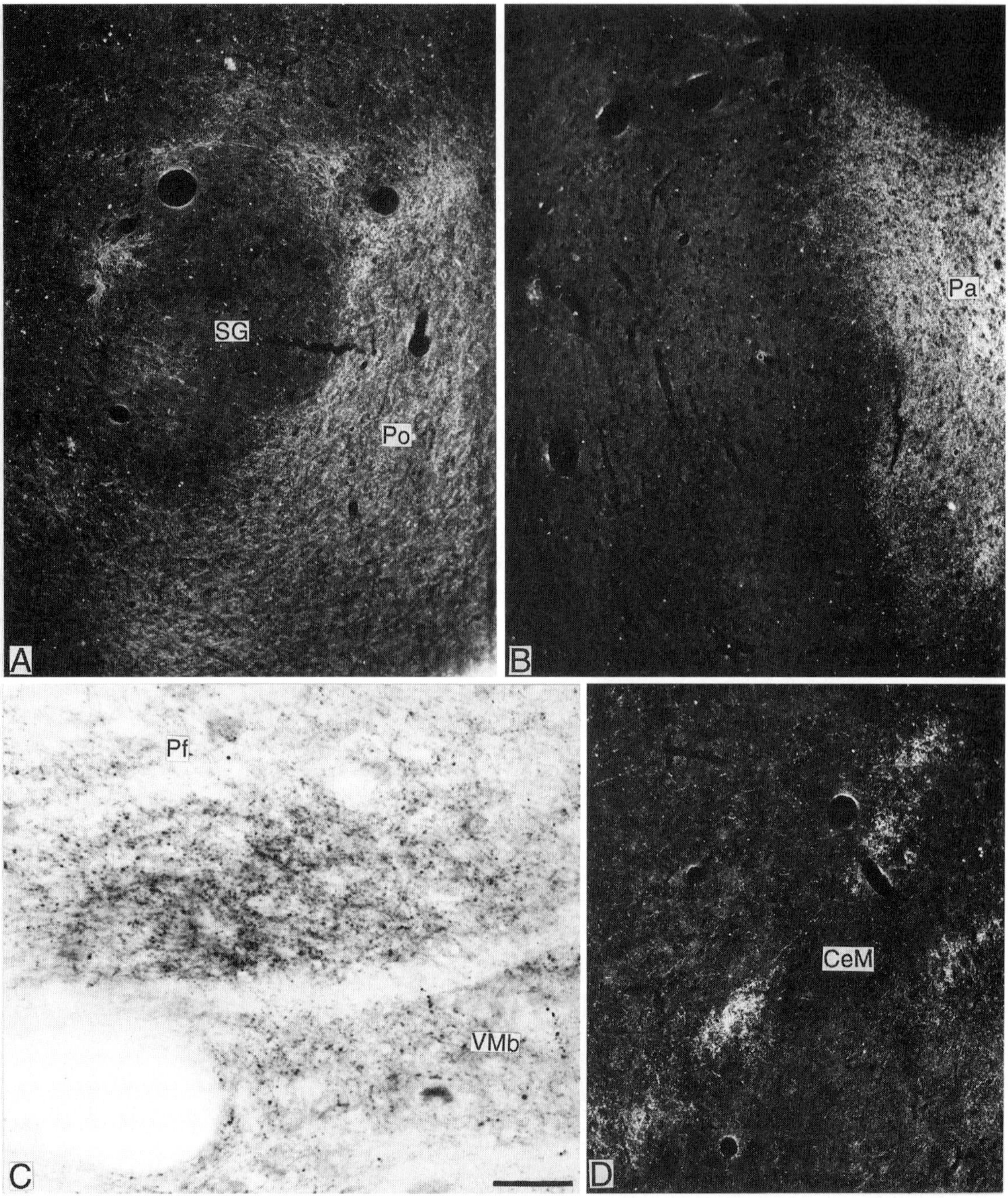

Fig. 58. Darkfield (A, B, D) and brightfield (C) photomicrographs showing tachykinin-(A, C) and CCK (B, D)-immunoreactive fibers in the posterior nucleus of a macaque monkey (A), the anterior paraventricular nuclei (B) of a macaque, in the submedial nucleus (C) of a human, and in the central medial nucleus of a macaque (D). Bar 250 μm. From material illustrated in Molinari et al. (1987) and Hirai and Jones (1989b).

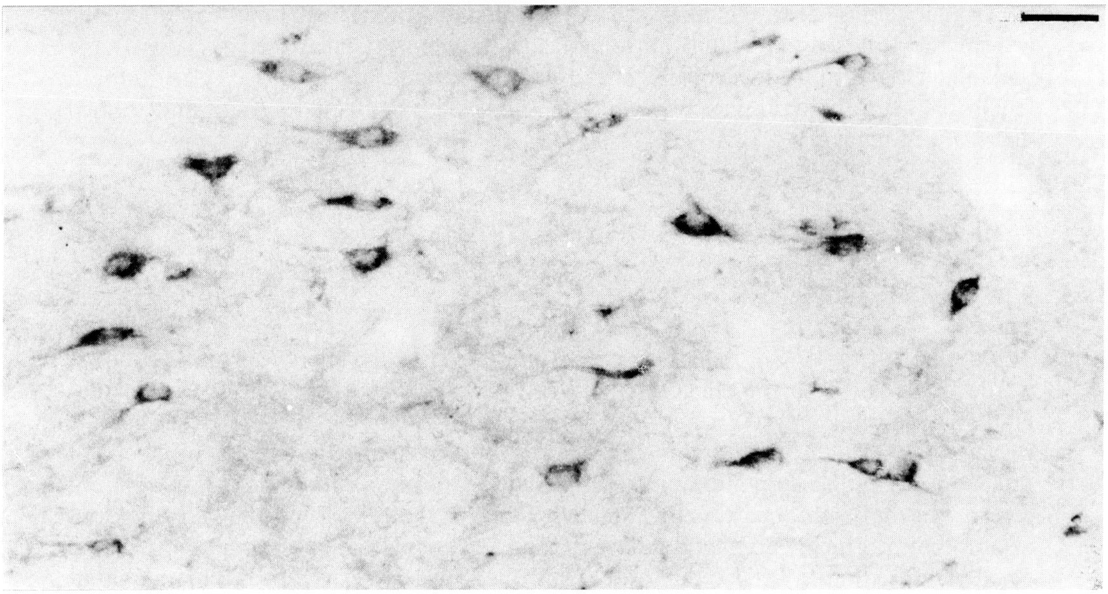

Fig. 59. Somatostatin immunoreactive cells in the reticular nucleus of a macaque monkey. Bar 50 μm. From material illustrated in Molinari et al. (1987)

for delta opiate receptors are reportedly absent from the thalamus of rats but mu receptor mRNAs are highly expressed. Kappa receptor mRNAs are weakly expressed and confined to cells in the epithalamus (Mansour et al., 1994).

Metenkephalin, when applied by iontophoresis to thalamic cells, has rather non-specific inhibitory effects (Hill et al., 1976).

8.6. HISTAMINE AND OTHER SUBSTANCES

A relatively small number of histamine immunoreactive fibers arising in the tubero-mamillary region of the hypothalamus, innervate the habenular, paraventricular, lateral posterior, medial geniculate, dorsal and ventral lateral geniculate nuclei and zona incerta of rodents and other species (Airaksinen and Panula, 1988; Airaksinen et al., 1989; Panula et al., 1989) and the lateral geniculate nuclei of cats (Uhlrich et al., 1993). The terminals of these fibers are similar to those formed by serotonin and noradrenergic fibers and form close membrane contacts without obvious membrane specializations. Autoradiography has demonstrated both H_1 and H_2 receptors in the thalamus of rats (Bouthenet et al., 1988; Ruat et al., 1990). When stimulated, these receptors result in a slow depolarization caused by suppression of the potassium leak current, I_{KL}, and similar to that elicited by stimulation of $\alpha 1$ noradrenergic, muscarinic cholinergic and metabotropic glutamate receptors (McCormick and Williamson, 1991) as described above. Adenosine which has not yet been demonstrated in fibers or terminals in the thalamus, tends to cause hyperpolarization of thalamic cells by increasing the postsynaptic potassium conductance, I_{KG} (Pape, 1992), and by acting upon metabotropic (A_1-purinergic) receptors. Nitric oxide, which is released from the terminals of cholinergic fibers in the thalamus following afferent stimulation (Do et al., 1994; Vincent and Kimura, 1992; Bickford et al., 1993), also enhances the hyperpolarization-activated cation current (Pape and Mager, 1992) and can affect the transmission of sensory signals throught the thalamus (Cudeiro et al., 1994). No thalamic cells are histochemically stainable for nitric oxide synthase, except for a few in the ventral lateral geniculate nucleus (Gabbott and Bacon, 1994a), and stained fibers are infrequent.

9. REGIONAL PROJECTIONS OF THE THALAMUS

9.1. DORSAL THALAMUS

The outflow of the dorsal thalamus, constituting most of the efferent projection of the whole thalamus, is to the telencephalon. The telencephalic targets of the dorsal thalamus include not only the cerebral cortex but also a number of basal telencephalic nuclei, of which the striatum is the largest, but substantial numbers of fibers of thalamic origin also terminate in the amygdala, substantia innominata and adjacent regions. In the cerebral cortex, thalamic fibers terminate not only in the neocortex but also in the entorhinal cortex and adjacent areas, including the hippocampal formation. There is no reason to believe that any area of neocortex lacks a dorsal thalamic input and this is likely also to be true of the palaeocortex and archicortex.

9.1.1. Projections to neocortical areas

Each dorsal thalamic nucleus, outside the intralaminar system, projects upon one or more neocortical areas (Figs. 60, 61). In monkeys, a projection to more than one area is the norm. Where the projection reaches two or more areas, separate populations of cells commonly form the projection to each area, with little or no collateralization of the thalamocortical axons. Other populations can have more widespread projections. These subpopulations of the cells in a nucleus are commonly identifiable by size, chemical properties and especially by location within the nucleus, Many examples have been given in the foregoing sections and others can be found in Section 10 which follows. The ventral posterior nucleus provides an example of submodality-specific relay cells located in different parts of the nucleus (Jones and Friedman, 1982) and projecting without branching to areas 3a, 3b, 1 or 2 of the somatosensory cortex (Jones, 1983a). These projections are formed by parvalbumin immunoreactive relay cells, the terminations of whose axons are found only in one or other of these somatosensory fields. The nucleus also possesses calbindin immunoreactive cells whose axons terminate in the somatosensory and/or motor areas and perhaps in other areas as well (Rausell et al., 1992). The dorsal lateral geniculate nucleus provides an example of lamina-specific locations of different functional classes relay cells: cells relaying

→

Fig. 60. Darkfield (A, B, D) and brightfield (C) photomicrographs from autoradiographs of sections through the postcentral gyrus of a macaque monkey (A, C, D) and of a squirrel monkey (B), showing labeling of terminations of thalamocortical fibers by anterograde transport of radioactive amino acids injected into the ventral posterior nucleus. Terminations are densest in areas 3a and 3b where they occupy layer IV and the adjacent part of layer III, with a subsidiary band of terminations at the junction of layers V and VI (A, B). In areas 1 and 2, terminations are less dense (A) and occupy the deeper half of layer III rather than layer IV, with a band of subsidiary terminations in layer I (C, D). Bars 500 μm (A), 100 μm (B-D). From material illustrated in Jones and Burton (1976a) and Jones and Friedman (1982).

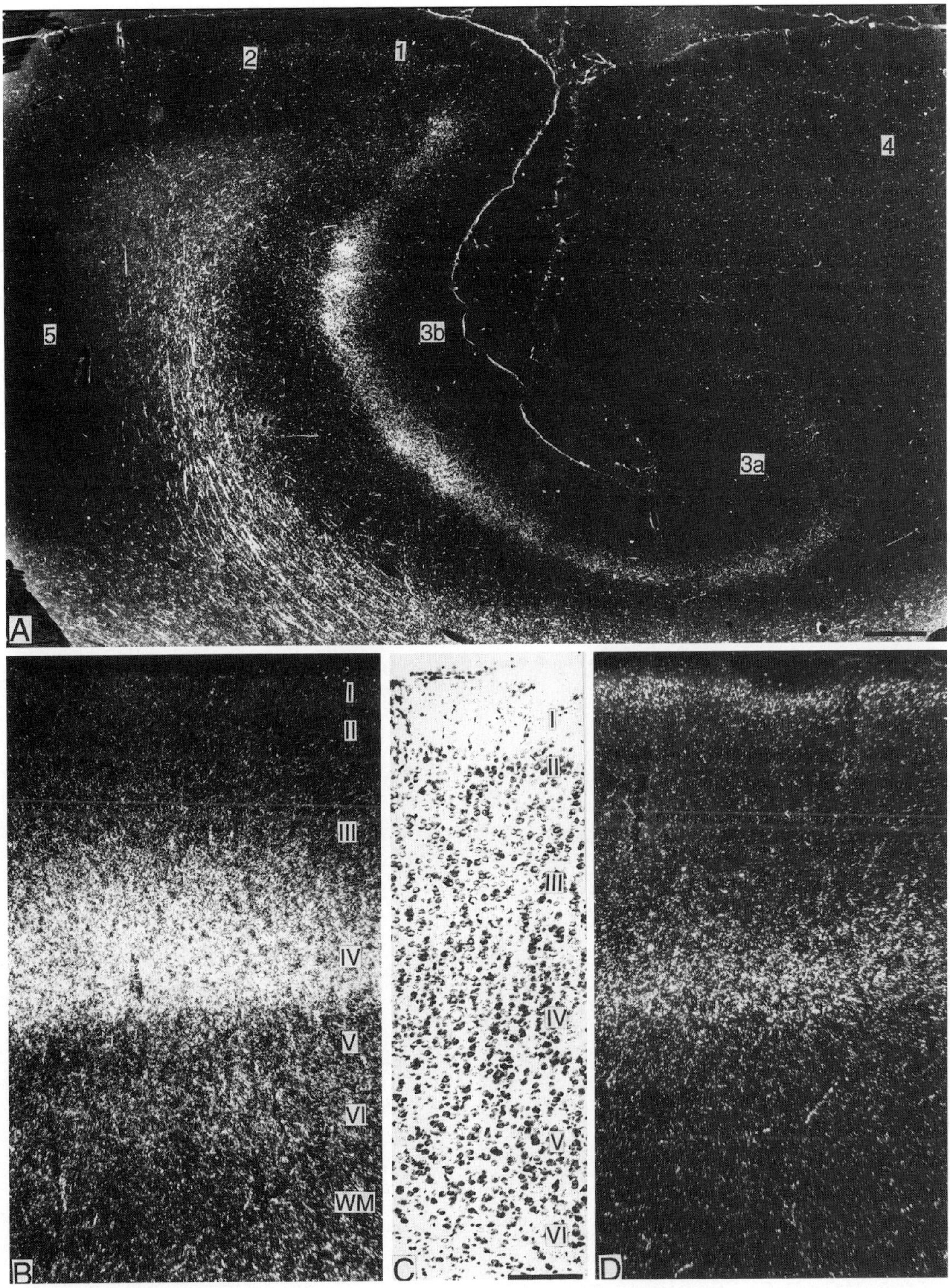
2
1
4
5
3b
3a
A
I
II
III
IV
V
VI
WM
B
I
II
III
IV
V
VI
C
D

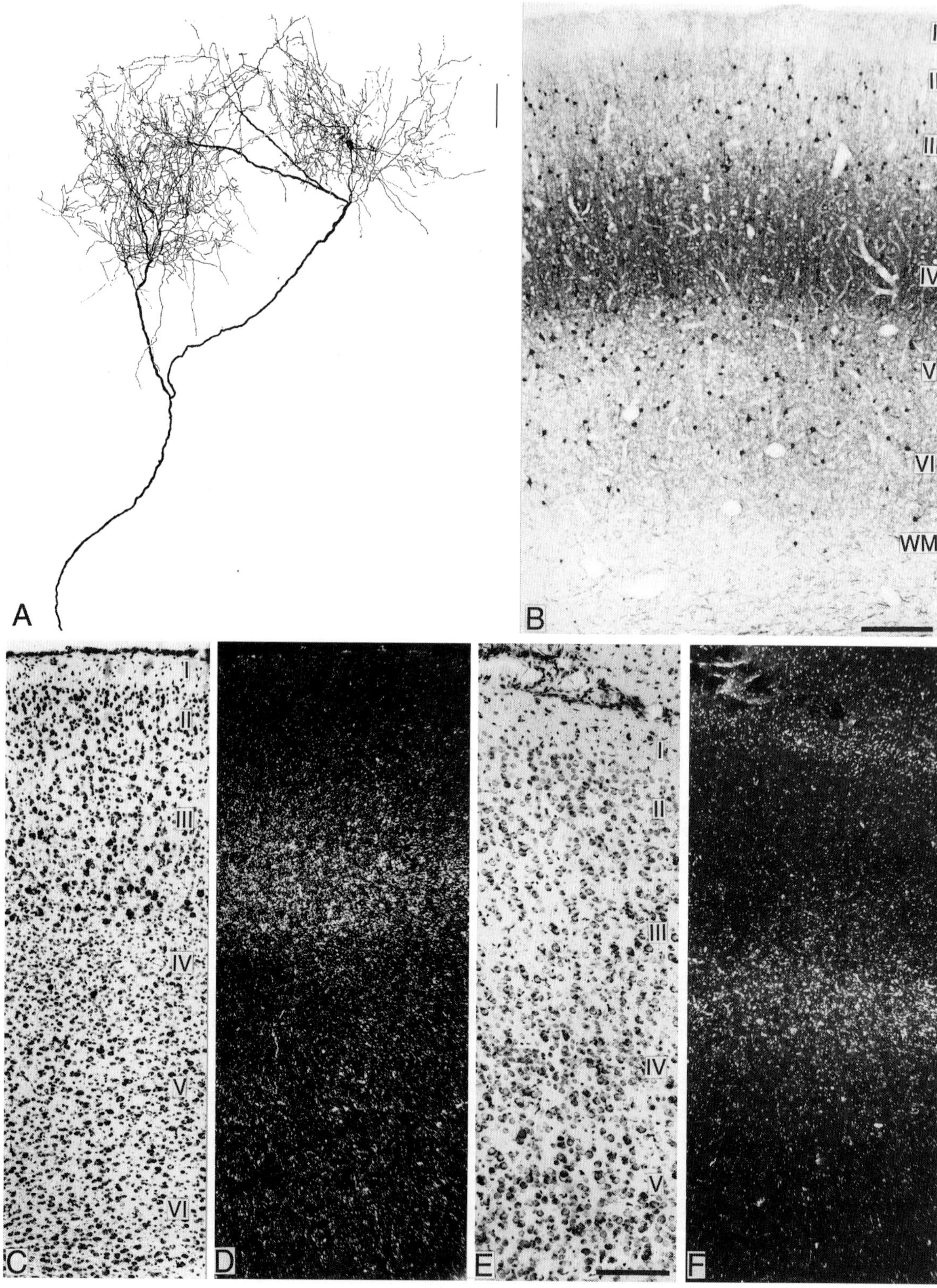
A
B
I
II
III
IV
V
VI
WM
C
D
E
F

Fig. 61. A: Terminal ramifications of a thalamocortical axon in area 3b of the somatosensory cortex of a macaque monkey; axon injected with horseradish peroxidase in the white matter beneath the cortex. Bar 50 μm. From unpublished work of J. DeFelipe and the author. B: Section through the primary auditory cortex of a macaque monkey, stained immunocytochemically for parvalbumin which shows the heavy terminations of parvalbumin immunoreactive thalamocortical fibers in layers IIIB and IV. Bar 100 μm. From material illustrated in Jones et al. (1995). C,D and E,F: Pairs of brightfield and darkfield photomicrographs from autoradiographs through the parietal cortex (area 5) (C, D) and agranular insular cortex (E, F) of squirrel monkeys, showing thalamocortical fiber terminations restricted to layer IIIB in area 5 and to layers IIIB and I in the insular cortex. Bar 100 μm. From material illustrated in Jones and Burton (1976a).

inputs from one class of retinal ganglion cell are restricted to the parvocellular layers (see Section 10.6), and those relaying inputs from a second class are restricted to the magnocellular layers. These two classes of relay cells have their particular chemical signatures, as described in Sections 6.2 and 6.3. A third class, the cells in the S and intercalated layers, provides an example not only of a third class of specifically located and chemically identifiable neurons (Section 6.2) but also of one that projects to areas beyond the striate area (Yukie and Iwai, 1981), the target of the axons of cells in the parvo- and magnocellular layers (Hubel and Wiesèl, 1972, 1977; Hendrickson et al., 1978). Similar examples of the kinds of subnuclear organization demonstrated by the ventral posterior and dorsal lateral geniculate nuclei can be found in virtually every other dorsal thalamic relay nucleus.

The dorsal lateral geniculate nucleus is perhaps the nucleus that comes closest to focusing all of its cortical projection upon a single neocortical area, the striate area (area 17), since the number of more widely projecting intercalated and S layer cells is quite small. Within the confines of other relay nuclei there are commonly two or more major populations of cells that project to different cortical areas.

From the point of view of the neocortex, the striate area may also furnish an example of an area that receives its sole thalamic input from a single dorsal thalamic nucleus, the dorsal lateral geniculate nucleus, although older reports indicate sparse projections from the inferior pulvinar nucleus as well (Rezak and Benevento, 1979; Ogren and Hendrickson, 1977). In the somatosensory cortex, each of its subareas receives its input from a subpopulation of cells in the ventral posterior nucleus. Areas on the orbital surface of the frontal lobe, similarly, receive their thalamic inputs from subpopulations of cells in the mediodorsal nucleus (Ray and Price, 1993; Section 10.4). Many areas, however, receive additional inputs from other dorsal thalamic nuclei. These inputs may come from the intralaminar nuclei (Jones and Leavitt, 1974; Jones, 1985a; Macchi and Bentivoglio, 1986; Royce et al., 1989; Smith and Parent, 1986; Darian-Smith et al., 1990b) and from the subpopulations of smaller, matrix-like cells described in Sections 6.2 and 10.7. In the striate area, these inputs are provided by the intercalated cells in the geniculate itself. It is still unclear if the striate area receives any input from the intralaminar nuclei in monkeys. Such an input is present in the *Galago* (Carey et al., 1979a,b; Raczkowski and Diamond, 1978b).

The presence of inputs to a cortical area from more than one thalamic nucleus has fed many theories about 'convergence' and parallel pathways for information flow, especially in the extrastriate visual areas. Most theories of this kind, however, fail to take into account that the projections from different nuclei converging on a single cortical field are fundamentally different in terms of quantity, character of the thalamic cells providing each projection, and especially in their laminar patterns of cortical termination. There is still, unfortunately, a dearth of quantitative information about this. Until it is forthcoming, it remains a valid view that each cortical area receives the greater part of its thalamic innervation from a nucleus or from a subpopulation of cells in a relay nucleus, the axons of which terminate in middle layers, and a subsidiary thalamic innervation from intralaminar cells and from cells in the same or a different relay nucleus, the axons of which terminate in superficial layers. To controvert this useful 'rule' it will be necessary to show that the axons of two large populations of relay cells located in different nuclei converge in the middle layers of a single cortical field. No examples of this have yet been forthcoming.

Quantification of a kind has been provided for the motor and premotor areas. There, on the basis of retrograde labeling studies, Darian-Smith et al. (1990b) estimated that at least 40% of the neurons of the VLp nucleus projected to area 4 while less than 15% projected to premotor areas. The trend is reversed for the VLa nucleus, many more cells of which project to the premotor and supplementary motor areas than to area 4. In these cases, it is likely that the quantitatively smaller groups of projecting cells are made up of the more diffusely projecting, small cells of the calbindin immunoreactive matrix in these nuclei (Rausell et al., 1992).

While recognizing that the presence of differently projecting populations of cells in probably every dorsal thalamic relay nucleus, especially when coupled with investigator bias, can result in the apportionment of different extents of cortex to each nucleus, a regional map of the cortical areas subtended by each of the major relay nuclei, based on the projections of their largest populations of relay cells, can probably be built up from maps such as those given for the ventral, mediodorsal, medial geniculate, LP/pulvinar and dorsal lateral geniculate nuclei in Section 10. Dorsal thalamic nuclei for which there are, unfortunately, few data on their cortical projections in primates, include the anterior, parataenial, medioventral (reuniens), limitans/suprageniculate, posterior nuclei and nuclei around the posterior pole of the ventral posterior nucleus. These are considered in the following section.

9.1.2. Nuclei projecting to limbic and insular regions

From studies in non-primates, it may be assumed that the nuclei of the anterior group, including the lateral dorsal, project mainly upon the cingulate, suprasplenial and retrosplenial areas, extending at least as far as the parasubiculum (see Jones, 1985a for a review and Swanson, 1987 for more recent information). The small amount of available information in monkeys confirms the projection to cingulate gyrus and retrosplenial region (Vogt et al., 1979; Baleydier and Mauguiére, 1980; Vogt, 1985). The three anterior nuclei and to a lesser extent the lateral dorsal nucleus contain retrogradely labeled cells when the hippocampal formation, particularly the subiculum and presubiculum, are injected with retrograde tracers (Amaral and Cowan, 1980).

In rats, the reuniens or medioventral nucleus provides the principal thalamic projection to the pre- and parasubculum and to the subiculum and field CA_1 of the hippocampus, as well as to the entorhinal and perirhinal areas. En route through the cingulum, fibers terminate in layer I of all the cingulate and retrosplenial areas (Herkenham, 1978; Wyss et al., 1979; Swanson et al., 1987). In monkeys retrograde labeling of cells occurs in the medioventral nucleus, particularly its anterior part, following injections of tracer in the hippocampal and entorhinal regions (Amaral and Cowan, 1980; Insausti et al., 1987), so the organization is undoubtedly similar to that found in the rat. The anterior and lateral dorsal nuclei do not appear to project to the entorhinal cortex in monkeys (Insausti et al., 1987). Labeled cells also occur in the medioventral nucleus after injections of tracer in the insula (Mesulam and Mufson,

1985), perhaps reflecting the continuity of the anterior part of the insula with the orbital and entorhinal areas. Midline components of the intralaminar system, especially the more anterior and dorsally located groups of cells, also provide input to the hippocampal, entorhinal and periamygdaloid regions (Amaral and Cowan, 1980; Insausti et al., 1987).

The parataenial nucleus is commonly reported to contain labeled cells after injections of anatomical tracers in many different parts of the forebrain. Probably in many instances the cells are in an adjacent part of the central lateral nucleus. Evidence of its projections in monkeys is scanty. In rats, it projects to prelimbic cortex (Jones and Leavitt, 1974) and/or the hippocampal formation (Wyss et al., 1979). The latter and projections to the entorhinal cortex have also been reported in monkeys (Amaral and Cowan, 1980). The paraventricular nuclei of the epithalamus are also commonly reported to contain retrogradely labeled cells after injections of tracers in the hippocampal and entorhinal areas. It is difficult to be sure that these are also not in the parataenial or central lateral nuclei.

The limitans/suprageniculate nucleus for reasons given in Sections 3 and 10.3, may best be considered a part of the intralaminar system. It has a clear cortical target in the granular insular cortex (Burton and Jones, 1976; Mesulam and Mufson, 1985) but cells can be retrogradely labeled in it from the superior parietal lobule (Pearson et al., 1978) and entorhinal cortex (Insausti et al., 1987) as well. The more anteriorly located, dysgranular insular field is the target of the ventral posterior inferior (VPI) and basal ventral medial (VMb) nuclei (Burton and Jones, 1976; Mesulam and Mufson, 1985). The VPI nucleus has been reported to provide the main thalamic input to the second somatosensory area (SII) of the cerebral cortex (Friedman and Murray, 1986) located in the parietal operculum. However, defining the thalamic inputs to SII presents a problem, mainly because of the relatively small amount of work that has been done since extensive microelectrode mapping of the lateral sulcus region revealed the presence of two or more representations of the body surface in what had previously been regarded as a single cortical field (reviewed in Burton, 1986 and Krubitzer et al., 1995). Early studies with anterograde labeling techniques suggested inputs from the core of VPL and VPM (Jones and Powell, 1970b; Burton and Jones, 1976). Later studies based on retrograde labeling highlighted VPI as possibly the sole source of inputs (Friedman and Murray, 1986) although this was not agreed to by all (Burton and Carlson, 1986). The difficulty in interpretation is mainly related to the intrusion of large, typical VPL cells into VPI and the similar interdigitation of small VPI cells into VPL (Jones, 1985a). Coupled with this is the fact that the large injections made to demonstrate thalamic projections to SII did not distinguish between the calbindin immunoreactive matrix (Section 10.7.5) which includes VPI and which projects diffusely to cortex, and the parvalbumin immunoreactive population of VP cells whose projection is more focused (Secions 6.2 and 10.1). Recent evidence from Krubitzer and Kaas (1992) indicates that projections to the three body representations that they describe in the parietal operculum of New World monkeys, are dominated by those from cytochrome oxidase (CO)-weak zones extending up from VPI between the CO-dense lobules of VPL proper and in the case of inputs to the face representations from the CO-weak zones similar to those found in macaque monkeys (Section 10.7 and 10.8) lying dorso-medial and posterior to VPM. It is likely that disagreements about the relative contributions of VPI and VPL/VPM cells to the inputs to SII stem from retrograde labeling of different proportions of superficially-projecting matrix (including VPI) cells and middle layer-projecting VP cells.

The VMb nucleus, commonly termed VPMpc in monkeys, contains neurons responding to gustatory stimuli, especially posteriorly (Pritchard et al., 1989) and modifications of gustatory sensation occur when it is involved in lesions of the ventral posterior nucleus (Blum et al., 1943; Patton et al., 1944; Reilly and Pritchard, 1995). Most neurons in the VMb nucleus are also reported to receive convergent tactile inputs from the tongue (Burton and Benjamin, 1971; Blomquist et al., 1962; Pritchard et al., 1989). The primary input to the nucleus arises from the lateral division of the nucleus of the solitary tract (Beckstead et al., 1980), a division in which inputs from the trigeminal, facial, glossopharyngeal and vagus nerves converge (Beckstead and Norgren, 1979). The gustatory area of the cerebral cortex to which VMb projects, is located in or adjacent to the frontal operculum in squirrel monkeys (Benjamin and Burton, 1968; Burton and Benjamin, 1971; Burton and Jones, 1976) and cynomolgus monkeys (Pritchard et al., 1986).

The posterior nucleus which is particularly associated with the terminations of spinothalamic fibers and with thalamic pain mechanisms (Section 10.8), appears to project to retroinsular areas in the parietal operculum (Burton and Jones, 1976), areas which are of interest in view of their reported association with painful syndromes or hypoalgesia when affected by lesions in humans (Biemond, 1956).

9.1.3. Projections of the intralaminar nuclei

These nuclei are considered in detail in Sections 4.3 and 10.3. Their positions as the principal (perhaps only) source of the dense thalamic projection upon the striatum (caudate nucleus, putamen, nucleus accumbens and areas linking these nuclei called the ventral striatum) is uncontested and has been well studied anatomically in monkeys (Section 10.3). Although never quantified, the size of the thalamostriatal projection must be very large, given the number of cells involved. The intralaminar nuclei also contain neurons which project to the cerebral cortex (Jones and Leavitt, 1974). Their numbers have never been quantified either but, judged from reports of experimental investigations, must be very many fewer than those projecting to the striatum. Unlike the thalamostriatal projection, the cortical projection of the intralaminar nuclei has received relatively little attention in monkeys or other primates, most knowledge about it stemming from work on cats (Macchi and Bentivoglio, 1986; Royce et al., 1989; Macchi, 1990; Minciacchi et al., 1993; Molinari et al., 1994). From those studies we learn that few or no intralaminar cells project by branched axons to both striatum and cortex, and that those projecting to the cortex, although tending not to respect the borders of cytoarchitectonic fields, project to relatively circumscribed areas so that individual intralaminar nuclei have regional cortical zones of projection. There may also be two classes of intralaminar cells that project to the cortex: one whose axon terminates in layer I and a second whose axon terminates in middle layers (Molinari et al., 1994). In monkeys, the lack of collateralization to striatum and cortex has been confirmed for the centre médian and parafascicular nuclei (Sadikot et al., 1992a).

For the monkey, and to a lesser extent, the *Galago*, information about the cortical projections of the intralaminar nuclei has come incidental to studies aimed at other aspects of the cerebral cortex or thalamus. Passing mention of retrograde labeling in an intralaminar nucleus after injections of anatomical tracers in one or other area of the cortex and the presence of a localized patch of corticothalamic terminals (which return to the location of the cells projecting to the cortex from which the corticothalamic fibers arise: Powell and Cowan, 1967; Jones et al., 1979), enables a picture of the organization of the intralaminar-cortical projection which is similar to that found in the cat, to be built up (see Fig. 72) (Jones and Leavitt, 1974; Strick, 1975; Darian-Smith et al., 1990a,b; Sadikot et al., 1992a; Leichnetz, 1986; Kievit and Kuypers, 1977; Vogt et al., 1987; Leíchnetz and Goldberg, 1988; Huerta et al., 1986; Huerta and Kaas, 1990; Stepniewska et al., 1994b). The central medial nucleus and the midline groups of intralaminar cells project to prefrontal, orbital and anterior cingulate regions, the paracentral and parts of the central lateral nucleus to the parietal, posterior cingulate and lateral temporal areas and the central lateral nucleus, especially its posteriorly-located larger cells to frontal eyefields, motor and somatosensory fields and possibly even to the visual cortex, although this is not confirmed. The most anterior and dorsal of the midline groups of cells, groups that are often confused with the parataenial and paraventricular nuclei, project to entorhinal, hippocampal and periamygdaloid areas (Amaral and Cowan, 1980; Insausti et al., 1987).

The centre médian nucleus projects to the motor cortex. The exact projection of the parafascicular nucleus is unclear but includes prefrontal, insular and adjacent areas. Auditory areas and adjacent fields do not receive fibers from the intralaminar nuclei proper, but the inputs from the limitans/suprageniculate and magnocellular medial geniculate nuclei may be regarded as equivalent to intralaminar-cortical connections.

9.1.4. Projections to the amygdala and adjacent regions

The lateral nucleus of the amygdala was reported to receive a rather heavy input from the medial nucleus of the pulvinar in an anterograde tracing study made on squirrel monkeys and macaque monkeys (Jones and Burton, 1976b) (Fig. 62). This was contested by Mehler (1980) in the discussion of a retrograde tracing study but no data were presented to support the claim, the pulvinar not being included in the sections presented. Thalamic nuclei that appear to have been consistently labeled in the study of Mehler were mainly those forming the midline extensions of the intralaminar nuclei and a small-celled region of the medioventral nucleus located just dorsal to the third ventricle and called by Mehler, the ventral reuniens nucleus, extending back into the midline region of the principal ventral medial nucleus. All of these nuclei also contain retrogradely labeled cells after injections of tracers in the substantia innominata (Russchen et al., 1985). The subparafascicular nucleus and adjacent parafascicular nucleus are also reported to project to the amygdala and substantia innominata in monkeys (Sadikot et al., 1992a). In other species, notably the cat, thalamic cells projecting to the amygdala have been identified in the posteromedial division of the lateral posterior-pulvinar complex (probably equivalent to the medial pulvinar nucleus of the monkey on topographic grounds) and in the adjacent limitans/suprageniculate nucleus, as well as in the midline groups of intralaminar cells (Russchen, 1982).

Many other cells projecting to the amygdala are located outside the confines of the dorsal thalamus in the subparafascicular, peripeduncular and adjacent nuclei at the midbrain/diencephalic junction (Jones et al., 1976; Mehler 1980; Russchen et al., 1987; Sadikot et al., 1992a). Cells with axons projecting to the hypothalamus or descending to the brainstem reticular formation are also located in these regions (Sadikot et al., 1992a) and may be the basis of reports of descending thalamic projections.

The centre médian and parafascicular nuclei are also reported to project to the internal and external segments of the globus pallidus and subthalamic nucleus (Sadikot et al., 1992a).

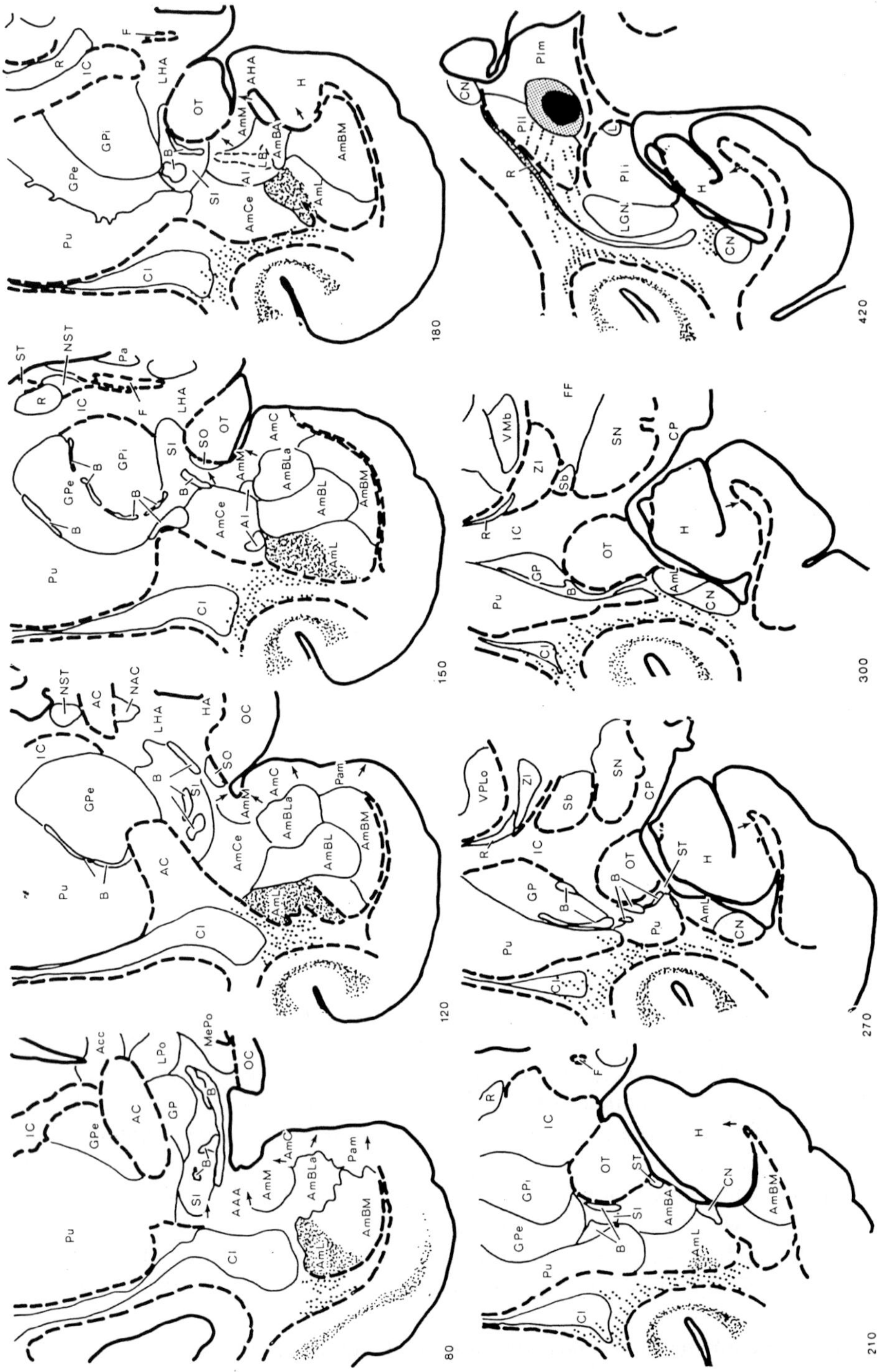
80
120
150
180
210
270
300
420
Pu
GPe
GPi
GP
IC
AC
Acc
LPo
MePo
OC
SI
B
AAA
AmM
AmC
AmBLa
AmBL
AmBM
AmCe
AmL
AmBA
Pam
Cl
NST
NAC
LHA
HA
SO
ST
R
Pa
F
OT
AHA
H
LB
Al
VPLo
ZI
Sb
SN
CP
CN
VMb
FF
Plm
Pll
Pli
LGN
L

Fig. 62. Fiber projections (dots) from the medial pulvinar nucleus (Plm) to the lateral nucleus of the amygdala (AmL) in a squirrel monkey. Demonstrated by autoradiography following injection of tritiated amino acids in the thalamus. Sections are in anterior (80) to posterior (420) order. From Jones and Burton (1976b).

9.2. VENTRAL THALAMUS

The connections of the reticular nucleus have been discussed in Sections 5.5 and 7.2. Its principal, perhaps only outflow is into the underlying dorsal thalamus. There are reports of this GABAergic connection in nonprimates crossing the midline and continuing into dorsal thalamic nuclei of the opposite side (Paré and Steriade, 1993; Battaglia et al., 1994). These may be seen as reflecting the role of the reticular nucleus in recruiting the whole thalamus into coherent oscillatory activity, which is likely to be bilateral. There are also reports in non-primates of descending projections from the reticular nucleus to brainstem centers. These are difficult to evaluate and have not been the subject of further investigation.

Unlike the reticular nucleus, the other components of the ventral thalamus, the zona incerta and ventral lateral geniculate nucleus, do not obviously receive terminals of axons arising in the dorsal thalamus. Both, however, receive terminations of axons emanating from the cerebral cortex, the zona incerta mainly from the somatic sensory and motor areas (Petras, 1965a-c; Jones and Powell, 1968; Wise and Jones, 1977; Künzle, 1978; Porter and White, 1983) and the ventral lateral geniculate nucleus from the visual areas (Nauta and Bücher, 1954; Giolli and Guthrie, 1969; Spatz, 1970; Abplanalp, 1970; Spatz and Tigges, 1973; Ogren and Hendrickson, 1976; Updyke, 1977; Giolli et al., 1978; Holländer et al., 1979; Graham et al., 1979; Conley and Friederich-Ecsy, 1993b). The principal sources of subcortical afferents to the zona incerta are the medial lemniscus, spinal cord, trigeminal nuclei and deep cerebellar nuclei (Mehler, 1966b; Lund and Webster, 1967a,b; Boivie, 1971a; Smith 1973; Faull and Carman, 1978; Basbaum and Fields, 1979; Erzurmulu and Killackey, 1980; Feldman and Kruger, 1980; Asanuma et al., 1983b; Berkley, 1983; Beitz, 1989; Carstens et al., 1990). Inputs from the hypothalamus are also found (Roger and Cadusseau, 1985; Swanson, 1987; Saper, 1990). Afferents to the ventral lateral geniculate nuclei come from the retina, pretectum and superior colliculus (Altman and Carpenter, 1961; Tarlov and Moore, 1966; Abplanalp, 1970; Rockel et al., 1972; Harting et al., 1973b; Graybiel, 1974, 1977; Takahashi et al., 1977; Hickey and Spear, 1976; Holländer and Sanides, 1976; Berman, 1977; Graham, 1977; Graybiel and Berson, 1980b; Weber et al., 1983; Holcombe and Guillery, 1984; Harman et al., 1990; Agarwala et al., 1992; Conley and Friederich-Ecsy, 1993b). In monkeys, apes and humans, the nucleus is bilaminar, the ventral lamina, adjacent to the dorsal lateral geniculate nucleus receiving a large contralateral retinal input and a smaller ipsilateral one. The more dorsal lamina is devoid of retinal inputs (Hendrickson, 1973; Babb, 1980; Balado and Franke, 1937; Hassler, 1959). Dense terminations of monoaminergic and cholinergic afferents from the brainstem are also concentrated in the nucleus (Section 8.4). Retinal inputs in cats are described as deriving mainly from small or W type retinal ganglion cells (Spear et al., 1977; Conley and Friederich-Ecsy, 1993a) and visually-responsive cells in the ventral lateral geniculate nucleus of cats have large, monocular receptive fields which lack the center-surround organization typical of cells in the dorsal lateral geniculate nucleus (Mathers and Mascitti, 1975; Spear et al., 1977; Hughes and Ater, 1977; Hale and Sefton, 1978; Sumitomo et al., 1979).

The major outflow of the ventral lateral geniculate nucleus is to hypothalamic nuclei involved in regulation of circadian rhythms (Card and Moore, 1994). This projection is made up of neuropeptide Y immunoreactive fibers. The nucleus also projects to other components of the ventral thalamus, notably the zona incerta, as well as commissurally to its fellow of the opposite side (Swanson et al., 1974). The zona incerta has diverse projections to sites as far distant as the spinal cord and some of its GABA cells project to the cerebral cortex, at least in rats (Lin et al., 1990; Nicolelis et al., 1992). Both it and the ventral lateral geniculate nucleus are reported to send fibers into the dorsal thalamus but only to the caudal intralaminar nuclei (Conley and Friederich-Ecsy, 1993b).

Apart from the GABAergic input to the rodent cerebral cortex (Nicolelis et al., 1992), efferent projections from the zona incerta have been described to the contralateral side of the spinal cord (Kuypers and Maisky, 1975; Nudo and Masterton, 1988), basal ganglia (Ricardo, 1981), superior colliclus, pretectum (Grofová et al., 1978; Ricardo 1980, 1981; Jones, 1985a), pons (Romanowski et al., 1985) and brainstem (Nicolelis et al., 1992). Other fibers arising in the ventral lateral geniculate nucleus terminate in the pretectum, superior colliculus and pontine nuclei (Hendrickson, 1973; Gaybiel, 1974a,b; Swanson et al., 1974; Ribak and Peters, 1975; Berman, 1977), and in some of the nuclei of the accessory optic system (Edwards et al., 1974; Swanson et al., 1974; Graybiel, 1974a,b; Ribak and Peters, 1975; Card and Moore, 1982; Abe and Rusak, 1992; Shinohara et al., 1993). The fibers to the hypothalamus end in the suprachiasmatic nuclei. The principal retinal input to the nucleus is derived from ganglion cells sensitive to luminance changes (Spear et al., 1977). Those connections, therefore, reflect the involvement of the ventral lateral geniculate nucleus in hypothalamic and brainstem circuitry underlying the maintenance of circadian rhythms which follow cycles of light and darkness. Whether all these connections are present in primates remains to be determined.

9.3. EPITHALAMUS

The epithalamus consist of the paraventricular and habenular nuclei. The former nuclei are all too commonly confused with the parataenial nucleus and with the dorsal groups of cells belonging to the intralaminar nuclei, so that they are frequently described as having connections more typical of the dorsal thalamus than of the epithalamus. The paraventricular nuclei can be distinguished histochemically by extremely dense acetyl cholinesterase staining (Figs. 7, 8) and immunocytochemically by a dense innervation by neuropeptide Y immunoreactive fibers, ascending from the region of the median eminence (Jones, 1985a; Molinari et al., 1987).

The outputs of the paraventricular nuclei have not been clearly distinguished from those of adjacent dorsal thalamic nuclei, especially the parataenial and dorsal components of the central lateral nucleus, but the amygdala, hippocampal formation, septum and nucleus accumbens are consistently mentioned. The principal inputs to the medial habenular nucleus arise in the septal region. Those to the magnocellular part of the lateral habenular nucleus arise in the globus pallidus and are GABAergic, while those ending in the parvocellular part arise in the basal forebrain and hypothalamus (Nauta and Mehler, 1966; Nauta, 1974, 1979a,b; Kim et al., 1976; Gottesfeld et al., 1981). Both habenular nuclei send their efferent fibrs via the habenulo-peduncular tract but only those arising in the medial habenular nucleus (which are cholinergic: Houser et al., 1983; Heckers et al., 1992), actually terminate in the interpeduncular nuclei. The fibers arising in the lateral habenular nucleus (which are non-cholinergic) leave the habenulo-peduncular tract in the ventral tegmental area and descend to terminate in the raphe and tegmental nuclei of the midbrain and in the pars compacta of the substantia nigra. Other fibers reach the lateral hypothalamic area (Herkenham and Nauta, 1979). These projections have not been examined in depth in primates.

10. THALAMIC NUCLEI AS COMPONENTS OF DEFINED FUNCTIONAL SYSTEMS

10.1. VENTRAL NUCLEI AND THE MOTOR SYSTEM

The major, large divisions of the ventral nuclei (VA, VLa, VLp and VPL/VPM) that are so clearly delineated by acetylcholinesterase- and Nissl-staining in the monkey and human thalamus, are of special significance (Figs. 7, 8, 63, 64). Experimental studies in monkeys show that these nuclei represent the principal thalamic relays for inputs from the pars reticulata of the substantia nigra, globus pallidus, deep cerebellar nuclei and medial lemniscus, respectively (Percheron, 1977; Tracey et al., 1980; De Vito and Anderson, 1982; Asanuma et al., 1983a,b; Alexander et al., 1986; Jones, 1987; Percheron et al., 1993, 1996; Rouiller et al., 1994; Sakai et al., 1996). For many years it was thought that there was substantial convergence of the last three of these pathways in the thalamus. This now seems unlikely, although the degree of segregation of the terminations of the pathways is still contested by some workers. There is virtually no overlap between the lemniscal and cerebellar terminations and that between the pallidal and cerebellar terminations is now regarded as small and may reflect no more than the protrusion of the fingers of VLp cells in among the islands of VLa cells (Section 3.2). The terminations of the inputs from the pars reticulata of the substantia nigra are perhaps less well-defined than the other three and there may be some overlap with the terminations of the pallidal input, but the nigral fiber terminations clearly have their principal focus in the VA and VMp nuclei (Goldman-Rakic et al., 1985; Ilinsky and Kultas-Ilinsky, 1987; Percheron et al., 1993, 1996). The greater differentiation of the borders between the nuclei of the ventral complex in the human thalamus may be a reflection of the increasing segregation of these inputs in comparison with the monkey.

Historically, delineations of the monkey and human ventral thalamic nuclei derive from different traditions, the first dominated by experimentation and the second by stereotaxic neurosurgery. Each has developed its own nomenclature, as pointed out in Section 1, and in the two nomenclatures have sown the seeds of much confusion. In monkeys, Walker (1938), mainly following Aronson and Papez (1934), divided the ventral nuclei into five. Anterolaterally, were the nuclei ventralis anterior (VA), ventralis lateralis (VL) and ventralis intermedius (VIM). Posteriorly, was the nucleus ventralis posterior, divided into medial (VPM) and lateral (VPL) subnuclei. VIM was characterized by the presence of extremely large cells and was regarded as the anterior part of the nucleus ventralis posterior. Medially, was a nucleus medialis ventralis (MV) which later came to be designated the nucleus ventralis medialis (VM).

Olszewski's (1952), in his atlas of the rhesus monkey thalamus (Table 3; Fig. 64) expanded on the parcellation of Walker, dividing VA into magnocellular (VAmc) and parvocellular (VA) subnuclei, calling VM a medial subnucleus (VLm) of a large ventral lateral nucleus (VL), the rest of which was divided into an anterior pars oralis

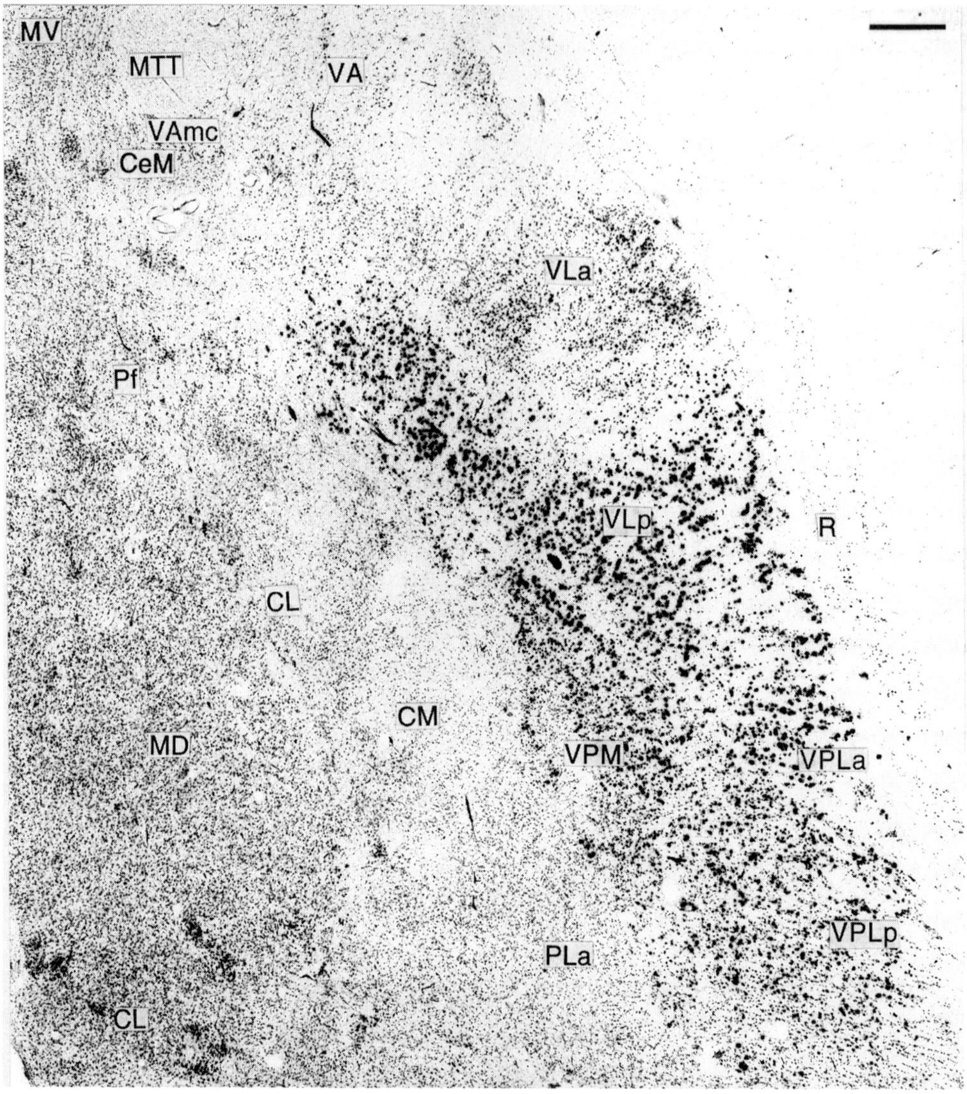

Fig. 63. Left Nissl-stained horizontal section through the ventral nuclei of the human thalamus, showing the distinctive cytoarchitecture of the various ventral nuclei. Note how the proprioceptive division of the ventral posterior nucleus (VPLa) consists of both large and medium sized cells which distinguishes it from the cerebellar relay nucleus (VLp) and from the cutaneous lemniscal relay nucleus (VPLp/VPM). Bar 1 mm. Right Nissl-stained horizontal section through the thalamus of a *M. fascicularis* at a level comparable to that through the human thalamus shown on the left. The cytoarchitecture of the ventral nuclei is similar to that of the human but a separate VPLa nucleus is not anatomically discernible. Bar 500 μm. From material illustrated in Hirai and Jones (1989a) and Asanuma et al. (1983a).

(VLo), an anteromedial part (area 'X'), a large posterodorsal pars caudalis (VLc), and a small posterodorsal pars postrema (VLps). Ventral and posterior to VLc, Olszewski recognized the very large celled VIM of Walker but called it the oral division of the ventral posterior lateral nucleus (VPLo), although remarking on the similarity of its cells to those of the subnuclei of VL. He included VPLo in the ventral posterior

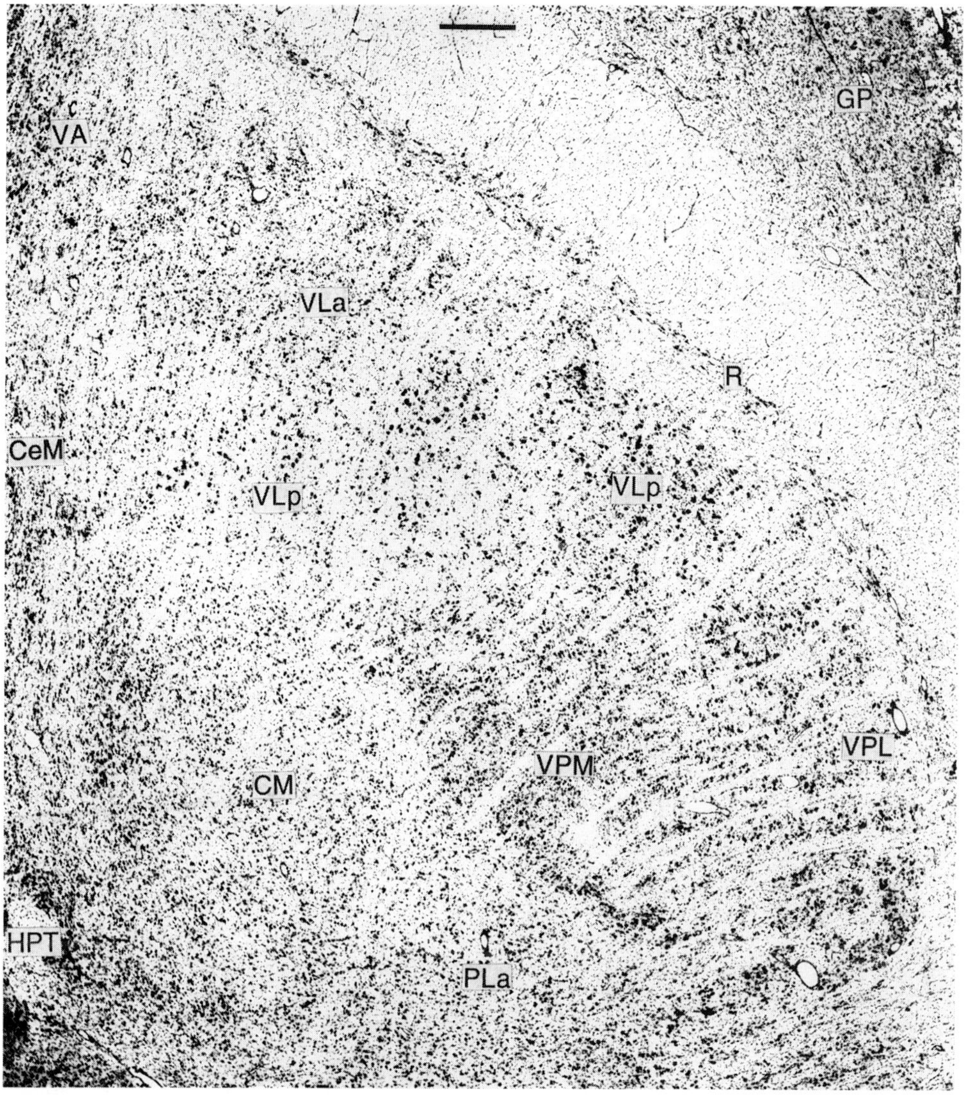

Fig. 63. (continued)

nucleus because he believed that it projected to the postcentral gyrus and formed part of the somatosensory relay. This has proven to be incorrect.

More recent delineations of the monkey ventral nuclei have been made in association with studies of the fiber connections of the nuclei. At the posterior region of the ventral nuclear mass, Tracey et al. (1980), Friedman and Jones (1981), Jones and Friedman (1982) and Asanuma et al. (1983a,b,c) concluded that the terminations of the medial lemniscus and cerebellothalamic pathways did not overlap, the cerebellar terminations coinciding with a zone of large, relatively sparsely distributed cells corresponding to Olszewski's VPLo, VLc and VLps nuclei and area X, and the lemniscal terminations with his caudal division of the ventral posterior lateral nucleus (VPLc) (Figs. 65, 66). Studies of Percheron (1977), Berkley (1983), Percheron et al. (1993,

TABLE 3. *A comparison of the nomenclatures used by various authors for the human and macaque ventral thalamic nuclei*

MONKEYS					HUMANS				
Walker (1938)	Olszewski (1952)	Ilinsky & Kultas-Ilinsky (1987)	Percheron et al. (1993)	Jones (1985) & Hirai & Jones (1989a)	Hassler (1959)	Dewulf (1971)	Van Buren & Borke (1972)	Walker (1982)	Ohye (1990)
VA	VApc	VApc	LRpo LO	**VA**	Dorso-oralis (D. o) internus (D. o. i) externus (D. o. e) Latero-polaris (L po).	v.a.	L po	Lpo	Lpo
VAmc	VAmc	VAmc	LRmc	**VAmc**	Latero-polaris magno-cellularis (L. po. mc)	v.o.m. (a)	L po (mc)		
VM	VLm	VM	LRvm	**VMp**	Ventro-oralis medialis (V. o. m)				
L	VLo	VAdc	LO	**VLa**	Ventro-oralis anterior (V. o. a) Ventro-oralis posterior (V. o. p)*	v.o.a.+ d.a.(p) v.o.p.	V o e(a) V o e(p)	Vo	Voe(a) Voe(p)
VL	Area X	VL	LIM	**VLp (anteromedial part)**	Ventro-oralis internus (V. o. i)	v.o.m. (p)	V o i	Vo	Voi
VL	VLc	VL	LI	**VLp (dorsal part)**	Dorso-intermedius externus (D. im. e) Dorso-intermedius internus (D. im. i) Dorso-intermedius superior (D. im. s)	d.a.(p)	D o	Vo	
LPpa	VLps	VL	LOd	**VLp (posterodorsal part)**	Dorso-intermedius externus magnocellularis (D. im. e. mc)				
VIM	VPLo	VL	LI/LIL	**VLp (ventral part)**	Ventro-intermedius internus (V. im. i) Ventro-intermedius externus (V. im. e) Zentro-lateralis intermedius internus (Z. im. i) Zentro-lateralis intermedius externus (Z. im. e)	v.im.	V im i V im e	VIM	Vim
VPL	VPLc	VPL	LCL	**VPLa**	Zentro-lateralis caudalis internus (Z. c. i) Zentro-lateralis caudalis externus (Z. c. e) Ventro-caudalis anterior externus (V.c.a.e)	v.p.l.	V c e	Vce	Vce
				VPLp	Ventro-caudalis posterior externus (V.c.p.e)				

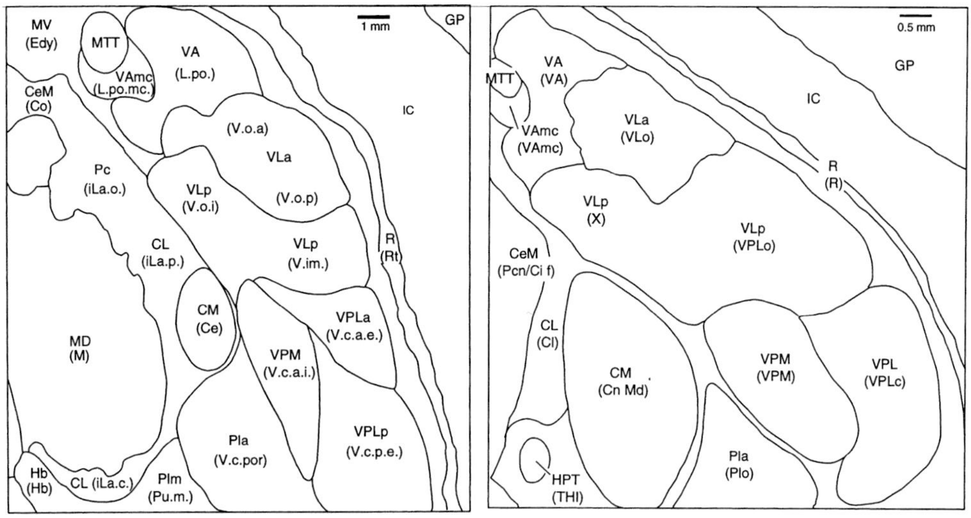

Fig. 64. Camera lucida drawings of the sections illustrated in Fig. 63. Human to left, monkey to right. These show the borders of the nuclei demonstrated cytoarchitectonically. In parentheses are the abbreviations for the names given the monkey nuclei by Olszewski (1952) and for those given the human nuclei by Hassler (1959).

1996), Rouiller et al. (1994) and Sakai et al. (1996) also confirm the cell sparse, large-celled region as the terminus of cerebellar fibers. Jones (1985a) later combined all of the large-celled, cerebellar-recipient nuclei in a single ventral lateral posterior nucleus (VLp). In the course of these investigations it was also found that the population of large cells forming the anterior pole and dorsal surface of Olszewski's VPLc and referred to in Section 3.2, received inputs specifically from muscle and other proprioceptive afferents and projected to areas 3a and 2 of the somatic sensory cortex (Fig. 65). The remainder of VPLc consisting of a more even mixture of large and small cells received inputs from cutaneous afferents and projected to areas 3b and 1 (Friedman and Jones, 1981; Jones and Friedman, 1982).

Anterior to the large celled zone of Asanuma et al. (1983a), the terminations of the cerebellothalamic fibers generally do not extend into Olszewski's VLo or VLm nuclei, although this was described in early studies (Nauta and Mehler, 1966; Kim et al., 1976). As pointed out above, small degrees of overlap of cerebellar terminations into VLo continue to be described (Asanuma et al., 1983 b,c; Percheron et al., 1993, 1996; Rouiller et al., 1994; Sakai et al., 1996). Nevertheless, Olszewski's VLo nucleus, has now been confirmed as the principal termination of pallidal afferents (Tracey et al., 1980; Anderson and De Vito 1981; Ilinsky and Kultas-Ilinsky 1987; Percheron et al., 1993; Arecchi-Bouchhioua et al., 1996) (Fig. 66). It was renamed the ventral lateral anterior nucleus (VLa) by Jones (1985a) and the densocellular ventral anterior nucleus by Ilinsky and Kultas-Ilinsky (1987). Ilinsky and Kultas-Ilinsky (1987) and Percheron et al. (1993, 1996) observed that the major site of terminations of afferents from the substantia nigra incorporated Olszewski's VA, VAmc and VLm nuclei. The last mentioned, was renamed the principal ventral medial nucleus (VMp) by Jones (1985a) and this term is used in the present account. For comparative purposes, the nomenclatures of Olszewski, of Ilinsky and Kultas-Ilinsky and of Percheron et al. are placed beside the current nomenclature in Table 3.

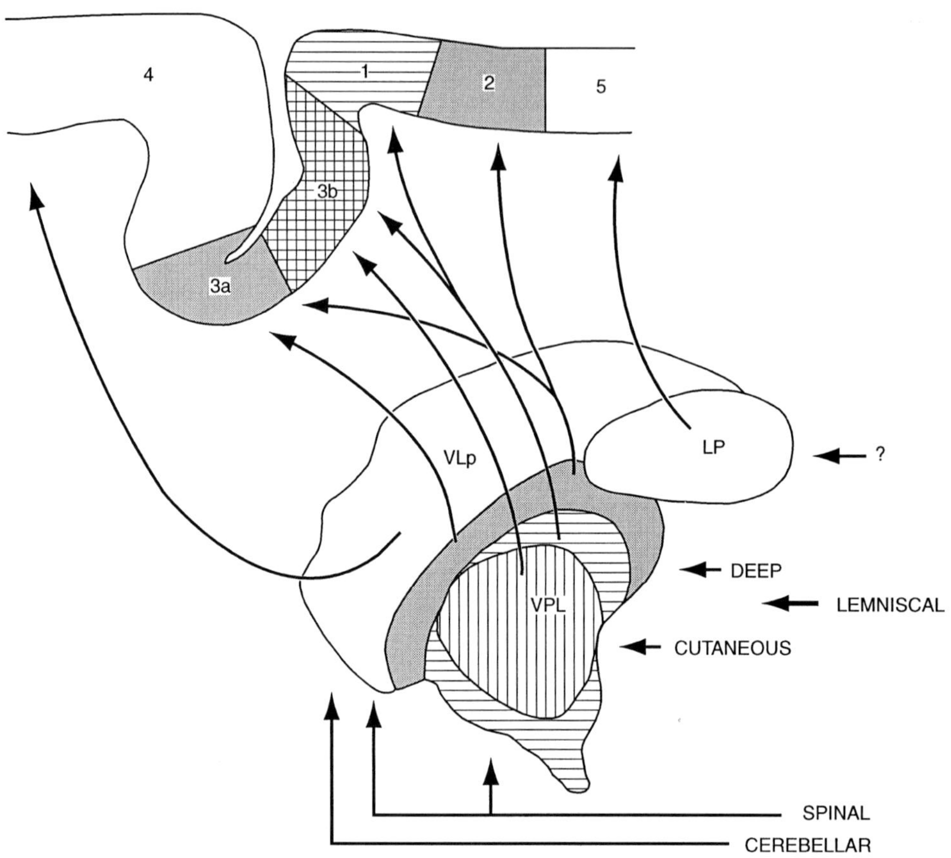

Fig. 65. Schematic figure, redrawn from Jones and Friedman (1982), showing the input-output connections of the ventral posterior nucleus (VPL) in monkeys. The anterodorsal, proprioceptive shell, receiving inputs from receptors in deep tissues, projects to areas 3a and 2, the primary muscle spindle component being focused on area 3a. The central core of the nucleus receives inputs from cutaneous receptors. The majority of its neurons project to area 3b. Others, mainly in the perimeter of the core, project to area 3b or 1. Spinal inputs terminate in VPL and in adjacent nuclei, including the VLp nucleus which is the principal target of afferents from the deep cerebellar nuclei and projects mainly to motor cortex, without overlap into area 3a.

In the human thalamus, the naming of the nuclei of the ventral nuclear mass derives from Hassler (1959) and has its roots in that of Vogt and Vogt (1941). Hassler divided the ventral nuclear mass into anterior ('polar'), dorsal, and ventral groups of nuclei (Fig. 64). Many of the nuclei were further subdivided into anterior ('oral'), posterior ('caudal'), medial ('internal'), lateral ('external') and intermediate subnuclei which were sometimes further subdivided. These divisions of divisions commonly led to the terms 'anterior' and 'oral', 'posterior' and 'caudal' or 'dorsal' and 'superior' being coupled in the name of a single nucleus (as, for example, in 'nucleus ventralis caudalis posterior externus'). Nuclei which Hassler thought formed transitions between major nuclear divisions were called 'zentro' nuclei. The resulting large array of lengthy names has little or no relationship to the list of names customarily used in experimental primates and can be a source of very great confusion to the uninitiated. Therefore, although taken up in the neurosurgical literature, little attempt has been made to use the names elsewhere.

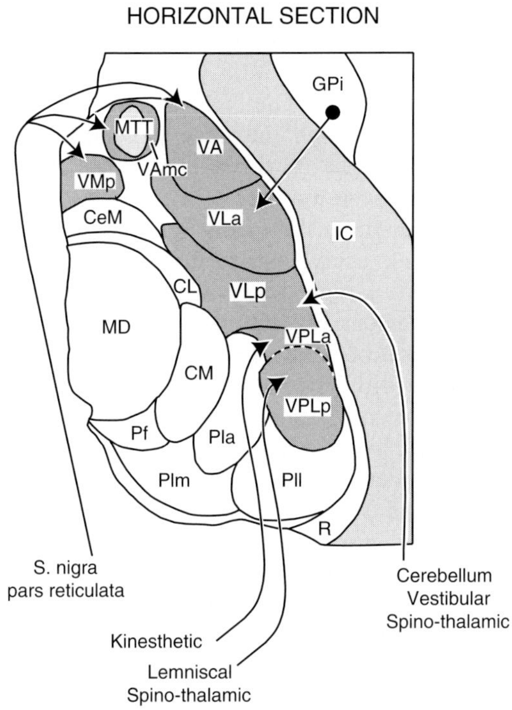

Fig. 66. Schematic horizontal section of the human thalamus, showing the principal sources of afferents to the ventral nuclei. From Macchi and Jones (1997).

Section 4.4, in which the structure of the human and monkey thalami were compared, makes it clear that it is possible to rationalize Hassler's nomenclature with that used in non-human primates, recognizing the equivalencies of nuclei in humans and other primates as one does so. Where the cyto- and chemoarchitecture of the nuclei across the species is virtually identical, recognition of equivalent ('homologous') nuclei is easy and the monkey term can be readily applied to the human nucleus. In the ventral nuclear group therefore, the VAmc nucleus of monkeys, with its large, densely staining cells and high acetylcholinesterase activity, clearly has its counterpart in the human in a nucleus with corresponding features that was called latero-polaris magnocellularis (L.po.mc) by Hassler.

VMp, a nucleus with perhaps the least well defined borders in the monkey thalamus, nevertheless occupies the same relative position and has a cyto- and chemoarchitecture essentially the same as the nucleus called in humans, ventro-oralis medialis (V.o.m) by Hassler.

There is disagreement about the exact position of the borders of VA in the monkey but all agree that VA is filled with dispersed, lightly stained, medium sized cells and has light-to moderate acetylcholinesterase staining. Thus, it is quite distinct from the immediately posterior VLa nucleus which is characterized by small, densely stained and closely packed neurons arranged in clusters and islands separated by cell-sparse regions, and showing relatively heavy acetylcholinesterase staining. Comparing the same general region of the human thalamus, VA should correspond to Hassler's nuclei dorso-oralis and latero-polaris (D.o and L.po), and VLa to his nucleus ventro-oralis (V.o) which he divided into anterior (V.o.a) and posterior (V.o.p) subnuclei. Hassler's

V.o.a is composed of islands of packed small cells which make its equivalence to VLa seem incontestable. His V.o.p is a mixed nucleus composed partly of packed small cells identical to those of his V.o.a and thus to VLa, but with medium-sized cells, similar to those of his V.im in its posterior part. The inclination of the present author is to see Hassler's V.o.p as the region in which the fingers of large cells of VLp intrude in among the small packed cells of VLa. This seems to reflect the interdigitation of VLa and VLp cells so clearly seen in the monkey.

VLp in monkeys is easily distinguished from VLa by its large, sparsely scattered, deeply staining cells and weak acetyl cholinesterase activity. On these grounds, a comparable nucleus of similarly large extent can be discerned in the human thalamus (Figs. 7, 8, 63, 64). In this extensive, cell sparse zone covered by the term, VLp, Hassler named a large number of subnuclei. Most of these can be equated on the grounds of topography alone with divisions made by Olszewski in the monkey. The antero-medial part of the human VLp, insinuated between the internal medullary lamina and VLa, corresponds to area X of Olszewski, and is called nucleus ventro-oralis internus (V.o.i) by Hassler. The dorsal part of the human VLp, which corresponds to Olszewski's VLc and VLps, is called nucleus dorso-intermedius (D.im) by Hassler. In some New World monkeys, this can show denser acetyl cholinesterase staining than the other divisions of VLp (Stepniewska et al., 1994a). The small posteriorly protruding tail, called VLps by Olszewski, can be equated with the subnucleus dorso-intermedius externus magnocellularis (D.im.e.mc) of Hassler. The ventral part of VLp, because of its very large cells, can only be equivalent to Olszewski's VPLo, and to what Hassler called the nucleus ventro-intermedius (V.im). This, he divided into medial (V.im.i) and lateral (V.im.e) subnuclei on the basis of differences in size between the medial and lateral parts of the nucleus. These may be meaningful in terms of the somatotopy that exists in the cerebellothalamic terminations (Asanuma et al., 1983b) and the evidence for somatotopy from stereotaxic interventions in this region in the human thalamus (Ohye et al., 1989; Lenz et al., 1988a,b). Hassler also added a nucleus zentro-lateralis intermedius (Z.im) which seems to represent a zone of transition between the very large cells of his V.im and the slightly smaller (although still large) cells of his D. im. That is, between the ventral and dorsal parts of VLp of the present account.

Immediately posterior to the zone of very large cells that make up the ventral part of VLp (Olszewski's VPLo and Hassler's V.im) in both monkeys and humans, is a region of mixed, very large and smaller cells in which acetyl cholinesterase staining is more dense than in VLp. This is the nucleus called VPLa in the human by Hirai and Jones (1989a). Its position relative to the largest cells of VLp makes it the same as Hassler's nucleus ventro-caudalis anterior externus (V.c.a.e). The same zone can be distinguished in the monkey where it represents the proprioceptive shell of the ventral posterior lateral nucleus (Jones and Friedman, 1982) but it is insufficiently well-defined to warrant separation off as a separate nucleus (but see Kaas et al., 1984, who in New World monkeys call it a ventral posterior superior nucleus). The nucleus called zentro-lateralis caudalis (Z.c) by Hassler and which extends posteriorly over the dorsal surface of the VPL nucleus, probably represents the dorsal extension of the proprioceptive shell found in monkeys. Posterior to VPLa, the cell population becomes more evenly mixed, the large cells becoming outnumbered by small and medium sized cells; the moderate density of acetyl cholinesterase staining is maintained. This is the VPLp nucleus, the cytoarchitecture of which corresponds exactly to the nucleus which Hassler called ventro-caudalis posterior externus (V.c.p.e). Medial to it, the clearly circumscribed

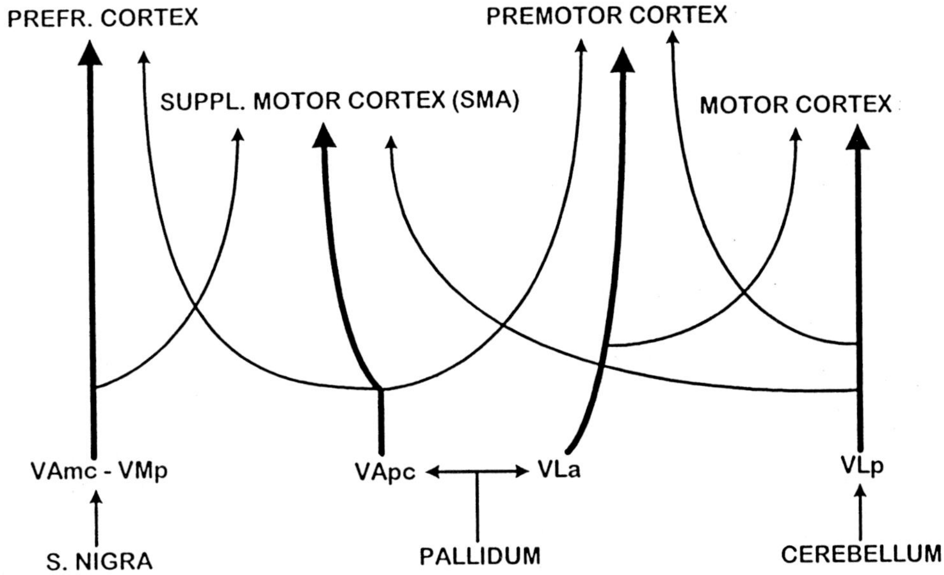

Fig. 67. Schematic figure, illustrating the principal (heavy lines) and subsidiary (thinner lines) sources of thalamic input to the motor, premotor and prefrontal (PREFR.) cortex of primates. From Macchi and Jones (1997).

nature of the VPM nucleus and its population of tightly packed, medium sized cells make it identical to Hassler's nucleus ventro-caudalis posterior internus (V.c.p.i).

Posterior and ventral to the VPL and VPM nuclei, are other components of the ventral nuclear group which are described and their names compared in sections 6.2, 10.7 and 10.8.

The lack of convergence in the terminations of the nigral, pallidal, cerebellar and lemniscal terminations makes the nuclei of the ventral complex relays for these parallel channels to the cerebral cortex. The degree of overlap in the projections of the nuclei to the cerebral cortex has not been clearly established and in the eyes of some remains controversial. Obviously, if the nuclei project without overlap to entirely separate cortical areas, the focus of convergence between the pathways is carried one step further away from the thalamus. Current evidence tends to support the view that the principal cortical projections of VA/VMp, VLa, VLp and VPL/VPM are focused on different cortical areas which may be loosely defined as prefrontal, premotor, motor and somatosensory, respectively (Fig. 67), but with overlap that is potentially quite substantial. The evidence as it stands, indicates that VA, VAmc and VMp project diffusely and to an extensive region of the neocortex concentrated in the prefrontal areas (Kievit and Kuypers, 1977; Goldman-Rakic and Porrino, 1985; Huerta et al., 1986; Huerta and Kaas, 1990; Kultas-Ilinsky and Illinsky, 1990; Barbas et al., 1991), but probably extending into the premotor cortex, including the supplementary motor area (Goldman-Rakic and Porrino, 1985; Percheron et al., 1986), frontal eye field (Huerta et al., 1986) and supplementary eye field (Huerta and Kaas, 1990), as well as into the anterior cingulate gyrus (Vogt et al., 1987) and even the posterior parietal cortex (Asanuma et al., 1985; Rouiller et al., 1994). The projections of VLa appear to be more focused and concentrated on the premotor and supplementary motor cortex but may extend substantially into the motor cortex (Kievit and Kuypers, 1977; Weisendanger and Weisendanger, 1985a; Ilinsky et al., 1985; Percheron et al., 1986; Strick, 1986; Ghosh et al., 1987; Darian-Smith et al., 1990b; Holsapple et al., 1991; Barbas et al., 1991; Matelli et al., 1989; Matelli and Luppino, 1993; Stepniewska et al., 1994b), but not into the prefrontal cortex (Ghosh et al., 1987; Huerta and Kaas, 1990).

The projection of VLp is focused on area 4 (Jones et al., 1979; Friedman and Jones, 1981; Matelli et al., 1989; Darian-Smith et al., 1990b; Matelli and Luppino, 1993; Stepniewska et al., 1994b) with a mediolateral topography that implies a body representation of face in the part of VLp equivalent to Olzewski's area X and forelimb and hindlimb successively more laterally (Asanuma et al., 1983b; Stepniewska et al., 1994b), but additional projections have been described to premotor (Kievit and Kuypers, 1977; Schell and Strick, 1984; Matelli et al., 1989; Darian-Smith et al., 1990; Matelli and Luppino, 1993) and supplementary motor areas (Matelli et al., 1989; Darian-Smith et al., 1990b; Matelli and Luppino, 1993; Rouiller et al., 1994), and to posterior parietal areas (Schmahmann and Pandya, 1990). Many of the reported differences in the published studies probably depend upon the manner in which the borders between cortical fields have been defined by the investigators, coupled with too great a reliance on atlases and a failure to recognize some of the subtleties in the boundaries between thalamic nuclei, for example the interdigitation of the cells of VLa and VLp.

Within the somatosensory relay, the VPL/VPM complex, parallel pathways to the cerebral cortex are also evident in the subdivision of at least VPL into an anterodorsal, proprioceptive shell and a cutaneous core, a division which is recognized by the division of the human VPL into cytoarchitectonically distinct VPLa and VPLp nuclei. In monkeys, the neurons of the anterodorsal shell (equivalent to VPLa) respond to inputs, including muscle spindle specific inputs, from deep tissues and project to areas 3a and 2 of the postcentral gyrus (Fig. 65). The neurons of the larger central core (equivalent to VPLp) respond to cutaneous stimuli and project to areas 3b and 1 (Poggio and Mountcastle, 1963; Friedman and Jones, 1981; Jones and Friedman, 1982; Jones et al., 1982; Jones, 1983b). A comparable division of VPM, although likely, has not yet been described.

In the human thalamus, the border between VPLa and the ventral, large celled part of VLp (Hassler's V.im) is a key landmark in placing stereotaxic lesions for the treatment of movement disorders, but it, too, has generated controversy. Single unit recording and microstimulation in the human thalamus in regions provisionally identified by atlas coordinates, with VPLa and VPLp (Hassler's V.c.a.e and V.c.p.e), and with VPM (Hassler's V.c.a.i), reveal nuronal responses to stimulation of localized peripheral receptive fields, or similarly localized peripheral paresthesiae, which follow a somatotopic order according to the part of the nucleus stimulated. A somatotopic body map can thus be demonstrated in VPM and VPL, as in monkeys, but it remains uncertain whether this encompasses both VPLa and VPLp (Hardy et al., 1980; Lenz et al., 1988a; Ohye and Narabayashi, 1979; Ohye et al., 1989). Some authors in single- and multi-unit recording studies considered that a single somatosensory map extended without interruption from V.c.a.e and V.c.p.e (VPL) into V.im.e (lateral VLp) and from V.c.a.i (VPM) into V.im.i (medial VLp) (Emmers and Tasker, 1970). Others (Ohye and Narabayashi, 1979; Ohye et al., 1989), recording with microelectrodes advanced into the ventral nuclei along an anterodorsal to posteroventral trajectory (Fig. 68), identified an anterior region in which neurons responded to deep stimuli and which they correlated with V.im (ventral VLp). This was immediately succeeded by a

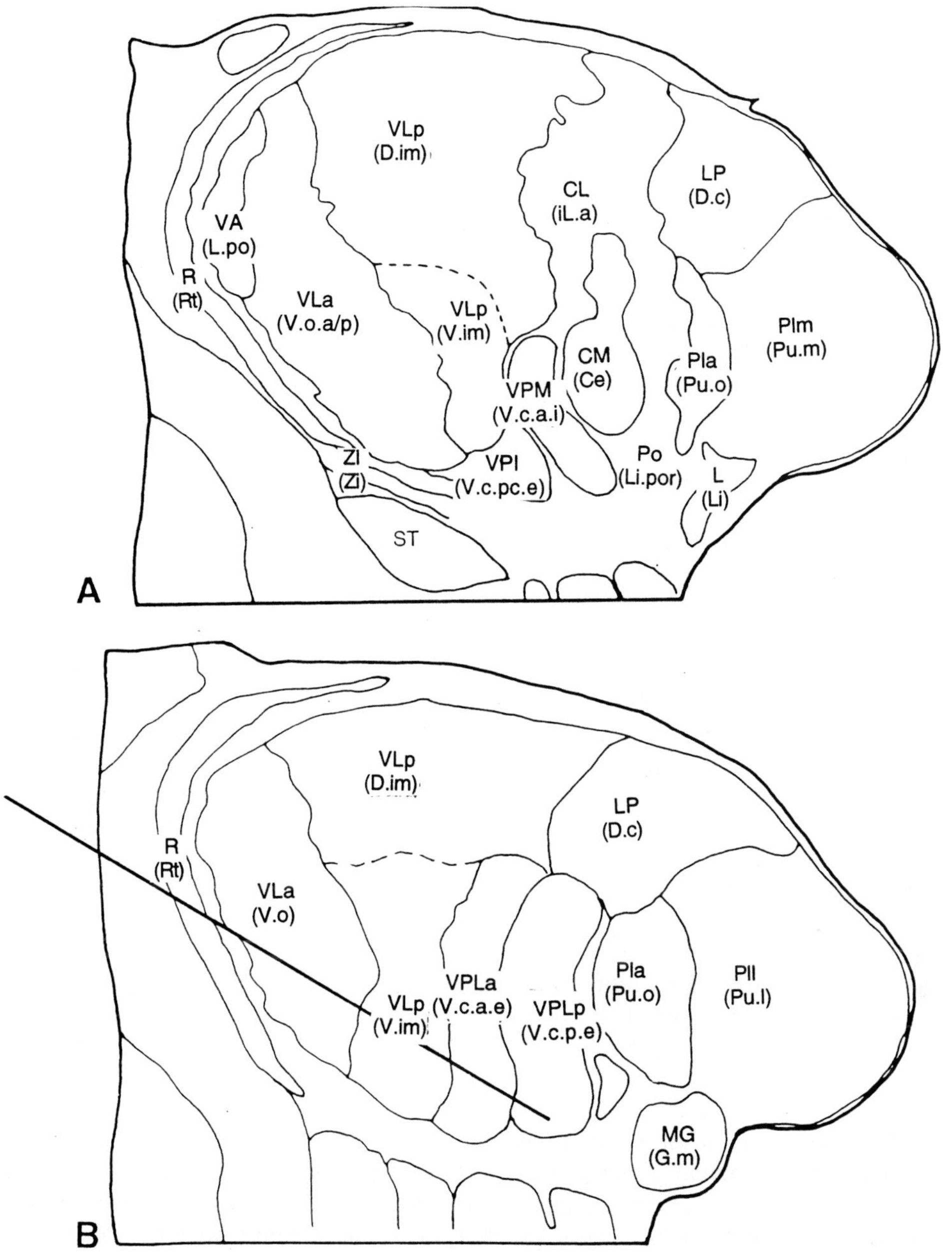

Fig. 68. Camera lucida drawings of parasagittal sections from a human thalamus (A is medial to B), showing the borders of the ventral nuclei named according to Hirai and Jones (1989a), with the nomenclature of Hassler in parenthesis. On B is drawn the trajectory of microelectrodes used in the majority of recent stereotaxic neurosurgical procedures aimed at the motor related nuclei of the human thalamus. From Macchi and Jones (1997).

region in which neurons responded to cutaneous stimuli and which they correlated with V.c.a.e and V.c.p.e plus V.c.a.i (VPLa and VPLp plus VPM). The knowledge derived from the studies of the monkey in this region immediately leads to the question whether the deep responses were actually in VPLa (equivalent to the proprioceptive shell of VPL in monkeys). In studies of Lenz et al. (1988), made with similarly oriented electrode penetrations, deep responses were found in a region described as the anterodorsal part of V.c, and thus probably equivalent to the anterodorsal shell of VPL of monkeys; cutaneous responses were encountered immediately posterior to this, in a region described as the core of V.c, and therefore probably equivalent to the central cutaneous core of VPL and VPM. V.im (ventral VLp) was described as unresponsive to somatosensory stimulation.

In anesthetized monkeys, neurons in VLp and VLa, unlike those in the proprioceptive anterodorsal shell of VPL, show little spontaneous activity and do not respond to peripheral stimuli, whereas the anterodorsal shell of VPL is exquisitely sensitive to vibratory stimuli and stimuli that stretch muscles (Poggio and Mountcastle, 1963; Friedman and Jones, 1981; Maendly et al., 1981). The shell receives only lemniscal afferents which do not overlap into VLp. VLp is the target of fibers from the deep cerebellar nuclei (Tracey et al., 1980; Asanuma et al., 1983b; Percheron et al., 1996). In unanesthertized monkeys, single unit recordings, putatively made, according to atlas coodinates, in the ventral part of VLp (Olszewski's VPLo) and from VLa (Olszewski's VLo) isolated many cells that responded to passive movements of joints (Anderson and Turner, 1991; Butler et al., 1992b; Vitek et al., 1994). In the absence of anesthesia, the neurons of VLp display quite high spontaneous firing rates, probably reflecting the tonic excitatory input from the cerebellum, and the majority respond to passive manipulations of joints and muscles, as well as to perturbations imposed on an actively moving limb (Butler et al., 1992a,b; Vitek et al., 1994). The inputs from the periphery arrive at relatively short latency (Butler et al., 1992b; Vitek et al., 1994) which in the posteroventral, large-celled part of the VLp, may reflect inputs via the spinothalamic tract rather than the cerebellum (Boivie 1979; Asanuma et al., 1983b; Berkley 1983).

Despite their rather obvious proprioceptive receptive fields, neurons in VLp generally do not discharge in relation to active movements performed by a monkey. By contrast, those in VLa, although not usually responding to passive manipulations of muscles and joints, tend preferentially to discharge during active movements (Vitek et al., 1994). Their spontaneous activity is low, probably reflecting the tonic inhibitory input from the globus pallidus.

In the thalamic regions in which stereotaxic lesions or chronic stimulation effectively eliminate tremor in humans, many neurons respond to passive movement of joints and microstimulation there can elicit sensations referred to deep tissues (Ohye, 1990; Lenz et al., 1994b; Benabid et al., 1996). In the eyes of many neurosurgeons, this area is synonymous with Hassler's V.im. The above discussion suggests that it may just as well be VPLa or a combination of the two nuclei. Both of these nuclei could show unusual or heightened neuronal responses in humans with tremor or dystonia on account of neuronal pathology (Lenz et al., 1988a,b; Lenz et al., 1990a,b; Vitek et al., 1990). This could further compound the difficulty of distinguishing between them. The posterior part of the region identified as Vim is regarded by some neurosurgeons as the best target for relieving tremor regardless of its type (Ohye et al., 1989, 1993; Ohye, 1990; Yamashiro and Tasker 1993). VPLa may, therefore, be the effective lesion site.

Stereotaxic lesions aimed at the V.o.p nucleus of Hassler (the region of interdigitation of VLa and VLp), are reported to be more effective than lesions aimed at V. im in reducing the rigidity although not the tremor of Parkinson's disease, while those aimed at V.o.i of Hassler (also part of VLa) are described as the most effective for alleviating dystonic syndromes (Hassler, 1982). These lesions are likely to interfere with thalamic nuclei receiving pallidal inputs while the more posterior, tremor-effective lesions are likely, as we have seen, to interfere with nuclei receiving proprioceptive and/or cerebellar fiber terminations.

10.2. PARCELLATION OF THE MEDIAL GENICULATE COMPLEX

The medial geniculate body of virtually all mammals is a complex of several nuclei. Three major nuclear subdivisions have been defined cytoarchitectonically in studies of New World and Old World monkeys (Jordan, 1973; Burton and Jones, 1976; Fitzpatrick and Imig, 1978; Molinari et al., 1995). They are termed the ventral, dorsal and magnocellular. The dorsal has at least two subdivisions: an anterodorsal, the cytoarchitecture of which resembles that of the ventral nucleus, and a posterodorsal which is clearly different and forms the posterior cap of the complex; it can be further divided in terms of its cortical projections.

The ventral nucleus is the primary auditory relay to the cerebral cortex (Walker, 1937b). In other species and to the extent that it has been studied in primates, its neurons show a high degree of frequency tuning and are excited reliably and at short latency; although influenced by inputs from the two ears, that from the contralateral ear is the dominant and most direct, both anatomically and physiologically (Aitkin and Webster, 1972; Semple and Aitkin, 1980; Allon, 1985; Calford and Aitkin, 1983; Aitkin et al., 1984). The inputs reach the nucleus fairly directly from the contralateral cochlear nuclei, with only one synaptic relay in the central nucleus of the inferior colliculus. The frequency specificity of the neurons forms the basis of a fine grain tonotopic organization in the nucleus (Aitkin and Webster, 1971, 1972; Gross et al., 1974; Calford and Webster, 1981; Allon et al., 1981; Imig and Morel, 1985b) and it projects to the auditory koniocortex in a manner that preserves this tonotopicity (Aitkin et al., 1988b; Morel et al., 1993; Morel and Kaas, 1992; Luethke et al., 1994; Molinari et al., 1995).

The dorsal and magnocellular nuclei receive auditory inputs that are less direct than those which reach the ventral nucleus. Various auditory brain stem nuclei contribute and their fibers converge in the dorsal and magnocellular nuclei. In the latter two nuclei, neuronal responses are more labile, frequency specificity is much less and tonotopic maps are less well-defined or absent (Aitkin, 1973; Irvine, 1980; Calford, 1983; Calford and Aitkin, 1983). The dorsal nuclei project mainly outside the primary auditory cortex. Different subnuclei appear to project to different functional areas surrounding the primary area. The magnocellular nucleus in addition to auditory inputs, receives afferent inputs from the somatosensory and vestibular systems (Poggio and Mountcastle, 1960; Blum et al., 1979a,b). It has more widespread projections than the other nuclei, including to areas well outside the auditory cortical areas, and the axons of many of its cells terminate specifically in layer I of the cortex (Jones and Burton, 1976; Hashikawa et al., 1995).

Because the borders of the nuclei in the medial geniculate complex are not perceived as clear-cut by many investigators, there has been little agreement about their relationships to the several fields of the auditory cortex which can be delimited by multiunit mapping in monkeys (Merzenich and Brugge, 1973; Imig et al., 1977; Aitkin et al., 1986; Luethke et al., 1989; Morel and Kaas, 1992; Morel et al. 1993; Kosaki et al., 1997). Most are agreed that the ventral nucleus forms the principal projection to the auditory koniocortex, but the extent of its projections to other fields is seen as zero by some but as extensive by others (Mesulam and Pandya, 1973; Burton and Jones, 1976; Fitzpatrick and Imig, 1978; Luethke et al., 1989; Morel and Kaas 1992; Morel et al.,

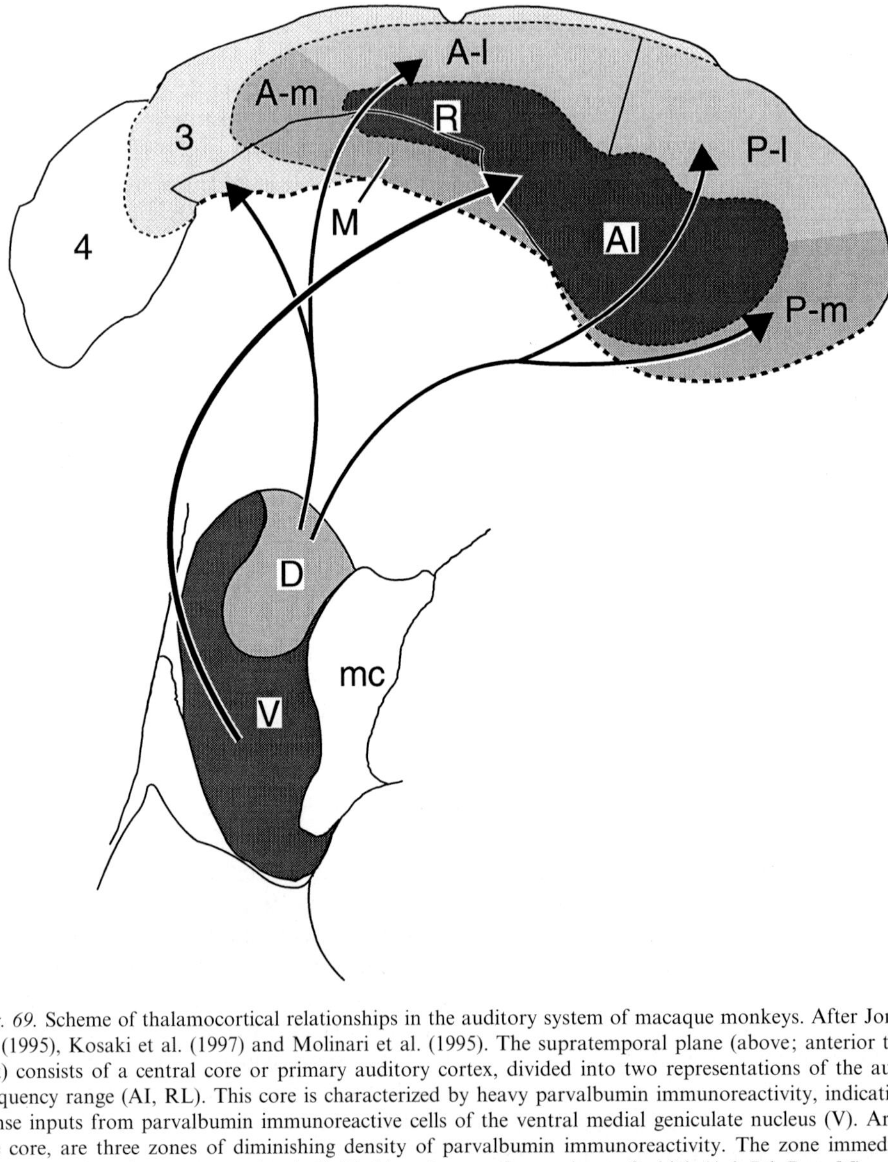

Fig. 69. Scheme of thalamocortical relationships in the auditory system of macaque monkeys. After Jones et al. (1995), Kosaki et al. (1997) and Molinari et al. (1995). The supratemporal plane (above; anterior to the left) consists of a central core or primary auditory cortex, divided into two representations of the audible frequency range (AI, RL). This core is characterized by heavy parvalbumin immunoreactivity, indicative of dense inputs from parvalbumin immunoreactive cells of the ventral medial geniculate nucleus (V). Around the core, are three zones of diminishing density of parvalbumin immunoreactivity. The zone immediately adjacent to the primary core region is divided into 5 areas, four at least of which (A-l, P-l, P-m, M) contain tonotopically organized representations. They and the outer zone (3), receive thalamic inputs characterized by strong contributions from calbindin immunoreactive cells in the dorsal nuclei (D) of the medial geniculate complex. The magnocellular nucleus (mc) has widespread projections to all fields.

1993; Molinari et al., 1995). There has been similar disagreement about the degree of specificity in the manner in which projections from the dorsal nuclei target particular fields, although most agree that their major projections are to areas around the primary koniocortical core area. All are agreed that the magnocellular nucleus has widespread projections to these and to other fields.

Another reason for discrepancy in the reports of different studies, apart from the difficulty of precisely locating the borders of the nuclei, may be the fact that these nuclei, like other relay nuclei of the dorsal thalamus in primates, contain two populations of relay cells: a larger, parvalbumin immunoreactive population which tends to project to a particular field or fields of the cerebral cortex and a second, less numerous, calbindin immunoreactive population of small cells that project diffusely to the cortex, unconstrained by architectonic borders (Avendaño et al., 1990; Hashikawa et al., 1991; Rausell and Jones, 1991a; Rausell et al., 1992a; Diamond et al., 1993; Molinari et al., 1995) (Fig. 69). Retrograde labeling of the second population following injection of a retrograde tracer in the cerebral cortex will tend to blur the precise pattern of projection of the major cell population.

Parvalbumin cells project to middle layers of the appropriate auditory cortical field and calbindin cells to superficial layers, as in other cortical areas. This difference, and particularly the relative density of parvalbumin fibers ending in each cortical area, imposes a unique chemoarchitecture on the various fields of the monkey auditory cortex (Jones et al., 1995). A central core region shows extremely dense parvalbumin immunoreactivity, especially in the thalamic fiber termination layers (IV and deep III). This core region is made up of two fields (AI and R), which can be delimited by tonotopic mapping (Morel et al., 1993; Kosaki et al., 1997). In these fields, neurons are highly frequency specific, and tuned at threshold to a single frequency. Both of the fields receive most of their thalamic input from cells in the ventral medial geniculate nucleus. Surrounding the central core but still located on the surface of the supratemporal plane, is a region of somewhat reduced, although still dense parvalbumin imunostaining in which at least four fields can be identified by differences in the character of the immunostaining and by tonotopic mapping. The responses of the cells, although sufficiently tuned to enable tonotopic maps to be made, are less specific and the cells respond to a range of frequences at threshold. The four fields so mapped are referred to as the anterolateral, posterolateral, posteromedial and medial fields (Kosaki et al., 1997). The posteromedial is divisible on physiological grounds into lateral and medial divisions. The chemoarchitectonic fields correspond to fields identified under different names in earlier studies (Merzenich and Brugge, 1973; Morel et al., 1993; Rauschecker et al., 1994). Surrounding these fields is a zone of much reduced parvalbumin immunostaining which extends out onto the superior temporal gyrus and in which neurons are less responsive to auditory stimuli and tonotopic maps cannot be revealed. Beyond this parvalbumin weak zone is a further zone, mainly located on the superior temporal gyrus and at the tip of the temporal pole in which parvalbumin fiber staining is virtually absent. Its neurons are not responsive to auditory stimuli. The fields in the zone immediately surrounding the primary core zone and some parts at least of the weak zones surrounding them in their turn, receive their thalamic inputs from the dorsal nuclei of the medial geniculate complex, with different subdivisions of these nuclei targeted on different cortical fields.

In the medial geniculate complex of monkeys, there is a strong complementarity in the distribution of parvalbumin and calbindin immunoreactive cells (Fig. 70). Although no nucleus is completely free of parvalbumin or calbindin cells, the relative

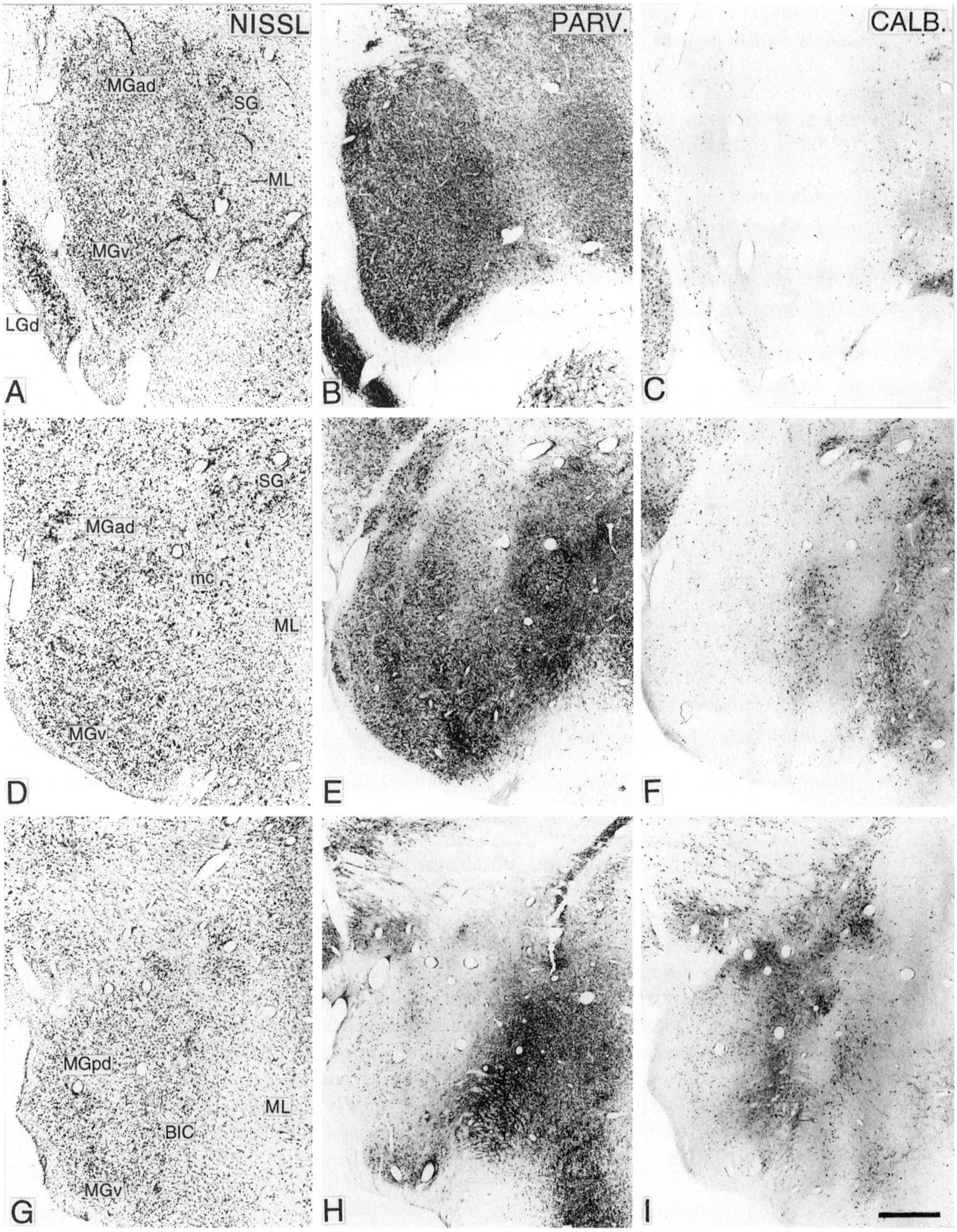
NISSL
PARV.
CALB.
MGad
SG
ML
MGv
LGd
A
B
C
SG
MGad
mc
ML
MGv
D
E
F
MGpd
ML
BIC
MGv
G
H
I

Fig. 70. Trios of adjacent frontal sections through anterior (A-C), middle (D-F) and posterior (G-I) levels of the medial geniculate complex of a Japanese monkey (*M. fuscata*), stained with thionin (NISSL) or immunocytochemically for parvalbumin (PARV.) or calbindin (CALB.). Note the enhanced numbers of parvalbumin cells in the ventral medial geniculate nucleus. These decline in number and are replaced by calbindin cells in the dorsal nuclei, especially in the posterior part of the posterodorsal nucleus (MGpd). Bar 1 mm. From material illustrated in Molinari et al. (1995).

numbers of the two cell types and the density of immunostaining of the neuropil give each of the nuclei of the complex a particular chemoarchitectonic signature. The ventral nucleus has a very high density of parvalbumin cells and very few calbindin cells; the anterodorsal nucleus has more calbindin cells but parvalbumin cells still predominate; the posterodorsal nucleus has approximately equal numbers of parvalbumin and calbindin cells but, unlike the ventral and anterodorsal nuclei, there is little parvalbumin immunostaining of the neuropil; in the posterior part of the posterodorsal nucleus, where it forms a cap to the posterior pole of the medial geniculate complex, calbindin cells predominate; the magnocellular nucleus is made up of domains in which clusters of one or other cell type predominate.

The differential immunostaining of the nuclei of the medial geniculate complex is correlated with their projections to the auditory cortical areas that display different densities of fiber immunostaining for parvalbumin (Fig. 69). The ventral nucleus in which parvalbumin cells predominate, projects primarily to the central core zone, i.e. fields AI and R, in which parvalbumin immunoreactivity is densest. Both of these fields receive a small input from the magnocellular nucleus, this input is mainly made up of parvalbumin immunoreactive fibers. Any small additional inputs from the anterodorsal and posterodorsal nuclei are invariably calbindin immunoreactive.

The zone of less dense parvalbumin immunostaining surrounding the two fields of the central core receives its thalamic input from the anterodorsal and posterodorsal nuclei. The anterodorsal nucleus has a much greater number of parvalbumin immunoreactive cells than the posterodorsal nucleus and projects mainly to the posteromedial field in which parvalbumin immunoreactivity in layer IV is denser than in the other fields (anteromedial, medial and posterolateral) of the surround. The posterodorsal nucleus has the fewest parvalbumin cells and projects mainly to the anterolateral, posterolateral and medial fields in which parvalbumin immunoreactivity in layer IV is weaker than in the posteromedial field. The posterior pole of the posterodorsal nucleus in which parvalbumin cells are virtually absent, projects to anterior regions of the temporal operculum in which parvalbumin immunoreactivity is at its weakest.

Fibers arising in the ventral medial geniculate nucleus and ending in fields AI and R, terminate in large numbers in layer IV and in the deep portion of layer III (layer IIIB), with a less dense lamina of terminations at the junction of layers V and VI (Fig. 71). Fibers arising in the dorsal nuclei and terminating in fields outside the central core zone, end mainly in layer IIIB. The terminal plexus in these areas, is less dense in comparison with that in the central core zone (Jones and Burton, 1976a; Hashikawa et al., 1995). Fibers arising in the magnocellular medial geniculate nucleus, appear to consist of two types with different laminar distribution patterns; one forms terminal plexuses in layers IV through IIIB which resemble those of fibers arising in the ventral nucleus; the other ends exclusively in layer I with the fibers extending over a wide region and showing relatively little branching (Fig. 71) (Hashikawa et al., 1995). Because the magnocellular nucleus contains clusters of parvalbumin and calbindin

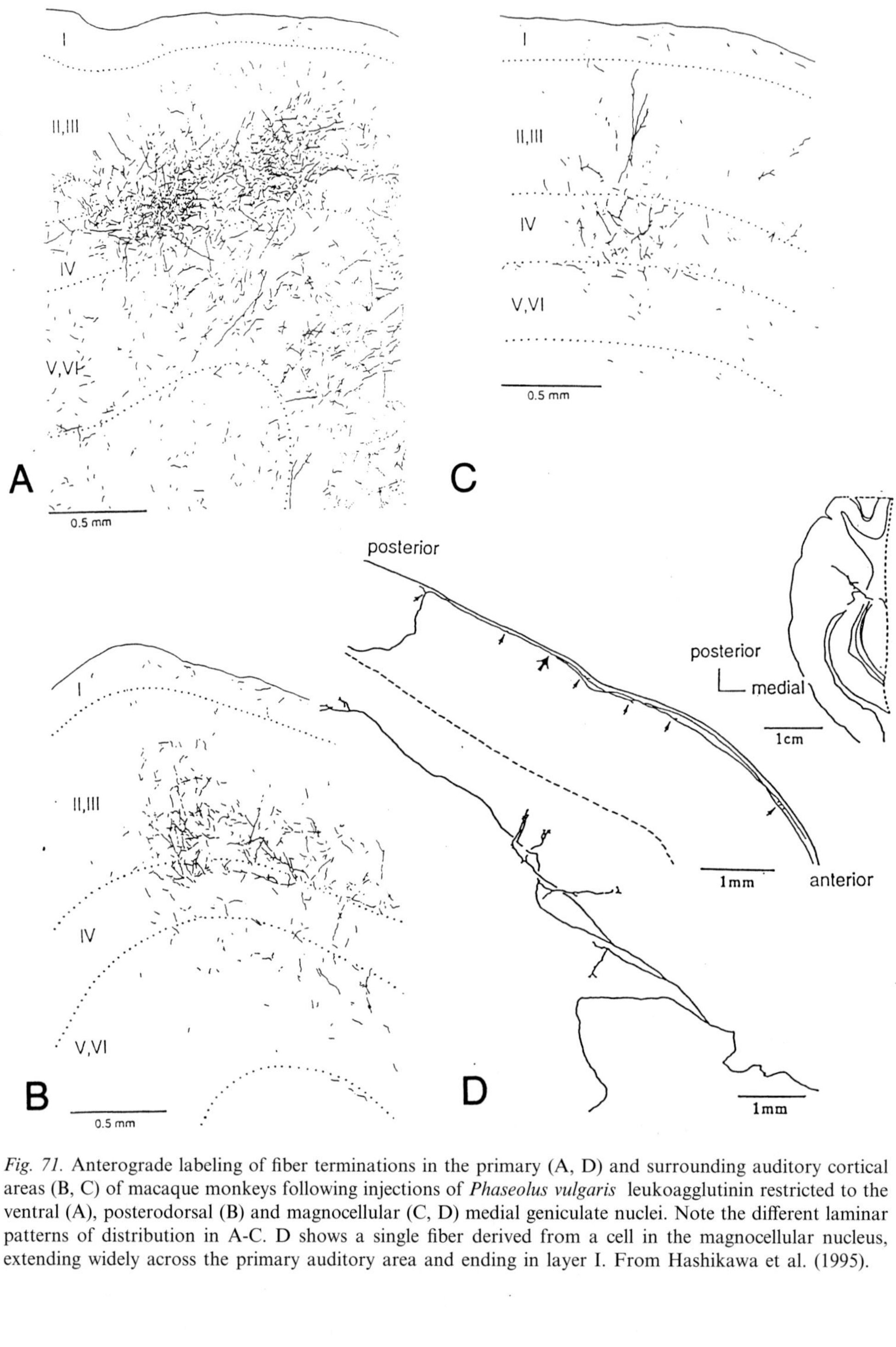

Fig. 71. Anterograde labeling of fiber terminations in the primary (A, D) and surrounding auditory cortical areas (B, C) of macaque monkeys following injections of *Phaseolus vulgaris* leukoagglutinin restricted to the ventral (A), posterodorsal (B) and magnocellular (C, D) medial geniculate nuclei. Note the different laminar patterns of distribution in A-C. D shows a single fiber derived from a cell in the magnocellular nucleus, extending widely across the primary auditory area and ending in layer I. From Hashikawa et al. (1995).

cells, it is likely that the parvalbumin neurons project to middle cortical layers and the calbindin cells to layer I (Hashikawa et al., 1991; Molinari et al., 1995).

The differential projections of the parvalbumin and calbindin cells in the medial geniculate complex of the monkey suggests that there are parallel pathways for auditory information transfer in the geniculocortical system. A parvalbumin pathway passes through the ventral nucleus to the primary core of the auditory cortex; others, also probably mostly parvalbumin positive, pass through the dorsal nuclei to surrounding areas. Added to these is a further parallel pathway which seems to dominate the posterodorsal projections but which is found in the projections of the other nuclei as well; this is made up of calbindin fibers directed to superficial layers of all areas.

The population of calbindin cells in any of the medial geniculate nuclei appears to form a background matrix to the complex, as in the ventral posterior and adjacent nuclei, and its cells project to the auditory cortex in a manner that is unconstrained by areal borders, the fibers terminating predominantly in superficial layers, including layer I (Hashikawa et al., 1991). When that field is injected with anatomical tracers, the diffusely projecting calbindin cells account for the presence of retrograde labeling in nuclei other than that which provides the predominant input to the field.

Parallel parvalbumin and calbindin pathways can also be discerned in the inputs to the nuclei of the medial geniculate complex. The ventral and anterodorsal nuclei receive the terminations of a densely parvalbumin immunoreactive pathway ascending in the brachium of the inferior colliculus from the central nucleus of the inferior colliculus (unpublished observations). The central nucleus provides the predominant afferent input to the ventral and anterodorsal medial geniculate nuclei, at least in non-primates (Kudo and Niimi, 1978, 1980; Andersen et al., 1980b; Calford and Aitkin, 1983; Aitkin et al., 1984; Jones, 1985a; Aitkin, 1986). The lateral lemniscus of the monkey is also densely parvalbumin immunostained (unpublished observations), suggesting that the primary auditory pathway from cochlear nuclei up to auditory cortex is parvalbumin positive. The posterodorsal nucleus, in which the population of calbindin cells is greatest and the parvalbumin population least, lacks parvalbumin staining of its neuropil and calbindin staining predominates instead, especially posteriorly. The fibers that contribute to this neuropil staining enter the medial geniculate complex through the peripeduncular nucleus from the lateral tegmental region. Their origins are undetermined but, by comparison with other species, they are likely to arise in the pericentral and/or dorsal nuclei of the inferior colliculus (Calford and Aitkin, 1983; Aitkin, 1986) and possibly in lower auditory brainstem nuclei as well (Irvine, 1986; Oliver, 1987; Casseday et al., 1989).

The medial geniculate complex of the Old World monkey has received little attention physiologically. Of the comparable nuclei in the cat, which have been extensively studied, the neurons of the ventral medial geniculate nucleus are characterized by short latency, and highly frequency-specific responses to tonal stimuli, while neurons in the dorsal nuclei have a much wider range of properties which include longer latency responses, little or no frequency tuning, lability or lack of responsiveness to pure tone stimuli and complex mixed properties (Aitkin and Webster, 1972; Calford, 1983; Calford and Webster, 1981; Irvine, 1986). A limited amount of work available from New World monkeys, suggests that the neuronal properties of the nuclei in primates are similar to those found in the cat (Allon and Wollberg, 1981; Allon et al., 1981). Therefore, the sharp frequency tuning of neurons in the AI and R fields of the auditory cortex undoubtedly reflects the fact that these areas receive their densest thalamocortical projections from the ventral nucleus (Molinari et al., 1995). The prop-

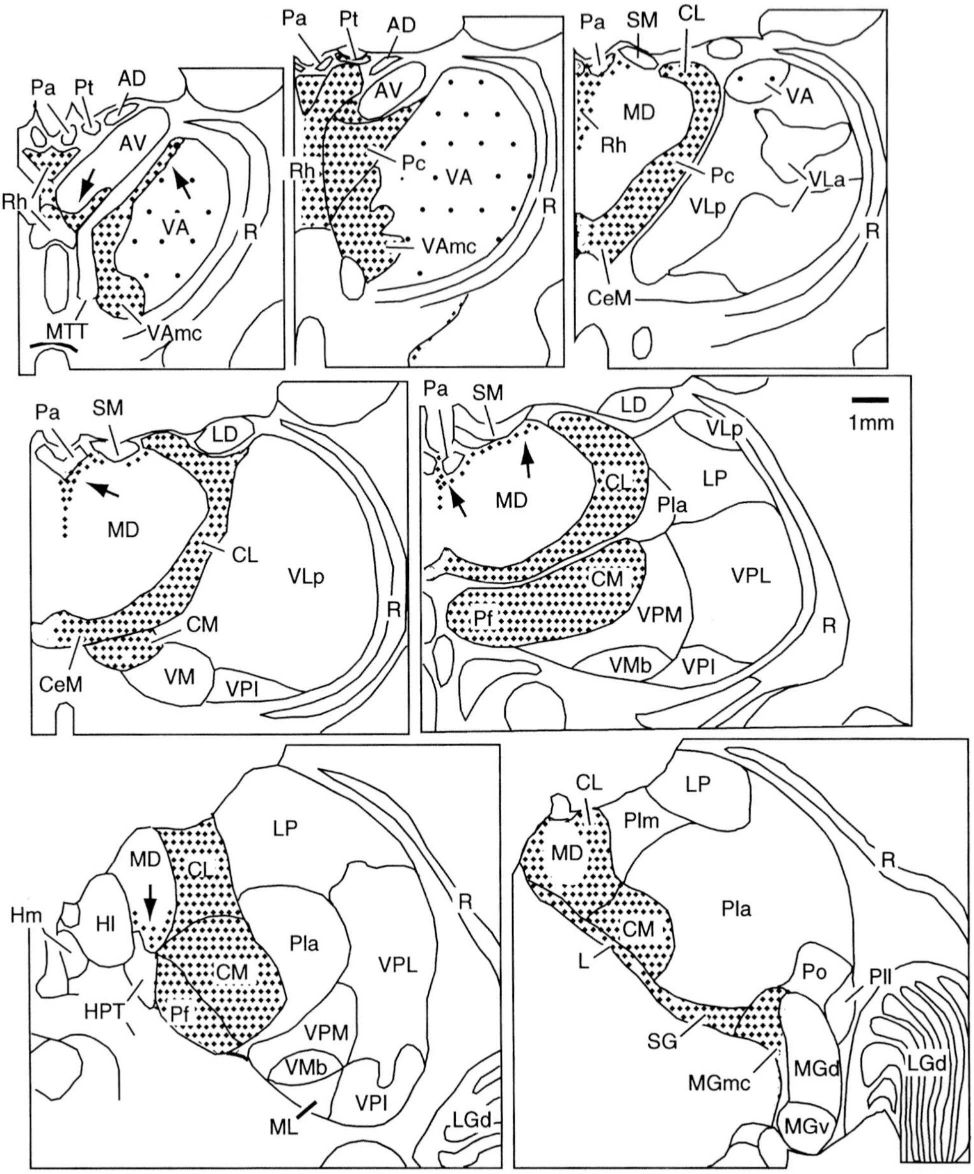

Fig. 72. A series of frontal sections in anterior (top left) to posterior (lower right) order, showing (dots) the distribution of cells projecting to the striatum in macaque monkeys, as determined from retrograde labeling experiments. Arrows indicate cells located outside the confines of the classical intralaminar nuclei. Modified from Jones (1989c).

erties of dorsal nuclei cells are reflected in the broad tuning and preference for complex sounds exhibited by neurons in the areas immediately surrounding R and AI (Rauschecker et al., 1994; Kosaki et al., 1997) which receive their principal thalamocortical inputs from the dorsal nuclei and are the targets of convergent corticocortical inputs from the AI and R fields (Jones et al., 1995).

10.3. PATHWAYS THROUGH THE INTRALAMINAR NUCLEI

10.3.1. Introduction

The intralaminar nuclei in both monkeys and humans can be readily identified by their dense histochemical staining for acetylcholinesterase (Figs. 7, 8). The unifying feature of the intralaminar nuclei, however, is the presence of a projection to the striatum (Jones, 1989) (Fig. 72). The presence of a striatal projection is justification for including nuclei that are not so obviously contained within the internal medullary lamina, for example the magnocellular ventral anterior and the suprageniculate/limitans nuclei, in the intralaminar system. Not all intralaminar cells, however, project to the striatum and some striatally-projecting cells may be found in other nuclei outside the boundaries of the acetyl cholinesterase stained complex. Furthermore, the intralaminar nuclei although traditionally not showing major retrograde degeneration after cortical lesions in monkeys (Powell and Cowan, 1967), nevertheless do have a significant projection to the cerebral cortex (Jones and Leavitt, 1974).

In humans, the posterior group of intralaminar nuclei (the centre médian and parafascicular) are traditionally associated with the thalamic projection to the putamen, since vascular lesions of the putamen led to retrograde cell loss in those nuclei (Le Gros Clark and Russell, 1939; Vogt and Vogt, 1941). The anterior group of intralaminar nuclei, on the basis of the distribution of retrograde degeneration subsequent to lesions, was traditionally related to the caudate nucleus (Powell and Cowan, 1956). The limbic part of the striatum, i.e. the nucleus accumbens and cell groups linking it to the other components of the striatum, from similar human studies, was thought to be the target of the paracentral nucleus of the anterior intralaminar group (Hassler, 1959; Hemphill et al., 1981). Recent studies in monkeys show much more overlap in the projections of the anterior and posterior groups of intralaminar nuclei within the divisions of the striatum (Fenelon et al., 1991; Smith and Parent, 1986; Nakano et al., 1990; Francois et al., 1991; Sadikot et al., 1992a). The principal organizing feature of the projection appears to be one in which intralaminar nuclei which are the recipients of corticothalamic fibers arising in a particular cortical area or areas, project to the part(s) of the striatum that receive corticostriatal fibers from the same cortical areas (Fig. 73). Hence, the striatal projections of the centre médian nucleus are focused on a part of the putamen dominated by corticostriatal fibers from the sensory-motor areas (Künzle, 1975, 1977; Jones et al., 1977; Malach and Graybiel, 1986; Flaherty and Graybiel, 1991, 1994) (the 'sensorimotor striatum') and those of the parafascicular nucleus are focused on a region of the caudate nucleus dominated by inputs from the prefrontal cortex (Künzle and Akert, 1977; Yeterian and Van Hoesen, 1978; Van Hoesen et al., 1981; Selemon and Goldman-Rakic, 1985) (the 'associative striatum'). The projections of the anterior group of intralaminar nuclei overlap from caudate nucleus into putamen and seem to be mostly associated with the territory receiving corticostriatal fibers from the parietotemporal and cingulate cortex. The full extent of their terminations has not been mapped in the striatum of primates but because the projections of the centre médian and parafascicular nuclei, together, fill most of the caudate nucleus and putamen, overlap can be expected to be considerable. The nucleus accumbens and adjacent regions (the 'limbic striatum') may receive their principal thalamic input from regions ventral to the parafascicular nucleus (Sadikot et al., 1992a) or from the midline groups of intralaminar nuclei (Gimenez-Amaya et al., 1995).

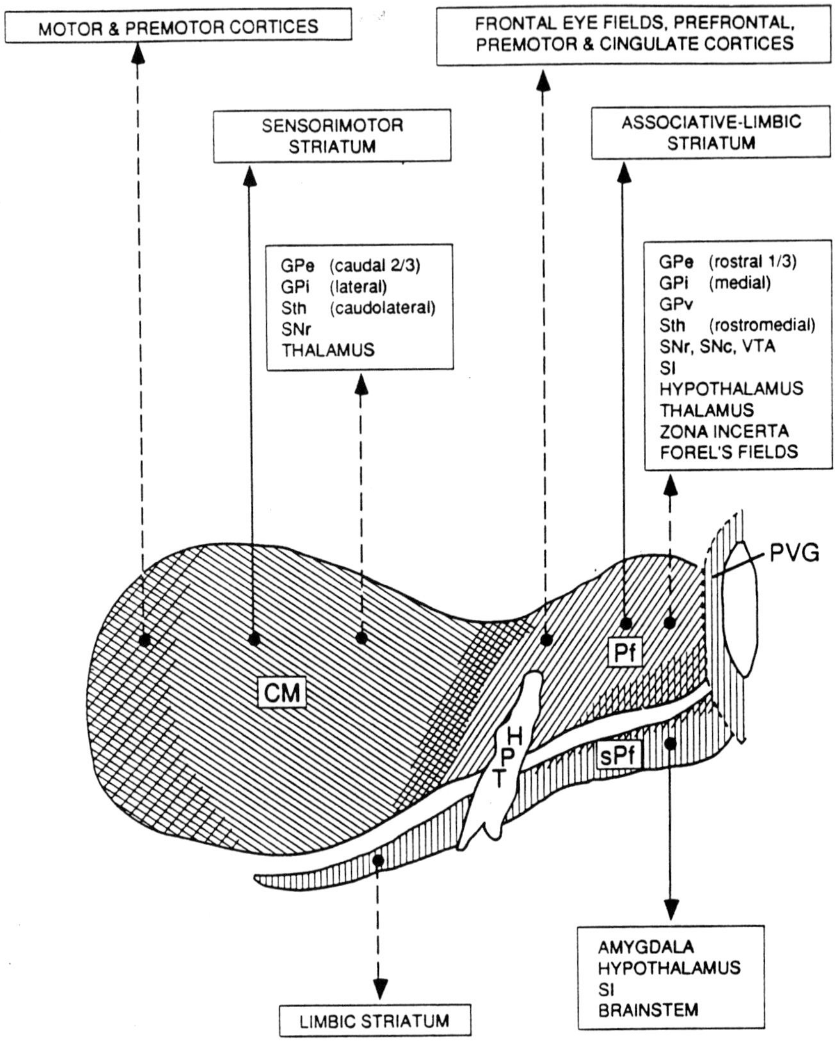

Fig. 73. Schematic view of the organization of striatal projections from the posterior intralaminar nuclei (CM, Pf) in monkeys. From Sadikot et al. (1992a).

Intralaminar neurons projecting to the striatum outnumber by far those projecting to the cerebral cortex and few intralaminar cells project by branched axons to both cortex and striatum, at least in non-primates (Steriade and Glenn, 1982; Royce, 1983; Macchi et al., 1984; Sadikot et al., 1992a). Thalamic fiber terminations in the striatum are focused in patches (Kalil, 1978; Royce, 1978a; Herkenham and Pert, 1981; Ragsdale and Graybiel, 1988, 1991; Sadikot et al., 1992b; Beckstead, 1984; Deschênes et al., 1996) that tend to coincide with zones of enhanced calbindin, acetyl cholinesterase and NADPH-diaphorase staining – the so-called matrix of the striatum, a region which also shows particular relationships to the terminations of afferents from the cerebral cortex and substantia nigra and to the locations of efferent cells (Reviewed in Selomon and Goldman-Rakic, 1985; Graybiel, 1990; Gimenez-Amaya and Graybiel, 1990, 1991).

In the matrix compartment, corticostriatal and thalamostriatal fibers and dopaminergic fibers from the pars compacta of the substantia nigra converge on the medium spiny relay cells (Kitai et al., 1976a,b; Kocsis et al., 1977; Van der Maelen and Kitai, 1980). In monkeys, the corticostriatal and dopaminergic fibers terminate in close proximity to one another, the terminals of the former being especially concentrated on the dendritic spines. Thalamostriatal synapses, which tend to be deconcentrated on dendritic shafts, have less intimate associations with the dopaminergic terminals (Smith et al., 1994). Both the cortical and thalamic inputs to the striatum are excitatory (Purpura and Malliani, 1967; Wilson et al., 1983) and glutamatergic (Divac et al., 1977; Reubi and Cuénod, 1979; McGeer et al., 1977; Hassler et al., 1982; Young and Bradford, 1986; Streit, 1980), but the closer association of the corticostriatal and dopaminergic innervation may have special significance in terms of modulation of cortical influences by dopamine. Disruptions of this relationship have been postulated as one of the key factors in the cognitive disturbances of schizophrenia (reviewed in Grace, 1991).

Other subcortical targets of projections from the intralaminar nuclei are reported to be the amygdala (Mehler et al., 1981; Royce and Mourey, 1985; Russchen, 1982; Sadikot et al., 1992a), substantia innominata (Russchen et al., 1985; Grove, 1988; Sadikot et al., 1992a), subthalamic nucleus (Royce and Mourey, 1985; Feger et al., 1994; Sadikot et al., 1992a), globus pallidus (Kincaid et al., 1991; Sadikot et al., 1992a) and claustrum (LeVay and Sherk, 1981; Kaufman and Rosenquist, 1985).

Retrogradely labeled cells have often been reported in nuclei outside the confines of the traditional intralaminar nuclei after injections of anatomical tracers in one or other part of the striatum. Commonly, the outlying cells ocupy 'midline' nuclei that represent unrecognized extensions of the central medial, rhomboid and central lateral nuclei. In other cases, the labeled cells occupy nuclei such as the so-called densocellular part of the mediodorsal nucleus, the magnocellular ventral anterior nucleus, the suprageniculate/limitans nucleus or the magnocellular medial geniculate nucleus, all of which can with justification be included in the intralaminar system (Heath and Jones, 1971b; Takada et al., 1985; Jones, 1985, 1989; Huntsman et al., 1996; Fenelon et al., 1991). Occasional cells, however, continue to be labeled in the principal ventral anterior nucleus and the pulvinar. Fig. 72 shows the full extent of the nuclei in which thalamostriatal cells are consistently demonstrated in monkeys.

10.3.2. Cortical projections of the intralaminar nuclei

The cortical projection of the intralaminar nuclei, mainly as the result of work on the mouse (Lorente de Nó, 1938), was thought to teriminate in layer I of the cortex and to be responsible for the recruiting response, a long latency, high voltage, repetitively waxing and waning, surface negative potential that can be elicited from large areas of the cerebral cortex following low frequency electrical stimulation in the intralaminar nuclei (Morison and Dempsey, 1942; Jasper, 1949, 1960). Available experimental methods, however, while demonstrating a substantial, topographically organized projection from the cortex *to* the intralaminar nuclei, were not able to identify an intralaminar-to-cortex projection (Powell and Cowan, 1967). The demonstration of a cortical projection in monkeys and other species came with the introduction of the retrograde tracing methods (Jones and Leavitt, 1974) and was soon confirmed (Kievit and Kuypers, 1975; Macchi et al., 1977; Itoh and Mizuno, 1977; Divac et al., 1977; Kennedy and Baleydier, 1977; Stanton et al., 1977; Macchi et al., 1984; Kasdon and

Jacobson, 1978; Miller and Benevento, 1979; Carey et al., 1979a,b; Hendry et al., 1979; Mizuno et al., 1981; Bentivoglio et al., 1981; Meyer and Albus, 1981; Niimi et al., 1981a,b, 1983; Tigges et al., 1982; Tong et al., 1982; Moran et al., 1982; Miyata and Sasaki, 1983; Royce, 1983; Brysch et al., 1984; Macchi et al., 1984, 1986; Royce and Mourey, 1985; Ullan, 1985; Cunningham and LeVay, 1986; Macchi and Bentivoglio, 1986; Smith and Parent, 1986; Royce et al., 1989; Macchi, 1990; Darian-Smith et al., 1990; Vogt and Sikes, 1990a,b; Gingold et al., 1991; Berendse and Groenewegen, 1991; Minciacchi et al., 1993).

Apart from GABAergic interneurons which form approximately 30% of the neuronal population of the intralaminar nuclei in monkeys (Hunt et al., 1991) (Fig. 16), the intralaminar nuclei consist of largely separate populations of striatally- and cortically-projecting cells. In other animals, the striatally-projecting group is the largest (Macchi and Bentivoglio, 1986) and few cells have branched axons to both striatum and cortex (Jinnai and Matsuda, 1981; Royce, 1983; Steriade and Glenn, 1982; Sadikot et al., 1992a). There is no reason to doubt that it is similar in primates. Emphasis has been given to the idea that the cortical projection terminates primarily in layer I of the cortex (e.g. Jones, 1975a; Glenn et al., 1982), but this has never been reliably confirmed in monkeys and in other species, at least, layer VI terminations have been described as well (Herkenham, 1978; Rieck and Carey, 1985; Rausell and Avendaño, 1985; Avendaño et al., 1990). The cortical projection is not particularly diffuse. Clusters of cells in the intralaminar nuclei project to relatively localized zones of cortex, and relatively few cells branch widely to innervate highly disparate regions (Steriade and Glenn, 1982; Berendse and Groenewegen, 1991; Macchi et al., 1984; Bullier et al., 1984; Kennedy and Bullier, 1985; Kaufman and Rosenquist, 1985; Molinari et al., 1986; Birnbacher and Albus, 1987; Deschênes et al., 1996). From work mainly in cats, the anterior group of intralaminar nuclei project to foci limited to a few functional areas in prefrontal, cingulate, parieto-temporal, entorhinal and prepiriform areas. The central medial nucleus provides most of the projection to the medial and basal areas of the cerebral cortex and the paracentral and central lateral nuclei provide most of the projection to lateral areas. In monkeys, this includes most of the superior and inferior parietal lobules (Pearson et al., 1978) as well as the entorhinal and periamydaloid cortex and hippocampus (Amaral and Cowan, 1980; Insausti et al., 1987). The central lateral and centre médian nuclei provide the inputs to the primary somatosensory and motor areas and the anterior parietal cortex but the projection to the somatosensory areas is weaker than that to the motor areas and much weaker than that to anterior parietal cortex (Royce and Mourey, 1985; Royce et al., 1989). The parafascicular nucleus seems to provide most of the intralaminar projection to deeply placed areas around the rhinal sulcus and to the cingulate areas (Jones and Leavitt, 1974; see also Royce and Mourey, 1985) although in monkeys, prefrontal and premotor areas may also be targets (Kievit and Kuypers, 1977; Vogt et al., 1987; Leichnetz, 1986; Huerta and Kaas, 1990; Huerta et al., 1986). Although all areas of the cerebral cortex probably receive an intralaminar projection, the density of this projection is less in the primary sensory areas, and intralaminar projections to primary visual cortex have been described by some workers but not by others. The principal auditory areas and the adjacent insular cortex do not receive a projection from the classical intralaminar nuclei and in these areas, the suprageniculate/limitans and possibly the magnocellular medial geniculate nuclei may provide comparable projections (Jones, 1985a, 1989; Hashikawa et al., 1995). There is some evidence in non-primates that within the intralaminar nuclei-to-cortex projection there may be two types of fiber, one terminating in the middle layers of the cortex and the other in layer I (Molinari et al., 1993, 1994). This is found in the magnocellular medial geniculate projection in monkeys (Hashikawa et al., 1995). The presence of cortically-projecting intralaminar cells with different conduction velocities (Steriade and Glenn, 1982) and with different chemical characteristics (see below) may be indicative of layer-specific projections.

10.3.3. Compartmental organization of the intralaminar nuclei

With the exception of the centre médian nucleus, all of the intralaminar nuclei of monkeys and humans are dominated by large, deeply staining cells. But they are not homogeneous. The central medial and parafascicular nuclei show little variation in cell size but others are more diverse. The central lateral nucleus consists of islands of small and medium sized cells anteriorly but has clusters of very large cells posteriorly. Similar variations can be seen in some of the other nuclei.

The cell populations of the anterior and posterior intralaminar nuclei in monkeys are distinguished by differences in calcium-binding protein immunoreactivity (Jones and Hendry, 1989; Molinari et al., 1993, 1994) (Fig. 24), and in the distribution of GABAergic cells (Hunt et al., 1991). Calbindin cells dominate the anterior group of intralaminar nuclei and the magnocellular ventral anterior and limitans/suprageniculate nuclei. There are pockets of parvalbumin cells in some of the nuclei. In the parafascicular and centre médian nuclei, however, all relay cells are parvalbumin immunoreactive and calbindin cells are excluded (Jones and Hendry, 1989). Since no GABA cells in the intralaminar nuclei of monkeys are immunoreactive for either of the calcium binding proteins, the parvalbumin and calbindin cells are all relay cells. There is no reason to believe, however, that the proteins uniquely specify cortically or striatally projecting relay cells.

The anterior group of intralaminar nuclei is dominated by inputs from the three deep cerebellar nuclei (Mehler, 1966a,b; Chan-Palay, 1977; Kalil, 1981; Sugimoto et al., 1981; Asanuma et al., 1983a,b). Other inputs come from the spinal cord (Mehler, 1962, 1966, 1969; Boivie, 1971, 1979; Jones and Burton, 1974; Burton and Craig, 1983; Mantyh, 1983), with additional inputs from the spinal trigeminal nuclei (Shigenaga et al., 1983), amygdala (Price and Amaral, 1981; Nauta, 1961), substantia nigra (Carpenter et al., 1976), superior colliculus and pretectal nuclei (Benevento et al., 1977; Graham, 1977; Faull and Mehler, 1978; Berman, 1977; Graham and Berman, 1981; Itoh, 1977; Partlow et al., 1977; Hendry et al., 1979; Beckstead et al., 1979; Harting et al., 1980; Parent and DeBellefeuille, 1983; Yamasaki and Krauthamer, 1980; Kaufman and Rosenquist, 1985; Yamasaki et al., 1986; Francois et al., 1991), vestibular and perihypoglossal nuclei (Kotchabhakdi et al., 1980; McGuinness and Krauthamer, 1980; Graybiel, 1977), rostral brainstem reticular formation (Edwards and De Olmos, 1976; Nauta and Kuypers, 1958; Comans and Snow, 1981; Kaufman and Rosenquist, 1985; Robertson and Feiner, 1982; Royce et al., 1991), periaqueductal gray matter (Hamilton, 1973; Edwards and deOlmos, 1976; Ruda, 1976; Cameron et al., 1995) and the cholinergic, serotoninergic and noradrenergic brainstem nuclei (Saper and Loewy, 1980; Bobillier et al., 1976; Jones and Moore, 1977; Parent and DeBellefeuille, 1983; Kaufman and Rosenquist, 1985; Peschanski and Besson, 1984; Paré et al., 1988; Lavoie and Parent, 1991).

Of the posterior group of intralaminar nuclei, the centre médian is primarily innervated by the internal segment of the globus pallidus, representing much of the output of the basal ganglia (Nauta and Mehler, 1966; Mehler, 1966; Kuo and Carpenter, 1973; Kim et al., 1976; Percheron, 1977; Hendry et al., 1979; Nauta, 1979; DeVito and Anderson, 1982; Edley and Graybiel, 1983; Filion and Harnois, 1978; Parent and DeBellefeuille, 1982, 1983; Ilinsky et al., 1985; Francois et al., 1988; Royce et al., 1991; Hazrati and Parent, 1991; Rouiller et al., 1994; Percheron et al., 1996). Many of the fibers terminating in the centre médian nucleus are branches of those projecting to the anterior ventral lateral nucleus (VLa) (Harnois and Filion, 1982; Parent and DeBellefeuille, 1982). The parafascicular nucleus seems to receive its input from brainstem sources (Royce et al., 1991; Robertson and Feiner, 1982; Graybiel, 1977; Edley and Graybiel, 1983; Paré et al., 1988; Peschanski and Besson, 1984), including the vestibular nuclei (Shiroyama et al., 1995).

The terminations of the afferent fiber systems in the anterior group of intralaminar nuclei, are patchy and the fibers end in relation to particular populations of intralaminar cells. For example, spinal terminations, although forming patches throughout most of the anterior intralaminar nuclei, are concentrated around the posterior, large cells of the central lateral nucleus (Mehler, 1966; Jones and Burton, 1974; Hendry et al., 1979; Molinari et al., 1986; Bentivoglio et al., 1988; Minciacchi et al., 1993). These large cells project mainly to the motor cortex (Jones and Leavitt, 1974; Molinari et al., 1986; Bentivoglio et al., 1988). Other cells, scattered among them and also presumably innervated by spinothalamic fibers, project to the caudate nucleus (Bentivoglio et al., 1988). Patches of cerebellar fiber terminations tend to interdigitate with those formed by spinal fibers and are concentrated on a different population of cells, especially anteriorly (Minciacchi et al., 1991). Hence a series of parallel channels through to the cortex may also be found in the intralaminar nuclei (Molinari et al., 1986, 1993, 1994; Velayos et al., 1989, 1993).

Corticothalamic fibers return to all the intralaminar nuclei (Powell and Cowan, 1967; Jones and Powell, 1971), the pattern of projection being such that cortical areas innervated by a particular intralaminar nucleus are the main sources of corticothalamic fibers to it. This projection, may only arise from layer V cells (see Section 7.1).

10.3.4. Intrathalamic connections of the intralaminar nuclei

Axons of intralaminar cells projecting to the cerebral cortex or striatum, give off collaterals to a part of the reticular nucleus en route (Jones, 1975b; Royce and Mourey, 1985; Sadikot et al., 1992a) and in rats, about 20% of cells in the parafascicular nucleus have intranuclear axon collaterals (Deschênes et al., 1996). There is no clear evidence for other intrathalamic connections formed by the intralaminar cells. However, fibers arising in the intralaminar nuclei commonly travel, for some distance through the internal medullary lamina and there are suggestions that they may give off terminals in other intralaminar nuclei as they do so (Nauta and Whitlock, 1954; Royce and Mourey, 1985; Beckstead, 1984; Mehler, 1966; Sadikot et al., 1992a).

10.3.5. Functional attributes of intralaminar cells

Intralaminar cells projecting to the striatum in the cat have significantly slower conduction velocities than those projecting to the cerebral cortex (Steriade and Glenn, 1982). The nuclei of the monkey have not been examined physiologically in any detail. In cats, cortically projecting cells in the posterior part of the central lateral nucleus possess a unique capacity to discharge trains of action potentials at very high rates –

more than 800–1000 Hz (Glenn and Steriade, 1982). This can occur when the cells are activated by stimulation of the midbrain reticular formation but also occurs in rhythmic bursts at 20–40 Hz during wakefulness and REM sleep (Steriade and Glenn, 1982; Steriade et al., 1993c). Enhanced regional blood flow occurs in these nuclei during selective attention (Kinomura et al., 1996). Recruitment of the intralaminar nuclei, during fast thalamocortical oscillations associated with the alert state, are thought to bind together the content and context of a sensory experience (Llinás and Ribary, 1993). Isolated, bilateral lesions of the intralaminar nuclei can be associated with prolonged hypersommolence (Façon et al., 1958; Castaigne et al., 1962).

Cells in the posterior part of the central lateral nucleus receiving spinothalamic inputs respond at high threshold to noxious stimuli (Giesler et al., 1981; Applebaum et al., 1979). Cells are reported to respond primarily to pinpricks, to sharp taps or to distension of viscera and have large receptive fields (Albe-Fessard and Kruger, 1962; Berkley et al., 1995). Other cells show convergence of auditory and visual inputs (Irvine, 1980; Schlag and Schlag-Rey, 1984; Grunwerg and Krauthamer, 1992).

10.4. PARCELLATION OF THE MEDIODORSAL NUCLEUS

The mediodorsal nucleus is by far the largest division of the medial group of nuclei in virtually all mammals. Of the other divisions, the parataenial nucleus is relatively large in prosimians, small in monkeys and difficult to discern in humans. The medioventral (reuniens) nucleus, separated from the mediodorsal nucleus by growth of the internal medullary lamina during development of the thalamus (Rose, 1942a), has become intimately associated with the ventral anterior and anterior nuclei, and its connections with the parahippocampal regions also bring it closer to the latter than to the mediodorsal nucleus. The mediodorsal nucleus in the three representative primates described here, is divided into three subnuclei: a medial, magnocellular nucleus, a parvocellular nucleus which virtually encircles the magnocellular nucleus, and a lateral, multiform nucleus (Olszewski, 1952; Hassler, 1959; Jones, 1985a) (Figs. 2, 3, 6, 74). The magnocellular nucleus, typically, has a dense myelinated fiber neuropil and is consequently sometimes referred to as the fibrous nucleus or division (Fig. 74 A). Adjacent to the multiform nucleus are the cells that are sometimes called a densocellular or 'paralaminar' division of the mediodorsal nucleus and which probably form part of the central lateral intralaminar nucleus (Olszewski, 1952; Jones, 1985a). The parvocellular nucleus has been subdivided into subsidiary divisions on the basis of thalamocortical connections in monkeys (Ray and Price, 1993).

Basal forebrain and olfactory structures provide the principal inputs to the mediodorsal nucleus. Fibers arising in different regions terminate in specific subdivisions. Olfactory inputs arising in the prepiriform cortex and adjacent regions such as the olfactory tubercle, are distributed to the magnocellular nucleus (Benjamin and Jackson, 1974; Yarita et al., 1980; Russchen et al., 1987). Neurons in the magnocellular nucleus show odor-specific responses (Tanabe et al., 1975b). The magnocellular division also receives fibers from the amygdala. GABAergic inputs reach all three subdivisions from the ventral pallidum and the pars reticulata of the substantia nigra. The parvocellular and multiform divisions also receive other inputs from the midbrain. These arise in the superior colliculus and midbrain tegmentum, in addition to the sources of cholinergic and monoaminergic fibers (Powell et al., 1965; Krettek and Price, 1977; Porrino et al., 1981; Aggleton and Mishkin, 1984; Ono and Niimi, 1986; Russchen et al., 1987; Velayos and Reinoso-Suárez, 1982; Mogenson et al.,

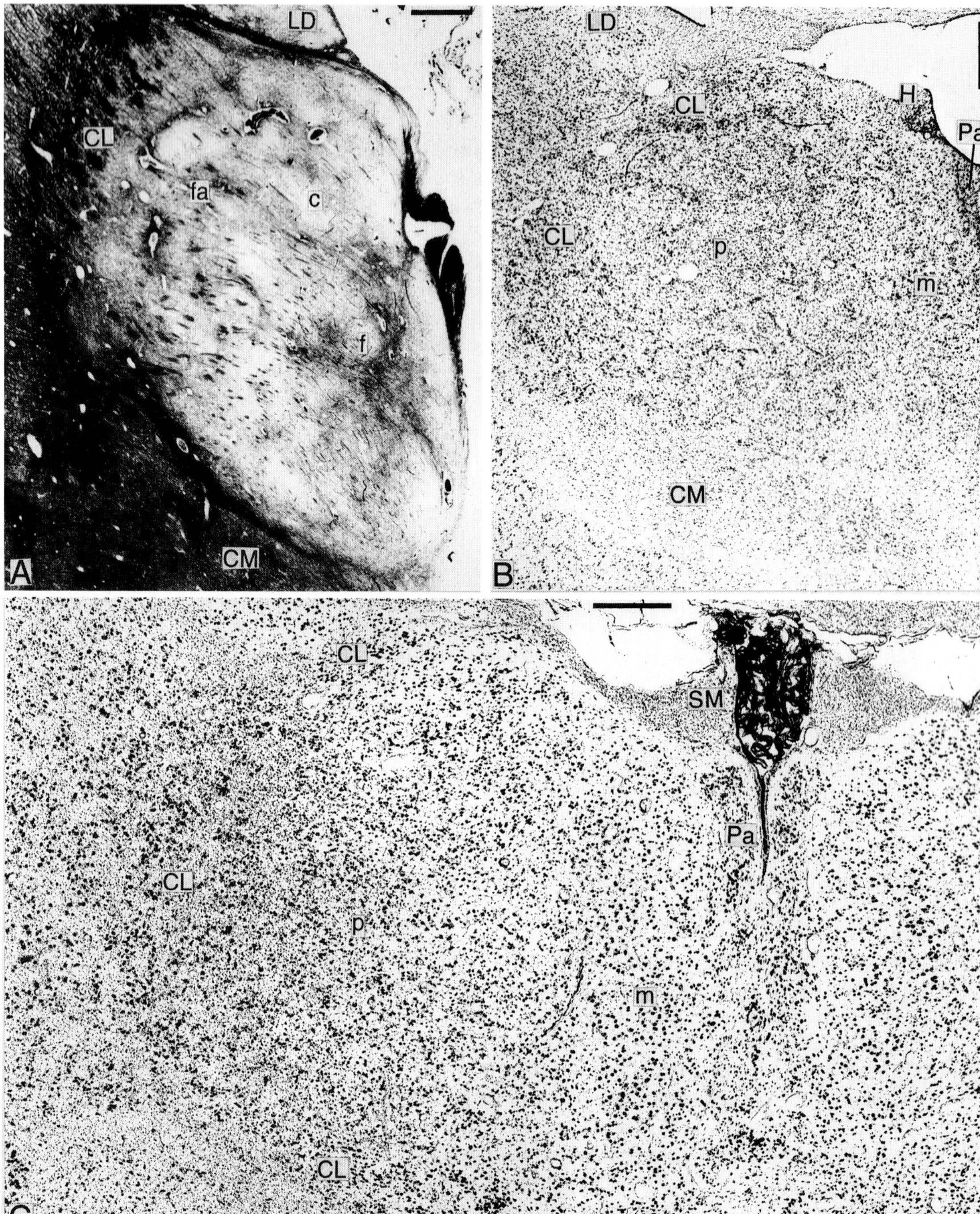
LD
CL
fa
c
f
CM
A
LD
CL
H
Pa
CL
p
m
CM
B
CL
SM
Pa
CL
p
m
CL
C

Fig. 74. Myelin- (A) and Nissl-stained (B, C) frontal sections through the mediodorsal nucleus of a human (A), a squirrel monkey (*S. sciureus*) (B) and a rhesus monkey (*M. mulatta*) (C), showing the magnocellular (m) and parvocellular (p) divisions and the encroaching cells of the central lateral nucleus (CL) which are sometimes regarded as a densocellular division of the nucleus. In A, the myelin-based divisions of Hassler (1959) are indicated. f (fibrosus) corresponds to the magnocellular division; c (caudalis) and fa (fasciculosis) together correspond to the parvocellular division; the region labeled CL is Hassler's paralamellar division. Bars 1 mm (A, C), 500 μm (B).

1987; Price and Slotnick, 1983; Hreib et al., 1988; Steriade et al., 1988; Cornwall and Phillipson, 1988; Groenewegen, 1988; Giguere and Goldman-Rakic, 1988; Kuroda and Price, 1991; Kuroda et al., 1992; Ray et al., 1992).

The cortical projection of the mediodorsal nucleus is targeted predominantly on the prefrontal granular cortex. Each subnucleus has a different area or areas of dominant projection (Figs. 75). The magnocellular nucleus, including the small-celled zone me-

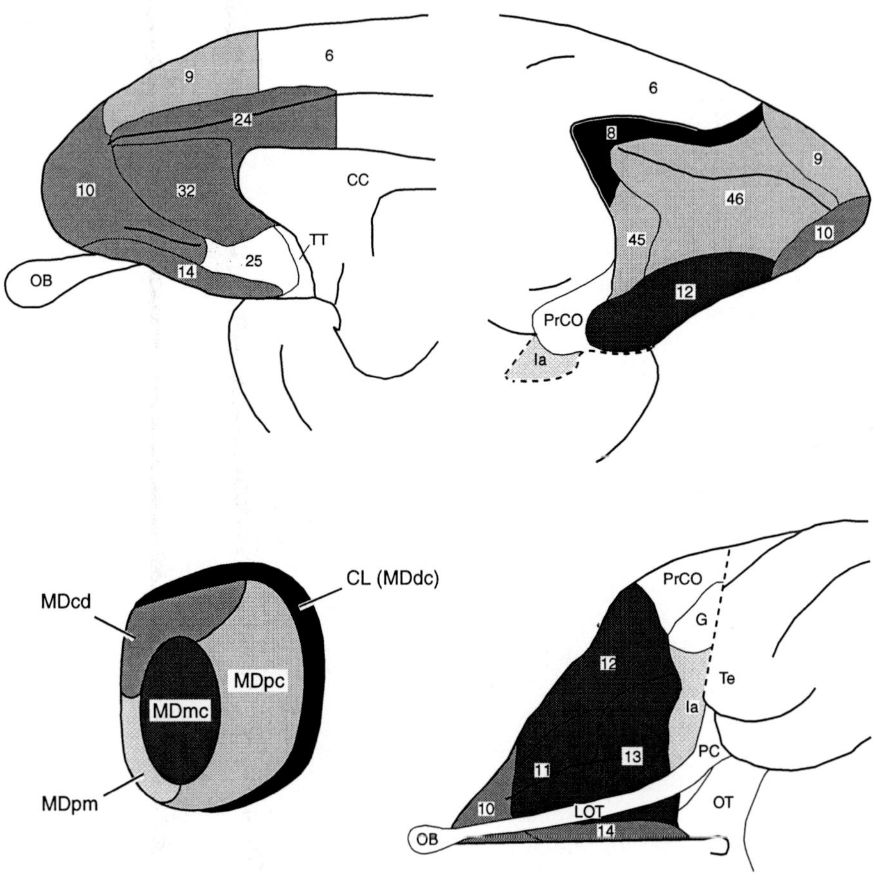

Fig. 75. Schematic frontal section (lower left) of the monkey mediodorsal nucleus and the topography of the projections of its various divisions on the prefrontal cortex and certain adjacent areas. Mainly after Ray and Price (1993). Numbers: areas of cerebral cortex; CC: corpus callosum; G: gustatory cortex; Ia: agranular insular cortex; LOT: lateral olfactory tract; MDcd, MDdc, MDmc, MDpc, MDpm: divisions of mediodorsal nucleus; OB: olfactory bulb; OT: optic tract; PC: piriform cortex; PrCO: precentral agranular field; Te: temporal cortex; TT: taenia tecta.

dial to it at the midline, projects primarily to orbital areas, the projection being somewhat more restricted than that of the parvocellular nucleus and including areas 12 and 13 and possibly areas 10 and 11, extending back to the agranular insular cortex. The projection of the parvocellular nucleus is much more widespread and includes lateral and medial areas of the frontal lobe, described as areas 9, 24, 32, 45, 46 and possibly others Its posterodorsal part projects to ventral medial areas abutting the orbital surface. The multiform nucleus projects to areas adjacent to the premotor cortex, including the frontal eye fields (Roberts and Akert, 1963; Akert, 1964; Naito et al., 1984; Barbas et al. 1991; Mufson and Mesulam, 1984; Barbas and Mesulam, 1981, 1985; Goldman-Rakic and Porrino, 1985; Huerta and Kaas, 1990; Siwek and Pandya, 1991; Morecraft et al., 1992; Ray and Price, 1993). Some of the projections to the frontal eye fields and connections reportedly to premotor cortex may in fact arise in the part of the central lateral nucleus commonly included as the densocellular division of the mediodorsal nucleus (see e.g. Huerta et al., 1986).

The areas in the orbital cortex to which the magnocellular nucleus projects, contain neurons that respond at relatively short latency to electrical stimulation of the olfactory bulb or olfactory cortex (Tanabe et al., 1974, 1975a,b; Takagi, 1979, 1980, 1986; Yarita et al., 1980; Naito et al., 1984), and respond to one or more odors presented to the nose of unanesthetized monkeys (Tanabe et al., 1975a,c; Yarita et al., 1980). At least two functional areas, based upon the recording of evoked potentials following stimulation at various sites along the central olfactory pathways, have been described in this region of the orbital cortex (Tanabe et al., 1975c; Yarita et al., 1980). Olfactory inputs may reach one of these by pathways that bypass the mediodorsal nucleus of the thalamus (Tanabe et al., 1975c). Bilateral ablation of these and adjacent areas in monkeys results in impairment of olfactory discrimination (Tanabe et al., 1975c).

The mediodorsal nucleus has been implicated in schizophrenia in humans in which disease it reportedly displays a more than 30% loss of cells (Pakkenberg, 1990). It is also a primary site of cellular atrophy in Korsakoff's psychosis, the atrophy reportedly affecting mainly the magnocellular division (Victor et al., 1971), although this has been contested (Mair et al., 1979). Traumatic and vascular lesions involving the mediodorsal nucleus in humans are usually reported to be associated with disturbances of recent memory similar to that found in Korsakoff's syndrome (Squire and Moore, 1979; Schott et al., 1980; Aggleton, 1991) and bilateral lesions affecting the magnocellular divisions in monkeys result in deficits in visual object oriented associative memory (Aggleton and Mishkin, 1983a,b; Zola-Morgan and Squire, 1985; Gower et al., 1985).

10.5. LATERAL POSTERIOR AND PULVINAR NUCLEI

These nuclei are a complex which has become particularly large in primates, undoubtedly in relation to the expansion of the parieto-temporal and occipital association cortex. The complex may also be more highly differentiated in primates in terms of its nuclear parcellation, perhaps a reflection of the increased functional differentiation of the large region of cortex to which it projects. In monkeys and humans a lateral posterior nucleus and four pulvinar nuclei (anterior, lateral, medial and inferior) can be identified with relative ease. In prosimians, although it is customary to identify only a lateral posterior and two divisions of the pulvinar (superior and inferior), as indicated in the description in Section 3.1 (Fig. 2), the inferior division may be divisable along lines similar to those of the monkey. The anterior pulvinar nucleus, although included in the pulvinar by convention, may have closer affinities with the ventral posterior nuclei and the somatosensory system and this is reflected in Hassler's delineation of the ventral posterior region of the human thalamus (see section 10.1).

The predominant input to the lateral posterior/pulvinar complex as a whole, arises in the visuomotor regions of the midbrain, especially the pretectum and superior colliculus (Harting et al., 1973, 1980; Graybiel, 1972, 1974a; Benevento and Fallon, 1975; Glendenning et al., 1975; Partlow et al., 1977; Graham, 1977; Berman, 1977; Berson and Graybiel, 1978; Gould et al., 1978; Itoh et al., 1979; Graybiel and Berson, 1980; Hughes, 1980; Weber and Harting, 1980; Marrocco et al., 1981; Benevento and Standage, 1983; Huerta and Harting, 1983; Lysakowski et al., 1986). But these inputs do not distribute homogeneously through all pulvinar nuclei or their subdivisions. There is, in addition, a small, direct retinal input to the inferior pulvinar nucleus (Campos Ortega et al., 1972; Tokunaga et al., 1981; Cowey et al., 1994). The cortical output of the whole complex is to that large region of parieto-temporo-occipital cortex intercalated between the primary somatosensory, auditory and visual areas (Burton and Jones, 1976; Benevento and Rezak, 1976; Baleydier and Mauguière, 1977, 1985; Mesulam et al., 1977; Ogren and Hendrickson, 1976; Benevento and Davis, 1977; Curcio and Harting, 1978; Jacobson and Trojanowski, 1976; Kasdon and Jacobson, 1978; Pearson et al., 1978; Symonds and Kaas, 1978; Mufson and Mesulam, 1984; Asanuma et al., 1985; Yeterian and Pandya, 1988, 1989; Schamamann and Pandya, 1990; Dick et al., 1991; Baleydier and Morel, 1992; Hardy and Lynch, 1992; Cavada

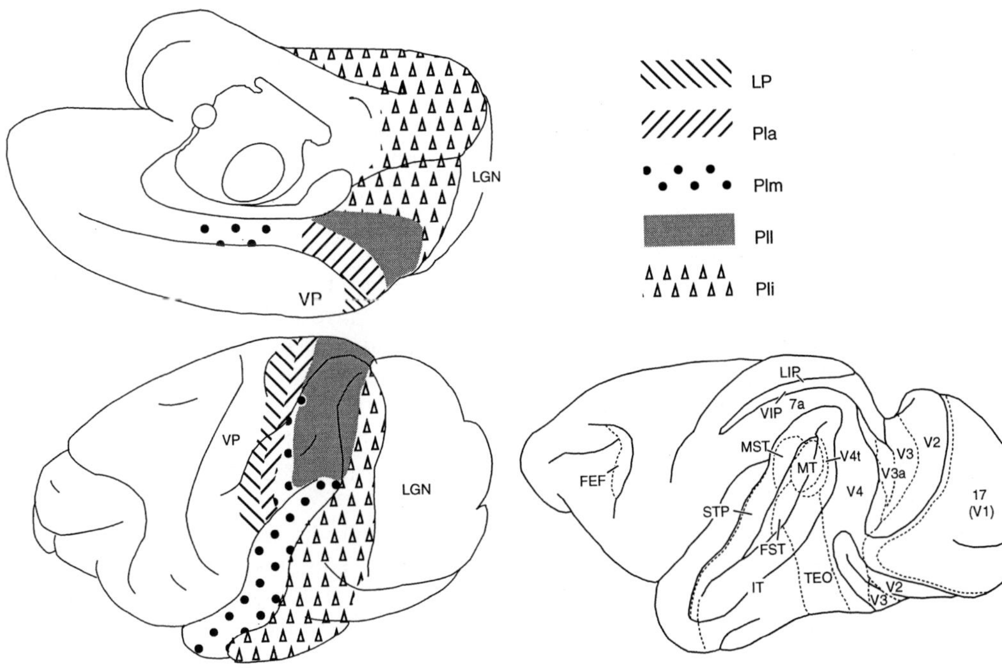

Fig. 76. Left Schematic view of the regional cortical projections of the lateral posterior (LP) and the four pulvinar nuclei (Pla, Pli, Pll, Plm) in monkeys. Modified from Jones (1985). LGN: dorsal lateral geniculate nucleus; VP: ventral posterior nucleus. Right Schematic view of the major, visually-related cortical fields within the region of projection of the pulvinar. The lunate, intraparietal and superior temporal sulci appear as though opened out. Redrawn from Rodman (1994). FEF: frontal eye field; FST: fundus of superior temporal sulcus; LIP: lateral intraparietal area; MST: medial superior temporal area; MT: middle temporal area; PO: parieto-occipital area; STP: superior temporal polysensory area; VIP: ventral intraparietal area; V1, V2, V3, V3a, V4, V4t: visual areas 1, 2, 3, 3a, 4 and transitional 4.

and Goldman-Rakic, 1993; Cusick et al., 1993; Steele and Weller, 1993). This includes the superior and inferior parietal lobules, the greater part of the temporal lobe and the prestriate cortex (Fig. 76), and to a lesser extent even the striate cortex itself.

There are numerous relay channels within this immense projection, based on the projections of individual nuclei and their subdivisions to different architectonic and functional fields. Few details are, however, available on account of the relatively little work that has been done in this area in monkeys. The inferior and lateral pulvinar nuclei receive their principal inputs from the superficial layers of the superior colliculus and large populations of their neurons are obviously visual in terms of their receptive field and stimulus-response properties (Gattass et al., 1978a,b; Bender, 1981, 1982, 1988; Acuña et al., 1983; Petersen et al., 1985; Robinson and Petersen, 1992). They project to prestriate and middle and inferotemporal visual areas. It is likely that there may be more than one channel based on projections of different thalamic cell populations to the different functional areas within the prestriate cortex but this has not been completely worked out (Baleydier and Mauguiére, 1977; Robinson et al., 1978; Baizer et al., 1991, 1993; Petersen et al., 1989; Webster et al., 1993; Baleydier and Morel, 1992; Cusick et al., 1993).

The inferior pulvinar nucleus contains at least four subdivisions based on connections (Lin and Kaas, 1979), and especially on differences in staining for cytochrome oxidase and the calcium binding proteins, calbindin and parvalbumin (Cusick et al., 1993). Each of these subdivisions may provide the predominant input to single areas in the large prestriate/inferotemporal region, with an emphasis on the so-called middle temporal and dorsolateral visual areas (Fig. 76) (Wall et al., 1982; Lin and Kaas, 1980; Maunsell and Van Essen, 1983; Standage and Benevento, 1983; Ungerleider et al., 1984; Cusick et al., 1993). However, current evidence suggests that each area may receive inputs from more than one subdivision and from parts of the lateral and medial pulvinar nuclei as well. Single cells may have axon branches to more than one visual area including the striate, prestriate and more anterior areas (Kennedy and Bullier, 1985; Lysakowski et al., 1988; Kaske et al., 1991).

The medial and lateral pulvinar nuclei also receive inputs from intermediate layers of the superior colliculus but large areas of the former have still not been charted in terms of afferent inputs. Neurons in the pulvinar outside the zones of termination of afferents from the superficial layers of the superior colliculus tend to be polysensory or have ill-defined properties (Richardson and Zorub, 1970; Mathers and Rapisardi, 1973; Chuong and Lindsley, 1973; Gattass et al., 1978; Benevento and Miller, 1981; Acuña et al., 1983, 1990; Martin-Rodriguez et al., 1982; Magariños-Ascone et al., 1988).

The inferior pulvinar nucleus in monkeys, undoubtedly as the result of its topographically organized input from the superior colliculus, contains one or more representations of the contralateral visual field (Bender, 1981, 1988; Allman et al., 1972; Gattass et al., 1978; Acuña et al., 1983; Petersen et al., 1985; Robinson and Petersen, 1992). The retinotopic maps are matched by a dense visuotopically organized corticothalamic projection from the primary visual cortex. The lateral pulvinar nucleus also contains two or more visual field representations, identified either by receptive field mapping (Bender, 1982) or by mapping the distribution of corticothalamic fibers from visual cortical areas which themselves contain visuotopic maps (Campos-Ortega and Hayhow, 1972; Ogren and Henderson, 1976; Rezak and Benevento, 1979; Ungerleider et al., 1983; Kennedy and Bullier, 1985; Cusick et al. 1993; Steele and Weller, 1993) (Fig. 77). Some of these maps overlap into the inferior pulvinar nucleus, but recent

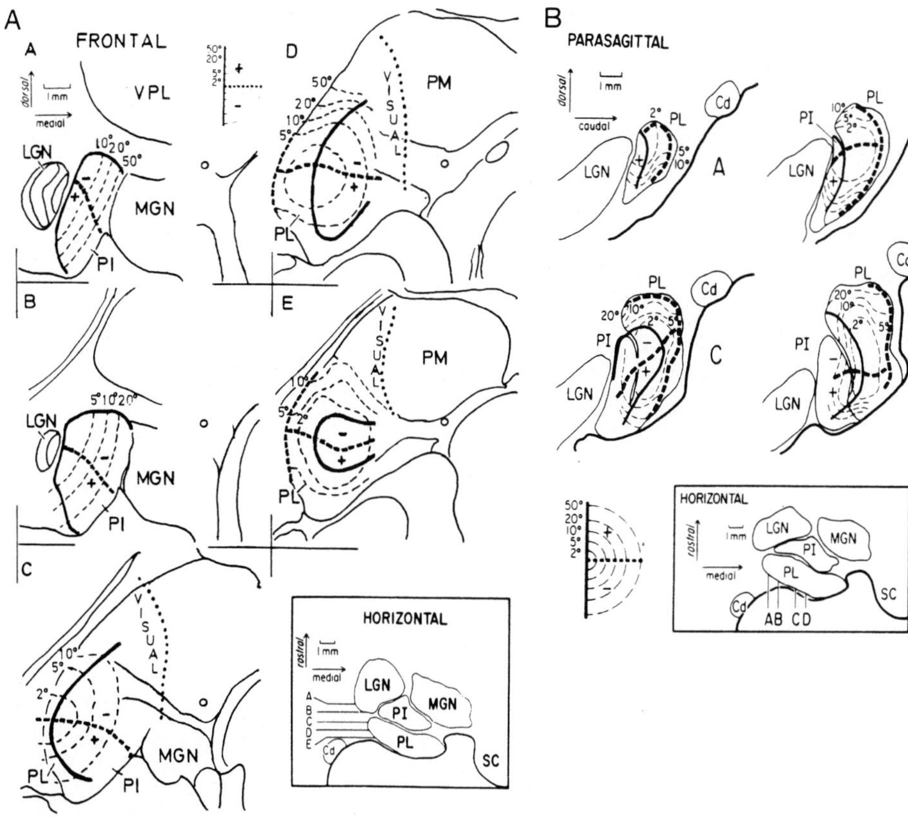

Fig. 77. Visual field maps obtained from single unit recordings in the pulvinar of monkeys, drawn on frontal (A) and parasagittal (B) sections through the posterior part of the thalamus. Redrawn from Bender (1981). Dashed line: horizontal meridian of visual field; continuous line: vertical meridian of visual field. PI: inferior pulvinar nucleus; Pl: lateral pulvinar nucleus.

studies based on immunocytochemical staining of the monkey pulvinar suggest that the traditional borders between the inferior and lateral pulvinar nuclei may need adjusting. This may bring nuclei into greater conformity with the locations of the visual maps than currently appears (Cusick et al., 1993; Guiterrez et al., 1994). The inferior pulvinar nucleus, unlike other nuclei rich in calbindin immunoreactive cells, contains two populations of such cells (Jones and Hendry, 1989), one relatively small and typical of calbindin cells in other nuclei, the second very large (Fig. 78). Both are non-GABA cells (Fig. 75 A, B). Their differential projections, if any, are unknown.

The lateral pulvinar nucleus projects mainly to posterior parts of the superior and inferior parietal lobules, the projection meeting and perhaps overlapping that of the medial pulvinar nucleus in the vicinity of the intraparietal sulcus (Hardy and Lynch, 1992). The medial pulvinar nucleus projects mainly to the superior temporal gyrus. The medial and lateral nuclei project to posterior parietal and superior temporal areas. The medial pulvinar nucleus, may also project to posterior cingulate cortex (Baleydier and Mauguiére, 1980, 1985; Vogt et al., 1979) and the parahippocampal region (Insausti et al., 1987; Mauguiére and Baleydier, 1978; Yeterian and Pandya, 1988; Mufson and Mesulam, 1984), the projection perhaps overlapping into the orbitofrontal cortex (Goldman-Rakic and Porrino, 1985; Yeterian and Pandya, 1988). Projections

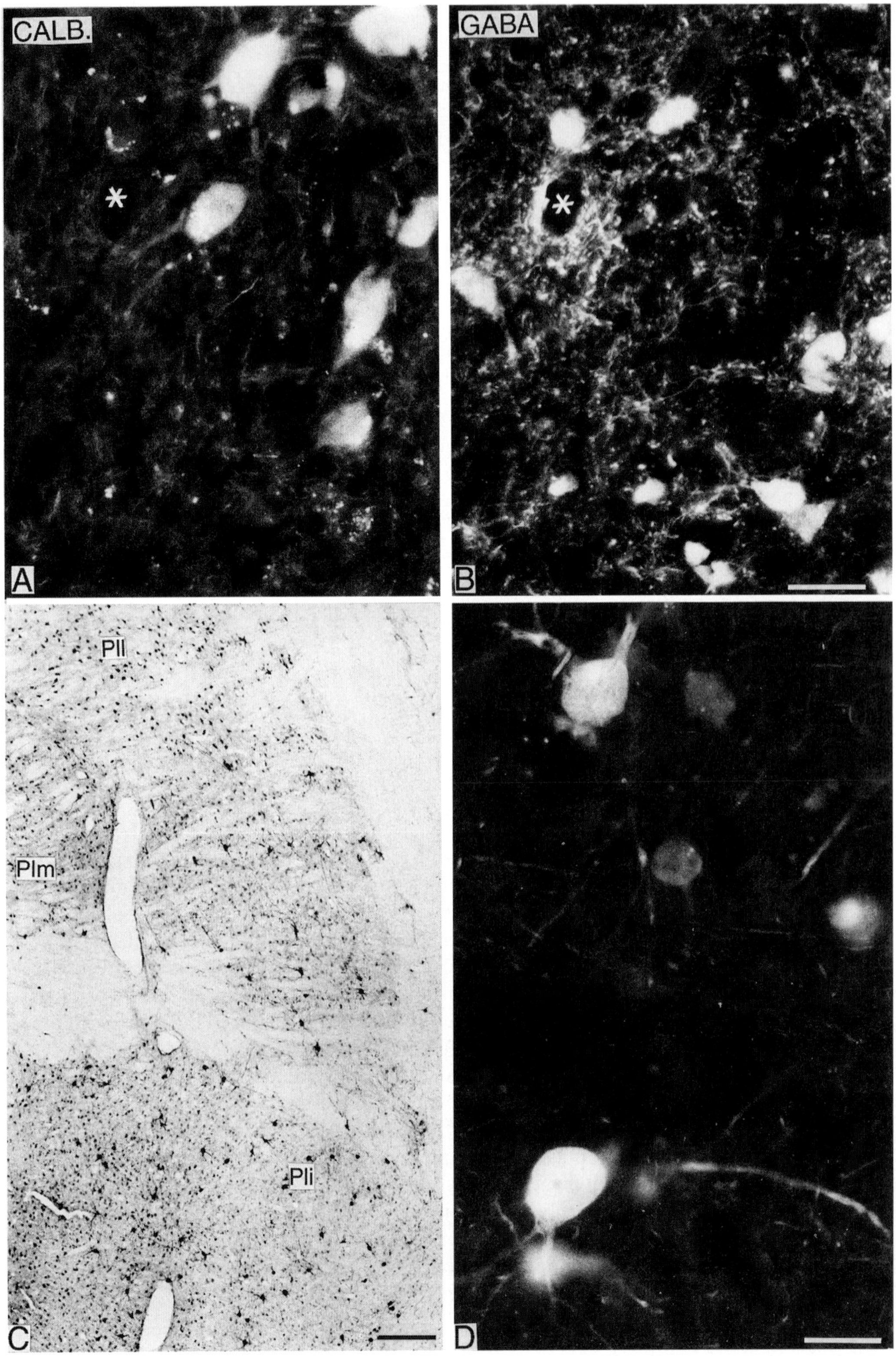
CALB.
GABA
A
B
Pll
Plm
Pli
C
D

Fig. 78. A, B: Fluorescence photomicrographs from the same microscopic field of the lateral pulvinar nucleus of a macaque monkey, showing calbindin immunoreactive cells labeled by fluorescein immunofluorescence and GABA immunoreactive cells labeled by rhodamine immunofluorescence. GABA is not found in the calbindin cells. Asterisk indicates same blood vessel. Bar 25 μm. C, D: Immunoperoxidase (C) and immunofluorescent (D) labeling of large and small cells for calbindin immunoreactivity in the inferior pulvinar nucleus of a macaque monkey. Bars 100 μm (C), 25 μm (D). From material illustrated in Jones and Hendry (1989).

←

to the frontal eye fields have been described from the medial pulvinar nucleus (Bos and Benevento, 1975; Trojanowski and Jacobson, 1976; Goldman-Rakic and Porrino, 1985; Huerta et al., 1986; Stanton et al., 1988; Bruce and Goldberg, 1985; Shook et al., 1991). In many instances these emanate from the part of the central lateral nucleus that extends over the posterior pole of the mediodorsal nucleus, misidentified as part of the pulvinar (Jones, 1985a).

The anterior pulvinar and lateral posterior nuclei project to anterior parietal association cortex (areas 5 and 7), the projection of the former overlapping into area 2 of the the somatosensory cortex and that of the latter into some of the extrastriate visual areas (Burton and Jones, 1976; Pearson et al., 1978; Jones et al., 1979; Graham et al., 1979; Pons and Kaas, 1985; Schamamann and Pandya, 1990; Cusick and Gould, 1990). Projections to the insula have also been described from the anterior pulvinar nucleus (Mesulam and Mufson, 1985). The subcortical inputs to these nuclei have not yet been fully defined. The lateral posterior nucleus seems to receive inputs from the superior colliculus. The anterior pulvinar nucleus receives some spinal and spinal trigeminal inputs (Rausell and Jones 1991b; Rausell et al., 1992) but these by no means fill the nucleus.

The pulvinar nuclei of prosimians and of some of the smaller New World monkeys, although divided up along lines rather different from those used in other primates, from the available evidence, seem to have connectional relationships with the cerebral cortex similar to those described in Old World monkeys (Glendenning et al., 1975; Symonds and Kaas, 1978; Carey et al., 1979; Graham et al., 1979; Raczkowski and Diamond, 1980; Norden et al., 1978; Lin and Kaas, 1979; Wall et al., 1982; Conley and Diamond, 1990). The manner of dividing up the cortex in the prosimians, makes it difficult to make precise correlations but the relative dispositions of the projections from the different pulvinar nuclei to the cortex seem identical. In the *Galago*, a part of the superior pulvinar nucleus, cytoarchitectonically comparable to the medial pulvinar nucleus of monkeys, for example, projects to cortex adjacent to the auditory areas and therefore equivalent to the superior temporal cortex. The posterolateral part of the inferior pulvinar nucleus of the *Galago* is the main site of terminations of fibers from the superficial layers of the superior colliculus (Raczkowski and Diamond, 1980). This region is, therefore, likely to correspond to the inferior pulvinar nucleus of monkeys and humans.

10.6. CELL-SPECIFIC STREAMS THROUGH THE DORSAL LATERAL GENICULATE NUCLEUS

10.6.1. Introduction

The dorsal lateral geniculate nucleus, one of the most distinctive nuclei of the primate thalamus, is a laminated structure with complete segregation of inputs from the two eyes, even in species in which the parvocellular mass is not histologically laminated (Section 3.2). By comparison with non-primates, the ipsilateral input is large and in humans is probably equal in size to the contralateral input. Two or three populations of anatomically and chemically distinct populations of cells are evident in prosimians, simians and the human; in the non-human primates, at least, they are functionally distinct and innervated by different classes of retinal ganglion cell. These represent relays in parallel pathways for the transfer of information from the different classes of retinal ganglion cell to the cerebral cortex. The probability that the human visual system is organized along lines similar to those of the macaque monkey, is high.

The cortical output of the majority of the relay cells in the nucleus is focused on the striate area (area 17). Here, the axons of the different classes of relay cell synaptically engage cortical cells in different layers. Smaller populations of geniculo-cortical relay cells of small size project more diffusely and their axons reach extrastriate areas as well. In the cortex, these axons terminate in superficial layers. In the visual cortex of Old World and some species of New World monkeys, the terminations of fibers arising in dorsal lateral geniculate laminae innervated by the contra- and ipsilateral eyes, do not overlap. Instead they terminate, primarily in layer IV, in a series of alternating left and right eye domains, each about 400–500 μm wide and referred to as ocular dominance columns or stripes (Hubel and Wiesel, 1968, 1969; Wiesel et al., 1974; LeVay et al., 1975, 1980, 1985; Florence et al., 1986; Hess and Edwards, 1987) (Fig. 79).

10.6.2. Ganglion cell-specific channel through the dorsal lateral geniculate nucleus

10.6.2.1. On-center and off-center cells

Axons arising from different classes of retinal ganglion cell have distinctive morphological features in cats (Bowling and Michael, 1980, 1984; Sur and Sherman, 1982; Garraghty et al., 1986; Hamos et al., 1987; Raczkowski et al., 1988; Weber et al., 1989) and comparable distinctions can be seen in monkeys (Michael and Bowling, 1982; Michael, 1988) and *Galagos* (Lachica et al., 1990). In monkeys, single ganglion cell axons end in one lamina only of the dorsal lateral geniculate nucleus (Michael and Bowling, 1982).

Virtually all neurons in the dorsal lateral geniculate nucleus have the typical center-surround receptive fields first described in the cat by Hubel and Wiesel (1961), and confirmed in the monkey by Wiesel and Hubel (1966). A center-surround receptive field is characterized by a central zone in which a flashed spot of light will elicit excitation ('on-center') or inhibition ('off-center') of the cell, and a concentric, antag-

Fig. 79. Upper Schematic representation of the layers of termination of fibers from the magnocellular (M1, M2), and parvocellular (P1, P2) layers and from the S and interlaminar (I) zones of the dorsal lateral geniculate nucleus in the primary visual cortex of the macaque monkey, together with recipient cortical cells and the projections from the recipient layers to other cortical layers. From Lund et al. (1994). Lower Cytochrome oxidase stained sections cut parallel to the surface of the visual cortex of a macaque monkey that had been subjected to monocular visual deprivation for several days prior to death. A passes mainly through layer III of area 17 (arrows indicate islands of layer IV) and shows the rows of blob-like periodicities characteristic of layers II and III. In places, every other row of blobs can be seen to be shrunken, indicating inputs from the deprived eye. B passes through area 18 and to a large extent through layer IVC of area 17. In area 18, alternating dark and light stripes are seen. Layer IVC of area 17 shows the alternating rows of pale, (deprived) and dark, (non-deprived) ocular dominance stripes which receive inputs from left and right eye layers of the dorsal lateral geniculate nucleus. Bar 1 mm.

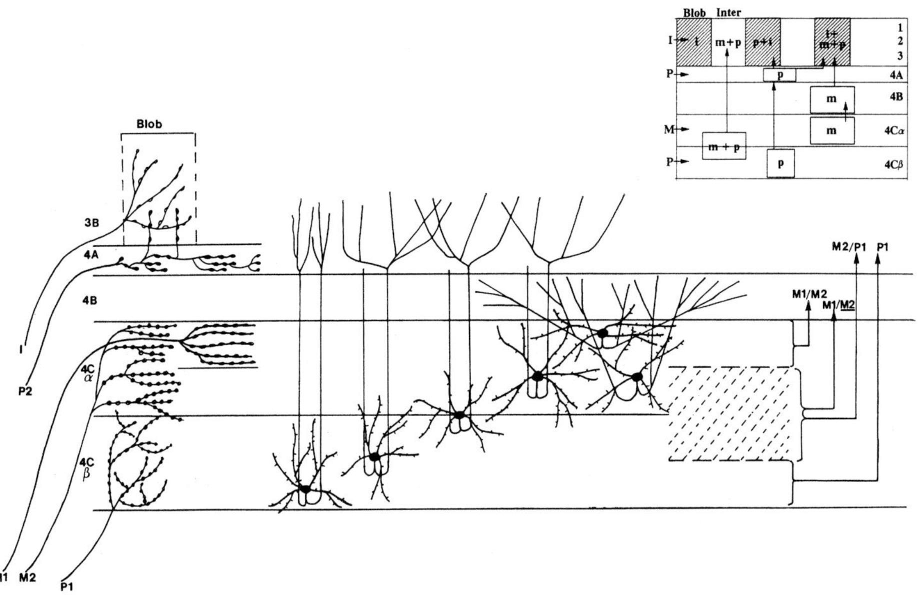
Blob Inter
I
m+p
p+i
i+ m+p
1
2
3
P
p
4A
m
4B
M
m
4Cα
m + p
P
p
4Cβ
Blob
3B
4A
4B
I
P2
4C α
4C β
M1 M2
P1
M2/P1 P1
M1/M2
M1/M2

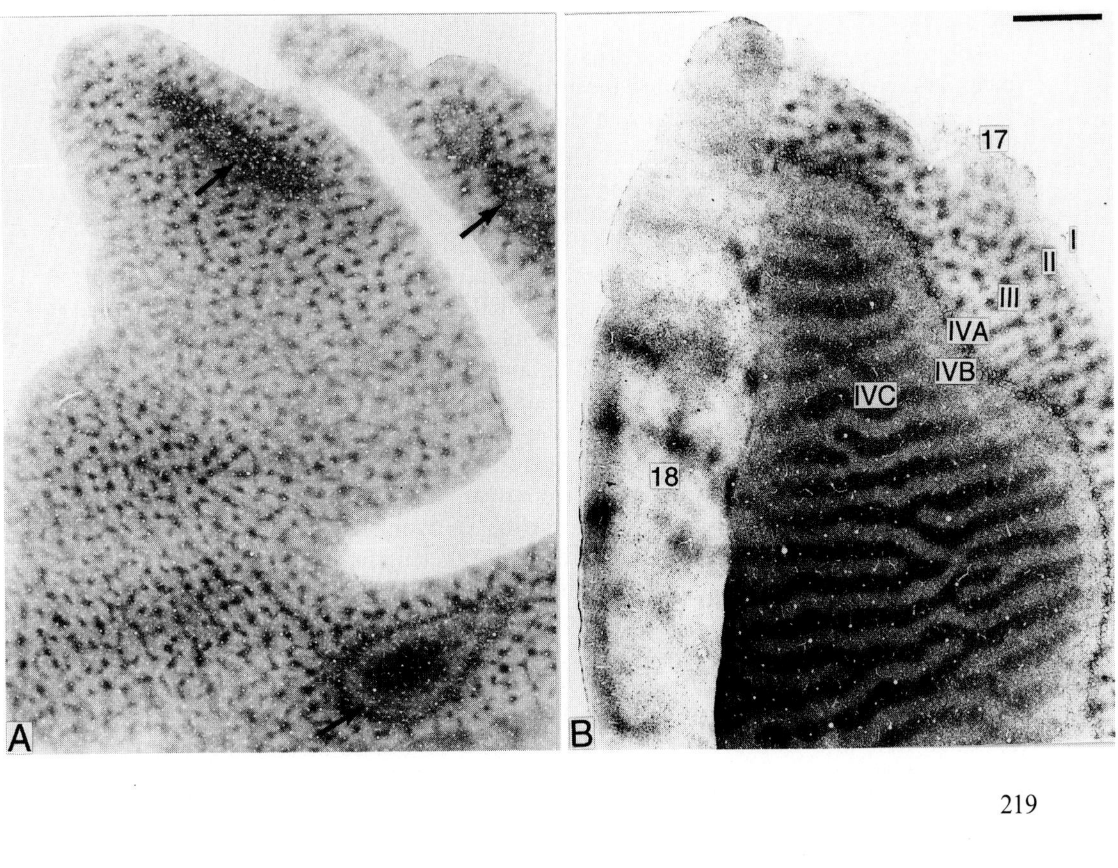
17
I
II
III
IVA
IVB
IVC
18
A
B

onistic surround in which flashed light stimuli will inhibit an on-center cell or excite an off-center cell. The center-surround antagonism is established in the ganglion cells of the retina (Kuffler, 1953) and the spatial characteristics of the receptive fields of most dorsal lateral geniculate cells are projected on to them with little change, by cell class specific connections from ganglion cell axon to the appropriate class of geniculate relay cell. The lateral geniculate cells tend to have smaller receptive field centers than in the retina and there is greater contrast between the effects of stimulating the center and surround. This reflects the action of intrageniculate GABAergic mechanisms (Sillito and Kemp, 1983).

Differential terminations of axons arising from on-center and off-center retinal ganglion cells in the dorsal lateral geniculate laminae result in the formation of on-center and off-center streams which are relayed separately through the dorsal lateral geniculate nucleus. In the Old World monkey, the majority of on-center and off-center cells occupy different parvocellular laminae (Schiller and Malpeli, 1978; Conway and Schiller, 1983; Michael, 1988), the on-center cells being found in laminae 5 and 6 and the off-center cells in laminae 3 and 4. The incomplete splitting of the two fundamental parvocellular layers of the geniculate (4 plus 6 and 3 plus 5) is, thus, a reflection of the segregation of the on-center and off-center streams. In the magnocellular layers, the on-center and off-center cells are not segregated in sublaminae. Infusion of DL-2-amino-4-phosphonobutyric acid into either the parvo- or magnocellular layers eliminates all responses of on-center cells, including the off-response from the antagonistic surround, leaving the responses of the off-center cells unaffected (Schiller, 1982).

10.6.2.2. The M and P channels through the dorsal lateral geniculate nucleus of monkeys

Virtually all ganglion cells in the retina of monkeys and presumably humans, have typical, antagonistic center-surround receptive fields (Wiesel and Hubel, 1966; Schiller and Malpeli, 1978). These can be divided into a number of classes on the basis of further functional criteria (De Monasterio and Gouras, 1975; De Monasterio, 1978). Two classes are predominant: (a) cells with large concentric receptive fields, the centers and surrounds of which receive summating inputs from red, green and possibly blue cones. These cells are therefore not color coded and have broad-band spectral sensitivity; they discharge transiently in response to changes in brightness and have rapidly conducting axons. Morphologically, they probably correspond to larger, so-called parasol ganglion cells (Polyak, 1941; Leventhal et al., 1981; Watanabe and Rodieck, 1989; Perry et al., 1984; Rodieck and Watanabe, 1993). They are now customarily called M or P-α cells (Perry et al., 1984; Shapley and Perry, 1986). They form about 10% of the ganglion cell population (Perry et al., 1984), depending on retinal eccentricity (Silveira and Perry, 1991). (b) cells with small center-surround receptive fields in which the center is determined by inputs from a single type of cone and the surround by another type (usually red and green form the antagonistic inputs). These cells are therefore color-coded, and are excited by certain wavelengths and inhibited by others. Their excitatory responses to chromatic stimuli at peak spectral sensitivity are usually sustained. They have relatively slowly conducting axons and have been correlated, with midget ganglion cells (Polyak, 1941). They are customarily called P or P-β cells (Shapley and Perry, 1986; Perry et al., 1984).

Other less well defined varieties of ganglion cell have been described physiologically (De Monasterio and Gouras, 1975; De Monasterio, 1978) and morphologically (Polyak, 1941; Perry and Cowey, 1984) in the primate retina. Some of these have small cell bodies and sparse radiating dendrites, and send their axons to the tectum rather than to the dorsal lateral geniculate nucleus (Leventhal et al., 1981; Perry et al., 1984). Based on considerations given below, it is tempting to suggest that these cells also project to the small-celled S and intercalated layers of the dorsal lateral geniculate nucleus. This has not been firmly established, although it is known that the S and intercalated layers do receive retinal inputs (Kaas et al., 1978).

The axons of the M and P classes of retinal ganglion cell terminate in segregated fashion among the laminae of the dorsal lateral geniculate nucleus, the M cells exclusively, in an eye dependent manner, in one or other of the magnocellular layers (1 or 2) and the P cells, in an eye dependent manner, in the parvocellular layers 3 and 5 or 4 and 6. The M and P properties are projected onto the geniculate relay cells with little decrement in properties and the magno- and parvocellular layers thus form parallel, broad-band and color-coded relay channels, respectively. In the magno- and parvocellular layers, relay cells take on the spatial or chromatic properties of the innervating ganglion cells (Wiesel and Hubel, 1966; Schiller and Malpeli, 1978; Kaplan and Shapley, 1982). Cells with broad-band spectral qualities thus predominate in the magnocellular layers and red-green, color opponent cells predominate in the parvocellular layers, reflecting inputs from the commonest form of color-coded retinal ganglion cell. The chief distinctions between the parvocellular and magnocellular layers and the P and M channels are, therefore, those of color opponency in the one and of broad band spectral sensitivity in the other.

Cells in the magnocellular layers, along with their broad band spectral sensitivity, display high contrast sensitivity. They are innervated by fast optic tract axons and generally show transient ('phasic') responses to visual stimuli. Cells in the parvocellular layers, along with their narrowly tuned spectral sensitivity, display low contrast sensitivity. They are innervated by slower optic tract axons and usually have tonic responses (Marrocco and Brown, 1975; Dreher et al., 1976; Sherman et al., 1976; Blakemore and Vital-Durand, 1981, 1986; Kaplan and Shapley, 1982; Marrocco et al., 1982; Derrington and Lennie, 1984; Shapley and Perry, 1986; Hubel and Livingstone, 1990).

Based on the immunocytochemical properties of cells in the dorsal lateral geniculate nucleus, as described in Section 6.2, we can state that both P and M cells are parvalbumin-containing while the M cells are distinguished by much higher levels of expression of the antigen recognized by the CAT 301 antibody (Hendry et al., 1988d). Neither set of cells is immunostained for calbindin or CAMII kinase-α (Fig. 80). These antigens are only expressed in the cells of the S and intercalated laminae (see below). Three other subunits of CAMII kinase (β, γ and δ) are expressed in cells of the principal laminae (Figs. 22, 23).

The channels formed by the chromatic, P system of the parvocellular laminae and the broadband, M system of the magnocellular laminae are maintained in the projections of these laminae to the striate cortex. Axons arising from cells in the parvocellular laminae terminate in a thick band in layer IVCβ, and in thinner bands in layer IVA and at the border of layers V and VI (Fig. 79). Axons of cells in the magnocellular laminae terminate in a single thick band in layer IVC α, immediately above the terminations from the parvocellular layers in IVCβ, and also at the border of layers V and VI (Hubel and Wiesel, 1972; Wiesel et al., 1974; Glendenning et al., 1976; Hubel et al., 1977; Kaas et al., 1976; Hendrickson et al., 1978; Livingstone and Hubel, 1982; Fitzpatrick et al., 1983; Blasdel and Lund, 1983a,b; Freund et al., 1989). In the visual

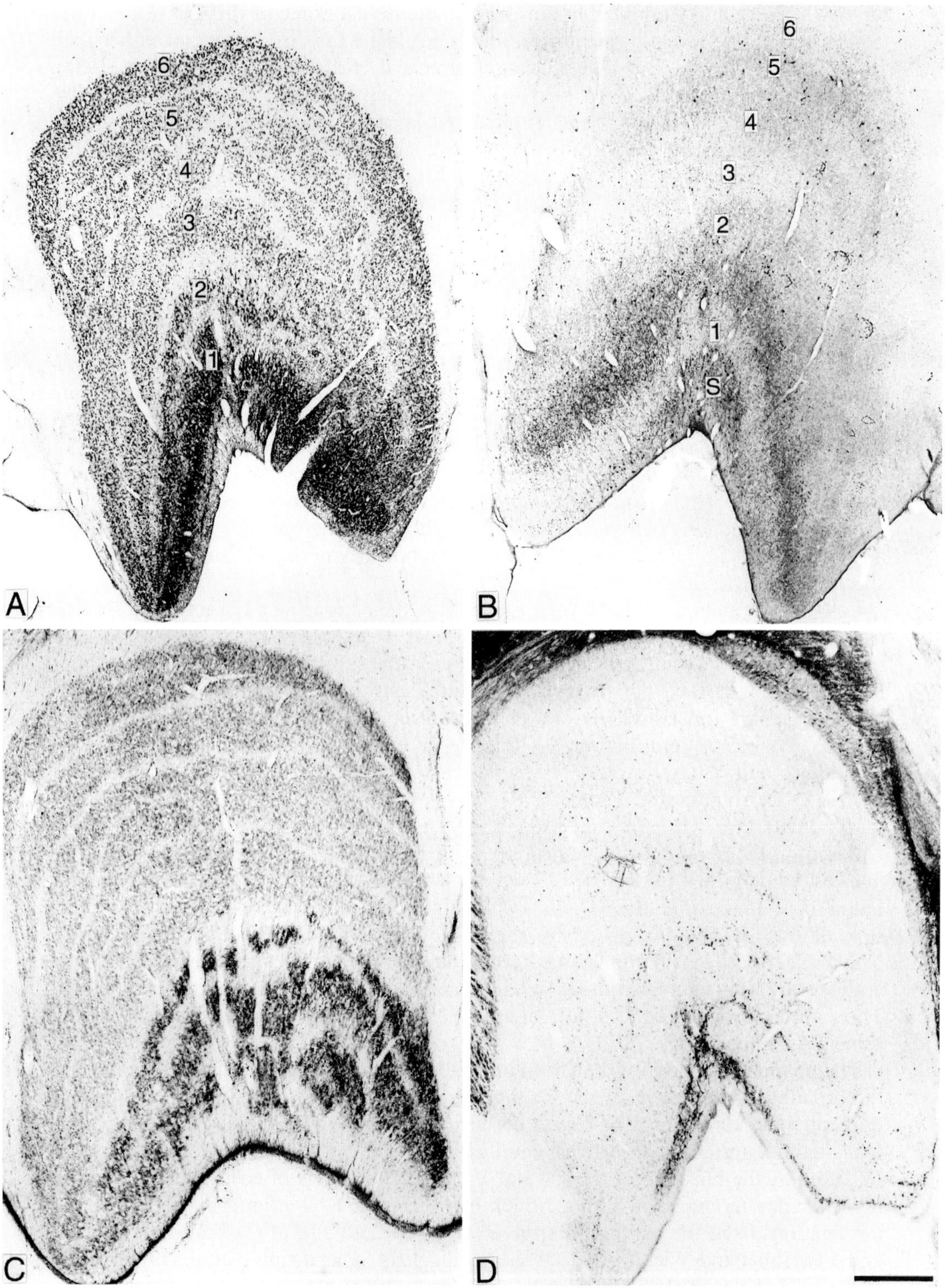
6
5
4
3
2
1
A
6
5
4
3
2
1
S
B
C
D

Fig. 80. Frontal sections of dorsal lateral geniculate nuclei from macaque monkeys, stained immunocytochemically for parvalbumin (A), calbindin (B), by the monoclonal antibody, CAT301 (C) or for alpha-type II calcium/calmodulin dependent protein kinase (D). In A and C, cells in the principal laminae are stained, with enhanced staining, especially in C of the magnocellular layers. In B and D, mainly cells in the S layers and interlaminar plexuses are stained. Bar 100 μm. From material illustrated in Jones and Hendry (1989), Hendry et al. (1988d) and Benson et al. (1991b).

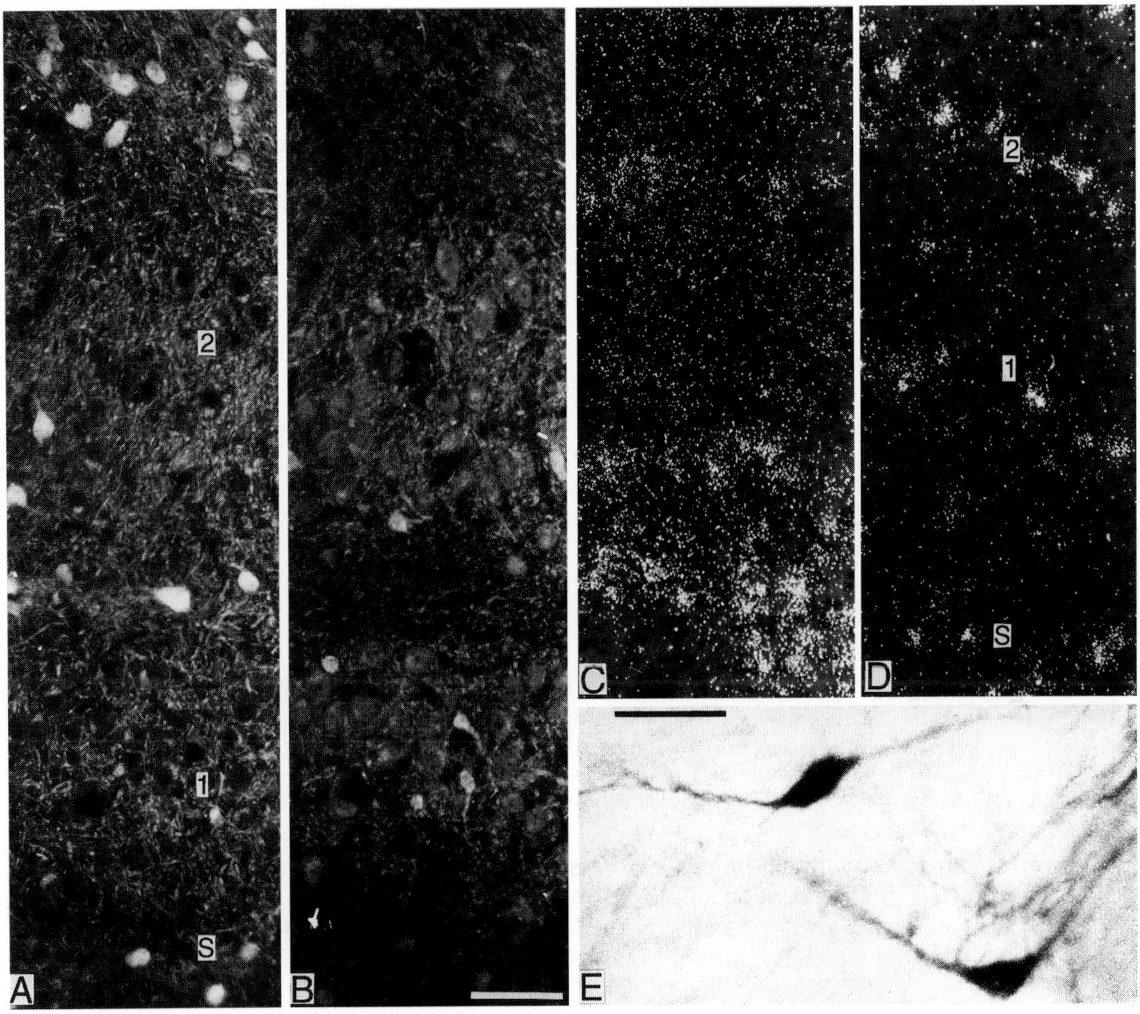

Fig. 81. A,B: Fluorescence photomicrographs from the same microscopic field showing (A) calbindin immunoreactivity in cells of the S layers and interlaminar zones between layers 1 and 2 and between layers 2 and 3 in the dorsal lateral geniculate nucleus of a macaque, with parvalbumin immunoreactive cells (B) confined to layers 1, 2 and 3. C,D: Matched dark field photomicrographs of autoradiographs, showing in situ hybridization of cRNA probes specific for alpha-type II calcium/calmodulin dependent protein kinase (CAMKII-α) (C) and 67kDa GAD (D) mRNAs. CAMKII-α mRNA is found in S layer and interlaminar cells, while GAD mRNA is found in cells of both the S and principal layers (1, 2). Bar 100 μm. E: CAMKII-α immunoreactive cells of the S layers. Bar 10 μm. From material illustrated in Jones and Hendry (1989) and Benson et al. (1991b).

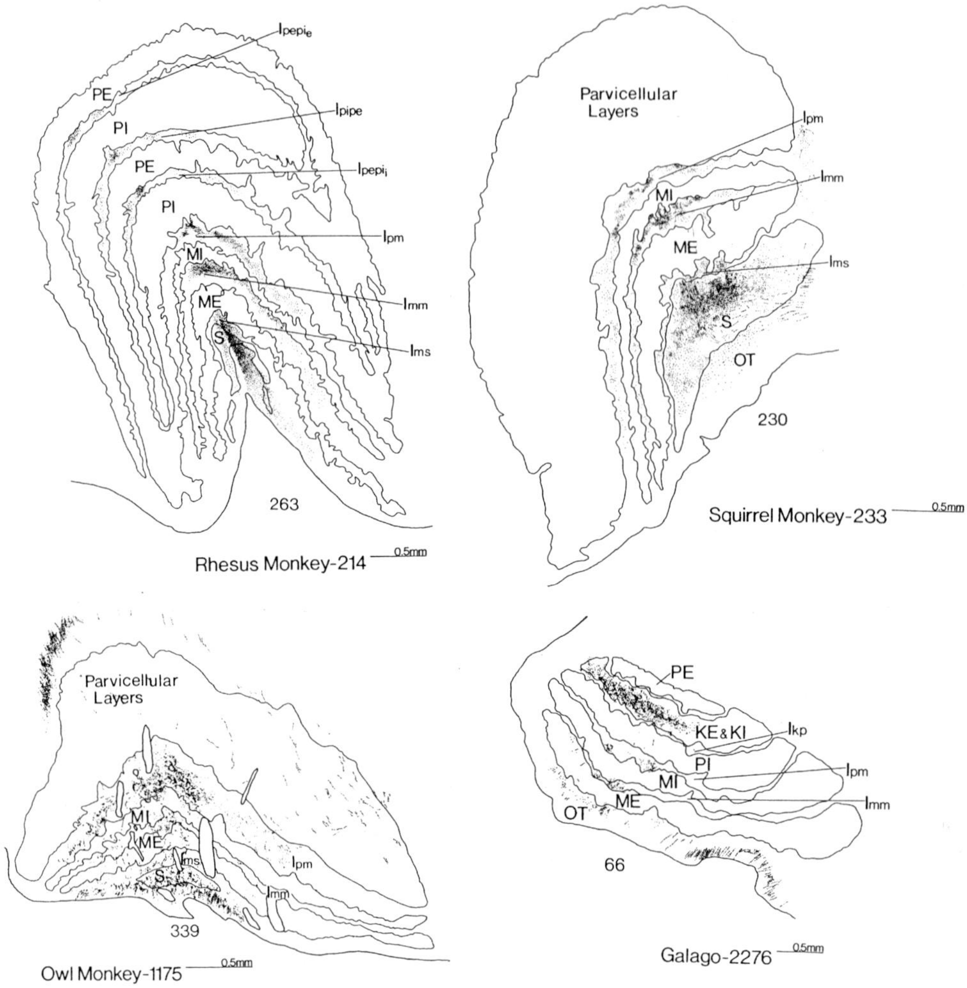

Fig. 82. Distribution of anterogradely labeled tecto-thalamic fibers in the S layers and interlaminar zones of the Old World, rhesus monkey (*Macaca mulatta.*), New World, squirrel monkey (*Saimiri sciureus*) and owl monkey (*Aotus trivirgatus*), and the prosimian *Galago crassicaudatus*. From Harting et al. (1991a).

cortex, further circuitry leading into the surrounding extrastriate areas and beyond may preserve the segregation of the M and the P channels (Shipp and Zeki, 1985; Livingstone and Hubel, 1987a,b; DeYoe and Van Essen, 1985; Hubel and Livingstone, 1990; Maunsell et al., 1990; Lund, 1994), although this view is not shared by all (Nealey and Maunsell, 1994; Ferrera et al., 1994).

A third, immunocytochemically and anatomically defined channel through the monkey dorsal lateral geniculate nucleus, is formed by the small cells of the S and inter-

→

Fig. 83. Frontal sections in anterior (A) to posterior (L) order through the ventral posterior medial (VPM) nucleus of a cynomolgus monkey (*M. fascicularis*), showing the cytochrome oxidase-stained aggregations that extend, rod-like, through the anteroposterior extent of the nucleus. One of the rods is indicated at all levels by an arrow. The cytochrome oxidase-weak matrix of VPM is indicated by s. Note how it expands posteriorly. Bar 250 μm. From material illustrated in Rausell and Jones (1991a).

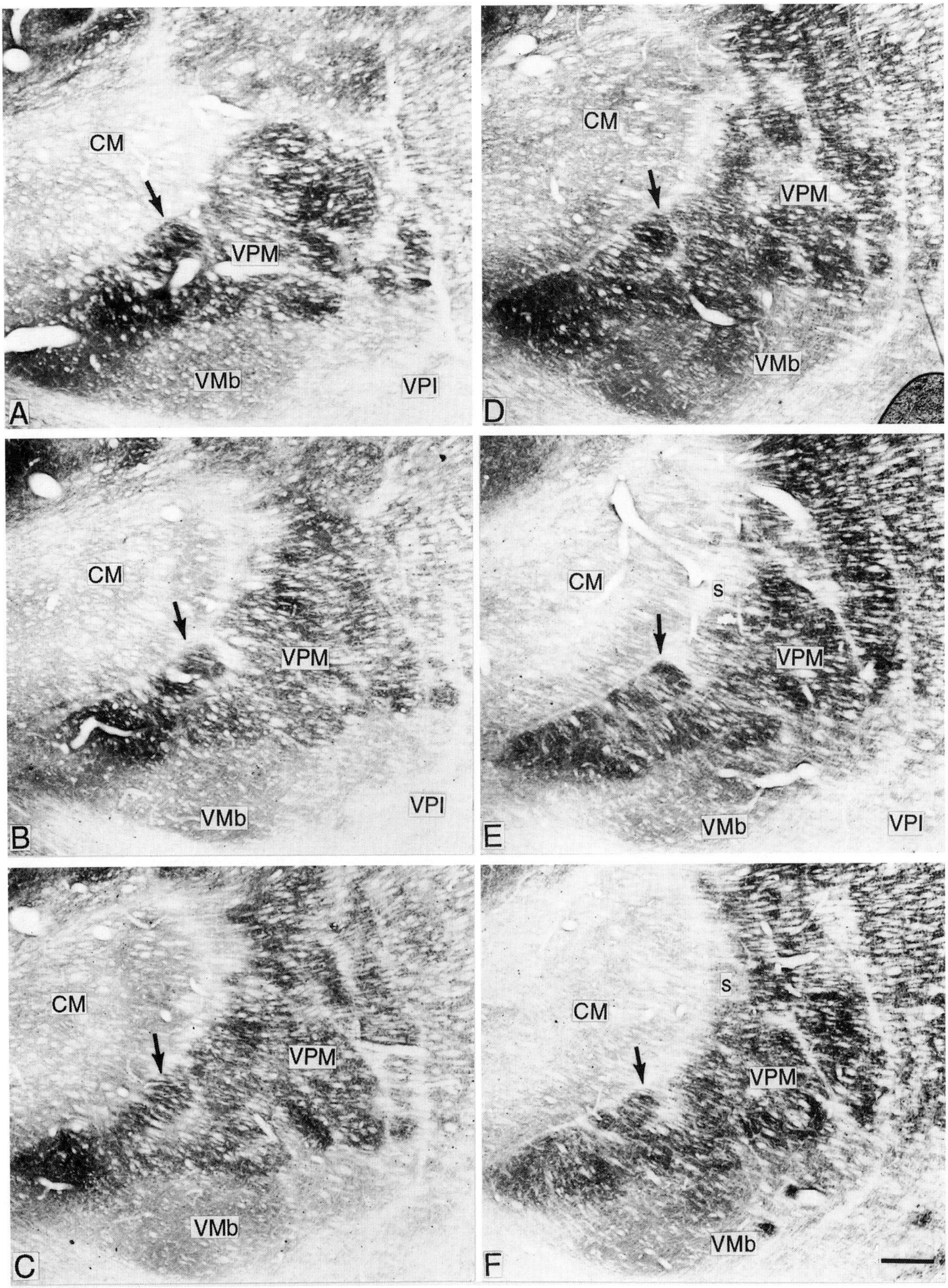

Fig. 83. (continued)

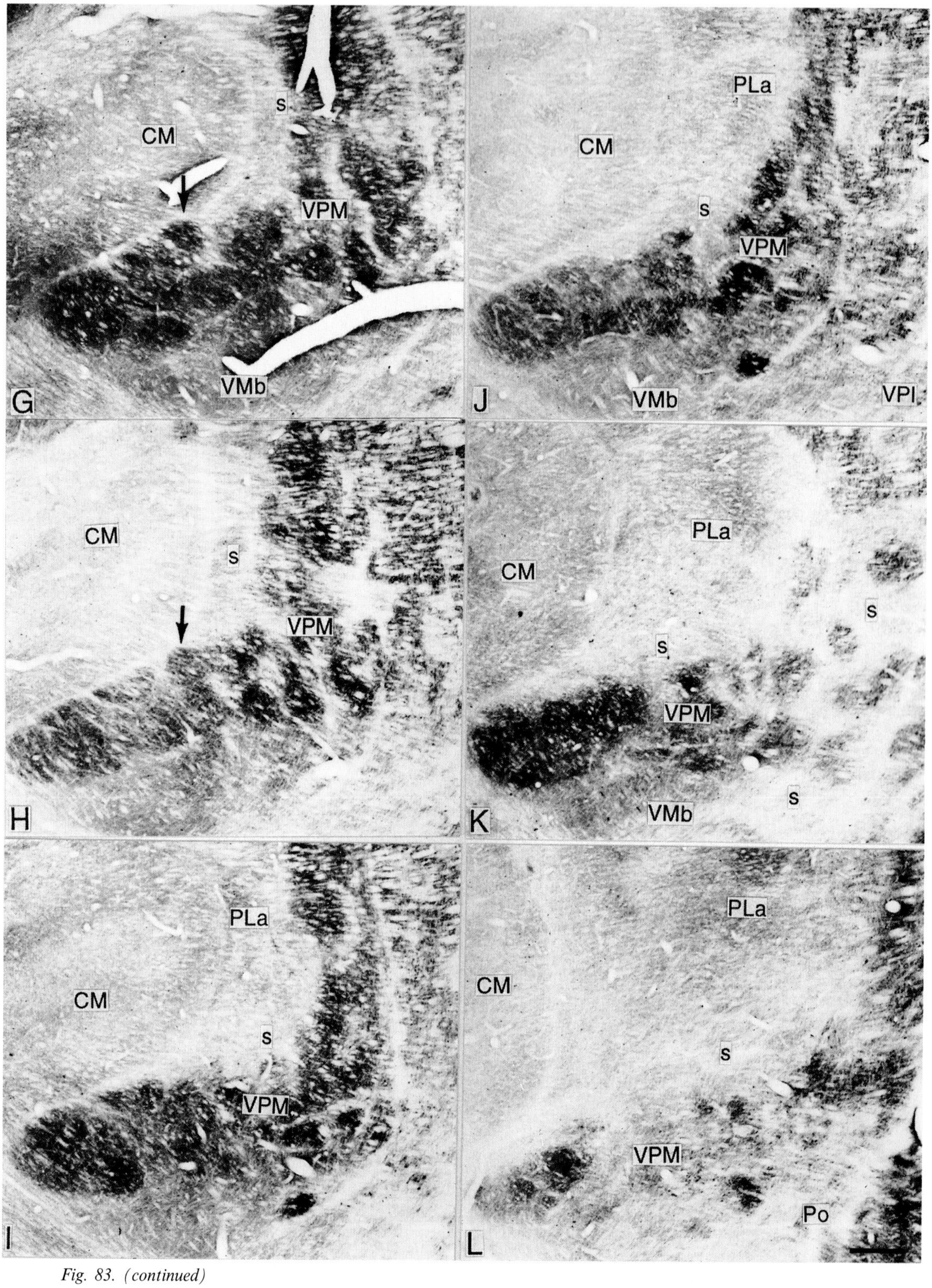

Fig. 83. (continued)

calated layers (Figs. 22, 80, 81). These are characterized by immunoreactivity for 28kDa calbindin, and for CAMII kinase-α (Jones, 1988; Jones and Hendry, 1989; Tighilet et al., 1998), and project only to superficial layers (I-III) of the visual cortex (Ogren and Hendrickson, 1976; Wong-Riley, 1974; Fitzpatrick et al., 1983; Livingstone and Hubel, 1982; Hendry and Yoshioka, 1994) (Fig. 79). These cells can be identified readily in humans and New World monkeys as well. The extent to which the S and intercalated layers are innervated by a specific class of retinal ganglion cell is not known. Anatomical tracing studies indicate that one of the principal sources of inputs to these cells is the tectum, especially the superficial layers of the superior colliculus (Benevento and Fallon, 1975; Kaas et al., 1978; Harting et al., 1978, 1980, 1986, 1991) (Fig. 82). Their projection to the cortex extends across areas 17 and 18 (Yukie and Iwai, 1981). The superficial projection and the innervation by both the tectum and retina has led to the cells of the S and intercalated layers being likened to the W cells of the cat dorsal lateral geniculate nucleus that are selectively innervated by sluggishly responding retinal ganglion cells with slowly conducting axons. There is no physiological justification for this analogy at the present time. The cells of the small-celled system may, in fact, form a matrix comparable to that described in the ventral and medial geniculate nuclei (Sections 10.3, 10.8), in this case extending through the dorsal lateral geniculate and inferior pulvinar nuclei (Tighilet et al., 1998) (Fig. 22E).

In the dorsal lateral geniculate nucleus of the *Galago*, the two small-celled layers (4 and 5) which are thought to be comparable to the interlaminar zones of the monkey (Fig. 4A), also stain specifically for calbindin and project to superficial layers of the visual cortex. The cells of the parvocellular laminae stain only for parvalbumin and project to middle layers, while the cells of the magnocellular laminae which also project to middle layers, stain for either calbindin or parvalbumin (Diamond et al., 1993; Johnson and Casagrande, 1995). Retinal ganglion cell inputs to the various layers of the *Galago* dorsal lateral geniculate nucleus suggest a division comparable to the X, Y and W cells of cats and may target specific layers, the W-like axons ending among the calbindin positive small cells (Irvin et al., 1986; Norton et al., 1988).

10.7. FUNCTIONAL COMPARTMENTS WITHIN THE VENTRAL POSTERIOR NUCLEI

10.7.1. Introduction

The ventral posterior nucleus of all primates is divided by a medullary lamina into medial (VPM) and lateral (VPL) subnuclei. These subnuclei are the termini of lemniscal fibers arising in the contralateral principal trigeminal and dorsal column nuclei respectively (reviewed in Jones 1985a; see also Jones et al., 1986b and Rausell et al., 1992). The body surface is represented systematically from medial to lateral, with the interior of the mouth represented most medially in VPM and the tail or ano-coccygeal region represented most laterally in VPL (Mountcastle and Henneman, 1949, 1952; Rose and Mountcastle, 1959; Poggio and Mountcastle, 1960, 1963; Pubols, 1968; Loe et al., 1977; Dykes et al., 1981; Jones and Friedman, 1982; Kaas et al., 1984). A part of the body such as a finger is represented as a curved lamella which extends through the dorsoventral and anteroposterior dimensions of VPL and as an elongated column through VPM. Each lamella or column represents the major region of terminations of axons arising from cells in the somatotopically related part of the principal trigeminal or dorsal column nuclei. Within VPL in many primates, including humans, a partial medullary lamina may separate the terminations of fibers carrying inputs from the hand and foot and in others, such as the *Galago*, further lamellae tend to separate off the representations of the individual digits.

10.7.2. Morphologically defined subunits in the trigeminal representation

The lamellar pattern of body representation is most obvious in VPL. In VPM, the representations of individual regions of the face, peri- and intraoral regions, although extending anteroposteriorly through the nucleus, are rod-like rather than lamellar. In

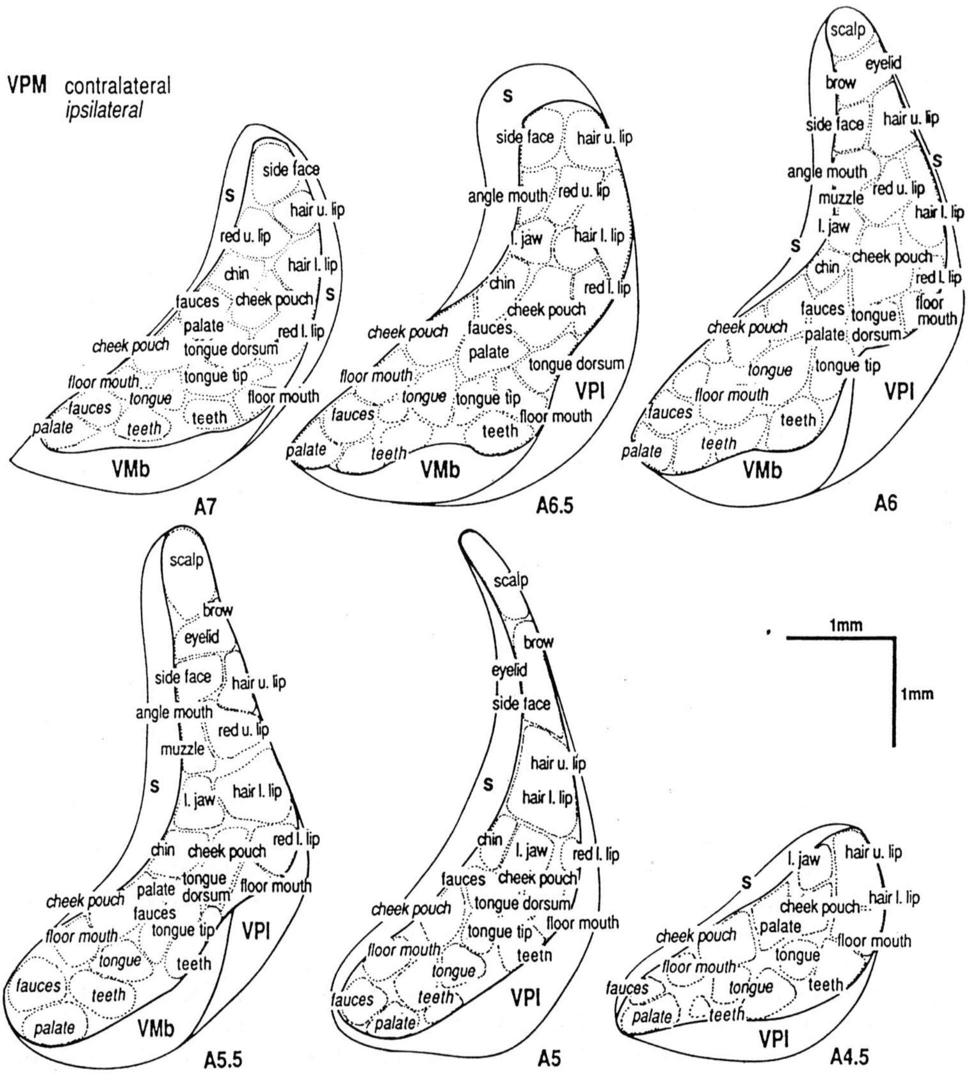

Fig. 84. Camera lucida drawings of frontal sections through the VPM and adjoining nuclei of a macaque monkey, in anterior (A7) to posterior (A4.5) order, showing the cytochrome oxidase stained rods (interrupted lines) and the representation of the contralateral (normal type) and ipsilateral (italic type) sides of the mouth, face and perioral regions, as derived from single and multiunit mapping. After Rausell and Jones (1991b).

monkeys, these rod-like units can be morphologically demonstrated by histochemical staining for cytochrome oxidase (CO) (Jones et al., 1986a,b; Rausell and Jones, 1991a; Fig. 83). The CO-stained periodicities and the aggregations of relay cells contained within them have been called 'rods' on account of their anteroposteriorly elongated shape. Their diameters are variable and measure from 200 to 800 μm. Single unit recordings of responses to peripheral stimuli show that all neurons throughout the antero-posterior extent of a rod have overlapping receptive fields on the same part of the face or on mucous membranes inside the mouth (Jones et al., 1986a; Rausell and Jones, 1991b) (Fig. 84). The relay cells of each rod project to a restricted focus in the somatosensory cortex (Rausell and Jones, 1991b; 1995).

The density of CO staining in the rods of VPM is determined by the large numbers of mitochondria in the large terminals of the trigeminal lemniscal fibers and the focal nature of the staining is a reflection of the manner in which trigeminal fibers bearing information from a single region of the body surface terminate on a restricted group of relay cells in VPM (Jones et al., 1986b; Rausell and Jones, 1991b). In VPL, rods are not visible but the CO staining pattern is often lamellar and within each lamella there is physiological evidence that relay cells are organized in rod-like domains comparable to those in VPM. If a microelectrode is introduced horizontally from behind and traverses VPL of a monkey from posterior to anterior, the neurons that it successively encounters over long distances all have overlapping receptive fields on the same part of the body surface (Jones and Friedman, 1982). When a microelectrode traverses VPL from dorsal to ventral, however, the successively encountered neurons within 100–200 μm steps show rapid shifts in the locations of their receptive fields to different parts of the body surface (Jones and Friedman, 1982). The results of microelectrode mapping in the three standard planes revealed that cells with overlapping receptive fields on the same part of the body surface are arranged in narrow, rod-like domains extending with a somewhat curving trajectory through the anteroposterior extent of VPL (Jones and Friedman, 1982; Jones et al., 1982). A lamella representing a finger, therefore, is made up of a series of rods representing parts of that finger.

The rod-like organization of neurons representing the same small part of the body in VPL of the monkey, as in VPM, is set up by the manner of deployment of medial lemniscal fiber terminations within the nucleus. The terminals of single medial lemniscal fibers are restricted anteroposteriorly (Jones, 1983a) but groups of fibers arising from clusters of cells in the trigeminal or dorsal column nuclei are grouped to form rod-like domains of terminations (Jones et al., 1982).

10.7.3. Modality-specific relay channels in VPL and VPM

Cells along a rod in VPL and VPM not only have overlapping receptive fields on the same part of the body, face or intraoral mucous membranes, but are also activated by stimuli that affect the same class of peripheral sensory receptor (Poggio and Mountcastle, 1963; Jones et al., 1982; Kaas et al., 1984). Neurons receiving inputs from receptors in hairy skin, those receiving inputs from glabrous skin and those receiving inputs from slowly- or rapidly-adapting mechanoreceptors are contained in different rods. A lamella representing a body part in VPL is therefore made up of multiple, elongated rod-like groupings of neurons each representing a small portion of that body part and receiving inputs from a specific class of peripheral mechanoreceptor. This is a basis for preservation of place and modality specificity in the relay of tactile information to the somatosensory cortex.

10.7.4. Modality segregation among subdivisions of the VPL nucleus

The VPL nucleus of monkeys is incompletely divided into zones in which neurons are selectively activated by receptors located in skin or in deep tissues such as muscles and joints (Friedman and Jones, 1981; Jones and Friedman, 1982). In humans (Hirai and Jones, 1989a; Section 4.4, 10.1), the divisions of the VPL are visible as distinct subnuclei. These were called VPLa and VPLp by Hirai and Jones and V.c.e.a and V.c.e.p by Hassler (1959) (Figs. 11–13, 64, 68).

In macaque monkeys, neurons responding specifically to movements of joints or manipulation of muscles and tendons are located in a thin, 300–500 μm thick shell that covers the anterior and dorsal surfaces of VPL (Friedman and Jones, 1981; Jones and Friedman, 1982) (Fig. 65). Neurons specifically activated by Group I muscle afferents are restricted to the anterior part of this shell (Maendly et al., 1981). Those in dorsal parts may be activated by other classes of muscle and joint receptor. The remaining larger division of VPL, forming a large core to the nucleus, contains neurons responding only to cutaneous and subcutaneous mechanoreceptors, organized in rods, as described above. In humans, the shell, i.e. the VPLa nucleus, is much thicker relative to the core (i.e. VPLp) than in monkeys. In humans, cells responding to movements of joints and/or deep pressure are found in the shell (in VPLa), anterodorsal to those responding to light cutaneous stimuli (in VPLp) (Lenz et al., 1988a,b).

Individual medial lemniscal afferents terminate in either the shell or core of VPL (Jones, 1983a), those ending in the core arising in the cuneate and gracile nuclei proper (Boivie, 1978) or the lateral cervical nucleus (Boivie, 1980). Those ending in the deep shell presumably arise in subdivisions of these nuclei in which group I afferent fibers relay (as in nucleus Z of the cat: Rosén, 1969; Landgren and Silfvenius, 1971; Rosén and Sjolund, 1973; Grant et al., 1973; Johansson and Silfvenius, 1977).

The segregation of deep and cutaneous inputs within subnuclei of VPL is a reflection of the fact that the ventral posterior nucleus relays somatosensory information to the cerebral cortex via receptor-specific, parallel pathways. This segregation is maintained in the projection of the subnuclei to the somatosensory cortex, for the relay cells in the shell and core project to different fields of the first somatosensory area (Friedman and Jones, 1981; Jones et al., 1982; Jones, 1983b). Cells in the anterior part of the 'deep' shell project to area 3a in which neurons respond specifically to muscle spindle primary afferents (Phillips et al., 1971; Wiesdanger, 1973; Lucier et al., 1975; Heath et al., 1976; Hore et al., 1976). Neurons in the dorsal part of the shell project to areas 3a and 2 (Fig. 65). Most neurons in area 2 respond to movements of joints (Powell and Mountcastle, 1959) and usually require complex stimuli (Iwamura and Tanaka, 1978; Iwamaura et al., 1993).

Neurons in the central part of the 'cutaneous' core project to area 3b, while more peripheral parts of the core project to areas 3b and 1. Individual relay cells of the core project to area 3b or area 1 only (Jones et al., 1979; Jones, 1983b). Neurons in area 1 respond mainly to rapidly-adapting peripheral mechanoreceptors while those in area 3b located in different patches of cortex respond either to slowly adapting or to rapidly adapting mechanoreceptors (Sur, 1980; Sur et al., 1980, 1984). This implies that VPL relay cells activated specifically by slowly- or rapidly-adapting mechanoreceptors relay this information separately through a further set of parallel channels to the somatosensory cortex (Fig. 85). A majority of the cells in VPL have rapidly-adapting responses and project to area 3b (Jones et al., 1982; Burton et al., 1995) giving a greater density of thalamocortical fiber terminations in area 3b than in the other cytoarchi-

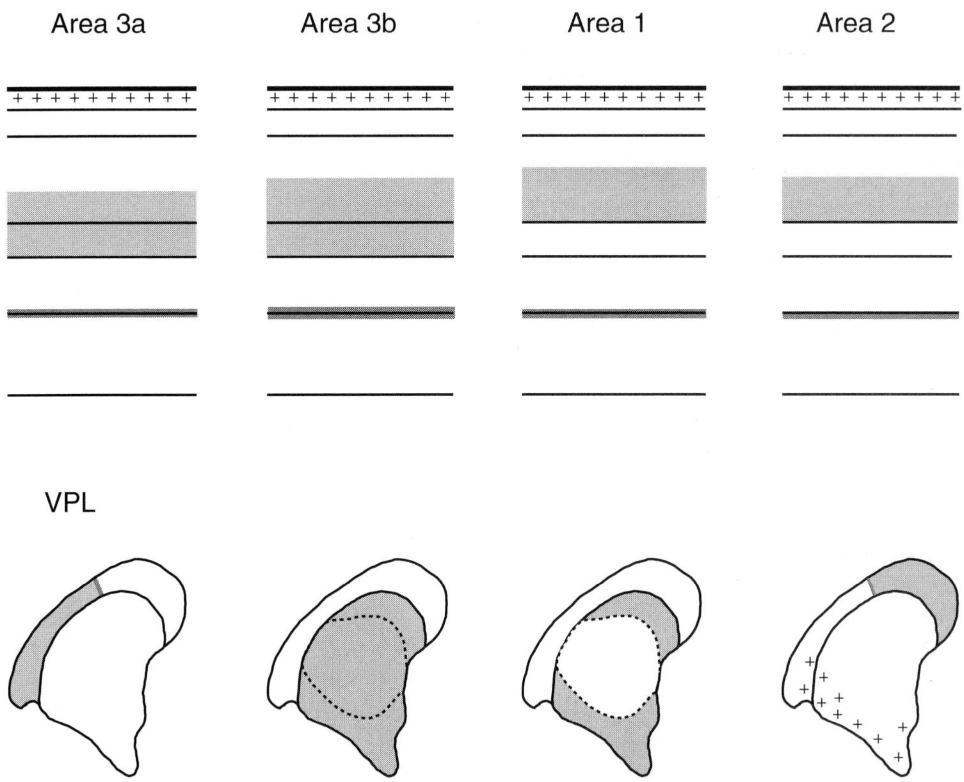

Fig. 85. Area- and layer-specific projections of cells in the ventral posterior nucleus of monkeys to the four fields of the postcentral gyrus. VPL is drawn in parasagittal section as in Fig. 65. Shaded areas in VPL indicate sources of projections to middle layers of each of the four cortical areas. Laminar terminations in cortex are indicated by similar shading. Note that terminations in areas 3a and 3b are in layer IV and deep layer III, while those in areas 1 and 2 are in deep layer III only'. A small projection is found to layer VI in each case. Crosses indicate population of smaller cells projecting to superficial layers (including layer I) in all areas. After Jones (1991).

tectonic areas of the somatosensory cortex (Jones and Powell, 1970; Jones and Burton, 1976). Fibers terminating in areas 3a or 3b end in a band in layer IV and the adjacent part of layer III, as well as in a thinner band at the junction of layers V and VI. Those ending in areas 1 or 2 end in the deep aspect of layer III and at the layer V/VI border (Fig. 85).

In VPM, neurons have the same types of receptor specificity as those in VPL, although neurons responding to manipulation of muscles and joints are rarely encountered and there is no obvious division of the nucleus into an anterodorsal deep shell and a central cutaneous core (Jones et al., 1982; Rausell and Jones, 1991a). Neurons receiving inputs from the mucous membranes of the lips, mouth and tongue tend to have rapidly adapting responses comparable to those found in the majority of neurons in VPL. In macaque monkeys (Jones et al., 1986b; Rausell and Jones, 1991a) there is a large representation of the ipsilateral sides of the lips, tongue and interior of the mouth in the medial part of VPM (Fig. 84) while that in VPM of squirrel monkeys is very small (Ganchrow and Mehler, 1986). The relative sizes of the ipsilateral representations are maintained in the ipsilateral representation in area 3b of the somatosensory cortex (Manger et al., 1995, 1996). It is probable that an expanded ipsilateral representation is found only in monkeys that possess cheek pouches (i.e. the *Cercopithecinae.*), with which a particular kind of food foraging and eating behavior is associated (Corlett and Lucas et al., 1990).

10.7.5. Histochemical compartments of the ventral posterior nucleus as relays for lemniscal and spinothalamic pathways

A different set of parallel pathways through the VPM and VPL nuclei is formed by the histochemically and immunocytochemically defined compartments described in Section 6.2. In monkeys (Jones et al., 1986a,b; Rausell and Jones 1991a; Rausell et al., 1992) and *Galagos* (Diamond et al., 1993) two compartments can be defined by histochemical and immunocytochemical staining. A matrix of small relay cells forming a background to both VPL and VPM, shows weak histochemical staining for cytochrome oxidase (CO) and its small relay cells are immunopositive for 28kDa calbindin, but immunonegative for parvalbumin (Figs. 24, 27). This matrix extends uninterruptedly from VPM and VPL into the VMb and VPI nuclei which form parts of the whole ventral posterior complex of nuclei, and into adjacent nuclei which include the anterior pulvinar, posterior, and ventral lateral. It forms a prominent strip (s) along the medial border of VPM (Figs. 26, 86). Superimposed on this matrix in VPL, VPM and VLp is a second compartment formed by mixed large and medium sized cells. This compartment stains densely for CO and its relay cells are immunopositive for parvalbumin, but immunonegative for calbindin. Both compartments contain GABAergic interneurons (Fig. 38). The CO-rich compartment of large, parvalbumin positive cells in VPM forms the rod-like aggregations of neurons with overlapping peripheral receptive fields described above. The CO- rich rods are separated from one another by the small-celled, calbindin positive matrix compartment which expands medially, ventrally and posteriorly. VPL is dominated by the CO-rich compartment and by clusters of large, parvalbumin positive cells. At intervals however, there are isolated patches of CO-weak matrix filled with calbindin-immunoreactive cells (Fig. 86 G-L).

The two compartments of VPL and VPM, like similar compartments in the medial geniculate complex of monkeys (Hashikawa et al., 1991; Molinari et al., 1995), contain relay cells that project differentially upon the cerebral cortex. The parvalbumin positive cells in the CO-rich compartment project to middle layers of the primary somatosensory cortex while the calbindin positive cells in the CO-weak matrix project to superficial layers of cortex, unconstrained by cytoarchitectonic borders. On the input side, the CO-rich compartment is dominated by terminations of medial lemniscal or principal trigeminal afferent fibers, while the CO-weak compartment is dominated by terminations of spinothalamic or spinal trigeminothalamic fibers (Rausell and Jones, 1991b; Rausell et al., 1992), including those specifically arising in the marginal zone of the caudal spinal trigeminal nucleus (Craig et al., 1994) and, thus, likely to be nociceptive-specific. The lamina-specific thalamocortical projections of the two compart-

Fig. 86. Trios of adjacent frontal sections through the ventral posterior nucleus of a macaque monkey, in anterior (A, B, C) to posterior (J, K, L) order, stained for cytochrome oxidase (CO), parvalbumin (PARV.) or calbindin (CALB.), showing the expansion of the CO-weak, parvalbumin-weak, calbindin-strong matrix zone (s) towards the posterior pole of the ventral nuclear complex where it merges with the posterior nucleus (Po). Note in D, E and G, H the CO/PARV.-weak zones in the interior of VPL. Bar 1 mm. From material illustrated in Rausell et al. (1992).

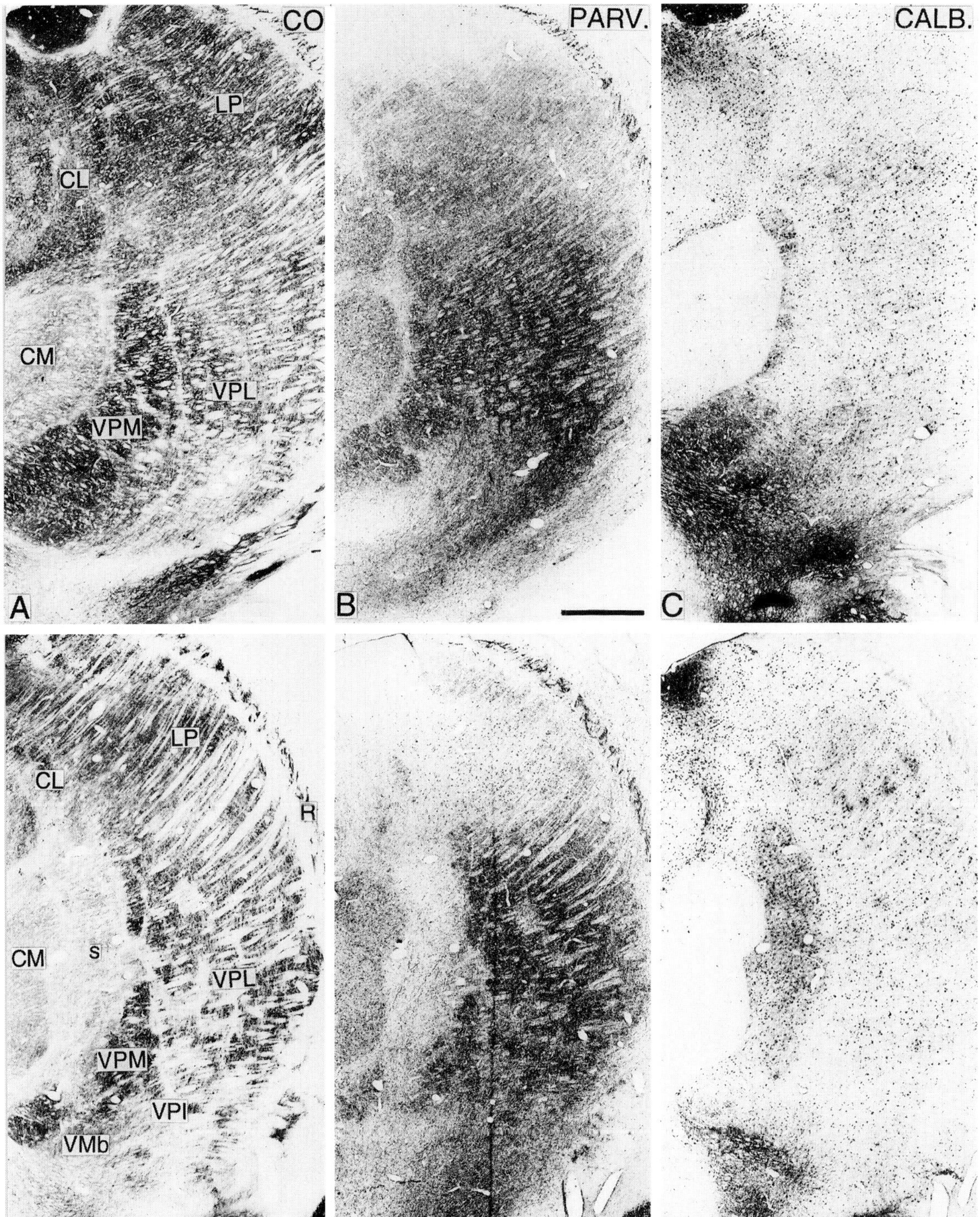

Fig. 86. (continued)

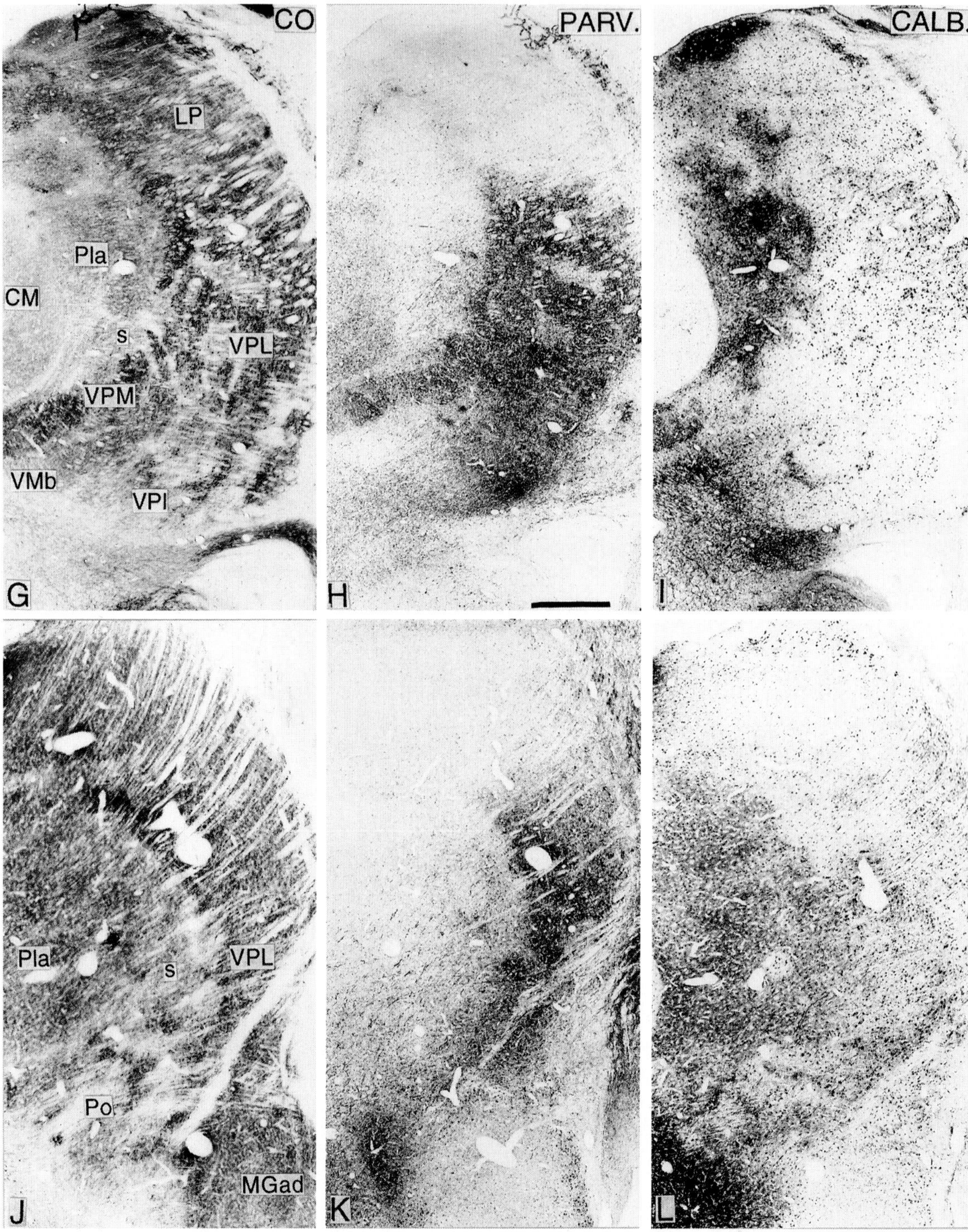

Fig. 86. (continued)

ments therefore form two further, parallel relay channels through the thalamus to the somatosensory cortex.

10.8. NUCLEI ASSOCIATED WITH THE PAIN SYSTEM

10.8.1. Introduction

Studies of the association of the thalamus with the phenomenon of pain have a long history and there is still no agreement as to the precise roles of the numerous thalamic nuclei in which fibers of the spinothalamic tract terminate or which have been related to pain mechanisms on other grounds. The distribution of the spinothalamic tract has been extensively studied in monkeys so that it is now possible to give a comprehensive description. Neurons receiving inputs from peripheral nociceptors have also been thoroughly characterized in the ventral posterior nucleus and to a lesser extent in other nuclei of monkeys and comparable information has been gathered from the human thalamus. With these data at our disposal, it is possible to draw a number of conclusions about the thalamic pain system in primates.

10.8.2. Distribution of the spinothalamic and spinal trigemino-thalamic pathways in monkeys

The distribution of these fibers in the monkey thalamus is widespread and, unlike the other major afferent pathways, not restricted to a single nucleus. The fiber distribution is also characteristically patchy, often with large gaps between isolated concentrations of terminals. The terminations of the fibers are, to a large extent, associated with regions of the thalamus in which cytochrome oxidase activity is low.

When labeled en masse, the main concentrations of spinothalamic fiber terminations are in the VPL, VLp and adjacent nuclei of the ventral posterior complex, notably VPI, as well as in the greater part of the central lateral nucleus and in the posterior nucleus (Mehler, 1969; Boivie, 1979; Berkley, 1983; Pearson and Haines, 1980; Burton and Craig, 1983; Asanuma et al., 1983b; Apkarian and Hodge, 1989; Mantyh, 1983; Ralston and Ralston, 1992; Rausell et al., 1992). However, isolated patches of terminations can be found scattered through other nuclei of the intralaminar system, including the parafascicular and centre médian nucleus, as well as in the lateral posterior and anterior pulvinar nucleus. The terminations in the central lateral nucleus are particularly concentrated around its large posterior cells, including those which form the so-called densocellular division of the mediodorsal nucleus. In the posterior nucleus in which there is also a dense concentration, the terminals spill over into the limitans-suprageniculate and magnocellular medial geniculate nuclei. In the VPL and VLp nuclei, the terminations are especially patchy, patches being found throughout the nucleus proper and in the small-celled (s) region that encapsulates the VPM nucleus and extends posteriorly to become continuous with the posterior nucleus.

Of special note is the invariable association of the spinothalamic fiber terminations with regions of reduced cytochrome oxidase (CO) activity (Fig. 87) (Rausell et al., 1992). These regions are further characterized by the presence of small neurons, all or a large proportion of which are immunoreactive for calbindin. The major concentrations of terminals in the central lateral and posterior nuclei are in cytochrome oxidase-weak, calbindin-rich regions. This is also true of the s region of the VPM nucleus and of virtually all adjacent nuclei such as VPI, VMb, anterior pulvinar, and limitans-

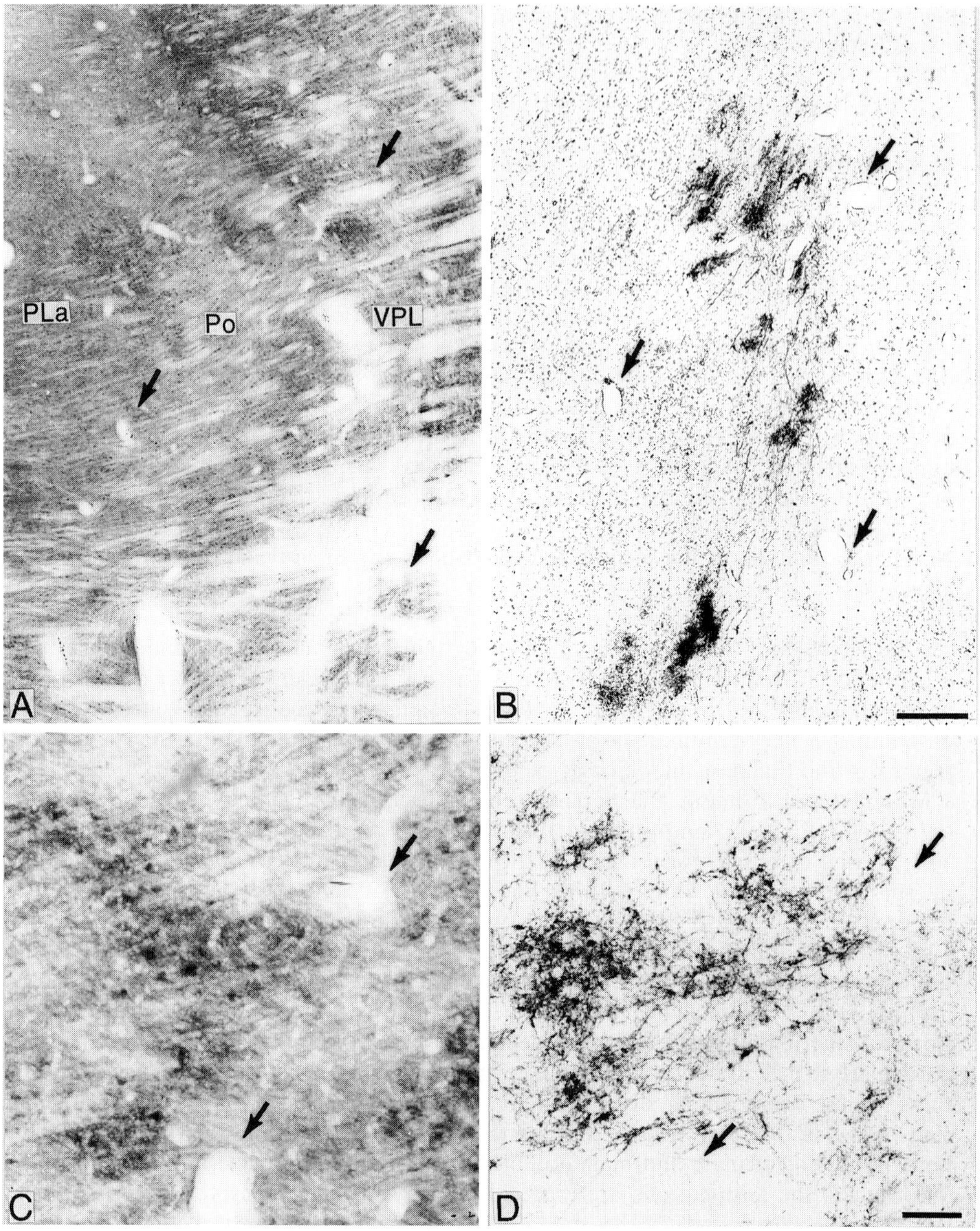

Fig. 87. Anterograde labeling of spinothalamic fiber terminations (B, D) with wheatgerm agglutinin-conjugated horseradish peroxidase, at the posterior pole (A, B) and in the interior (C, D) of VPL in a macaque monkey. A and C, from adjacent sections stained for cytochrome oxidase (arrows indicate same blood vessels), show that fiber terminations tend to be concentrated in CO-weak regions. Bars 1 mm (A, B), 100 μm (C, D). From material illustrated in Rausell et al. (1992).

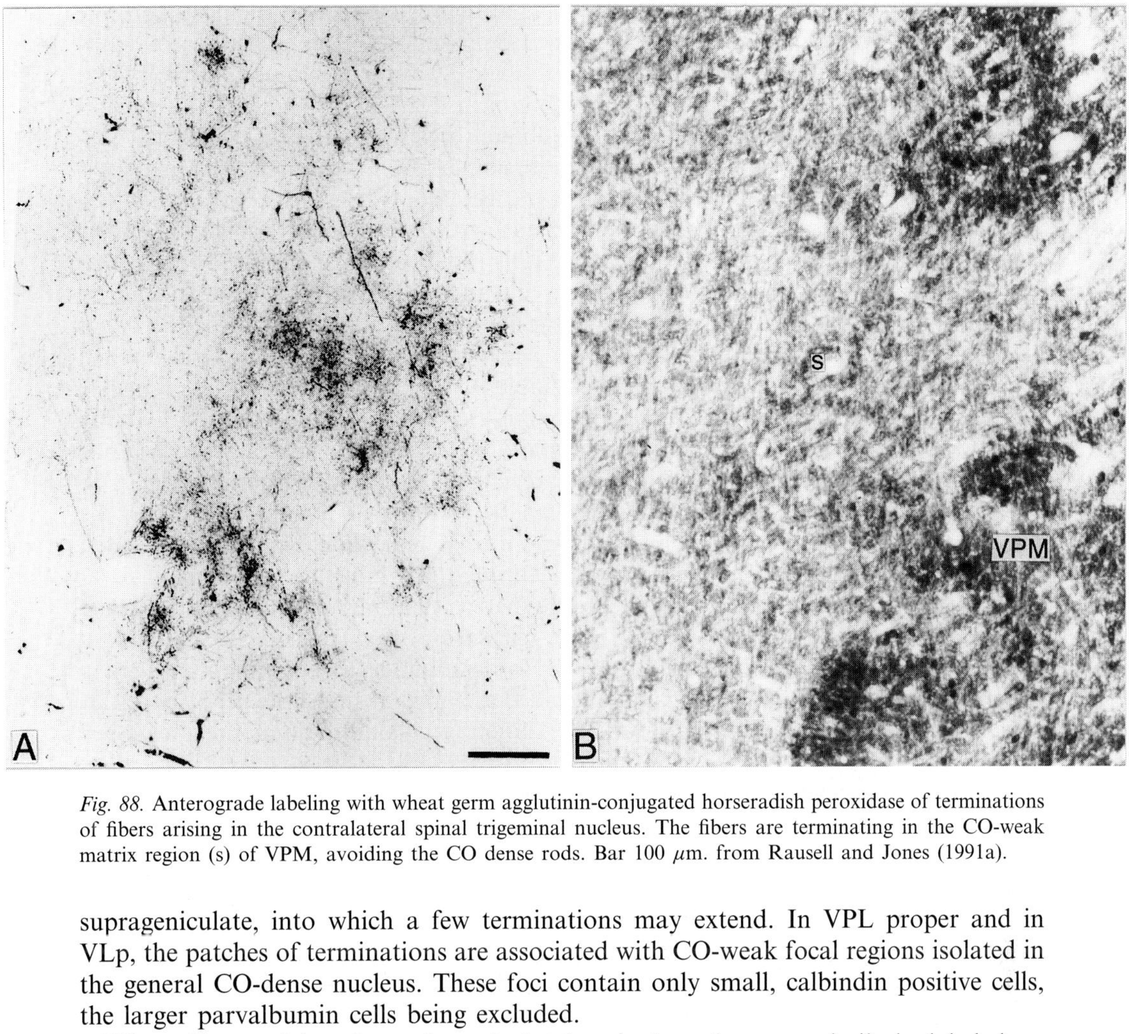

Fig. 88. Anterograde labeling with wheat germ agglutinin-conjugated horseradish peroxidase of terminations of fibers arising in the contralateral spinal trigeminal nucleus. The fibers are terminating in the CO-weak matrix region (s) of VPM, avoiding the CO dense rods. Bar 100 μm. from Rausell and Jones (1991a).

suprageniculate, into which a few terminations may extend. In VPL proper and in VLp, the patches of terminations are associated with CO-weak focal regions isolated in the general CO-dense nucleus. These foci contain only small, calbindin positive cells, the larger parvalbumin cells being excluded.

When fibers arising from the spinal trigeminal nucleus are similarly labeled en masse, their terminations can be identified scattered in many of the same nuclei as the spinothalamic fibers (Rausell and Jones, 1991b) (Fig. 87). The principal concentration of terminations is in the small-celled s region that lies between the VPM and anterior pulvinar nuclei but from there, other foci extend in the small celled matrix between the CO-dense rods of VPM as well as into the Pla and posterior nuclei. There is a particularly dense focus of terminations related to the posteromedial tip of VPM. Other terminations are found in the posterior part of the central lateral nucleus. The association of the spinal trigemino-thalamic terminations with CO-weak regions in which small cells predominate, is as marked as in the case of the spinothalamic terminations. The same regions are also made up of a predominance of calbindin-immunoreactive cells. The focus of terminations at the posteromedial tip of VPM is associated with a particularly dense concentration of calbindin immunoreactive cells and fibers (Fig. 27 E). The fibers terminating here are thought to be derived specifically from the marginal layer of the caudal spinal trigeminal nucleus (Craig et al., 1994). The spinothalamic and spinal trigeminothalamic fibers are not themselves calbindin immunoreactive (Rausell et al., 1992). Looking at the region around the posterior pole of the ventral posterior nucleus in CO-stained preparations, especially when matched with adjacent sections stained immunocytochemically for parvalbumin or calbindin (Figs. 87, 88), it is impossible to escape the conclusion that the pole of the nucleus is embedded in a small-celled, CO-weak, calbindin positive matrix in which the majority of spinothalamic and spinal trigeminothalamic fibers terminate. This matrix is difficult to delineate cytoarchitectionically, as reflected in the inconsistency with which it is represented in atlases. The posterior nucleus, for example, is included in the suprageniculate nucleus by Olszewski (1952). The rather prominent anterior expansion of this between the VPM and Pla nuclei is included in VPM by Olszewski and by Rausell and Jones (1991a) and in the posterior nucleus ('Pom') by Boivie (1978). Together, these two small celled, CO-weak, calbindin positive regions form a kind of capsule to the VPM nucleus from which fingers extend into VPM, forming a matrix between the CO-dense, large-celled, parvalbumin-rich rods. A similar matrix is found in VPL and VLp. Whenever the matrix is particularly evident, foci of spinothalamic and spinal trigeminothalamic terminations are invariably found. Posteriorly, the CO-weak, small-celled, calbindin-rich domain can be pictured as a kind of hilus to the thalamus, interposed between the pulvinar and medial geniculate complex at the point at which the thalamus meets the midbrain. Through this hilus, the major pain pathways enter the thalamus and terminate in the matrix. A more medially directed group of fibers separates from those entering the matrix and is distributed to the central lateral and submedial nuclei. In these, the association of the terminations with a matrix comparable to that seen in and around the ventral posterior nuclei is less clear. Nevertheless, the terminations are found in relation to calbindin immunoreactive cells.

10.8.3. High threshold neurons in the ventral posterior nucleus

It has been known for some time that a certain percentage of neurons in the ventral posterior nucleus of monkeys have receptive fields that are much larger than those of neurons typically activated by low threshold mechanoreceptors and that these receptive fields depend upon inputs from the anterolateral pathways of the spinal cord (Perl and Whitlock 1961; Whitlock and Perl, 1961; Pollin and Albe-Fessard, 1979).

Ventral posterior neurons with stimulus-response properties very similar to those of wide dynamic range, spinothalamic tract neurons in the dorsal horn of the spinal cord are few in comparison with those responding only to low threshold mechanoreceptors (Willis et al., 1974; Appelbaum et al., 1979; Kenshalo et al., 1979, 1980; Chung et al., 1986; Ferrington et al., 1987; Surmeier et al., 1986, 1988). The wide dynamic range neurons in the thalamus have large receptive fields and usually respond in a graded manner to both innocuous and noxious mechanical stimuli and commonly to noxious thermal stimuli as well. The number of cells with these properties observed in VPL and VPM of unanesthetized monkeys is small (Casey and Morrow, 1983; Bushnell and Duncan, 1987; Chandler et al., 1992; Morrow and Casey, 1992; Bushnell et al., 1993; Tremblay et al., 1993; Duncan et al., 1993; Apkarian and Shi, 1994). Larger numbers may be encountered however if electrical stimulation of Aβ and Aδ fibers is used as a search stimulus (Chung et al., 1986).

Apart from the wide dynamic range cells, a small number of neurons responding specifically to noxious mechanical or thermal stimuli have also been reported in VPL of monkeys (Kenshalo et al., 1980; Chung et al., 1986; Bushnell et al., 1993; Apkarian and Shi, 1994). The receptive fields of these high threshold cells are small so that the neurons are capable of relaying information relevant to the localization of a noxious stimulus.

Ventral posterior cells responding to noxious thermal and/or mechanical stimuli in monkeys are clustered and the clusters are dispersed (Kenshalo et al., 1979, 1980; Chung et al., 1986). This is in keeping with the dispersed, patchy nature of the small islands of matrix in which spinothalamic fibers terminate in VPL. In VPM, at least, the cells tend to occupy regions of the nucleus comparable to the matrix zone of small calbindin positive neurons, especially around the medial perimeter of VPM where the matrix is at its most obvious (Bushnell et al., 1993; Bushnell and Duncan 1987; Tremblay et al., 1993; Duncan et al., 1993). High threshold neurons in the VPI nucleus (Apkarian and Shi, 1994) are also in a calbindin positive zone.

In humans, wide dynamic range and nociceptive-specific cells can be recorded in the vicinity of the VPL nucleus (Lenz et al., 1994b) and painful sensations including those of noxious heat, are projected to peripheral sites by microstimulation in regions posteroinferior to VPL (Lenz et al., 1993, 1994a). Usually, only nonpainful paresthesiae and non-painful thermal sensations are elicited by microstimulation in the core of VPL. These results suggest that nociceptive specific cells are concentrated around the borders of the ventral posterior nucleus in the human as well in the monkey.

It is interesting that continuous retrograde labeling of cells occurs in the VPI and VMb nuclei and in the medial (s) region of the VPM nucleus after injections of anatomical tracers in the insular cortex (Mesulam and Mufson, 1985; Mufson and Mesulam, 1985), especially in view of the association of the insular and opercular cortex with painful or hypoalgesic syndromes in humans (Biemond, 1956), and the association of these thalamic regions with the terminations of spinothalamic fibers. The posterior nucleus is a source of projections to retroinsular areas of cortex (Jones and Burton, 1976) which have also been implicated in painful conditions or with the alleviation of pain in humans (Biemond, 1956; Lende et al., 1971).

10.8.4. The ventrocaudal thalamic region in the human thalamus

The ill-defined posterior nucleus/ventral posterior matrix region has its equivalent in humans among a group of nuclei that Hassler (1959) recognized in the vicinity of the posterior pole of the ventral posterior nucleus. Most of these nuclei can be identified, from the cytological descriptions of Hassler, with nuclei traditionally given other names in monkeys (Figs. 11–13, Table 2). Hassler's nucleus ventrocaudalis parvocellularis (V.c. pc) is divided into medial (V.c. pc. i) and lateral (V.c. pc. e) subnuclei which are clearly the equivalents of the basal ventral medial (VMb) and ventral posterior inferior (VPI) nuclei respectively. Posterior to and merging with these two nuclei is an ill-defined region which Hassler called porta thalami, seeing it as the gate through which lemniscal and spinothalamic fibers enter the thalamus. The region lies between the posterior pole of VPLp (Hassler's V.c.p.e) and the suprageniculate part of the limitans/suprageniculate nucleus which Hassler called the nucleus limitans portae (Li. por.). This region of transition between the two nuclei is in a position comparable to the posterior nucleus of the present account, although Hassler's description of the cytoarchitecture makes it impossible to exclude adjacent parts of the anterior pulvinar nucleus from it. Hassler calls the region of transition, nucleus ventralis caudalis portae (V.c. por.).

On the basis of myelinated fiber loss in the human brain following lesions of the spinothalamic tract, and clinical experience following stimulation or electrolytic lesions of the region posterior to VPLp, Hassler and Riechert (1959; Hassler 1959) considered that the principal terminations of the pathway bringing pain and temperature messages to the thalamus were 'particularly basal in the *ventrocaudalis parvocellularis*', i.e. in VPI and VMb of the present account. Hassler implies that VMb is the terminus of spinal trigeminothalamic fibers and VPI of spinothalamic fibers. The experimental evidence obtained from monkeys and described above would suggest that Hassler's nucleus ventrocaudalis portae (i.e. Po) is a more likely termination site, although terminals may extend into the other two nuclei.

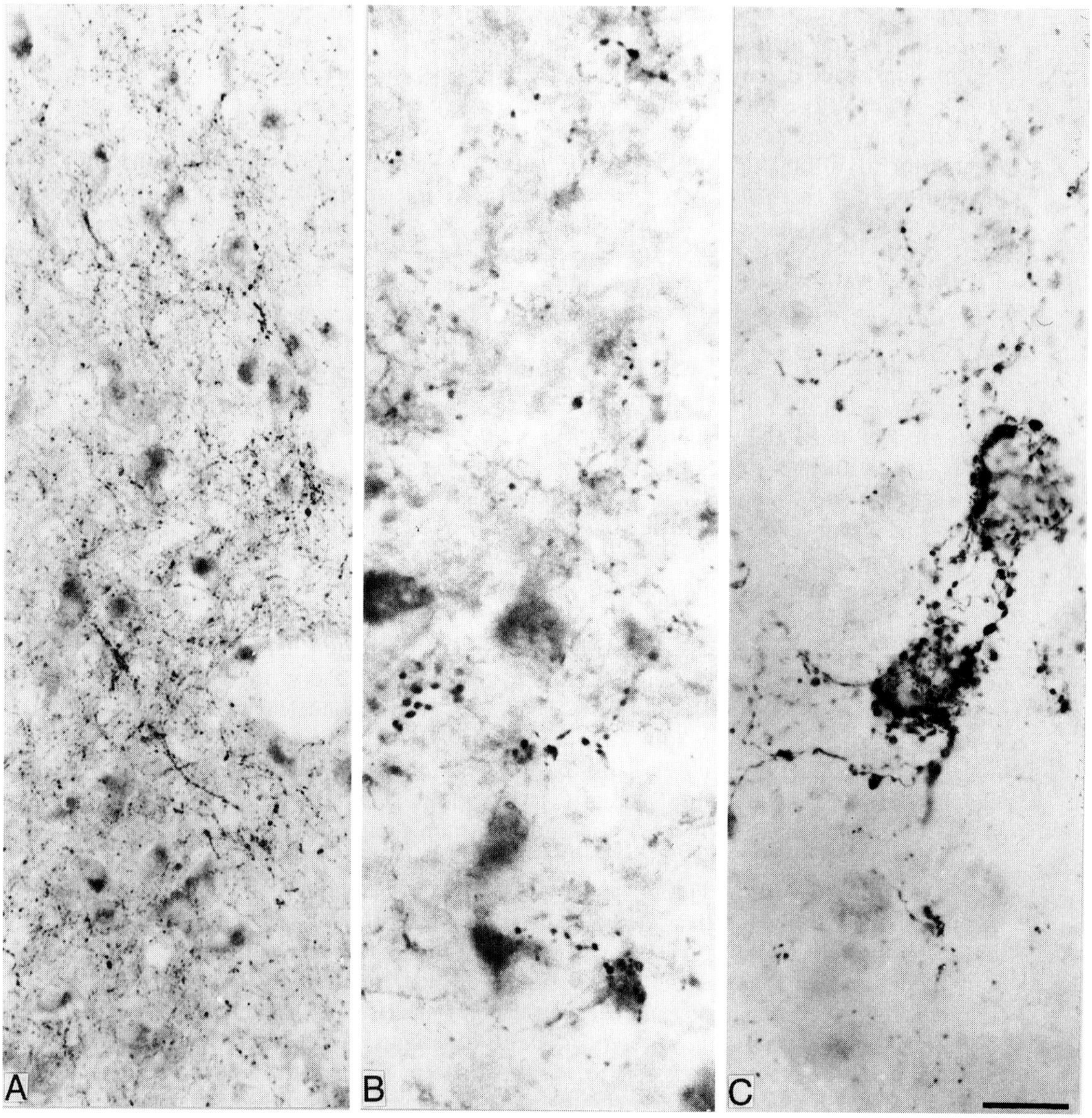

Fig. 89. Tachykinin immunoreactive fibers and terminals in the central lateral (A), anteroventral (B) and mediodorsal (C) nuclei of a human thalamus. Bar 500 μm (A), 20 μm (B, C). From material illustrated in Hirai and Jones (1989b).

In studies of the distribution of degenerating spinothalamic tract fibers in postmortem human brains, Mehler (1966) observed fiber ramifications in VPI, VMb and Po, as well as in VPL. The ramifications were clustered in a manner resembling the terminations of spinothalamic fibers in monkeys. The Po (V.c. por.) nucleus in humans is also filled with ramifications of tachykinin and enkephalin immunoreactive fibers (Hirai and Jones 1959b; Figs. 89, 90). These may arise in the spinal cord.

Single unit recording in humans reveals that the majority of neurons responding to noxious heat are located 'posteroinferior' to the main region of cutaneous responses which corresponds to the core region of the VPL (VPLp) (Lenz et al., 1993). Thermal/pain sensations can be elicited by microstimulation in this region (Hassler and Riechert 1959; Halliday and Logue 1972; Dostrovsky et al., 1991; Lenz et al., 1993). Cells responding to painful mechanical stimuli can be recorded, however, throughout VPLp (Lenz et al., 1994b). Visceral pain can also be elicited from stimulation in this general territory (Davis et al., 1995; Lenz et al., 1994a).

Cells responding to noxious mechanical and thermal stimuli have also been recorded in regions of the human thalamus probably corresponding to the central lateral nucleus, although sometimes identified under other names, (Ishijima et al., 1975; Jeanmonod et al., 1993; Rinaldi et al., 1991; Tsubokawa and Moriyasu, 1975). Stimulation in this region will provoke painful sensations (Sano, 1977; Rinaldi et al., 1991).

A number of different thalamic regions have been lesioned in humans in efforts to relieve intractable pain. The commonest regions subjected to stereotaxic thalamotomy are the general region of VPL, posteromedial regions in the vicinity of the posterior central lateral nucleus or posterior-inferior regions in the vicinity of the Po nucleus in an effort to interrupt the spinothalamic tract or some part of it (Wycis and Spiegel, 1962; Helfant et al., 1965; Spiegel and Wyais, 1966; Albe-Fessard et al., 1970; Hageman and De Grood, 1970; Voris and Whistler, 1975; Colombo, 1984; Frank et al., 1987; Laitinen, 1988; Tasker, 1989; Bosch, 1991). All seem to be effective in the short term but pain commonly returns. Anesthesia, paresthesia and loss of proprioception accompanying lesions in VPL are, however, often permanent.

10.8.5. The nucleus submedius

The nucleus submedius, which has traditionally been regarded as part of the medial group of thalamic nuclei (M. Rose, 1935), achieved some prominence in the recent past as a result of indications that it received potentially nociceptive-specific inputs from the spinal cord and caudal medulla oblongata (Craig and Burton, 1981; Burton and Craig, 1983; Mantyh, 1983; Craig, 1987). In New World monkeys, it is a rather insignificant group of dispersed cells slung beneath the internal medullary lamina close to the anterior pole of the centre médian nucleus (Emmers and Akert, 1963; Jones, 1985a). In humans, its position is not very clear, although Hirai and Jones (1989b) suggested that it might be represented by a small group of cells, richly innervated by enkephalin and tachykinin-immunoreactive fibers displaced ventrally beneath the enlarged centre médian nucleus (Fig. 3H-J, 58). The cortical projections of the nucleus submedius have not been described in monkeys or other primates. In cats, it targets an area of orbital frontal cortex deep in the presylvian sulcus (Craig et al., 1982). For recent studies the reader should refer to Ericson et al. (1995) and in rats to Kawakita et al. (1993).

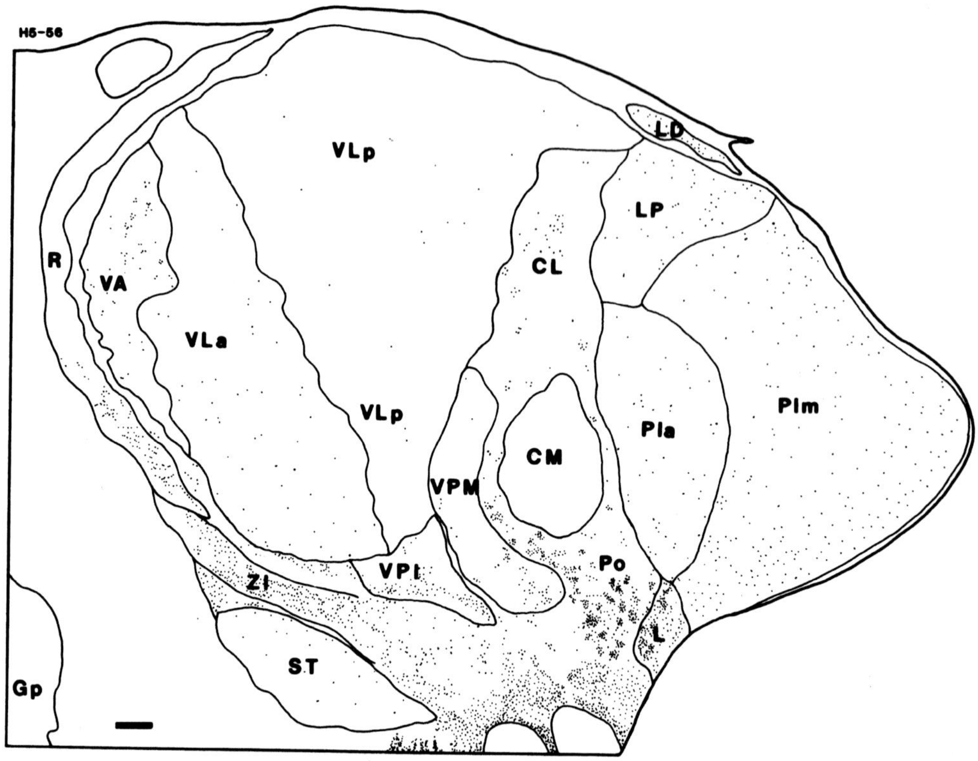
H5-56
LD
VLp
LP
CL
R
VA
VLa
Plm
VLp
Pla
CM
VPM
Po
VPl
ZI
L
ST
Gp

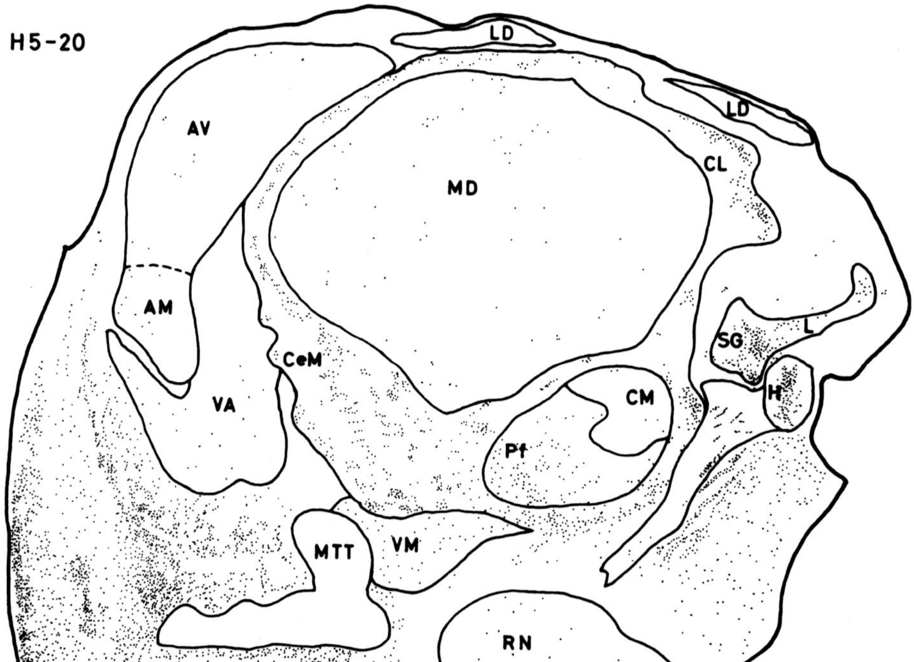
H5-20
LD
LD
AV
MD
CL
AM
L
SG
CeM
H
VA
CM
Pf
MTT
VM
RN

Fig. 90. Camera lucida drawings from parasagittal sections of a human thalamus (A is lateral to B), showing the distribution of tachykinin (A) and metenkephalin immunoreactive fibers (B) in relation to the posterior and intralaminar nuclei. From Hirai and Jones (1989b).

←

11. SUMMARY

This account covers the cytoarchitecture and chemoarchitecture of the thalamus in representative prosimians and monkeys and in the human. Parcellations of the thalamus on the basis of histochemical staining patterns serve to confirm and extend those made on the basis of cyto- and myeloarchitecture alone, and help to reinforce the conviction that the thalamus of all primates is fundamentally similar. The major differences between prosimians and simians are in the relative reduction of the anterior nuclei and expansion of the ventral lateral, posterior intralaminar and pulvinar nuclei in the latter family. In humans, the enlargement of the thalamus parallels that of the cerebral cortex and is especially associated with expansions of the ventral lateral, posterior intralaminar and pulvinar nuclei. The nuclear divisions of the human thalamus are, nevertheless, fundamentally similar to those of the Old World monkey thalamus so that a common nomenclature, despite historical precedent to the contrary, is not only feasible but desirable in order to facilitate the exchange of experimental and clinical information.

The basic circuitry of the thalamus is built up on the basis of glutamatergic thalamocortical relay neurons, divisible into subclasses with different immunocytochemical signatures, and two sets of inhibitory, GABAergic interneurons – those intrinsic to the relay nuclei and those in the reticular nucleus. Differential expression of calcium binding proteins is a major distinguishing feature of relay neurons. Those containing 28kDa calbindin form a small-celled matrix in many thalamic nuclei and project relatively diffusely to superficial layers of the cerebral cortex. Those expressing parvalbumin form the topographically-organized projection to middle layers.

The afferent innervation of the thalamus comes from a variety of nucleus-specific input pathways. The nucleus- and subnucleus-specific terminations of these pathways form the basis of parallel, segregated channels through the thalamus to the cerebral cortex. These parallel pathways are especially evident in the motor and auditory nuclei of the thalamus. In the somatosensory and visual relay nuclei, further functional channels within a major thalamocortical projection are formed by subpopulations of neurons innervated by specific classes of cutaneous mechanoreceptor or retinal ganglion cell.

The excitatory corticothalamic projection, especially acting in concert with the inhibitory reticular nucleus, is a major factor in inducing collective oscillations of large ensembles of thalamic neurons which underlie changes in conscious state. The synapses made by subcortical, cortical and reticular nucleus inputs to the dorsal thalamus are associated with the expression of a wide variety of genes encoding $GABA_A$ receptor subunits and glutamate receptors. Many of these show nucleus-specific patterns of expression. The thalamus is also innervated in a relatively non-specific manner by cholinergic, noradrenergic, serotoninergic and peptidergic fibers of brainstem origin but by few or no dopaminergic fibers. Many of these inputs elicit effects upon the same set of membrane conductances while others have more specific effects related to interactions with particular classes of receptor.

The dorsal thalamus projects not only to the cerebral cortex, including the neocortex, palaeocortex and archicortex, but also to the striatum and amygdala and related basal telencephalic structures. The cortical projection is based upon populations of cells, often with particular morphological and chemical identities, that project in an areal-specific or more diffuse manner to the cortex, their axons commonly terminating in different cortical layers. The striatal projection arises from the intralaminar nuclei and from certain other nuclei whose chemistry is identical to that of the intralaminar nuclei and should probably be included in the intralaminar system.

Many aspects of thalamic function, such as the relay of sensory information and the relay of information in systems that do not form components of obvious sensory systems, are reflected in the organization of the specific relay nuclei of the dorsal thalamus. These are considered individually in Section 10. Other, more generalized functions of the thalamus, such as its role in state dependent activities of the whole forebrain, are best considered in relation to the intrinsic and transmitter-defined circuitry of the thalamus. These are recurrent themes throughout the whole chapter.

12. ACKNOWLEDGMENTS

Personal work reported in the text was supported by grant numbers NS 21377, NS 30101, NS 22317, EY 07193 and MH 44188 from the National Institutes of Health, United States Public Health Service, and by Human Brain Project grant, number MH/DA 52154, jointly funded by the National Institute of Mental Health, the National Institute on Drug Abuse and the National Aeronautics and Space Administration. I thank Drs. D.L. Benson, E. Dell'Anna, S.H.C. Hendry, T. Hirai, C.N. Honda, C.A. Hunt, G.W. Huntley, M.M. Huntsman, P.J. Isackson, M.E. Leggio, X.B. Liu, G. Macchi, M. Molinari, E. Rausell, H.D. Schwark, B. Tighilet and R.A. Warren for their contributions and Mr. Phong Nguyen, Mr. Clyde King and Ms. Hao Truong for technical help.

13. ABBREVIATIONS

AD	Anterodorsal nucleus
AHA	Anterior hypothalamic area
AM	Anteromedial nucleus
Am	Amygdala
Arc	Arcuate nucleus of hypothalamus
AV	Anteroventral nucleus
CeM	Central medial nucleus
CL	Central lateral nucleus
CM	Centre médian nucleus
CN (Cd)	Caudate nucleus
DMH	Dorsomedial hypothalamic area
F	Fornix
FF	Field of Forel
Gpe	Globus pallidus (external division)
Gpi	Globus pallidus (internal division)
H	Habenular nuclei
Hl	Lateral habenular nucleus

Hm	Medial habenular nucleus
HPT	Habenulo-peduncular tract
IC	Internal capsule
IML	Internal medullary lamina
Is	Interstitial nucleus of Cajal
L	Limitans nucleus
LD	Lateral dorsal nucleus
LGd	Dorsal lateral geniculate nucleus
LGv	Ventral lateral geniculate nucleus
LMN	Lateral mamillary nucleus
LP	Lateral posterior nucleus
L-SG	Limitans-suprageniculate nucleus
MD	Mediodorsal nucleus
MDl	Mediodorsal nucleus, lateral divisions
MDm	Mediodorsal nucleus, magnocellular division
MG	Medial geniculate complex
MGad	Anterodorsal medial geniculate nucleus
MGd	Dorsal medial geniculate nuclei
MGmc	Magnocellular medial geniculate nucleus
(mc)	Magnocellular medial geniculate nucleus
MGpd	Posterodorsal medial geniculate nucleus
MGv	Ventral medial geniculate nucleus
ML	Medial lemniscus
MMN	Medial mamillary nucleus
MTT	Mamillothalamic tract
MV	Medioventral (reuniens) nucleus
OT	Optic tract
Pa	Paraventricular nuclei
PC	Pericentral gray matter
Pc	Paracentral nucleus
Pf	Parafascicular nucleus
Pla	Anterior pulvinar nucleus
Pli	Inferior pulvinar nucleus
Pll	Lateral pulvinar nucleus
Pls	Superior pulvinar nucleus
Po	Posterior nucleus
Prg	Pregeniculate nucleus
PT	Pretectum
Pt	parataenial nucleus
Pu	Putamen
R	Reticular nucleus
Re	Reuniens (medioventral) nucleus
Rh	Rhomboid nucleus
RN	Red nucleus
SB (ST)	Subthalamic nucleus
SC	Superior colliculus
SG	Suprageniculate nucleus
SI	Substantia innominata
SM	Stria medullaris thalami

SN	Substantia nigra
SNc	Pars compacta of substantia nigra
SNr	Pars reticulata of substantia nigra
Sm	Submedial nucleus
Spf	Subparafascicular nucleus
ST	Stria terminalis
TF	Thalamic fasciculus
Tu	Tuberal nucleus
VA	Ventral anterior nucleus
VAmc	Magnocellular division of ventral anterior nucleus
VLa	Anterior ventral lateral nucleus
VLp	Posterior ventral lateral nucleus
VM	Ventral medial nucleus
VMp	Principal ventral medial nucleus
VMb	Basal ventral medial nucleus
VMH	Ventromedial hypothalamic nuclei
VPI	Ventral posterior inferior nucleus
VPM	Ventral posterior medial nucleus
VPL	Ventral posterior lateral nucleus
VPLa	Anterior ventral posterior lateral nucleus
VPLp	Posterior ventral lateral posterior nucleus
ZI	Zona incerta

14. REFERENCES

Abe H, Rusak B (1992) Stimulation of the hamster ventral lateral geniculate nucleus induces Fos-like immunoreactivity in suprachiasmatic nucleus cells. *Neurosci Lett 148*:185–189.

Abplanalp P (1970) Some subcortical connections of the visual system in tree shrews and squirrels. *Brain Behav Evol 3*:155–168.

Acuña C, Cudeiro J, Gonzalez F, Alonso JM, Perez R (1990) Lateral-posterior and pulvinar reaching cells-comparison with parietal area 5a: a study in behaving *Macaca nemestrina* monkeys. *Exp Brain Res 82*:158–166.

Acuña C, Gonzalez F, Dominguez R (1983) Sensorimotor unit activity related to intention in the pulvinar of behaving Cebus apella monkeys. *Exp Brain Res 52*:411–422.

Agarwala S, Gunluk AE, May JG, III, Petry HM (1992) Immunohistochemical organization of the ventral lateral geniculate nucleus in the ground squirrel. *J Comp Neurol 318*:255–266.

Aggleton JP, Mishkin M (1983a) Visual recognition impairment following medial thalamic lesions in monkeys. *Neuropsychologia 21*:189–197.

Aggleton JP, Mishkin M (1983b) Memory impairments following restricted medial thalamic lesions in monkeys. *Exp Brain Res 52*:199–209.

Aggleton JP, Mishkin M (1984) Projections of the amygdala to the thalamus in the cynomolgus monkey. *J Comp Neurol 222*:56–68.

Agmon A, Yang LT, Jones EG, O'Dowd DK (1995) Topological precision in the thalamic projection to neonatal mouse barrel cortex. *J Neurosci 15*:549–561

Ahlsén G, Lindström S (1982a) Excitation of perigeniculate neurones via axon collaterals of principal cells. *Brain Res 236*:477–481.

Ahlsén G, Lindström S (1982b) Mutual inhibition between perigeniculate neurones. *Brain Res 236*:482–486.

Ahlsén G, Lindström S, Lo FS (1982) Functional distinction of perigeniculate and thalamic reticular neurons in the cat. *Exp Brain Res 46*:118–26.

Ahlsén G, Lindström S, Lo F-S (1984) Inhibition from the brain stem of inhibitory interneurones of the cat's dorsal lateral geniculate nucleus. *J Physiol Lond 347*:593–609.

Ahlsén G, Lindström S, Lo F-S (1985) Interaction between inhibitory pathways to principal cells in the lateral geniculate nucleus of the cat. *Exp Brain Res 58*:134–143.

Ahlsén G, Lindström S, Sybirska E (1978) Subcortical axon collaterals of principal cells in the lateral geniculate body of the cat. *Brain Res 156*:106–109.

Albe-Fessard D, Condes-Lara M, Sanderson P (1983) The focal tonic cortical control of intralaminar thalamic neurons may involve a cortico-thalamic loop. *Acta Morphol Hung 31*:9–26.

Albrecht D, Muller R, Zippel U, Gabriel HJ, Henklein P, Davidowa H (1994) Cholecystokinin-induced activity changes of dorsal lateral geniculate neurons in rats. *Neurosci 59*:953–60.

Albrecht D, Quaschling U, Zippel U, Davidowa H (1996) Effects of dopamine on neurons of the lateral geniculate nucleus: an iontophoretic study. *Synapse 23*:70–78.

Airaksinen MS, Flugge G, Fuchs E, Panula P (1989) Histaminergic system in the tree shrew brain. *J Comp Neurol 286*:289–310.

Airaksinen MS, Panula P (1988) The histaminergic system in the guinea pig central nervous system: an immunocytochemical mapping study using an antiserum against histamine. *J Comp Neurol 273*:163–186.

Aitkin L (1986) The Auditory Midbrain. Clifton: Humana Press.

Aitkin LM (1973) Medial geniculate body of the cat: responses to tonal stimuli of neurons in medial division. *J Neurophysiol 36*:275–283.

Aitkin LM, Gates GR, Phillips SC (1984) Responses of neurons in inferior colliculus to variations in sound-source azimuth. *J Neurophysiol 52*:1–17.

Aitkin LM, Kudo M, Irvine DRF (1988b) Connections of the primary auditory auditory cortex in the common marmoset, *Callithrix jacchus*, jacchus. *J Comp Neurol 269*:235–248.

Aitkin LM, Merzenich MM, Irvine DRF, Clarey JC, Nelson JE (1986) Frequency representation in auditory cortex of the common marmoset (*Callithrix jacchus*, jacchus). *J Comp Neurol 252*:175–185.

Aitkin LM, Webster WR (1971) Tonotopic organization in the medial geniculate body of the cat. *Brain Res 26*:402–405.

Aitkin LM, Webster WR (1972) Medial geniculate body of the cat: organizations and responses to tonal stimuli of neurons in ventral division. *J Neurophysiol 35*:365–380.

Akert K (1964) Comparative anatomy of the frontal cortex and thalamocortical connections. In: The Frontal Granular Cortex and Behavior (Warner JM, Albert K eds), pp 372–394. New York: McGraw Hill.

Albe-Fessard D, Dondey M, Nicolaidis S, Le Beau J (1970) Remarks concerning the effect of diencephalic lesions on pain and sensitivity with special reference to lemniscally mediated control of noxious afferences. *Confinia Neurologica 32*:174–184.

Albe-Fessard D, Kruger L (1962) Duality of unit discharges from cat centrum medianumus in response to natural and electrical stimulation. *J Neurophysiol 25*:3–20.

Albrecht D, Zippel U, Henklein P (1996) Inhibitory effects of cholecystokinin on rat's geniculate activity can be blocked by GABA-antagonists. *Neurosci Lett 197*:69–71.

Alexander GE, DeLong MR, Strick PL (1986) Parallel organization of functionally integrated circuits linking basal ganglia and cortex. *Ann Rev Neurosci 9*:357–382.

Allman JM, Kaas JH, Lane RH, Miezin FM (1972) A representation of the visual field in the inferior nucleus of the pulvinar in the owl monkey (*Aotus Trivirgatus*). *Brain Res 40*:291–302.

Allon N, Yeshurun Y (1985) Functional organization of the medial geniculate body's subdivisions of the awake squirrel monkey. *Brain Res 360*:75–82.

Allon N, Yeshurun Y, Wollberg Z (1981) Responses of single cells in the medial geniculate body of awake squirrel monkeys. *Exp Brain Res 41*:222–232.

Amadeo A, Arcelli P, Spreafico R, de Biasi S (1995) Ultrastructural immunolocalization of muscarinic acetylcholine receptor in the dorsal thalamus of rat. *Neurosci Lett 184*:161–164.

Amaral DG, Cowan WM (1980) Subcortical afferents to the hippocampal formation in the monkey. *J Comp Neurol 189*:573–591.

Andersen RA, Knight PL, Merzenich MM (1980) The thalamocortical and corticothalamic connections of AI, AII and the anterior auditory field (AAF) in the cat: Evidence for two largely segregated systems of connections. *J Comp Neurol 194*:663–701.

Andersen RA, Roth GL, Aitkin LM, Merzenich MM (1980c) The efferent projections of the central nucleus and the pericentral nucleus of the inferior colliculus in the cat. *J Comp Neurol 194*:649 662.

Anderson ME, Turner RS (1991) Activity of Neurons in cerebellar-receiving and pallidal-receiving areas of the thalamus of the behaving monkey. *J Neurophyiol 66*:879–893.

Apkarian AV, Hodge CJ (1989) A dorsolateral spinothalamic tract in macaque monkey. *Pain 37*:323–333.

Apkarian AV, Shi T (1994) Squirrel monkey lateral thalamus. I. Somatic nociresponsive neurons and their relation to spinothalamic terminals. *J Neurosci 14*:6779–6795.

Appelbaum AE, Leonard RB, Kenshalo DR, Martin RF, Willis WD (1979) Nuclei in which functionally identified spinothalamic tract neurons terminate. *J Comp Neur 188*:575–586.

Araki M, McGeer PL, McGeer EG (1985) Striatonigral and pallidonigral pathways studied by a combination of retrograde horseradish peroxidase tracing and a pharmacohistochemical method for gamma-aminobutyric acid transaminase. *Brain Res 331*:17–24.

Arikuni T, Sakai M, Kubota K (1983) Columnar aggregation of prefrontal and anterior cingulate cortical cells projecting to the thalamic mediodorsal nucleus in the monkey. *J Comp Neurol 220*:116–125.

Arecchi-Bouchhioua P, Yelnik J, François C, Percheron G, Tandé D (1997) Three-dimensional morphology and distribution of pallidal axons projecting to both the lateral region of the thalamus and the central complex in promates. *Brain Res 754*:311–314.

Aronson LR, Papez JW (1934) Thalamic nuclei of Pithecus (Macacus) rhesus: II. Dorsal Thalamus. *Archs Neurol Psychiat Chicago 32*:27–44.

Asanuma C (1989a) Axonal arborizations of a magnocellular basal nucleus input and their relation to neurons in the thalamic reticular nucleus of rats. *Proc Natl Acad Sci USA 86*:4746–4750.

Asanuma C (1994) GABAergic and pallidal terminals in the thalamic reticular nucleus of squirrel monkeys. *Exp Brain Res 101*:439–451.

Asanuma C, Andersen RA, Cowan WM (1985) The thalamic relations of the caudal inferior lobule and the lateral prefrontal cortex in monkeys; divergent cortical projections from cell clusters in the medial pulvinar nucleus. *J Comp Neurol 241*:357–381.

Asanuma C, Porter LL (1990) Light and electron microscopic evidence for a GABAergic projection from the caudal basal forebrain to the thalamic reticular nucleus in rats. *J Comp Neurol 302*:159–172.

Asanuma C, Thach WT, Jones EG (1983a) Cytoarchitectomic delineation of the ventral lateral thalamic region in monkeys. *Brain Res Rev 5*:219–235.

Asanuma C, Thach WT, Jones EG (1983b) Distribution of cerebellar terminations and their relation to other afferent termination in the ventral lateral thalamic region of the monkey. *Brain Res Rev 5*:237–265.

Asanuma C, Thach WT, Jones EG (1983c) Anatomical evidence for segregated focal groupings of efferent cells and their terminal ramifications in the cerebellothalamic pathways. *Brain Res Rev 5*:267–297.

Atweh SF, Kuhar MJ (1977a) Autoradiographic localization of opiate receptors in rat brain II. The brain stem. *Brain Res 129*:1–12.

Atweh SF, Kuhar MJ (1977b) Autoradiographic localization of opiate receptors in rat brain III. The telencephalon. *Brain Res 134*:393–406.

Avendaño C, Stepniewska I, Rausell E, Reinoso-Suárez F (1990) Segregation and heterogeneity of thalamic cell populations projecting to superficial layers of posterior parietal cortex: a retrograde tracer study in cat and monkey. *Neurosci 39*:547–560.

Azmitia EC, Segal M (1978) An autoradiographic analysis of the differential ascending projections of the dorsal and median raphe nuclei in the rat. *J Comp Neurol 179*:641–667.

Babb RS (1980) The pregeniculate nucleus of the monkey (*Macaca mulatta*). 1. A study at the light microscopy level. *J Comp Neurol 190*:651–672.

Baizer JS, Desimone R, Ungerleider LG (1993) Comparison of subcortical connections of inferior temporal and temporal parietal cortex in monkeys. *Vis Neurosci 10*:59–72.

Baizer JS, Ungerleider LG, Desimone R (1991) Organization of visual inputs to the inferior temporal and posterior parietal cortex in macaques. *J Neurosci 11*:168–190.

Bajo VM, Rouiller EM, Welker E, Clarke S, Villa AEP, de Ribaupierre Y, de Ribaupierre F (1995) Morphology and spatial distribution of corticothalamic terminals originating from the cat auditory cortex. *Hear Res 83*:161–174.

Baker FH, Malpeli JG (1977) Effects of cryogenic blockade of visual cortex on the responses of lateral gericulate neurons in the monkey. *Exp Brain Res 29*:433–444.

Bal T, von Krosigk M, McCormick DA (1995a) Synaptic and membrane mechanisms underlying synchronized oscillations in the ferret lateral geniculate nucleus in vitro. *J Physiol Lond 483*:641–663.

Bal T, McCormick DA (1993a) Ionic mechanisms of rhythmic burst firing and tonic activity in the nucleus reticularis thalami, a mammalian pacemaker. *J Physiol Lond 468*:669–691.

Bal T, McCormick DA (1993b) Mechanisms of oscillatory activity in guinea-pig nucleus reticularis thalami in vitro: a mammalian pacemaker. *J Physiol Lond 468*:669–691.

Bal T, McCormick DA (1995) A mechanism for the waning of thalamic spindle and 'absence-seizure like' oscillations in vitro. *Soc Neurosci Abstr 21*:11

Bal T, von Krosigk M, McCormick DA (1995b) Role of the ferret perigeniculate nucleus in the generation of synchronized oscillations in vitro. *J Physiol Lond 483*:665–685.

Balado M, Franke E (1937) Das Corpus Geniculatum Externum. Berlin: Springer

Baleydier C, Mauguière F (1977) Pulvinar-latero posterior afferents to cortical area 7 in monkeys demonstrated by horseradish peroxidase tracing technique. *Exp Brain Res 27*:501–508.

Baleydier C, Mauguière F (1980) The duality of the cingulate gyrus in monkey. Neuroanatomical study and functional hypothesis. *Brain 103*:525–554.

Baleydier C, Mauguière F (1985) Anatomical evidence for medial pulvinar connections with posterior

cingulate cortex, the retrosplenial area, and the posterior parahippocampal gyrus in monkeys. *J Comp Neurol 232*:219–228.
Baleydier C, Mauguière F (1985) Pulvinar-lateroposterior afferents to cortical area 7 in monkeys demonstrated by horseradish peroxidase tracing technique. *Exp Brain Res 232*:219–228.
Baleydier C, Morel A (1992) Segregated thalamocortical pathways to inferior parietal and inferotemporal cortex in macaque monkeys. *Vis Neurosci 8*:391–405.
Barbas H, Haswell Henion TH, Dermon CR (1991) Diverse thalamic projections to the prefrontal cortex in the rhesus monkey. *J Comp Neurol 313*:65–94.
Barbas H, Mesulam M-M (1981) Organization of afferent input to subdivisions of area 8 in the rhesus monkey. *J Comp Neurol 200*:407–431.
Basbaum AI, Fields HL (1979) The origin of descending pathways in the dorsalateral funiculus of the spinal cord of the cat and rat: futher studies on the anatomy of pain modulation. *J Comp Neurol 187*:513–531.
Battaglia G, Lizier C, Colacitti C, Princivalle A, Spreafico R (1994) A reticuloreticular commissural pathway in the rat thalamus. *J Comp Neurol 347*:127–138.
Baughman RW, Gilbert CD (1981) Aspartate and glutamate as possible neurotransmitters in the visual cortex. *J Neurosci 1*:427–439.
Beaton JA, Stemsrud K, Monaghan DT (1992) Identification of a novel *N*-methyl-D-aspartate receptor population in the rat medial thalamus. *J Neurochem 59*:754–757.
Beckstead RM (1984) The thalamostriatal projection in the cat. *J Comp Neurol 223*:313–346.
Beckstead RM, Domesick VB, Nauta WJH (1979) Efferent connections of the substantia nigra and vertral tegmental area in the rat. *Brain Res 175*:191–218.
Beckstead RM, Morse JR, Norgren R (1980) The nucleus of the solitary tract in the monkey: projections to the thalamus and brain stem nuclei. *J Comp Neurol 190*:259–282.
Beckstead RM, Norgren R (1979) An autoradiographic examination of the central distribution of the trigeminal, facial, glossopharyngeal, and vagal nerves in the monkey. *J Comp Neurol 184*:455–472.
Ben-Ari Y, Dingledine R, Kanazawa I, Kelly JS (1976) Inhibitory effects of acetylcholine on neurones in the feline nucleus reticularis thalami. *J Physiol Lond 261*:647–671.
Benabid AL, Pollack P, Gao D, Hoffman D, Limousin P, Gay E, Payen I, Benazzouz A (1996) Chronic electrical stimulation of the ventralis intermedius nucleus of the thalamus as a treatment of movement disorders. *J Neurosurg 84*:203–214.
Benavides J, Savaki HE, Malgouris C, Laplace C, Daniel M, Begassat F, Desban M, Uzan A, Dubroeucq MC, Renault C (1984) Autoradiographic localization of peripheral benzodiazepine binding sites in the cat brain with [3H]PK 11195. *Brain Res Bull 13*:69–77.
Bender DB (1981) Retinotopic organization of macaque pulvinar. *J Neurophysiol 46*:672–693.
Bender DB (1982) Receptive field properties of neurons in the macaque inferior pulvinar. *J Neurophysiol 48*:1–17.
Bender DB (1988) Electrophysiological and behavioral experiments on the primate pulvinar. In: Progress in Brain Research (Hicks TP, Benedek G eds), vol. 75, pp 55–64. New York: Elsevier.
Benevento LA, Davis B (1977) Topographical projections of the prestriate cortex to the pulvinar nuclei in the macaque: an autoradiographic study. *Exp Brain Res 30*:405–424.
Benevento LA, Fallon JH (1975a) The ascending projections of the superior colliculus in the rhesus monkey (*Macaca mulatta*). *J Comp Neurol 160*:339–361.
Benevento LA, Fallon JH (1975b) The projection of occipital cortex to the dorsal lateral geniculate nucleus in the rhesus monkey (*Macaca Mulatta*). *Exp Neurol 46*:409–417.
Benevento LA, Miller J (1981) Visual responses of single neurons in the caudal lateral pulvinar of the macaque monkey. *J Neurosci 11*:1268–1278.
Benevento LA, Rezak M (1976) The cortical projections of the inferior pulvinar and adjacent lateral pulvinar in the rhesus monkey. (*Macaca mulatta*): an autoradiographic study. *Brain Res 108*:1–24.
Benevento LA, Rezak M, Santos-Anderson R (1977) An autoradiographic study of the projections of the pretectum in the rhesus monkey (*Macaca mulatta*): evidence for sensorimotor links to the thalamus and oculomotor nuclei. *Brain Res 127*:197–218.
Benevento LA, Standage GP (1983) The organization of projections of the retinorecipient and nonretinorecipient nuclei of the pretectal complex and layers of the superior colliculus to the lateral pulvinar and medial pulvinar in the macaque monkey. *J Comp Neurol 217*:307–336.
Benjamin RM, Burton H (1968) Projection of taste nerve afferents to anterior opercular-insular cortex in squirrel monkey (*Saimiri sciureus*). *Brain Res 7*:221–231.
Benjamin RM, Jackson JC (1974) Unit discharges in the mediodorsal nucleus of the squirrel monkey evoked by electrical stimulation of the olfactory bulb. *Brain Res 75*:181–191.
Benson DL, Isackson PJ, Gall CM, Jones EG (1991a) Differential effects of monocular deprivation on

glutamic acid decarboxylase and type II calcium-cadmodulin-dependent protein kinase gene expression in the adult monkey visual cortex. *J Neurosci 11*:31–47.

Benson DL, Isackson PJ, Hendry SHC, Jones EG (1991b) Differential gene expression for glutamic acid decarboxylase and type II calcium-calmodulin-dependent protein kinase in basal ganglia, thalamus and hypothalamus of the monkey. *J Neurosci 11*:1540–1564.

Benson DL, Isackson PJ, Gall CM, Jones EG (1992) Contrasting patterns in the localization of glutamic acid decarboxylase and Ca2+/calmodulin protein kinase gene expression in the rat central nervous system. *Neurosci 46*:825–850.

Bentivoglio M, Chen S, Peng Z-C, Bertini G, Ringstedt T, Persson H (1993) Thalamus, neurotrophins and their receptors. In: Thalamic Networks for Relay and Modulation (Minciacchi D, Molinari M, Macchi G, Jones EG eds), pp 309–320. New York: Pergamon.

Bentivoglio M, Macchi G, Albanese A (1981) The cortical projections of the thalamic intralaminar nuclei, as studied in cat and rat with the multiple fluorescent retrograde tracing technique. *Neurosci Lett 26*:5–10.

Bentivoglio M, Minciacchi D, Molinari M, Granato A, Spreafico R, Macchi G (1988) The intrinsic and extrinsic organization of the thalamic intralaminar nuclei. In: Cellular Thalamic Mechanisms (Bentivoglio M, Spreafico R eds), pp 221–238. Amsterdam: Elsevier

Bentivoglio M, Spreafico R, Alvarez-Bolado G, Sanchez MP, Fairén A (1991a) Differential expression of the GABA-A receptor complex in the dorsal thalamus and reticular nucleus: An immunohistochemical study in the adult and developing rat. *Eur J Neurosci 3*:118–125.

Bentivoglio M, Spreafico R, Minciacchi D, Macchi G (1991b) GABAergic interneurons and neuropil of the intralaminar thalamus: an immunohistochemical study in the rat and the cat, with notes in the monkey. *Exp Brain Res 87*:85–95.

Berardi N, Morrone MC (1984a) Development of γ-aminobutyric acid mediated inhibition of X cells of the cat lateral geniculate nucleus. *J Physiol Lond 357*:525–537.

Berardi N, Morrone MC (1984b) The role of γ-aminobutyric acid mediated inhibition in the response properties of cat lateral geniculate nucleus neurones. *J Physiol Lond 357*:505–523.

Berendse HW, Groenewegen HJ (1991) Restricted cortical termination fields of the midline and intralaminar thalamic nuclei in the rat. *Neurosci 42*:73–102.

Berkley KJ (1983) Spatial relationships between the terminations of sensory and motor pathways in the rostral brainstem of cats and monkeys.II. Cerebellar projections compared with those of the ascending somatic sensory pathways in lateral diencephalon. *J Comp Neurol 220*:229–251.

Berkley KJ, Benoist J-MG, M., Guilbaud G (1995) Responses of neurons in the caudal intralaminar thalamic complex of the rat to stimulation of the uterus, vagina, cervix, colon and skin. *Brain Res 695*:92–95.

Berman N (1977) Connections of the pretectum in the cat. *J Comp Neurol 174*:227–254.

Berson DM, Graybiel AM (1978) Parallel thalamic zones in the LP-pulvinar complex of the cat identified by their afferent and efferent connections. *Brain Res 147*:139–148.

Bickford ME, Gunluk AE, Guido W, Sherman SM (1993) Evidence that cholinergic axons from the parabrachial region of the brainstem are the exclusive source of nitric oxide in the lateral geniculate nucleus of the cat. *J Comp Neurol 334*:410–430.

Bickford ME, Günlük AE, Van Horn SC, Sherman SM (1994) GABAergic projection from the basal forebrain to the visual sector of the thalamic reticular nucleus in the cat. *J Comp Neurol 348*:481–510.

Biegon A, Rainbow TC, McEwen BS (1982) Quantitative autoradiography of serotonin receptors in the rat brain. *Brain Res 242*:197–204.

Biemond A (1955) The conduction of pain above the level of the thalamus opticus. *Arch Neurol Psychiatr 231*–244.

Birnbacher D, Albus K (1987) Divergence of single axons in afferent projections to the cat's visual cortical areas 17, 18, and 19: a parametric study. *J Comp Neurol 261*:543–561.

Blakemore C, Vital-Durand F (1981) Postnatal development of the monkey's visual system. *Ciba Foundation Symposium 86*:152–171.

Blakemore C, Vital-Durand F (1986) Organization and post-natal development of the monkey's lateral geniculate nucleus. *J Physiol Lond 380*:453–491.

Blasco I, Alvarez FJ, Villalba RM, Solano ML, Martinez-Murillo R, Rodrigo J (1989) Light and electron microscopic study of galanin-immunoreactive nerve fibers in the rat posterior thalamus. *J Comp Neurol 283*:1–12.

Blasdel GG, Lund JS (1983a) Termination of afferent axons in macaque striate cortex. *J Neurosci 3*:1389–1413.

Blomquist AJ, Benjamin M, Emmers R (1962) Thalamic localization of afferents from the tongue in squirrel monkey. *J Comp Neurol 118*:77–87.

Bloomfield SA, Sherman SM (1988) Postsynaptic potentials recorded in neurons of the cat's lateral geniculate nucleus following electrical stimulation of the optic chiasm. *J Neurophysiol 60*:1924–1945.

Bloomfield SA, Sherman SM (1989) Dendritic current flow in relay cells and interneurons of the cat's lateral geniculate nucleus. *Proc Natl Acad Sci USA 86*:3911–3914.

Blum M, Ruch TC, Walker AE (1943) Localisation of taste in the thalamus of *Macaca mulatta. Yale J Biol Med 16*:175–191.

Blum PS, Abraham LD, Gilman S (1979a) Vestibular, auditory, and somatic input to the posterior thalamus of the cat. *Exp Brain Res 34*:1–10.

Blum PS, Day MJ, Carpenter MB, Gilman S (1979b) Thalamic components of the ascending vestibular system. *Exp Neurol 64*:587–603.

Bobillier P, Sequin S, Petityean F, Salvert D, Touret M, Jouvet M (1976) The raphe nuclei of the cat brain stem: a topographical atlas of their efferent projections as revealed by autoradiography. *Brain Res 113*:449–486.

Boivie J (1971a) The termination in the thalamus and the zona incerta of fibres from the dorsal column nuclei (DCN) in the cat. An experimental study with silver impregnation methods. *Brain Res 28*:459–490.

Boivie J (1971b) The termination of the spinothalamic tract in the cat An experimental study with silver impregnation methods. *Exp Brain Res 112*:331–353.

Boivie J (1978) Anatomical observations on the dorsal column nuclei, their thalamic projection and the cytoarchitecture of some somatosensory thalamic nuclei in the monkey. *J Comp Neurol 178*:17–48.

Boivie J (1979) An anatomical reinvestigation of the termination of the spinothalamic tract in the monkey. *J Comp Neurol 186*:343–370.

Boivie J (1980) Thalamic projections from lateral cervical nucleus in monkey. A degeneration study. *Brain Res 198*:13–26.

Bonnnano G, Raiteri M (1993) Multiple $GABA_B$ receptors. *Trends Pharmacol Sci 14*:259–261.

Bormann J, Feigenspan A (1995) $GABA_C$ receptors. *Trends Neurosci 18*:515–519.

Bos J, Benevento LA (1975) Projections of the medial pulvinar to orbital cortex and frontal eye fields in the rhesus monkey. *Exp Neurol 49*:487–496.

Bosch DA (1991) Stereotactic rostral mesencephalotomy in cancer pain and deafferentation pain. A series of 40 cases with follow-up results. *J Neurosurg 75*:747–751.

Bouras C, Magistretti PJ, Morrison JH, Constantinidis J (1987) An immunohistochemical study of pro-somatostatin-derived peptides in the human brain. *Neurosci 22*:781–800.

Bouras C, Vallet PG, Dobrinov H, de St.-Hilaire S, Constantinidis J (1986) Substance P neuronal cell bodies in the human brain: complete mapping by immunohistoflourescence. *Neurosci Lett 69*:31–36.

Bourassa J, Deschênes M (1995) Corticothalamic projections from the primary visual cortex in rats: A single fiber study using bicytin as an anterograde tracer. *Neurosci 66*:253–263.

Bourassa J, Pinault D, Deschênes M (1995) Corticothalamic projections from the cortical barrel field to the somatosensory thalamus in rats: A single-fibre study using biocytin as an anterograde tracer. *Eur J Neurosci 7*:19–30.

Bouthenet ML, Ruat M, Sales N, Garbarg M, Schwartz JC (1988) A detailed mapping of histamine H_1 receptors in guinea-pig central nervous system established by autoradiography with $[^{125}I]$ iodobolpyramine. *Neurosci 26*:553–600.

Bowery N (1989) $GABA_B$ receptors and their significance in mamalian pharmacology. *Trends Pharmacol Sci 10*:401–407.

Bowery NG, Hudson AL, Price GW (1987) $GABA_A$ and $GABA_B$ receptor site distribution in the rat central nervous system. *Neurosci 20*:365–384.

Bowery NG, Knott C, Moratalla R, Pratt GD (1990) $GABA_B$ receptors and their heterogeneity. advances in biochemical psychopharmacology 46:127–139.

Bowling DB, Michael CR (1980) Projection patterns of single physiologically characterized optic tract fibres in cat. *Nature 286*:899–902.

Bowling DB, Michael CR (1984) Terminal patterns of single, physiologically characterized optic tract fibers in the cat's lateral geniculate nucleus. *J Neurosci 4*:198–216.

Boyson SJ, McGonigle P, Molinoff PB (1986) Quantitative autoradiographic localization of the D1 and D2 subtypes of dopamine receptors in rat brain. *J Neurosci 6*:3177–3188.

Broide RS, O'Connor LT, Smith MA, Smith JAM, Leslie FM (1995) Developmental expression of a7 neuronal nicotinic receptor messenger RNA in rat sensory cortex and thalamus. *Neurosci 67*:83–94.

Broman J (1994) Neurotransmitters in subcortical somatosensory pathways. *Anat Embryol 189*:181–214.

Broman J, Ottersen OP (1992) Cervicothalamic tract terminals are enriched in glutamate-like immunoreactivity: An electron microscope double-labeling study in the cat. *J Neurosci 12*:204–221.

Bruce CJ, Goldberg ME (1985) Primate frontal eye fields. I. Single neurons discharging before saccades. *J Neurophysiol 53*:603–635.

Bruinvels AT, Palacios JM, Hoyer D (1993) 5-hydroxytryptamine$_1$ recognition sites in rat brain: heterogeneity of non-5-hydroxytryptamine$_{1A/1C}$ binding sites revealed by quantitative receptor autoradiography. *Neurosci 53*:465–474.

Brysch I, Brysch W, Creutzfeldt O, Hayes NL, Schlingensiepen KH (1984) The second, intralaminar thalamo-cortical projection system. *Anat Embryol 169*:111–118.

Buckley NJ, Bonner TI, Brann MR (1988) Localization of a family of muscarinic receptor mRNAs in rat brain. *J Neurosci 8*:4646–4652.

Buller AL, Larson HC, Schneider BE, Beaton JA, Morrisett RA, Monaghan DT (1994) The molecular basis of NMDA receptor subtypes: native receptor diversity is predicted by subunit composition. *J Neurosci 14*:5471–5484.

Bullier J, Kennedy H, Salinger W (1984) Bifurcation of subcortical afferents tovisual areas 17,18, and 19 in the cat cortex. *J Comp Neurol 228*:309–328.

Bullier J, Norton TT (1979) X and Y relay cells in cat lateral geniculate nucleus: quantiative analysis of receptive-field properties and classification. *J Neurophysiol 42*:244–273.

Burchfiel JL, Duffy FH (1974) Corticofugal influence upon cat thalamic ventrobasal complex. *Brain Res 70*:395–411.

Burgunder J-M, Young WS (1990) Cortical neurons expressing the cholescystokinin gene in the rat: distribution in the adult brain, ontogeny and some of their projections. *J Comp Neurol 300*:26–46.

Burgunder J-M, Young WS (1992) Expression of cholecystokinin and somatostatin genes in the human thalamus. *J Comp Neurol 324*:14–22.

Burgunder J-M, Young WS, III (1988) The distribution of thalamic projection neurons containing cholecystokinin messenger RNA, using in situ hybridization histochemistry and retrograde labelling. *Brain Res 404*:221–89

Burgunder J-M, Young WS, III (1990) Ontogeny of tyrosine hydroxylase and cholecystokinin gene expression in the rat mesencephalon. *Dev Brain Res 52*:85–94.

Burton H (1984) Corticothalamic connections from the second somatosensory area and neighboring regions in the lateral sulcus of macaque monkeys. *Brain Res 2*:368–372.

Burton H (1986) Second somatosensory cortex and related areas. In: Cerebral Cortex Vol. 5. Sensory Motor Areas and Aspects of Cortical Connectivity (Jones EG, Peters A eds), pp 31–98. New York: Plenum.

Burton H, Benjamin RM (1971) Central projections of the gustatory system. *Handbook Sensory Physiol 4*:149–164.

Burton H, Carlson M (1986) Second somatic sensory cortical area (SII) in a prosimian primate, *Galago crassicaudatus*. *J Comp Neurol 247*:200–220.

Burton H, Craig AD (1983) Spinothalamic projections in cat, raccoon and monkey: a study based on anterograde transport of horseradish peroxidase. In: Somatosensory Integration in the Thalamus (Macchi G, Rustioni A, Spreafico A eds), pp 1–25. Amsterdam: Elsevier.

Burton H, Fabri M, Alloway K (1995) Cortical areas within the lateral sulcus connected to cutaneous representations in areas 3b and 1: a revised interpretation of the second somatosensory area in Macaque monkeys. *J Comp Neurol 355*:539–562.

Burton H, Jones EG (1976) The posterior thalamic region and its cortical projection in New World and Old World monkeys. *J Comp Neurol 168*:249–301.

Bushnell MC, Duncan GH (1987) Mechanical response properties of ventroposterior medial thalamic neurons in the alert monkey. *Exp Brain Res 67*:603–614.

Bushnell MC, Duncan DH, Tremblay N (1993) Thalamic VPM nucleus in the behaving monkey. I. Multimodal and discriminative properties of thermosensitive neurons. *J Neurophysiol 69*:739–752.

Butler EG, Horne MK, Hawkins NJ (1992a) The activity of monkey thalamic and motor cortical neurones in a skilled, ballistic movement. *J Physiol Lond 445*:25–48.

Butler EG, Horne MK, Rawson JA (1992b) Sensory characteristics of monkey thalamic and motor cortex neurones. *J Physiol Lond 445*:1–24.

Buzsaki G, Kennedy B, Solt VB, Ziegler M (1991) Noradrenergic control of thalamic oscillation: the role of α-2 receptors. *Eur J Neurosci 3*:222–229.

Cajal SR (1904) Textura del Sistema Nervioso del Hombre y de los Vertebrados. Vol. 2, parts 1 and 2. Madrid: Moya.

Calford MB (1983) The parcellation of the medial geniculate body of the cat defined by the auditory response of single units. *J Neurosci 3*:2350–2364.

Calford MB, Aitkin LM (1983) Ascending projections to the medial geniculate body of the cat: evidence for multiple, parallel auditory pathways through thalamus. *J Neurosci 3*:2365–2380.

Calford MB, Webster WR (1981) Auditory representation within principal division of cat medial geniculate body: an electrophysiological study. *J Neurophysiol 45*:1013–1028.

Cameron AA, Khan IA, Westlund KN, Cliffer K, D., Willis WD (1995) The efferent projections of the periaqueductal gray in the rat: a *Phaseolus vulgaris*-leucoagglutinin study. I. Ascending projections. *J Comp Neurol 351*:568–584.

Campbell MJ, Pandya DN (1994) Medial geniculate body chemoarchitectonic subdivisions. *Neurosci Abstr 20*:977

Campos-Ortega JA, Hayhow WR, de Clüver PF (1972) A note on the problem of retinal projections to the inferior pulvinar of primates. *Brain Res 22*:126–130.

Card JP, Moore RY (1982) Ventral lateral geniculate nucleus efferents to the rat suprachiasmatic nucleus exhibit avian pancreatic polypeptide-like immunoreactivity. *J Comp Neurol 206*:390–396.

Carey RG, Fitzpatrick D, Diamond IT (1979a) Thalamic projections to layer 1 of striate cortex shown by retrograde transport of horseradish peroxidase. *Science 203*:556–558.

Carey RG, Fitzpatrick D, Diamond IT (1979b) Layer I of striate cortex of *Tupaia glis* and *Galago senegalensis*: projections from thalamus and claustrum revealed by retrograde transport of horseradish peroxidase. *J Comp Neurol 186*:393–438.

Carpenter MB, Nakano K, Kim R (1976) Nigrothalamic projections in the monkey demonstrated by autoradiographic technics. *J Comp Neurol 165*:401–415.

Carstens E, Leah J, Lechner J, Zimmermann M (1990) Demonstration of extensive brainstem projections to medial and lateral thalamus and hypothalamus in the rat. *Neurosci 35*:609–626.

Casagrande VA (1994) A third parallel visual pathway to primate area V1. *Trends Neurosci 17*:305–310.

Casey KL, Morrow TJ (1983) Ventral posterior thalamic neurons differentially responsive to noxious stimulation of the awake monkey. *Science 221*:675–677.

Casseday JH, Kobler JB, Isbey SF, Covey E (1989) Central acoustic tract in an echolocating bat: an extralemniscal auditory pathway to the thalamus. *J Comp Neurol 287*:247–259.

Catania MV, Landwehrmeyer GB, Testa CM, Standaert DG, Penney JB, Jr., Young AB (1994) Metabotropic glutamate receptors are differentially regulated during development. *Neurosci 61*:481–496.

Catsman-Berrevoets CE, Kuypers HGJM (1978) Differential laminar distribution of corticothalamic neurons projecting to the VL and the center median. An HRP study in the cynomolgus monkey. *Brain Res 154*:359–365.

Cavada C, Goldman-Rakic PS (1993) Multiple visual areas in the posterior parietal cortex of primates. *Progress in Brain Research 95*:123–137.

Chalmers DT, Watson SJ (1991) Comparative anatomical distribution of 5-HT 1A receptor mRNA and 5-HT 1A binding in rat brain-a combined in situ hybridisation/in vitro receptor autoradiographic study. *Brain Res 561*:51–60.

Chan-Palay V (1977) Cerebellar Dentate Nucleus. Organization, Cytology and Transmitters. Berlin: Springer.

Chandler MJ, Hobbs SF, Fu Q-G, Kenshalo DR, Blair RW, Foreman RD (1992) Responses of neurons in ventroposterolateral nucleus of primate thalamus to urinary bladder distension. *Brain Res 571*:26–34.

Chen S, Bentivoglio M (1993) Nerve growth factor receptor-containing cholinergic neurons of the basal forebrain project to the thalamic reticular nucleus in the rat. *Brain Res 606*:207–212.

Chmielowska J, Carvell GE, Simons DJ (1989) Spatial organization of thalamocortical and corticothalamic projection systems in the rat SmI barrel cortex. *J Comp Neurol 285*:325–338.

Chung JM, Lee KH, Surmeier DJ, Sorkin LS, Kim J, Willis WD (1986a) Response characteristics of neurons in the ventral posterior lateral nucleus of the monkey thalamus. *J Neurophysiol 56*:370–390.

Chuong CH, Lindsley DB (1973) Polysensory responses and sensory interaction in pulvinar and related posterio-lateral thalamic nuclei in cat. Electroencephalogr *Clin Neurophysiol 34*:265–280.

Ciabarra AM, Sullivan JM, Gahn LG, Pecht G, Heinemann S, Severino KA (1995) Cloning and characterization of χ-1: a developmentally regulated member of a novel class of the ionotropic glutamate receptor family. *J Neurosci 15*:6498–6508.

Cicirata F, Angaut P, Serapide MF, Panto MR (1990) Functional organization of the direct and indirect projection via the reticularis thalami nuclear complex from the motor cortex to the thalamic nucleus ventralis lateralis. *Exp Brain Res 79*:325–337.

Cimino M, Marini P, Fornasari D, Cattabeni F, Clementi F (1992) Distribution of nicotinic receptors in cynomolgus monkey brain and ganglia: localization of alpha3 subunit mRNA, alpha-bungarotoxin and nicotine binding sites. *Neurosci 51*:77–86.

Clark AS, Schwartz ML, Goldman-Rakic PS (1989) GABA-immunoreactive neurons in the mediodorsal nucleus of the monkey thalamus. *J Chem Neuroanat 2*:259–267.

Cleland BG, Dubin MW, Levick WR (1971) Sustained and transient neurones in the cat's retina and lateral geniculate nucleus. *J Physiol Lond 217*:473–496.

Coffield JA, Miletic V (1987) Immunoreactive enkephalin is contained within some trigeminal and spinal neurons projecting to the rat medial thalamus. *Brain Res 425*:380–383.

Colombo F (1984) Somatosensory-evoked potentials after mesencephalic tractotomy for pain syndromes. Neuroradiologic and clinical correlations. *Surgical Neurology 21*:453–458.

Colonnier M, Guillery RW (1964) Synaptic organization in the lateral geniculate nucleus of the monkey. *Z Zellforsch 62*:333–355.

Comans PE, Snow PJ (1981) Ascending projections to nucleus parafascicularis of the cat. *Brain Res 230*:337–341.

Conley M, Birecree E, Casagrande VA (1985) Neuronal classes and their relation to functional and laminar organization of the lateral geniculate nucleus: a Golgi study of the prosimian primate, *Galago crassicaudatus*. *J Comp Neurol 242*:561–583.

Conley M, Diamond IT (1990) Organization of the visual sector of the thalamic reticular nucleus in Galago. Evidence that the dorsal lateral geniculate and pulvinar nuclei occupy separate, parallel tiers. *Eur J Neurosci 2*:211–226.

Conley M, Friederich-Ecsy B (1993a) Functional organization of the ventral lateral geniculate complex of the tree shrew (*Tupaia belangeri*): I. Nuclear subdivisions and retinal projections. *J Comp Neurol 328*:1–20.

Conley M, Friederich-Ecsy B (1993b) Functional organization of the ventral lateral geniculate complex of the tree shrew (*Tupaia belangeri*): II. Connections with the cortex, thalamus, and brainstem. *J Comp Neurol 328*:21–42.

Conley M, Kupersmith AC, Diamond IT (1991a) The organization of projections from subdivisions of the auditory cortex and thalamus to the auditory sector of the thalamic reticular nucleus in *Galago*. *Eur J Neurosci 3*:1089–1103.

Conley M, Schmechel DE, Diamond IT (1991b) Differential distribution of somatostatin-like immunoreactivity in the visual sector of the thalamic reticular nucleus in Galago. *Eur J Neurosci 3*:237–242.

Conrad LCA, Leonard CM, Pfaff DW (1974) Connections of the median and dorsal raphe nuclei in the rat: an autoradiographic and degeneration study. *J Comp Neurol 156*:179–205.

Consolazione A, Priestley JV, Cuello AC (1984) Serotonin-containing projections to the thalamus in the rat revealed by a horseradish peroxidase and peroxidase antiperoxidase double-staining technique. *Brain Res 322*:233–244.

Contreras D, Steriade M (1995) Cellular basis of EEG slow rhythms: a study of dynamic corticothalamic relationships. *J Neurosci 15*:604–622.

Contreras D, Steriade M (1996) Spindle oscillation in cats: the role of corticothalamic feedback in a thalamically generated rhythm. *J Physiol Lond 490*:159–179.

Conway JL, Schiller PH (1983) Laminar organization of tree shrew dorsal lateral geniculate nucleus. *J Neurophysiol 50*:1330–1342.

Corlett RT, Lucas PW (1990) Alternative seed-handling strategies in primates: seed-splitting by long-tailed macaques (*Macaca fascicularis*). *Oecologia 82*:166–171.

Cornwall J, Cooper JD, Phillipson OT (1990) Projections to the rostral reticular thalamic nucleus in the rat. *Exp Brain Res 80*:157–171.

Cornwall J, Phillipson OT (1988a) Afferent projections to the dorsal thalamus of the rat as shown by retrograde lectin transport: I. The medio-dorsal nucleus. *Neurosci 24*:1035–1050.

Cornwall J, Phillipson OT (1988b) Mediodorsal and reticular thalamic nuclei receive colateral axons from prefrontal cortex and laterodorsal tegmental nucleus in the rat. *Neurosci Lett 88*:121–126.

Cortes R, Palacios JM (1986) Muscarinic cholinergic receptor subtypes in the rat brain: I Quantitative autoradiographic studies. *Brain Res 362*:227–238.

Coulter DA, Huguenard JR, Prince DA (1990a) Differential effects of petit mal anticonvulsants and convulsants on thalamic neurones: calcium current reduction. *Brit J Pharmacol 100*:800–806.

Coulter DA, Huguenard JR, Prince DA (1990b) Differential effects of petit mal anticonvulsants and convulsants on thalamic neurones: GABA current blockade. *Brit J Pharmacol 100*:807–813.

Coveñas R, Alonso JR, Tramu G (1989) Immunocytochemical study of angiotensin II cell bodies in the rat thalamus. *Brain Res 481*:185–189.

Cowey A, Stoerig P, Bannister M (1994) Retinal ganglion cells labelled from the pulvinar nucleus in macaque monkeys. *Neurosci 61*:691–706.

Cox CL, Huguenard JR, Prince DA (1995) Cholecystokinin depolarizes rat thalamic reticular neurons by suppressing a K^+ conductance. *J Neurophysiol 74*:990–1000.

Cox CL, Huguenard JR, Prince DA (1996) Heterogenous axonal arborizations of rat thalamic neurons in the ventrobasal nucleus. *J Comp Neurol 366*:416–430.

Cox CL, Huguenard JR, Prince DA (1997) Peptidergic modulation of intrathalamic circuit activity *in vitro*: Actions of cholecystokinin. *J Neurosci 17*:70–82.

Crabtree JW (1992) The somatotopic organization within the rabbit's thalamic reticular nucleus. *Eur J Neurosci 4*:1343–1351.

Crabtree JW (1996) Organization in the somatosensory sector of the cat's thalamic reticular nucleus. *J Comp Neurol 366*:207–222.

Crabtree JW, Killackey HP (1989) The topographic organization and axis of projection within the visual sector of the rabbit's thalamic reticular nucleus. *Eur J Neurosci 1*:94–109.

Craig AD (1987) Medial thalamus and nociception: the nucleus submedius. In: Thalamus and Pain (Besson J-M, Guilbaud G, Peschanski M eds), pp 227–244. Amsterdam: Excerpta Medica.

Craig AD, Burton H (1981) Spinal and medullary lamina I projection to nucleus submedius in medial thalamus: a possible pain center. *J Neurophysiol 45*:443–466.

Craig AD, Bushnell MC, Zhang E-T, Blomqvist A (1994) A thalamic nucleus specific for pain and temperature sensation. *Nature 372*:770–773.

Craig AD, Dostrovsky JO (1991) Thermoreceptive lamina I trigeminothalamic neurons project to the nucleus submedius in the cat. *Exp Brain Res 85*:470–474.

Craig AD, Wiegand SJ, Price JL (1982) The thalamo-cortical projection of the nucleus submedius in the cat. *J Comp Neurol 206*:28–48.

Cropper EC, Eisenman JS, Azmitia EC (1984) An immunocytochemical study of the serotonergic innervation of the thalamus of the rat. *J Comp Neurol 224*:38–50.

Crouch RL (1934) The nuclear configuration of the thalamus of Macacus rhesus. *J Comp Neurol 59*:451–485.

Crunelli V, Haby M, Jassik-Gerschenfeld D, Leresche N, Pirchio M (1988) Cl^-- and K^+-dependent inhibitory postsynaptiac potentials evoked by interneurones of the rat lateral geniculate nucleus. *J Physiol Lond 399*:153–176.

Crunelli V, Kelly JS, Leresche N, Pirchio M (1987a) On the excitatory post-synaptic potential evoked by stimulation of the optic tract in the rat lateral geniculate nucleus. *J Physiol Lond 384*:603–618.

Crunelli V, Kelly JS, Leresche N, Pirchio M (1987b) The ventral and dorsal lateral geniculate nucleus of the rat: intracellular recordings in vitro. *J Physiol Lond 384*:587–601.

Crunelli V, Leresche N, Parnavelas JG (1987c) Membrane properties of morphologically identified X and Y cells in the lateral geniculate nucleus of the cat in vitro. *J Physiol Lond 390*:243–256.

Castaigne P, Buge A, Escourolle R, Masson (1962) Ramollissement pedonculaire médian tegmentothalamique avec optalmoplégie et hypersomnie. *Rev Neurol (Paris) 106*:357–367.

Cucchiaro JB, Bickford ME, Sherman SM (1991a) A GABAergic projection from the pretectum to the dorsal lateral geniculate nucleus in the cat. *Neurosci 41*:213–226.

Cucchiaro JB, Uhlrich DJ, Sherman SM (1991b) Electron-microscopic analysis of synaptic input from the perigeniculate nucleus to the A-laminae of the lateral geniculate nucleus in cats. *J Comp Neurol 310*:316–336.

Cucchiaro JB, Uhlrich DJ, Sherman SM (1993) Ultrastructure of synapses from the pretectum in the A-laminae of cat's lateral geniculate nucleus. *J Comp Neurol 334*:618–630.

Cudeiro J, Rivadulla C, Rodriguez R, Martinez-Conde S, Acuña C, Alonso JM (1994) Modulatory influence of putative inhibitors of nitric oxide synthesis on visual processing in the cat lateral geniculate nucleus. *J Neurophysiol 71*:146–149.

Cudeiro J, Sillito AM (1995) Spatial frequency tuning of orientation-discontinuity-sensitive corticofugal feedback to the cat lateral geniculate nucleus. *J Neurophysiol 490*:481–492.

Cunningham ET, Jr., LeVay S (1986) Laminar and synaptic organization of the projection from the thalamic nucleus centralis to primary visual cortex in the cat. *J Comp Neurol 254*:65–77.

Curcio CA, Harting JK (1978) Organization of pulvinar afferents to area 18 in the squirrel monkey: evidence for stripes. *Brain Res 143*:155–161.

Curró Dossi R, Nuñez A, Steriade M (1992a) Electrophysiology of a slow (0.5–4 Hz) intrinsic oscillation of cat thalamocortical neurones in vivo. *J Physiol Lond· 447*:215–234.

Curró Dossi R, Paré D, Steriade M (1991) Short-lasting nicotinic and long-lasting muscarinic depolarizing responses of thalamocortical neurons to stimulation of mesopontine cholinergic nuclei. *J Neurophysiol 65*:393–406.

Curró Dossi R, Paré D, Steriade M (1992b) Various types of inhibitory postsynaptic potentials in anterior thalamic cells are differentially altered by simulation of laterodorsal tegmental cholinergic nucleus. *Neurosci 47*:279–290.

Cusick CG, Gould HJ (1990) Connections between area 3b of the somatosensory cortex and subdivision of ventroposterior nuclear complex and the anterior pulvinar nucleus in squirrel monkeys. *J Comp Neurol 292*:83–102.

Cusick CG, Scripter JL, Darensbourg JG, Weber JT (1993) Chemoarchitectonic subdivision of the visual pulvinar in monkeys and their connectional relations with the middle temporal and rostral dorsolateral visual areas, MT and DLr. *J Comp Neurol 336*:1–30.

Darian-Smith C, Darian-Smith I, Cheema SS (1990a) Thalamic projections to sensorimotor cortex in the newborn macaque. *J Comp Neurol 299*:47–63.

Darian-Smith C, Darina-Smith I, Cheema SS (1990b) Thalamic projections to sensorimotor cortex in the macaque monkey: use of multiple retrograde fluorescent tracers. *J Comp Neurol 299(1)*:17–46.

Davis KD, Tasker RR, Kiss ZHT, Hutchison WD, Dostrovsky JO (1995) Visceral pain evoked by thalamic microstimulation in humans. *NeuroReprt 6*:369–374.

Dawson TM, Gehlert DR, McCabe RT, Barnett A, Wamsley JK (1986) D-1 dopamine receptors in the rat brain: a quantitative autoradiographic analysis. *J Neurosci 6*:2352–2365.

Dawson DR, Killackey HP (1985) Distinguishing topography and somatotopy in the thalamocortical projections of the developing rat. *Dev Brain Res 17*:309–313.

de Biasi S, Amadeo A, Spreafico R, Rustioni A (1994a) Enrichment of glutamate immunoreactivity in lemniscal terminals in the ventroposterolateral thalamic nucleus of the rat: An immunogold and WGA-HRP study. *Anat Rec 240*:131–140.

de Biasi S, Arcelli P, Spreafico R (1994b) Parvalbumin imunoreactivity in the thalamus of guinea pig: light and electron microscopic correlation with gamma-aminobutyric acid immunoreactivity. *J Comp Neurol 348*:556–569.

de Lima A, Montero VM, Singer W (1985) The cholinergic innervation of the visual thalamus: an EM immunocytochemical study. *Exp Brain Res 59*:206–212.

de Lima AD, Singer W (1987a) The brainstem projection to the lateral geniculate nucleus in the cat: identification of cholinergic and monoaminergic elements. *J Comp Neurol 259*:92–121.

de Lima AD, Singer W (1987b) The serotoninergic fibers in the dorsal lateral geniculate nucleus of the cat: distribution and synaptic connections demonstrated with immunocytochemistry. *J Comp Neurol 258*:339–351.

de Monasterio FM (1978) Center and surround mechanisms of opponent-color X and Y ganglion cells of retina of macaques. *J Neurophysiol 41*:1418–1434.

de Monasterio FM, Gouras P (1975) Functional properties of ganglion cells of the rhesus monkey retina. *J Physiol Lond 251*:167–195.

De Vito JL, Anderson ME (1982) An autoradiographic study of efferent connections of the globus pallidus in *Macaca mulatta*. *Exp Brain Res 46*:107–117.

Del Toro ED, Juiz JM, Peng X, Lindström J, Criado M (1994) Immunocytochemical Localization of the a7 Subunit of the nicotinic acetylcholine receptor in the rat central nervous system. *J Comp Neurol 349*:325–342.

DeLorey TM, Olsen RW (1992) Gamma-aminobutyric $acid_A$ receptor structure and function. *J Biol Chem 267*:16747–16750.

Demeulemeester H, Vandesande F, Orban GA, Heizmann CW, Pochet R (1989) Calbindin D-28K and parvalbumin immunoreactivity is confined to two separate neuronal subpopulations in the cat visual cortex, whereas partial coexistence is shown in the dorsal lateral geniculate nucleus. *Neurosci Lett 99*:6–11.

Deniau JM, Lackner D, Feger J (1978) Effect of substantia nigra stimulation on identified neurons in the VL-VA thalamic complex: comparison between intact and chronically decorticated cats. *Brain Res 145*:27–36.

Derrington AM, Lennie P (1984) Spatial and temporal contrast sensitivities of neurons in lateral geniculate nucleus of macaque. *J Physiol Lond 357*:219–240.

Deschênes M, Bourassa J, Doan VD, Parent A (1996) A single-cell study of the axonal projections arising from the posterior intralaminar thalamic nuclei in the rat. *Eur J Neurosci 8*:329–343.

Deschênes M, Hu B (1990a) Electrophysiology and pharmacology of the corticothalamic input to lateral thalamic nuclei: an intracellular study in the cat. *Eur J Neurosci 2*:140–152.

Deschênes M, Hu B (1990b) Membrane resistance increase induced in thalamic neurons by stimulation of brainstem cholinergic afferents. *Brain Res 513*:339–342.

Deschênes M, Madariaga-Domich A, Steriade M (1985) Dendrodendritic synapses in the cat reticularis thalami nucleus: a structural basis for thalamic spindle synchronization. *Brain Res 334*:165–168.

Deschênes M, Paradis M, Roy JP, Steriade M (1984) Electrophysiology of neurons of lateral thalamic nuclei in cat: Resting properties and burst discharges. *J Neurophysiol 51*:1196–1219.

Deshmukh PP, Yamamura HI, Woods I, Nelson DL (1983) Computer-assisted autoradiographic localization of subtypes of serotonin1 receptors in rat brain. *Brain Res 288*:338–343.

Destexhe A, Sejnowski TJ (1995) G protein activation kinetics and spillover of γ-aminobutyric acid may account for differences between inhibitory responses in the hippocampus and thalamus. *Proc Natl Acad Sci USA 92*:9515–9519.

Deutch AY, Holliday J, Roth RH, Chun LLY, Hawrot E (1987) Immunohistochemical localization of a neuronal nicotinic acetylcholine receptor in mammalian brain. *Proc Natl Acad Sci USA 84*:8697–8701.

DeVito JL, Anderson ME (1982) An autoradiographic study of efferent connections of the globus pallidus in *Macaca mulatta*. *Exp Brain Res 46*:107–117.

Dewulf A (1971) Anatomy of the Normal Human Thalamus. Topometry and Standardized Nomenclature. Amsterdam: Elsevier.

DeYoe EA, Van Essen D (1985) Segregation of efferent connections and receptive field properties in visual area V2 of the macaque. *Nature 317*:58–61.

Di Chiara G, Porceddu ML, Morelli M, Mulas ML, Gessa GL (1979) Evidence for a GABAergic projection from the substantia nigra to the ventromedial thalamus and to the superior colliculus of the rat. *Brain Res 2*:273–284.

Diamond IT, Fitzpatrick D, Conley M (1992) A projection from the parabigeminal nucleus to the pulvinar nucleus in *Galago*. *J Comp Neurol 316*:375–382.

Diamond IT, Fitzpatrick D, Schmechel D (1993) Calcium binding proteins distinguish large and small cells of the ventral posterior and lateral geniculate nuclei of the prosimian G*alago* and the tree shrew (*Tupaia belangeri*). *Proc Natl Acad Sci USA 90*:1425–1429.

Diamond ME, Armstrong-James M, Ebner FF (1992a) Somatic sensory responses in the rostral sector of the posterior group (POm) and in the ventral posterior medial nucleus (VPM) of the rat thalamus. *J Comp Neurol 318*:462–476.

Diamond ME, Armstrong-James M, Budway MJ, Ebner FF (1992b) Somatic sensory responses in the rostral sector of the posterior group (POm) and in the ventral posterior medial nucleus (VPM) of the rat thalamus: dependence on the barrel field cortex. *J Comp Neurol 319*:66–84

Diamond ME, Armstrong-James M, Ebner FF (1993) Experience-dependent plasticity in adult rat barrel cortex. *Proc Natl Acad Sci USA 90*:2082–2086.

Dick A, Kaske A, Creutzfeldt OD (1991) Topographical and topological organization of the thalamocortical projection to the striate and prestriate cortex in the marmoset (*Callithrix jaccus*). *Exp Brain Res 84*:233–253.

Dineley-Miller K, Patrick J (1992) Gene transcripts for the nicotinic acetylcholine receptor subunit, Beta4, are distributed in multiple areas of the rat central nervous system. *Mol Brain Res 16*:339–344.

Dingledine R, Kelly JS (1977) Brain stem stimulation and the acetylcholine-evoked inhibition ofneurones in the feline nucleus reticularis thalami. *J Physiol Lond 271*:135–154.

Divac I, LaVail JH, Rakic P, Winston KR (1977) Heterogeneous afferents to the inferior parietal lobule of the rhesus monkey revealed by the retrograde transport method. *Brain Res 123*:197–208.

Do K-Q, Binns KE, Salt TE (1994) Release of the nitric oxide precursor, arginine, from the thalamus upon sensory afferent stimulation, and its effect on thalamic neurons *in vivo*. *Neurosci 60*:581–586.

Doble A, Malgouris C, Daniel N, Imbault F, Basbaum A, Uzan A, Gueremy C, Le Fur G (1987) Labelling of peripheral-type benzodiazepine binding sites in humanbrain with [3H]PK 11195: anatomical and subcellular distribution. *Brain Res Bull 18*:49–61.

Dostrovsky JO, Davis KD, Kawakita K (1991) Central mechanisms of vascular headaches. *Canadi J Physiol Pharmacol 69*:652–658.

Dreher B, Fukada Y, Rodieck RW (1976) Identification, classification and anatomical segregation of cells with X-like and Y-like properties in the lateral geniculate nucleus of Old-World primates. *J Physiol Lond 258*:433–452.

Dubin MW, Cleland BG (1977) Organization of visual inputs to interneurons of lateral geniculate nucleus of the cat. *J Neurophysiol 40*:410–427.

Dubin W, Cleland BG (1977) Organization of visual inputs to interneurons of lateral geniculate nucleus of the cat. *J Neurophysiol 40*:410–427.

Dubois A, Savasta M, Curet O, Scatton B (1986) Autoradiographic distribution of the D1 agonist [3H] SKF 38393, in the rat brain and spinal cord. Comparison with the distribution of D2 dopamine receptors. *Neurosci 19*:125–137.

Dumont Y, Fournier A, St-Pierre S, Quirion R (1993) Comparative characterization and autoradiographic distribution of neuropeptide Y receptor subtypes in the rat brain. *J Neurosci 13*:73–86.

Duncan GH, Bushnell MC, Oliveras J-L, Bastrash N, Tremblay N (1993) Thalamic VPM nucleus in the behaving monkey. III. Effects of reversible inactivation by lidocaine on thermal and mechanical discrimination. *J Neurophysiol 70*:2086–2096.

Duvoisin RM, Deneris ES, Patrick J, Heinemann S (1989) The functional diversity of the neuronal nicotinic acetylcholine receptors is increased by a novel subunit: beta 4. *Neuron 3*:487–496.

Dykes RW, Rasmusson DD, Hoeltzell PB (1980) Organization of primary somatosensory cortex in the cat. *J Neurophysiol 43*:1527–1546.

Dykes RW, Sur M, Merzenich MM, Kaas JH, Nelson RJ (1981) Regional segregation of neurons responding to quickly adapting, slowly adapting, deep and pacinian receptors within thalamic ventroposterior inferior nuclei in the squirrel monkey (*Saimiri sciureus*). *Neurosci 6*:1687–1692.

Eaton SA, Birse EF, Wharton B, Udvarhelyi PM, Watkins JC, Salt TE (1993) Mediation of thalamic sensory responses in vivo by ACPD-activated excitatory amino acid receptors. *Eur J Neurosci 5*:186–189.

Eaton SA, Salt TE (1989) Modulatory effects of serotonin on excitatory amino acid responses and sensory synaptic transmission in the ventribasal nucleus. *Neurosci 33*:285–292.

Eaton SA, Salt TE (1996) Role of N-Methyl-D-Aspartate and metabotropic glutamate receptors in corticothalamic excitatory postsynaptic potentials in vivo. *Neurosci 73*:1–5.

Edley SM, Graybiel AM (1983) The afferent and efferent connection of the feline nucleus tegmenti pedunculopontinus, pars compacta. *J Comp Neurol 217*:187–215.

Edwards SB, Rosenquist AC, Palmer LA (1974) An autoradiographic study of ventral lateral geniculate projections in the cat. *Brain Res 72*:282–287.

Edwards SB, de Olmos JS (1976) Autoradiographic studies of the projections of the midbrain reticular formation: ascending projections of nucleus cuneiformis. *J Comp Neurol 165*:417–431.

Emmers R, Akert K (1963) A Stereotaxic Atlas of the Brain of the Squirrel Monkey (*Samiri sciureus*). Madison: Univ. Wis. Press.

Emmers R, Tasker RR (1970) The Human Somesthetic Thalamus, with Maps for Physiological Target Localization During Stereotaxic Neurosurgery. New York: Raven.

Emri Z, Turner JP, Crunelli V (1996b) Tonic activation of presynaptic $GABA_B$ receptors on thalamic sensory afferents. *Neurosci 72*:689–698.

Enz R, Brandstatter JH, Hartveit E, Wassle H, Bormann J (1995) Expression of GABA receptor ρ-1 and ρ-2 subunits in the retina and brain of the rat. *Eur J Neurosci 7*:1495–1501.

Ericson A-C, Blomqvist A, Craig AD, Otterson OP, Broman J (1995) Evidence for glutamate as neurotransmitter in trigemino- and spinothalamic tract terminals in the nucleus submedius of cats. *Eur J Neurosci 7*:305–317.

Erzurumlu RS, Killackey HP (1980) Diencephalic projections of the subnucleus interpolaris of the brainstem trigeminal complex in the rat. *Neurosci 5*:1891–1902.

Esclapez M, Tillakaratne NJK, Kaufman DL, Tobin AJ, Houser CR (1994) Comparative localization of two forms of glutamic acid decarboxylase and their mRNAs in rat brain supports the concept of functional differences between the forms. *J Neurosci 14*:1834–1855.

Eysel UT, Pape H-C, Schayck RV (1986) Excitatory and differential disinhibitory actions of acetylcholine in the lateral geniculate nucleus of the cat. *J Physiol Lond 370*:233–254.

Façon E, Steriade M, Wertheim N (1958) Hypersomnie prolongeé engendreé par des lesions bilaterales du systeme activateur medial. Les syndrome thrombotique de la bifurcation du tronc basilaire. *Rev Neurol 98*:117–133.

Faull RLM, Carman JB (1978) The cerebellofugal projections in the brachium conjunctium of the rat. I. The contralateral ascending pathway. *J Comp Neurol 178*:495–518.

Faull RLM, Mehler WR (1978) The cells of origin of nigrotectal, nigrothalamic and nigrostriatal projections in the rat. *Neurosci 3*:989–1002.

Feger J, Bevan M, Crossman AR (1994) The projections from the parafascicular thalamic nucleus to the subthalamic nucleus and the striatum arise from separate neuronal populations: a comparison with the corticostriatal and corticosubthalamic efferents in a retrograde fluorescent double-labelling study. *Neurosci 60*:125–132.

Feig S, Harting JK (1994) Ultrastructural studies of the primate lateral geniculate nucleus: morphology and spatial relationships of axon terminals arising fFrom the retina, visual cortex (Area 17), superior colliculus, parabigeminal nucleus, and pretectum of *Galago crassicaudatus*. *J Comp Neurol 343*:17–34.

Feigenspan A, Wässle H, Bormann J (1993) Pharmacology of GABA receptor Cl^- channels in rat retinal bipolar cells. *Nature 361*:159–162.

Feldman SG, Kruger L (1980) An axonal transport study of the ascending projection of medial lemniscal neurons in the rat. *J Comp Neurol 192*:427–454.

Fenelon G, Francois C, Percheron G, Yelnik J (1991) Topographic distribution of the neurons of the central complex (centre médian-parafascicular complex) and of other thalamic neurons projecting to the striatum in macaques. *Neurosci 45*:495–510.

Ferrera VP, Nealey TA, Maunsell JHR (1994a) Responses in macaque visual area V4 following inactivation of the parvocellular and magnocellular LGN pathways. *J Neurosci 14*:2080–2097.

Ferrera VP, Rudolph KK, Maunsell JHR (1994b) Responses of neurons in the parietal and temporal visual pathways during a motion task. *J Neurosci 14*:6171–6186.

Ferrington DG, Sorkin LS, Willis WD, Jr. (1987) Responses of spinothalamic tract cells in the superficial dorsal horn of the primate lumbar spinal cord. *J Physiol Lond 388*:681–703.

Filion M, Harnois C (1978) A compariosn of projections of entopeduncular neurons to the thalamus, the midbrain and the habenula in the cat. *J Comp Neurol 181*:763–780.

FitzGibbon T, Tevah LV, Sefton AJ (1995) Connections between the reticular nucleus of the thalamus and pulvinar-lateralis posterior complex: a WGA-HRP study. *J Comp Neurol 363*:489–504.
Fitzpatrick D, Conley M, Luppino G, Matelli M, Diamond IT (1988) Cholinergic projections from the midbrain reticular formations and the parabigeminal nucleus to the lateral nucleus in the tree shrew. *J Comp Neurol 272*:43–67.
Fitzpatrick D, Diamond IT, Raczkowski D (1989) Cholinergic and monoaminergic innervation of the cat's thalamus: comparison of the lateral geniculate nucleus with other principal sensory nuclei. *J Comp Neurol 288*:647–675.
Fitzpatrick D, Itoh K, Diamond IT (1983a) The laminar organization of the lateral geniculate body and the striate cortex in the squirrel monkey (*Saimiri sciureus*). *J Neurosci 3*:673–702.
Fitzpatrick D, Penny GR, Schmechel DE (1984) Glutamic acid decarboxylase-immunoreactive neurons and terminals in the lateral geniculate nucleus of the cat. *J Neurosci 4*:1809–1829.
Fitzpatrick D, Raczkowski D (1991) The localization of neurotransmitters in the geniculostriate pathway. In: Neuroanatomy of the Visual Pathways and their Development (Dreher B, Robinson SR eds), pp 278–301. Boca Raton: CRC Press.
Fitzpatrick K, Imig TJ (1978) Projections of auditory cortex upon the thalamus and midbrain in the owl monkey. *J Comp Neurol 177*:437–556.
Flaherty AW, Graybiel AM (1991) Corticostriatal transformations in the primate somatosensory system. Projections from physiologically mapped body-part representations. *J Neurophysiol 66*:1249–1263.
Flaherty AW, Graybiel AM (1994) Input-output organization of the sensorimotor striatum in the squirrel monkey. *J Neurosci 14*:599–610.
Florence SL, Conley M, Casagrande VA (1986) Ocular dominance columns and retinal projections in new world spider monkeys (*Ateles ater*). *J Comp Neurol 243*:234–248.
Fonnum F, Grofová I, Rinvik E (1978) Origin and distribution of glutamate decarboxylase in the nucleus subthalamicus of the cat. *Brain Res 153*:370–374.
Fonnum F, Storm-Mathisen J, Divac I (1981) Biochemical evidence for glutamate as neurotransmitter in corticostriatal and corticothalamic fibres in rat brain. *Neurosci 6*:863–874.
Fosse VM, Kolstad J, Fonnum F (1986) A bioluminescence method for the measurement of L-glutamate: applications to the study of changes in the release of L-glutamate from lateral geniculate nucleus and superior coliculus after visual cortex ablation in rats. *J Neurochem 47*:340–349.
Francesconi W, Muller CM, Singer W (1988) Cholinergic mechansisms in the reticular control of transmission in the cat lateral geniculate nucleus. *J Neurophysiol 59*:1690–1718.
Francois C, Percheron G, Parent A, Sadikot AF, Fenelon G, Yelnik J (1991) Topography of the projection from the central complex of the thalamus to the sensorimotor striatal territory in monkeys. *J Comp Neurol 305*:17–34.
Francois C, Percheron G, Yelnik J, Tande D (1988a) A topographic study of the course of nigral axons and of the distribution of pallidal axonal endings in the centre médian-parafascicular complex of macaques. *Brain Res 473*:181–186.
Frank F, Fabrizi AP, Gaist G, Weigel K, Mundinger F (1987) Sterotactic mesencephalotomy versus multiple thalamotomies in the treatment of chronic cancer pain syndromes. *Appl Neurophysiol 50*:314–318.
Freeman AS, Chiodo LA, Lentz SI, Wade K, Bannon MJ (1991) Release of cholecystokinin from rat midbrain slices and modulatory effect of D2DA receptor stimulation. *Brain Res. 555*:281–287.
French CR, Sefton AJ, Mackay-Sim A (1985) The inhibitory role of the visually responsive region of the thalamic reticular nucleus in the rat. *Exp Brain Res 57*:471–479.
Freund TF, Martin KAC, Soltesz I, Somogyi P, Whitteridge D (1989) Arborisation pattern and postsynaptic targets of physiologically identified thalamocortical afferents in striate cortex of the macaque monkey. *J Comp Neurol 289*:315–336.
Friedemann M (1911) Die Cytoarchitektonik des Zwischenhirns der Cercopitheken mit besonderer Berücksichtigung des Thalamus opticus. *J Psychol Neurol Lpz 18*:309–378.
Friedlander MJ, Lin C-S, Stanford LR, Sherman SM (1981) Morphology of functionally identified neurons in lateral geniculate nucleus of cat. *J Neurophysiol 46*:80–129.
Friedman DP, Jones EG (1980) Focal projection of electrophysiologiclaly defined groupings of thalamic cells on the monkey somatic sensory cortex. *Brain Res 1*:249–252.
Friedman DP, Jones EG (1981) Thalamic input to areas 3a and 2 in monkey. *J Neurophysiol 45*:59–85.
Friedman DP, Murray EA (1986) Thalamic connectivity of the second somatosensory area and neighboring somatosensory fields of the lateral sulcus of the macaque. *J Comp Neurol 252*:348–373.
Frigyesi T, Rinvik E, Yahr MD (Eds) (1972) Corticothalamic Projections and Sensorimotor Activities. New York: Raven Press.
Frigyesi TC (1972) Intracellular recordings from neurons in the dorsolateral thalamic reticular nucleus during capsular,basal ganglia and midline thalamic stimulation. *Brain Res 48*:157–172.

Frigyesi TL, Schwartz R (1972) Cortical control of thalamic sensorimotor relay activities in the cat and the squirrel monkey. In: Corticothalamic Projections and Sensorimotor Activities (Frigyesi T, Rinvik E, Yahr MD eds), pp 161–195. New York: Raven Press.

Fritschy JM, Mohler H (1995) $GABA_A$-receptor heterogeneity in the adult rat brain: differential regional and cellular distribution of seven major subunits. *J Comp Neurol 359*:154–194.

Fuchs JL, Schwark HD (1993) Distribution of [3H]QNB and [125I]alpha-Bungarotoxin binding and acetylcholinesterase activity in visual system and hippocampal structures of eleven mammalian species. *J Comp Neurol 329*:427–437.

Funke K, Eysel UT, FitzGibbon T (1991) Retinogeniculate transmission by NMDA and non-NMDA receptors in the cat. *Brain Res 547*:229–238.

Funke K, Eysel UT (1993) Modulatory effects of acetylcholine, serotonin and noradrenaline on the activity of cat perigeniculate neurons. *Exp Brain Res 95*:409–420.

Fusco M, Polato P, Vantini G, Cavicchioli L, Bentivoglio M, Leon A (1991) Nerve growth factor differentially modulates the expression of its receptor within the CNS. *J Comp Neurol 312*:477–491.

Gabbott PLA, Bacon SJ (1994a) An oriented framework of neuronal processes in the ventral lateral geniculate nucleus of the rat demonstrated by NADPH diaphorase histochemistry and GABA immunocytochemistry. *Neurosci 60*:417–440.

Gabbott PLA, Bacon SJ (1994b) Two ypes of interneuron in the dorsal lateral geniculate nucleus of the rat: A aombined NADPH diaphorase histochemical and GABA immunocytochemical study. *J Comp Neurol 350*:281–301.

Gall CM, Lauterborn J, Burks D, Seroogy K (1987) Colocalization of enkephalin and cholecystokinin in discrete areas of the rat brain. *Brain Res 403*:403–408.

Ganchrow D, Mehler WR (1986) Dual efferent projections of the trigeminal principal sensory nucleus to the thalamic ventroposteromedial nucleus in the squirrel monkey. *Neurosci Lett 68*:181–186.

Garraghty PE, Sur M, Sherman SM (1986) Morphology of retinogeniculate X and Y axon arbors in monocularly enucleated cats. *J Comp Neurol 251*:198–215.

Gattass R, Oswaldo-Cruz E, Sousa APB (1978a) Visuotopic organization of the cebus pulvinar: a double representation of the contralateral hemifield. *Brain Res 152*:1–16.

Gattass R, Sousa APB, Oswaldo-Cruz E (1978b) Single unit response type in the pulvinar of the cebus monkey to multisensory stimulation. *Brain Res 158*:75–88.

Geisert EE, Langsetmo A, Spear PD (1981) Influence of the cortico-geniculate pathway on response properties of cat lateral geniculate neurons. *Brain Res 208*:409–415.

Ghosh S, Brinkman C, Porter R (1987) A quantitative study of the distribution of neurons projecting to the precentral motor cortex in the monkey (*M. fascicularis*). *J Comp Neurol 259*:424–444.

Giesler GJ, Jr., Yezierski RP, Gerhart RP, Willis WD (1981) Spinothalamic tract neurons that project to medial and/or lateral thalamic nuclei: evidence for a physiologically novel population of spinal cord neurons. *J Neurophysiol 46*:1285–1308.

Giguére M, Goldman-Rakic PS (1988) Mediodorsal nucleus: areal, laminar, and tangential distribution of afferents and efferents in the frontal lobe of rhesus monkeys. *J Comp Neurol 277*:195–213.

Gilbert CD, Kelly JP (1975) The projections of cells in different layers of the cat's visual cortex. *J Comp Neurol 163*:81–105.

Gimenez-Amaya JM, Graybiel AM (1990) Compartmental origins of the striatopallidal projection in the primate. *Neurosci 34*:111–126.

Gimenez-Amaya J-M, Graybiel AM (1991) Modular organization of projection neurons in the matrix compartment of the primate striatum. *J Neurosci 11*:779–791.

Gimenez-Amaya JM, McFarland NR, De Las Heras S, Haber SN (1995) Organization of thalamic projections to the ventral striatum in the primate. *J Comp Neurol 354*:127–149.

Gingold SI, Greenspan JD, Apkarian A (1991) Anatomic evidence of nociceptive inputs to primary somatosensory cortex: relationship between spinothalamic terminals and thalamocortical cells in squirrel monkeys. *J Comp Neurol 308*:467–490.

Giolli RA, Guthrie MD (1969) the primary optic projections in the rabbit: an experimental degeneration study. *J Comp Neurol 136*:99–126.

Giolli RA, Towns LC, Takahashi TT, Karamanlidis AN, Williams DD (1978) An autoradiographic study of the projections of visual cortical area to the thalamus, pretectum and superior colliculus of the rabbit. *J Comp Neurol 180*:743–792.

Giüffrida R, Palmeri A, Ricca G, Sapienza S (1988) Cortical control on exteroceptive relay thalamic neurons in the cat. *Pflügers Arch 412*:S27

Giuffrida R, Rustioni A (1988) Glutamate and aspartate immunoreactivity in corticothalamic neurons of rat. In: Cellular Thalamic Mechanisms (Bentivoglio M, Spreafico R eds), pp 311–320. Amsterdam, Elsevier.

Glendenning KK, Hall JA, Diamond IT, Hall WC (1975) The pulvinar nucleus of *Galago senegalensis*. *J Comp Neurol 161*:419–458.
Glendenning KK, Kofron EA, Diamond IT (1976) Laminar organization of projections of the lateral geniculate nucleus to the striate cortex in *Galago*. *Brain Res 105*:538–547.
Glenn LL, Hada J, Roy JP, Steriade M (1982) Anterograde tracer and field potential analysis of the neocortical layer I projection from nucleus ventralis medialis of the thalamus in cat. *Neurosci 7*:1861–1878.
Glenn LL, Steriade M (1982) Discharge rate and excitability of cortically projecting intralaminar thalamic neurons during working and sleep states. *J Neurosci 2*:1387–1404.
Gloor P, Fariello RG (1988) Generalized epilepsy: some of its cellular mechanisms differ from those of focal epilepsy. *Trends Neurosci 11*:63–68.
Godfraind JM (1975) Micro-electrophoretic studies in the cat pulvinar region: effect of acetylcholine. *Exp Brain Res 22*:243–254.
Godfraind JM (1978) Acetylcholine and somatically evoked inhibition on perigeniculate neurons in the cat. *Brit J Pharmacol 63*:295–302.
Goldman D, Duneris E, Luyten W, Kochhar A, Patrick J, Heinemann S (1987) Members of a nicotinic acetylcholine receptor gene family are expressed in different regions of the mammalian central nervous system. *Cell 48*:965–973.
Goldman-Rakic PS, Ilinsky IA, Jounadet ML (1985) Organization of the nigrothalamocortical system in the rhesus monkey. *J Comp Neurol 236*:315–330.
Goldman-Rakic PS, Porrino LJ (1985) The primate mediodorsal (MD) nucleus and its projection to the frontal lobe. *J Comp Neurol 242*:535–560.
Golshani P, Warren RA Jones EG (1998) Progression of change in NMDA, non-NMDA and metabotropic glutamate receptor function during development of the corticothalamic synapse. *J. Neurophysiol.* in press.
Gottesfeld Z, Brandon C, Wu J-Y (1981) Immunocytochemistry of glutamate decarboxylase in the deafferented habenula. *Brain Res 208*:181–186.
Gottschaldt K-M, Vahle-Hinz C, Hicks TP (1983a) Electrophysiological and micropharmacological studies on mechanisms of input-output transformation in single neurones of the somato-sensory thalamus. In: Somatosensory Integration in the Thalamus (Macchi G, Rustioni A, Spreafico R eds), pp 199–216. Amsterdam: Elsevier.
Gottschaldt K-M, Vahle-Hinz C, Young DW (1983b) The extraction of specific stimulus information out of complex afferent signals in subcortical relay stations of the cat's sinus hair system. *Fortschrift Zool 28*:113–128.
Gould HJ, Hall WC, Ebner FF (1978) Connections of visual cortex in the hedgehog (*Paraechinus hypomelas*) I. thalamacortical projections. *J Comp Neurol 177*:445–472.
Gower EC, Kasdon DL, Jacobsen S, Butters NM (1985) Memory dificits in monkeys with mediodorsal lesions. *Neurosci Abstr 11*:831
Grace AA (1991) Phasic versus tonic dopamine release and the modulation of dopamine system responsivity: a hypothesis for the etiology of schizophrenia. *Neurosci 41*:1–24.
Graham J (1977) An autoradiographic study of the efferent connections of the superior colliculus in the cat. *J Comp Neurol 173*:629–654.
Graham J, Linn CS, Kaas JH (1979) Subcortical projections of six visual cortical areas in the owl monkey. *Aotus trivirgatus*. *J Comp Neurol 187*:557–580.
Graham J, Berman N (1981a) Origins of the projections of the superior colliculus to the dorsal lateral geniculate nucleus and the pulvinar in the rabbit. *Neurosci Lett 26*:101–106.
Graham J, Berman N (1981b) Origins of pretectal and tectal projections to the central lateral nucleus in the cat. *Neurosci Lett 26*:209–214.
Grant G, Boivie J, Silfvenius H (1973) Course and termination of fibres from the nucleus z of the medulla oblongata. An experimental light microscopical study in the cat. *Brain Res 55*:55–70.
Graybiel AM (1972a) Some ascending connections of the pulvinar and nucleus lateralis posterior of the thalamus in the cat. *Brain Res 44*:99–125.
Graybiel AM (1972b) Some extrageniculate visual pathways in the cat. *Invest Ophthalmol 11*:322–332.
Graybiel AM (1974a) Studies on the anatomical organization of posterior association cortex. In: The Neurosciences: Third Study Program (Schmitt FO, Worden FG eds), pp 205–214. Cambridge, Mass: MIT Press.
Graybiel AM (1974b) Visuo-cerebellar and cerebello-visual connections involving the ventral lateral geniculate nucleus. *Exp Brain Res 20*:303–306.
Graybiel AM (1977) Direct and indirect oculomotor pathways of the brainstem: an autoradiographic study of the pentine reticular formation in the cat. *J Comp Neurol 175*:37–78.

Graybiel AM (1990) Neurotransmitters and neuromodulators in the basal ganglia. *Trends Neurosci 13*:245–254.
Graybiel AM, Berson DM (1980a) Histochemical identification and afferent connections of subdivisions in the lateralis posterior-pulvinar complex and related thalamic nuclei in the cat. *Neurosci 5*:1175–1238.
Graybiel AM, Berson DM (1980b) Autoradiographic evidence for a projection from the pretectal nucleus of the optic tract to the dorsal lateral geniculate complex in the cat. *Brain Res 195*:1–12.
Graybiel AM, Elde RP (1983) Somatostatin-like immunoreactivity characterizes neurons of the nucleus reticularis thalami in the cat and monkey. *J Neurosci 3*:1308–1321.
Groenewegen HJ (1988) Organization of the afferent connections of the mediodorsal thalamic nucleus in the rat, related to the mediodorsal-prefrontal topography. *Neurosci 24;*2:379–432.
Grofová I, Ottersen OP, Rinvik E (1978) Mesencepahlic and diencephalic afferents to the superior colliculus and periaqueductal gray substance demonstrated by retrograde axonal transport pf horseradish peroxidase in the cat. *Brain Res 146*:205–220.
Gross NB, Lifschitz WS, Anderson DJ (1974) Tonotopic organization of the auditory thalamus of the squirrel monkey (*Saimiri sciureus*). *Brain Res 65*:323–332.
Grove EA (1988a) Efferent connections of the substantia innominata in the rat. *J Comp Neurol 277*:347–364.
Grove EA (1988b) Neural associations of the substantia innominata in the rat: afferent connections. *J Comp Neurol 277*:315–346.
Grunwerg BS, Krauthamer GM (1992) Sensory responses of intralaminar thalamic neurons activated by the superior colliculus. *Exp Brain Res 88*:541
Guillery RW (1966) A study of Golgi preparations from the dorsal lateral geniculate nucleus of the adult cat. *J Comp Neurol 128*:21–50.
Guillery RW (1969a) The organization of synaptic interconnections in the laminae of the dorsal lateral geniculate nucleus of the cat. *Z Zellforsh 96*:1–38.
Guillery RW (1969b) A quantitative study of synaptic interconnections in the dorsal lateral geniculate nucleus of the cat. *Z Zellforsch 96*:39–48.
Guillery RW, Adrian HO, Woolsey CN, Rose JE (1966) Activation of somatosensory areas I and II of cat's cerebral cortex by focal stimulation of the ventrobasal complex. In: The Thalamus (Purpura, Yahr MD eds), pp 197–206. New York: Columbia Univ. Press.
Guillery RW, Colonnier M (1970) Synaptic patterns in the dorsal lateral geniculate nucleus of the monkey. *Z Zellforsch 103*:90–108.
Guillery RW, Kaas JH (1971) A study of normal and congenitally abnormal retinogeniculate projections in cats. *J Comp Neurol 143*:73–100.
Gutierrez C, Yaun A, Cusick CG (1994) Calbindin and SMI-32 immunostaining delineate subdivisions of the visual pulvinar in macaque monkeys. *Soc Neurosci Abstr 24*:770
Guyon A, Leresche N (1995) Modulation by different $GABA_B$ receptor types of voltage-activated calcium currents in rat thalamocortical neurones. *J Physiol Lond 485*:29–42.
Hageman R, De Grood MP (1970) Experiences with stereotactic treatment of intractable pain. *Psychiatr Neurol Neurochir 73*:113–133.
Hale PT, Sefton AJ (1978) A comparison of the visual and electrical response properties of cells in dorsal and ventral lateral geniculate nuclei. *Brain Res 153*:591–595.
Hale PT, Sefton AJ, Bauer LA, Cottee LJ (1982) Interrelations of the rat's thalamic reticular and dorsal lateral geniculate nuclei. *Exp Brain Res 45*:217–229.
Hallanger AE, Levey AI, Lee HJ, Rye DB, Wainer BH (1987) The origins of cholinergic and other subcortical afferents to the thalamus in the rat. *J Comp Neurol 262*:105–124.
Halliday AM, Logue V (1972) Painful sensations evoked by electrical stimulation in the thalamus. In: Neurophysiology Studied in Man. Somjen GG (Ed) pp 221–230 Amsterdam, Excerpta Medica.
Hamilton BL (1973) Projections of the nuclei of the periaqueductal gray matter in the cat. *J Comp Neurol 152*:45–58.
Hamos JE, Van Horn SC, Raczkowski D, Sherman SM (1987) Synaptic circuits involving an individual retinogeniculate axon in the cat. *J Comp Neurol 259*:165–192.
Hamos JE, Van Horn SC, Raczkowski D, Uhlrich DJ, Sherman SM (1985) Synaptic connectivity of a local circuit neuron in lateral geniculate nucleus of the cat. *Nature 317*:618–621.
Harding BN (1971) Dendro-dendritic synapses, including reciprocal synapses, in the ventrolateral nucleus of the monkey thalamus. *Brain Res 34*:181–185.
Harding BN (1972) An electron microscopic study of afferent fibre connexions to the monkey thalamus from motor cortex and basal ganglia. *J Anat 111*:503–504.
Harding BN (1973a) An ultrastructural study of the centre median and ventrolateral thalamic nuclei of the monkey. *Brain Res 54*:335–340.

Harding BN (1973b) An ultrastructural study of the termination of afferent fibres within the ventrolateral and centre median nuclei of the monkey thalamus. *Brain Res 54*:341–346.
Harding BN, Powell TPS (1977) An electron microscopic study of the centre-median and ventrolateral nuclei of the thalamus in the monkey. *Phil Trans R Soc London, B 279*:357–412.
Hardy SGP, Lynch JC (1992) The spatial distribution of pulvinar neurons that project to two subregions of the inferior parietal lobule in the macaque. *Cerebral Cortex 2*:217–230.
Hardy TL, Bertrand G, Thompson CJ (1980) Position and organization of thalamic cellular activity during diencephalic recording. *Appl Neurophysiol 43*:18–27.
Harman AM, Coleman L-A, Beazley LD (1990) Retinofugal projections in a marsupial, *Tarsipes rostratus* (honey possum). *Brain Behav Evol 36*:30–38.
Harris RM (1987) Axon collateral in the thalamic reticular nucleus from thalamocortical neurons of the rat ventrobasal thalamus. *J Comp Neurol 258*:397–406.
Harting JK, Casagrande VA, Weber JT (1978) The projection of the primate superior colliculus upon the dorsal lateral geniculate nucleus: autoradiographic demonstration of interlaminar distribution of tectogeniculate axons. *Brain Res 150*:593–599.
Harting JK, Diamond IT, Hall WC (1973a) Anterograde degeneration study of the cortical projections of the lateral geniculate and pulvinar nuclei in the tree shrew (*Tupaia glis*). *J Comp Neurol 150*:393–440.
Harting JK, Hall WC, Diamond IT, Martin GF (1973b) Anterograde degeneration study of the superior colliculus in *Tupaia glis*: evidence for a subdivision between superficial and deep layers. *J Comp Neurol 148*:361–386.
Harting JK, Hashikawa T, Van Lieshout D (1986) Laminar distribution of tectal, parabigeminal and pretectal inputs to the primate dorsal lateral geniculate nucleus: connectional studies in *Galago crassicaudatus*. *Brain Res 366*:358–363.
Harting JK, Huerta MF, Frankfurter AJ, Strominger NL, Royce GJ (1980) Ascending pathways from the monkey superior colliculus: an autoradiographic analysis. *J Comp Neurol 192*:853–882.
Harting JK, Huerta MF, Hashikawa T, Van Lieshout DP (1991a) Projection of the mammalian superior colliculus upon the dorsal lateral geniculate nucleus: organization of tectogeniculate pathways in nineteen species. *J Comp Neurol 304*:275–306.
Harting JK, Van Lieshout DP, Feig S (1991b) Connectional studies of the primate lateral geniculate nucleus: distribution of axons arising from the thalamic reticular nucleus of Galago crassicaudatus. *J Comp Neurol 310*:411–427.
Harting JK, Van Lieshout DP, Hashikawa T, Weber JT (1991c) The parabigeminogeniculate projection: Connectional studies in eight mammals. *J Comp Neurol 305*:559–581.
Hartveit E, Heggelund P (1990) Neurotransmitter receptors mediating excitatory input to cells in the cat lateral geniculate nucleus. II. Nonlagged cells. *J Neurophysiol 63*:1361–1372.
Harvey AR (1980a) The afferent connexions and laminar distribution of cells in area 18 of the cat. *J Physiol Lond 302*:483–505.
Harvey AR (1980b) A physiological analysis of subcortical and commissural projections of areas 17 and 18 in the cat. *J Physiol Lond 302*:507–534.
Hashikawa T, Molinari M, Rausell E, Jones EG (1995) Patchy and laminar teminations of medial geniculate axons in monkey auditory cortex. *J Comp Neurol 362*:195–207
Hashikawa T, Rausell E, Molinari M, Jones EG (1991) Parvalbumin- and calbindin-containing neurons in the monkey medial geniculate complex: differential distribution and cortical layer specific projections. *Brain Res 544*:335–341.
Hassler R (1959) Anatomy of the thalamus. In: Introduction to Stereotaxis with an Atlas of the Human Brain (Schaltenbrand G, Bailey P eds), pp 230–290. New York: Thieme.
Hassler R (1982) Architectonic organization of the thalamic nuclei. In: Stereotaxy of the Human Brain (Schaltenbrand G, Walker AE eds), pp 140–180. Stuttgart: Thieme.
Hassler R, Haug P, Nitsch C, Kim JS, Paik K (1982) Effect of motor and premotor cortex ablation on concentrations of amino acids, monoamines, and acetylcholine and on the ultrastructure in rat striatum. A confirmation of glutamate as the specific cortico-striatal transmitter. *J Neurochem 38*:1087–1098.
Hassler R, Riechert T (1959b) Klinische und anatomische Befunde bei stereotaktischen Schmerzoperationen im Thalamus. *Arch Psychiat Nervenkr 200*:93–122.
Havton LA, Ohara PT (1993) Quantitative analyses of intracellularly characterized and labeled thalamocortical projection neurons in the ventrobasal complex of primates. *J Comp Neurol 336*:135–150.
Havton LA, Ohara PT (1994) Cell body and dendritic tree size of intracellularly labeled thalamocortical projection neurons in the ventrobasal complex of cat. *Brain Res 651*:76–84.
Hazlett JC, Dutta CR, Fox CA (1976) The neurons in the centromedian-parafascicular complex of the monkey (*macaca mulatta*): a Golgi study. *J Comp Neurol 168*:41–74.

Hazrati L-N, Parent A (1991) Projection from the external pallidum to the reticular thalamic nucleus in the squirrel monkey. *Brain Res 550*:142–146.

He J (1997b) Modulatory effects of regional cortical activation on the onset responses of the cat medial geniculate neurons. *J Neurophysiol 77*:896–908.

Heath CJ, Hore J, Phillips CG (1976) Inputs from low threshold muscle and cutaneous afferents of hand and forearm to areas 3a and 3b of baboon's cerebral cortex. *J Physiol Lond 257*:199–227.

Heath CJ, Jones EG (1971) An experimental study of ascending connections from the posterior group of thalamic nuclei in the cat. *J Comp Neurol 141*:397–426.

Heckers S, Geula C, Mesulam M-M (1992) Cholinergic innervation of the human thalamus: dual origin and differential nuclear distribution. *J Comp Neurol 325*:68–82.

Helfant MH, Leksell L, Strang RR (1965) Experiences with intractable pain treated by stereotaxic mesencephalotomy. *Acta Chir Scand 129*:573–580.

Hemphill M, Holm G, Crutcher M, Delong M, Hedreen J (1981) Afferent connections of the nucleus accumbens in the monkey. In: The Neurobiology of the Nucleus Accumbens (Chronister RB, Defrance JF eds), pp 75–81. Brunswick: Haer Institute.

Hendrickson A (1973) The pregeniculate nucleus of the monkey. *Anat Rec 175*:341

Hendrickson A, Wilson ME, Toyne MJ (1970) The distribution of optic nerve fibers in *Macaca mulatta. Brain Res 23*:425–427.

Hendrickson AE, Wilson JR, Ogren MP (1978) The neuroanatomical organization of pathways between the dorsal lateral geniculate nucleus and visual cortex in old world and new world primates. *J Comp Neurol 182*:123–135.

Hendry SHC, Hockfield SJ, McKay RDG, Jones EG (1983) Features of the organization of macaque monkey visual cortex demonstrated with monoclonal antibody, CAT301. *Neurosci Abstr 9*:39

Hendry SHC, Jones EG (1988b) Activity-dependent regulation of GABA expression in the visual cortex of adult monkeys. *Neuron 1*:701–712.

Hendry SHC, Jones EG, Graham J (1979) Thalamic relay nuclei for cerebellar and certain related fiber system in the cat. *J Comp Neurol 185*:679–714.

Hendry SHC, Jones EG, Hockfield S, McKay RDG (1988a) Neuronal populations stained with the monoclonal antibody cat-301 in the mammalian cerebral cortex and thalamus. *J Neurosci 8*:518–542.

Hendry SHC, Yoshioka T (1994) A neurochemically distinct third channel in the macaque dorsal lateral geniculate nucleus. *Science 264*:575–577.

Herkenham M (1978) The connections of the nucleus reuniens thalami: evidence for a direct thalamo-hippocampal pathway in the ra. *J Comp Neurol 177*:589–610.

Herkenham M (1987) Mismatches between neurotransmitter and receptor localizations in brain: observations and implications. *Neurosci 23*:1–38.

Herkenham M, McLean S (1988) The anatomical relationship of opioid peptides and opiate receptors in the hippocammpi of four rodent species. *NIDA Res Monogr 82*:33–47.

Herkenham M, McLean S, Moon SL, Pert CB (1988) Autoradiography of receptor distributions in thalamus suggests and endocrine function for neurotransmitters: evidence from comparative anatomy of the opiate system. In: Cellular Thalamic Mechanisms (Bentivoglio M, Spreafico R eds), pp 417–434. Amsterdam Elsevier..

Herkenham M, Nauta WJH (1979) Efferent connections of the habenular nuclei in the rat. *J Comp Neurol 187*:19–48.

Herkenham M, Pert CB (1981) Mosaic distribution of opiate receptors, parafascicular projections and acetylcholinesterase in rat striatum. *Nature 291*:415–417.

Hersch SM, Gutekunst CA, Rees HD, Heilman CJ, Levey AI (1994) Distribution of m1-m4 muscarinic receptor proteins in the rat striatum: light and electron microscopic immunocytochemistry using subtype-specific antibodies. *J Neurosci 14*:3351–63.

Hess DT, Edwards MA (1987) Anatomical demonstration of ocular segregation in the retinogeniculocortical pathway of the new world capuchin monkey (*Cebus apella*). *J Comp Neurol 264*:409–420.

Hickey TL, Spear PD (1976) Retinogeniculate projections in hooded and albino rats: an autoradiographic study. *Exp Brain Res 24*:523–529.

Hicks TP, Kaneko T, Metherate R, Oka J-L, Stark CA (1991) Amino acids as transmitters of synaptic excitation in neocortical sensory processes. *Can J Physiol Pharmacol 69*:1099–1114.

Hicks TP, Metherate R, Landry P, Dykes RW (1986) Bicuculline-induced alterations of response properties in functionally identified ventroposterior thalamic neurones. *Exp Brain Res 63*:248–264.

Hill RG, Pepper CM, Mitchell JF (1976) Depression of nociceptive and other neurones in the brain by iontophoretically applied met-enkephalin. *Nature, Lond 262*:604–606.

Hirai T, Jones EG (1989a) A new parcellation of the human thalamus on the basis of histochemical staining. *Brain Res Rev 14*:1–34.

Hirai T, Jones EG (1989b) Distribution of tachykinin- and enkephalin-immunoreactive fibers in the human thalamus. *Brain Res Rev 14*:35–52.

Hockfield S, McKay RD (1983) A surface antigen expressed by a subset of neurons in the vertebrate central nervous system. *Proc Natl Acad Sci U.SA 80*:5758–61.

Hockfield S, McKay RD (1985) Identification of major cell classes in the developing mammalian nervous system. *J Neurosci 5*:3310–28.

Hockfield S, McKay RDG, Hendry SHC, Jones EG (1983) A surface antigen that identifies ocular dominance columns in the visual cortex and lamina features of the lateral geniculate nucleus. Cold Spring Harbor Symp Quant Biol 4:877–889.

Hoffman KP, Stone J, Sherman SM (1972) Relay of receptive-field properties in dorsal lateral geniculate nucleus of the cat. *J Neurophysiol 35*:518–531.

Holcombe V, Guillery RW (1984) The organization of retinal maps within the dorsal and ventral lateral geniculate nuclei of the rabbit. *J Comp Neurol 225*:469–491.

Holländer H, Sanides D (1976) The retinal projection to the ventral part of the lateral geniculate nucleus. An experimental study with silver-impregnation methods and axoplasmic protein tracing. *Exp Brain Res 26*:329–342.

Holländer H, Tietze J, Distel H (1979) An autoradiographic study of the subcortical projections of the rabbit striate cortex in the adult and during postnatal development. *J Comp Neurol 184*:783–794.

Hollmann M, Heinemann S (1994) Cloned glutamate receptors. *Ann Rev Neurosci 17*:31–108.

Holsapple JW, Preston JB, Strick PL (1991) The origin of thalamic inputs to the 'hand' representation in the primary motor cortex. *J Neurosci 11*:2644–2654.

Hoogland PV, Welker E, Van der Loos H (1987) Organization of the projections from barrel cortex to thalamus in mice studied with *Phaseolus vulgaris*-leucoagglutinin and HRP. *Exp Brain Res 68*:73–87.

Hoogland PV, Welker E, Van der Loos H, Wouterland FG (1988) The organization and structure of the thalamic afferents from the barrel cortex in the mouse: a PHA-L study. In: Cellular Thalamic Mechanisms (Bentivoglio M, Spreafico R eds), pp 151–162. Elsevier.

Hoover DB, Muth EA, Jacobowitz DM (1978) A mapping of the distribution of acetylcholine, choline acetyltransferase and acetylcholinesterase in discrete areas of the brain. *Brain Res 153*:295–306.

Hore J, Preston JB, Durkovic RG, Cheney PD (1976) Responses of cortical neurons (areas 3a and 4) to ramp stretch of hindlimb muscles in the baboon. *J Neurophysiol 39*:484–500.

Houser CR, Crawford GD, Barber RP, Salvaterra PM, Vaughn JE (1983) Organization and morphological characteristics of cholinergic neurons: an immunocytochemical study with a monoclonal antibody to choline acetyltransferase. *Brain Res 266*:97–119.

Houser CR, Phelps PE, Vaughn JE (1988) Cholinergic innervation of the rat thalamus as demonstrated by the immunocytochemical localization of the choline acetyltransferase. In: Cellular Thalamic Mechanisms (Bentivoglio M, Spreafico R Eds) pp 387–398. Amsterdam: Elsevier.

Houser CR, Vaughn JE, Barber RP, Roberts E (1980) GABA neurons are the major cell type of the nucleus reticularis thalami. *Brain Res 200*:341–354.

Hreib KK, Rosene DL, Moss MB (1988) Basal forebrain efferents to the medial dorsal thalamic nucleus in the rhesus monkey. *J Comp Neurol 277*:365–390.

Hu B, Senatorov V, Mooney D (1994) Lemniscal and non-lemniscal synaptic transmission in rat auditory thalamus. *J Physiol Lond 479*:217–231.

Hu B, Steriade M, Deschênes M (1989a) The effects of brainstem peribrachial stimulation on neurons of the lateral geniculate nucleus. *Neurosci 31*:13–25.

Hu B, Steriade M, Deschênes M (1989b) The effects of brainstem peribrachial stimulation on perigeniculate neurons: the blockage of spindle waves. *Neurosci 31*:1–13.

Hu B, Steriade M, Deschênes M (1989c) The cellular mechanism of thalamic ponto-geniculo-occipital waves. *Neurosci 31*:25–36.

Huang Q, Zhou D, Chase K, Gusella JF, Aronin N, DiFiglia M (1992) Immunohistochemical localization of the D1 dopamine receptor in rat brain reveals its axonal transport, pre- and postsynaptic localization, and prevalence in the basal ganglia, limbic system, and thalamic reticular nucleus. *Proc Natl Acad Sci USA 89*:11988 92.

Hubel DH, Livingstone MS (1990) Color and contrast sensitivity in the lateral geniculate body and primary visual cortex of the macaque monkey. *J Neurosci 10*:2223–2237.

Hubel DH, Wiesel TN (1961) Integrative action in the cat's lateral geniculate body. *J Physiol Lond 155*:385–398.

Hubel DH, Wiesel TN (1968) Receptive fields and functional architecture of monkey striate cortex. *J Physiol Lond 195*:215–243.

Hubel DH, Wiesel TN (1969) Anatomical demonstration of columns in the monkey striate cortex. *Nature 221*:747–750.

Hubel DH, Wiesel TN (1972) Laminar and columnar distribution of geniculo-cortical fibers in the macaque monkey. *J Comp Neurol 146*:421–450.
Hubel DH, Wiesel TN (1977) Functional architecture of macaque monkey visual cortex. *Proc R Soc Lond B 198*:1–59.
Hubel DH, Wiesel TN, LeVay S (1977) Plasticity of ocular dominance columns in monkey striate cortex. *Phil Trans R Soc London, B 278*:377–409.
Huerta MF, Krubitzer LA, Kaas JH (1986) Frontal eye field as defined by intracortical microstimulation in squirrel monkeys, owl, monkeys, and macaque monkeys: I. subcortical connections. *J Comp Neurol 253*:415–439.
Huerta MF, Harting JK (1983) Sublamination within the superficial gray layer of the squirrel monkey: an analysis of the tectopulvinar projection using anterograde and retrograde transport methods. *Brain Res 261*:119–126.
Huerta MF, Kaas JH (1990) Supplementary eye field as defined by intracortical microstimulation: connections in macaques. *J Comp Neurol 293*:299–330.
Hughes A (1980) Directional units in the rat optic nerve. *Brain Res 202*:196–200.
Hughes CP, Ater SB (1977) Receptive field properties in the ventral lateral geniculate nucleus of the cat. *Brain Res 132*:163–166.
Hughes HC (1980) Efferent organization of the cat pulvinar complex with a note on bilateral claustrocortical and reticulocortical connections. *J Comp Neurol 193*:937–963.
Hughes HC, Mullikin WH (1984) Brainstem afferents to the lateral geniculate nucleus of the cat. *Exp Brain Res 54*:253–258.
Huguenard JR, Prince DA (1994a) Intrathalamic rhythmicity studied *in vitro*: nominal T-current modulation causes robust antioscillatory effects. *J Neurosci 14*:5485–5502.
Huguenard JR, Prince DA (1994b) Clonazepam suppresses $GABA_B$-mediated inhibition in thalamic relay neurons through effects in nucleus reticularis. *J Neurophysiol 71*:2576–2581.
Hull EM (1968) Corticofugal influence in the macaque lateral geniculate nucleus. *Vision Research 8*:1285–1298.
Hunt CA, Pang DZ, Jones EG (1991) Distribution and density of GABA cells in intralaminar and adjacent nuclei of monkey thalamus. *Neurosci 43*:185–196.
Hunt CA, Seroogy KB, Gall CM, Jones EG (1987) Cholecystokinin innervation of rat thalamus, including fibers to ventroposterolateral nucleus from dorsal column nuclei. *Brain Res 426*:257–269.
Hunt S, Schmidt J (1978) Some observations on the binding patterns of α-bungarotoxin in the central nervous system of the rat. *Brain Res 157*:213–232.
Huntsman MM, Jones EG (1993) Distribution of GABA-A receptor subunit mRNAs in monkey thalamus. *Neurosci Abstr 19*:132
Huntsman MM, Leggio MG, Jones EG (1995) Expression patterns and deprivation effects on $GABA_A$ receptor subunit and GAD mRNAs in monkey lateral geniculate nucleus. *J Comp Neurol 352*:235–247.
Huntsman MM, Leggio MG, Jones EG (1996) Nucleus-specific expression of $GABA_A$ receptor subunit mRNAs in monkey thalamus. *J Neurosci 16*:3571–3589.
Ilinsky IA, Jouandet ML, Goldman-Rakic PS (1985) Organization of the nigrothalamocortical system in the rhesus monkey. *J Comp Neurol 236*:315–330.
Ilinsky IA, Kultas-Ilinsky K (1982) Stereotactic surgery in the rhesus monkey (*Macaca mulatta*) utilizing intracerebral landmarks. *Appl. Neurophysiol 45*:563–572.
Ilinsky IA, Kultas-Ilinsky K (1984) An autoradiographic study of topographical relationships between pallidal and cerebellar projections to the cat thalamus. *Exp Brain Res 54*:95–106.
Ilinsky IA, Kultas-Ilinsky K (1987) Sagittal cytoarchitectonic maps of the *Macaca mulatta* thalamus with a revised nomenclature of the motor-related nuclei validated by observations on their connectivity. *J Comp Neurol 262*:331–364.
Ilinsky IA, Kultas-Ilinsky K (1990a) Fine structure of the magnocellular subdivision of the ventral anterior thalamic nucleus (VAmc) of *macaca mulatta*: I. Cells types and synaptology. *J Comp Neurol 294*:455–478.
Ilinsky IA, Kultas-Ilinsky K (1990b) Fine structure of the magnocellular subdivision of the ventral anterior thalamic nucleus (VAmc) of *Macaca mulatta*: II. Organization of nigrothalamic afferents as revealed with EM autoradiography. *J Comp Neurol 294*:479–489.
Ilinsky IA, Kultas-Ilinsky K (1990c) Fine structure of the magnocellular subdivision of the ventral anterior thalamic nucleus (VAmc) of *Macaca mulatta*: I. Cell types and synaptology. *J Comp Neurol 294*:455–478.
Ilinsky IA, Tourtellotte WG, Kultas-Ilinsky K (1993) anatomical distinctions between the two basal ganglia afferent territories in the primate motor thalamus. *Stereotact Funct Neurosurg 60*:62–69.
Imig TJ, Morel A (1985a) Tonotopic organization in ventral nucleus of medial geniculate body in the cat. *J Neurophysiol 53*:309–340.

Imig TJ, Morel A (1985b) Tonotopic organization in lateral part of posterior group of thalamic nuclei in the cat. *J Neurophysiol 53*:836–851.

Imig TJ, Ruggero MA, Kitzes LM, Javel E, Brugge JF (1977a) Organization of auditory cortex in the owl monkey (*Aotus trigvirgatus*). *J Comp Neurol 171*:111–128.

Inagaki S, Kubota Y, Shinoda K, Kawai Y, Tohyama M (1983) Neurotensin containing pathway from the endopiriform nucleus and the adjacent prepiriform cortex to the dorsomedial thalamic nucleus in the rat. *Brain Res 260*:143–146.

Insausti R, Amaral DG, Cowan WM (1987) The entorhinal cortex of the monkey: II. cortical afferents. *J Comp Neurol 264*:396–408.

Irvin GE, Norton TT, Sesma MA, Casagrande VA (1986) W-like response properties on interlaminar zone cells in the lateral geniculate nucleus of a primate (*Galago crassicaudatus*). *Brain Res 362*:254–270.

Irvine DRF (1980) Acoustic properties of neurons in posteromedial thalamus of cat. *J Neurophysiol 43*:395–408.

Irvine DRF (1986) The Auditory Brainstem. Berlin: Springer.

Ishijima B, Yoshimasu N, Fukushima T, Hori T, Sekino H, Sano K (1975) Nociceptive neurons in the human thalamus. *Confin Neurol 37*:99–106.

Itoh K (1977) Efferent projections of the pretectum in the cat. *Exp Brain Res 30*:89–105.

Itoh K, Mizuno N (1977) Topographical arrangement of thalamocortical neurons in the centrolateral nucleus (CL) of the cat, with special reference to a spino-thalamo-motor cortical path through the CL. *Exp Brain Res 30*:471–480.

Itoh K, Misuno N, Sugimoto T, Nomura S, Nakamura Y, Konishi A (1979) A cerebello-pulvino-cortical and a retino-pulvino-cortical pathways in the cat as revealed by the use of the anterograde and retrograde transport of horseradish peroxidase. *J Comp Neurol 187*:349–357.

Iwamura Y, Tanaka M (1978a) Postcentral neurons in hand region of area 2: their possible role in the form discrimination of tactile objects. *Brain Res 150*:662–666.

Iwamura Y, Tanaka M, Sakamoto M, Hikosaka O (1993) Rostrocaudal gradients in the neuronal receptive field complexity in the finger region of the alert monkey's postcentral gyrus. *Exp Brain Res 92*:360–368.

Jacobson S, Trojanowski JQ (1975a) Corticothalamic neurons and thalamocortical terminal fields: an investigation in rat using horseradish peroxidase and autoradiography. *Brain Res 85*:385–401.

Jacobson S, Trojanowski JQ (1975b) Amygdaloid projections to prefrontal granular cortex in rhesus monkey demonstrated with horseradish peroxidase. *Brain Res 100*:132–139.

Jacobson SJ, Trojanowski JQ (1975c) Corticothalamic neurons and thalamocortical terminals fields: an investigation in rat using horseradish peroxidase and autoradiography. *Brain Res 85*:385–401.

Jahnsen H, Llinás R (1984) Electrophysiological studies of guinea-pig thalamic neurones: An in vitro study. *J Physiol Lond 349*:205–226.

Jasper HH (1949) Diffuse projection systems: the integrative activity of the thalamic reticular system. *Electroenceph Clin Neurophysiol 1*:405–420.

Jasper HH (1960) Unspecific thalamocortical relations. In: Handbook of Physiology Section I Neurophysiology Volume 2 (Field J, Magoun HW, Hall VE eds), pp 1307–1321. Washington, D.C. American Physiological Society.

Jeanmonod D, Magnin M, Morel A (1993) Thalamus and neurogenic pain: physiological, anatomical and clinical data. *NeuroReprt 4*:475–478.

Jennes I, Stumph WE, Kalivas PW (1980) Neurotensin: topographical distribution in rat brain by immunohistochemistry. *J Comp Neurol 210*:211–224.

Jinnai K, Matsuda Y (1981) Thalamocaudate projection neurons with a branching axon to the cerebral motor cortex. *Neurosci Lett 26*:95–100.

Johansson H, Silfvenius H (1977) Input from ipsilateral proprio- and exteroceptive hind limb afferents to nucleus Z of the cat medulla oblongata. *J Physiol Lond 265*:371–393.

Johnson JK, Casagrande VA (1995) Distibution of calcium-binding proteins within the parallel visual pathways of a primate (*Galago crassicaudatus*). *J Comp Neurol 356*:238–260.

Johnson RR, Burkhalter A (1992) Evidence for excitatory amino acid neurotransmitters in the geniculocortical pathway and local projections within rat primary visual cortex. *Exp Brain Res 89*:20–30.

Johnston GA (1994a) GABA receptors: as complex as ABC? *Clin Exp Pharmacol and Physiol 21*:521–526.

Johnston GA (1994b) $GABA_C$ receptors. *Prog Brain Res 100*:61–65.

Jones BE, Moore RY (1977) Ascending projections of the locus coeruleus in the rat. II Autoradiographic study. *Brain Res 127*:23–53.

Jones EG (1975a) Possible determinants of the degree of retrograde neuronal labeling with horseradish peroxidase. *Brain Res 85*:249–253.

Jones EG (1975b) Some aspects of the organization of the thalamic reticular complex. *J Comp Neurol 162*:285–307.

Jones EG (1975c) Lamination and differential distribution of thalamic afferents in the sensory-motor cortex of the squirrel monkey. *J Comp Neurol 160*:167–204.
Jones EG (1983) The thalamus. In: Chemical Neuroanatomy (Emson PC ed), pp 257–293. New York: Raven.
Jones EG (1983a) Lack of collateral thalamocortical projections to fields of the first somatic sensory cortex in monkeys. *Exp Brain Res 52*:375–384.
Jones EG (1983b) Distribution patterns of individual medial lemniscal axons in thalamic ventobasal complex of monkeys. *J Comp Neurol 215*:1–16.
Jones EG (1983c) The thalamus. In: Chemical Neuroanatomy Emson PC Ed. Raven New York pp. 257–293.
Jones EG (1983d) Identification and classification of intrinsic circuit elements in the neocortex. In: Dynamic Aspects of Neocortical Organization (Edelman G, Cowan WM, Gall WE eds), New York: John Wiley.
Jones EG (1984) Laminar distribution of cortical efferent cells. In: Cerebral Cortex Volume 1 Cellular Components of the Cerebral Cortex (Jones EG, Peters A eds), pp 521–553. New York: Plenum.
Jones EG (1985) The Thalamus. New York: Plenum.
Jones EG (1987) Ascending inputs to, and internal organization of, cortical motor areas. In: CIBA Foundation Symposium, 132, Motor Areas of the Cerebral Cortex (Porter R ed), pp 21–39. Chichester: Wiley.
Jones EG (1987) Immunocytochemical studies on thalamic afferent transmitters. In: Thalamus and Pain (Besson JM, Guilbaud G, Peschanski M eds), pp 83–109. Amsterdam: Elsevier.
Jones EG (1988) Modern views of cellular thalamic mechanisms. In: Cellular Thalamic Mechanisms. (Bentivoglio M, Spreafico R eds), pp 1–22. Amsterdam: Elsevier.
Jones EG (1989) Defining the thalamic intralaminar nuclei in primates. In: Neurologia e Scienze di Base. Scritti in Onore di Giorgio Macchi. (Gainotti G, Bentivoglio M, Bergonzi P, Ferro FM eds), pp 161–193. Milan: Universita Cattolica del Sacro Cuore.
Jones EG, Burton H (1976a) Areal differences in the laminar distribution of thalamic afferents in cortical fields of the insular, parietal and temporal regions of primates. *J Comp Neurol 168*:197–248.
Jones EG, Burton H (1976b) A projection from the medial pulvinar to the amygdala in primates. *Brain Res 104*:142–147.
Jones EG, Burton H, Saper CB, Swanson L (1976) Midbrain, diencephalic and cortical relationships of the basal nucleus of Meynert and related structures in primates. *J Comp Neurol 167*:385–420.
Jones EG, Coulter JD, Burton H, Porter R (1977) Cells of origin and terminal distribution of corticostriatal fibers arising in the sensory-motor cortex of monkeys. *J Comp Neurol 173*:53–80.
Jones EG, Dell'Anna ME, Molinari M, Rausell E, Hashikawa T (1995) Subdivisions of macaque monkey auditory cortex revealed by calcium binding protein immunoreactivity. *J Comp Neurol 362*:153–170.
Jones EG, Friedman DP (1982) Projection pattern of functional components of thalamic ventrobasal complex on monkey somatosensory cortex. *J Neurophysiol 48*:521–544.
Jones EG, Friedman DP, Hendry SHC (1982) Thalamic basis of place and modality-specific columns in monkey somatosensory cortex: a correlative anatomical and physiological study. *J Neurophysiol 48*:545–568.
Jones EG, Hendry SHC (1989) Differential calcium binding protein immunoreactivity distinguishes classes of relay neurons in monkey thalamic nuclei. *Eur J Neurosci 1*:222–246.
Jones EG, Hendry SHC, Brandon C (1986a) Cytochrome oxidase staining reveals functional organization of monkey somatosensory thalamus. *Exp Brain Res 62*:438–442.
Jones EG, Huntley GW, Benson DL (1994) Alpha calcium/calmodulin-dependent protein kinase II selectively expressed in a subpopulation of excitatory neurons in monkey sensory-motor cortex: comparison with GAD-67 expression. *J Neurosci 14*:611–629.
Jones EG, Leavitt RY (1974) Retrograde axonal transport and the demonstration of non-specific projections to the cerebral cortex and striatum from thalamic intralaminar nuclei in the rat, cat and monkey. *J Comp Neurol 154*:349–378.
Jones EG, Powell TPS (1968) The projection of the somatic sensory cortex upon the thalamus in the cat. *Brain Res 10*:369–391.
Jones EG, Powell TPS (1969a) The cortical projection of the ventroposterior nucleus of the thalamus in the cat. *Brain Res 13*:298–318.
Jones EG, Powell TPS (1969b) Electron microscopy of synaptic glomeruli in the thalamic relay nuclei of the cat. *Proc R Soc Lond B 121*:153–171.
Jones EG, Powell TPS (1969c) An electron microscopic study of the mode of termination of corticothalamic fibers in the thalamic relay nuclei of the cat. *Proc R Soc Lond B 172*:173–185.
Jones EG, Powell TPS (1970a) An electron microscopic study of the laminar pattern and mode of termination of the afferent fibre pathways to the somatic sensory cortex. *Phil Trans Roy Soc 257*:45–62.

Jones EG, Powell TPS (1970b) Connexions of the somatic sensory cortex of the rhesus monkey III. Thalamic connexions. *Brain 93*:37–56.
Jones EG, Powell TPS (1971) An analysis of the posterior group of the thalamic nuclei on the basis of its afferent connections. *J Comp Neur 143*:185–216.
Jones EG, Schwark HD, Callahan PJ (1986b) Extent of the ipsilateral representation in the ventral posterior medial nucleus of the monkey thalamus. *Exp Brain Res 63*:310–320.
Jones HE, Sillito AM (1992) The action of the putative neurotransmitters *N*-acetylaspartylglutamate and L-homocysteate in cat dorsal lateral geniculate nucleus. *J Neurophysiol 68*:663–672.
Jones HE, Sillito AM (1994a) Directional asymmetries in the length-response profiles of cells in the feline dorsal lateral geniculate nucleus. *J Physiol Lond 479*:475–486.
Jones HE, Sillito AM (1994b) The length-response properties of cells in the felilne perigeniculate nucleus. *Eur J Neurosci 6*:1199–1204.
Jones EG, Wise SP (1977) Size, laminar and columnar distribution of efferent cells in the sensory-motor cortex of monkeys. *J Comp Neurol 175*:391–438.
Jones EG, Wise SP, Coulter JD (1979) Differential thalamic relationships of sensory motor and parietal cortical fields in monkeys. *J Comp Neurol 183*:833–882.
Jordan H (1973) The structure of the medial geniculate nucleus (MGN): a cyto-and myeloarchitectonic study in the squirrel monkey. *J Comp Neurol 148*:469–480.
Jourdain A, Semba K, Fibiger HC (1989) Basal forebrain and mesopontine tegmental projections to the reticular thalamic nucleus: an axonal collateralization and immunohistochemical study in the rat. *Brain Res 505*:55–65.
Juhasz G, Emri Z, Kekesi KA, Salfay O, Crunelli V (1994) Blockade of thalamic $GABA_B$ receptors decreases EEG synchronization. *Neurosci Lett 172*:155–158.
Julius D, MacDermott AB, Axel RMW, Jessel TM (1988) Molecular characterization of a functional cDNA encoding the serotonin 1c receptor. *Science 241*:558–564.
Kaas JH, Huerta MF, Weber JT, Harting JK (1978) Patterns of retinal terminations and laminar organization of the lateral geniculate nucleus of primates. *J Comp Neurol 182*:517–554.
Kaas JH, Lin CS, Casagrande VA (1976) The relay of ipsilateral and contralateral retinal input from the lateral geniculate nucleus to striate cortex in the owl monkey: a transneuronal transport study. *Brain Res 106*:371–378.
Kaas JH, Nelson RJ, Sur M, Dykes RW, Merzenich MM (1984) The somatotopic organization of the ventroposterior thalamus of the squirrel monkey. *Saimiri sciureus. J Comp Neurol 226*:111–140.
Kalil K (1978) Patch-like termination of thalamic fibers in the putamen of the rhesus monkey: an autoradiographic study. *Brain Res 140*:333–339.
Kalil K (1981) Projections of the cerebellar and dorsal column nuclei upon the thalamus of the rhesus monkey. *J Comp Neurol 195*:25–50.
Kalil RE, Chase R (1970) Corticofugal influence on activity of lateral geniculate neurons in the cat. *J Neurophysiol 33*:459–474.
Kanagasuntheram R, Krishnamurti A (1970) The termination of optic fibers in the lateral geniculate nucleus of some primates. *Brain Res 17*:129–132.
Kanagasuntheram R, Krishnamurti A, Wong WC (1969) Observations on the lamination of the lateral geniculate body in some primates. *Brain Res 14*:623–631.
Kaneko T, Tashiro K, Sugimoto T, Konishi A, Mizuno N (1985) Identification of thalamic neurons with vasoactive intestinal polypeptide-like immunoreactivity in the rat. *Brain Res 347*:390–393.
Kaplan E, Shapley RM (1982) X and Y cells in lateral geniculate nucleus of macaque monkeys. *J Physiol Lond 330*:125–143.
Kasdon DL, Jacobson S (1978) The thalamic afferents to the inferior parietal lobule of the rhesus monkey. *J Comp Neurol 177*:685–706.
Kaske A, Dick A, Creutzfeldt OD (1991) The local domain for divergence of subcortical afferents to the striate and extrastriate visual cortex in the common marmoset (*Callithrix jacchus*): a multiple labelling study. *Exp Brain Res 84*:254–265.
Kaufman EFS, Rosenquist AC (1985) Efferent projections of the thalamic intralaminar nuclei in the cat. *Brain Res 335*:257–279.
Kawakita K, Dostrovsky JO, Tang JS, Chiang CY (1993) Responses of neurons in the rat thalamic nucleus submedius to cutaneous, muscle and visceral nociceptive stimuli. *Pain 55*:327–338.
Kayama Y (1985) Ascending, descending and local control of neuronal activity in the rat lateral geniculate nucleus. *Vision Res 25*:339–347.
Kayama Y, Negi T, Sugitani M, Iwama K (1982) Effects of locus coeruleus stimulation on neuronal activities of dorsal lateral geniculate nucleus and perigeniculate reticular nucleus of the rat. *Neurosci 7*:655–666.

Kayama Y, Shosaku A, Doty RW (1984) Cryogenic blockade of the visual cortico-thalamic projection in the rat. *Exp Brain Res 54*:157–165.

Kayama Y, Sumitomo I, Ogawa T (1986a) Does the ascending cholinergic projection inhibit or excite neurons in the rat thalamic reticular nucleus? *J Neurophysiol 56*:1310–1320.

Kayama Y, Takagi M, Ogawa T (1986b) Cholinergic influence of the laterodorsal tegmental nucleus on neuronal activity in the rat lateral geniculate nucleus. *J Neurophysiol 56*:1297–1309.

Kechagias S, Broman J (1995) Immunocytochemical evidence for vesicular storage of glutamate in cat spinocervical and cervicothalamic tract terminals. *Brain Res 675*:316–320.

Keinänen K, Wisden W, Sommer B, Werner P, Herb A, Verdoorn A, Sakmann B, Seeburg PH (1990) A family of AMPA-selective glutamate receptors. *Science 249*:556–560.

Kemp JA, Sillito AM (1981) The nature of excitatory transmitter mediating S and Y cell inputs to the cat dLGN. *J Physiol Lond 323*:377–391.

Kennedy H, Baleydier C (1977) Direct projections from thalamic intralaminar nuclei to extra-striate visual cortex in the cat traced with horseradish peroxidase. *Exp Brain Res 28*:133–139.

Kennedy H, Bullier J (1985) A double-labeling investigation of the afferent connectivity to cortical areas V1 and V2 of the macaque monkey. *J Neurosci 5*:2815–2830.

Kenshalo DR, Leonard RB, Chung JM, Willis WD (1979) Responses of primate spinothalamic neurons to graded and to repeated noxious heat stimuli. *J Neurophysiol 42*:1370–1389.

Kenshalo DR, Jr., Giesler GJ, Jr., Leonard RB, Willis WD (1980) Responses of neurons in primate ventral posterior lateral nucleus to noxious stimuli. *J Neurophysiol 43*:1594–1614.

Khachaturian H, Lewis ME, Watson SJ (1983) Enkephalin systems in diencephalon and brainstem of the rat. *J Comp Neurol 220*:310–320.

Kharazia VN, Weinberg RJ (1994) Glutamate in thalamic fibers terminating in layer IV of primary sensory cortex. *J Neurosci 14*:6021–6032.

Khrestchatisky M, MacLennan AJ, Chiang M-Y, Xu W, Jackson MB, Brecha N, Sternini C, Olsen RW, Tobin AJ (1989) A novel a subunit in rat brain $GABA_A$ receptors. *Neuron 3*:745–753.

Kia HK, Miquel M-C, Brisorgueil M-J, Daval G, Riad M, El Mestikawy S, Hamon M, Verge D (1996) Immunocytochemical localization of $serotonin_{1A}$ receptors in the rat central nervous system. *J Comp Neurol 365*:289–305.

Kievit J, Kuypers HGJM (1975) Subcortical afferents to the frontal lobe in the rhesus monkey studied by means of retrograde transport of horseradish peroxidase. *Brain Res 85*:261–266.

Kievit J, Kuypers HGJM (1977) Organization of the thalamo-cortical connexions to the frontal lobe in the rhesus monkey. *Exp Brain Res 29*:299–322.

Kilpatrick IC, Starr MS, Fletcher A, James TA, Macleod NK (1980) Evidence for a GABAergic nigrothalamic pathway in the rat. I Behavioral and biochemical studies. *Exp Brain Res 40*:45–54.

Kim R, Nakano K, Carpenter MB, Jayaraman A (1976a) Projections of the globus pallidus and adjacent structures: an autoradiographic study in the monkey. *J Comp Neurol 169*:263–290.

Kincaid AE, Penney JB, Jr., Young AB, Newman SW (1991) The globus pallidus receives a projection from the parafascicular nucleus in the rat. *Brain Res 553*:18–26.

Kinomura S, Larsson J, Gulyas B, Roland PE (1996) Activation by attention of the human reticular formation and thalamic intralaminar nuclei. *Science 271*:512–515.

Kitai ST, Kocsis JD, Preston RJ, Sugimori M (1976a) Monosynaptic inputs to caudate neurons identified by intracellular injection of horseradish peroxidase. *Brain Res 109*:601–605.

Kitai ST, Kocsis JD, Wood J (1976b) Origin and characteristics of the cortico-caudate afferents: an anatomical and electrophysiological study. *Brain Res 118*:137–141.

Kocsis JD, Sugimori M, Kitai ST (1977) Convergence of excitatory synaptic inputs to caudate spiny neurons. *Brain Res 124*:403–414.

Kölliker A von (1896) Handbuch der Gewebelehre des Menschen, Sixth edition, Vol. 2 Nervensystem des Menschen und der Thiere. Leipzig: Engelmann.

Kosaki H, Hashikawa T, He J, Jones EG (1997) Tonotopic organization of auditory cortical fields delineated by parvalbumin immunoreactivity in macaque monkeys. *J Comp Neurol* 386:304–316.

Kotchabhakdi N, Rinvik E, Yingchareon Y, Walberg F (1980) Afferent projections to the thalamus from the perihypoglossal nuclei. *Brain Res 187*:457–461.

Krettek JE, Price JL (1977) Projections from the amygdaloid complex to the cerebral cortex and thalamus in the rat and cat. *J Comp Neurol 172*:687–722.

Krnjević K (1974) Chemical nature of synaptic transmission in vertebrates. *Physiol Rev 54*:418–540.

Kromer LF, Moore RY (1980) A study of the organization of the locus coerulus projections to the lateral geniculate nuclei in the albino rat. *Neurosci 5*:255–271.

Krubitzer L, Clarey J, Tweedale R, Elston G, Calford M (1995) A redefinition of somatosensory areas in the lateral sulcus of macaque monkeys. *J Neurosci 15*:3821–3839.

Krubitzer LA, Kaas JH (1992) The somatosensory thalamus of monkeys: cortical connections and a redefinition of nuclei in marmosets. *J Comp Neurol 319*:123

Kruger L, Mantyh PW, Sternini C, Brecha NC, Mantyh CR (1988a) Calcitonin gene-related peptide (CGRP) in the rat central nervous system: patterns of immunoreactivity and receptor binding sites. *Brain Res 463*:223–244.

Kruger L, Sternini C, Brecha N, Mantyh PW (1988b) Distribution of calcitonin gene-related peptide immunoreactivity in relation to the rat central somatosensory projection. *J Comp Neurol 273*:149–162.

Kubo T, Bujo H, Akiba I, Nakai J, Mishina M, Numa S (1988) Location of a region of the muscarinic acetylcholine receptor involved in selective effector couping. *Febs Lett 241*:119–125.

Kudo M, Niimi K (1978) Ascending projections of the inferior colliculus onto the medial geniculate body in the cat studied by anterograde and retrograde tracing techniques. *Brain Res 155*:113–117.

Kudo M, Niimi K (1980) Ascending projections of the inferior colliculus in the cat: An autoradiographic study. *J Comp Neurol 191*:545–566.

Kuffler S (1953) Discharge patterns and functional organization of mammalian retina. *J Neurophysiol 16*:37–68.

Kultas-Ilinsky K, Fogarty JD, Hughes B, Ilinsky IA (1988) Distribution and binding parameters of GABA and benzodiazepine receptors in the cat motor thalamus and adjacent nuclear groups. *Brain Res 459*:1–16.

Kultas-Ilinsky K, Ilinsky IA (1991) Fine structure of the ventral lateral nucleus (VL) of the *Macaca mulatta* thalamus: cell types and synaptology. *J Comp Neurol 314*:319–349.

Kultas-Ilinsky K, Ilinsky IA, Massopust LC, Young PA, Smith KR (1978) Nigrothalamic pathway in the cat demonstrated by autoradiographic and electron microscopy. *Exp Brain Res 33*:481–492.

Kultas-Ilinsky K, Ilinsky I, Warton S, Smith KR (1983) Fine structure of nigral and pallidal afferents in the thalamus: An EM autoradiography study in the cat. *J Comp Neurol 216*:390–405.

Kultas-Ilinsky K, Ribak CE, Peterson GM, Oertel WH (1985) A description of the GABAergic neurons and axon terminals in the motor nuclei of the cat thalamus. *J Neurosci 5*:1346–1369.

Kultas-Ilinsky K, Yi H, Ilinsky IA (1995) Nucleus reticularis thalami input to the anterior thalamic nuclei in the monkey: a light and electron microscopic study. *Neurosci Lett 186*:25–28.

Künzle H (1975) Bilateral projections from precentral motor cortex to the putamen and other parts of the basal ganglia. An autoradiographic study in macaca fascicularis. *Brain Res 88*:195–210.

Künzle H (1977) Projections from the primary somatosensory cortex to basal ganglia and thalamus in the monkey. *Exp Brain Res 30*:481–493.

Künzle H (1978) An autoradiographic analysis of the efferent connections from premotor and adjacent prefrontal regions (areas 6 and 9) in *Macaca fascicularis. Brain Behav Evol 15*:185–234.

Künzle H, Akert K (1977b) Efferent connections of cortical area 8 (frontal eye field) in *Macaca fascicularis*. A reinvestigation using autoradiographic techinques. *J Comp Neurol 173*:147–164.

Kuo J-S, Carpenter MB (1973) Organization of pallidothalamic projections in the rhesus monkey. *J Comp Neurol 151*:201–236.

Kuroda M, Lopez-Mascaraque L, Price JL (1992a) Neuronal and synaptic composition of the mediodorsal thalamic nucleus in the rat: a light and electron microscopic Golgi study. *J Comp Neurol 326*:61–81.

Kuroda M, Murakami K, Kishi K, Price JL (1992b) Distribution of the piriform cortical terminals to cells in the central segment of the mediodorsal thalamic nucleus of the rat. *Brain Res 595*:159–163.

Kuroda M, Price JL (1991a) Synaptic organization of projections from basal forebrain structures to the mediodorsal thalamic nucleus of the rat. *J Comp Neurol 303*:513–533.

Kuroda M, Price JL (1991b) Ultrastructure and synaptic organization of axon terminals from brainstem structures to the mediodorsal thalamic nucleus of the rat. *J Comp Neurol 313*:539–552.

Kuroda M, Price JL, Ray JP (1991) An ultrastructural study of neurotensin-like immunoreactive terminals in the mediodorsal thalamic nucleus of the rat. *Brain Res 548*:353–357.

Kuypers HGJM, Maisky VA (1975) Retrograde axonal transport of horseradish peroxidase from spinal cord to brain stem cell groups in the cat. *Neurosci Lett 1*:9–14.

Kvale I, Fonnum F (1983) The effects of unilateral neonatal removal of visual cortex on transmitter parameters in the adult superior colliculus and lateral geniculate body. *Dev Brain Res 11*:261–266.

Kwon YH, Esguerra M, Sur M (1991) NMDA and Non-NMDA receptors mediate visual responses of neurons in the cat's lateral geniculate nucleus. *J Neurophysiol 66*:414–428.

Lachica EA, Casagrande VA (1992) Direct W-like geniculate projections to the cytochrome oxidase (CO) blobs in primate visual cortex: axon morphology. *J Comp Neurol 319*:141

Lachica EA, Crooks MW, Casagrande VA (1990) Effects of monocular deprivation on the mophology of retinogeniculate axon arbors in a primate. *J Comp Neurol 296*:303–323.

Laitinen LV (1977) Anterior pulvinotomy in the treatment of intractable pain. In: Neurosurgical Treatment in Psychiatry, Pain and Epilepsy. Sweet WH, Obrador S, Martinez-Rodriguez JC eds), pp 669–672. Baltimore: University Park Press.

Laitinen LV (1988) Mesencephalotomy and thalamotomy for chronic pain. In: Modern Stereotactic Neurosurgery. (Lunsford LD ed), pp 269–271. Boston: Nijhoff.

LaMotte CC, Snowman A, Pert CB, Snyder SN (1978) Opiate receptor binding in rhesus monkey brain: association with limbic structures. *Brain Res 155*:374–379.

Lanaud P, Popovici T, Normand E, Lemoine C, Bloch B, Roques BP (1989) Distribution of CCK mRNA in particular regions (hippocampus, periaqueductal grey and thalamus) of the rat by in situ hybridization. *Neurosci Lett 104*:38–42.

Landgren S, Silfvenius H (1971) Nucleus Z, the medullary relay in the projection path to the cerebral cortex of group I muscle afferents from the cat's hind limb. *J Physiol Lond 218*:551–571.

Lane RH, Allman JM, Kaas JH (1971) Representation of the visual field in the superior colliculus of the grey squirrel (*Sciurus carolinensis*) and the tree shrew (*Tupaia glis*). *Brain Res 26*:277–292.

Lavoie B, Parent A (1991) Serotoninergic innervation of the thalamus in the primate: an immunohistochemical study. *J Comp Neurol 312*:1–18.

Le Gros Clark WE, Russell WR (1939) Observations on the efferent connexions of the centre median nucleus. *J Anat Lond 73*:255–262.

Lechner J, Leah JD, Zimmerman M (1993) Brainstem peptidergic neurons projecting to the medial and lateral thalamus and zona incerta in the rat. *Brain Res 603*:47–56.

Lee BB, Creutzfeldt OD, Elepfandt A (1979) The responses of magno-and parvocellular cells of the monkey's lateral geniculate body to moving stimuli. *Exp Brain Res 35*:547–557.

Lee K, McCormick DA (1995) Acetylcholine excites GABAergic neurons of the ferret perigeniculate nucleus through nicotinic receptors. *J Neurophysiol 73*:2123–2128.

Lee SM, Friedberg MH, Ebner FF (1994a) The role of GABA-mediated inhibition in the rat ventral posterior medial thalamus. I. assessment of receptive field changes following thalamic reticular nucleus lesions. *J Neurophysiol 71*:1702–1715.

Lee SM, Friedberg MH, Ebner FF (1994b) The role of GABA-mediated inhibition in the rat ventral posterior medial thalamus. II. differential effects of $GABA_A$ and $GABA_B$ receptor antagonists on responses of VPM neurons. *J Neurophysiol 71*:1716–1726.

Leger L, Sakai K, Salvert D, Touret M, Jouvet M (1975) Delineation of dorsal lateral geniculate afferents from the cat brain stem as visualised by the horseradish peroxidase techniques. *Brain Res 93*:490–496.

Leichnetz GR (1986) Afferent and efferent connections of the dorsolateral precentral gyrus (area 4, hand/arm region) in the macaque monkey, with comparisons to area 8. *J Comp Neurol 254*:460–492.

Leichnetz GR, Goldberg ME (1988) Higher centers concerned with eye movement and visual attention: cerebral cortex and thalamus. *Rev Oculomotor Res 2*:365–429.

Lende RA, Kirsch WM, Druckman R (1971) Relief of facial pain after combined removal of precentral and postcentral cortex. *J Neurosurg 34*:537–543.

Lenz FA, Dostrovsky JO, Tasker RR, Yamashiro K, Kwan HC, Murphy JT (1988a) Single-unit analysis of the human ventral thalamic nuclear group: somatosensory responses. *J Neurophysiol 59*:299–316.

Lenz FA, Tasker RR, Kwan HC, Schnider S, Kwong R, Murayama Y, Dostrovsky JO, Murphy JT (1988b) Single unit analysis of the human ventral thalamic nuclear group: correlation of thalamic 'tremor' cells with the 3–6Hz component of Parkinsonian tremor. *J Neurosci 8*:754–764.

Lenz FA, Gracely RH, Hope EJ, Baker FH, Rowland LH, Doughkerty PM, Richardson RT (1994a) The sensation of angina can be evoked by stimulation of the human thalamus. *Pain 59*:119–126.

Lenz FA, Kwan HC, Martin R, Tasker R, Richardson RT, Dostrovsky JO (1994b) Characteristics of somatotopic organization and spontaneous neuronal activity in the region of the thalamic principal sensory nucleus in patients with spinal cord transection. *J Neurophysiol 72*:1570–1587.

Lenz FA, Kwan HC, Martin RL, Tasker RR, Dostrovsky JO, Lenz YE (1994c) Single unit analysis of the human ventral thalamic nuclear group. Tremor-related activity in functionally identified cells. *Brain 117*:531–543.

Lenz FA, Kwan HC, Dostrovsky JO, Tasker RR, Murphy JT, Lenz YE (1990a) Single unit analysis of the human ventral thalamic nuclear group. Activity correlated with movement. *Brain 113*:1795–1821.

Lenz FA, Martin R, Kwan HC, Tasker RR, Dostrovsky JO (1990b) Thalamic single-unit activity occurring in patients with hemidystonia. *Stereotact Funct Neurosurg 54–55*:159–162.

Lenz FA, Seike M, Richardson RT, Lin YC, Baker FH, Khoja I, Jaeger CJ, Gracely RH (1993) Thermal and pain sensations evoked by microstimulation in the area of human ventrocaudal nucleus. *J Neurophysiol 70*:200–212.

LeVay S, Connolly M, Houde J, Van Essen DC (1985) The complete pattern of ocular dominance stripes in the striate cortex and visual field of the macaque monkey. *J Neurosci 5*:486–501.

LeVay S, Hubel DH, Wiesel TN (1975) The pattern of ocular dominance columns in macaque visual cortex revealed by a reduced silver stain. *J Comp Neurol 159*:559–576.

LeVay S, Wiesel TN, Hubel DH (1980) The development of ocular dominance columns in normal and visually deprived monkeys. *J Comp Neurol 191*:1–51.

Leventhal AG, Rodieck RW, Dreher B (1981) Retinal ganglion cell classes in the Old World monkey: morphology and central projections. *Science 213*:1139–1142.

Levesque D, Diaz J, Pilon C, Martres MP, Giros B, Souil E, Schott D, Morgat JL, Schwartz JC, Sokoloff P (1992) Identification, characterization, and localization of the dopamine D3 receptor in rat brain using 7-[3H]hydroxy-N,N-di-n-propyl-2-aminotetralin. *Proc Natl Acad Sci USA 89*:8155–8159.

Levey AI, Hallanger AE, Wainer BH (1987) Choline acetyltransferase immunoreactivity in the rat thalamus. *J Comp Neurol 257*:317–332.

Levey AI, Hersch SM, Rye DB, Sunahara RK, Niznik HB, Kitt CA, Price DL, Maggio R, Brann MR, Ciliax BJ (1993) Localization of D_1 and D_2 dopamine receptors in brain with subtype-specific antibodies. *Proc Natl Acad Sci USA 90*:8861–8865.

Lin C-S, Kaas JH (1979) The inferior pulvinar complex in owl monkeys: architectonic subdivisions and patterns of input from superior colliculus and subdivisions of visual cortex. *J Comp Neurol 187*:655–677.

Lin SC, Kaas JH (1980) Projections from the medial nucleus of the inferior pulvinar complex to the middle temporal area of the visual cortex. *Neurosci 5*:2219–2228.

Lin C-S, Nicolelis MAL, Schneider JS, Chapin JK (1990) A major direct GABAergic pathway from zona incerta to neocortex. *Science 248*:1553–1556.

Lindvall O, Björklund A (1974) The organization of the ascending catecholamine neuron system in the rat brain as revealed by the glyoxylic acid fluorescence method. *Acta Physiol Scand Suppl 412*:1–48.

Liu F-Y, Qiao J-T, Dafny N (1993) Cerebellar stimulation modulates thalamic noxious-evoked responses. *Brain Res Bull 30*:529–534.

Liu X-B, Honda CN, Jones EG (1995a) Distribution of four types of synapse on physiologically identified relay neurons in the ventral posterior thalamic nucleus of the cat. *J Comp Neurol 352*:69–91.

Liu X-B, Jones EG (1991) The fine structure of serotonin and tyrosine hydroxylase immunoreactive terminals in the ventral posterior thalamic nucleus of cat and monkey. *Exp Brain Res 85*:507–518.

Liu X-B, Jones EG (1992) Simultaneous demonstration of serotonin-immunoreactive terminals and GABAergic neurons in the VPL nucleus of the cat thalamus. *Eur J Neurosci 4*:1065–1078.

Liu X-B, Jones EG (1996) Localization of alpha type II calcium calmodulin-dependent protein kinase at glutamatergic but not gamma-aminobutyric acid (GABAergic) synapses in thalamus and cerebral cortex. *Proc Natl Acad Sci USA 93*:7332–7336.

Liu X-B, Jones EG, Huntley GW, Molinari M (1989) Tachykinin immunoreactivity in terminals of trigeminal afferent fibers in adult and fetal monkey thalamus. *Exp Brain Res 78*:479–488.

Liu X-B, Warren RA, Jones EG (1995b) Synaptic distribution of afferents from reticular nucleus in ventroposterior nucleus of cat thalamus. *J Comp Neurol 352*:187–202.

Livingstone MS, Hubel DH (1982) Thalamic inputs to cytochrome oxidase-rich regions in monkey visual cortex. *Proc Natl Acad Sci USA 79*:6098–6101.

Livingstone MS, Hubel DH (1987a) Connections between layer 4B of area 17 and thick cytochrome oxidase stripes of area 18 in the squirrel monkey. *J Neurosci 7*:3371–3377.

Livingstone MS, Hubel DH (1987b) Psychophysical evidence for separate channels for the perception of form, color, movement, and depth. *J Neurosci 7*:3416–3468.

Llinás R, Ribary U (1993) Coherent 40-Hz oscillation characterizes dream state in humans. *Proc Natl Acad Sci USA 90*:2078–2081.

Loe PR, Whitsel BL, Dreyer DA, Metz CB (1977) Body representaiton in ventrobasal thalamus of macaque: a single-unit analysis. *J Neurophysiol 40*:1339–1355.

London ED, Waller SB, Wamsley JK (1985) Autoradiographic localization of [3H] nicotine binding sites in the rat brain. *Neurosci Lett 53*:179–184.

Lorén I, Alumets J, Hakanson R, Sundler F (1979a) Distribution of gastrin and CCK-like peptides in rat brain. *Histochem 59*:249–257.

Lorén I, Emson PC, Fahrenkrug J, Bjorklund A, Alumets J, Hakanson R, Sundler F (1979b) Distribution of vasoactive intestinal polupeptide in the rat and mouse brain. *Neurosci 4*:1953–1976.

Lorente de Nó R (1938) Cerebral cortex: architectonics, intra-cortical connections. In: Physiology of the Nervous System (Fulton JR ed), pp 274–301. New York: Oxford Univ. Pres.

Lozsádi DA (1995) Organization of connections between the thalamic reticular and the anterior thalamic nuclei in the rat. *J Comp Neurol 358*:233–246.

Lübke J (1993) Morphology of neurons in the thalamus reticular nucleus (TRN) of mammals as revealed by intracellular injections into fixed brain slices. *J Comp Neurol 329*:458–471.

Lucier GE, Rüegg DC, Wiesendanger M (1975) Responses of neurones in motor cortex and in area 3A to controlled stratches of forelimb muscles in cebus monkeys. *J Physiol Lond 251*:833–853.

Lüddens H, Wisden W (1991) Function and pharmacology of multiple $GABA_A$ receptor subunits. *Trends Pharmacol Sci 12*:49–51.

Luethke LE, Krubitzer LA, Kaas JH (1989) Connections of primary auditory cortex in the New World monkey, *Saguinus*. *J Comp Neurol 285*:487–513.

Lund JS, Lund RD, Hendrickson AE, Bunt AH, Fuchs AF (1975) The origin of efferent pathways from the primary visual cortex, Area 17, of the macaque monkey as shown by retrograde transport of horseradish peroxidase. *J Comp Neurol 164*:287–304.

Lund JS, Yoshioka T, Levitt JB (1994) Substrates for interlaminar connections in area VI of macaque monkey cerebral cortex. In: Cerebral Cortex Volume 10 Primary Visual Cortex in Primates (Peters A, Rockland KS eds), pp 37–60. New York and London: Plenum Press.

Lund RD, Webster KE (1967a) Thalamic afferents from the dorsal column nuclei. An experimental anatomical study in the rat. *J Comp Neurol 130*:301–312.

Lund RD, Webster KE (1967b) Thalamic afferents from the spinal cord and trigeminal nuclei. An experimental anatomical study in the rat. *J Comp Neurol 130*:313–328.

Lysakowski A, Standage GP, Benevento LA (1986) Histochemical and architectonic differentiation of zones of pretectal and collicular inputs to the pulvinar and dorsal lateral geniculate nuclei in the macaque. *J Comp Neurol 250*:431–448.

Lysakowski A, Standage GP, Benevento LA (1988) An investigation of collateral projections of the dorsal lateral geniculate nucleus and other subcortical structures to the cortical areas V1 and V4 in the macaque monkey: a double label retrograde tracer study. *Exp Brain Res 69*:651–661.

Macchi G (1990) The intralaminar system and sleep. In: The Diencephalon and Sleep (Mancia M, Marini G eds), pp 211–229. New York: Raven Press.

Macchi G, Bentivoglio M (1986) The thalamic intralaminar nuclei and the cerebral cortex. In: Cerebral Cortex Volume 5 Sensory-Motor Cortex and Aspects of Cortical Connectivity (Jones EG, Peters A eds), pp 355–401. New York: Plenum.

Macchi G, Bentivoglio M, D'Atena C, Rossin P, Tempesta E (1977) The cortical projection of the thalamic intralaminar nuclei restudied by means of the HRP retrograde axonal transport. *Neurosci Lett 4*:121–126.

Macchi G, Bentivoglio M, Minciacchi D, Molinari M (1986) Organization des connexions thalamiques. *Rev Neurol Paris 142*:267–282.

Macchi G, Bentivoglio M, Molinari M, Minciacchi D (1984) The thalamo-caudate versus thalamo-cortical projections as studied in the cat with fluorescent retrograde double labeling. *Exp Brain Res 54*:225–239.

Macchi G, Jones EG (1996) Towards an agreement on terminology of nuclear and subnuclear divisions of the motor thalamus. *J Neurosurg 86*:670–685.

MacDonald RL, Olsen RW (1994) $GABA_A$ receptor channels. *Annu Rev Neurosci 17*:569–602.

Mackay-Sim A, Sefton AJ, Martin PR (1983) Subcortical projections to lateral geniculate and thalamic reticular nuclei. *J Comp Neurol 213*:24–35.

MacLeod NK, James TA, Kilpatrick IC, Starr MS (1980) Evidence for a GABAergic nigrothalamic pathway in the rat. II Electrophysiological studies. *Exp Brain Res 40*:55–63

Maendly R, Rüegg DG, Wiesendanger M, Wiesendanger R, Lagowska J, Hess B (1981) Thalamic relay for group I muscle afferents of forelimb nerves in the monkey. *J Neurophysiol 46*:901–917.

Magariños-Ascone C, Buno WJ, Garcia-Austt E (1988) Monkey pulvinar units related to motor activity and sensory responses. *Brain Res 445*:30–38.

Magnusson KR, Clements JR, Larson AA, Madl JE, Beitz AJ (1987) Localization of glutamate in trigeminothalamic projection neurons: a combined retrograde transport-immunohistochemical study. *Somatosens Motor Res 4*:177–190.

Mair WGP, Warrington EK, Weiskrartz L (1979) Memory disorder in Korsakoff's psychosis, a neuropathological and neuropsychological investigation of two cases. *Brain 102*:749–783.

Malach R, Graybiel AM (1986) Mosaic architecture of the somatic sensory-recipient sector of the cat's striatum. *J Neurosci 6*:3436–3458.

Manger PR, Woods TM, Jones EG (1995) Representation of the face and intraoral structures in area 3b of the squirrel monkey (Saimiri sciureus) somatosensory cortex, with special reference to the ipsilateral representation. *J Comp Neurol 362*:597–607.

Manger PR, Woods TM, Jones EG (1996) Representation of face and intra-oral structures in area 3b of macaque monkey somatosensory cortex. *J Comp Neurol 371*:513–521.

Mansour A, Fox CA, Burke S, Meng F, Thompson RC, Akil H, Watson SJ (1994) Mu, Delta, and Kapoa Opioid receptor mRNA expression in the rat CNS: An in situ hybridization study. *J Comp Neurol 350*:412–438.

Mansour A, Khachaturian H, Lewis ME, Akil H, Watson SJ (1987) Autoradiographic differentiation of mu, delta, and kappa opioid receptors in the rat forebrain and midbrain. *J Neurosci 7*:2445–2464.

Mansour A, Meador-Woodruff JH, Bunzow JR, Civelli O, Akil H, Watson SJ (1990) Localization of dopamine D2 receptor mRNA and D1 and D2 receptor binding in the rat brain and pituitary: an in situ hybridization-receptor autoradiographic analysis. *J Neurosci 10*:2587–2600.

Mantyh PW (1983) The spinothalamic tract in the primate: a re-examination using wheatgerm agglutinin conjugated to horseradish peroxidase. *Neurosci 9*:847–862.

Mantyh PW, Kemp A (1983) The distribution of putative neurotransmitters in the lateral geniculate nucleus of the rat. *Brain Res 288*:344–348.

Marrocco RT, Brown JB (1975) Correlation of receptive field properties of monkey LGN cells with the conduction velocity of retinal afferent input. *Brain Res 92*:137–144.

Marrocco RT, McClurkin JW (1985) Evidence for spatial structure in the cortical input to the monkey lateral geniculate nucleus. *Exp Brain Res 59*:50–56.

Marrocco RT, McClurkin JW, Alkire MT (1996) The influence of the visual cortex of the spatiotemporal response properties of lateral geniculate nucleus cells. *Brain Res 737*:110–118.

Marrocco RT, McClurkin JW, Young RA (1981) Spatial properties of superior colliculers cells projecting to the inferior pulvinar and parabigeninal nucleus of the monkey. *Brain Res 222*:150–154.

Marrocco RT, McClurkin JW, Young RA (1982) Modulation of lateral geniculate nucleus cell responsiveness by visual activation of the corticogeniculate pathway. *J Neurosci 2*:256–263.

Martin LJ, Blackstone CD, Huganir RL, Price DL (1992) Cellular localization of a metabotropic glutamate receptor in rat brain. *Neuron 9*:259–270.

Martin LJ, Blackstone CD, Levey AI, Huganir RL, Price DL (1993) AMPA glutamate receptor subunits are differentially distributed in rat brain. *Neurosci 53*:327–358.

Martin-Rodriguez JG, Buño WJ, García-Austt E (1982) Human pulvinar units, spontaneous activity and sensory-motor influences. *Electroencephalogr Clin Neurophysiol 54*:388–398.

Mash DC, Potter LT (1986) Autoradiographic localization of M1 and M2 muscarinic receptors in the rat brain. *Neurosci 19*:551–564.

Mason A, Ilinsky IA, Beck S, Kultas-Ilinsky K (1996) Reevaluation of synaptic relationships of cerebellar terminals in the ventral lateral nucleus of the rhesus monkey thalamus based on serial section analysis and three-dimensional reconstruction. *Exp Brain Res 109*:219–239.

Matelli M, Luppino G (1993a) Cortical projections of motor thalamus. In: Thalamic Networks for Relay and Modulation (Minciacchi D, Molinari M, Macchi G, Jones EG eds), pp 165–174. Oxford: Pergamon.

Matelli M, Luppino G, Fogassi L, Rizzolatti G (1989) Thalamic input to inferior area 6 and area 4 in the Macaque monkey. *J Comp Neurol 280*:468–488.

Mathers LH (1972a) Ultrastructure of the pulvinar of the squirrel monkey. *J Comp Neurol 146*:15–42.

Mathers LH (1972b) The synaptic organization of the cortical projection to the pulvinar of the squirrel monkey. *J Comp Neurol 146*:43–60.

Mathers LH, Mascetti GG (1975) Electrophysiological and morphological properties of neurons in the ventral lateral geniculate nucleus of the rabbit. *Exp Neurol 46*:506–520.

Mathers LH, Rapisardi SC (1973) Visual and somatosensory receptive fields of neurons in the squirrel monkey pulvinar. *Brain Res 64*:65–83.

Matute C, Martinez-Millan l (1985) Selective retrograde labeling in some afferents to the rabbit lateral geniculate nucleus following injections of tritiated neurotransmitter-related compounds. *Neurosci Lett 53*:9–14.

Mauguière F, Baleydier C (1978) Topographical organization of medial pulvinar neurons sending fibres to Brodmann's areas 7, 21, and 22 in the monkey. *Exp Brain Res 31*:605–607.

Maunsell JHR, Nealey TA, DePriest DD (1990) Magnocellular and parvocellular contributions to responses in the middle temporal visual area (MT) of the Macaque monkey. *J Neurosci 10*:3323–3334.

Maunsell JHR, Van Essen DC (1983a) Functional properties of neurons in middle temporal visual area of the macaque monkey. I. Selectivity for stimulus, direction, speed and orientation. *J Neurophysiol 49*:1127–1147.

Maunsell JHR, Van Essen DC (1983b) Functional properties of neurons in middle temporal area of the macaque monkey. II. Binocular interactions and sensitivity to binocular disparity. *J Neurophysiol 49*:1148–116

McClurkin JW, Marrocco RT (1984) Visual cortical input alters spatial tuning in monkey lateral geniculate nucleus cells. *J Physiol Lond 348*:135–152.

McClurkin JW, Optican LM, Richmond BJ (1994) Cortical feedback increases visual information transmitted by monkey parvocellular lateral geniculate nucleus neurons. *Vis Neurosci 11*:601–617.

McCormick DA (1992a) Neurotransmitter actions in the thalamus and cerebral cortex and their role in neuromodualtion of thalamocortical activity. *Prog Neurobiol 39*:337–388.

McCormick DA (1992b) Cellular mechanisms underlying cholinergic and noradrenergic modulation of neuronal firing mode in the cat and guinea pig dorsal lateral geniculate nucleus. *J Neurosci 12*:278–289.

McCormick DA, Pape H-C (1988) Acetylcholine inhibits identified interneurons in the cat lateral geniculate nucleus. *Nature 334*:246–248.
McCormick DA, Pape H-C (1990a) Noradrenergic and serotonergic modulation of a hyperpolarization-activated cation current in thalamic relay neurons. *J Physiol Lond 431*:319–342.
McCormick DA, Pape HC (1990b) Properties of a hyperpolarization-activated cation current and its role in rhythmic oscillation in thalmic relay neurones. *J Physiol Lond 431*:291–318.
McCormick DA, Prince DA (1986) Acetycholine induces burst firing in thalamic reticular neurones by activating a potassium conductance. *Nature 319*:402–405.
McCormick DA, Prince DA (1987) Actions of acetylcholine in the guinea-pig and cat medial and lateral geniculate nuclei, in vitro. *J Physiol Lond 392*:147–165.
McCormick DA, Prince DA (1988) Noradrenergic modulation of firing pattern in guinea pig and cat thalamic neurons. *J Neurophysiol 59*;3:978–996.
McCormick DA, von Krosigk M (1992) Corticothalamic activation modulates thalamic firing through glutamate 'metabotropic' receptors. *Proc Natl Acad Sci USA 89*:2774–2778.
McCormick DA, Wang Z (1991) Serotonin and noradrenaline excite GABAergic neurones of the guinea-pig and cat thalamic reticular nucleus. *J Physiol Lond 442*:235–255.
McCormick DA, Williamson A (1991) Modulation of neuronal firing mode in cat and guinea pig LGNd by histamine: possible cellular mechanisms of histaminergic control of arousal. *J Neurosci 11*:3188–3199.
McGeer PL, McGeer EG, Scherer U, Singh K (1977) A glutamatergic cortico-striatal path. *Brain Res 128*:369–373.
McGuinness CM, Krauthamer GM (1980) The afferent projections to the centrum medianim of the cat as demonstrated by retrograde transport of horseradish peroxidase. *Brain Res 184*:255–270.
McKay RDF, Hockfield S (1983) A surface antigen expressed by a subset of neurons in the vertebrate central nervous system. *Proc Natl Acad Sci USA 80*:5758–5761.
Meador-Woodruff JH, Mansour A, Bunzow JR, Van Tol HH, Watson SJ, Civelli O (1989) Distribution of D2 dopamine receptor mRNA in rat brain. *Proc Natl Acad Sci USA 86*:7625–7628.
Meador-Woodruff JH, Mansour A, Grandy DK, Damask SP, Civelli O, Watson SJ (1992) Distribution of D5 dopamine receptor mRNA in rat brain. *Neurosci Lett 145*:209–212.
Meguro H, Mori H, Araki K, Kushiya E, Kutsuwada T, Yamazaki M, Kumanishi T, Arakawa M, Sakimura K, Mishina M (1992) Functional characterization of a heteromeric NMDA receptor channel expressed from cloned cDNAs. *Nature 357*:70–4.
Mehler WR (1962) The anatomy of the so-called pain tract in man: An analysis of the course and distribution of the ascending fibers of the fasciculus anterolateralis. In: Basic Research in Paraplegia (French JD, Porter RW eds), pp 26–55. Springfield, Ill: CC Thomas.
Mehler WR (1966a) Further notes on the center median nucleus of Luys. In: The Thalamus (Purpura DP, Yahr MD eds), pp 109–122. New York: Columbia University Press.
Mehler WR (1966b) Some observations on secondary ascending afferent systems in the central nervous system. In: Pain (Knighton RS, Dumke PR eds), pp 11–32. Boston: Little, Brown and Co.
Mehler WR (1966c) The posterior thalamic region in man. *Confin Neurol 27*:18–29.
Mehler WR (1969) Some neurological species differences–a posteriori. *Ann NY Acad Sci 167*:424–468.
Mehler WR (1980) Subcortical afferent connections of the amygdala in the monkey. *J Comp Neurol 190*:733–762.
Meibach RC, Maayani S, Green JP (1980) characterization and radioautography of [3H]LSD binding by rat brain slices. *Eur J Pharmacol 67*:371–382.
Melander T, Hokfelt T, Rokaeus A (1986) Distribution of galaninlike immunoreactivity in the rat central nervous system. *J Comp Neurol 248*:475–517.
Mengod G, Martinez-Mir MI, Vilaro MT, Palacios JM (1989) Localization of the mRNA for the dopamine D2 receptor in the rat brain by in situ hybridization histochemistry. *Proc Natl Acad Sci USA 86*:8560–8564.
Merzenich MM, Brugge JF (1973) Representation of the cochlear partition on the superior temporal plane of the macaque monkey. *Brain Res 50*:275–296.
Mesulam M-M, Mufson EJ (1985) The insula of Reil in man and monkey: Architectonics, connectivity, and function. In: Cerebral Cortex Volume 4 Association and Auditory Cortices (Peters A, Jones EG eds), pp 179–228. New York and London: Plenum Press.
Mesulam M-M, Mufson EJ, Wainer BH, Levey AI (1983) Central cholinergic pathways in the rat: an overview based on an alternative nomenclature (Ch1-Ch6). *Neurosci 10*:1185–1202.
Mesulam M-M, Pandya DN (1973) The projections of the medial geniculate complex within the sylvian fissure of the rhesus monkey. *Brain Res 60*:315–333.
Mesulam M-M, Van Hoesen GW, Pandya DN, Geschwind N (1977) Limbic and sensory connections of the

inferior parietal lobule (area PG) in the rhesus monkey: a study with a new method for horseradish peroxidase. *Brain Res 136*:393–414.
Meyer G, Albus K (1981) Topography and cortical projections of morphologically identified neurons in the visual thalamus of the cat. *J Comp Neurol 201*:353–374.
Miceli MO, Van Der Kooy D, Post CA, Della-Fera MA, Baile CA (1919a) Differential distributions of cholecystokinin in hamster and rat forebrain. *Brain Res 402*:318–330.
Michael CR (1988) Retinal afferent arborization patterns, dendritic field orientations, and the segregation of function inn the lateral geniculate nucleus of the monkey. *Proc Natl Acad Sci USA 85*:4914–4918.
Michael CR, Bowling DB (1982) Projection patterns of single physiologically characterized optic tract fibres in cat. *Nature 286*:899–902.
Miller JW, Benevento LA (1979) Demonstration of a direct projection from the intralaminar central lateral nucleus to the primary visual cortex. *Neurosci Lett 14*:229–234.
Minciacchi D, Granato A, Antonini A, Sbriccoli A, Macchi G (1991a) A procedure for the simultaneous visualization of two anterograde and different retrograde fluorescent tracers. Application to the study of the afferent-efferent organization of thalamic anterior intralaminar nuclei. *J Neurosci Meth 38*:183–191.
Minciacchi D, Granato A, Santarelli M, Macchi G (1993) Different weights of subcortico-cortical projections upon primary sensory areas: the thalamic anterior intralaminar system. In: Thalamic Networks for Relay and Modulation (Minciacchi D, Molinari M, Macchi G, Jones EG eds), pp 209–228. Oxford: Pergamon.
Miyata M, Sasaki K (1983) HRP studies on thalamocortical neurons related to the cerebellocerebral projection in the monkey. *Brain Res 274*:213–224.
Mizuno N, Uchida K, Nomura S, Nakamura Y, Sugimoto T, Uemura-Sumi M (1981) Extrageniculate projections to the visual cortex in the macaque monkey: an HRP study. *Brain Res 212*:454–459.
Mogenson GJ, Ciriello J, Garland J, Wu M (1987) Ventral pallidum projections to mediodorsal nucleus of the thalamus: an anatomical and electrophysiological investigation in the rat. *Brain Res 404*:221–230.
Mohler H, Knoflach F, Paysan J, Motejlek K, Benke D, Luscher B, Fritschy JM (1995) Heterogeneity of $GABA_A$-receptors: cell-specific expression, pharmacology, and regulation. *Neurochem Res 20*:631–636.
Molinari M, Bentivoglio M, Minciacchi D, Granato A, Macchi G (1986) Spinal afferents and cortical efferents of the anterior intralaminar nuclei: an anterograde-retrograde tracing study. *Neurosci Lett 72*:258–264.
Molinari M, Dell'Anna ME, Rausell E, Leggio MG, Hashikawa T, Jones EG (1995) Auditory thalamocortical pathways defined in monkeys by calcium binding protein immunoreactivity. *J Comp Neurol 362*:171–194.
Molinari M, Hendry SHC, Jones EG (1987) Distributions of certain neuropeptides in the primate thalamus. *Brain Res 426*:270–289.
Molinari M, Leggio MG, Dell'Anna ME, Gianetti S, Macchi G (1993) Structural evidence in favour of a relay function for the anterior intralaminar nuclei. In: Thalamic Networks for Relay and Modulation (Minciacci D, Molinari M, Macchi G, Jones EG eds), pp 197–208. Oxford: Pergamon.
Molinari M, Leggio MG, Dell'Anna ME, Giannetti S, Macchi G (1994) Chemical compartmentation and relationships between calcium-binding protein immunoreactivity and layer-specific cortical and caudate-projecting cells in the anterior intralaminar nuclei of the cat. *Eur J Neurosci 6*:299–312.
Molineaux SM, Jessell TM, Axel R, Julius D (1989) 5-HT1c receptor is a prominent serotonin receptor subtype in the central nervous system. *Proc Natl Acad Sci USA 86*:6793–6797.
Molnar E, Baude A, Richmond SA, Patel PB, Somogyi P, McIlhinney RAJ (1993) Biochemical and immunocytochemical characterization of antipeptide antibodies to a cloned GluR1 glutamate receptor subunit: cellular and subcellular distribution in the rat forebrain. *Neurosci 53*:307–326.
Molotchnikoff S, Richard D, Lachapelle P (1980) Influence of the visual cortex upon receptive field organization of lateral geniculate cells in rabbits. *Brain Res 193*:383–399.
Molotchnikoff S, Tremblay F, Lepore F (1984) The role of the visual cortex in response properties of lateral geniculate cells in rats. *Exp Brain Res 53*:223–232.
Monaghan DT, Cotman CW (1982) The distribution of [3H]kainic acid binding sites in rat CNS as determined by autoradiography. *Brain Res 252*:91–100.
Monaghan DT, Cotman CW (1985) Distribution of N-Methyl-D-aspartate-sensitive L-[^{3}H]glutamate-binding sites in rat brain. *J Neurosci 5*:2909–2919.
Monaghan DT, Olverman HJ, Nguyen L, Watkins JC, Cotman CW (1988) Two classes of N-methyl-D-aspartate recognition sites: differential distribution and differential regulation by glycine. *Proc Natl Acad Sci USA 85*:9836–9840.
Montero VM (1986a) Localization of γ-aminobutyric acid (GABA) in type 3 cells and demonstration of their source to F2 terminals in the cat lateral geniculate nucleus: A Golgi-electron-microscopic GABA-immunocytochemical study. *J Comp Neurol 254*:228–245.
Montero VM (1986b) The interneuronal nature of GABAergic neurons in the lateral geniculate nucleus of

the rhesus monkey: a combined HRP and GABA immunohistochemical study. *Exp Brain Res 64*:615–622.

Montero VM (1987) Ultrastructural identification of synaptic terminals from the axon of type 3 interneurons in the cat lateral geniculate nucleus. *J Comp Neurol 264*:268–283.

Montero VM (1988) Ultrastructural identification of synaptic terminals from collateral axons of geniculo-cortical relay cells in the perigeniculate nucleus of the cat. *Brain Res 261*:1–8.

Montero VM (1989a) The GABA-immunoreactive neurons in the interlaminar regions of the cat lateral geniculate nucleus: light and electron microscopic observations. *Exp Brain Res 75*:497–512.

Montero VM (1989b) Ultrastructure identification of synaptic terminals from cortical axons and from collateral axons of geniculo-cortical relay cells in the perigeniculate nucleus of the cat. *Exp Brain Res 75*:65–72.

Montero VM (1990) Quantitative immunogold analysis reveals high glutamate levels in synaptic terminals of retino-geniculate, cortico- geniculate, and geniculo-cortical axons in the cat. *Visual Neurosci 4*:437–443.

Montero VM (1991) A quantitative study of synaptic contacts on interneurons and relay cells of the cat lateral geniculate nucleus. *Exp Brain Res 86*:257–70.

Montero VM (1994) Quantitative immunogold evidence for enrichment of glutamate but not aspartate in synaptic terminals of retino-geniculate, geniculo-cortical, and cortico-geniculate axons in the cat. *Visual Neurosci 11*:675–681.

Montero VM, Guillery RW, Woolsey CN (1977) Retinotopic organization within the thalamic reticular nucleus demonstrated by a double label autoradiographic technique. *Brain Res 138*:407–421.

Montero VM, Scott GL (1981) Synaptic terminals in the dorsal lateral geniculate nucleus from neurons of the thalamic reticular nucleus: a light and electron microscopic autoradiographic study. *Neurosci 6*:2561–2578.

Montero VM, Singer W (1985) Ultrastructural identification of somata and neural processes immunoreactive to antibodies against glutamic acid decarboxylase (GAD) in the dorsal lateral geniculate nucleus of the cat. *Exp Brain Res 59*:151–165.

Montero VM, Wenthold RJ (1989) Quantitative immunogold analysis reveals high glutamate levels in retinal and cortical synaptic terminals in the lateral geniculate nucleus of the macaque. *Neurosci 31*:639–647.

Montero VM, Zempel J (1986) The proportion and size of GABA-immunoreactive neurons in the magno-cellular and parvocellular layers of the lateral geniculate nucleus of the rhesus monkey. *Exp Brain Res 62*:215–223.

Monyer H, Burnashev N, Laurie DJ, Sakmann B, Seeburg PH (1994) Developmental and regional expression in the rat brain and functional properties of four NMDA receptors. *Neuron 12*:529–540.

Monyer H, Sprengel R, Herb A, Higuchi M, Lomeli H, Burnashev N, Sakmann B, Seeburg PH (1992) Heteromeric NMDA receptors: molecular and functional distinction of subtypes. *Science 256*:1217–1221.

Moore RY, Bloom FE (1978) Central catecholamine neuron system: Anatomy and physiology of the dopamine system. *Ann Rev Neurosci 1*:129–169.

Moore RY, Bloom FE (1979) Central catecholamine neuron system: Anatomy and physiology of the norephinephrine and epinephrine systems. *Ann Rev Neurosci 2*:113–168.

Moore RY, Card JP (1994) Intergeniculate leaflet: an anatomically and functionally distinct subdivsion of the lateral geniculate complex. *J Comp Neurol 344*:403–430.

Moran A, Avendaño C, Reinoso-Suárez F (1982) Thalamic afferents to the motor cortex in the cat. A horseradish peroxidase study. *Neurosci Lett 33*:229–233.

Morecraft RJ, Geula C, Mesulam M-M (1992) Cytoarchitecture and neural afferents of orbitofrontal cortex in the brain of the monkey. *J Comp Neurol 323*:341–358.

Morel A, Garraghty PE, Kaas JH (1993) Tonotopic organization, architectonic fields, and connections of auditory cortex in macaque monkeys. *J Comp Neurol 335*:437–459.

Morel A, Kaas JH (1992) Subdivisions and connections of auditory cortex in owl monkeys. *J Comp Neurol 318*:27–63.

Morison RS, Dempsey EW (1942) A study of thalamo-cortical relations. *Am J Physiol 135*:281–292.

Morita T, Sakimura K, Kushiya E, Yamazaki M, Meguro H, Araki K, Abe T, Mori KJ, Mishina M (1992) Cloning and functional expression of a cDNA encoding the mouse Beta2 subunit of the kainate-selective glutamate receptor channel. *Mol Brain Res 14*:143–146.

Morris BJ (1989) Neuronal localisation of neuropeptide Y gene expression in rat brain. *J Comp Neurol* 358–368.

Morris BJ, Herz A (1986) Autoradiographic localization in rat brain of kappa opiate binding sites labelled by [3H] bremazocine. *Neurosci* 839–846.

Morrison JH, Foote SL (1986) Noradrenergic and serotoninergic innervation of cortical, thalamic, and tectal visual structures in old and new world monkeys. *J Comp Neurol 243*:117–138.

Morrow AL, Battaglia G, Norman AB, Creese I (1985) Identification of subtypes of [^{3}H] prazosin-labelled a_1 receptor binding sites in rat brain. *Eur J Pharmacol 109*:285–287.

Morrow AL, Creese I (1986) Characterization of a_1-adrenergic receptor subtypes in rat brain: reevaluation of [^{3}H] WB4104 and [^{3}H] prazosin binding. *Mol Pharmacol 29*:321–330.

Morrow TJ, Casey KL (1992) State-related modulation of thalamic somatosensory responses in the awake monkey. *J Neurophysiol 67*:305–317.

Mountcastle V, Henneman E (1949) Pattern of tactile representation in thalamus of cat. *J Neurophysiol 12*:85–100.

Mountcastle VB, Henneman E (1952) The representation of tactile sensibility in the thalamus of the monkey. *J Comp Neurol 97*:409–439.

Mufson EJ, Mesulam M-M (1984) Thalamic connections of the insula in the rhesus monkey and comments on the paralimbic connectivity of the medial pulvinar nucleus. *J Comp Neurol 227*:109–120.

Mulle C, Madariaga A, Deschênes M (1986) Morphology and electrophysiological properties of reticularis thalami neurons in cat: In vivo study of a thalamic pacemaker. *J Neurosci 6*:2134–2145.

Muller F, O'Rahilly R (1988) The development of the human brain, including the longitudinal zoning in the diecephalon at stage 15. *Anat Embryol 179*:55–71.

Muñoz A, Huntsman MM, Jones EG (1998) Nucleus- and cell-specific localization of $GABA_B$ receptor mRNAs in monkey thalamus. *J Comp Neurol* (in press)

Murphy PC, Sillito AM (1987a) Corticofugal feedback influences the generation of length tuning in the visual pathway. *Nature 329*:727–729.

Murphy PC, Sillito AM (1989a) The binocular input to cells in the feline dorsal lateral geniculate nucleus (dLGN). *J Physiol Lond 415*:393–408.

Murphy PC, Sillito AM (1996) Functional morphology of feedback pathways from area 17 of the cat visual cortex to the lateral geniculate nucleus. *J Neurosci 16*:1–11.

Murphy PC, Uhlrich DJ, Tamamaki N, Sherman SM (1994) Brain-stem modulation of the response properties of cells in the cat's perigeniculate nucleus. *Visual Neurosci 11*:781–791.

Mushiake H, Shosaku A, Kayama Y (1984) Inhibition of thalamic ventrobasal complex neurons by glutamate infusion into the thalamic reticular nucleus in rats. *J Neurosci Res 12*:93–100.

Nagy JI, Carter DA, Lehman J, Fibiger HC (1978) Evidence for GABA containing projection from the entopedurcular nucleus to the lateral hebenula in the rat. *Brain Res 145*:360–364.

Naito J, Kawamura K, Takagi SF (1984) An HRP study of neuronal pathways to neocortical olfactory areas in monkeys. *Neurosci Res* 19–33.

Najlerahim A, Barton AJ, Harrison PJ, Heffernan J, Pearson RC (1989a) Messenger RNA encoding the D2 dopaminergic receptor detected by in situ hybridization histochemistry in rat brain. *Febs Letts* 255:335–339.

Nakai Y, Takaori S (1974) Influence of norepinephrine containing neurons derived from the locus coeruleus on lateral geniculate neuronal activities of cats. *Brain Res 71*:47–60.

Nakanishi S (1994) Metabotropic glutamate receptors: synaptic transmission, modulation, and plasticity. *Neuron 5*:1031–1037.

Nakano K, Hasegawa Y, Tokushige A, Nakagawa S, Kayahara T, Mizuno N (1990) Topographical projections from the thalamus, subthalamic nucleus and pendunculopontine tegmental nucleus to the striatum in the Japanese monkey, *Macaca fuscata*. *Brain Res 537*:54–68.

Nauta HJW (1974) Evidence of a pallidohabenular pathway in the cat. *J Comp Neurol 156*:19–28.

Nauta HJW (1979) Projections of the pallidal complex: an autoradiographic study in the cat. *Neurosci 4*:1853–11873.

Nauta WJH (1961) Fibre degeneration following lesions of the amygdaloid complex in the monkey. *J Anat 95*:515–531.

Nauta WJH, Bücher VM (1954) Efferent connections of the striate cortex in the albino rat. *J Comp Neurol 180*:257–285.

Nauta WJH, Kuypers HGJM (1958) Some ascending pathways in the brain stem reticular formation. In: Reticular Formations in the Brain (Jasper HH, Proctor LD, Knighton RS, Noshay WC eds), Boston: Little Brown.

Nauta WJH, Mehler WR (1966) Projections of the lentiform nucleus in the monkey. *Brain Res 1*:3–42.

Nauta WJH, Whitlock DG (1954) An anatomical analysis of the non-specific thalamic projection system. In: Brain Mechanisms and Consciousness (Delafresnaye JF, Thomas CC eds), Oxford: Springfield, Blackwell.

Nealey TA, Maunsell JHR (1994) Magnocellular and parvocellular contributions to the responses of neurons in macaque striate cortex. *J Neurosci 14*:2069–2079.

Neki A, Ohishi H, Kaneko T, Shigemoto R, Nakanishi N, Mizuno N (1996) Pre- and post synaptic localization of a metabotropic glutamate receptor, mGluR2, in the rat brain: an immunohistochemical study with a monoclonal antibody. *Neurosci Lett 202*:197–200.

Nicolelis MAL, Chapin JK, Lin RCS (1992) Somatotopic maps within the zona incerta relay parallel GABAergic somatosensory pathways to the neocortex, superior colliculus, and brainstem. *Brain Res 577*:134–141.

Niimi K, Matsuoka H, Aisaka T, Okada Y (1981a) Thalamic afferents to the prefrontal cortex in the cat traced with horseradish peroxidase. *J Hirnforsch 22*:221–241.

Niimi K, Matsuoka H, Yamazaki Y, Matsumoto H (1981b) Thalamic afferent to the visual cortex in the cat studied by retrograde transport of horseradish peroxidase. *Brain Behav Evol 18*:114–139.

Niimi K, Matsuoka H, Yamazaki Y, Katayama T (1983) Thalamic afferents to the anterior and middle suprasylvian gyri in the cat traced with horseradish peroxidase. *J Hirnforsch 24*:173–187.

Nishiyama K, Kwak S, Murayama S, Kanazawa I (1995) Substance P is a possible neurotransmitter in the rat spinothalamic tract. *Neurosci Res 21*:261–266.

Norden JJ, Lin CS, Kaas JH (1978) Subcortical projections of the dorsomedial visual area (DM) of visual association cortex in the owl monkey, *Aotus trivirgatus*. *Exp Brain Res 32*:321–334.

Norita M, Katoh Y (1987) The GABAergic neurons and axon terminals in the lateralis medialis-suprageniculate nuclear complex of the cat: GABA-immunocytochemical and WGA-HRP studies by light and electron microscopy. *J Comp Neurol 263*:54–67.

Norton TT, Casagrande VA, Irvin GE, Sesma MA, Petry HM (1988) Contrast-sensitivity functions of W-, X-, and Y-like relay cells in the lateral geniculate of bush baby, *Galago crassicaudatus*. *J Neurophysiol 59*:1639–1656.

Nothias F, Onteniente B, Roudier F, Peschanksi M (1988a) Immunocytochemical study of serotoninergic and noradrenergic innervation of the ventrobasal complex of the rat thalamus. *Neurosci Lett 95*:59–63.

Nothias F, Peschanski M, Besson J-M (1988b) Somatotopic reciprocal connections between the somatosensory cortex and the thalamic Po nucleus in the rat. *Brain Res 447*;1:169–174.

Nudo RJ, Masterton RB (1988) Descending pathways to the spinal cord: a comparative study of 22 mammals. *J Comp Neurol 277*:53–79.

Oertel WH, Graybiel AM, Mugnaini E, Elde RP, Schmechel DE, Kopin IJ (1983) Coexistence of glutamic acid decarboxylase and somatostatin-like immunoreactivity in neurons of the feline nucleus reticularis thalami. *J Neurosci 3*:1322–1332.

Ogawa R, Itoh K, Kaneko T, Mizuno N (1989) Co-existence of vasoactive intestinal polypeptide (VIP)- and cholecystokinin (CCK)-like immunoreactivities in thalamocortical neuron in the ventrolateral nucleus of the rat. *Brain Res 490*:152–156.

Ogden TE (1960) Cortical control of thalamic somatosensory relay nuclei. *Electroencephalogr Clin Neurophysiol 12*:621–634.

Ogren M, Hendrickson A (1976) Pathways between striate cortex and subcortical regions in *Macaca mulatta* and *Saimiri sciureus*: evidence for a reciprocal pulvinar connection. *Exp Neurol 53*:780–800.

Ogren MP, Hendrickson AE (1977) The distribution of pulvinar terminals in visual areas 17 and 18 of the monkey. *Brain Res 137*:343–350.

Ohara PT, Chazal G, Ralston HJ, III (1989) Ultrastructural analysis of GABA-immunoreactive elements in the monkey thalamic ventrobasal complex. *J Comp Neurol 283*:541–558.

Ohara PT, Lieberman AR (1981) Thalamic reticular nucleus: anatomical evidence that cortico-reticular axons establish monosynaptic contact with reticulo-geniculate projection cells. *Brain Res 207*:153–156.

Ohara PT, Lieberman AR (1985) The thalamic reticular nucleus of the adult rat: experimental anatomical studies. *J Neurocytol 14*:365–411.

Ohara PT, Lieberman AR, Hunt SP, Wu JY (1983) Neural elements containing glutamic acid decarboxylase (GAD) in the dorsal lateral geniculate nucleus of the rat: immunohistochemical studies by light and electronmicroscopy. *Neurosci 8*:189–212.

Ohara PT, Sefton AJ, Lieberman AR (1980) Mode of termination of afferents from the thalamic reticular nucleus in the dorsal lateral geniculate nucleus of the rat. *Brain Res 197*:503–506.

Ohishi H, Shigemoto R, Nakanishi S, Mizuno N (1993a) Distribution of the mRNA for a metabotropic glutamate receptor (mGluR3) in the rat brain: an in situ hybridization study. *J Comp Neurol 335*:252–266.

Ohishi H, Shigemoto R, Nakanishi S, Mizuno N (1993b) Distribution of the messenger RNA for a metabotropic glutamate receptor, mGluR2, in the central nervous system of the rat. *Neurosci 53*:1009–1018.

Ohye C (1990) Thalamus. In: The Human Nervous System (Paxinos G ed), pp 439–468. San Diego: Academic Press.

Ohye C, Imai S, Nakajima H, Shibazaki T, Hirai T (1979) Experimental study of spontaneous postural tremor induced by a more successful tremor-producing procedure in the monkey. *Adv Neurol 24*:83–91.

Ohye C, Narabayashi H (1979) Physiological study of presumed ventralis intermedius neurons in the human thalamus. *J Neurosurg 50*:290–297.

Ohye C, Shibazaki T, Hirai T, Wada H, Hirato M, Kawashima Y (1989) Further physiological observations on the ventralis intermedius neurons in the human thalamus. *J Neurophysiol 61*:488–500.

Ohye C, Shibazaki T, Hirai T, Kawashima Y, Hirato M, Matsumura M (1993) Tremor-mediating thalamic zone studied in humans and in monkeys. *Stereotact Funct Neurosurg 60*:136–145.

Ohye C, Shibazaki T, Hirato M, Kawashima Y, Matsumura M (1990) Strategy of selective VIM thalamotomy guided by microrecording. *Stereotact Funct Neurosurg 54–55*:186–191.

Ojima H (1994) Terminal morphology and distribution of corticothalamic fibers originating from layers 5 and 6 of cat primary auditory cortex. *Cerebral Cortex 4*:646–663.

Ojima H, Murakami K, Kishi K (1996) Dual termination modes of corticothalamic fibers originating from pyramids of layers 5 and 6 in cat visual cortical area 17. *Neurosci Lett 208*:57–60.

Okamoto N, Hori S, Akazawa C, Hayashi Y, Shigemoto R, Mizuno N, Nakanishi S (1994) Molecular characterization of a new metabotropic glutamate receptor mGluR7 coupled to inhibitory cyclic AMP signal transduction. *J Biol Chem 269*:1231–1236.

Oliver DL (1987) Projections to the inferior colliculus from the anteroventral cochlear nucleus in the cat: possible substrates for binaural interaction. *J Comp Neurol 264*:24–46.

Olivier A, Parent A, Poirier LJ (1970) Identification of the thalamic nuclei on the basis of their cholinesterase content in the monkey. *J Anat 106*:37–50.

Olsen JF, Suga N (1991) Combination-sensitive neurons in the medial geniculate body of the mustached bat: encoding of target range information. *J Neurophysiol 65*:1275–1296.

Olsen RW, Bureau MH, Endo S, Smith G (1991) The $GABA_A$ receptor family in the mammalian brain. *Neurochem Res 16*:317–25.

Olsen RW, McCabe RT, Wamsley JK (1990) GABAA receptor subtypes: autoradiographic comparison of GABA, benzodiazepine, and convulsant binding sites in the rat central nervous system. *J Chem Neuroanat 3*:59–76.

Olsen RW, Szamraj O, Houser CR (1987) [3H]AMPA binding to glutamate receptor subpopulations in rat brain. *Brain Res 2*:243–254.

Olsen RW, Tobin AJ (1990) Molecular biology of GABA A receptors. *FASEB J. 4*:1–12.

Olszewski J (1952) Thalamus of the *Macaca mulatta*: an Atlas for Use with the Stereotaxic Instrument. S Karger, Basel

Ono K, Niimi K (1986) Afferent projections to the thalamic mediodorsal nucleus in the cat studied by retrograde and anterograde axonal transport of horseradish peroxidase. *J Hirnforsch 27*:597–610.

Page KJ, Everitt BJ (1995) The distribution of neurons coexpressing immunoreactibity to AMPA-sensitive glutamate receptor subtypes (GluR1–4) and nerve growth factor receptor in the rat basal forebrain. *Eur J Neurosci 7*:1022–1033.

Pakkenberg B (1990) Pronounced reduction of total neuron number in mediodorsal thalamic nucleus and nucleus accumbens in shizophrenics. *Arch Gen Psychiatry 47*:1023–1028.

Palacios JM, Kuhar MJ (1981) Neurotensin receptors are located on dopamine-containing neurones in rat midbrain. *Nature 294*:587–588.

Palacios JM, Wamsley JK, Kuhar MJ (1981) High affinity GABA receptors autoradiographic localization. *Brain Res 222*:285–308.

Panula P, Pirvola U, Auvinen S, Airaksinen MS (1989) Histamine-immunoreactive nerve fibers in the rat brain. *Neurosci 28*:585–610.

Panula P, Yang HY, Costa E (1982) Neuronal location of the bomesin-like immunoreactivity in the central nervous system of the rat. *Regul Pept 4*:275–283.

Papadopoulos GC, Parnavelas JG (1990a) Distribution and synaptic organization of dopaminergic axons in the lateral geniculate nucleus of the rat. *J Comp Neurol 294*:356–361.

Papadopoulos GC, Parnavelas JG (1990b) Distribution and synaptic organization of serotoninergic and noradrenergic axons in the lateral geniculate nucleus of the rat. *J Comp Neurol 294*:345–355.

Pape H-C (1992) Adenosine promotes burst activity in guinea-pig geniculocortical neurones through two different ionic mechanisms. *J Physiol Lond 447*:729–753.

Pape H-C, Eysel UT (1986) Binocular interactions in the lateral geniculate nucleus of the cat: GABAergic inhibition reduced by dominant afferent activity. *Exp Brain Res 61*:265–271.

Pape H-C, Mager R (1992) Nitric oxide controls oscillatory activity in thalamocortical neurons. *Neuron 9*:441–448.

Pape HC, McCormick DA (1989) Noradrenaline and serotonin selectively modulate thalamic burst firing by enhancing a hyperpolarization-activated cation current. *Nature 340*:715–718.

Pape H-C, McCormick DA (1995) Electrophysical and pharmacological properties of interneurons in the cat dorsal lateral geniculate nucelus. *Neurosci 68*:1105–1126.

Paré D, Dossi RC, Steriade M (1991) Three types of inhibitory postsynaptic potentials generated by interneurons in the anterior thalamic complex of cat. *J Neurosci 66*:1190–1204.

Paré D, Smith Y, Parent A, Steriade M (1988) Projections of brainstem core cholinergic and non-cholinergic neurons of cat to intralaminar and reticular thalamic nuclei. *Neurosci 25*:69–86.

Paré D, Steriade M (1993) The reticular thalamic nucleus projects to the contralateral dorsal thalamus in macaque monkey. *Neurosci Letts 154*:96–100.

Parent A, DeBellefeuille L (1982) Organization of efferent projections from the internal segment of globus pallidus in primate as revealed by fluorescence retrograde labeling method. *Brain Res 245*:201–213.

Parent A, DeBellefeuille L (1983) Pallidointralaminar and pallidonigral projections in primate as studied by retrogrde double labeling method. *Brain Res 278*:11–28.

Parent A, Descarries L, Beaudet A (1981) Organization of ascending serotonin systems in the adult rat brain A radioautographic study after intraventricular administration of [3H] 5-hydroxtryptamine. *Neurosci 6*:115–138.

Parent A, Mackey A, de Bellefeuille L (1983) The subcortical afferents to caudate nucleus and putamen in primate: a fluorescence retrograde double labeling study. *Neurosci 10*:1137–1150.

Parent A, Paré D, Smith Y, Steriade M (1988) Basal forebrain cholinergic and noncholinergic projections to the thalamus and brainstem in cats and monkeys. *J Comp Neurol 277*:281–301.

Partlow GD, Colonnier M, Szabo J (1977a) Thalamic projections of the superior colliculus in the rhesus monkey, *Macaca mulatta*. A light and electron microscopic study. *J Comp Neurol 171*:285–318.

Pasik P, Pasik T, Hámori J (1976) Synapses between interneurons in the lateral geniculate nucleus of monkeys. *Exp Brain Res 25*:1–13.

Pasik P, Pasik T, Hámori J, Szentágothai J (1973) Golgi type II interneurons in the neuronal circuit of the monkey lateral geniculate nucleus. *Exp Brain Res 17*:18–34.

Pasik P, Pasik T, Holstein GR (1988) Serotonin-immunoreactivity in the monkey lateral geniculate nucleus. *Exp Brain Res 69*:662–666.

Patton HD, Ruch TC, Walker AE (1944) Experimental hypogeusia from Horsely Clarke lesions of the thalami in *Macaca mulatta. J Neurophysiol 7*:171–184.

Pazos A, Palacios JM (1985) Quantitative autoradiographic mapping of serotonin receptors in the rat brain. I. Serotonin-1 receptors. *Brain Res 346*:205–230.

Pearson JC, Garfunkel DA (1987) Evidence for thalamic projections from external cuneate nucleus, cell groups Z and X, and the mesencephalic nucleus of the trigeminal nerve in squirrel monkey. *Neurosci Lett 74*:125–131.

Pearson JC, Haines DE (1980a) Somatosensory thalamus of a prosimian primate (*Galago senegalensis*). I. Configuration of nuclei and termination of spinothalamic fibers. *J Comp Neurol 190*:533–558.

Pearson JC, Haines DE (1980b) Somatosensory thalamus of a prosimian primate (*Galago senegalensis*). II. An HRP and Golgi study of the ventral posterolateral nucleus (VPL). *J Comp Neurol 190*:559–580.

Pearson JC, Norris JR, Phelps CH (1984) The cytoarchitecture and some efferent projections of the centromedian-parafascicular complex in the lesser bushbaby (*Galago senegalensis*). *J Comp Neurol 225*:554–568.

Pearson RCA, Brodal P, Powell TPS (1978) The projection of the thalamus upon the parietal lobe in the monkey. *Brain Res 144*:143–148.

Pearson RCA, Neal JW, Powell TPS (1987) Increase in immunohistochemical staining of GABAergic axons in the superior colliculus and thalamus of the rat following damage of the ipsilateral striatum and frontal cortex. *Brain Res 412*:352–356.

Pecci-Saavedra J, Vaccarezza OL, Reader TA (1968) Ultrastructure of cells and synapses in the parvocellular portion of the cebus monkey lateral geniculate nucleus. *Z Zellforsch 89*:462–477.

Pelaprat D, Peschanski M, Broer Y, Besson JM, Roques BP (1987) Postsynaptic receptors for cholecystokinin in the thalamic reticular nucleus: a possible modulatory system for sensory transmission. *Neurosci Lett* 80:16–20.

Penney JB, Jr., Young AB (1983) Speculations on the functional anatomy of basal ganglia disorders. *Annu Rev Neurosci 6*:73–94.

Penny GR, Conley M, Schmechel DE, Diamond IT (1984) The distribution of glutamic acid decarboxylase immunoreactivity in the diencephalon of the opossum and rabbit. *J Comp Neurol 228*:38–56.

Penny GR, Fitzpatrick D, Schmechel DE, Diamond IT (1983) glutamatic acid decarboxylase-immunoreactive neurons and horseradish peroxidase-labeled projection neurons in the ventral posterior nucleus of the cat and Galago senegalensis. *J Neurosci 3*:1868–1887.

Percheron G (1977) The thalamic territory of cerebellar afferents and the lateral region of the thalamus of the macaque in stereotaxic ventricular coordinates. *J Hirnforsch 18*:375–400.

Percheron G, Francois C, Talbi B, Meder JF, Fenelon G, Yelnik J (1993) The primate motor thalamus analysed with reference to subcortical afferent territories. *Stereotact Funct Neurosurg 60*:32–41.

Percheron G, Francois C, Talbi B, Yelnik J, Fenelon G (1996) The primate motor thalamus. *Brain Res Rev 22*:93–181.

Percheron G, Francois C, Yelnik J (1986) Relations between the basal ganglia and the thalamus of the primate. New morphologic data. New physiopathologic interpretations. *Rev Neurol Paris 142*:337–353.
Perl ER, Whitlock DG (1961) Somatic stimuli exciting spinothalamic projections to thalamic neurons in cat and monkey. *Exp Neurol 3*:256–296.
Perry VH, Oehler R, Cowey A (1984) Retinal ganglion cells that project to the dorsal lateral geniculate nucleus in the macaque monkey. *Neurosci 12*:1101–1124.
Persohn E, Malherbe P, Richards JG (1992) Comparative molecular neuroanatomy of cloned $GABA_A$ receptor subunits in the rat CNS. *J Comp Neurol 326*:193–216.
Pert CB, Kuhar MJ, Snyder SH (1976) Autoradiographic localization of opiate receptor in rat brain. *Proc Natl Acad Sci USA 73*:3729–3733.
Peschanski M, Besson JM (1984) Diencephalic connections of the raphe nuclei of the rat brain-stem: an anatomical study with reference to the somatosensory system. *J Comp Neurol 224*:503–534.
Peschanski M, Pelaprat D, Besson JM, Broer Y, Roques BP (1987) Postsynaptic receptors for cholecystokinin in the thalamic reticular nucleus: a possible modulatory system for sensory transmission. *Neurosci Lett 80*:16–20.
Petersen SE, Robinson DL, Currie JN (1989) Influences of lesions of parietal cortex on visual spatial attention in humans. *Exp Brain Res 76*:267–280.
Petersen SE, Robinson DL, Keys W (1985) Pulvinar nuclei of the behaving rhesus monkey: visual responses and their modulation. *J Neurophysiol 54*:867–886.
Petralia RS, Wang Y-X, Wenthold RJ (1994a) The NMDA receptor subunits NR2A and NR2B show histological and ultrastructural localization patterns similar to those of NR1. *J Neurosci 14*:6102–6120.
Petralia RS, Wang Y-X, Wenthold RJ (1994b) Histological and ultrastructural localization of the kainate receptor subunits, KA2 and GluR6/7, in the rat nervous system using selective antipeptide antibodies. *J Comp Neurol 349*:85–110.
Petralia RS, Wenthold RJ (1992) Light and electron immunocytochemical localization of AMPA-selective glutamate receptors in the rat brain. *J Comp Neurol 318*:329–354.
Petras JM (1965a) Some fiber connections of the precentral and postcentral cortex with the basal ganglia, thalamus and subthalamus. *Trans Am Neurol 90*:274–275.
Petras JM (1965b) Fiber degeneration in the basal ganglia and diencephalon following lesions in the precentral and postcentral cortex of the monkey (*Macaca mulatta*) with additional observation in the chimpanzee. In: Abstracts of Eighth International Anatomical Congress, Wiesbaden
Phillips CG, Powell TPS, Wiesendanger M (1971) Projection from low threshold muscle afferents of hand and forearm to area 3 a of baboon's cortex. *J Physiol Lond 217*:419–446.
Phillis JW (1970) The Pharmacology of Synapses. Oxford: Pergamon Press.
Phillis JW, Tēbēcis AK, York DH (1967a) A study of cholinoceptive cells in the lateral geniculate nucleus. *J Physiol 192*:695–713.
Phillis JW, Tēbēcis AK, York DH (1967b) A study of cholinoceptive cells in the lateral geniculate nucleus. *J Physiol 192*:695–713.
Pieribone VA, Nicholas AP, Dagerlind A, Hökfelt T (1994) Distribution of alpha 1 adrenoceptors in rat brain revealed by in situ hybridization experiments utilizing subtype-specific probes. *J Neurosci 14*:4252–68.
Pinault D, Bourassa J, Deschênes M (1995a) Thalamic reticular input to the rat visual thalamus: a single fiber study using biocytin as an anterograde tracer. *Brain Res 670*:147–152.
Pinault D, Bourassa J, Deschênes M (1995b) The axonal arborization of single thalamic reticular neurons in the somatosensory thalamus of the rat. *Eur J Neurosci 7*:31–40.
Pinault D, Deschênes M (1992a) Muscarinic inhibition of reticular thalamic cells by basal forebrain neurones. *NeuroReprt 3*:1101–1104.
Pinault D, Deschênes M (1992b) Control of 40-Hz firing of reticular thalamic cells by neurotransmitters. *Neurosci 51*:259–268.
Pinault D, Deschênes M (1992c) The origin of rhythmic fast subthreshold depolarizations in thalamic relay cells of rats under urethane anasthesia. *Brain Res 595*:295–300.
Pinault D, Deschênes M (1992d) Voltage-dependent 40-Hz oscillations in rat reticular thalamic neurons in vivo. *Neurosci 51*:245–258.
Pioro EP, Cuello AC (1990) Distribution of nerve growth factor receptor-like immunoreactivity in the adult rat central nervous system. Effect of colchicine and correlation with the cholinergic system. I. Forebrain. *Neurosci 34*:57–88.
Poggio GF, Mountcastle VB (1960) A study of the functional contributions of the lemniscal and spinothalamic systems to somatic sensibility. *Bull Johns Hopkins Hosp 106*:266–316.
Poggio GF, Mountcastle VB (1963) The functional properties of ventrobasal thalamic neurons studied in unanesthetized monkeys. *J Neurophysiol 26*:775–806.

Pollin B, Albe-Fessard D (1979) Organization of somatic thalamus in monkeys with and without section of dorsal spinal tracts. *Brain Res 173*:431–449.

Pollin B, Rokyta R (1982) Somatotopic organization of nucleus reticularis thalami in chronic awake cats and monkeys. *Brain Res 250*:211–221.

Polyak SL (1941) The Retina. Chicago: Univ. Chicago Press.

Pompeiano M, Palacios JM, Mengod G (1992) Distribution and cellular localization of mRNA coding for 5-HT1A receptor in the rat brain: correlation with receptor binding. *J Neurosci 12*:440–453.

Pons TP, Kaas JH (1985) Connections of area 2 of somatosensory cortex with the anterior pulvinar and subdivisions of the ventroposterior complex in macaque monkeys. *J Comp Neurol 240*:16–36.

Porrino LJ, Crane AM, Goldman-Rakic PS (1981) Direct and indirect pathways from the aymgdala to the frontal lobe in rhesus monkeys. *J Comp Neurol 198*:121–126.

Porter LL, White EL (1983) Afferent and efferent pathways of the vibrissal region of primary motor cortex in the mouse. *J Comp Neurol 214*:279–289.

Powell TPS, Cowan WM (1956) A study of thalamo-striate relations in the monkey. *Brain 79*:364–389.

Powell TPS, Cowan WM (1967) The interpretation of the degenerative changes in the intralaminar nuclei of the thalamus. *J Neurol Neurosurg Psychiatr 30*:140–153.

Powell TPS, Cowan WM, Raisman G (1965) The central olfactory connexions. *J Anat 99*:791–813.

Powell TPS, Mountcastle VB (1959) Some aspects of the functional organization of the cortex of the postcentral gyrus of the monkey: a correlation of findings obtained in a single unit analysis with cytoarchitecture. *Bull Johns Hopkins Hosp 105*:133–162.

Price JL, Slotnick BM (1983) Dual olfactory representation in the rat thalamus and anatomical and dectrophysiological study. *J Comp Neurol 215*:63–77.

Price JL, Amaral DG (1981) An autoradiographic study of the projections of the central nucleus of the monkey amygdala. *J Neurosci 1*:1242–1259.

Pritchard TC, Hamilton RB, Morse JR, Norgren R (1986) Projections of thalamic gustatory and lingual areas in the monkey, *Macaca fascicularis*. *J Comp Neurol 244*:213–229.

Pritchard TC, Hamilton RB, Norgren R (1989) Neural coding of gustatory information in the thalamus of *Macaca mulatta*. *J Neurophysiol 61*:1–14.

Pubols LM (1968) Somatic sensory representation in the thalamic ventrobasal complex of the spider monkey (*Ateles*). *Brain Behav Evol 1*:305–323.

Purpura DP, Malliani A (1967) Synaptic potentials and discharge characteristics of caudate neurons activated by thalamic stimulation. *Brain Res 6*:325–340.

Quinlan JT, Phillips MI (1981) Immunoreactivity for an anagiotensin II-like peptide in the human brain. *Brain Res 205*:212–218.

Raczkowski D, Diamond IT (1978) Connections of the striate cortex in *Galago senegalensis*. *Brain Res 144*:383–388.

Raczkowski D, Diamond IT (1980) Cortical connections of the pulvinar nucleus in Galago. *J Comp Neurol 193*:1–40.

Raczkowski D, Hamos JE, Sherman SM (1988) Synaptic circuitry of physiologically identified W-cells in the cat's dorsal lateral geniculate nucleus. *J Neurosci 8*:31–48.

Raczkowski D, Fitzpatrick D (1989) Organization of cholinergic synapses in the cat's dorsal lateral geniculate and perigeniculate nuclei. *J Comp Neurol 288*:676–690.

Raczkowski D, Rosenquist AC (1983) Connections of the multiple visual cortical areas with lateral posterior-pulvinar complex and adjacent thalamic nuclei in the cat. *J Neurosci 3*:1912–1942.

Ragsdale CW, Graybiel AM (1988) Multiple patterns of thalamostriatal innervation in the cat. In: Cellular Thalamic Mechanisms (Bentivoglio M, Spreafico R eds), pp 261–270. Amsterdam: Elsevier

Ragsdale CW, Graybiel AM (1991) Compartmental organization of the thalamostriatal connection in the cat. *J Comp Neurol 311*:134–167.

Rainbow TC, Parsons B, Wolfe BB (1984) quantitative autoradiography of beta 1- and beta 2-adrenergic receptors in rat brain. *Proc Natl Acad Sci USA 81*:1585–1589.

Rainbow TC, Wieczorek CM, Halpain S (1984) Quantitative autoradiography of binding sites for [3H]AMPA, a structural analogue of glutamic acid. *Brain Res 309*:173–7.

Ralston HJ (1971a) Evidence for presynaptic dendrites and a proposal for their mechanism of action. *Nature 230*:585–587.

Ralston HJ, Ralston DD (1992) The primate dorsal spinothalamic tract: evidence for a specific termination in the posterior nuclei (Po/SG) of the thalamus. *Pain 48*:107–118.

Ralston HJ, III, Ohara PT, Ralston DD, Chazal G (1988) The neuronal and synaptic organization of the cat and primate somataosensory thalamus. In: Cellular Thalamic Mechanisms (Bentivoglio M, Spreafico R eds), pp 127–141. Amsterdam: Elsevier

Ralston HJ, III, Ralston DD (1994) Medial lemniscal and spinal projections to the macaque thalamus: An

electron microscopic study of differing GABAergic circuitry serving thalamic somatosensory mechanisms. *J Neurosci 14*:2485–2502.
Ramoa AS, McCormick DA (1994a) Developmental changes in electrophysiological properties of LGNd neurons during reorganization of retinogeniculate connections. *J Neurosci 14*:2089–2097.
Ramoa AS, McCormick DA (1994b) Ehanced activation of NMDA receptorresponses at the immature retinogeniculate synapse. *J Neurosci 14*:2098–2105.
Rapisarda C, Palmeri A, Saplenza S (1992) Cortical modulation of thalamo-cortical neurons relaying exteroceptive information: a microstimulation study in the guinea pig. *Exp Brain Res 88*:140–150.
Rauschecker JP, Tian B, Hauser M (1995) Processing of complex sounds in the macaque nonprimary auditory cortex. *Science 5207*:111–114.
Rausell E, Avendaño C (1985) Thalamocortical neurons projecting to superficial and to deep layers in parietal, frontal and prefrontal regions in the cat. *Brain Res 347*:159–165.
Rausell E, Bae CS, Viñuela A, Huntley GW, Jones EG (1992a) Calbindin and parvalbumin cells in monkey VPL thalamic nucleus: distribution, laminar cortical projections, and relations to spinothalamic terminations. *J Neurosci 12*:4088–4111.
Rausell E, Cusick CG, Taub E, Jones EG (1992b) Chronic deafferentation in monkeys differentially affects nociceptive and nonnociceptive pathways distinguished by specific calcium-binding proteins and down-regulates gamma- aminobutyric acid type A receptors at thalamic levels. *Proc Natl Acad Sci USA 89*:2571–2575.
Rausell E, Jones EG (1991a) Chemically distinct compartments of the thalamic VPM nucleus in monkeys relay principal and spinal trigeminal pathways to different layers of the somatosensory cortex. *J Neurosci 11*:226–237.
Rausell E, Jones EG (1991b) Histochemical and immunocytochemical compartments of the thalamic VPM nucleus in monkeys and their relationship to the representational map. *J Neurosci 11*:210–225.
Rausell E, Jones EG (1995) Extent of intracortical arborization of thalamocortical axons as a determinant of representational plasticity in monkey somartic sensory cortex. *J Neurosci 15*:4270–4288
Ray JP, Price JL (1990) Postnatal changes in the density and distribution of neurotensin-like immunoreactive fibers in the mediodorsal nucleus of the thalamus in the rat. *J Comp Neurol 292*:269–282
Ray JP, Price JL (1992) The organization of the thalamocortical connections of the mediodorsal thalamic nucleus in the rat, related to the ventral forebrain-prefrontal cortex topography. *J Comp Neurol 323*:167–197.
Ray JP, Price JL (1993) The organization of projections from the mediodorsal nucleus of the thalamus to orbital and medial prefrontal cortex in Macaque monkeys. *J Comp Neurol 337*:1–31.
Ray JP, Russchen FT, Fuller TA, Price JL (1992) Sources of presumptive glutamatergic/aspartatergic afferents to the mediodorsal nucleus of the thalamus in the rat. *J Comp Neurol 320*:435–456.
Reilly S, Pritchard TC (1995) The effect of thalamic lesions on primate taste preference. *Exp Neurol 135*:56–66.
Reubi JC, Cuénod M (1979) Glutamate release in vitro from corticostriatal terminals. *Brain Res 176*:185–188.
Rezak M, Benevento LA (1979) A comparison of the organization of the projections of the dorsal lateral geniculate nucleus, the inferior pulvinar and adjacent lateral pulvinar to primary visual cortex (area 17) in the macaque monkey. *Brain Res 167*:19–40.
Ribak CE, Peters A (1975) An autoradiographic study of the projections from the lateral geniculate body of the rat. *Brain Res 92*:341–368.
Ricardo JA (1980) Efferent connections of the subthalamic region in the rat. I. The subthalamic nucleus of Luys. *Brain Res 202*:257–271.
Ricardo JA (1981) Efferent connections of the subthalamic region in the rat. II. The zona incerta. *Brain Res 214*:43–60.
Richard D, Gioanni Y, Kitsikis A, Buser P (1975) A study of geniculate unit activity during cryogenic blockade of the primary visual cortex in the cat. *Exp Brain Res 22*:235–242.
Richardson DE, Zorub DS (1970) Sensory function of the pulvinar. *Confin Neurol 32*:165–173.
Rieck RW, Carey RG (1985) Organization of the rostral thalamus in the rat: evidence for connections to layer I of visual cortex. *J Comp Neurol 234*:137–154.
Rinaldi PC, Young RF, Albe-Fessard D, Chodakiewitz J (1991) Spontaneous neuronal hyperactivity in the medial and intralaminar thalamic nuclei of patients with deafferentation pain. *J Neurosurg 74*:415–421.
Rinvik E, Ottersen OP, Storm-Mathisen J (1987) Gamma-aminobutyrate-like immunoreactivity in the thalamus of the cat. *Neurosci 21*:781–806.
Roberts TS, Akert K (1963) Insular and opercular cortex and its thalamic projection in *Macaca mulatta*. *Schweiz Arch Neurol Neurochir Psychiatr 92*:1–43.

Robertson RT, Feiner AR (1982) Diencephalic projections from the pontine reticular formation: antoradiographic studies in the cat. *Brain Res 239*:3–16.

Robinson DL, Goldberg ME, Stanton GB (1978) Parietal association cortex in the primate: sensory mechanisms and behavioral modulations. *J Neurophysiol 41*:910–932.

Robinson DL, Petersen SE (1992) The pulvinar and visual salience. *Trends Neurosci 15*:127–132.

Rodieck RW, Watanabe M (1993) Survey of the morphology of macaque retinal ganglion cells that project to the pretectum, superior colliculus, and parvicellular laminae of the lateral geniculate nucleus. *J Comp Neurol 338*:289–303.

Rodman, HR (1994) Development of inferior temporal cortex in the monkey. *Cerebral Cortex 4*:485–498.

Rogawski MA, Aghajanian GK (1980a) Norepinephrine and serotonin: opposite effects on the activity of lateral geniculate neurons evoked by optic pathway stimulation. *Exp Neurol 69*:678–694.

Rogawski MA, Aghajanian GK (1980b) Activation of lateral geniculate neurons by norepinephrine: mediation by an α-drenergic receptor. *Brain Res 182*:345–360.

Rogawski MA, Aghajanian GK (1980c) Modulation of lateral geniculate excitability by noradrenaline microiontophoresis or locus coeruleus stimulation. *Nature 287*:731–734.

Rogawski MA, Aghajanian GK (1982) Activation of lateral geniculate neurons by locus coeruleus or dorsal noradrenergic bundle stimulation: selective blockade by the alpha1-adrenoreceptor antagonist prazosin. *Brain Res 250*:31–40.

Roger M, Cadusseau J (1985) Afferents to the zona incerta in the rat: a combined retrograde and anterograde study. *J Comp Neurol 241*:480–492.

Romanowski CAJ, Mitchell IJ, Crossman AR (1985) The organisation of the efferent projections of the zona incerta. *J Anat 143*:75–95.

Rosa MG, Pettigrew JD, Cooper HM (1996) Unusual pattern of retinogeniculate projections in the Controversial primate, Tarsius. *Brain Behav Evol 48*:121–129.

Rose JE (1942a) The ontogenetic development of the rabbit's diencephalon. *J Comp Neurol 77*:61–129.

Rose JE (1942b) The thalamus of the sheep: cellular and fibrous structure and comparison with pig, rabbit and cat. *J Comp Neurol 77*:469–523.

Rose JE, Mountcastle VB (1952) The thalamic tactile region in rabbit and cat. *J Comp Neurol 97*:441–490.

Rosén I (1969) Afferent connexions to group I activated cells in the main cuneate nucleus of the cat. *J Physiol Lond 205*:209–236.

Rosén I, Sjolund B (1973) Organization of group I activated cells in the main and external cuneate nuclei of the cat: identification of muscle receptors. *Exp Brain Res 16*:221–237.

Rotter A, Birdsall NJM, Burgen ASV, Field PM, Hulme EC, Raisman G (1979) Muscarinic receptors in the central nervous system of the rat. I Technique for autoradiographic localization of the binding of [3H] propylbenzilycholine mustard and its distribution in the forebrain. *Brain Res 1*:141–166.

Rouiller EM, Colomb E, Capt M, de Ribaupierre F (1985) Projections of the reticular complex of the thalamus onto physiologically characterized regions of the medial geniculate body. *Neurosci Lett 53*:227–232.

Rouiller EM, de Ribaupierre F (1985) Origin of afferents to physiologically defined regions of the medial geniculate body of the cat: ventral and dorsal divisions. *Hear Res 19*:97–114.

Rouiller EM, de Ribaupierre F (1990) Arborization of corticothalamic axons in the auditory thalamus of the cat: a PHA-L tracing study. *Neurosci Lett 108*:29–35.

Rouiller EM, Liang F, Babalian A, Moret V, Wiesendanger M (1994) Cerebellothalamocortical and pallidothalamocortical projections to the primary and supplementary motor cortical areas: A multiple tracing study in macaque monkeys. *J Comp Neurol 345*:185–213.

Rouiller EM, Liang FY, Moret V, Wiesendanger M (1991) Patterns of corticothalamic terminations following injection of *Phaseolus vulgaris* leukoagglutinin (PHA-L) in the sensorimotor cortex of the rat. *Neurosci Lett 125*:93–97.

Rouiller EM, Welker E (1991) Morphology of corticothalamic terminals arising from the auditory cortex of the rat: A *Phaseolus vulgaris*-leucoagglutinin (PHA-L) tracing study. *Hear Res 56*:179–190.

Royce GJ (1978a) Autoradiographic evidence for a discontinuous projection to the caudate nucleus from the centromedian nucleus in the cat. *Brain Res 146*:145–150.

Royce GJ (1978b) Cells of origin of subcortical afferents to the caudate nucleus: a horseradish peroxidase study. *Brain Res 153*:465–476.

Royce GJ (1982) Cells of origin of corticothalamic projections upon the centromedian and parafascicular nuclei in the cat. *Brain Res 258*:11–22.

Royce GJ (1983) Cortical neurons with collateral projections to both the caudate nucleus and the centromedian-parafascicular thalamic complex: a fluorescent retrograde double labeling study in the cat. *Exp Brain Res 50*:157–165.

Royce GJ, Bromley S, Gracco C, Beckstead RM (1989) Thalamocortical connections of the rostral intralaminar nuclei: an autoradiographic analysis in the cat. *J Comp Neurol 288*:555–582.
Royce GJ, Bromley S, Gracco C (1991) Subcortical projections to the centromedian and parafascicular thalamic nuclei in the cat. *J Comp Neurol 306*:129–155.
Royce GJ, Mourey RJ (1985) Efferent connections of the centromedian and parafascicular thalamic nuclei: an autoradiographic investigation in the cat. *J Comp Neurol 235*:277–300.
Ruat M, Traiffort E, Bouthenet ML, Schwartz JC, Hirschfield J, Buschauer A, Schunack W (1990) Reversible and irreversible labeling and autoradiographic localization of cerebral histamine H_2 receptor using {^{125}I] iodinated probes. *Proc Natl Acad Sci USA 87*:1658–1662.
Russchen FT (1982) Amygdalopetal projections in the cat. II. Subcortical afferent connections. A study with retrograde tracing techinques. *J Comp Neurol 207*:157–176.
Russchen FT, Amaral DG, Price JL (1987) The afferent input to the magnocellular division of the mediodorsal thalamic nucleus in the monkey, *Macaca fascicularis. J Comp Neurol 256*:175–210.
Russchen FT, Bakst I, Amaral DG, Price JL (1985) The amygdalostriatal projections in the monkey. An anterograde tracing study. *Brain Res 329*:241–257.
Russchen FT, Amaral DG, Price JL (1987) The afferent input to the magnocellular division of the mediodorsal thalamic nucleus in the monkey, *Macaca fascicularis. J Comp Neurol 256*:175–210.
Rustioni A, Battaglia G, de Biasi S, Giuffrida R (1988) Neuromediators in somatosensory thalamus: an immunocytochemical overview. In: Cellular Thalamic Mechanisms (Bentivoglio M, Spreafico R eds), pp 271–296. Amsterdam: Elsevier
Rustioni A, Schmechel DE, Spreafico R, Cheema S, Cuénod M (1983) Excitatory and inhibitory amino acid putative neurotransmitters in the ventralis posterior complex: an autoradiographic and immunocytochemical study in rats and cats. In: Somatosensory Integration in the Thalamus (Macchi G, Rustioni A, Spreafico R eds), Amsterdam: Elsevier.
Rye DB, Saper CB, Lee HJ, Wainer BH (1987) Pedunculopontine tegmental nucleus of the rat: cytoarchitecture, cytochemistry, and some extrapyramidal connections of the mesopontine tegmentum. *J Comp Neurol 259*:483–528.
Sadikot AF, Parent A, Francois C (1992a) Efferent connections of the centromedian and parafascicular thalamic nuclei in the squirrel monkey: A PHA-L study of subcortical projections. *J Comp Neurol 315*:137–159.
Sadikot AF, Parent A, Smith Y, Bolam JP (1992b) Efferent connections of the centromedian and parafascicular thalamic nuclei in the squirrel monkey: A light and electron microscopic study of the thalamostriatal projection in relation to striatal heterogeneity. *J Comp Neurol 320*:228–242.
Saini KD, Garey LJ (1981) Morphology of neurons in the lateral geniculate nucleus of the monkey–a Golgi study. *Exp Brain Res 42*:235–248.
Sakai ST, Inase M, Tanji J (1996) Comparison of cerebellothalamic and pallidothalamic projections in the monkey (*Macaca fuscata*): a double anterograde labeling study. *J Comp Neurol 368*:215–228.
Sakurai SY, Penney JB, Young AB (1993) Regionally distinct N-methyl-D-aspartate receptors distinguished by quantitative autoradiography of [3H]MK-801 binding in rat brain. *J Neurochem 60*:1344–53.
Sales N, Martres MP, Bouthenet ML, Schwartz JC (1989) Ontogeny of dopaminergic D-2 receptors in the rat nervous system: characterization and detailed autoradiographic mapping with [125I]iodosulpride. *Neurosci 28*:673–700.
Salt TE (1986a) Mediation of thalamic sensory input by both NMDA receptors and non-NMDA receptors. *Nature 322*:263–265.
Salt TE (1986b) Physiologically evoked inhibition mediated by y-amino-butyric acid (GABA) in the rat ventrobasal thalamus (VB). *J Physiol Lond 382*:91.
Salt TE (1987) Excitatory amino acid receptors and synaptic transmission in the rat ventrobasal thalamus. *J Physiol Lond 391*:499–510.
Salt TE (1989) Gamma-aminobutyric acid and afferent inhibition in the cat and rat ventrobasal thalamus. *Neurosci 28*:17–26.
Salt TE, Eaton SA (1989) Function of non-NMDA receptors and NMDA receptors in synaptic responses to natural somatosensory stimulation in the ventrobasal thalamus. *Exp Brain Res 77*:646–652.
Salt TE, Eaton SA (1990a) Postsynaptic potentials evoked in ventrobasal thalamus neurones by natural sensory stimuli. *Neurosci Lett 114*:295–299.
Salt TE, Eaton SA (1991a) Sensory excitatory postsynaptic potentials mediated by NMDA and non-NMDA receptors in the thalamus in vivo. *Eur J Neurosci 3*:296–300.
Salt TE, Eaton SA (1991b) Excitatory actions of the metabotropic excitatory amino acid receptor agonist, trans-1-aminocyclopentane-1,3-dicarboxylate (t-ACPD), on rat thalamic neurons In Vivo. *Eur J Neurosci 3*:1104–1111.

Salt TE, Eaton SA (1996) Functions of ionotropic and metabotropic glutamate receptors in sensory transmission in the mammalian thalamus. *Prog Neurobiol 48*:55–72

Sano K (1977) Intralaminar thalamotomy (thalamolaminotomy) and postero-medial hypothalamotomy in the treatment of intractable pain. *Prog Neuro Surg 8*:50–103.

Saper CB (1990) Hypothalamus. In: The Human Nervous System (Paxinos, G ed), pp 389–413. San Diego: Academic Press.

Saper CB, Loewy AF (1980) Efferent connections of the parabrachial nucleus in the rat. *Brain Res 197*:291–318.

Saporta S, Kruger L (1977) The organization of thalamocortical relay neurons in the rat ventrobasal complex studied by the retrograde transport of horseradish peroxidase. *J Comp Neurol 174*:187–208.

Sato K, Kiyama H, Tohyama M (1993) The differential expression patterns of messenger RNAs encoding non-N-methyl-D-aspartate glutamate receptor subunits (GluR1–4) in the rat brain. *Neurosci 52*:515–540.

Saugstad JA, Kinzie JM, Mulvihill ER, Segerson TP, Westbrook GL (1994) Cloning and expression of a new member of the L-2-amino-4-phosphobutyric acid-sensitive class of metabotropic glutamate receptors. *Mol Pharmacol 45*:367–372.

Savasta M, Dubois A, Scatton B (1986) Autoradiographic localization of D1 dopamine receptors in the rat brain with [3H]SCH 23390. *Brain Res 375*:291–301.

Savasta M, Palacios JM, Mengod G (1988) Regional localization of the mRNA coding for the neuropetide cholecystokinin in the rat brain studied by in situhybridization. *Neurosci Lett 93*:132–138.

Scharfman HE, Lu S-M, Guido W, Adams PR, Sherman SM (1990) N-methyl-D-aspartate receptors contribute to excitatory postsynaptic potentials of cat lateral geniculate neurons recorded in thalamic slices. *Proc Natl Acad Sci USA 87*:4548–4552.

Scheibel ME, Scheibel A (1966) Patterns of organization in specific and nonspecific thalamic fields. In: The Thalamus (Purpura DP, Yahr MD eds), pp 13–46. New York: Columbia Univ. Press.

Scheibel ME, Scheibel AB (1966a) The organization of the nucleus reticularis thalami a Golgi study. *Brain Res 1*:43–62.

Scheibel ME, Scheibel AB (1966b) The organization of the ventral anterior nucleus of the thalamus A Golgi study. *Brain Res 1*:250–268.

Schell GR, Strick PL (1984) The origin of thalamic inputs to the arcuate premotor and supplementary motor areas. *J Neurosci 4*:539–560.

Schiller PH (1982) Central connections of the retinal on and off pathways. *Nature 297*:580

Schiller PH, Malpeli JG (1978) Functional specificity of lateral geniculate nucleus laminae of the rhesus monkey. *J Neurophysiol 41*:788–797.

Schlag J, Schlag-Rey M (1984) Visuomotor functions of central thalamus in monkey. II. Unit activity related to visual events, targeting, and fixation. *J Neurophysiol 51*:1175–1195.

Schmahmann JD, Pandya DN (1990) Anatomical investigation of projections from thalamus to posterior parietal cortex in the rhesus monkey: a WGA-HRP and fluorescent tracer study. *J Comp Neurol 295*:299–326.

Schott B, Mauguière F, Lawrent B, Serderat O, Fischer C (1980) L'amnesie thalamique. *Rev Neurol 136*:117–130.

Schulz DW, Loring RH, Aizenman E, Zigmond RE (1991) Autoradiographic localization of putative nicotinic receptors in the rat brain using $^{125}I_I$-Neuronal bungarotoxin. *J Neurosci 11*:287–297.

Schwartz M, Dekker JJ, Goldman-Rakic PS (1991) Dual mode of corticothalamic synaptic termination in the mediodorsal nucleus of the rhesus monkey. *J Comp Neurol 309*:289–304.

Segal M, Dudai Y, Amsterdam A (1978) Distribution of an α-bungarotoxin-binding cholinergic nicotinic receptor in rat brain. *Brain Res 148*:105–120.

Selemon LD, Goldman-Rakic PS (1985) Longitudinal topography and interdigitation of corticostriatal projections in the rhesus monkey. *J Neurosci 5*:776–794.

Semple MN, Aitkin LM (1980) Physiology of pathway from dorsal cochlear nucleus to inferior colliculus revealed by electrical and auditory stimulation. *Exp Brain Res 41*:19–28.

Senatorov VV, Trudeau VL, Hu B (1995) Expression of cholecystokinin mRNA in corticothalamic projecting neurons: a combined fluorescence in situ hybridization and retrograde tracing study in the ventrolateral thalamus of the rat. *Mol Brain Res 3*:87–96.

Shapley R, Perry VH (1986) Cat and monkey retinal ganglion cells and their visual functional roles. *Trends Neurosci 9*:229–235.

Shaw C, Cynader M (1986) Laminar distribution of receptors in monkey (*Macaca fascicularis*) geniculostriate system. *J Comp Neurol 248*:301–312.

Sherman SM (1988) Functional organization of the cat's lateral geniculate nucleus. In: Cellular Thalamic Mechanisms (Bentivoglio M, Spreafico R eds), pp 163–183. Amsterdam: Elsevier.

Sherman SM, Wilson JR, Kaas JH, Webb SV (1976) X- and Y-cells in the dorsal lateral geniculate nucleus of the owl monkey (*Aotus trivirgatus*). *Science 192*:475–476.

Shigemoto R, Nakanishi S, Mizuno N (1992) Distribution of the mRNA for a metabotropic glutamate receptor (mGluR1) in the central nervous system: An in situ hybridization study in adult and developing rat. *J Comp Neurol 322*:1221–1135.

Shigenaga Y, Nakatani Z, Nishimori T, Suemune S, Kuroda R, Matano S (1983) The cells of onigin of cat trigeminothalamic projections: especially in the caudal medulla. *Brain Res 277*:201–222.

Shimazu H, Yanagisawa N, Garoutte B (1965) Cortico-pyramidal influences on thalamic somatosensory transmission in the cat. *Jpn J Physiol 15*:101–124.

Shin HC, Chapin JK (1990) Mapping the effects of motor cortex stimulation on somatosensory relay neurons in the rat thalamus: direct responses and afferent modulation. *Brain Res Bull 24*:257–265.

Shinohara K, Tominaga K, Fukuhara C, Otori Y, Inouye S-IT (1993) Processing of photic information within the intergeniculate leaflet of the lateral geniculate body: assessed by neuropeptide Y immunoreactivity in the suprachiasmatic nucleus of rats. *Neurosci 56*:813–822.

Shipp S, Zeki S (1985) Segregation of pathways leading from area V2 to areas V4 and V5 of macaque monkey visual cortex. *Nature 315*:322–325.

Shiroyama T, Kayahara T, Yasui Y, Nomura J, Nakano K (1995) The vestibular nuclei of the rat project to the lateral part of the thalamic parafascicular nucleus (centromedian nucleus in primates). *Brain Res 704*:130–134.

Shook BL, Schlag-Rey M, Schlag J (1991) Primate supplementary eye field. II. Comparative aspects of connections with the thalamus, corpus striatum, and related forebrain nuclei. *J Comp Neurol 307*:562–583.

Shosaku A (1985) A comparison of receptive field properties of vibrissa neurons between the rat thalamic reticular and ventrobasal nuclei. *Brain Res 347*:36–40.

Shosaku A, Kayama Y, Sumitomo I (1984) Somatotopic organization in the rat thalamic reticular nucleus. *Brain Res 311*:57–64.

Shosaku A, Kayama Y, Sumitomo I, Sugitani M, Iwama K (1989) Analysis of recurrent inhibitory circuit in rat thalamus: Neurophysiology of the thalamic reticular nucleus. *Prog Neurobiol 32*:77–102.

Shosaku A, Sumitomo I (1983) Auditory neurons in the rat thalamic nucleus. *Exp Brain Res 49*:432–442.

Sidibé M, Beavan M, Bolam JP (1997) Efferent connections of the internal globus pallidus in the squirrel monkey: I. topography and synaptic organization of the pallidothalamic projection. *J Comp Neurol 382*:323–347.

Sieghart W (1991) Heterogeneity of $GABA_A$-benzodiazepine receptors. *Biochem Soc Trans 19*:129–132.

Sieghart W (1992) $GABA_A$ receptors: ligand-gated Cl- ion channels modulated by multiple drug-binding sites. *Trends Pharmacol Sci13*:446–50.

Sillito AM, Cudeiro J, Murphy PC (1993) Orientation sensitive elements in the corticofugal influence on centre-surround interactions in the dorsal lateral geniculate nucleus. *Exp Brain Res 93*:6–16.

Sillito AM, Jones HE, Gerstein GL, West DC (1994) Feature-linked synchronization of thalamic relay cell firing induced by feedback from the visual cortex. *Nature 369*:479–482.

Sillito AM, Kemp JA (1983a) Cholinergic modulation of the functional organization of the cat visual cortex. *Brain Res 289*:143–156.

Sillito AM, Kemp JA (1983b) The influence of GABAergic inhibitory processes on the receptive field structure of X and Y cells in cat dorsal lateral geniculate nucleus (dLGN). *Brain Res 277*:55–62.

Sillito AM, Kemp JA, Barardi N (1983) The cholinergic influence on the function of the cat dorsal lateral geniculate nucleus (dLGN). *Brain Res 280*:299–308.

Sillito AM, Murphy PC, Salt TE (1990a) The contribution of the non-N-methyl-D-aspartate group of excitatory amino acid receptors to retinogeniculate transmission in the cat. *Neurosci 34*:273–280.

Sillito AM, Murphy PC, Salt TE, Moody CI (1990b) Dependence of retinogeniculate transmission in cat on NMDA receptors. *J Neurophysiol 63*:347–355.

Silveira LCL, Perry VH (1991) The topography of magnocellular projecting ganglion cells (M-ganglion cells) in the primate retina. *Neurosci 40*:217–238.

Sim LJ, Joseph SA (1992) Serotonin and substance P afferents to parafascicular and central medial nuclei. *Peptides 13*:171–176.

Singer W (1977a) Effects of monocular deprivation on excitatory and inhibitory pathways in cat striate cortex. *Exp Brain Res 30*:25–41.

Singer W (1977b) Control of thalamic transmission by corticofugal and ascending reticular pathways in the visual. *Physiol Rev 57*:386–420.

Siwek D, Pandya D (1991) Prefrontal Projections to the mediodorsal nucleus of the thalamus in the rhesus monkey. *J Comp Neurol 312*:509–524.

Skofitsch G, Jacobowitz DM (1985) Galanin-like immunoreactivity in capsaicin sensitive sensory neurons and ganglia. *Brain Res Bull 15*:191–195.

Smith ML, Hale BD, Booze RM (1994) Calbindin-D28k immunoreactivity within the cholinergic and GABAergic projection neurons of the basal forebrain. *Exp Neurol 130*:230–236.

Smith RL (1973) The ascending fiber projections from the principal sensory trigeminal nucleus in the rat. *J Comp Neurol 148*:423–446.

Smith Y, Bennett BD, Bolam JP, Parent A, Sadikot AF (1994) Synaptic relationships between dopaminergic afferents and cortical or thalamic input in the sensorimotor territory of the striatum in monkey. *J Comp Neurol 344*:1–19.

Smith Y, Paré D, Deschênes M, Parent A, Steriade M (1988) Cholinergic and non-cholinergic projections from the upper brainstem core to the visual thalamus in the cat. *Exp Brain Res 70*;1:166–180.

Smith Y, Parent A (1986) Differential connections of caudate nucleus and putamen in the squirrel monkey (*Saimiri sciureus*). *Neurosci 18*:347–371.

Smith Y, Seguela P, Parent A (1987) Distribution of GABA-immunoreactive neurons in the thalamus of the squirrel monkey (*Saimiri sciureus*). *Neurosci 22*:579–591.

So YT, Shapley R (1981) Spatial tuning of cells in and around lateral geniculate nucleus of the cat: X and Y relay cells and perigeniculate interneurons. *J Neurophysiol 45*:107–120.

Sofroniew MV, Priestley JV, Consolazione A, Eckenstein F, Cuello AC (1985) Cholinergic projections from the midbrain and pons to the thalamus in the rat, identified by combined retrograde tracing and choline acetyltransferase immunohistochemistry. *Brain Res 329*:213–223.

Soltesz I, Crunelli V (1992a) A role for low-frequency, rhythmic synaptic potentials in the synchronization of cat thalamocortical cells. *J Physiol Lond 457*:257–276.

Soltesz I, Crunelli V (1992b) $GABA_A$ and pre- and post-synaptic $GABA_B$ receptor-mediated responses in the lateral geniculate nucleus. In: Progress in Brain Research vol. 90 (Mize RR, Mare RE, Sillito AM eds), pp 151–169. Amsterdam: Elsevier.

Soltesz I, Haby M, Leresche N, Crunelli V (1988) The $GABA_B$ antagonist phaclofen inhibits the late K^+-dependent IPSP in cat and rat thalamic and hippocampal neurones. *Brain Res 448*:351–354.

Soltesz I, Lightowler S, Leresche N, Crunelli V (1989a) Optic tract stimulation evokes GABAA but not $GABA_B$ IPSPs in the rat ventral lateral geniculate nucleus. *Brain Res 479*:49–55.

Soltesz I, Lightowler S, Leresche N, Crunelli V (1989b) On the properties and origin of the $GABA_B$ inhibitory postsynaptic potential recorded in morphologically identified projection cells of the cat dorsal lateral geniculate nucleus. *Neurosci 33*:23–33.

Sommer B, Keinänen K, Verdoorn TA, Wisden W, Burnashev N, Herb A, Kohler M, Takagi T, Sakmann B, Seeburg PH (1990) Flip and flop: a cell-specific functional switch in glutamate-operated channels of the CNS. *Science 249*:1580–1585.

Spatz WB, Tigges J (1973) Studies on the visual area MT in primates. II. Projection fibers to subcortical stuctures. *Brain Res 61*:374–378.

Spatz WB, Tigges J, Tigges M (1970) Subcortical projections, cortical associations, and some intrinsic interlaminar connections of the striate cortex in squirrel monkey (Saimiri). *J Comp Neurol 140*:155–174.

Spear PD, Smith DC, Williams LL (1977) Visual receptive-field properties of single neurons in cat's ventral lateral geniculate nucleus. *J Neurophysiol 40*:390–409.

Spiegel EA, Wycis HT, Szekely EG, Gildenberg PL (1966) Medial and basal thalamotomy in so-called intractable pain. In: Pain (Knighton RS, Dumke PR eds), pp 503–517. London: Churchill.

Spreafico R, Amadeo A, Angoscini P, Panzica F, Battagilia G (1993a) Branching projections from mesopontine nuclei to the nucleus reticularis and related thalamic nuclei: a double labelling study in the rat. *J Comp Neurol 336*:481–492.

Spreafico R, de Biasi S, Amadeo A, De Blas AL (1993b) $GABA_A$-receptor immunoreactivity in the rat dorsal thalamus: An ultrastructural investigation. *Neurosci Lett 158*:232–236.

Spreafico R, Frassoni C, Arcelli P, Battaglia G, Wenthold RJ, de Biasi S (1994a) Distribution of AMPA selective glutamate receptors in the thalamus of adult rats and during postnatal development. A light and ultrastructural immunocytochemical study. *Dev Brain Res 82*:231–244.

Spreafico R, Frassoni C, Arcelli P, de Biasi S (1994b) GABAergic interneurons in the somatosensory thalamus of the guinea-pig: a llight and ultrastructural immunocytochemical investigation. *Neurosci 59*:961–974.

Spreafico R, Frassoni C, Regondi MC, Arcelli P, de Biasi S (1993c) Interneurons in the mammalian thalamus: a marker of species? In: Thalamic Networks for Relay and Modulation (Minciacchi D, Molinari M, Macchi G, Jones EG eds), Oxford: Pergamon.

Spreafico R, Schmechel DE, Ellis LC, Rustioni A (1983) Cortical relay neurons and interneurons in the N ventralis posterolateralis of cats: a horseradish peroxidase, electron microscopic, Golgi and immunocytochemical study. *Neurosci 9*:491–510.

Squire LR, Moore RY (1979) Dorsal thalamic lesion in a noted case of human memory dysfunction. *Ann Neurol 6*:503–506.

Standage GP, Benevento LA (1983) The organization of connections between the pulvinar and visual area MT in the macaque monkey. *Brain Res 262*:288–294.

Stanford LR, Friedlander MJ, Sherman SM (1981) Morphology of physiologically identified W cells in the C-laminae of the cat's lateral geniculate nucleus. *J Neurosci 1*:578–584.

Stanford LR, Friedlander MJ, Sherman MS (1983) Morphological and physiological properties of geniculate W-cells of the cat: a comparison with X- and Y-cells. *J Neurophysiol 50*:582–608.

Stanton GB, Cruce WLR, Goldberg ME, Robinson DL (1977) Some ipsilateral projections to areas PF and PG of the inferior parietal lobule in monkeys. *Neurosci Lett 6*:243–250.

Stanton GB, Goldberg ME, Bruce CJ (1988) Frontal eye field efferents in the macaque monkey: I. subcortical pathways and topography of striatal and thalamic terminal fields. *J Comp Neurol 271*:473–492.

Starr MS, Kilpatrick IC (1981) Distribution of gamma-aminobutyrate in the rat thalamus: specific decreases in thalamic gamma-aminobutyrate following lesion or electrical stimulation of the substantia nigra. *Neurosci 6*:1095–1104.

Steele GE, Weller RE (1993) Subcortical connections of subdivisions of inferior temporal cortex in squirrel monkeys. *Vis Neurosci 10*:563–583.

Steinbusch HWM (1981) Distribution of serotonin-immunoreactivity in the central nervous system of the rat-cell bodies and terminals. *Neurosci 6*:557–618.

Stephan H, Frahm H, Baron G (1981) New and revised data on volumes of brain structures in insectivores and primates. *Folia Primatol 35*:1–29.

Stepniewska I, Preuss TM, Kaas JH (1994a) Thalamic connections of the primary motor cortex (M1) of owl monkeys. *J Comp Neurol 349*:558–582.

Stepniewska I, Preuss TM, Kaas JH (1994b) Architectonic subdivisions of the motor thalamus of owl monkeys: Nissl, acetylcholinesterase, and cytochrome oxidase patterns. *J Comp Neurol 349*:536–557.

Steriade M, Amzica F, Nuñez A (1993a) Cholinergic and noradrenergic modulation of the slow (approx. 0.3 Hz) oscillation in neocortical cells. *J Neurophysiol 70*:1385–1400.

Steriade M, Apostol V, Oakson G (1971) Control of unitary activities in cerebellothalamic pathways during wakefulness and synchronized sleep. *J Neurophysiol 34*:384–413.

Steriade M, Contreras D (1995) Relations between cortical and thalamic cellular events during transition from sleep patterns to paroxysmal activity. *J Neurosci 15*:623–642.

Steriade M, Contreras D, Curró Dossi R, Nuñez A (1993b) The Slow (<1 Hz) oscillation in reticular thalamic and thalamocortical neurons: scenario of sleep rhythm generation in interacting thalamic and neocortical networks. *J Neurosci 13*:3284–3299.

Steriade M, Curró Dossi R, Contreras D (1993c) Electrophysiological properties of intralaminar thalamocortical cells discharging rhythmic (approximately 40 Hz) spikebursts at approximately 1000 Hz during waking and rapid eye movement sleep. *Neurosci 56*:1–10.

Steriade M, Datta S, Paré D, Oakson G, Dossi RC (1990a) Neuronal activities in brain-stem cholinergic nuclei related to tonic activation processes in thalamocortical systems. *J Neurosci 10*:241–259.

Steriade M, Deschênes M (1984) The thalamus as a neuronal oscillator. *Brain Res Rev 8*:1–63.

Steriade M, Deschênes M, Domich L, Mulle C (1985) Abolition of spindle oscillations in thalamic neurons disconnected from nucleus reticularis thalami. *J Neurophysiol 54*:1473–1497.

Steriade M, Deschênes M, Oakson G (1987a) The deafferented reticular thalamic nucleus generates spindle rhythmicity. *J Neurophysiol 57*:260–273.

Steriade M, Domich L, Oakson G (1986) Reticularis thalami neurons revisited: activity changes during shifts in states of vigilance. *J Neurosci 6*:68–81.

Steriade M, Glenn LL (1982) Neocortical and caudate projections of intralaminar thalamic neurons and their synaptic excitation from midbrain reticular core. *J Neurophysiol 48*:352–371.

Steriade M, Gloor P, Llinás RR, Lopes da Silva FH, Mesulam M-M (1990b) Basic mechanisms of cerebral rhythmic activities. *Electroencephalogr Clin Neurophysiol 76*:481–508.

Steriade M, Jones EG, Llinás RR (1990c) Thalamic Oscillations and Signalling. J Wiley & Sons

Steriade M, Jones EG, McCormick DA (1997a) Thalamus. Vol 1 Oxford Pergamon

Steriade M, McCarley RW (1990) Brainstem control of wakefulness and sleep. New York: Plenum Press.

Steriade M, McCormick DA, Sejnowski TJ (1993d) Thalamocortical oscillations in the sleeping and aroused brain. *Science 262*:679–685.

Steriade M, Nuñez A, Amzica F (1993e) Intracellular analysis of relations between the slow (<1Hz) neocortical oscillation and other sleep rhythms of the elctroencephalogram. *J Neurosci 13*:3266–3283.

Steriade M, Nuñez A, Amzica F (1993f) A Novel Slow (<1 Hz) Oscillation of neocortical neurons *in vivo*: depolarizing and hyperpolarizing components. *J Neurosci 13*:3252–3265.

Steriade M, Paré D, Bouhassira D, Deschênes M, Oakson G (1989) Phasic activation of lateral geniculate and perigeniculate thalamic neurons during sleep with ponto-geniculo-occipital waves. *J Neurosci 9*:2214–2229.

Steriade M, Paré D, Datta S, Oakson G, Dossi RC (1990d) Different cellular types in mesopontine cholinergic nuclei related to ponto-geniculo-occipital waves. *J Neurosci 10*:2560–2579.

Steriade M, Paré D, Parent A, Smith Y (1988) Projections of cholinergic and noncholinergic neurons of the brainstem core to relay and asociational thalamic nuclei in the cat and macaque monkey. *Neurosci 25*:47–67.

Steriade M, Parent A, Hada J (1984) Thalamic projections of nucleus reticularis thalami: a study using retrograde transport of horseradish peroxidase and double fluorescent tracers. *J Comp Neurol 229*:531–547.

Steriade M, Parent A, Paré D, Smith Y (1987b) Cholinergic and non-cholinergic neurons of cat basal forebrain project to reticular and mediodorsal thalamic nuclei. *Brain Res 408*:372–376.

Steriade M, Wyzinski P (1972) Cortically elicited activities in thalamic reticularis neurons. *Brain Res 42*:514–520.

Steriade M, Wyzinski P, Apostol V (1972) Corticofugal projections governing rhythmic thalamic activity. In: Corticothalamic Projections and Sensorimotor Activities (Frigyesi T, Rinvik E, Yahr MD eds), pp 221–272. New York: Raven Press.

Streit P (1980) Selective retrograde labeling indicating the transmitter of neuronal pathways. *J Comp Neurol 191*:429–464.

Strick PL (1975) Multiple sources of thalamic input to the primate motor cortex. *Brain Res 88*:372–377.

Strick PL (1986) The organization of thalamic inputs to the 'premotor' areas. *Prog Brain Res 64*:99–109.

Sucher NJ, Akbarian S, Chi CL, Leclerc CL, Awobuluyi M, Deitcher DL, Wu MK, Yuan JP, Jones EG, Lipton SA (1995) Developmental and regional expression pattern of a novel NMDA receptor-like subunit (NMDAR-L) in the rodent brain. *J Neuosci 15*:6509–6520.

Sugimoto T, Itoh K, Yasui Y, Kaneko T, Mizuno N (1985) Coexistence in neuropeptides in projection neurons of the thalamus in the cat. *Brain Res 347*:381–384.

Sugimoto T, Mizuno N, Itoh K (1981) An autoradiographic study on the terminal distribution of cerebellothalamic fibers in the cat. *Brain Res 215*:29–48.

Sugitani M (1979) Electrophysiological and sensory properties of the thalamic reticular neurones related to somatic sensation in rats. *J Physiol Lond 290*:79–96.

Sumitomo I, Hsiao C-F, Fukuda Y (1988a) Two types of thalamic reticular cells in relation to the two visual thalamocortical systems in the rat. *Brain Res 446*;2:354–362.

Sumitomo I, Sugitaini M, Fukuda Y, Iwama K (1979) Properties of cells responding to visual stimuli in the rat ventral lateral geniculate nucleus. *Exp Neurol 66*:721–736.

Sumitomo I, Takahashi Y, Kayama Y, Ogawa T (1988b) Burst discharges associated with phasic hyperpolarizing oscillations of rat ventrobasal relay neurons. *Brain Res 447*:376–379.

Sur M (1980) Receptive fields of neurons in areas 3b and 1 of somatosensory cortex in the monkey. *Brain Res 198*:465–471.

Sur M, Merzenich MM, Kaas JH (1980) Magnification, receptive-field area, and 'hypercolumn' size in areas 3b and 1 of somatosensory cortex in owl monkeys. *J Neurophysiol 44*:295–311.

Sur M, Sherman SM (1982) Retinogeniculate terminations in cats: morphological differences between X and Y cell axons. *Science 218*:389–391.

Sur M, Wall JT, Kaas JH (1984) Modular distribution of neurons with slowly adapting and rapidly adapting responses in area 3b of somatosensory cortex in monkeys. *J Neurophysiol 51*:724–744.

Surmeier DJ, Honda CN, Willis WD (1986a) Responses of primate spinothalamic neurons to noxious thermal stimulation of glabrous and hairy skin. *J Neurophysiol 56*:328–250.

Surmeier DJ, Honda CN, Willis WD (1986b) Temporal features of the responses of primate spinothalamic neurons to noxious thermal stimulation of hairy and glabrous skin. *J Neurophysiol 56*:351–369.

Surmeier DJ, Honda CN, Willis WD (1988) Natural groupings of primate spinothalamic neurons based on cutaneous stimulation. Physiological and anatomical features. *J Neurophysiol 59*:833–860.

Swanson LW (1987) The Hypothalamus. In: Handbook of Chemical Neuroanatomy Volume 5, Integrated Systems of the CNS, Part I Hypothalamus, Amygdala, Retina. (Björklund A, Hökfelt T, Swanson LW eds), pp 1–277. Amsterdam: Elsevier.

Swanson LW, Cowan WM, Jones EG (1974) An autoradiographic study of the efferent connections of the ventral lateral geniculate nucleus in the albino rat and cat. *J Comp Neurol 156*:143–164.

Swanson LW, Hartman BK (1975) The central adrenergic system: an immunofluorescence study of the location of cell bodies and their efferent connections in the rat utilizing dopamine-B-hydroxylase as a marker. *J Comp Neurol 163*:467–506.

Swanson LW, Simmons DM, Whiting PJ, Lindstrom J (1987) Immunohistochemical localization of neuronal nicotinic receptors in the rodent central nervous system. *J Neurosci 7*:3334–3342.

Symonds LL, Kaas JH (1978) Connections of striate cortex in the prosimian, *Galago senegalensis*. *J Comp Neurol 181*:477–512.

Symonds LL, Rosenquist AC, Edwards SB, Palmer LA (1981) Projections of the pulvinar lateral posterior complex to visual cortical areas in the cat. *Neurosci 6*:1995–2020.

Tai Y, Yi H, Ilinsky IA, Kultas-Ilinsky K (1995) Nucleus reticularis thalami connections with the mediodorsal thalamic nucleus: a light and electron microscopic study in the monkey. *Brain Res Bull 38*:475–488.

Takada M, Itoh K, Yasui Y, Sugimoto T, Mizuno N (1985) Topographical projections from the posterior thalamic regions on the striatum in the cat, with reference to possible tecto-thalamo-striatal connections. *Exp Brain Res 60*:385–396.

Takagi SF (1979) Dual systems for sensory olfactory processing in higher primates. *Trends Neurosci 12*:313–315.

Tanabe T, Iino M, Ooshima Y, Takagi SF (1974) An olfactory area in the frontal lobe. *Brain Res 80*:127–130.

Tanabe T, Iino M, Takagi SF (1975a) Discrimination of odors in olfactory bulb, pyriform-amygdaloid areas, and orbitofrontal cortex of the monkey. *J Neurophysiol 38*:1284–1296.

Tanabe T, Yarita H, Iino M, Ooshima Y, Takagi SF (1975b) An olfactory projection area in orbitofrontal cortex of the monkey. *J Neurophysiol 38*:1269–1283.

Tanabe Y, Nomura A, Masu M, Shigemoto R, Mizuno N, Nakanishi S (1993) Signal transduction, pharmacological properties, and expression patterns of two rat metabotropic glutamate receptors, mGluR3 and mGluR4. *J Neurosci 13*:1372–8.

Tarlov EC, Moore RY (1966) The tecto-thalamic connections in the brain of the rabbit. *J Comp Neurol 126*:403–421.

Tasker RR (1988) Anatomical correlation of detailed sensory mapping of the human thalamus. *Confin Neurol 34*:184

Tasker RR (1989) Stereotactic surgery. In: Textbook of Pain (Wall PD, Melzack R Eds), pp 840–855. New York: Churchill-Livingstone.

Tēbēcis AK (1972a) Cholinergic and non-cholinergic transmission in the medial geniculate nucleus of the cat. *J Physiol Lond 226*:153–172.

Tēbēcis AK (1972b) Cholinergic and non-cholinergic transmission in the medial geniculate nucleus of the cat. *J Physiol Lond 226*:153–172.

Thomson AM (1988a) Inhibitory postsynaptic potentials evoked in thalamic neurons by stimulation of the reticularis nucleus evoke slow spikes in isolated rat brain slices–I. *Neurosci 25*:491–502.

Thomson AM (1988b) Biphasic responses of thalamic neurons to GABA in isolated rat brain slices–II. *Neurosci 25*:503–512.

Tiberi M, Jarvie KR, Silvia C, Falardeau P, Gingrich JA, Godinot N, Bertrand L, Yang-Feng TL, Fremeau RT Jr, Caron MG (1991) Cloning, molecular characterization, and chromosomal assignment of a gene encoding a second D1 dopamine receptor subtype: differential expression pattern in rat brain compared with the D1A receptor. *Proc Natl Acad Sci USA 88*:7491–5.

Tigges J, Bos J, Tigges M (1977) An autoradiographic investigation of the subcortical visual system in chimpanzee. *J Comp Neurol 172*:367–80.

Tigges J, Tigges M, Cross NA, McBride RL, Letbetter WD, Anschel S (1982) Subcortical structures projecting to visual cortical areas in squirrel monkey. *J Comp Neurol 209*:29–40.

Tighelet B, Huntsman MM, Hashikawa T, Murray KD, Isackson PJ, Jones EG (1998) Cell-specific expression of type II calcium/calmodulin dependent protein kinase isoforms and glutamate receptors in normal and visually deprived lateral geniculate nucleus of monkeys. *J Comp Neurol 390*:278–296.

Tokunaga A, Akert K, Garey LJ, Otani K (1981) Primary and secondary subcortical projections of the monkey visual system: an autoradiographic study. *Brain Res 214*:137–143.

Tömböl T, Bentivoglio M, Macchi G (1990) Neuronal cell types in the thalamuc intralaminar central lateral nucleus of the cat. *Exp Brain Res 81*:491–499.

Tracey DJ, Asanuma C, Jones EG, Porter R (1980) Thalamic relay to motor cortex: afferent pathways from brain stem, cerebellum, and spinal cord in monkeys. *J Neurophysiol 44*:532–554.

Tremblay N, Bushnell MC, Duncan DH (1993) Thalamic VPM nucleus in the behaving monkey. II. Response to air-puff stimulation during discrimination and attention tasks. *J Neurophysiol 69*:753–763.

Trojanowski JQ, Jacobson S (1976) Areal and laminar distribution of some pulvinar cortical efferents in rhesus monkey. *J Comp Neurol 169*:371–391.

Tseng G-F, Royce GJ (1986) A Golgi and ultrastructural analysis of the centromediam nucleus of the cat. *J Comp Neurol 245*:359–378.

Tsubokawa T, Moriyasu N (1975) Follow-up results of centre median thalamotomy for relief of intractable pain. A method of evaluating the effectiveness during operation. *Confin Neurol 37*:280–284.

Tsumoto T, Nakamura S, Iwana K (1975) Pyramidal tract control over cutaneous and kinestesic sensory transmission in the cat thalamus. *Exp Brain Res 22*:281–294.

Tsumoto T, Creutzfeldt OD, Legendy CR (1978) Functional organization of the corticofugal system from visual cortex to lateral geniculate nucleus in the cat. (With an appendix on geniculocortical mono-synaptic connections.) *Exp Brain Res 32*:345–364.

Tsumoto T (1990) Excitatory amino acid transmitters and their receptors in neural circuits of the cerebral neocortex. *Neurosci Res 9*:79–102.

Turgeon SM, Albin RL (1994) Postnatal ontogeny of $GABA_B$ binding in rat brain. *Neurosci 62*:601–614.

Ueki A, Uno M, Anderson M, Yoshida M (1977b) Monosynaptic inhibition of thalamic neurons produced by stimulation of substantia nigra. *Experientia 33*:1480–1481.

Ulrich D, Huguenard JR (1995) Purinergic inhibition of GABA and glutamate release in the thalamus: implications for thalamic network activity. *Neuron 15*:909–18.

Uhlrich DJ, Cucchiaro JB, Sherman SM (1988) The projection of individual axons from the parabrachial region of the brain stem to the dorsal lateral geniculate nucleus in the cat. *J Neurosci 8*:4565–4575.

Uhlrich DJ, Cucchiaro JB, Humphrey AL, Sherman SM (1991) Morphology and axonal projection patterns of individual neurons in the cat perigeniculate nucleus. *J Neurophysiol 65*:1528–1541.

Uhlrich DJ, Manning KA, Pienkowski TP (1993) The histaminergic innervation of the lateral geniculate complex in the cat. Vis *Neurosci 10*:225–235.

Ullan J (1985) Cortical topography of thalamic intralaminar nuclei. *Brain Res 328*:333–340.

Ulrich D, Huguenard JR (1995) Purinergic inhibition of GABA and glutamate release in the thalamus: implications for thalamic network activity. *Neuron 15*:909–918.

Ulrich D, Huguenard JR (1996) $GABA_B$ receptor-mediated responses in GABAergic projection neurones of rat nucleus reticularis thalami in vitro. *J Physiol Lond 493*:845–854.

Ungerleider LG, Desimone R, Galkin TW, Mishkin M (1984) Subcortical projections of area MT in the macaque. *J Comp Neurol 223*:368–386.

Ungerleider LG, Galkin TW, Mishkin M (1983) Visuotopic organization of projections from striate cortex to inferior and lateral pulvinar in rhesus monkey. *J Comp Neurol 217*:137–157.

Uno M, Ozawa N, Yoshida M (1978) The mode of pallido-thalamic transmission investigated with intracellular recording from cat thalamus. *Exp Brain Res 33*:493–507.

Uno M, Yoshida M (1975a) Monosynaptic inhibition of thalamic neurons produced by stimulation of the pallidal nucleus in rats. *Brain Res 99*:377–380.

Updyke BV (1977) Topographic organization of the projections from cortical areas 17, 18, and 19 onto the thalamus, pretectum and superior colliculus in the cat. *J Comp Neur 173*:81–122.

Van Hoesen GW, Yeterian EH, Lavizzo-Mourey R (1981) Widespread corticostriate projections from temporal cortex of the rhesus monkey. *J Comp Neurol 199*:205–219.

Van der Maelen CP, Kitai ST (1980) Intracellular analysis of synaptic potentials in rat neostriatum following stimulation of the cerebral cortex, thalamus, and substantia nigra. *Brain Res Bull 5*:725–733.

Varela FJ, Singer W (1987) Neuronal dynamics in the visual corticothalamic pathway revealed through binocular rivalry. *Exp Brain Res 66*:10–20.

Velayos JL, Casa-Puig R, Reinoso-Suárez F (1993a) Laminar organization of the cortical projections to the intralaminar and medial thalamic nuclei in the cat. In: Thalamic Networks for Relay and Modulation (Minciacchi D, Molinari M, Jones EG eds), pp 185–196. Oxford: Pergamon.

Velayos JL, Jimenez-Castellanos J, Jr., Reinoso-Suárez F (1989) Topographical organization of the projections from the reticular thalamic nucleus to the intralaminar and medial thalamic nuclei in the cat. *J Comp Neurol 279*:457–469.

Velayos JL, Reinoso-Suárez F (1982) Topographic organization of the brainstem afferents to the mediodorsal thalamic nucleus. *J Comp Neurol 206*:17–27.

Victor M, Adams R, Collins G (1971) The Wernicke-Korsakoff Syndrome. Philadelphia: Davis.

Vilaro MT, Wiederhold KH, Palacios JM, Mengod G (1992a) Muscarinic M2 receptor mRNA expression and receptor binding in cholinergic and non-cholinergic cells in the rat brain: a correlative study using in situ hybridization histochemistry and receptor autoradiography. *Neurosci 47*:367–393.

Vincent SR, Kimura H (1992) Histochemical mapping of nitric oxide synthase in the rat brain. *Neurosci 46*:755–784.

Vincent SR, Reiner PB (1987) The immunohistochemical localization of choline acetyltransferase in the cat brain. *Brain Res Bull 18*:371–415.

Vidnyanszky Z, Hámori J (1994) Quantitative electron microscopic analysis of synaptic input from cortical areas 17 and 18 to the dorsal lateral geniculate nucleus in cats. *J Comp Neurol 349*:259–68.

Vitek JL, Ashe J, DeLong MR (1990) Altered somatosensory response properties of neurons in the 'motor' thalamus of MPTP treated parkinsonian monkeys. *Neurosci 16*:425

Vitek JL, Ashe J, DeLong MR, Alexander GE (1994) Physiological properties and somatotopic organization of the primate motor thalamus. *J Neurophysiol 71*:1498–1513.

Vogt BA (1985) Cingulate cortex. In: Cerebral Cortex Vol 4 Association and Auditory Cortices (Peters A, Jones EG eds), pp 89–149. New York: Plenum Press.

Vogt BA, Pandya DN, Rosene DL (1987) Cingulate cortex of the rhesus monkey: I. cytoarchitecture and thalamic afferents. *J Comp Neurol 262*:256–270.

Vogt BA, Rosene DL, Pandya DN (1979) Thalamic and cortical afferents differentiate anterior from posterior cingulate cortex in the monkey. *Science 204*:205–207.

Vogt BA, Sikes RW (1990) Lateral magnocellular thalamic nucleus in rabbits: architecture and projections to cingulate cortex. *J Comp Neurol 299(1)*:64–74.

Vogt C (1909) La myeloarchitecture du thalamus du cercopithéque. *J Psychol Neurol 12*:285–324.

Vogt C, Vogt O (1941) Thalamusstudien I-III: I. Zur Einführung. II. Homogenität und Grenzgestaltung der Grisea des Thalamus. III. Das Griseum centrale (Centrum medianum Luys). *J Psychol Neurol 50*:32–154.

Voigt MM, Uhl FR (1988a) Preprocholecystokinin mRNA in rat brain: regional expression includes thalamus. *Brain Res 464*:247–253.

von Krosigk M, Bal T, McCormick DA (1993) Cellular mechanisms of a synchronized oscillation in the thalamus. *Science 261*:361–364.

Voris HC, Whisler WW (1975) Results of stereotaxic surgery for intractable pain. *Confin Neurol 37*:86–96.

Wada E, McKinnon D, Heinemann S, Patrick J, Swanson LW (1990) The distribution of mRNA encoded by a new member of the neuronal nicotinic acetylcholine receptor gene family ($\alpha 5$) in the rat central nervous system. *Brain Res 526*:45–53.

Wada E, Wada K, Boulter J, Deneris E, Heinemann S, Patrick J, Swanson LW (1989) Distribution of Alpha2, Alpha3, Alpha4 and Beta2 neuronal nicotinic receptor subunit mRNAs in the central nervous system: a hybridization histochemical study in the rat. *J Comp Neurol 284*:314–335.

Wahle P, Albus K (1985) Cholecystokinin octapeptide-like immunoreactive material in neurons of the intralaminar nuclei of the cat's thalamus. *Brain Res 327*:348–353.

Walker AE (1937a) A note on the thalamic nuclei of *Macaca mulatta*. *J Comp Neurol 66*:145–155.

Walker AE (1937b) The projection of the medial geniculate body to the cerebral cortex in the macaque monkey. *J Anat 17*:319–331.

Walker AE (1938) The Primate Thalamus. Chicago: Univ. Chicago Press.

Wall JT, Symonds LL, Kaas JH (1982) Cortical and subcortical projections of the middle temporal area (MT) and adjacent cortex in Galago. *J Comp Neurol 211*:193–214.

Wamsley JK, Gehlert DR, Roeske WR, Yamamura HI (1984) Muscarinic antagonist binding site heterogeneity as evidenced by autoradiography after direct labeling with [3H]-QNB and [3H]-pirenzepine. *Life Sci 34*:1395–402.

Wamsley JK, Young WS, III, Kuhar MJ (1980a) Immunohistochemical localization of enkephalin in rat forebrain. *Brain Res 190*:153–174.

Wamsley JK, Zarbin M, Birdsall N, Kuhar MJ (1980b) Muscarinic cholinergic receptors: autoradiographic localization of high and low affinity agonist binding sites. *Brain Res 200*:1–12.

Wamsley JK, Zarbin MA, Young WS, Kuhar MJ (1982) Distribution of opiate receptors in the monkey brain: an autoradiographic study. *Neurosci 7*:595–613.

Warden MK, Young WS, III (1988) Distribution of cells containing mRNAs encoding substance P and neurokinin B in the rat central nervous system. *J Comp Neurol 272*:90–113.

Warren RA, Agmon A, Jones EG (1994a) Oscillatory synaptic interactions between ventroposterior and reticular neurons in mouse thalamus in vitro. *J Neurophysiol 72*:1993–2003.

Warren RA, Golshani P, Jones EG (1997) $GABA_B$ Receptor-mediated inhibition in developing mouse ventral posterior thalamic nucleus. *J Neurophysiol 28*:549–553.

Warren RA, Jones EG (1994) Glutamate activation of cat thalamic reticular nucleus: effects on response properties of ventroposterior neurons. *Exp Brain Res 100*:215–226.

Warren RA, Jones EG (1997) Maturation of neuronal form and function in a mouse thalamo-cortical circuit. *J Neurosci 17*:277–295.

Waszak M (1974) Firing pattern of neurons in the rostral and ventral part of nucleus reticularis thalami during EEG-spindles. *Exp Neurol 43*:38–59.

Watanabe M, Inoue Y, Sakimura K, Mishina M (1993) Distinct distributions of five N-Methyl-D-Aspartate receptor channel subunit mRNAs in the forebrain. *J Comp Neurol 338*:377–390.

Watanabe M, Rodieck RW (1989) Parasol and midget ganglion cells of the primate retina. *J Comp Neurol 289*:434–454.

Weber AJ, Kalil RE (1983) The percentage of interneurons in the dorsal lateral geniculate nucleus of cat and observations on several variables that affect the sensitivity of horseradish peroxidase as a retrograde marker. *J Comp Neurol 220*:336–346.

Weber AJ, Kalil RE, Behans M (1989) Synaptic connections between corticogeniculate axons and interneurones in the dorsal lateral geniculate nucleus of cat. *J Comp Neurol 289*:156–164.

Weber E, Barchas JD (1983) Immunohistochemical distribution of dynorphin-B in rat brain Relation to dynorphin-A and alpha-neo-endorphin systems. *Proc Natl Acad Sci USA 80*:1125–1129.

Weber JT, Harting JK (1980) The efferent projections of the pretectal complex: an autoradiographic and horseradish peroxidase analysis. *Brain Res 194*:1–28.

Weber JT, Huerta MF, Kaas JH, Harting JK (1983) The projections of the lateral geniculate nucleus of the squirrel monkey: studies of the interlaminar zones and the S layers. *J Comp Neurol 213*:135–145.

Webster MJ, Bachevalier J, Ungerleider LG (1993) Subcortical connections of inferior temporal areas TE and TEO in macaque monkeys. *J Comp Neurol 335*:73–91.

Weiner DM, Levey AI, Brann MR (1991a) Expression of muscarinic acetylcholine and dopamine receptor mRNAs in rat basal ganglia. *Proc Natl Acad Sci USA 87*:7050–7054.

Welker E, Hoogland PV, Van der Loos H (1988) Organization of feedback and feedforward projections of the barrel cortex: a PHA-L study in the mouse. *Exp Brain Res 73*:411–435.

Westlund KN, Sorkin LS, Ferrington DG, Carlton SM, Willcockson HH, Willis WD (1990b) Serotoninergic and noradrenergic projections to the ventral posterolateral nucleus of the monkey thalamus. *J Comp Neurol 295*:197–207.

Whitlock DG, Perl ER (1961) Thalamic projections of spinothalamic pathways in monkey. *Exp Neurol 3*:240–255.

Widdowson PS (1993) Quantitative receptor autoradiography demonstrates a differential distribution of neuropeptide-Y Y1 and Y2 receptor subtypes in human and rat brain. *Brain Res 631*:27–38.

Wieland HA, Luddens H, Seeburg PH (1992) Molecular determinants in $GABA_A$/BZ receptor subtypes. *Adv Biochem Psychopharmacol 47*:29–40.

Wiesel TN, Hubel DH (1966) Spatial and chromatic interactions in the lateral geniculate body of the rhesus monkey. *J Neurophysiol 29*:1115–1156.

Wiesel TN, Hubel DH, Lam DMK (1974) Autoradiographic demonstration of ocular-dominance columns in the monkey striate cortex by means of transneuronal transport. *Brain Res 79*:273–279.

Wiesendanger M (1973) Input from muscle and cutaneous nerves of the hand and forearm to neurones of the precentral gyrus of baboons and monkeys. *J Physiol Lond 228*:203–219.

Wiesendanger R, Wiesendanger M (1985) The thalamic connections with medial area 6 (supplementary motor cortex) in the monkey (*Macaca fascicularis*). *Exp Brain Res 59*:91–104.

Williams MN, Zahm DS, Jacquin MF (1994) Differential foci and synaptic organization of the principal and spinal trigeminal projections to the thalamus in the rat. *Eur J Neurosci 6*:429–453.

Williams SR, Turner JP, Crunelli V (1994) Pharmacological and electrophysiological properties of local circuit interneurones in brain slices of the rat dorsal lateral geniculate nucleus. *Soc Neurosci Abstr 20*:8

Williamson AM, Ohara PT, Ralston HJ (1993) Electron microscopic evidence that cortical terminals make direct contact onto cells of the thalamic reticular nucleus in the monkey. *Brain Res 631*:175–179.

Williamson AM, Ohara PT, Ralston DD, Milroy AM, Ralston HJ, III (1994) Analysis of gamma-aminobutyric acidergic synaptic contacts in the thalamic reticular nucleus of the monkey. *J Comp Neurol 349*:182–192.

Williamson AM, Ralston HJ (1993) Fine structure of calcitonin gene-related peptide immunoreactive synaptic contacts in the thalamus of the rat. *J Comp Neurol 328*:130–144.

Willis WD, Jr. (1988a) Physiology of pain perception. In: Biowarning System in the Brain (Takagi H, Oomura Y, Ito M, Otsuka M eds), pp 69–85. University of Tokyo Press.

Willis WD, Jr. (1988b) Nociceptive neurons in the primate ventral posterior lateral (VPL) nucleus. In: Cellular Thalamic Mechanisms (Bentivoglio M, Spreafico R eds), pp 77–92. Elsevier.

Willis WD, Trevino DL, Coulter JD, Maunz RA (1974) Responses of primate spinothalamic tract neurons to natural stimulation of hindlimb. *J Neurophysiol 37*:358–372.

Wilson CJ, Chang HT, Kitai ST (1983) Origins of post synaptic potentials evoked in spiny neostriatal projection neurons by thalamic stimulation in the rat. *Exp Brain Res 51*:217–226.

Wilson JR (1989) Synaptic organization of individual neurons in the macaque lateral geniculate nucleus. *J Neurosci 9*:2931–2953.

Wilson JR, Friedlander MJ, Sherman SM (1984) Fine structural morphology of identified X- and Y-cells in the cat's lateral geniculate nucleus. *Proc R Soc Lond B 221*:411–436.

Wilson JR, Hendrickson AE (1981) Neuronal and synaptic structure of the dorsal lateral geniculate nucleus in normal and monocularly deprived *Macaca* monkeys. *J Comp Neurol 197*:517–540.

Wilson JS, Hull CD, Buchwald NA (1983) Intracellular studies of the convergence of sensory input on caudate neurons of cat. *Brain Res 270*:197–208.

Wilson KM, Minneman KP (1989) Regional variations in α_1-adrenergic receptors subtypes in rat brain. *J Neurochem 53*:1782–1786.

Winer J, A., Larue DT (1987) Patterns of reciprocity in auditory thalamocortical and corticothalamic connections: study with horseradish peroxidase and autoradiographic methods in the rat medial geniculate body. *J Comp Neurol 257*:282–315.

Winfield, DA (1980) The synaptic organization of glomeruli in the magnocellular and parvocellular laminae of the lateral geniculate nucleus in the monkey. *Brain Res 198*:55–62.

Winfield DA, Lieberman AR, Webster KE (1974) An electron microscopical study of the projection from the primary somatosensory cortex to the posterolateral complex of the rat thalamus. *J Anat 118*:372–377.

Wisden W, Laurie DJ, Monyer H, Seeburg PH (1992) The distribution of 13 $GABA_A$ receptor subunit mRNAs in the rat brain. I. telencephalon, dicephalon, mesencephalon. *J Neurosci 12*:1040–1062.

Wisden W, Seeburg PH (1992a) $GABA_A$ receptor channels: from subunits to functional entities. *Curr Op Neurobiol 2*:263–269.

Wise SP, Jones EG (1977) Cells of origin and terminal distribution of descending projections of the rat somatic sensory cortex. *J Comp Neurol 175*:129–158.

Wong-Riley MTT (1970) Somato-dendritic and dendro-dendritic synapses in the squirrel monkey lateral geniculate nucleus. *Brain Res 20*:135–139.

Wong-Riley MTT (1972a) Neuronal and synaptic organization of normal dorsal lateral geniculate nucleus of the squirrel monkey *Saimiri sciureus*. *J Comp Neurol 144*:25–59.

Wong-Riley MTT (1972b) Changes in the dorsal lateral geniculate nucleus of the squirrel monkey after unilateral ablation of the visual cortex. *J Comp Neurol 146*:519–548.

Wong-Riley MTT (1974) Demonstration of geniculocortical and callosal projection neurons in the squirrel monkey by means of retrograde axonal transport of horseradish peroxidase. *Brain Res 79*:267–272.

Woolf NJ, Butcher LL (1986) Cholinergic systems in the rat brain: projections from the pontomesencephalic tegmentum to the thalamus, tectum, basal ganglia and basal forebrain. *Brain Res Bull 16*:603–637.

Wycis MT, Spiegel EA (1962) Long-range results in the treatment of intractable pain by stereotoxic and midbrain surgery. *J Neurosurg 19*:101–107.

Wyss JM, Swanson LW, Cowan WM (1979) A study of subcortical afferents to the hippocampal formation in the rat. *Neurosci 4*:463–476.

Xiong M, Van Horn SC, Godwin DW, Erisir A, Sherman SM (1996) Localization of metabotropic glutamate receptors in monkey LGN and VPL. *Neurosci Abstr 22*:1606

Yamamoto T, Noda T, Samejima A, Oka H (1988) Electrophysiological and morphological features of thalamic neurons with special reference to the cerebellar and pallidal inputs. In: Cellular Thalamic Mechanisms (Bentivoglio M, Spreafico R eds), pp 239–260. Amsterdam: Elsevier.

Yamasaki DSG, Krauthamer GM (1990) Somatosensory neurons projecting from the superior colliculus to the intralaminar thalamus in the rat. *Brain Res 523*:188–194.

Yamasaki DSG, Krauthamer G, Rhoades RW (1986) Superior collicular projection to intralaminar thalamus in rat. *Brain Res 378*:223–233.

Yamashiro K, Tasker RR (1993) Stereotactic thalamotomy for dystonic patients. *Stereotact Funct Neurosurg 60*:81–85.

Yarita H, Iino M, Tanabe T, Kogure S, Takagai SF (1980) A transthalamic olfactory pathway to orbitofrontal cortex in the monkey. *J Neurophysiol 43*:69–85.

Yen C-T, Conley M, Jones EG (1985a) Morphological and functional types of neurons in cat ventral posterior thalamic nucleus. *J Neurosci 5*:1316–1338.

Yen C-T, Conley M, Hendry SHC, Jones EG (1985b) The morphology of physiologically identified GABAergic neurons in the somatic sensory part of the thalamic reticular nucleus in the cat. *J Neurosci 5*:2254

Yeterian EH, Pandya DN (1988) Corticothalamic connections of paralimbic regions in the rhesus monkey. *J Comp Neurol 269*:130–146.

Yeterian EH, Pandya DN (1989) Thalamic connections of the cortex of the superior temporal sulcus in the thesus monkey. *J Comp Neurol 282*:80–97.

Yeterian EH, Pandya DN (1994) Laminar origin of striatal and thalamic projections of the prefrontal cortex in rhesus monkeys. *Exp Brain Res 99*:383–399.

Yeterian EH, Van Hoesen GW (1978) Cortico-striate projections in the rhesus monkey: the organization of certain cortico-caudate connections. *Brain Res 139*:43–63.

Yoshida M (1981) The GABAergic systems and the role of the basal ganglia in motor control. In: GABA and the Basal Ganglia (Di Chiara G, Gessa GL eds), pp 37–52. New York.

Young AB, Bromberg MB, Penney JB (1981) Decreased glutamate uptake in subcortical areas deafferented by sensorimotor cortical ablation in the cat. *J Neurosci 1*:241–249.

Young AM, Bradford HF (1986) Excitatory amino acid neurotransmitters in the corticostriate pathway: studies using intracerebral microddialysis in vivo. *J Neurochem 47*:1399–1404.

Young KA, Wilcox RE (1991) Characterization of D2 receptors and dopamine levels in the thalamus of the rat. *Life Science 48*:1845–1852.

Young WS, Alheid GF, Heimer L (1984) The ventral pallidal projection to the mediodorsal thalamus: a study with fluorescent retrograde tracers and immunohistofluorescence. *J Neurosci 4*:1626–1638.

Young WS, III, Kuhar MJ (1980) Noradrenergic α_1 and α_2 receptors: light microscopic autoradiographic localization. *Proc Natl Acad Sci USA 77*:1696–1700.

Young WS, III, Kuhar MJ (1981) Neurotensin receptor localization by light microscopic autoradiography in rat brain. *Brain Res 206*:273–286.

Yuan B, Morrow TJ, Casey KL (1985) Responsiveness of ventrobasal thalamic neurons after suppression of SI cortex in the anaesthetized rat. *J Neurosci 5*:2971–2978.

Yuan B, Morrow TJ, Casey KL (1986) Corticofugal influences of S1 cortex on ventrobasal thalamic neurons in the awake rat. *J Neurosci 6*:3611–3617.

Yukie M, Iwai E (1981) Direct projection from the dorsal lateral geniculate nucleus to the prestriate cortex in macaque monkeys. *J Comp Neurol 201*:1–14.

Zahm DS, Zaborszky L, Alheid GF, Heimer L (1987) The ventral striatopallidothalamic projection: II. The ventral pallidothalamic link. *J Comp Neurol 255*:592–605.

Zarbin MA, Innis RB, Wamsley JK, Snyder SH, Kuhar MJ (1983) Autoradiographic localization of cholecystokinis receptors in rodent brain. *J Neurosci 3*:877–906.

Zola-Morgan S, Squire LR (1985) Medial temporal lesions in monkeys impair memory on a variety of tasks sensitive to human amnesia. *Behav Neurosci 99*:22–34.

CHAPTER II

Neurochemical organization of the primate visual cortex

J.H. MORRISON, P.R. HOF AND G.W. HUNTLEY

1. INTRODUCTION

Morphology has traditionally been the primary determinant of cell type in the cerebral cortex (e.g. Cajal 1899a,b), and remains the principal basis for distinguishing fundamental classes of neocortical neurons. Additionally, however, the wealth of information gleaned from the application of modern, sophisticated neuroanatomical and neurophysiological techniques indicates that connectivity, intrinsic biophysical membrane properties and neurochemical features can also be viewed as distinguishing characteristics of cell type. In attempting to construct a comprehensive scheme of visual information flow through the cortical circuitry, what emerges when all of these characteristics of cell type are considered together is a complex cellular organization in which distinct neuronal classes occupy different positions in the microcircuitry underlying the processing of visual information. These unique functional roles are often reflected in the neurochemical phenotype of each neuronal class, and it is this aspect of cell type which will be the principal focus of the following review. In what follows, emphasis will be placed on integrating the particular neurochemical feature under consideration into the context of regional and laminar location, morphologic characteristics, and connectivity. Throughout this chapter, the primary visual cortex (area 17, striate cortex, area V1) of primates will serve as the focus of neurochemical specificity in cortical neuronal organization, connectivity and function; however, variations in chemically-specified neural circuits and cell classes across different areas and species will also be discussed.

2. GENERAL ORGANIZATION OF PRIMATE VISUAL CORTEX

2.1. REGIONAL SPECIALIZATION OF VISUAL CORTEX

It was appreciated early on in the parcellation of neocortex that the three major areas of occipital cortex as defined by Brodmann (areas 17, 18, and 19) were all primarily visual in function, with Brodmann's area 17 having a particularly well-defined role as primary visual cortex (Brodmann 1909). However, as more sophisticated anatomic, physiologic, and behavioral techniques were applied to the primate neocortex, it became apparent that much of the parietal and temporal neocortex that was assumed to

Handbook of Chemical Neuroanatomy, Vol. 14: The Primate Nervous System, Part II
F.E. Bloom, A. Björklund and T. Hökfelt, editors

be multimodal or modality-independent association cortex was actually closely linked to a single modality: vision (Van Essen 1985). Furthermore, regions that were traditionally thought to be uniform based on cytoarchitecture, such as area 19, have been shown to consist of several distinct, separable visual areas when subjected to modern anatomic and physiologic techniques (Van Essen 1985).

It is now clear that the primate visual system includes a large number of cortical regions that are involved in integration of visual stimuli into an accurate, cohesive, and dynamic representation of the subject's visual environment. As such, visual information processing in the neocortex involves both localized functional specializations, and distributed processing that simultaneously involves multiple regions. Information processing in the monkey visual system is distributed along well defined sets of connections linking the primary (area V1) to secondary (area V2) visual cortex, and from there to higher level association cortical areas located in the occipital, parietal and temporal lobes (Rockland and Pandya 1979; Stone et al. 1979; Lennie 1980; Tigges et al. 1981; Mishkin et al. 1983; Maunsell and Newsome 1987; DeYoe and Van Essen 1988; Livingstone and Hubel 1988; Zeki and Shipp 1988; Morel and Bullier 1990; Felleman and Van Essen 1991; DeYoe et al. 1994; Ungerleider 1995). The notion of both distributed information processing along multiple cortical pathways and the existence of a hierarchical distribution of visual areas, responsible for particular aspects of visual stimulus analysis, arises from the observation that specific components of a visual stimulus are segregated into submodalities such as form, motion and color which are analyzed within restricted and separate cortical areas (Mishkin et al. 1983; Maunsell and Newsome 1987; DeYoe and Van Essen 1988; Livingstone and Hubel 1988; Zeki and Shipp 1988; Morel and Bullier 1990; Felleman and Van Essen 1991; DeYoe et al. 1994; Ungerleider 1995). These areas are linked, such that there exist divergent pathways projecting toward cortical regions located in the temporal lobe, primarily subserving form and color analysis, or to parietal areas more involved in visuomotor functions and spatial information (detailed below; DeYoe and Van Essen 1988; Livingstone and Hubel 1988; Baizer et al. 1991). Physiologic recordings in behaving animals and correlative anatomic studies have revealed the existence more than 25 cortical visual fields (for review see Van Essen et al. 1990, 1992; Felleman and Van Essen 1991; see Fig. 1).

Using classical staining methods such as Nissl and myelin stains, definite anatomic criteria can be applied in a reliable manner for only a limited number of cortical fields. For instance, area V1 displays well defined cyto- and myeloarchitectonic patterns that differentiate it from area V2. Similarly, the heavily myelinated zone in the depth of the posterior portion of the superior temporal sulcus corresponds to the cortical area known as MT (or V5), but such patterns may be less reliable in the case of other cortical areas. However, detailed anatomical characterization of cortical areas can be reliably achieved using immunohistochemical techniques to analyze neurochemical features specific to certain groups of neurons. Cortical regions can therefore be defined according to the distribution and density of select neuronal populations exhibiting a specific neurochemical phenotype. In this context, the analysis of regional variations in the distribution of specific neurochemical markers represents a helpful means to define precisely the anatomic characteristics and topographic boundaries of visual areas that possibly correspond to different hierarchical steps of information processing, and to provide quantitative measures of their specific neuronal composition. Also, the combination of tract-tracing with immunohistochemistry allows for the identification of reliable neurochemical, connectional, and morphological criteria that can be used to

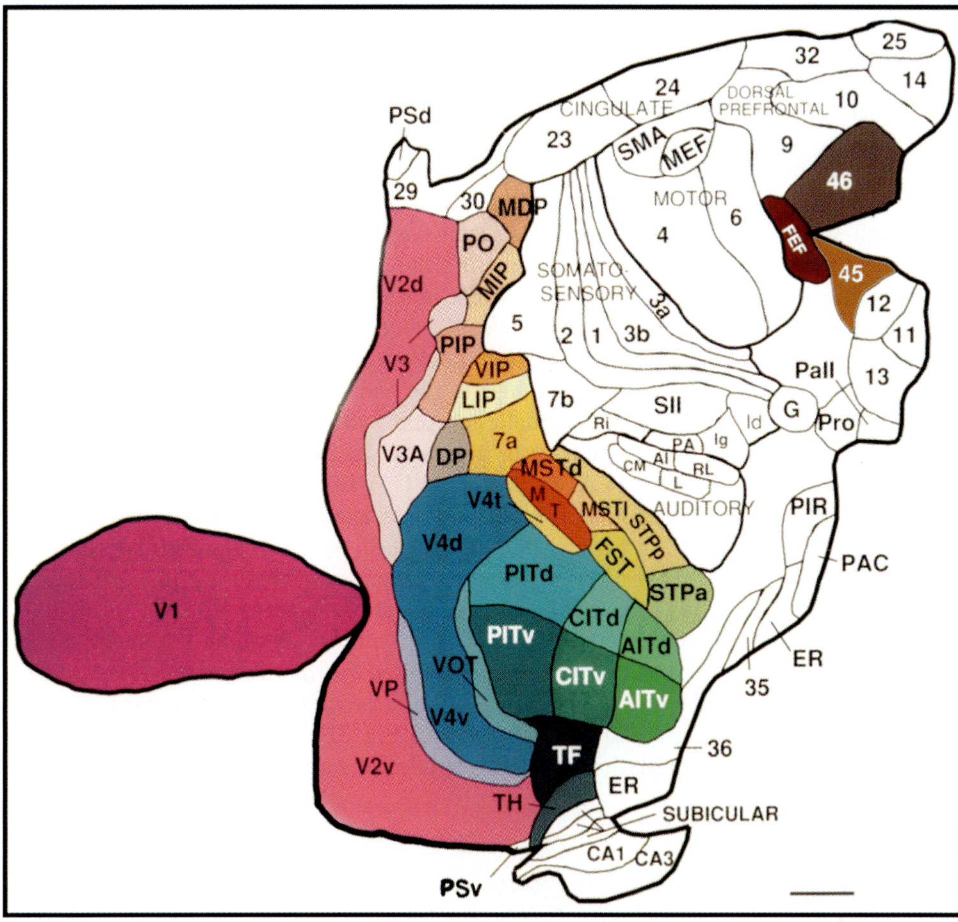

Fig. 1. Map of cortical areas in the macaque monkey highlighting those that are primarily visual in function (see Felleman and Van Essen 1991 for details). The visual areas of neocortex are designated by colors, while areas corresponding to other Brodmann areas or various functional systems (e.g., auditory, somatosensory, and motor) are in black and white. The upper left figure is a dorsolateral view of the cortical surface, and the lower left is a mid-sagittal view. The large panel on the right represents a flattened image where the sulci have been unfolded to obtain a more complete view of areas that are buried in sulci. The color scheme is such that visual areas of the occipital lobe have purple, blue, and reddish hues, those of the parietal lobe have yellow, orange and light brown hues, visual areas in the temporal lobe have green hues, and relevant areas of frontal lobe have brown hues. Area V1 stands out at the caudal end as the large purple area. Much of the cellular and synaptic material that will be presented in this chapter will be based on data from area V1, with the other visual areas discussed primarily as they relate to regional comparisons and corticocortical connectivity. There is a total of 32 visual areas designated in this diagram. These regional designations reflect, as much as possible, cytoarchitectonic, connectional, and physiologic properties based on a very broad array of experimental analyses. This diagram is particularly useful to make the point that approximately 40% of the cortical surface is visual in function, and that it is highly parcellated in respect to distinct areas. Scale bar = 1 cm. Adapted from Felleman and Van Essen (1991).

define the particular phenotype of different subsets of corticocortically projecting neurons (Campbell et al. 1991; Carmichael and Price 1994; Hof et al. 1995a, 1996; Nimchinsky et al. 1996), or other cortical efferents that are particularly well represented in a given region. A fairly detailed example of this approach to parcellation based on the localization of neurofilament protein will be presented later in this chapter.

2.2. LAMINAR ORGANIZATION

As is the case for all neocortical areas, V1 and other visual areas are generally divided into six layers of varying thickness that proceed in order from the pia mater to the subcortical white matter according to a scheme which basically derives from Brodmann (Brodmann 1905), with layer I occupying the most superficial position and layer VI occupying a position adjacent to the white matter. In all visual areas, layer I is a thin, neuron-poor layer, whereas the adjacent layer II is densely packed with small neurons. Layer III is marked by less densely packed, larger neurons that often appear to form parallel rows oriented perpendicular to the pial surface. The subjacent layer IV marks a return to densely packed small neurons, with area V1 exhibiting further specialization of layer IV, particularly in primates. In primate area V1, layer IV is subdivided into the sublayers IVA, IVB, and IVC, with IVC further divided into a superficial IVCα and a deeper IVCβ (Valverde 1985; Lund 1981). While this standard scheme is based on the cytoarchitectonic appearance of Nissl-stained materials, the laminar organization of visual cortical areas in general, and area V1 in particular, is also reflected in fiber stains and various chemically-specific stains (Fig. 2). In area V1, layer IVA is a very thin band of tightly packed cells, most of which are spiny stellate cells. Layer IVB is a layer that contains a distinct band of myelinated fibers, corresponding to the external band of Baillarger (the stria of Gennari), as well as fairly widely dispersed large, pyramidal-like neurons (see Fig. 2D). Layer IVC is very densely packed with a majority population of spiny stellate cells and fewer numbers of aspiny

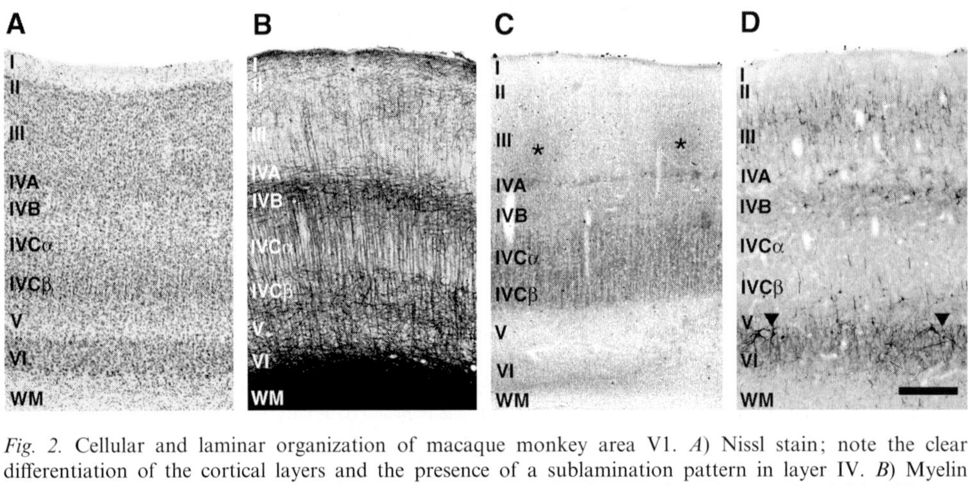

Fig. 2. Cellular and laminar organization of macaque monkey area V1. *A*) Nissl stain; note the clear differentiation of the cortical layers and the presence of a sublamination pattern in layer IV. *B*) Myelin stain; there is a very high density of fibers in layers I, IVB (stria of Gennari), and VI, and a visible difference between sublayers IVCα and IVCβ. *C*) Cytochrome oxidase staining; note the high level of cytochrome oxidase labeling in layers IVA, IVC, and VI as well as the typical blobs in layer III (*asterisks*). *D*) Neurofilament protein immunoreactivity (see also Figs 53–55); there is a characteristic paucity of labeled neurons in layer IVC, while layers IVA and V display isolated pyramidal neurons; layers III and VI contain a higher density of neurofilament protein-immunoreactive neurons. Layer IVB is characterized by the presence of clusters of large modified pyramidal neurons, and the border between layers V and VI contains groups of large Meynert cells (*arrowheads*). Layers are indicated by Roman numerals; WM, white matter. Scale bar (on *D*) = 150 μm.

interneurons. Layer V is similar to layer III in that it is marked by large widely spaced pyramidal cells, and is subdivided into sublayers VA and VB, based on differences in density. Layer VI is also subdivided into an upper, cell-dense layer VIA and a sparser, deeper layer VIB. The primary visual cortex is distinctive in that it contains the giant Meynert cells at the border of layers V and VI (Fig. 2D).

Not only do the layers provide a structural framework that helps to designate location in the cortex, but they also relate importantly to several aspects of cortical organization in regard to afferent termination, efferent origin and cell type distribution (Valverde 1985), and as will be described in this chapter, chemically-specific patterns as well. For example, the projection from the dorsal lateral geniculate nucleus (LGN) to VI terminates most heavily in layers IVA and IVC, however, cortical efferents are also segregated by layer in that projections from area V1 back to LGN originate in layer VI, whereas those to the brainstem (e.g., superior colliculus) originate in layer V (Jones 1984). Corticocortical projections interconnecting visual areas originate and terminate in multiple layers that can be separated into subsets with a high degree of laminar specificity that relates to cortical hierarchies (see below). In addition, as discussed below, cortical cell types display preferential laminar distribution. In these respects, V1 is a particularly good example of the degree to which lamination can be linked to segregation of inputs, outputs, and intrinsic neurons, all of which are reflected in patterns of chemical specificity.

2.3. CELL TYPES IN PRIMATE VISUAL CORTEX

Primate visual cortex, like all neocortical areas, is composed of two fundamental classes of neurons: those that bear dendritic spines (spiny neurons), and those which are completely or mostly aspiny. Spiny neurons are themselves divided into two types, pyramidal cells and spiny stellate cells. Pyramidal neurons form the majority of the cells in neocortex, constituting approximately 70–80% of the total neuronal population. Such neurons occupy all layers, with the exception of layer I, and, in primary visual cortex, are also mostly absent from layers IVA and IVC. Pyramidal cells are regarded as the major output neuron type of the cerebral cortex, with most supplying a principal axon which leaves the immediate cortical area to terminate in other cortical or subcortical regions (Fig. 3), and a variable number of vertically and horizontally-directed axon collaterals which provide a rich network of local connections linking different layers as well as more distant, functionally-distinct columns. Spiny stellate cells, in contrast, are short-axon cells located exclusively in layers IVA and IVC, and provide ascending, vertically-oriented axons which target superficial layers (Mates and Lund 1983; Saint Marie and Peters 1985). Functionally, the spiny class of neuron forms exclusively asymmetric synapses, and is excitatory (e.g. Feldman 1984).

The other fundamental class of neocortical neuron – the aspiny cell class – is composed of a rich array of morphologically and neurochemically diverse neurons, which is detailed below (Fig. 4). However, in general, such neurons are non-pyramidal neurons with smooth or sparsely-spiny dendrites and axons which form relatively local ramifications. Most form exclusively symmetric synapses, and are inhibitory in function (Houser et al. 1984). In monkey primary visual cortex, they constitute approximately 20% of the total neuronal population, and are present in all layers, even extending into the subcortical white matter (Beaulieu et al. 1992; Fitzpatrick et al. 1987; Hendry et al. 1987).

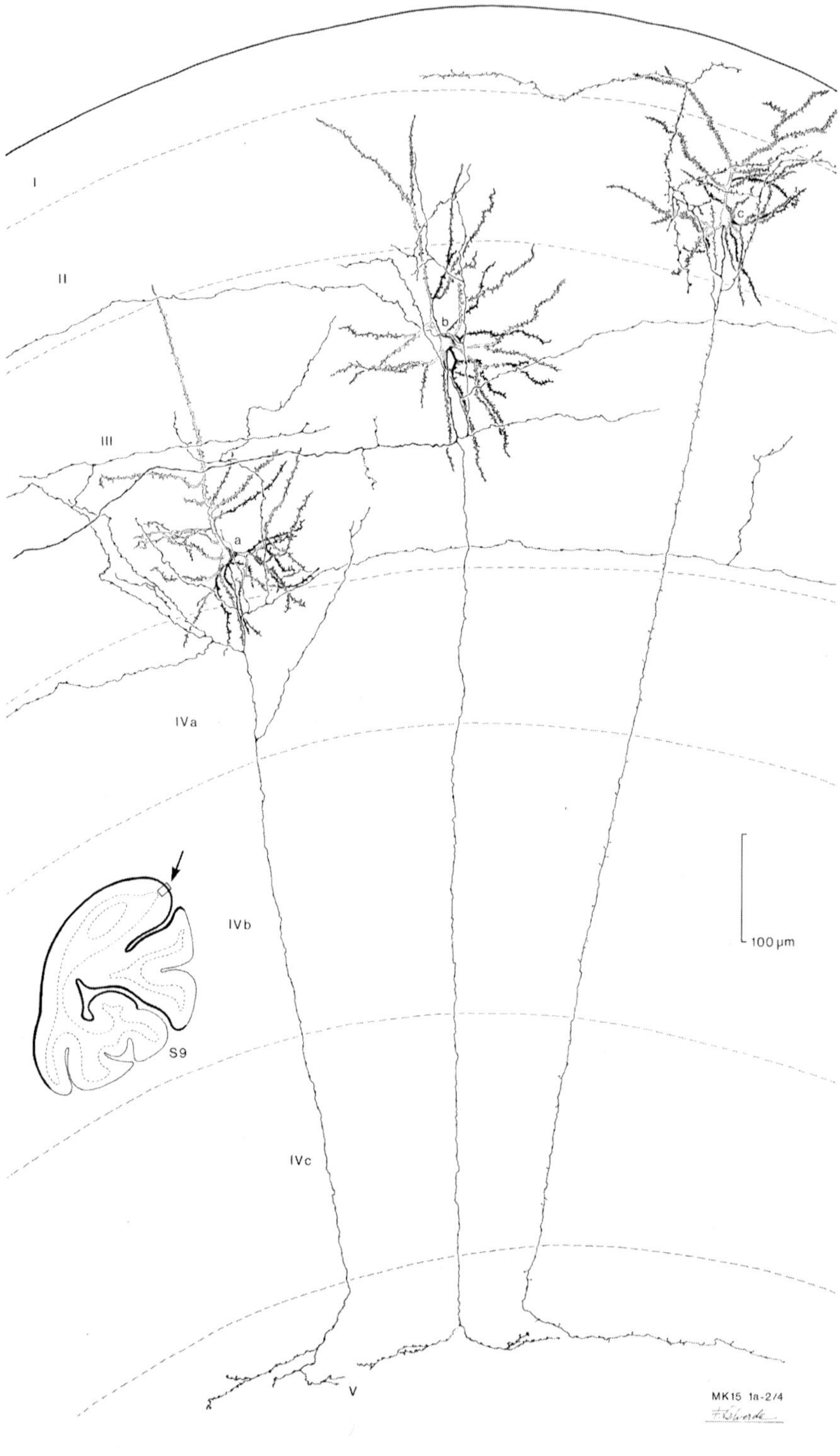
I
II
III
IVa
IVb
IVc
V
a
b
c
S9
100 μm
MK15 1a-2/4

Fig. 3. Camera lucida drawings of three examples of pyramidal cells in primary visual cortex of rhesus monkey. Note that examples *A* and *B* are in layer III, and *C* (far right) is in layer II. The diagram of the coronal section on the left notes the portion of primary visual cortex from which the material was taken. All three have the relatively classic morphology of a pronounced apical dendrite, an axon that exits from the cortical gray matter, and several recurrent collaterals that extend for a mm or more in the horizontal plane (*A, B*). The drawings are based on a Golgi stain. Reproduced from Valverde (1985).

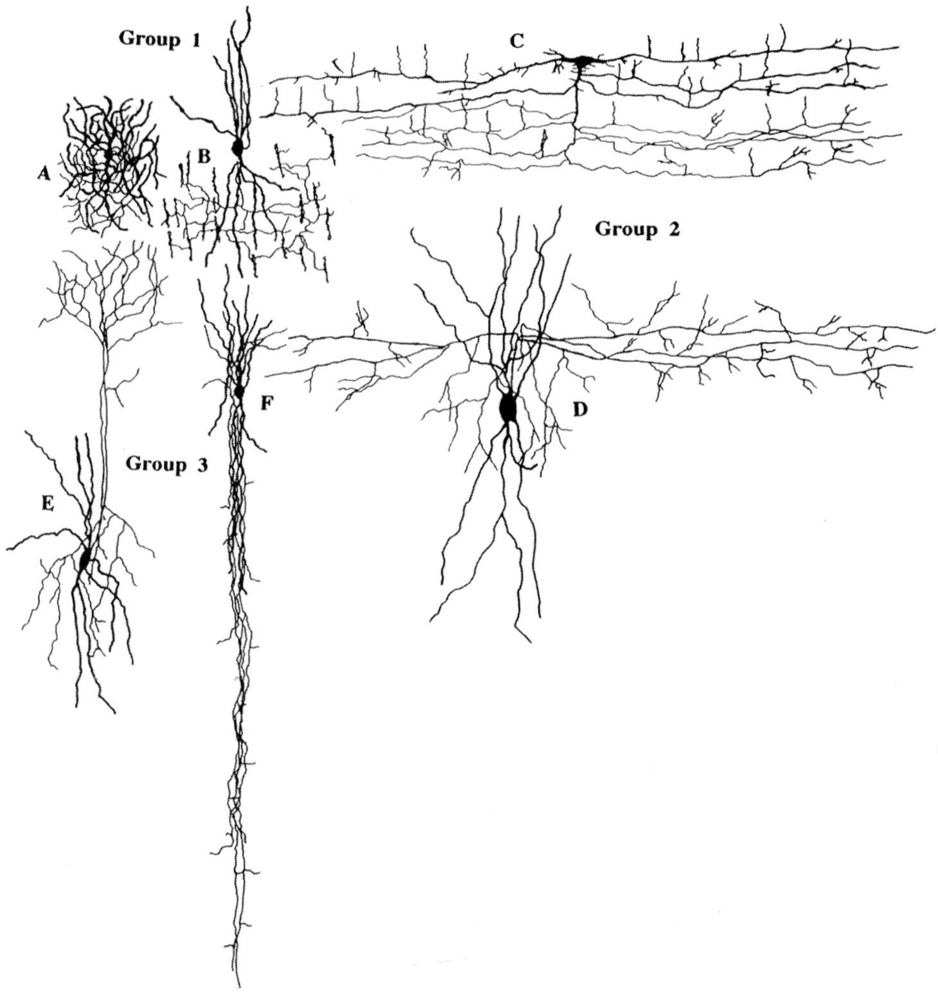

Fig. 4. Major classes of aspiny non-pyramidal cells in the primate cerebral cortex as seen in a Golgi preparation. Group 1 is represented by cells *A* and *B*, with *A* being a neurogliaform cell, and *B* a chandelier cell. Both of these neuron types form very local connections. Group 2 consists of neurons with long horizontal axon collaterals, such as *C*, a Cajal-Retzius neuron, generally found in layer I, and a large basket cell (*D*). Group 3 consists of neurons that form vertical connections, and is represented by a Martinotti cell with an ascending axonal arbor (*E*), and a double bouquet cell with both ascending and descending axonal arbor that can extend for up to a mm or more in the radial domain (*E*). All of these neurons utilize GABA as their neurotransmitter, as well several peptides, and in addition, calcium-binding proteins are colocalized in different combinations of these mrophologic classes (see details in text). Reproduced from Jones et al. (1994).

3. NEUROTRANSMITTER LINKS TO CORTICAL CIRCUITS: OVERVIEW

Visual cortical circuitry is composed of afferent fiber systems of extrinsic origin, and local projections provided by intrinsic pyramidal and non-pyramidal neurons. Collectively, these circuits account for three, broad functional effects: rapid excitation, inhibition, and combinations of more slowly-developing depolarization and hyperpolarization which are often described as neuromodulatory. To a certain degree, each circuit type can be linked to a broad class of neurotransmitter.

The key excitatory circuits of the visual cortex are provided by thalamocortical afferents, corticocortical and callosal afferents, the horizontal axon collaterals of intrinsic pyramidal neurons and the ascending axons of spiny stellate neurons. Each of these projections forms focused aggregations of axon terminals (e.g. Jones 1981), which, particularly in the case of geniculocortical afferents, is critical for maintaining the point-to-point connectivity necessary for preserving precise, visuotopically organized information. These excitatory circuits use the fast-acting, excitatory amino acids glutamate and aspartate as their neurotransmitter (Streit 1984), and possibly also *N*-acetylaspartylglutamate (NAAG), an endogenous dipeptide which interacts with glutamate receptors, meets most of the criteria for a neurotransmitter (Blakely and Coyle 1988), and has been localized immunocytochemically to geniculocortical terminals in cats (Tieman et al. 1987; Xing and Tieman 1993) and monkeys (Tieman et al. 1991).

The inhibitory circuits of the visual cortex are provided by the population of intrinsic interneurons which synthesize and release the inhibitory neurotransmitter γ-aminobutyric acid (GABA; Houser et al. 1984). GABAergic neurons are composed of a wide variety of morphologically and neurochemically diverse subpopulations (Fig. 4), which is detailed below (see section 4).

The neuromodulatory circuits are represented by the system of more diffusely projecting, widespread afferent fibers which use one of several monoamines as a neurotransmitter. For the present purposes, acetylcholine will be considered with the monoamines since the cholinergic projection shares the organizational characteristics of the monoamine projections, although chemically it is not a monoamine. Given the broad spatial domain of these projections and the relatively long time course of monoamine actions, these systems appear to be ideally suited for a role in regulation of activities that involve the cohesive activity of large areas of neocortex, such as vigilance, attention, affective state changes, and mood (Foote and Morrison 1987). Their patterns of termination in primate visual cortex are described below (see section 6).

While these links between circuit type and transmitter class are helpful in a broad sense, they are greatly oversimplified because they imply that a given neurotransmitter will have a unitary cellular effect. However, in reality, a high degree of receptor diversity exists which is reflected by the existence of multiple receptor subtypes for a given neurotransmitter, and, of these, some are composed of heteromeric combinations of distinct receptor subunits. Since the different receptor subtypes have different functional properties, the complete characterization of a neocortical circuit will require precise identity of the subunit components of the postsynaptic receptor as well as the neurotransmitter employed. While the precise molecular characterization of the postsynaptic component of a chemically-identified circuit is in its most primitive state in regard to neocortical circuitry, recent developments in the molecular neurobiology of receptors and availability of subunit specific reagents have rendered such an analysis feasible, particularly for glutamate and $GABA_A$ receptors (Morrison et al. 1996; Huntley et al. 1994b; Huntsman et al. 1994).

4. EXCITATORY CIRCUITS

4.1. INTRODUCTION

Glutamate and aspartate are the major excitatory amino acid neurotransmitters in the central nervous system in general (Krnjevic and Phillis 1963), and area V1 in particular (Streit 1984). With respect to localization, glutamate and aspartate have been linked to key circuits in visual cortex by studying the uptake and transport of tritiated aspartate and glutamate (Baughman and Gilbert 1981; Kisvárday et al. 1987; Barbaresi et al. 1987), and, more recently, immunocytochemistry (Dori et al. 1992; Carder and Hendry 1994). While excellent antibodies have been developed against glutamate and aspartate (Ottersen and Storm-Mathisen 1987; Hepler et al. 1988), the fact that these amino acids are also represented in metabolic, non-neurotransmitter pools can make interpretation difficult. However, in a variety of species including rat, cat and monkey, the vast majority of the labeled neurons are pyramidal neurons in layers II to VI (Conti et al. 1987, 1988; Dori et al. 1989; Dori et al. 1992; Carder and Hendry 1994) although some layer IV non-pyramidal neurons are also labeled (Conti et al. 1987). In monkey area V1, none of the layer IV glutamate-containing neuronal somata also display GABA immunoreactivity, indicating that such neurons likely represent the excitatory spiny stellate cells (Carder and Hendry 1994). Transport studies have also demonstrated glutamate or aspartate immunoreactivity or uptake of tritiated amino acids in corticospinal neurons (Baughman and Gilbert 1981; Giuffrida and Rustioni 1987); corticostriatal and corticopontine neurons (Dinopoulos et al. 1991; Dori et al. 1992); corticocortically and callosally projecting neurons (Barbaresi et al. 1987; Conti et al. 1988); and thalamocortical neurons and their terminals (Ottersen et al. 1983; Montero 1990; Johnson and Burkhalter 1992; Carder and Hendry 1994; Kharazia and Weinberg 1994). In addition, by employing uptake of tritiated aspartate as a means to visualize excitatory circuits, Kisvárday et al. (1989) and Johnson and Burkhalter (1992) demonstrated the existence of very extensive horizontal networks of excitatory amino acid-containing pyramidal cell axon collaterals in monkey and rat visual cortex. At the ultrastructural level, most glutamate-immunoreactive synapses in neocortex are asymmetrical (DeFelipe et al. 1988; Conti et al. 1989; Kharazia and Weinberg 1994), and thus presumably excitatory (Gray 1969). In addition, Donoghue et al. (1985) used an antibody to localize glutaminase, a metabolic enzyme for glutamate, which was also present preferentially in pyramidal cells in primate visual cortex, particularly in layers V and VI. These data taken together support the notion that glutamate and aspartate are the major excitatory neurotransmitters employed by pyramidal and spiny stellate cells for both local and corticofugal projections, as well as the major afferent thalamocortical terminations.

4.2. THALAMOCORTICAL PROJECTIONS

Area V1 receives its principal thalamic input from the LGN (Hubel and Wiesel 1972), which projects essentially only to area V1 in primates (Benevento and Standage 1982; Bullier and Kennedy 1983), and a lesser projection from the inferior pulvinar nucleus (Benevento and Rezak 1976; Ogren and Hendrickson 1977). Visual association regions in the temporal and parietal regions receive a projection from the pulvinar, a thalamic nucleus that is particularly well developed in primates and serves as the key thalamic link in the retino-tecto-pulvinar-extrastriate system which plays a critical role in visual attention (Robinson and Petersen 1992).

The projection from the LGN terminates most densely in layer IV in a highly topographic, point to point fashion so as to represent the visual field with very high acuity, and in most details is quite similar across primates (see Casagrande and Kaas 1994 for review). In Old World monkeys, the projection from the magnocellular layers of the LGN terminates primarily in layer IVCα, while that from the parvocellular layers terminates primarily in layers IVA and IVCβ (Hubel and Wiesel 1972; Hendrickson et al. 1978; Blasdel and Lund 1983; Freund et al. 1989). There are also minor collateral projections from both parvo- and magnocellular layers of the LGN to layer VI. In addition, direct geniculocortical projections arising from small thalamic relay neurons in the intercalated and S-layers of the LGN terminate in the cytochrome oxidase blobs of layers II/III and in layer I (Hendrickson et al. 1978; Livingstone and Hubel 1982; Weber et al. 1983; Lachica and Casagrande 1992). The most detailed information on the innervation patterns of LGN axons in macaque area V1 comes from the work of Blasdel and Lund (1983) and Freund et al. (1989) where they carefully reconstructed HRP-filled, physiologically-characterized LGN axons. Freund et al. (1989) followed the anatomic analysis to the ultrastructural level as well. Both studies demonstrated the extraordinary laminar and radial precision of the geniculocortical terminal patterns. For example, the magno- and parvocellular projections to layer IV tended to terminate largely, if not exclusively, in one particular sublamina of layer IV (Figs 5, 6). In addition, the parvocellular projections to layer IVCβ (Fig. 5) terminate in very circumscribed, narrow radial domains, approximately one-sixth the surface area of magnocellular terminations in layer IVCα (Fig. 6). However, the synaptic targets of the magnocellular projections to layer IVCα and the parvocellular projections to layer IVCβ are similar, with the vast majority of inputs on spines and shafts of spine-bearing dendrites, and a very small but potentially significant portion forming synapses with GABAergic somata and dendrites (Freund et al. 1989). Thus, a high level of precision and specificity is reflected in the geniculocortical projection to area V1 at the laminar, radial, and synaptic levels of resolution.

4.3. CORTICOCORTICAL PROJECTIONS

Area V1 and other visual areas are the targets of a complex system of interconnections and inputs from other cortical areas (reviewed in Casagrande and Kaas 1994). In fact, the interplay between the information content of the thalamocortical and corticocortical afferents to a given region virtually defines that region's function. Whereas area V1 is dominated by geniculocortical input, visual association areas in the temporal lobe are sites of extensive convergence from numerous other cortical areas and have a less dominant input from the thalamus than area V1.

In a simplistic sense, the visual cortex follows a strategy of segregation, regionally dedicated information processing, and integration through complex regional interconnections. The complex system of corticocortical connections allows for integration of visual information that has been largely segregated beginning in the retina and proceeding through the LGN to area V1. This material will be summarized only briefly here, since there have been several extensive and lucid reviews written on this topic over the last fifteen years (Mishkin et al. 1983; Livingstone and Hubel 1988; Zeki and Shipp 1988; Desimone and Ungerleider 1989; Fellman and Van Essen 1991; Merigan and Maunsell 1993; Van Essen and Gallant 1994; Ungerleider 1995).

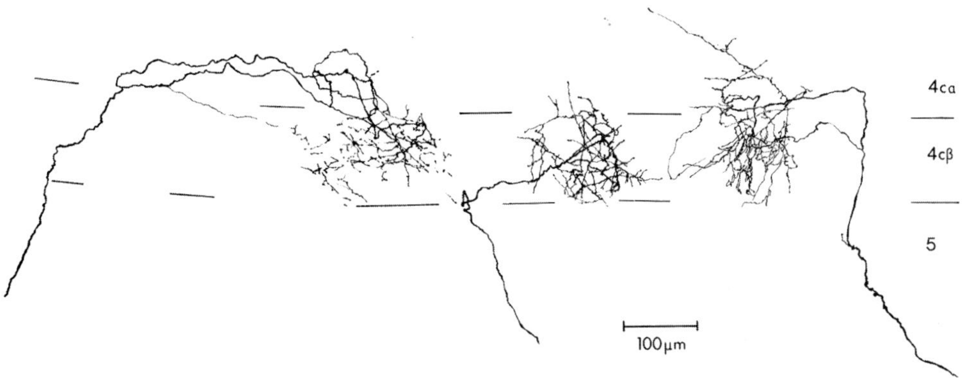

Fig. 5. Geniculocortical terminations in layer IVCβ, of macaque monkey area V1. Three camera lucida drawings are shown, each of which was filled with HRP. The filling was from white matter injections of HRP. These axons are typical of the other axons that were found arborized in this layer in this study by Blasdel and Lund (1983). As is apparent from these reconstructions, they are highly stereotyped with axonal fields that are roughly circular and generally confined to less than 200 μm spread laterally. Since layer IVCβ is know to receive its primary geniculocortical afferents from the parvocellular layers, it is believed that these axons represent parvocellular afferents. These three terminations demonstrate the level of precision that the geniculocortical projections to striate cortex achieve both in respect to laminar termination and limited spread in the horizontal plane. Reproduced from Blasdel and Lund (1983).

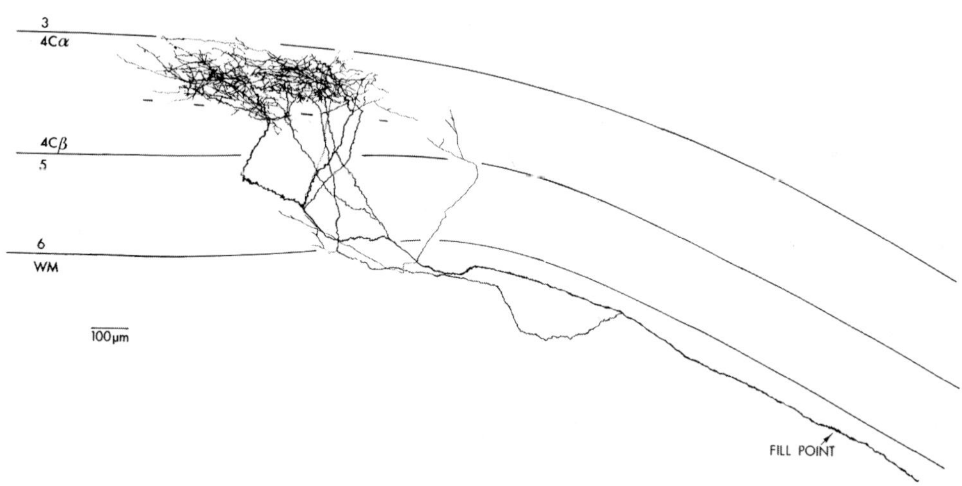

Fig. 6. Terminal arborization of an LGN axon terminating in layer IVCα of primary visual cortex in the macaque monkey. This axon was recorded intracellularly and filled with HRP. This axon had physiological properties of a monocular ON-center receptive field, and gave a highly transient ON response when tested with a flashing spot of light. It also had low contrast threshold and displayed no sensitivity to color. The terminal arbor was reconstructed through the use of computer-assisted microscopy. The terminals in this arborization occupy a surface area of 0.4 mm^2, or approximately 1.2 mm in length and 0.3 mm width. The authors suggest that this band corresponds to the ocular dominance territory of one eye. Note the contrasting characteristics of this arborization pattern with that in Figure 5 terminating in layer IVCβ. This presumably represents the termination of a projection from a neuron in the magnocellular layers of the LGN. Reproduced from Blasdel and Lund (1983).

Ganglion cells in the retina are divided into two major classes based primarily on the size: the magnocellular (M) or α class, and the parvocellular (P) or β class. These ganglion cells also differ in their response to visual stimuli. These differences in response properties as well as spatial segregation are largely maintained in the LGN where the α cells project primarily to the magnocellular layers (layers 1 and 2), and the β cells project primarily to the parvocellular layers (layers 3–6). In studies where neurons have been recorded in the LGN, the parvocellular neurons are color selective, with high spatial resolution, relatively low temporal resolution, and low contrast sensitivity (Livingstone and Hubel 1988). In contrast, magnocellular neurons show little color selectivity, have high contrast sensitivity and high temporal resolution, but relatively low spatial resolution (Van Essen and Gallant 1994). These properties reflect a segregation of information into one system that is designed primarily for visuospatial and motion analysis (the M pathway), and one system that is more clearly linked to color and form analysis (the P pathway) (Mishkin et al. 1983; Desimone and Ungerleider 1989; Van Essen and Gallant 1994; Ungerleider 1995). These patterns of connectivity and related functional characteristics are illustrated in Figure 7. This segregation is largely maintained in the terminal patterns in area V1 (Fig. 7), although this segregation is not complete. In fact, Sawatari and Callaway (1996) recently demonstrated a convergence of M and P pathways on layer IVB neurons in macaque primary visual cortex. In addition, these two parallel pathways are essentially split into three pathways by their terminal patterns in areas V1 and V2. This is best reflected by the segregation of afferents that is in register with the patterns of cytochrome oxidase staining that are most clearly apparent in tangential sections (Fig. 8; Wong-Riley 1994). The cytochrome oxidase patterns in area V1 have been referred to as puffs or blobs, particularly in layer III (Fig. 8), where there are circular areas of high cytochrome oxidase intensity surrounded by less intense labeling, referred to respectively as blobs and interblobs (Livingstone and Hubel 1988). In addition, layer IVC stands out as having a uniformly dense labeling pattern while IVA is moderately dense with a honeycomb-like pattern in cytochrome oxidase preparations (Wong-Riley 1994). It has been demonstrated (Livingstone and Hubel 1988) that neurons within the

---

Fig. 7. Diagram representing the hierarchical organization of parallel processing streams in the macaque monkey visual system. This diagram illustrates several principles of visual information processing from the retina and the LGN through primary visual cortex and onto the major corticocortical streams. Boxes represent individual cortical visual areas at the high levels of the hierarchy (e.g., areas PP, PIT, etc.), as well as layers or intraregional compartments (e.g., blob, interblob, etc.) within areas V1, V2 or the LGN at the lower levels of the hierarchy. The thick solid lines represent major connections between areas, most of which are reciprocal, and the thinner lines represent more minor interconnections. The subcortical streams originating in the retina and projecting to the lateral geniculate nucleus include the M, K and P streams (gray, yellow and pink, respectively). Cortical streams at the early and intermediate stages include the MD or magno-dominated stream, BD or blob-dominated stream, and ID or interblob- dominated stream (red, green and blue, respectively). The PP complex (posterier parietal) is shown in orange as a high level area in the MD stream, and the IT complex includes posterior inferotemporal areas (PIT) as well as the central and interior areas (CIT and AIT) which are components of the BD and ID streams. It is clear from this diagram that spatial relationships represented by properties such as orientation or direction of movement are represented primarily in the stream seen on the left of the diagram, and colors and patterns are represented more in the streams represented on the right side of the diagram with increasingly complex response properties represented at the highest levels of both hierarchies. The anatomic information present in the diagram is based largely on information sited in Felleman and Van Essen 1991, and Van Essen and De Yoe 1994. The physiological specializations represented by icons and explained at the bottom of the diagram are based on information that is reviewed in Felleman and Van Essen 1987, De Yoe and Van Essen 1988, and Van Essen and Gallant 1994. Reproduced from Van Essen and Gallant (1994).

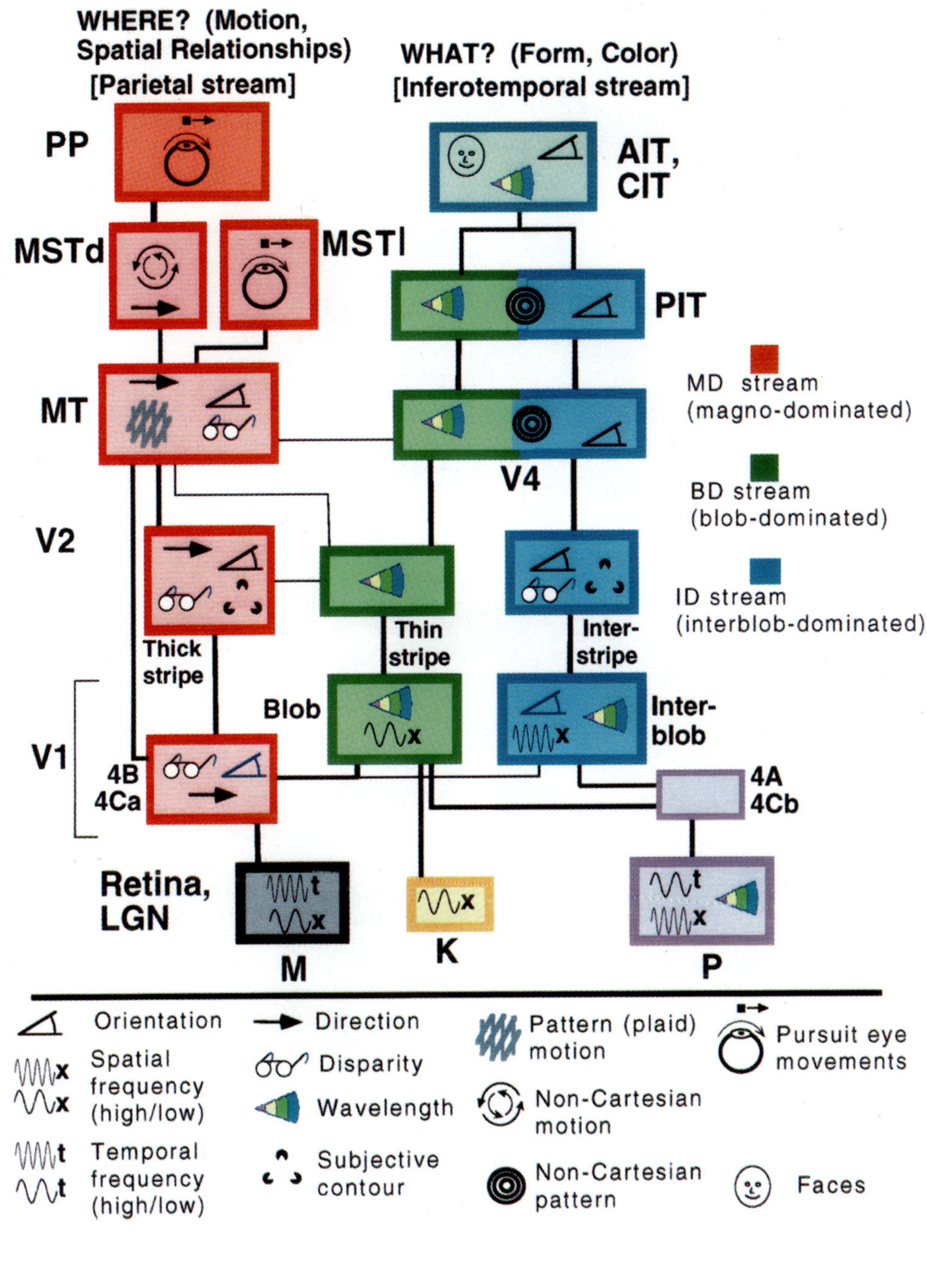

blobs of area V1 are particularly color sensitive. The cytochrome oxidase pattern that is apparent in area V2 is one of stripes, with cytochrome oxidase density variations appearing as a densely labeled thick stripe, a cytochrome oxidase intense thin stripe, and a region in between of low intensity referred to as the interstripe region (see Fig. 7). The M pathway proceeds from area V1 to area V2, terminating primarily in the thick stripes, whereas the P pathway from area V1 to area V2 is subdivided in area V2

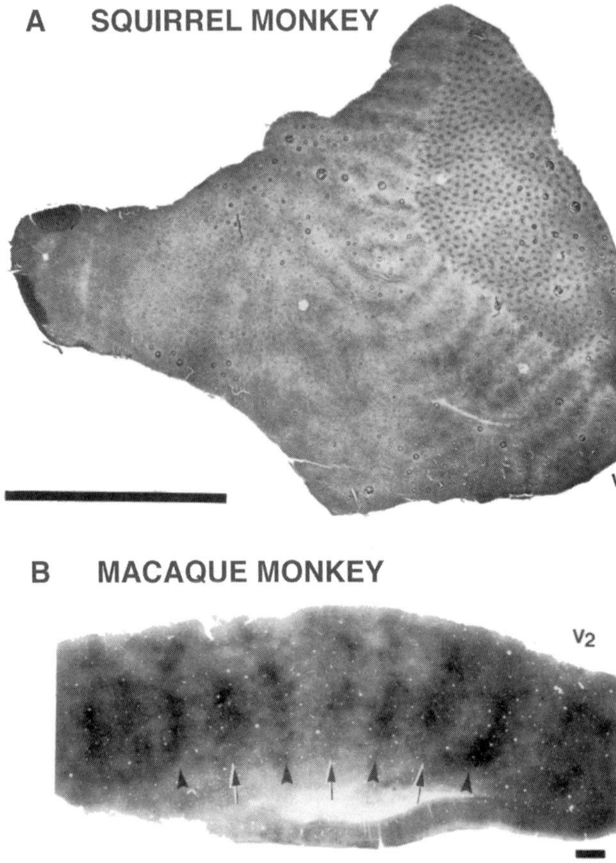

Fig. 8. Illustrations from squirrel monkey (*A*) and macaque monkey (*B*) demonstrating the intraregional patterns of areas V1 and V2 following cytochrome oxidase staining. *A*) A flat-mounted section from the lateral surface of the squirrel monkey occipital cortex stained for cytochrome oxidase (CO). Area V1 is shown to the right, and area V2 to the left. Note that area V1 contains the typical array of spots or 'blobs' and the immediately adjacent portion of area V2 is noted by parallel stripes. This section was taken from layer III. Scale bar = 1 cm. *B*) This is a tangential section through macaque monkey areaV2 reacted for CO. Note the CO rich zones that form the thick bands (*arrows*), and those that form the thin stripes (*arrowheads*), as well as the CO-poor interstripe regions. These are the patterns referred to in Figure 7 as thin stripes, thick stripes and interstripes. Scale bar = 1 mm. Reproduced from Wong-Riley (1994).

with the blobs projecting to the thin stripe, and the interblob regions projecting into the interstripe. This scheme maintains some segregation of color and form analysis in area V2. Information passes sequentially from area V2 to higher level regions (Fig. 7). There are also projections that skip levels of the hierarchy, such as the magnocellular projections directly from area V1 to area MT. Area MT, an area in the caudal portion of the superior temporal sulcus prominently involved in motion detection receives a direct projection from layer IVB and the Meynert cells of area V1.

Within this scheme, areas V1 and V2 can be seen as maintaining a segregation of information in respect to form and color analysis vs. motion and visuospatial analysis that was established in the retina and LGN. In their role as segregators, areas V1 and V2 furnish corticocortical projections that send information relevant to each set of functions along one or the other stream, with the P pathway leading into the temporal lobe, and the M pathway leading into the parietal lobe. However, even this fairly complex description of these parallel hierarchical pathways is clearly oversimplified (see Fig. 9 for a more complete corticocortical scheme). It has been estimated by Van Essen and Gallant (1994), that there are in fact thirty two distinct areas in the macaque cerebral cortex that are visual in nature and there are more than three hundred corticocortical pathways interconnecting these regions. Based on physiologic and connectivity data, these authors have proposed a topographic map of the macaque monkey visual cortex reflecting its parcellation into different functional compartments and outlining possible hierarchical levels and converging pathways between the occipitoparietal and occipitotemporal streams (Felleman and Van Essen 1991). Such parcellation offers a conceptual framework to investigate morphologic criteria to accurately describe the visual areas. The position of a given cortical region in one of the hierarchical schemes is also reflected by the laminar organization of its corticocortical interconnections with other visual regions within the same hierarchical stream (Felleman and Van Essen 1991). The vast majority of these pathways are reciprocal. A projection characterized as forward (e.g., from area V2 to areas V3 or V4) emanates from layer III and terminates primarily in layers III and IV, whereas a feedback projection (e.g., from area V4 to area V2) emanates primarily from layers V and VI, as well as III, and terminates primarily in layers I, II and V. While these highly specified reciprocal projections maintain the two or three hierarchical streams that process visual information, there is extensive opportunity for crosstalk between these streams by corticocortical connections that proceed, for instance, from area MT, an important region in the M pathway, to area V4, an important region in the P pathway. Thus while a high degree of regional segregation of information processing exists in cortical visual systems, extensive corticocortical networks are in place that establish a distributed system from which a cohesive visual world emerges (Fig. 9).

4.4. HORIZONTAL AXON COLLATERALS OF PYRAMIDAL NEURONS

As already alluded to, the axon collaterals of most pyramidal cells form both vertically-oriented, recurrent branches which synapse in layers superficial or deep to their origin, but which remain primarily within the same narrow columnar domain of the parent soma (Fitzpatrick et al. 1985), as well as a separate, longer-range, horizontally-oriented set of projections (Gilbert and Wiesel 1989; Rockland and Lund 1983; Das and Gilbert 1995; Fig. 10).

One of the major functions of the more vertically oriented recurrent collaterals may be to interconnect layers III and V, the two major output layers (Lund and Boothe 1975; Fitzpatrick et al. 1985; Lund et al. 1994). The synaptic targets of the recurrent collaterals have been characterized in detail in one particular case by Kisvárday and colleagues (Kisvárday et al. 1986) for layer III pyramidal cells in cat striate cortex, where 95% of the synaptic targets of the recurrent axon terminals were other pyramidal cells. This applied to both the vertical and distant horizontal projections. In addition, the majority of these synapses were on spines, although shafts are also targets of recurrent collaterals. McGuire and colleagues (1991) carried out a similar analysis directed at the horizontal pyramidal cell collateral terminations in macaque primary visual cortex and also found that the vast majority formed synapses with the spines of other pyramidal cells, but smooth stellate cells (i.e. GABAergic interneurons) were also contacted. These authors hypothesized that such horizontal interconnections might

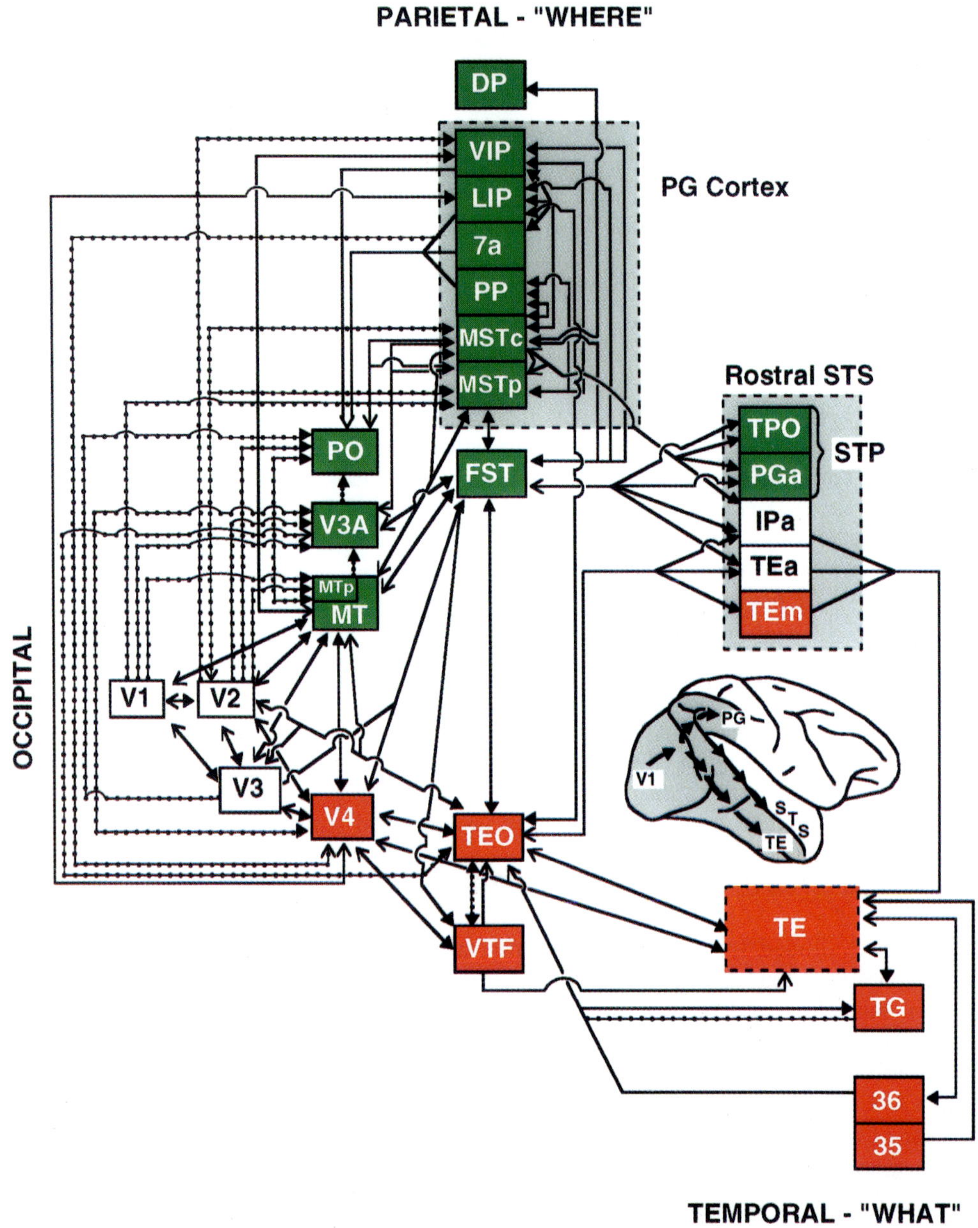

Fig. 9. Schematic representation of visual processing pathways in the macaque monkey. Solid lines indicate connections from both central and peripheral visual field representations. Dotted lines show connections from peripheral visual field representations only. The red boxes indicate visual areas in the occipito-temporal pathway that relate principally to object vision ('temporal what'), and green boxes indicate areas in the occipito-parietal pathway that subserve primarily spatial vision ('parietal where'). The white boxes represent areas that contain elements related to both functional streams. The extent of the regions included in the diagram are shown by the shaded area in the view of the macaque monkey brain. See list for abbreviations. Adapted from Ungerleider (1995).

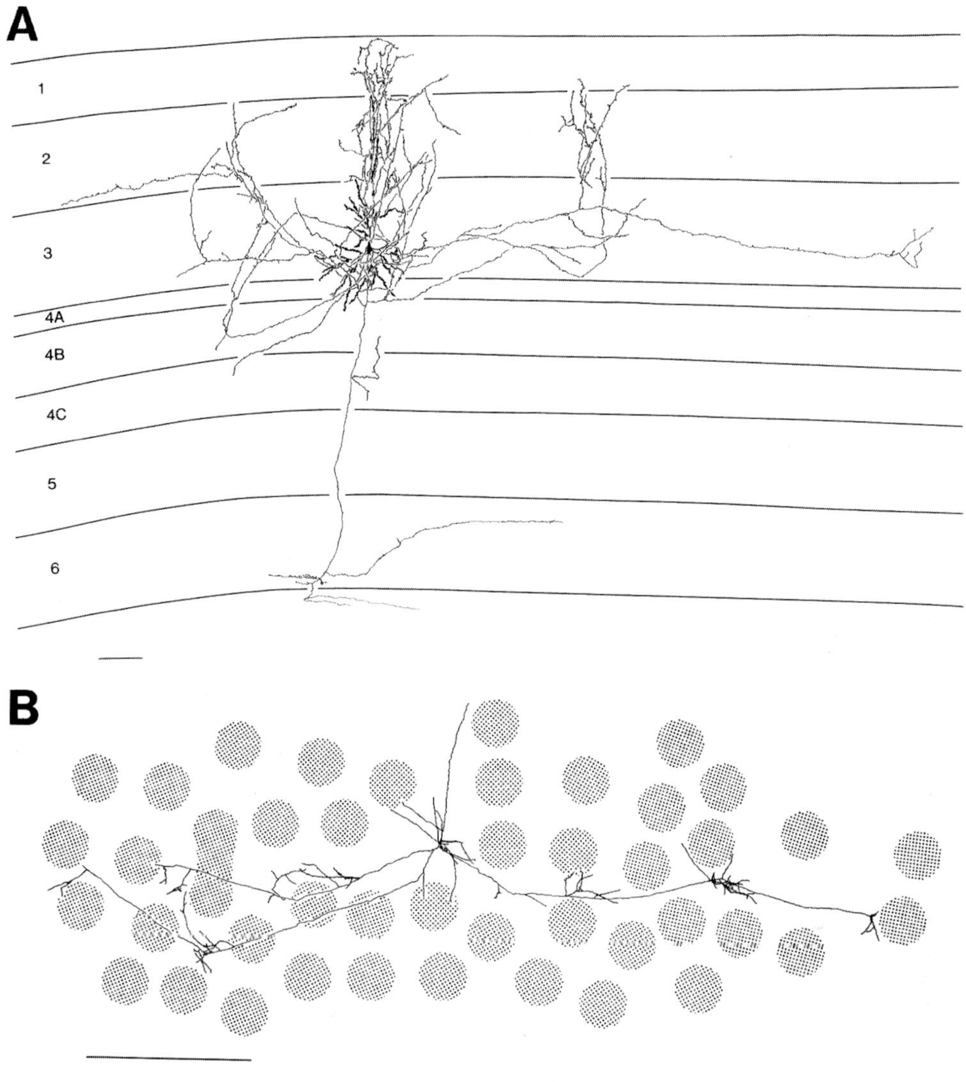

Fig. 10. Camera lucida drawings of intracellularly filled pyramidal neurons in macaque monkey primary visual cortex. Note the very extensive axonal arbor in the horizontal plane. The neuron in *A* is a layer III pyramidal neuron with an axon that extends for more than 4 mm parallel to the cortical surface. The dendrite is visualized as the thicker, darker lines in the center, and the axons are seen as the finer plexus that extends throughout the diagram. *B* is the axon of a different pyramidal neuron seen at a surface view. This neuron extended for approximately 6 mm parallel to the cortical surface. It is shown superimposed on the gray patches representing the cytochrome oxidase blobs. This gives a sense of perspective and shows that cells receiving input from axons such as these would be capable of integrating information over a large area of cortex represented by multiple receptive fields and larger than the receptive fields of this individual neuron. Reproduced from Gilbert (1992).

serve as the substrate for the interactions between the receptive field center and surround in VI.

Functionally, long distance excitatory horizontal collaterals appear to innervate other pyramidal cells that have the same orientation selectivity in primary visual cortex, but are in a different hypercolumn and in so doing skip over intervening regions (Matsubara et al. 1985; Gilbert and Wiesel 1989). Thus, it appears that horizontal intrinsic connections are likely to play a role in integrating the activity of columnar modules with similar functional properties and/or inhibiting columns of dissimilar function (reviewed in Gilbert 1992; Fig 10).

4.5. SPINY STELLATE NEURONS

The other major excitatory input to pyramidal cells of intrinsic origin is that provided by spiny stellate cells (reviewed in Lund 1984). Spiny stellate cells are found primarily, if not exclusively in layers IVA and IVC of primary visual cortex. The spiny stellate cell of primate area V1 is a small or mid-size multipolar neuron whose dendrites and axons tend to arborize extensively within layer IV (Fig. 11), as well as provide crucial excitatory links to superficial layers (Lund 1984; Saint Marie and Peters 1985; Lund et al. 1994). These neurons resemble pyramidal cells in that they have a high density of spines on their dendrites. The major feature that distinguishes them from pyramidal cells is that they lack an apical dendrite (Lund 1984). The excitatory spiny stellate cell of layer IV also resembles the pyramidal cell in the fact that it provides asymmetrical synapses which are excitatory (Saint Marie and Peters 1985), and as is the case for the pyramidal cell, these neurons are thought to use glutamate as their neurotransmitter. These asymmetrical synapses have been identified on spines as well as on shafts of both smooth and spiny dendrites.

While the dendritic arbor of these neurons tends to be local, it can vary from a primarily radial orientation to one that is more horizontal. The relatively restricted dendritic arbor of these neurons presumably reflects the fact that they are high resolution neurons that sample afferents to a very restricted region of V1. The dendrites rarely leave layer IV in which the cell body resides (Lund 1984; Saint Marie and Peters 1985; Lund et al. 1994), and they receive asymmetric synapses on their spines and shafts.

The axon of the spiny stellate cell is primarily radial in orientation and appears to play an important role in forming radial links among the subdivisions of layer IV and from layer IV to III, V, and VI. In some respects the axonal arbor mirrors the vertical plexus of recurrent collaterals, however, they are more radially restricted than recurrent collaterals of pyramidal cells (Lund et al. 1994). It is clear that spiny stellate cells receive a major input from the LGN (Freund et al. 1985). Given its finely tuned axonal distribution, it appears to function as a high fidelity translator of thalamic inputs, maintaining strict topographic organization and in so doing setting up initial vertical links of information transfer within sensory areas. Thus, the spiny stellate cell of area V1 appears to be specialized for reception and local dissemination of glutamatergic geniculocortical inputs (Valverde 1985). Apparently both pyramidal cells and aspiny non-pyramidal cells receive such radially limited inputs from spiny stellate neurons, suggesting that this is likely to be a key interneuron in setting up the excitatory component of the cortical column in area V1. The vertical connections with layer III, particularly from layer IVA, provide important input to the cytochrome oxidase-rich blobs, and thus represent a crucial early radial link in information proces-

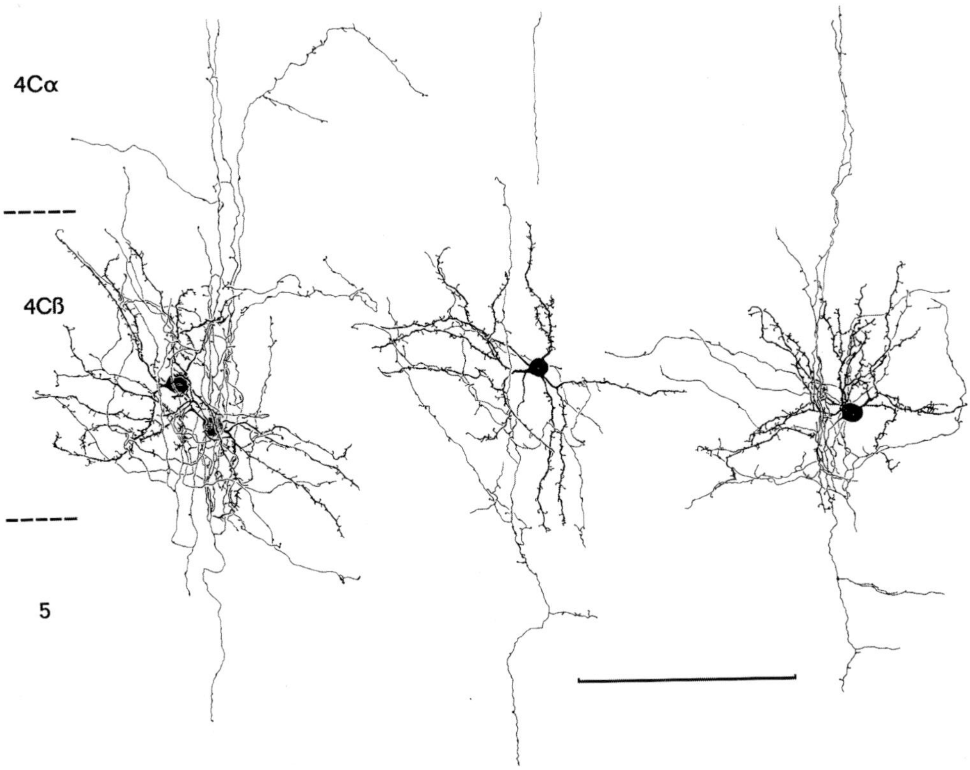

Fig. 11. Three examples of camera lucida drawings of Golgi impregnated spiny stellate neurons in layer IVCβ of monkey primary visual cortex. Note the very limited dendritic arbor confined to layer IVCβ and the extension of the axons into layer V and IVCα within a narrow radial domain. The panel on the left actually includes the arborization of two intermingled spiny stellate cells. Scale bar equals 100 μm demonstrating that the total arbor of the spiny stellate cells is not much more than 100 μm in the horizontal dimension. Reproduced from Saint Marie and Peters (1985).

sing. While the spiny stellate cell may serve as a major excitatory intrinsic link from the thalamus, there is extensive opportunity for inhibition of these neurons by the clutch cell, and thus, even the initial excitatory thalamic link in the neocortex is heavily regulated by GABAergic interneurons.

4.6. GLUTAMATE RECEPTORS IN VISUAL CORTEX

While the neurotransmitter nature of the excitatory connections may be relatively uniform on the pre-synaptic side, it is likely to be extremely complicated on the postsynaptic side due to the variable numbers of glutamate receptor subtypes and their constituent subunits. It is quite possible that the activity of different classes of convergent excitatory inputs (e.g., thalamocortical vs. corticocortical projections) is mediated by different combinations of glutamate receptor subtypes, each composed of differing combinations of subunits, a possibility for which there is a precedent in CA3 of primate hippocampus (Siegel et al. 1995). Therefore, while it is known that the major excitatory afferents to visual areas originate in the LGN, pulvinar, and other regions of cortex, there is much less known about the precise nature of the receptor component of these circuits and, particularly, the degree to which postsynaptic specificity of glutamate receptor proteins is the defining characteristic of a glutamatergic circuit.

The excitatory synaptic activity conveyed by glutamatergic circuits is mediated by three pharmacologically defined subtypes of ionotropic glutamate receptors: AMPA/low-affinity kainate (AMPA receptors), high-affinity kainate (AMPA/kainate receptors), and NMDA receptors. Each of the pharmacologically defined GluR classes are composed of multimeric assemblies of subunit proteins, details of which have been summarized in several recent reviews (Hollmann and Heinemann 1994; Sommer and Seeburg 1992; Nakanishi 1992). Briefly, the established subunit families are: GluR1–4 (or A-D) comprising the AMPA receptors; GluR5–7 and the kainate binding proteins KA1 and KA2 comprising the KA receptors; and subunits NMDAR1 and NMDAR2A-D comprising the NMDA receptors (Hollmann and Heinemann 1994; Kalb 1995). Clearly, while the diversity and specificity of the subunit composition of the GluR complex has profound functional implications, the precise subunit composition of AMPA and/or KA GluRs at any specific postsynaptic site in visual cortex remains largely unknown.

In monkey neocortex, including primary visual cortex, quantitative colocalization studies suggest that distinct pyramidal and non-pyramidal cell classes can be subdivided by GluR expression patterns (Huntley et al. 1994a,b; Vickers et al. 1993; Fig. 12). At least two distinct subpopulations of NMDAR1-immunoreactive pyramidal cells can be identified on the basis of differential non-NMDA receptor subunit immunoreactivity: one which possesses AMPA subunits GluR2/3 and kainate subunits GluR5/6/7 (Fig. 12), while the other possesses GluR2/3, but lacks GluR5/6/7. These data begin to suggest that particular classes of neocortical pyramidal neurons display different excitatory amino acid receptor subunit profiles.

Neocortical GABAergic interneurons can be subdivided into largely non-overlapping subsets as defined by the selective distribution of three calcium binding proteins; parvalbumin, calbindin, and calretinin (see below, section 5.5.2). Thus, double labeling studies using subunit specific GluR antibodies and antibodies to the calcium-binding proteins are particularly useful to obtain data on the relative representation of different GluRs in GABAergic cells in general, as well as in chemically-defined subsets. With respect to NMDAR1 and GABAergic interneurons in macaque area V1, studies demonstrate that while NMDAR1 is present in virtually all parvalbumin-containing GABAergic interneurons, NMDAR1 is less broadly distributed in calretinin-immunoreactive neurons (Huntley et al. 1994a). The situation with AMPA subunits is more complex, since the presence or absence of GluR2 has different functional implications than for GluRs1, 2, and 4. While AMPA receptors that possess GluR2 are not significantly permeable to calcium, those that contain any combination of GluRs 1, 3, and 4 but lack GluR2 are fairly permeable to calcium (Hollmann et al. 1991; Sommer et al. 1991; Hollmann and Heinemann 1994). This is particularly relevant in hippocampus and neocortex, because electrophysiological analyses combined with RT-PCR analyses of mRNA levels have suggested that pyramidal cells contain AMPA receptors that are relatively impermeable to calcium, whereas non-pyramidal GABAergic neurons contain calcium permeable AMPA receptors, and this is matched by a higher GluR2:GluR1 mRNA ratio in pyramidal cells as compared to non-pyramidal cells (Jonas et al. 1994; Geiger et al. 1995). Initial double labeling studies suggested that GluR2/3 were poorly represented in GABAergic interneurons in both primate hippocampus and neocortex (Vickers et al. 1993; Leranth et al. 1996), however, more recent

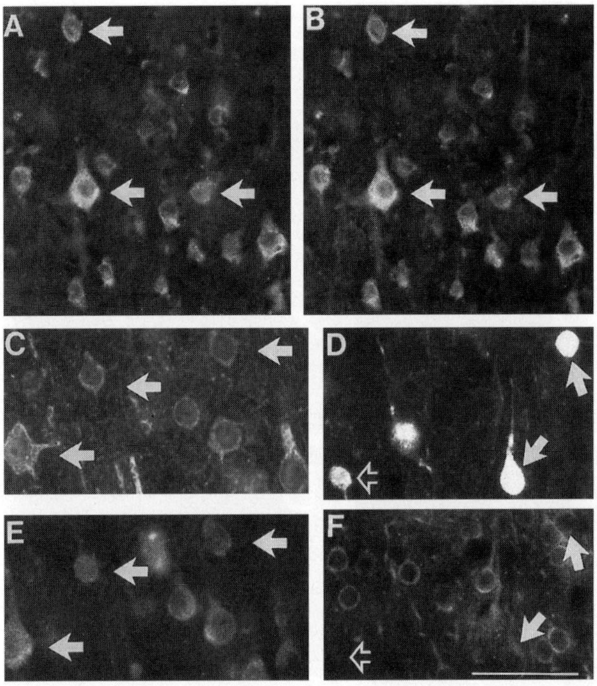

Fig. 12. Pairs of fluorescence micrographs taken from single sections of macaque monkey neocortex showing colocalized combinations of NMDAR1, non-NMDA receptor subunits, and calcium binding proteins. *A, B*) section through layer III of primary somatosensory cortex showing cells that are immunoreactive for both GluR2/3 (designated by *arrows* in *A*), and NMDAR1 (designated by *arrows* in *B*). *C* and *E* represent the same section through layer III of primary motor cortex showing cells immunoreactive for both GluR5/6/7 (designated by *arrows* in *C*), and GluR2/3 (designated by *arrows* in *E*). *D* and *F* represent the same section showing cells in layer IVCβ of area V1 labeled for parvalbumin, solid arrows in *D*. There are also double labeled for NMDAR1 (*solid arrows* in *F*). One neuron labeled in *D* (*open arrow*) is not immunoreactive for NMDAR1 in *F* (*open arrow*). Scale bar = 50 μm. Reproduced from Huntley et al. (1994).

studies employing a highly selective and sensitive GluR2-specific monoclonal antibody have demonstrated that virtually all parvalbumin and calbindin-containing GABAergic interneurons in rodent hippocampus and somatosensory cortex (Vissavajjhala et al. 1996), as well as in macaque primary visual cortex (Hof et al. 1996; Fig. 13), in fact contain GluR2. However, preliminary ultrastructural analyses with this antibody suggest that individual AMPA synapses on GABAergic interneurons contain less GluR2-immunoreactive signal than do AMPA-mediated synapses on pyramidal cells (He and Morrison, unpublished observations), which may explain the relatively higher calcium permeability of these AMPA synapses on GABAergic interneurons as compared to those on pyramidal cells. These data are instructive in the degree to which they point out the need for high resolution data with subunit selective antibodies in order to draw precise linkage between an identified circuit or cell class and the molecular or functional attributes of glutamatergic circuits.

Regional and laminar patterns of subunit-identified cells and synapses suggest that different subunits are also differentially represented in mediating the activities of distinct excitatory pathways (Huntley et al. 1993; Vickers et al 1993; Huntley et al. 1994a,b). For example, the density of GluR5/6/7 immunoreactive neurons is greatest

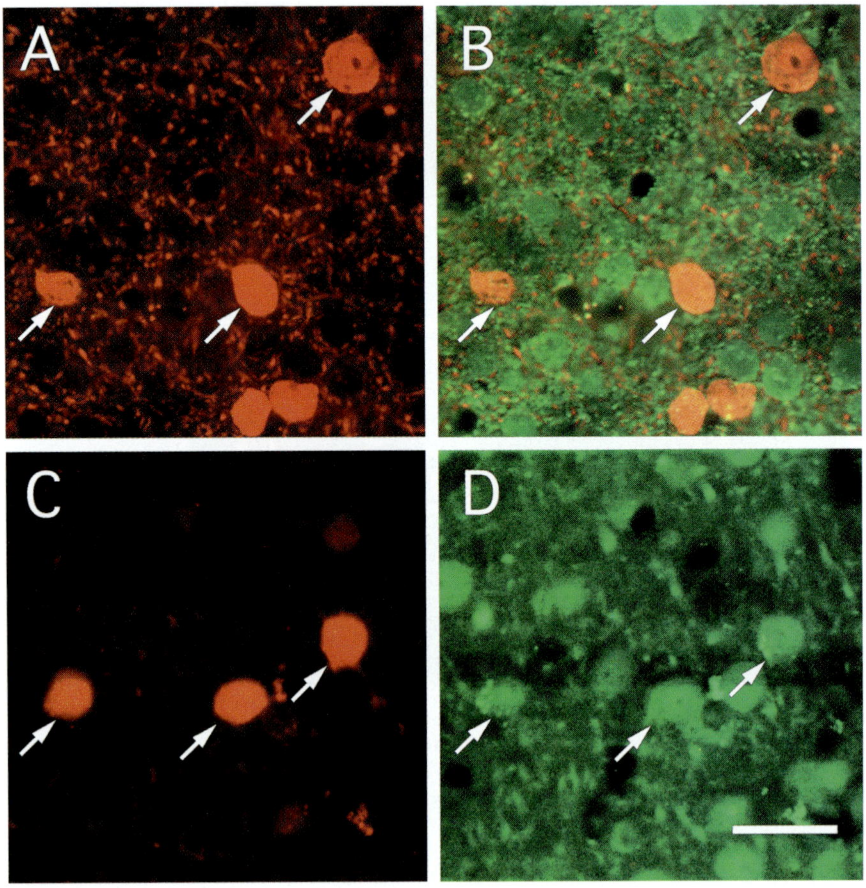

Fig. 13. Examples of the laminar distribution and colocalization of GluR2- (*A, C*) and parvalbumin- (PV; *B, D*) immunoreactive neurons in the macaque monkey area V1. Note that the vast majority of neurons appear to contain GluR2 (*A*). *C, D*) Double labeling at the confocal microscopy level shows that the vast majority of PV-immunoreactive neurons (*D*) also contain GluR2-immunoreactivity (*C*). Some double labeled neurons are marked by *white arrows*. Scale bar 25 μm.

in higher-order association areas of the prefrontal cortex in comparison with primary visual cortex (Huntley et al. 1993). Additionally, across prefrontal, temporal, motor and sensory areas of monkey neocortex, GluR5/6/7 subunit-immunoreactivity is densest in the principal cortical layers which furnish and receive corticocortical, commissural and local excitatory projections, while relatively sparse in layers which receive a large share of thalamocortical afferents. This is particularly evident in layer IV of areas V1 and V2 (Huntley et al. 1993; Fig. 14). Recent data by Carder (1997) demonstrate that the AMPA receptor subunits GluR1, GluR2/3/4c and GluR4 have a preferential distribution in three bands of dense labeling corresponding to layers IVA, IVC and VI. In layer IVA the staining has honeycomb pattern and in layer IVC it displays a lattice pattern. In addition the cytochrome oxidase rich blobs in layers I-II and V-VI are intensely stained, suggesting a high degree of correlation between direct geniculocortical inputs distribution in area V1 and AMPA receptor localization (Carder 1997).

In contrast, NMDAR1-immunoreactive cells and synapses are densely distributed

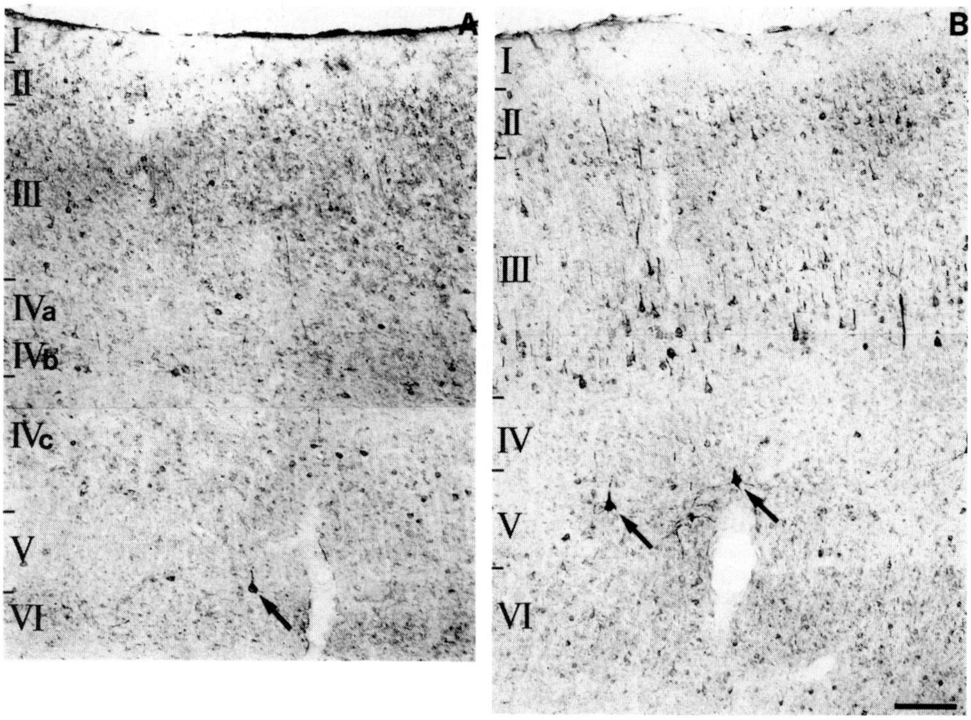

Fig. 14. Distribution of kainate subunits GluR5/6/7 in areas V1 (*A*), and V2 (*B*) of macaque monkey. The overall staining intensity is higher in area V2 than in area V1. The arrows in both *A* and *B* note relatively intensely stained neurons located at the layer V-VI border in each area. Scale bar = 100 μm. Reproduced from Huntley et al. (1993).

throughout all layers of rat (Aoki et al. 1994; Petralia et al. 1994) and monkey visual cortex, including layer IVC in monkey area V1 (Fig. 15; Huntley et al. 1994a). The presence of such high levels of NMDAR1 subunits in layer IV of area V1 in adult monkeys was unexpected, given extensive electrophysiological data suggesting that NMDA receptors do not mediate thalamocortical transmission in adult cortex (Fox et al. 1989). During development, however, NMDA receptors appear to participate in monosynaptic thalamocortical activity (Fox et al. 1989). The change in the role of NMDA receptors in mediating thalamocortical activity may be accounted for in part by developmental changes in the expression or distribution of NMDA receptors, since in developing rodent brain, mRNAs encoding different NMDA receptor subunits undergo significant changes in expression levels and distribution over time (Monyer et al. 1994). Nevertheless, given the significant percentages of NMDAR1-labeled synapses in layers IV to VI of adult monkey cortex (Huntley et al. 1994a; Fig. 16, Table 1), it may be the case that under normal conditions, various modulatory influences, particularly GABAergic inputs, render NMDA receptors in granular and infragranular layers functionally quiescent (Huntley et al. 1994b). In this context it will be important to determine whether or not the NMDAR1-positive synapses in layer IV of area V1 mediate geniculocortical input and are thus in a position to be 'unmasked'.

In addition to ionotropic glutamate receptor subtypes, recent studies also point to differential distribution and variable effects of certain metabotropic glutamate receptor

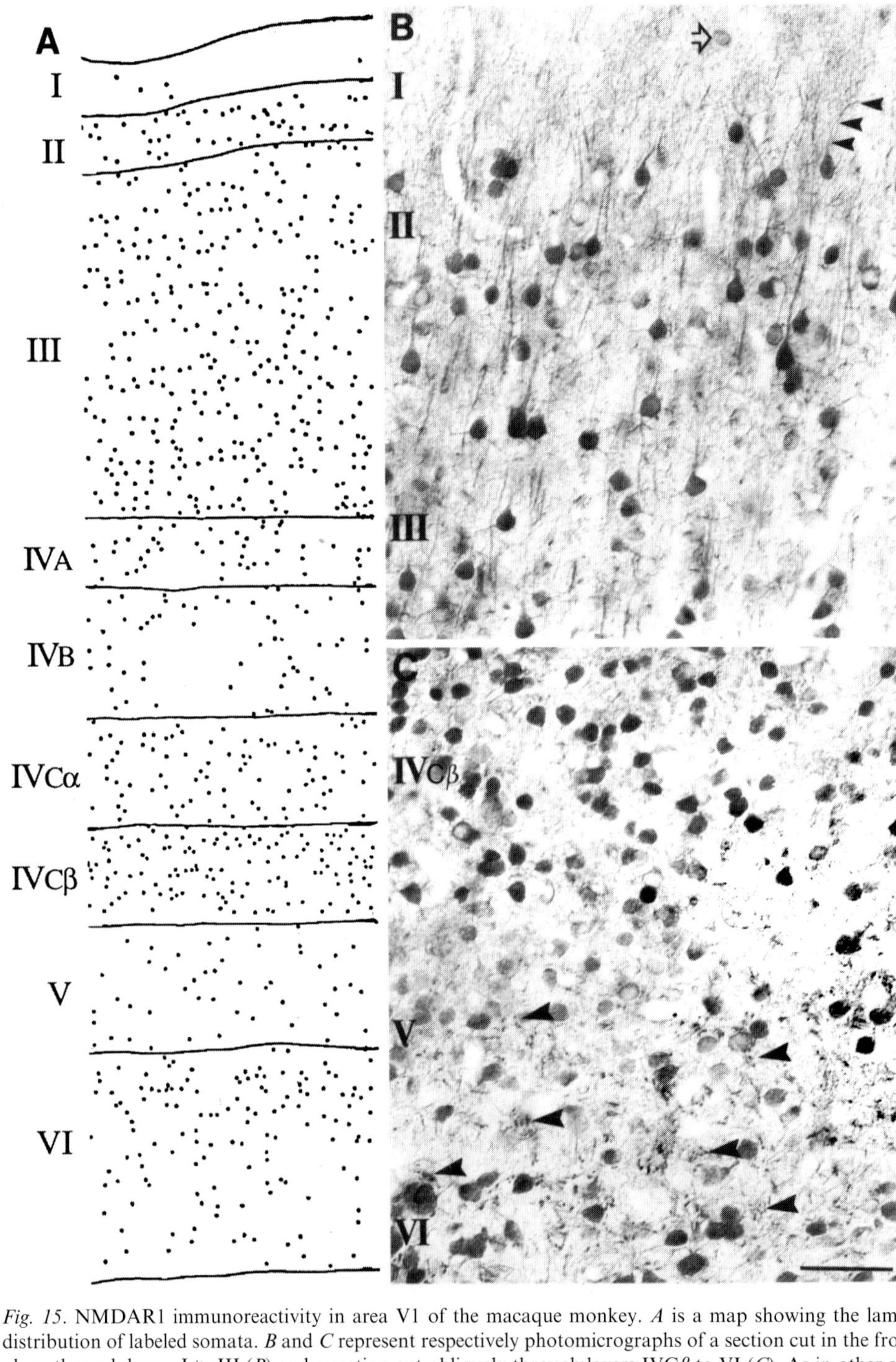

Fig. 15. NMDAR1 immunoreactivity in area V1 of the macaque monkey. *A* is a map showing the laminar distribution of labeled somata. *B* and *C* represent respectively photomicrographs of a section cut in the frontal plane through layers I to III (*B*) and a section cut obliquely through layers IVCβ to VI (*C*). As in other areas labeled apical dendrites could be followed from the pyramidal cell somata in layer II into layer I (*arrowheads* in *B*). Also a dendritic plexus is apparent in layer I as well as small faintly labeled somata (*open arrow* in *B*). In obliquely cut sections through the deep layers small clusters of cut dendrites were observed in layers V and VI (*arrowheads* in *C*). Scale bar = 50 μm. Reproduced from Huntley et al. (1994).

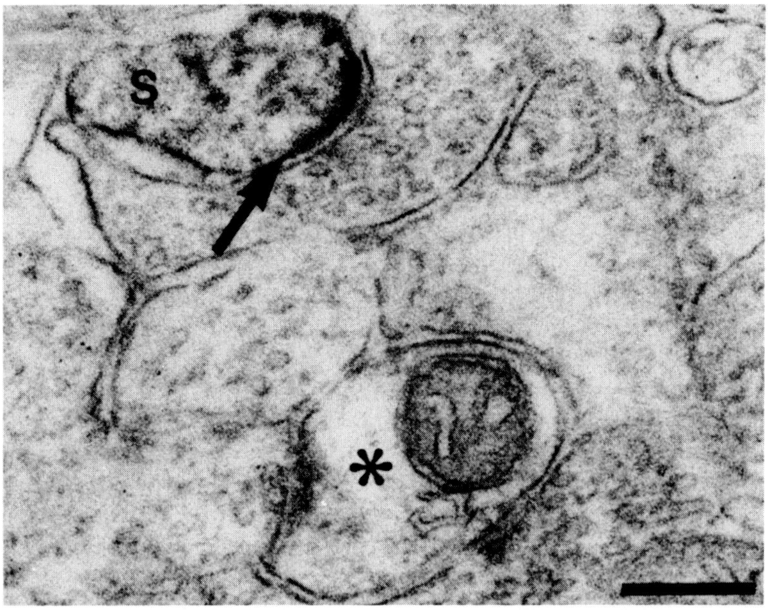

Fig. 16. Ultrastructural localization of NMDAR1 immunoreactivity in layer IVCβ of area V1 in the macaque monkey. This is an illustration of diaminobenzidine-labeled immunostained ultrastructural material. Note the immunopositive dendritic spine noted by an *S* and an *arrow* pointing to the synaptic specialization in the upper left. Note also the unlabeled dendritic shaft noted by an *asterisk* in the lower part of the figure. Scale bar = 0.2 μm. Reproduced from Huntley et al. (1994).

(mGluR) subtypes in the neocortex (Wang and McCormick 1993; Baskys 1994). In the visual cortex most available data have been obtained from rodents and cats (Sledezcek et al. 1993; Kato 1994; Reid et al. 1995; Reid and Daw 1997), where agonists of mGluRs have both excitatory and inhibitory effects on neuronal activity. Single unit recordings from cat area 17 demonstrated that mGluR dependent increases in spontaneous activity were observed in most cases in the infragranular layers and decreases in the supragranular layers (Reid et al. 1995), indicating that mGluRs have both facilitatory and depressive effects that appear to be cell type- and layer-dependent. This physiologic observation is also paralleled by the fact that immunohistochemical studies have shown that in the cat visual cortex, different mGluR subtypes have a

TABLE 1. *Percentages of NMDAR1-immunoreactive synapses in primary motor (M1), somatosensory (S1) and visual (V1) cortex of the macaque monkey*

Layers	Areas		
	M1	S1	V1
II-III	46.0 (52/113)	26.4 (83/314)	44.3 (144/325)
IV	–	45.6 (68/149)	34.5 (91/264)
V-VI	40.8 (53/130)	37.4 (92/246)	19.2 (53/276)

Values represent percentages of total asymmetric synapses that were NMDAR1-immunoreactive counted from randomly chosen microscopic fields in the indicated layers and cortical areas. Raw numbers are shown in parentheses. Layers are indicated by Roman numerals. The data from area S1 were obtained through area 3b. In layer IV of area V1, data were collected in layer IVCβ only. Adapted from Huntley et al. (1994).

different laminar distribution. Thus, mGluR1α is moderately present in layer I to IV but is very dense in layers V and VI, while mGluR2 and 3 have a nearly opposite distribution, and mGluR5 is most dense in layer V (Reid et al. 1995; Reid and Daw 1997), suggesting that mGluRs may influence visual function in a manner that is equally complex but distinct from that of ionotropic receptors.

5. INHIBITORY CIRCUITS: GABAERGIC INTERNEURONS

5.1. INTRODUCTION

Inhibition is a major component of the response properties of cortical neurons (e.g. Creutzfeldt and Ito 1968; Innocenti and Fiore 1974), and is critical for the establishment or fine-tuning of a variety of receptive field properties of visual cortical neurons which include orientation, directional specificity, stereopsis and end-stopping (Sillito 1975, 1977, 1979; Poggio and Fischer 1977; Tsumoto et al. 1979; Ferster 1988; Wörgötter and Eysel 1991). Functional inhibition in visual and other cortical areas is mediated by a population of intrinsic neurons which synthesize and release GABA, recognized as the principal inhibitory neurotransmitter of the CNS (Krnjevic 1974, 1987).

5.2. PROPORTION AND LAMINAR DISTRIBUTION OF GABAERGIC NEURONS AND TERMINALS IN PRIMATE AREA V1

The primary visual cortex contains the greatest density of GABAergic neurons in comparison with other cortical areas (Hendry et al. 1987). However, given the greater overall cellular density of this cortical region (Rockel et al. 1980; O'Kusky and Colonnier 1982), the proportion of all neurons which displays GABA immunoreactivity is somewhat less (20%) than the proportions typical of other areas (25–30%) (Fitzpatrick et al. 1987; Hendry et al. 1987; Beaulieu et al. 1992). At the synaptic level, GABA-immunopositive synapses account for approximately 17% of all synapses in monkey V1 (Beaulieu et al. 1992).

GABAergic neurons and terminals are present throughout all layers of primate area V1 (Fig. 17), and are also found in the subjacent white matter. However, both GABA-containing somata and terminals are unevenly distributed across layers, with the greatest concentrations found in layers which receive the terminations of thalamocortical axons (Fig. 17B; Hendrickson et al. 1981; Fitzpatrick et al. 1987; Hendry et al. 1987). Numerically, the greatest concentrations of GABA-immunoreactive neurons are found in supragranular layers, which contain approximately 40–50% of the total number of GABAergic cells, and in layers IVA and IVC, which contain approximately 35% of the total number of GABAergic cells. The proportion of total neurons which are GABAergic also varies by layer, ranging from 85–95% in layer I, to 15–25% in other layers (Fitzpatrick et al. 1987; Hendry et al. 1987; Beaulieu et al. 1992). The uneven distribution of GABAergic cells across layers is also evident in studies localizing glutamic acid decarboxylase (GAD), the enzyme which synthesizes GABA, by immunocytochemical detection of the GAD protein (Fitzpatrick et al. 1987) or by *in situ* hybridization of probes specific for GAD mRNA (Figs 17C, 18; Benson et al. 1991a). However, GAD localization, particularly that by *in situ* hybridization methods, reveals a further sublaminar organization in which sublayer IVCα contains two distinct sub-

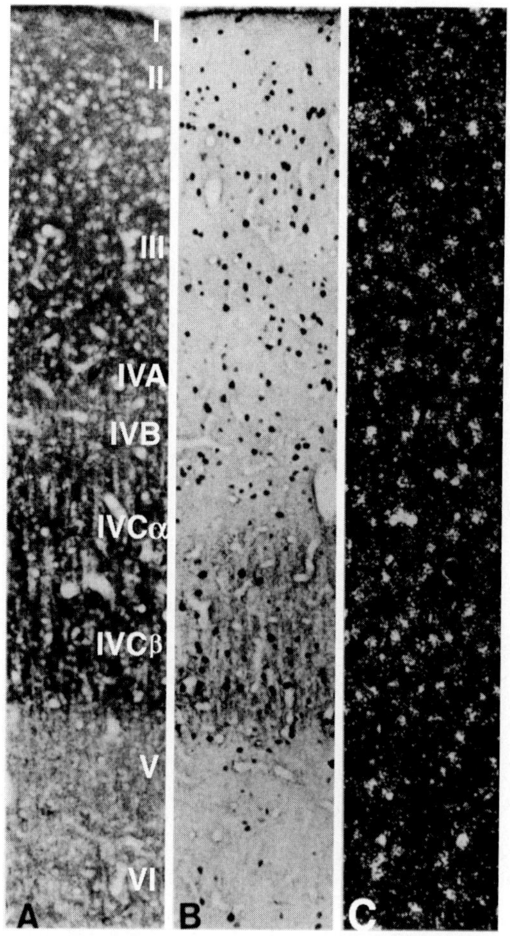

Fig. 17. Distribution of GABA neurons in macaque monkey area V1. Photomicrographs of a section processed histochemically for cytochrome oxidase activity (*A*); section processed immunocytochemically for GABA (*B*); and section processed by *in situ* hybridization of autoradiographically localized cRNA probe to GAD mRNA (*C*). Modified from Jones (1993).

layers, while IVCβ and layer VI contain three distinct sublayers (Fig. 18; Benson et al. 1991a).

While GABAergic cell somata and terminals are distributed unevenly *across* layers, they are uniformly distributed *within* layers, displaying no variations in density or distribution which correlate with eye dominance columns or functionally distinct compartments revealed by differential intensity patterns of cytochrome oxidase activity (Fig. 19; Hendrickson et al. 1981; Fitzpatrick et al. 1987; Hendry et al. 1987; Beaulieu et al. 1992). Such cytochrome oxidase patterns include the intensely stained patches and the surrounding, lightly stained inter-patch regions most apparent in superficial layers (Fig. 19B), as well as an irregular honeycomb pattern in layer IVA, in which intensely stained walls surround weakly stained hollows (Horton and Hubel 1981; Humphrey and Hendrickson 1983; Horton 1984; Wong-Riley and Carroll 1984). In both cases, regions of intense cytochrome oxidase activity (patches in layers II/III and honeycomb walls of layer IVA) correspond precisely to regions of geniculocortical

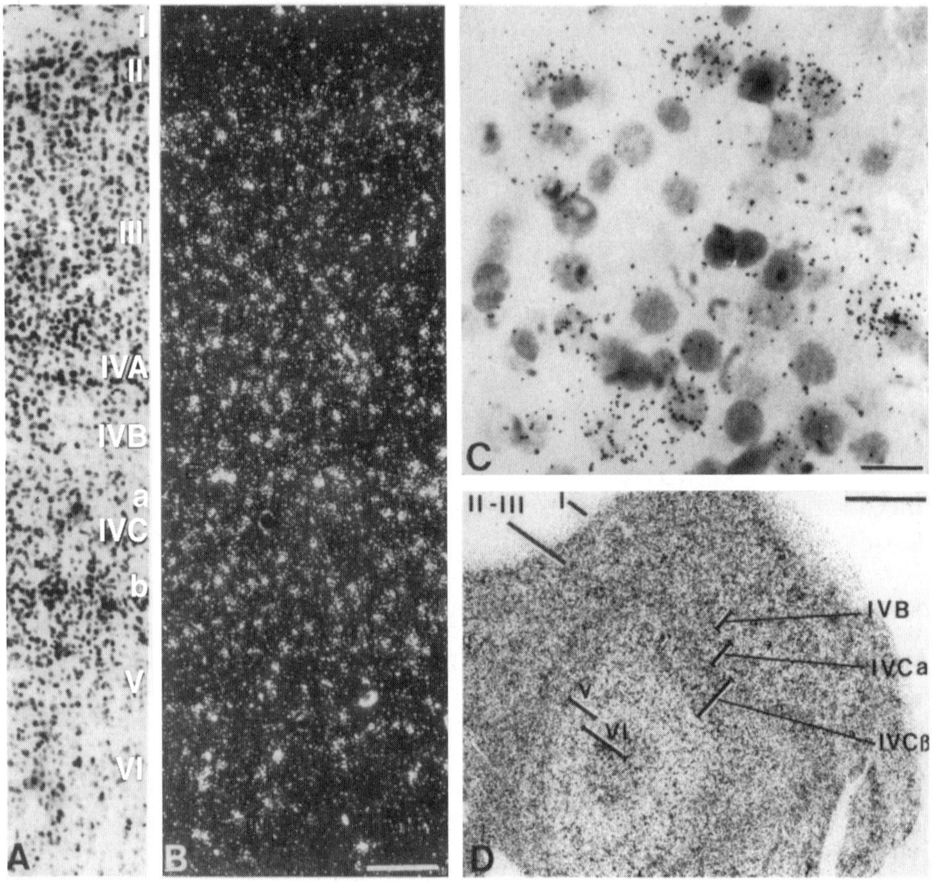

Fig. 18. Laminar distribution of GAD mRNA-expressing cells in macaque monkey area V1 labeled by *in situ* hybridization. *A, B*) Photomicrographs of matched pair of sections, one stained with thionin to indicate lamination (*A*), the other processed autoradiographically to visualize GAD mRNA-containing neurons (*B*). *C*) High-power photomicrograph demonstrating silver-grain clusters overlying pale Nissl-stained nuclei characteristic of neurons. *D*) Film autoradiograph of flattened, tangentially-cut section demonstrating sublaminar hybridization patterns. Scale bars = 100 μm (*A, B*), 15 μm (*C*). Reproduced from Jones et al. 1994, after Benson et al. (1991a).

terminations (Hendrickson et al. 1978; Livingstone and Hubel 1982; Blasdel and Lund 1983; Fitzpatrick et al. 1983; Itaya et al. 1984). When terminals are labeled with anti-GAD antisera, they appear enhanced within the layer III patches, but this effect is due to an overall larger terminal size within the patches in comparison with those occupying interpatch regions (Hendrickson et al. 1981; Fitzpatrick et al. 1987), and not an actual difference in terminal number or density (Beaulieu et al. 1992).

5.3. GENERAL FEATURES OF GABAERGIC NEURONS: MORPHOLOGY AND SYNAPTIC TARGETS

GABAergic neurons can be distinguished by several common morphological features. Such neurons belong to the class of aspiny or sparsely spiny, non-pyramidal cells (Ribak 1978; Houser et al. 1983b), characterized by dendrites that are smooth or

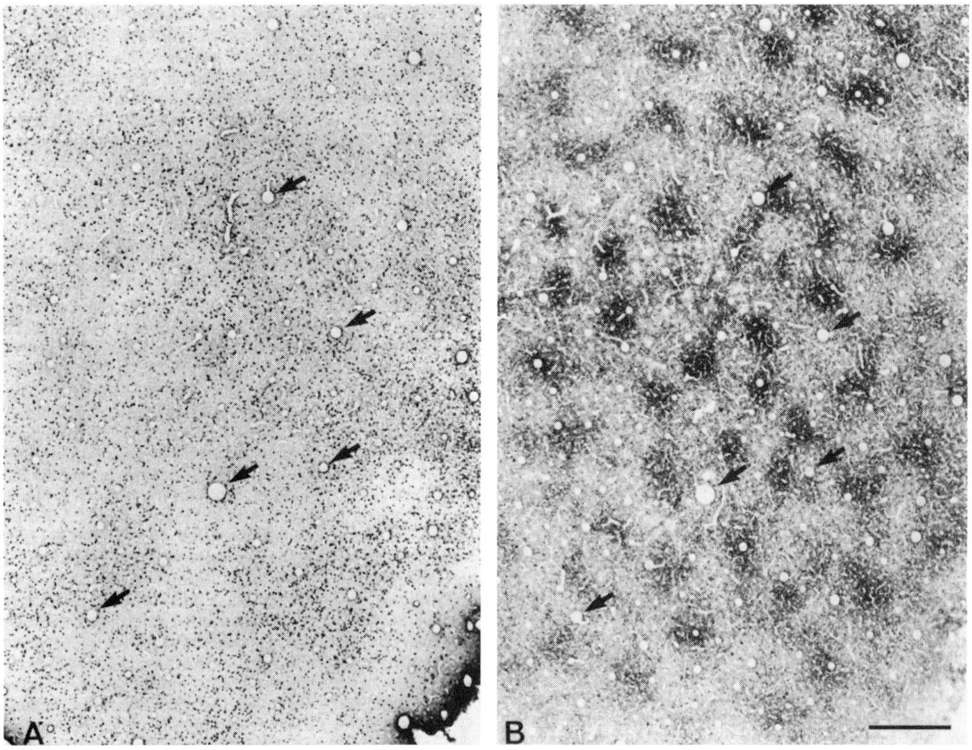

Fig. 19. Photomicrographs of adjacent sections cut tangentially through layer III of macaque monkey area V1, stained immunocytochemically for GABA (*A*) or processed histochemically for cytochrome oxidase activity (*B*). *Arrows* indicate matching blood vessels in both photomicrographs. There is no consistent relationship between the distribution of GABAergic neurons and the puffs of intense cytochrome oxidase activity. Reproduced from Hendry et al. (1987).

lightly beaded, and axons which form relatively local ramifications, mostly in the vicinity of their parent soma (Fig. 20; Cajal 1899b; Valverde 1971; Lund 1973; Lund et al. 1979; Fairén and Valverde 1980; Kisvárday et al. 1986). GABAergic neurons can also be characterized by their synaptic organization: their somata receive both asymmetric and symmetric synapses, while their axons form exclusively symmetric synapses with target cells (Somogyi and Soltész 1986). A quantitative estimate of the cell regions targeted by GABA-immunopositive synapses in monkey V1 indicates that nearly 66% of all GABAergic synapses target dendritic shafts, while approximately 25% target dendritic spines, 12% target neuronal somata and less than 0.1% target axon initial segments (Beaulieu et al. 1992). While such general morphological characteristics serve to distinguish GABAergic cells from the other major class of cortical neuron–the spiny class, which includes both pyramidal and spiny stellate cells–GABAergic neurons are themselves a markedly heterogeneous group, exhibiting a large degree of morphological and neurochemical diversity.

5.4. MORPHOLOGICALLY-DEFINED SUBTYPES OF GABA NEURONS

Based principally on unique patterns of axon terminations, there are four major classes of GABAergic interneuron which have been identified in monkey area V1: basket

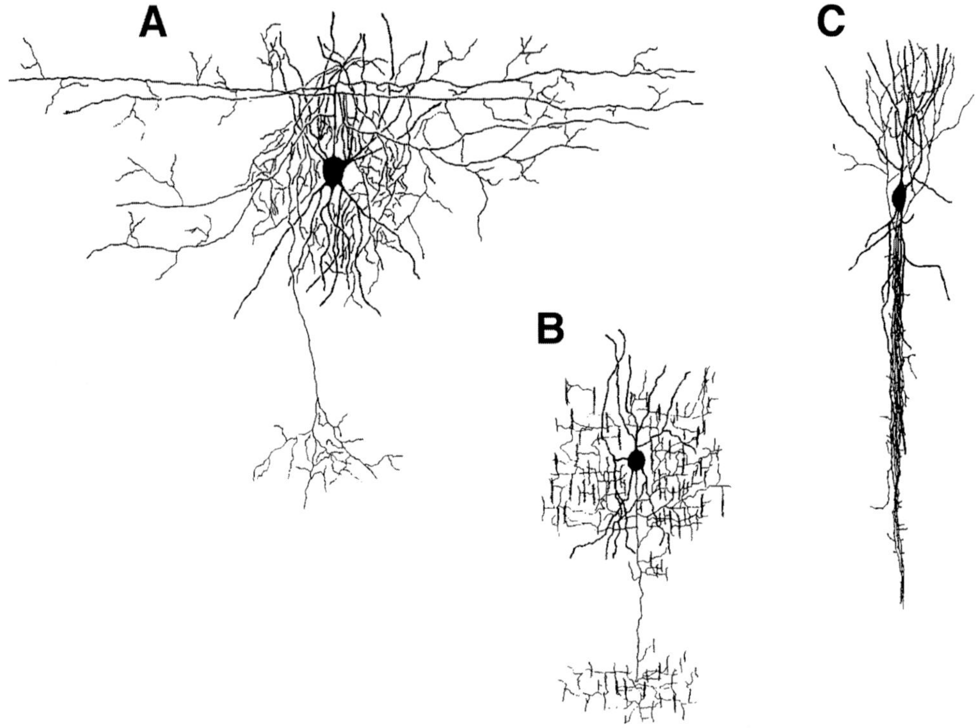

Fig. 20. Drawings, based on Golgi-impregnated neurons, of the major morphological types of GABAergic interneurons. *A*) Classic basket cell; *B*) chandelier cell; *C*) double bouquet cell. Modified from DeFelipe (1992).

cells, chandelier cells, double-bouquet cells and clutch cells (Fig. 20). Details of their organization and identification as GABA-containing have been supported or inferred by additional studies of area V1 in other species, or other cortical areas in monkeys, and these studies will, by necessity, be referred to. In addition to those GABA neurons which fall into one of the above classes, there are many others present in primate area V1 for which solid links between chemical identity and morphological categorization are more conjectural. However, they probably belong to one of the many other varieties of morphologically defined aspiny nonpyramidal cells which have been described in visual and other cortical areas from Golgi studies (e.g. Cajal 1899b; Lund 1973; Jones 1975).

5.4.1. Basket cells

Basket cells give rise to myelinated axons which form pericellular clusters of terminations around pyramidal cell somata and their proximal apical and basilar dendrites (both shafts and spines; Figs 20A, 21), and possibly around the somata and proximal dendrites of other non-pyramidal neurons as well (Marin-Padilla 1969; Jones 1975; Freund et al. 1983; Somogyi et al. 1983b; DeFelipe et al. 1986; Somogyi and Soltész 1986; Lund 1987; Lund et al. 1988). Such clusters of terminations form a local plexus, typically very close to the parent soma, or found in layers above or below that in which the parent soma is found, as well as a distant plexus, which arises from axon collaterals which can extend for longer distances (up to several millimeters) in horizontally-directed traverses (Marin-Padilla 1969; Jones 1975; Somogyi et al. 1983b; DeFelipe et al. 1986; Kisvárday et al. 1987; Lund 1987; Lund et al. 1988). The relatively long range of such horizontally-oriented collaterals is supported by autoradiographic studies of the tangential distribution of neurons accumulating ^{3}H-GABA or related molecules following small injections into cat or monkey striate cortex or other cortical areas in monkeys (Hendry and Jones 1981; Somogyi et al. 1983b; Kisvárday et al. 1987; Kritzer et al. 1992), and suggests that basket cells are positioned to mediate inhibition across functionally distinct columns or compartments. The somata of large basket cells can approach 20 μm in diameter in sensory-motor cortex (Marin-Padilla 1969; Jones 1975; DeFelipe et al. 1986), although smaller versions are present in area V1 (e.g. Lund et al. 1988). The multipolar dendrites are also extensive, and tend to align themselves vertically. In monkey area V1, GABA-immunoreactive neurons with the morphological characteristics of classic basket neurons have been identified (Fitzpatrick et al. 1987), and comparative studies in area V1 of other species or of other areas confirm that symmetric axosomatic synapses are almost invariably GABA- or GAD-immunoreactive (Ribak 1978; Freund et al. 1983; Hendry et al. 1983a; Somogyi and Soltész 1986; Beaulieu and Somogyi 1990; Fariñas and DeFelipe 1991a). Although details of the precise laminar distributions of basket cells are more limited, they appear to be present in layers II, III, IVB, V and VI in area V1 of monkeys (Lund 1987; Lund et al. 1988) and humans (Cajal 1899b).

5.4.2. Chandelier cells

Chandelier cells give rise to axon terminations which form short, vertically oriented rows of boutons resembling candlesticks (Fig. 20B). The configurations of the dendritic and axonal fields of chandelier cells are variable. However, the synaptic targets of chandelier cells are very specific: synapses formed by such neurons target exclusively the axon initial segments of pyramidal neurons and some spiny stellate cells (Fig. 21), and thus they are variably referred to as axo-axonic cells (Jones and Powell 1969; Szentágothai and Arbib 1974; Jones 1975; Somogyi 1977, 1979; Fairén and Valverde 1980; Peters et al. 1982; Somogyi et al. 1982; Freund et al. 1983; DeFelipe et al. 1985; Lund 1987; Werner et al. 1989; Peters and Harriman 1990). A single chandelier cell axon can give rise to up to several hundred rows of boutons, where any given row of boutons can represent all of the synapses a single pyramidal cell axon initial segment receives, or can be one of a set of converging rows in which the others are contributed by one or two other chandelier cells (e.g. Freund et al. 1983; DeFelipe et al. 1985). The axon initial segments of non-pyramidal neurons mostly do not receive the terminations of chandelier cell axons, although an exception to this may be the large basket cells of monkey sensory-motor cortex (DeFelipe et al. 1986). The somata of chandelier cells appear to be present throughout layers III to VI of monkey striate cortex, possibly with the exception of layer IVCβ (Lund 1987), but the laminar density of chandelier cell terminals in primate area V1 is probably variable based on generalizations made from studies of other species and areas. Quantitative studies of chandelier cell terminations on identified projection neurons in cat visual cortex have shown, for example, that the axon initial segments of callosally-projecting and corticocortically projecting pyramidal neurons of layer III receive a greater density of synaptic contacts in comparison with the density of those on the axon initial segments of corticothalamic neurons of layer VI (Fariñas and DeFelipe 1991b), confirming previous indications

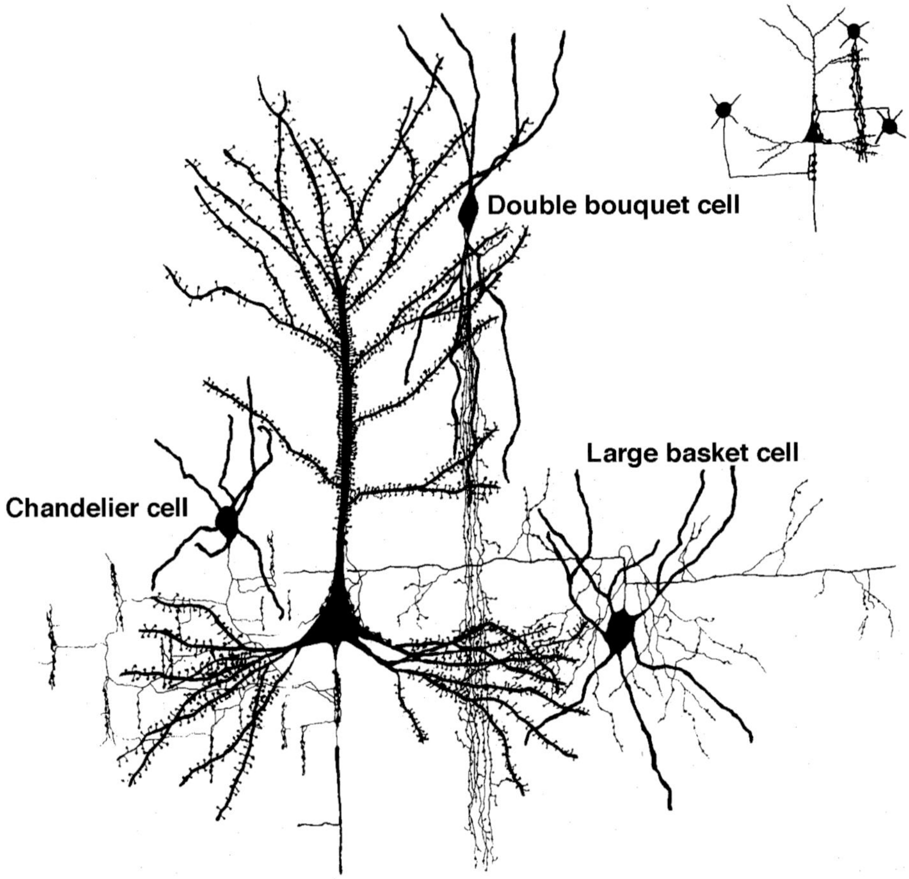

Fig. 21. Schematic drawing showing location and patterns of synaptic innervation of a representative spiny pyramidal neuron by chandelier, double bouquet and large basket cells. The inset is a simplified cartoon of the synaptic relationships of such neurons to emphasize that each of these morphologically distinct types of GABAergic interneurons innervates a distinct region of the pyramidal cell. Reproduced from DeFelipe and Fariñas (1992).

that pyramidal cells in superficial layers are more densely innervated by chandelier cells than those in deeper layers (Sloper and Powell 1979; DeFelipe et al. 1985). The GABAergic nature of chandelier cells has been verified by combined Golgi-impregnation and immunocytochemistry, or correlative light- and electron-microscopy of immunocytochemically prepared material, from many cortical areas of different species (Peters et al. 1982; Freund et al. 1983; DeFelipe et al. 1985; Somogyi et al. 1985).

5.4.3. Double bouquet cells

Double bouquet cells possess bitufted dendritic fields and characteristically give rise to axon collaterals which form tightly intertwined, vertically aligned bundles spanning layers II through V (Figs 20C, 21, 22; Somogyi and Cowey 1981; Cajal, 1899a,b; DeFelipe and Jones 1992). The somata of double bouquet cells are concentrated in layers II and III (Fig. 22A). The axon bundles they give rise to are quite numerous and thus they are a major source of inhibitory synaptic input. In tangential sections through layer III of monkey area V2, double bouquet cell axons identified by their immunoreactivity for calbindin or tachykinins form a repeating, microcolumnar structure of regularly spaced bundles (DeFelipe and Jones 1992). From serial reconstructions of Golgi-impregnated double bouquet cells in layer III of monkey area V1, the target region of the synapses formed by double bouquet cells includes both dendritic shafts and dendritic spines of pyramidal cells (Fig. 22C-E), particularly those forming oblique side branches of the apical dendrite (Somogyi and Cowey 1981), which has been confirmed by studies of double bouquet cell axon terminations in other monkey cortical areas (de Lima and Morrison 1989; DeFelipe et al. 1989a, 1990). The identity of these neurons as GABAergic has been shown directly by GABA immunocytochemistry in monkey area 3b (DeFelipe and Jones 1992), and more indirectly in monkey area V1 and other areas by autoradiographic studies showing the strong verticality of silver grains linking retrogradely-labeled somata in layers II and III with more deeply placed injections of ^{3}H-GABA (Somogyi et al. 1981, 1983a; DeFelipe and Jones 1985). Such vertically-oriented strings of silver grains undoubtedly represent the axon bundles of double bouquet cells linking superficial and deep layers.

5.4.4. Clutch cells

Clutch cells have been described in the visual cortices of cat, monkey and human and are characterized by an elongated dendritic field and a dense plexus of axon collaterals, both of which ramify very locally in relationship to their parent soma, and are mostly contained within layer IV (Fig. 23; Cajal 1899b; Valverde 1971; Lund 1973; Mates and Lund 1983; Kisvárday et al. 1986, 1990). The synaptic targets of these cells appears to be mostly the dendritic shafts and spines, and to a lesser extent the somata, of spiny stellate cells. The GABAergic nature of clutch cells has been verified by combined Golgi impregnation and postembedding GABA immunocytochemistry in the visual cortices of human and monkey (Kisvárday et al. 1986, 1990).

5.5. NEUROCHEMICALLY-DEFINED SUBTYPES OF GABAERGIC NEURONS

A variety of neuropeptides, intracellular proteins and other neurochemicals have been colocalized with GABAergic neurons to yield a diverse array of chemically heterogeneous subpopulations. Such subpopulations often occupy distinct laminar and/or compartmental positions, suggesting that each may play a specialized functional role within the cortical circuitry.

5.5.1. Neuropeptides

GABAergic neurons in monkey area V1 have been colocalized with tachykinins, somatostatin, neuropeptide Y (NPY), and cholecystokinin octapeptide (CCK) by immunocytochemical methods. Numerically, however, the numbers of GABA neurons in cat or monkey visual cortex which do not contain any of the known peptides far exceed those which do (Hendry et al. 1984b; Jones and Hendry 1986; Demeulemeester et al. 1988).

GABA/tachykinin-immunoreactive cells are small multipolar neurons whose somata and the axon plexuses they give rise to occupy two principal strata in monkey area V1:

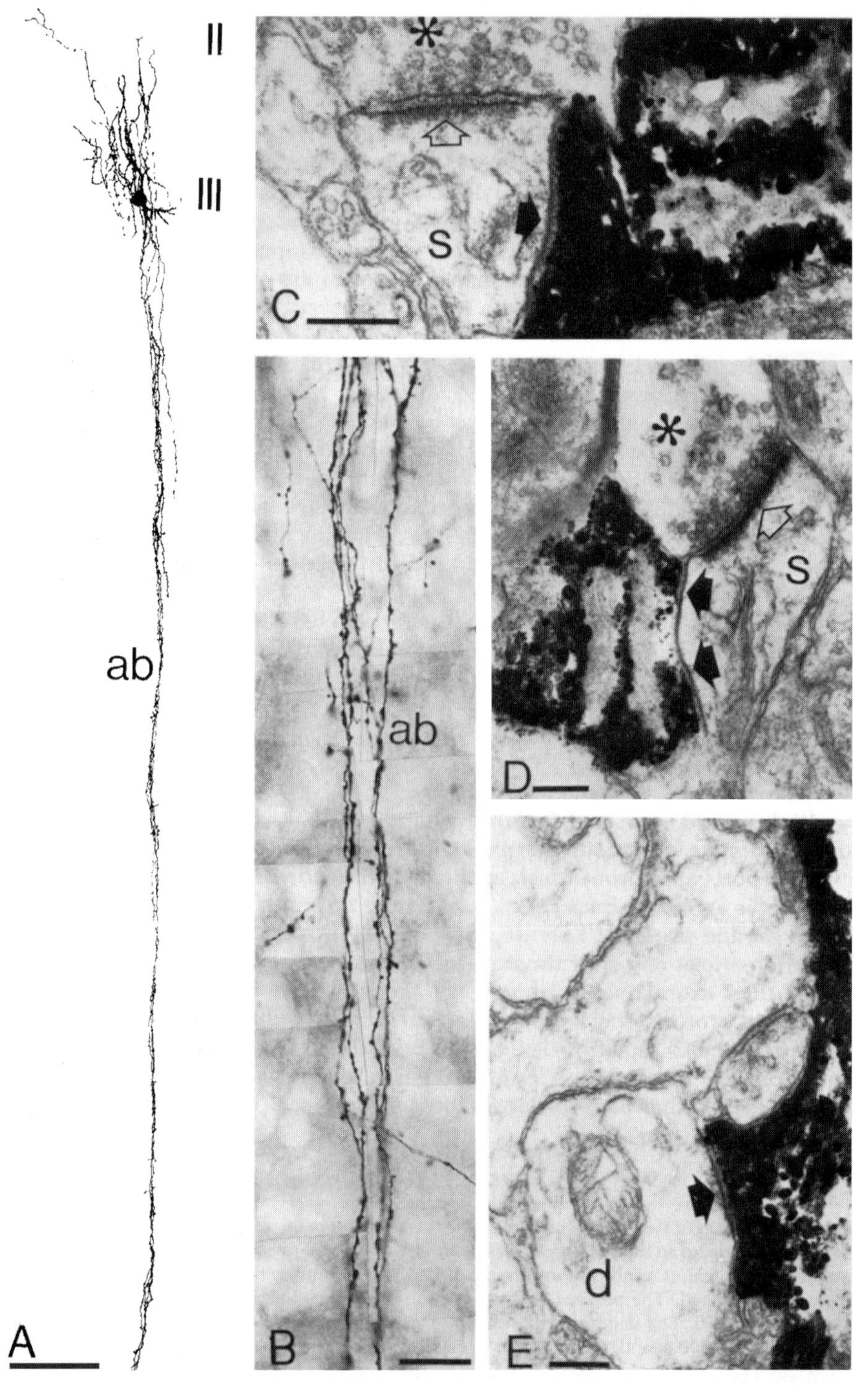
II
III
ab
A
*
S
C
ab
B
*
S
D
d
E

Fig. 22. Laminar distribution and synaptic organization of Golgi-impregnated double bouquet cells in macaque monkey area V1. *A*) Drawing of a double bouquet neuron, indicating the laminar position of the soma and dendrites in layer III, with long, vertically-oriented axonal bundle spanning layers II through V. *B*) Photomicrograph of Golgi-impregnated double bouquet cell axon bundle (*ab*). Numerous boutons are visible, some of which are shown at the ultrastructural level in *C-E*. Such boutons form symmetrical synapses (*solid arrows*) with dendritic spines (*s*) or shafts (*d*). The spines also receive unlabeled, asymmetrical synapses (*open arrows*) characterized by the presence of clear, round vesicles in the axon terminal (*asterisks*). Scale bars = 100 μm (*A*), 20 μm (*B*), 0.2 μm (*C-E*). Reproduced from Somogyi and Cowey (1984).

one of which is located in superficial layers (II and III), the other mostly confined to layer IV (Fig. 24A; Hendry et al. 1988a; Jones et al. 1988). In both strata, the distribution of somata is relatively uniform, however, in layers II and III, the concentration of tachykinin-immunoreactive puncta is greater within rows which correspond to rows of cytochrome-oxidase patches (Hendry et al. 1988a). In comparison with GABAergic neurons containing other peptides, the GABA/tachykinin-immunoreactive population is by far the most numerous, where in layer IVC it has been estimated that approximately one-half of the GABAergic neurons also contain tachykinin immunoreactivity. The morphologically-defined types of neuron that display GABA/tachykinin immunoreactivity are incompletely established: in monkey area V2, tachykinin-immunoreactive double-bouquet cells with their characteristic axon bundles have been described (Fig. 25), but these are rare in area 17 (DeFelipe et al. 1990). A second population of larger tachykinin-immunoreactive neurons are found in superficial and deep layers of monkey striate cortex; however, these larger peptide-containing neurons are not GABA-immunoreactive (Jones et al. 1988).

GABA neurons displaying somatostatin or NPY immunoreactivity are large multipolar cells present throughout both superficial and deep layers, and also extend into the subjacent white matter (Hendry et al. 1984a,b; Campbell et al. 1987; Huntley et al.

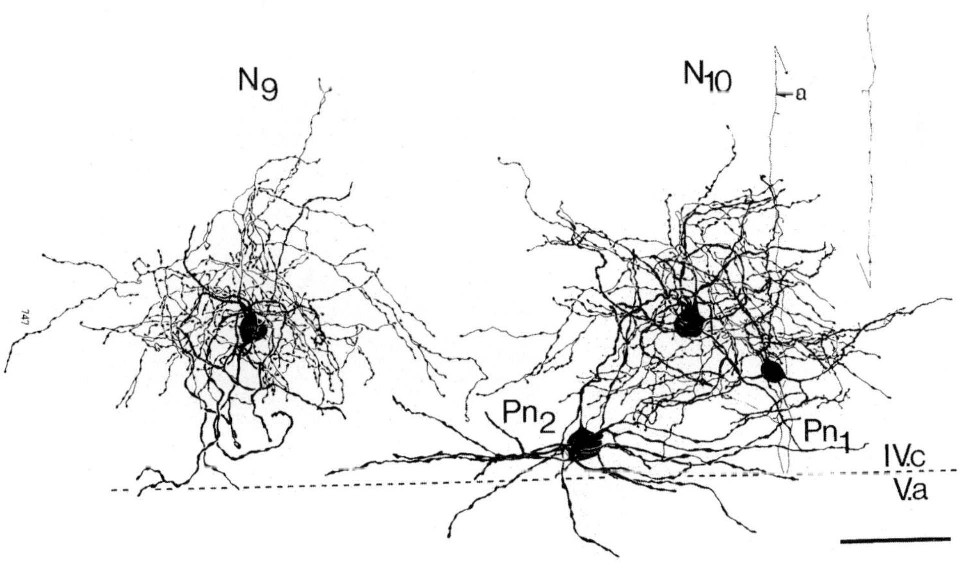

Fig. 23. Laminar and morphological features of GABA-immunoreactive clutch cells in macaque monkey area V1. Drawing showing two clutch cells (*N9* and *N10*) in layer IVC. Clutch cell *N10* forms synapses with a spiny stellate neuron (*Pn1*, with ascending axon, *a*) and with an aspiny neuron (*Pn2*), which was also GABA-immunoreactive. Scale bar = 50 μm. Reproduced from Kisvárday et al. (1986).

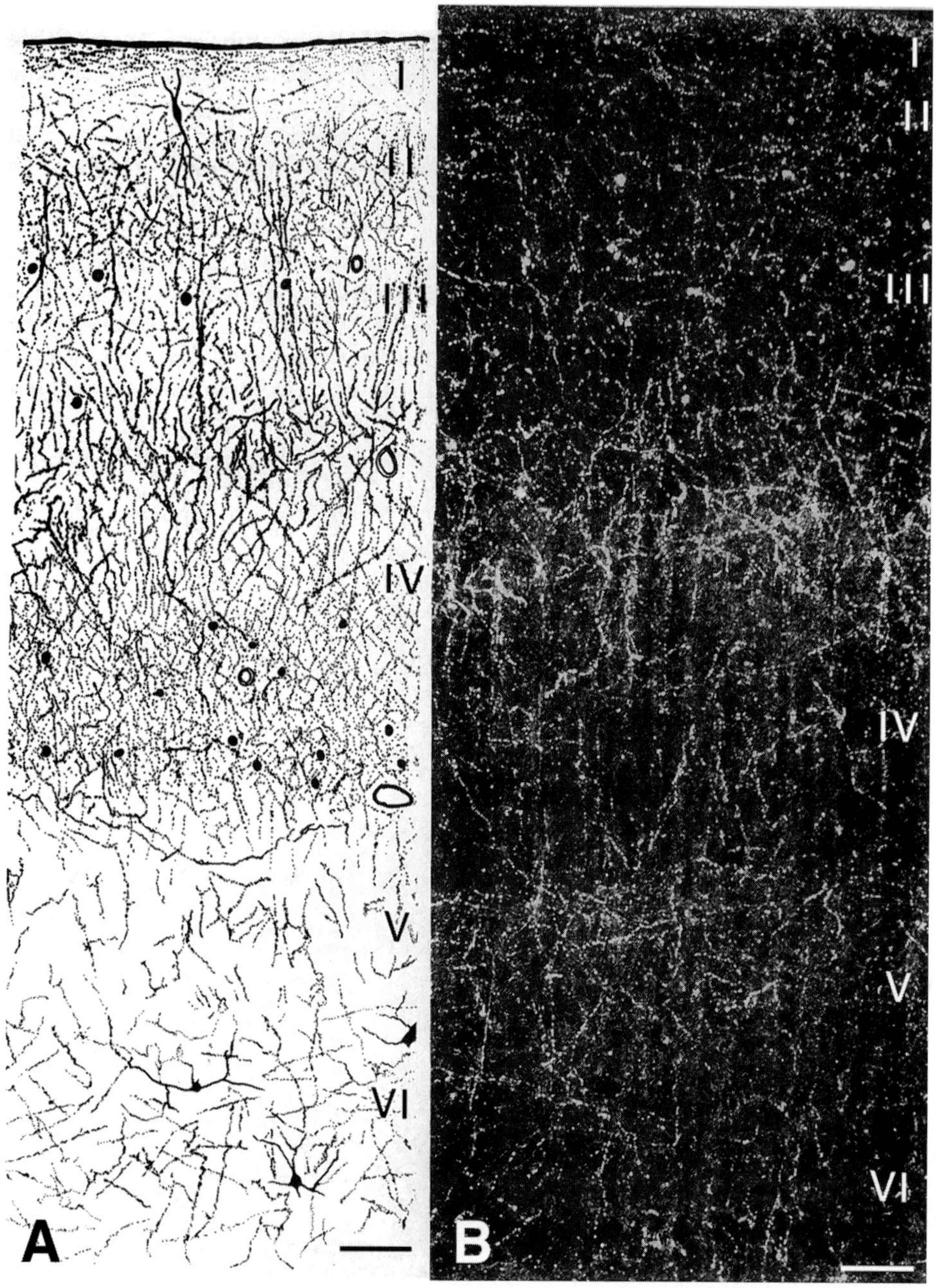

Fig. 24. Distribution of (*A*) tachykinin and (*B*) NPY immunoreactivity in macaque monkey area V1. Camera lucida drawing (*A*) and darkfield photomicrograph (*B*) demonstrate characteristic laminar distribution patterns of labeled somata and fiber plexuses. Scale bars = 100 μm (*A*), 50 μm (*B*). Reproduced from Jones (1990).

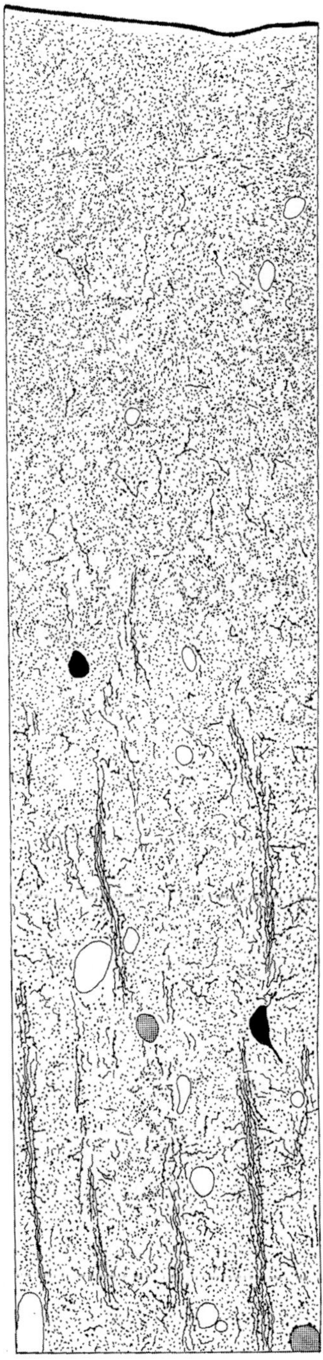

Fig. 25. Camera lucida drawing of tachykinin immunoreactivity spanning layers I to III in macaque monkey area V2. Immunolabeled elements include intensely stained somata (*solid profiles*); more lightly immunostained somata (*shaded profiles*); puncta (*stipple*) and axon bundles, characteristic of double bouquet cells. Reproduced from DeFelipe et al. (1990).

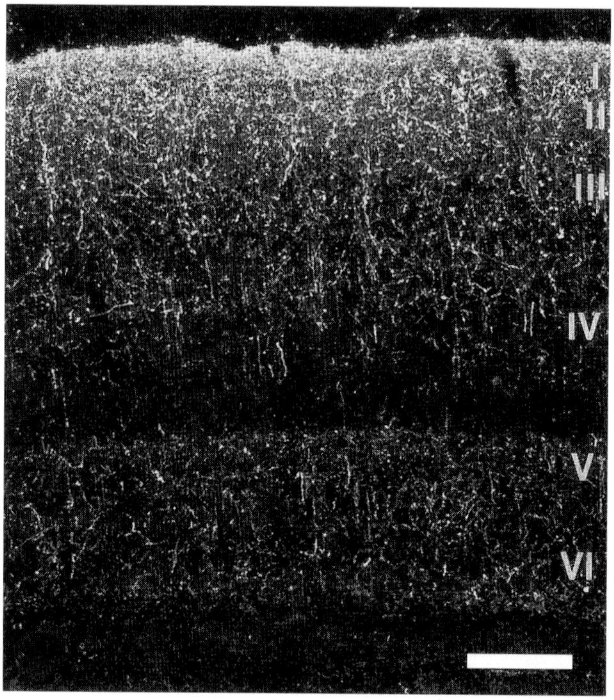

Fig. 26. Darkfield photomicrograph through macaque monkey area V1 showing characteristic laminar distribution of somatostatin-immunoreactive fiber plexuses. Scale bar = 300 μm. Reproduced from de Lima and Morrison (1989).

1988; de Lima and Morrison 1989; Kuljis and Rakic 1989a,b; Tigges et al. 1989). GABA/somatostatin and GABA/NPY-containing neurons give rise to several plexuses of axon terminations in monkey area V1 (Figs 24B, 26): a superficial (layer I) and deep (layer VI and white matter) set; and a set of middle-layer ones, occupying the lower half of layer III through IVB, and the upper portion of layer V. Such GABA/peptide-containing somata and their axon plexuses are largely absent from layer IVC, while those GABA neurons displaying NPY immunoreactivity are unevenly distributed in layer III, where they are concentrated in regions which surround the cytochrome oxidase patches (Kuljis and Rakic 1989b). Often, somatostatin and NPY are colocalized, suggesting that many GABA neurons contain both peptides (Hendry et al. 1984a). A few somatostatin-immunoreactive axon bundles characteristic of double bouquet cells are found in area V1, but these are more numerous in occipital and temporal association areas (Fig. 27A; de Lima and Morrison 1989). The existence of somatostatin-immunoreactive double bouquet cells is further supported by combined light and electron microscopic analysis in visual association cortex demonstrating ascending bundles of immunoreactive fibers making multiple symmetrical synaptic contacts on both small dendritic shafts and spines, with the vast majority (e.g. 83%, in supragranular layers) on small shafts (Fig. 11B, C). The morphological subtyping of other NPY- and somatostatin-containing GABAergic cells remains incomplete.

Small numbers of GABAergic neurons also contain CCK immunoreactivity in monkey striate cortex (Hendry et al. 1983b, 1984b). Such neurons are concentrated in superficial layers (II and III), and are also found in deeper layers as well but in general

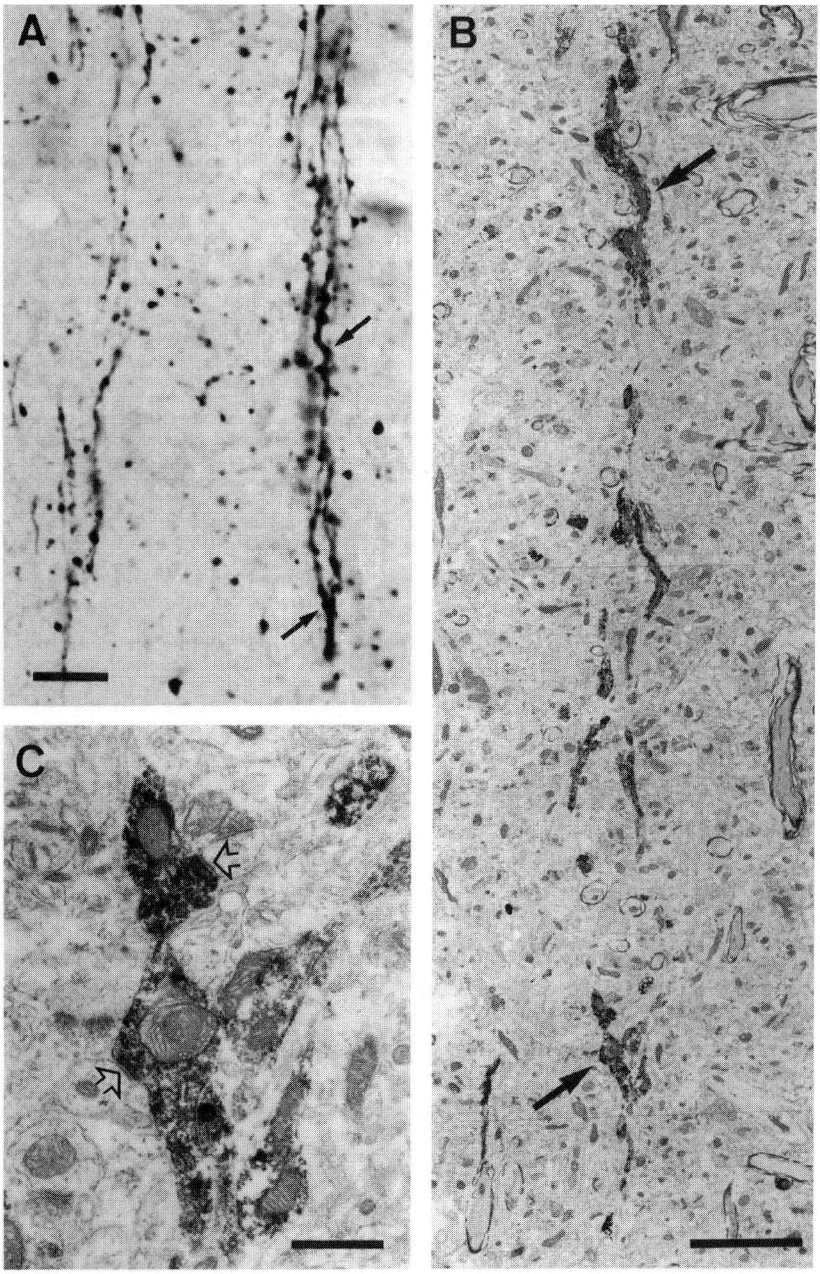

Fig. 27. Synaptic organization of somatostatin-immunoreactive double bouquet cell axons in macaque monkey temporal association neocortex. *A*) Semi thin section showing characteristic, radially-oriented axon bundles with numerous varicosities. Arrows indicate region shown at the ultrastructural level in the correlative electron-micrograph (*B*), with the region indicated by the lower arrow in *B* shown at higher-power in *C*. Two somatostatin-immunoreactive symmetrical synapses are shown (*open arrows, C*). The upper one forms a synaptic contact with a dendritic spine and the lower one with a dendritic shaft. Scale bars = 10 μm (*A*), 4 μm (*B*), and 1 μm (*C*). Reproduced from de Lima and Morrison (1989).

tend to avoid layer IV. GABA/CCK-immunoreactive neurons in monkey area V1 conform to a general multipolar or bitufted morphological class, but in cat area 17, it has been suggested that some GABA/CCK-immunoreactive cells are double bouquet and small basket cells (Freund et al. 1986).

A population of small cells containing vasoactive intestinal polypeptide (VIP) precursor mRNAs has been identified in superficial layers of monkey area V1 using *in situ* hybridization histochemistry (Benson et al. 1991b), but these cells have so far eluded detection by immunocytochemical methods for VIP protein localization. Thus, it is unclear whether they represent subpopulations of GABAergic cells, as some VIP-immunoreactive cells in rat cortex do (Kubota and Jones 1992).

A subpopulation of chandelier neurons has also been identified by immunoreactivity for corticotropin releasing factor (CRF) in monkey prefrontal and occipital association areas (Lewis and Lund 1990). However, it is unknown whether CRF-immunoreactive chandelier neurons are present in primate area V1.

5.5.2. Calcium-binding proteins

Immunoreactivities for the calcium-binding proteins calbindin (28 kDa, vitamin D-dependent; CB), parvalbumin (PV) and calretinin (CR) have been used to identify subpopulations of GABA neurons in primate area V1. While such subpopulations identified by calcium-binding protein immunoreactivity are mostly non-overlapping, and together account for the vast majority of all GABA neurons, small numbers of cells occasionally possess immunoreactivities for two calcium-binding proteins, while a few GABA cells – particularly in layer III – display no calcium-binding protein immunoreactivity (Hendry et al. 1989; van Brederode et al. 1990; Carder et al. 1996).

CB immunoreactivity has been examined in detail in the primary visual cortex of two species of monkeys (Old World macaque monkeys and New World squirrel monkeys), in bushbabies (*Galago*, a prosimian primate), and human, where a number of similarities as well as differences in organization emerge across primate species (Celio et al. 1986; Hendry et al. 1989; van Brederode et al. 1990; Hendrickson et al. 1991; Hendry and Carder 1993; Johnson and Casagrande 1995; Yoshioka and Hendry 1995). In the primary visual cortices of macaque monkeys and humans, intensely CB-immunoreactive GABAergic neurons are present throughout layers I to VI, with a greater concentration of somata and processes in layers II-III, IVB-IVCα and V (Fig. 28A). In all layers with the exception of layers III and IVA, such neurons appear evenly distributed when viewed in flattened sections cut tangential to the pial surface. However, in layer III, the intense CB-immunoreactive GABAergic neurons display an irregular pattern consisting of an intensely immunostained matrix surrounding more lightly immunostained circular regions. A comparison of immunostained sections with adjacent ones processed histochemically for cytochrome oxidase activity demonstrates that the intensely CB-immunostained matrix corresponds to the cytochrome oxidase-poor interpatch regions, while the lightly immunostained circular regions correspond to the cytochrome-oxidase dense patches. In layer IVA of humans and macaques, CB-immunoreactive neurons preferentially occupy the walls of the cytochrome-oxidase intense lattice, and are absent from the cytochrome-oxidase pale hollows. The only notable difference in CB immunoreactivity between the visual cortices of human and macaque monkey is the presence in humans of a much greater number of lightly CB-immunoreactive neurons, particularly in layer IVA (Hendry and Carder 1993). Colocalization studies in visual and other areas of monkey and human cortex have shown

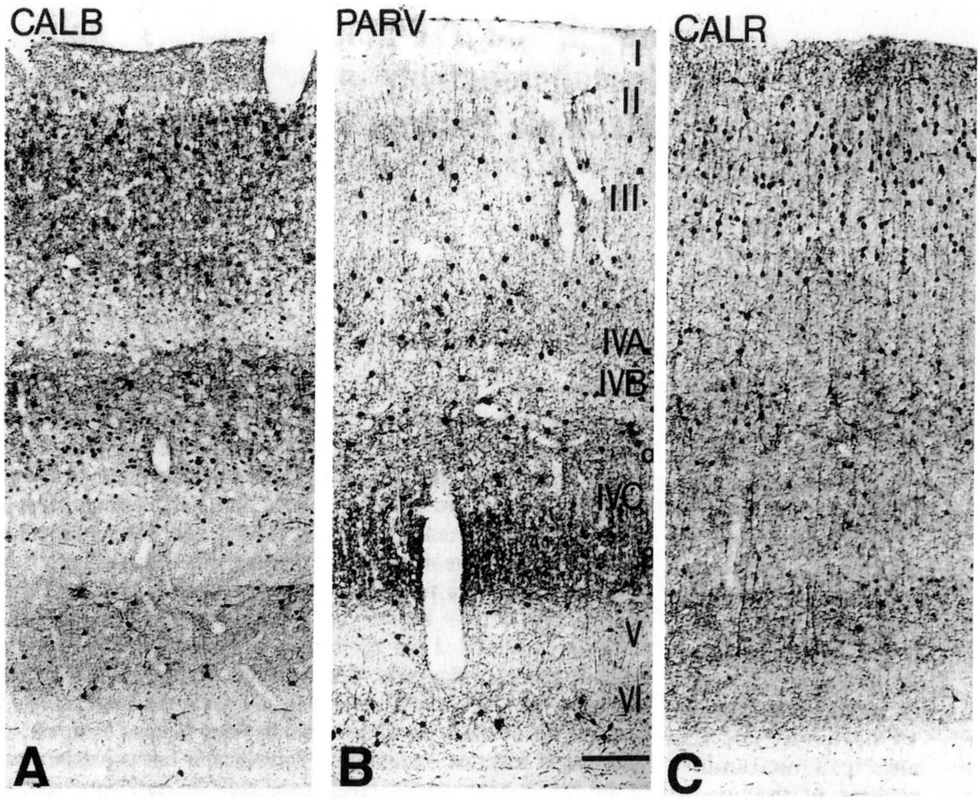

Fig. 28. Brightfield photomicrographs of sections through macaque monkey area V1 demonstrating laminar distribution patterns of calbindin (*A*), parvalbumin (*B*) or calretinin (*C*) immunoreactive somata and fibers. Scale bar = 250 μm. Reproduced from Carder et al. (1996).

that such lightly labeled CB neurons lack GABA immunoreactivity, and thus probably represent a population of pyramidal or spiny stellate cells (DeFelipe and Jones 1992; Hayes and Lewis 1992; Hendry and Carder 1993).

In the primary visual cortex of squirrel monkeys, the laminar distribution of CB-immunoreactive GABA neurons and their irregular layer III pattern is similar to that described for macaques. However, the irregular staining pattern extends into layers II, IVB, V and VI, where an intensely immunostained matrix surrounds the cytochrome oxidase-dense patches in each of these layers (Celio et al. 1986).

In comparison to the patterns described for macaque monkeys and humans, the laminar and compartmental distribution of CB immunoreactivity differs in the Prosimian bushbabies (Johnson and Casagrande 1995). Calbindin-containing somata are concentrated in layers II and IIIA, with much lower densities in other layers. In addition, there is no obvious correlation between CB immunoreactivity and regions of intense or pale cytochrome oxidase staining in superficial layers.

Morphologically, many of the CB-immunoreactive GABA cells, particularly in superficial layers of visual and other monkey cortical areas, form dense CB-immunoreactive bundles of axons which are characteristic of double bouquet cells (DeFelipe et al. 1989, 1990).

The distribution of PV immunoreactivity in monkey and human area V1 is largely complementary to that of CB immunoreactivity (Fig. 28B; Hendry et al. 1989; Blümcke et al. 1990, 1994; Hendrickson et al. 1991; Spatz et al. 1994; Johnson and Casagrande 1995; Cao et al. 1996; Carder et al. 1996). In monkey area V1, all PV-immunoreactive somata are also GABA-immunoreactive. Parvalbumin-immunoreactive somata and terminals are present throughout layers II-VI, but show greater concentrations in layers III, IVA, IVC – particularly sublayer IVCβ – and VI (Fig. 28B). In layer III, the distribution of PV immunostaining is inhomogeneous, with more intensely stained patches corresponding to the cytochrome oxidase-dense patches. Immunolabeled somata, however, are distributed throughout both patch and interpatch regions, and thus intensity differences in immunoreactivity are due largely to differences in terminal labeling. This overall laminar pattern is largely similar in the primary visual cortices of bushbabies (Johnson and Casagrande 1995), New World marmoset monkeys (Spatz et al. 1994), and humans, although in the latter, the enhanced terminal labeling in layer IV is mostly restricted to layer IVC (Blümcke et al. 1990). In layer IVA of monkeys and humans, the distribution of PV-immunoreactive somata is similar to that described for CB immunoreactivity: PV-positive somata preferentially occupy the thin walls of the cytochrome oxidase intense lattice. In relationship to morphologically-defined cell types, some PV immunoreactive neurons in visual and other cortical areas of monkeys and humans have all of the features of classic large basket cells, while others possess characteristics of chandelier cells (DeFelipe et al. 1989b, 1993; Hendry et al. 1989; Lewis and Lund 1990; Akil and Lewis 1992a,b; Williams et al. 1992; Fonseca et al. 1993; Anderson et al. 1995). Not all basket or chandelier cells are labeled by PV immunoreactivity, however. The remaining PV-immunoreactive neurons are undefined in strict morphological terms.

The laminar pattern of CR immunoreactivity is largely similar to that of CB immunoreactivity, with greater concentrations of immunostained somata and processes in layers II-III, IVA-IVCα, and V (Fig. 28C; Carder et al. 1996). The distribution of CR-immunoreactive somata is homogeneous across layers III, indicating no preferential parcellation to patch or interpatch regions. Most CR-immunoreactive neurons are also GABA-containing, although a small number lack GABA immunoreactivity. Morphologically, some CR-immunostained neurons possess bundles of radially-oriented processes, indicating the likelihood that such neurons are double-bouquet cells, while others form basket-like terminations around unstained pyramidal cell somata, suggesting that these are basket cells (Carder et al. 1996).

5.5.3. Cell-surface molecules

Subsets of GABAergic neurons can be identified by the expression of different cell-surface proteoglycans which can be visualized with monoclonal antibodies or the binding of plant lectins. One such monoclonal antibody, Cat-301, labels a diverse set of neurons in monkey area V1, the large majority of which are GABAergic cells (Fig. 29; Hendry et al. 1984a, 1988b; Hockfield et al. 1984; DeYoe et al. 1990). The GABA/Cat-301-containing cells are distributed in most layers of V1, but form two prominent bands, one in layer IVB and the other in layer VI. In tangential sections through the deeper half of layer III, Cat-301 immunostaining is inhomogeneous, and characterized by dense aggregations of stained somata which correspond to the core regions of the cytochrome oxidase-intense patches. In contrast, the cytochrome oxidase-intense patches which extend into upper layer III and layer II bear no consistent

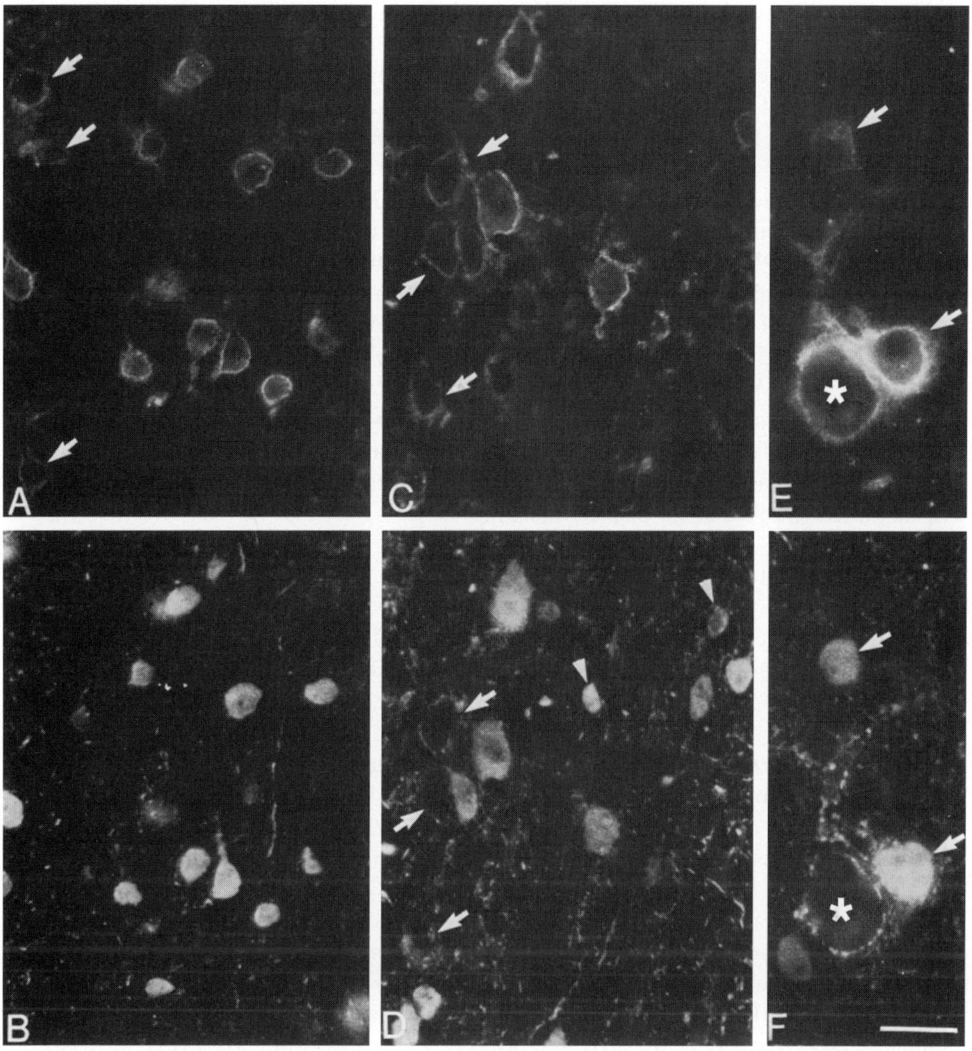

Fig. 29. Pairs of matched sections labeled by immunofluorescence for Cat-301 (*A, C, E*) or GABA (*B, D, F*) in macaque monkey area V1. *A, B*) Section through deep layer III. Most small Cat-301-immunolabeled cells (A) are also immunoreactive for GABA (*B*), but some (*arrows*) are not. *C, D*) Section through layer IVB. Some large Cat-301-immunoreactive cells (*C, arrows*) are not also immunolabeled for GABA (*D, arrows*), but are surrounded by GABA-immunoreactive puncta. Some small GABA-immunoreactive neurons (*D, arrowheads*) are not also Cat-301-immunoreactive. *E, F*) Section through layer VI. *Arrows* denote small neurons immunoreactive for Cat-301 (*E*) and GABA (*F*), while *asterisk* denotes Cat-301-positive cell which is not also GABA-immunoreactive, but is surrounded by GABAergic boutons. Scale bar = 25 μm (A-D), 15 μm (E,F). Reproduced from Hendry et al. (1988).

relationship with Cat-301-immunostained somata. The correlation between regions of intense cytochrome oxidase activity and aggregations of Cat-301 somata is also evident in area V2, where bands of immunostaining line up with bands of intense enyzme activity (Hendry et al. 1988b; DeYoe et al. 1990). Based on these cortical laminar and compartmental patterns, and the predominant Cat-301 staining of thalamic relay neurons in the magnocellular layers of the LGN, it has been suggested that within the monkey central visual system, Cat-301 staining preferentially – but not exclusively – identifies neurons with certain common functional properties (Hendry et al. 1984a, 1988b; Hockfield et al. 1984; DeYoe et al. 1990). The morphological types of GABA neurons which Cat-301-labeled cells represent are diverse, but some possess features of large basket cells. In addition, virtually none are also immunoreactive for the known neuropeptides.

Another example of a distinct subpopulation of GABAergic cells identified by cell-surface glycoproteins is one revealed by binding of the plant lectin *Vicia villosa* (VVA) in monkey area V1 (Mulligan et al. 1989). Similar to Cat-301 neurons, VVA-labeled neurons are virtually all GABA neurons, and this population accounts for approximately 30% of all GABAergic cells. VVA/GABA-containing neurons are found in most layers, but they form a prominent band which spans the deeper half of layer III through layer IVCα, and are mostly absent from layers I and IVCβ. Unlike Cat-301 staining, however, VVA/GABA-containing neurons are homogeneously distributed across differentially stained cytochrome oxidase compartments in superficial layers. VVA/GABA-containing neurons are also a morphologically heterogeneous group, with some resembling large basket neurons.

5.6. DISTRIBUTION OF GABA RECEPTORS

The laminar and compartmental distributions of $GABA_A$ receptors in monkey area V1 are similar to the distribution patterns displayed by GABA neurons and terminals, and therefore also appear enriched in layers receiving geniculocortical inputs. Autoradiographic studies of ^{3}H-muscimol or ^{3}H-flunitrazepam binding reveal dense concentrations of GABA/benzodiazepine binding sites in layers II-III, IVA, IVC and VI, with sublayer IVCβ showing a greater density of ligand binding in comparison with sublayer IVCα (Shaw and Cynader 1986; Rakic et al. 1988; Hendry et al. 1990; Shaw et al. 1991). The general laminar distribution pattern revealed by ligand binding is closely matched by the distribution patterns of $\alpha 1$, $\beta 2$, and $\gamma 2$ subunit proteins and mRNAs, but with additional compartmental distinctions evident (see Fig. 30; Hendry et al. 1990, 1994; Huntsman et al. 1994). For example, immunocytochemical studies reveal that each of these subunits is concentrated within the intense cytochrome oxidase patches of layers II-III, and forms a honeycomb lattice in layers IVA and IVC in which intensely stained walls surround weakly stained hollows. In layers IVB, V and VI, which also contain cytochrome oxidase-intense patches, receptor subunit staining is uniform (Hendry et al. 1990, 1994). In contrast to the distribution of $\alpha 1$, $\beta 2$, and $\gamma 2$ subunits, *in situ* hybridization studies suggest that other α subunit ($\alpha 2$, $\alpha 4$, and $\alpha 5$) and β subunit ($\beta 1$) mRNAs are present in weak to moderate levels in all layers but layer VI and the subjacent white matter, where they are densely distributed (Fig. 30; Huntsman et al. 1994). This pattern shows little correspondence to those revealed by immunocytochemical or autoradiographic methods.

In human primary visual cortex, the distribution of $GABA_A$ receptors – visualized either by autoradiographic localization of ^{3}H-flunitrazepam binding sites or by the immunocytochemical localization of $\alpha 1$, $\beta 2/3$, and $\gamma 2$ subunits – is virtually identical to that found in macaque monkeys (Zezula et al. 1988; Hendry et al. 1994).

$GABA_B$ receptors have been localized in monkey area V1 by the autoradiographic detection of ^{3}H-baclofen binding, which reveals a rather diffuse distribution of receptors throughout all layers with a slightly lower concentration in middle layers (Shaw et al. 1991).

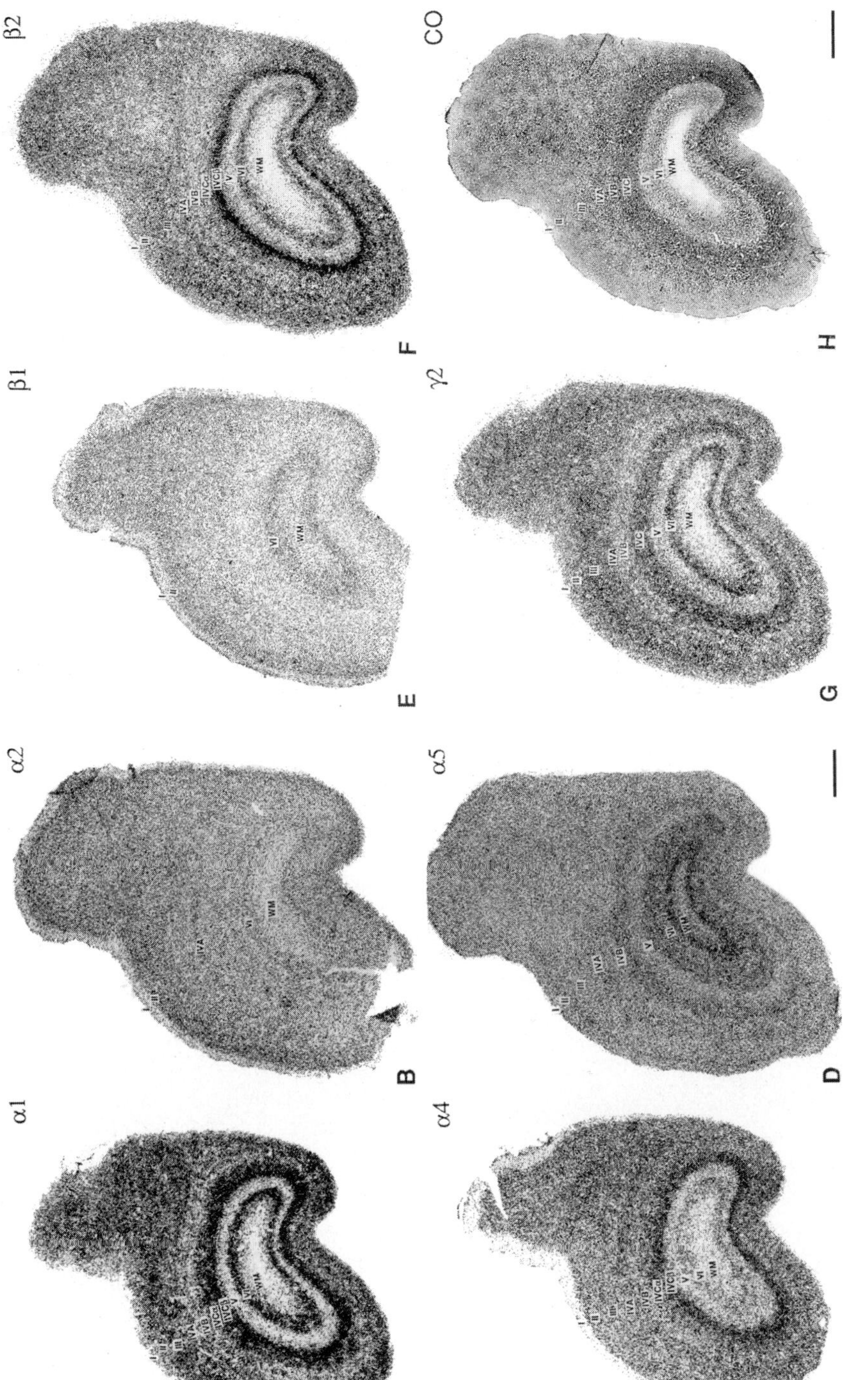

Fig. 30. Laminar and compartmental distribution of $GABA_A$-receptor subunit mRNAs in macaque monkey area V1. Adjacent, flattened sections processed by *in situ* hybridization histochemistry using cRNA probes for the $GABA_A$-receptor subunit, shown in photomicrographs of film autoradiograms (*A-G*), or processed for cytochrome oxidase activity (*H*). Time of exposure to film was identical for all subunits. Scale bar = 1 mm. Reproduced from Huntsman et al. (1994).

5.7. DEVELOPMENT OF GABA NEURONS, RECEPTORS AND RELATED NEUROCHEMICALS IN PRIMATE AREA V1

An adult-like cytoarchitecture of monkey area V1 emerges by the beginning of the last third of the 165-day gestational period in macaque monkeys, including a sharply defined area V1/V2 border and all of the sublayers of layer IVC (Fig. 31; Rakic 1982; Huntley et al. 1988). Neurochemically-defined populations of neurons, including those identified by GABA, neuropeptide and calcium-binding protein immunoreactivity, as well as $GABA_A$ receptors, are present at very early stages, but undergo considerable changes in density and distribution over the last third of gestation which, in some cases, extend into the early postnatal period.

5.7.1. Development of GABA neurons

GABA neurons can be detected as early as E41, where they are sparsely scattered across the entire telencephalic wall (Schwartz and Meinecke 1992). By E110, significant numbers of GABA-immunoreactive somata are concentrated within the marginal zone and the subjacent white matter (subplate), and are also evident within cell-dense layers II to VI (Fig. 32A; Huntley et al. 1988). By E155, the overall density of GABAergic neurons within layers II to VI has increased, with a relatively greater concentration in layer IVC evident, while the density of those present within the white matter has declined (Fig. 32A; Huntley et al. 1988). The final adult-like distribution pattern becomes evident during the first postnatal week (Schwartz and Meinecke 1992). The general developmental progression of GABAergic cell distribution and density leading to the adult pattern is similar in human visual cortex (Yan et al. 1992).

5.7.2. Development of GABA receptors

$GABA_A$ receptors localized autoradiographically by binding of ^{3}H-muscimol or ^{3}H-flunitrazepam are also present from very early stages (at least by E61) in develop-

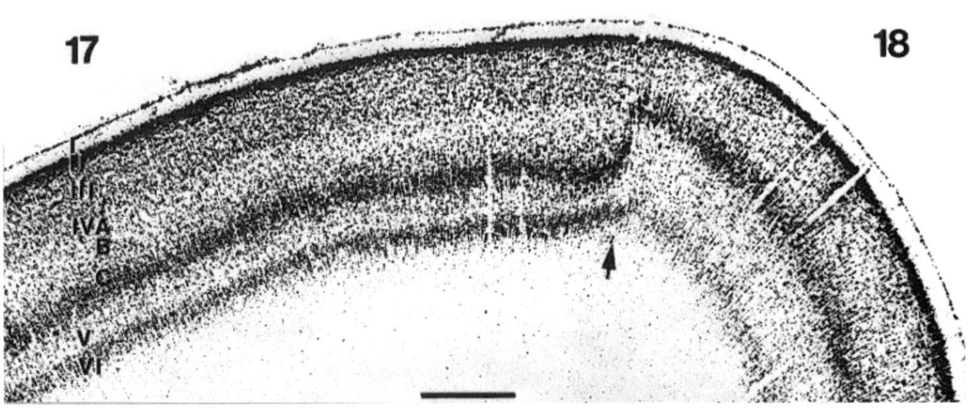

Fig. 31. Photomicrograph of thionin-stained section through the area V1/V2 border region taken from an E110 macaque monkey fetus. Adult-like cytoarchitectonic features, including all sublaminae of layer IV, and a sharp V1/V2 border (*arrow*), are already evident by this stage of development. Scale bar = 500 μm. Reproduced from Huntley et al. (1988).

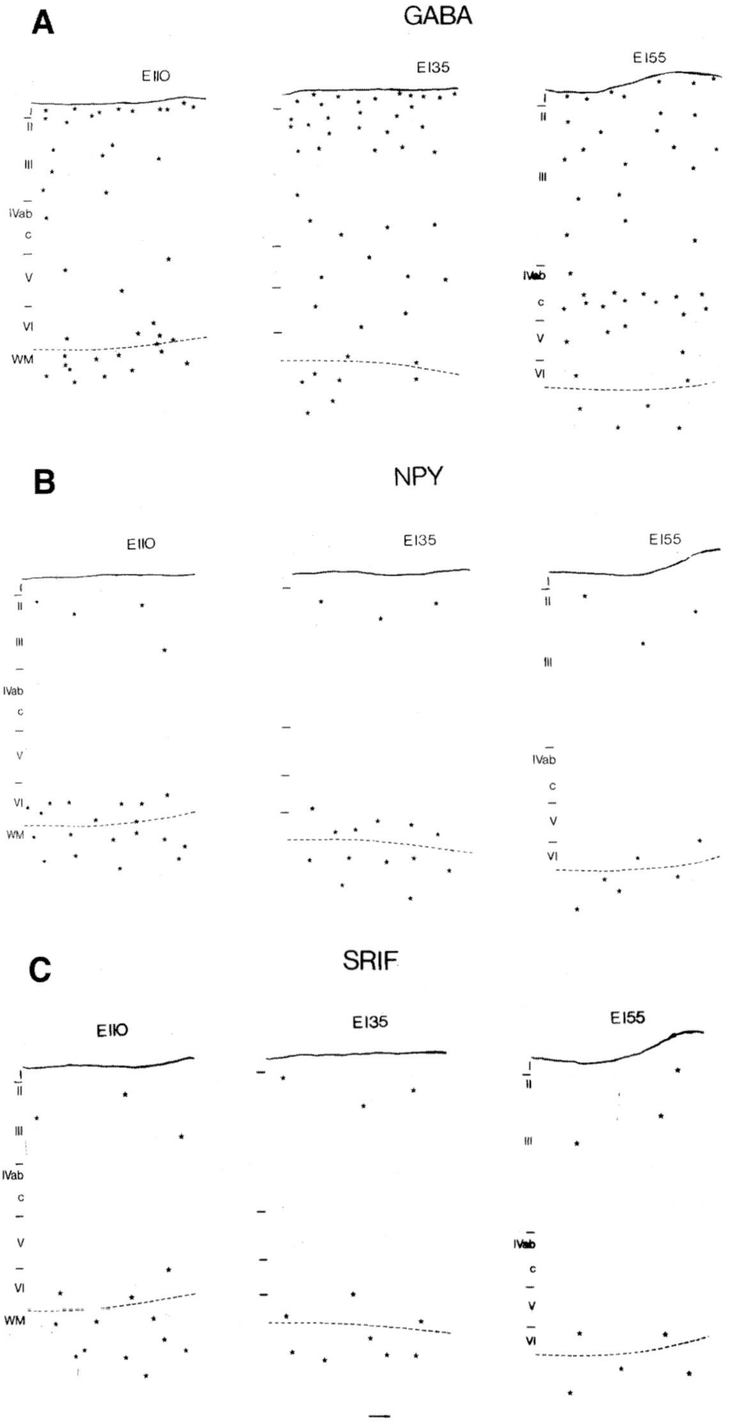

Fig. 32. Camera lucida drawings of sections through area V1 taken from macaque monkey fetuses aged E110, E135 and E155, stained immunocytochemically for GABA (*A*), NPY (*B*), or somatostatin (*C*). *Stars* indicate immunoreactive somata. Scale bar = 100 μm. Modified from Huntley et al. (1988).

ing macaque area V1 (Shaw et al. 1991), matching the presence of GABAergic neurons. The greatest densities of binding sites are found within the marginal and subplate zones, with less dense binding in the cortical plate. With increasing fetal age, an adult-like laminar pattern is gradually established by birth. However, after birth, the density of muscimol binding sites increases dramatically in all layers, while the density of flunitrazepam binding sites decreases in superficial and deep layers, thus leaving layer IVC with the densest binding. The final adult-like density of both binding sites emerges very late, by 8.5 years of life (Shaw et al. 1991). Immunocytochemical and *in situ* hybridization studies of the distribution of individual $GABA_A$ receptor subunits during development of monkey visual cortex are consistent with the patterns suggested by ligand binding (Meinecke and Rakic 1992; Hornung and Fritschy 1996). These studies indicate, in addition, a developmental switch in subunit composition whereby $\alpha 2$ subunit immunoreactivity, which dominates during fetal development, is replaced near birth by the appearance of $\alpha 1$ subunit immunoreactivity (Hornung and Fritschy 1996).

^{3}H-baclofen binding sites indicating $GABA_B$ receptors are also present during development of monkey area V1, but do not appear until E126 (Shaw et al. 1991). At this time, the laminar distribution and density is already adultlike, remaining relatively unchanged with increasing age through adulthood.

5.7.3. Development of neuropeptides

As in the adult, populations of NPY-, somatostatin-, tachykinin-, and CCK-containing cells can also be identified in fetal monkey area V1 from early stages (Figs 32, 33; Huntley et al. 1988). Double labeling studies indicate that some are GABA-immunoreactive, but many do not contain GABA. NPY- and somatostatin-immunoreactive neurons and fibers are densely packed in the subjacent white matter by E110 (Fig. 34), and are sparsely distributed in layers II to VI. With increasing age, their numbers decline in the white matter, leaving an adult-like pattern of neuron distribution by late fetal stages (Figs 32B,C, 34; Huntley et al. 1988; Yamashita et al. 1989). The density and distribution of NPY-immunoreactive fiber plexuses, however, continues to change well into postnatal life, reaching an adult pattern between 7 and 13 months of age (Tigges et al. 1989).

Tachykinin-immunoreactive neurons display a relatively restricted, but transient, distribution during development of monkey area V1 (Fig. 33A; Huntley et al. 1988). By E110, SP-immunoreactive somata are densely packed in a thin band corresponding to layer V. Such neurons, particularly by E135, give rise to thick, radially ascending processes which give rise to a dense fiber plexus in layers II/III and I (Fig. 35). This pattern remains at least through late fetal stages; it is not known when the adult-like pattern emerges but it most likely occurs postnatally.

CCK-immunoreactive neurons are mostly absent from monkey area V1 at E110 (Fig. 35B); their density increases with age as do fiber the plexuses they give rise to (Huntley et al. 1988). By E135, a transient population of stained somata are present in layer V, but these disappear by later fetal ages. In addition, long immunoreactive fibers are present deep within the white matter, and can be followed into the overlying cortex by E135. Such fibers, which may be of extrinsic origin, are much denser in area V1 than in area V2, and readily delineate this areal border by late fetal stages (Fig. 36). It is unknown when the adult-like pattern emerges.

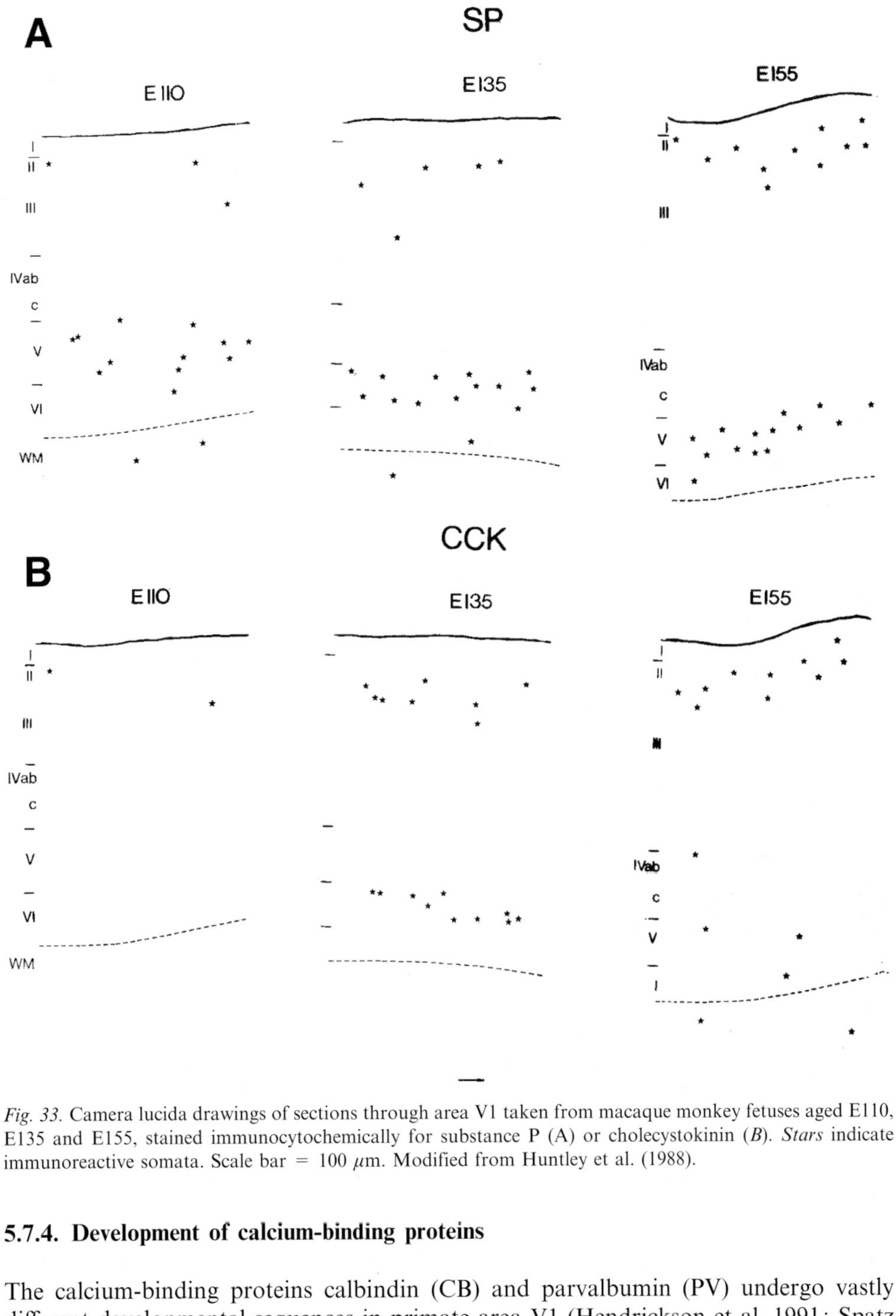

Fig. 33. Camera lucida drawings of sections through area V1 taken from macaque monkey fetuses aged E110, E135 and E155, stained immunocytochemically for substance P (A) or cholecystokinin (*B*). *Stars* indicate immunoreactive somata. Scale bar = 100 μm. Modified from Huntley et al. (1988).

5.7.4. Development of calcium-binding proteins

The calcium-binding proteins calbindin (CB) and parvalbumin (PV) undergo vastly different developmental sequences in primate area V1 (Hendrickson et al. 1991; Spatz et al. 1994; Cao et al. 1996; Condé et al. 1996). In layer I of macaque monkeys and humans, CB-immunopositive Cajal-Retzius neurons are found which also display PV-immunoreactivity for a brief period before finally disappearing by birth (Fig. 37;

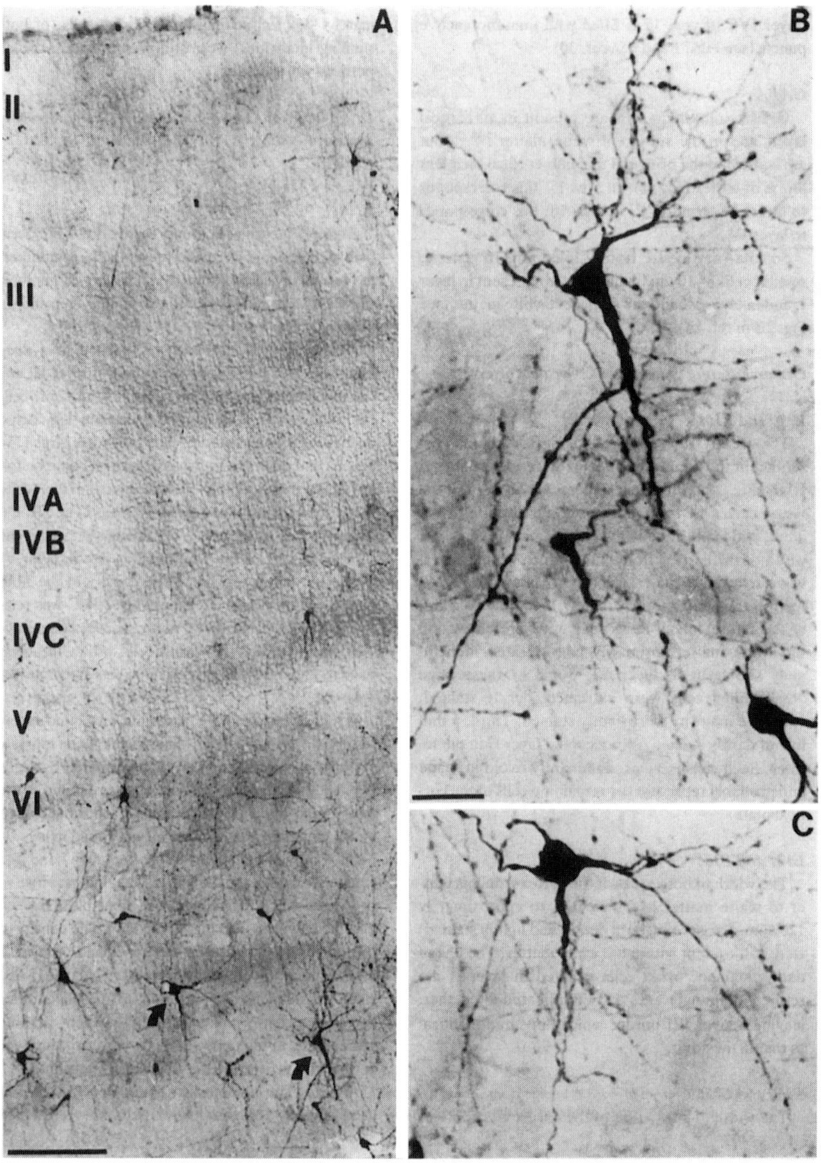

Fig. 34. Laminar distribution and morphological features of NPY-immunoreactive somata and fibers in developing macaque monkey area V1 (E110). Immunoreactive neurons in subplate zone, subjacent to layer VI, are particularly large and give rise to numerous beaded processes. *Arrows* in *A* indicate cells shown at higher power in *B* and *C*. Scale bar = 150 μm (*A*), and 25 μm (*B, C*). Modified from Huntley et al. (1988).

Huntley and Jones 1990; Yan et al. 1995, 1997). In deeper layers, CB-immunoreactive neurons, including both pyramidal and non-pyramidal cells, are concentrated in layers IV-VI by E113 in monkeys. With increasing age up through birth, the number and staining intensity of such neurons throughout layers II-VI increase dramatically, but then exhibit a decline by 9–12 postnatal weeks. Between 20–36 postnatal weeks, there is a second wave of increasing density and intensity of CB-immunostaining affecting

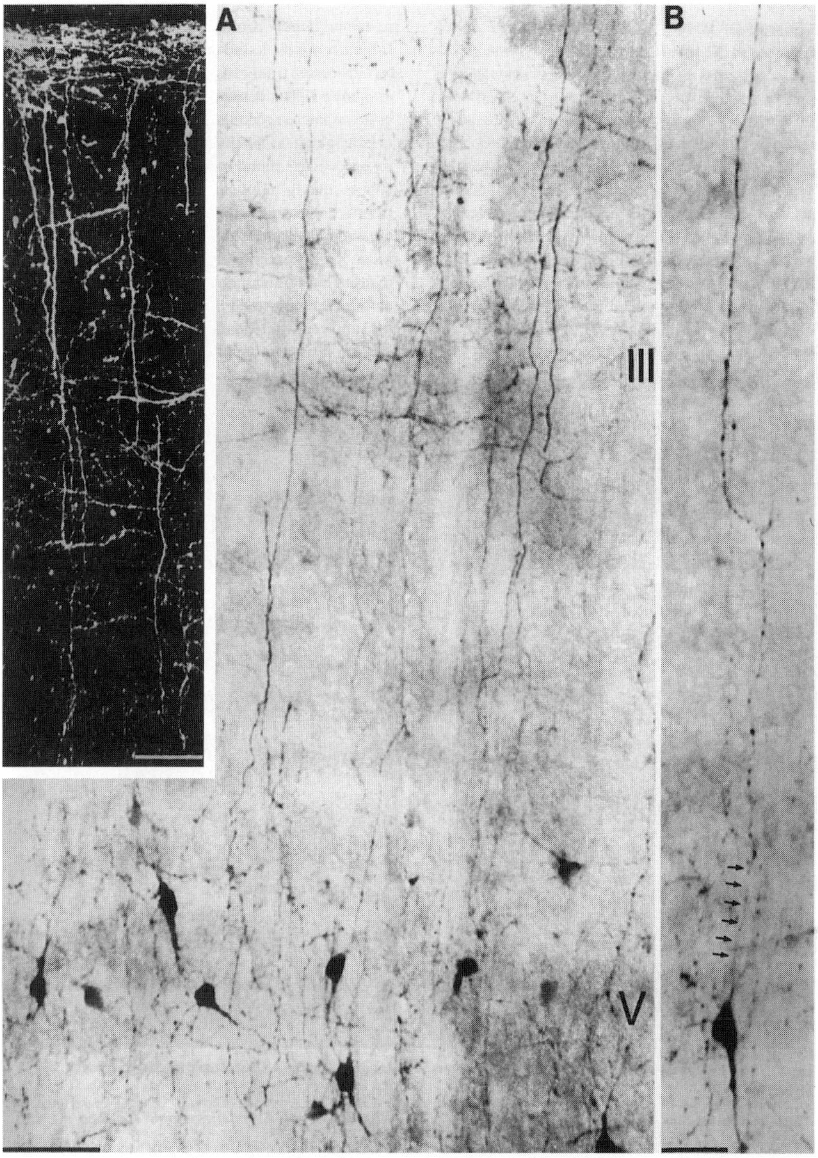

Fig. 35. Photomicrographs of sections through area V1 of a macaque monkey at embryonic day E135. By these stages of development, immunoreactive somata are concentrated transiently in layer V, and give rise to ascending, radially-oriented beaded fibers (*arrows in B*) which form plexuses in superficial layers and layer I (*inset*, darkfield photomicrograph). Scale bars = 50 μm (*A*), 100 μm (*inset*), 25 μm (*C*). Reproduced from Huntley et al. (1988).

mostly superficial layers, which then slowly declines to adult intensity levels and distribution patterns by the second year of life (Hendrickson et al. 1991).

In human area V1, CB-immunoreactive somata and fibers are present from very early stages (by 15 gestational weeks), but are restricted to marginal and intermediate zones, avoiding the undifferentiated cortical plate. By 24 weeks, CB-immunoreactive

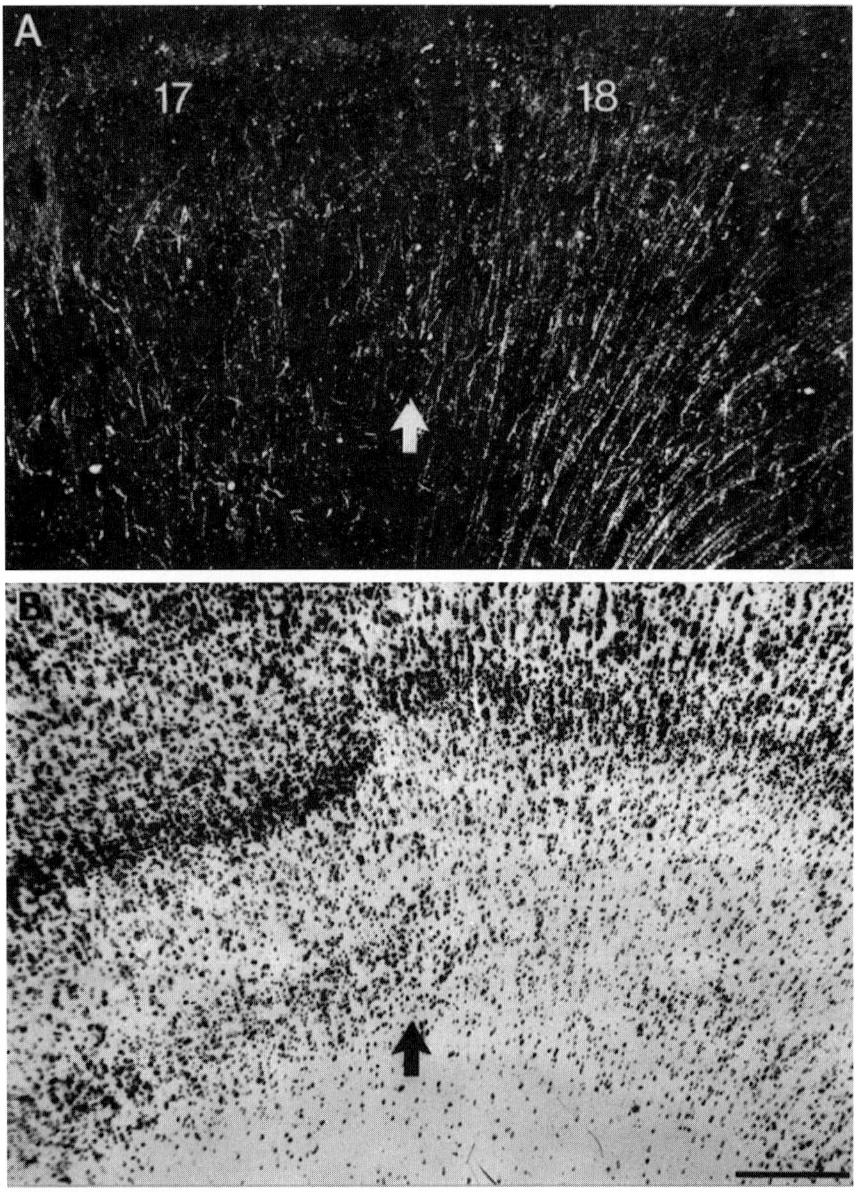

Fig. 36. Distribution of cholecystokinin-immunoreactive fibers at the area V1/V2 border region in an E155 macaque monkey fetus. Photomicrographs of immunoreactive fibers (*A*, darkfield) shows much denser innervation of area V2 in comparison with area V1; the areal border is indicated by the arrows in (A) and the adjacent thionin-stained section (B). Scale bar = 250 μm. Reproduced from Huntley et al. (1988).

somata and fibers are present mostly in deep layers, increasing in number and intensity throughout the rest of the cortical layers, with higher concentrations in subdivisions of layer IV, over subsequent weeks in an 'inside-out' pattern (Yan et al. 1997).

In contrast, PV-immunostaining in layers II-VI of V1 is sparse at mid-fetal periods, becoming dense only at very late fetal and early neonatal periods in macaque monkeys

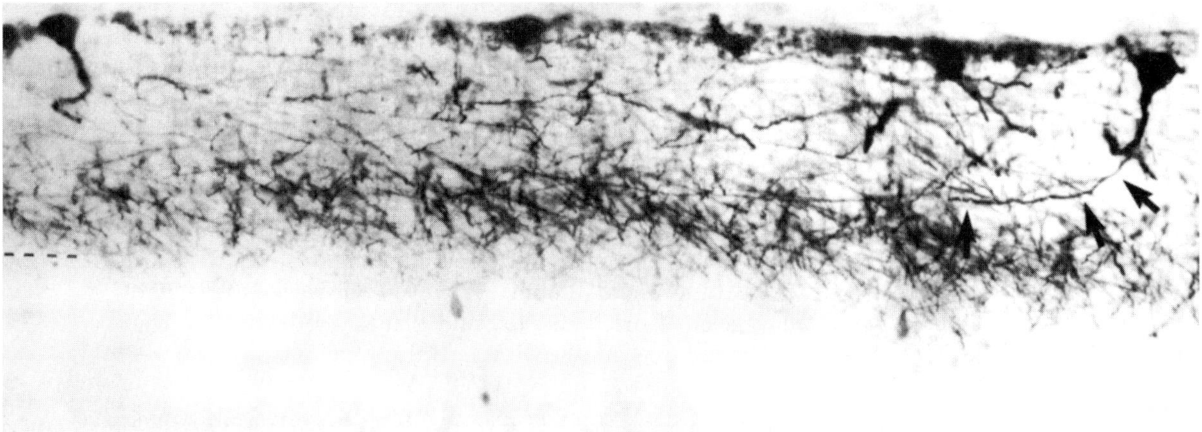

Fig. 37. Photomicrograph of section taken through layer I of area V1 taken from an E121 macaque monkey fetus, and stained immunocytochemically for calbindin. Numerous labeled Cajal-Retzius neuron somata are present, characterized by descending process (*arrows*), which give rise to dense immunoreactive fiber plexus in deep layer I. The dotted line indicates the layer I/II border. Modified from Huntley and Jones (1990).

(Hendrickson et al. 1991; Condé et al. 1996). Immunostained somata and processes become evident first in deep layers, where presumed basket and chandelier cells can first be visualized, and layer IVCα, but are mostly absent from IVCβ through 6 postnatal weeks. Between 6–12 postnatal weeks, the density of the PV-immunoreactive neuropil and numbers of stained somata increase rapidly in layer IVCβ, eventually exceeding that of IVCα by the end of this period. The density and distribution of PV immunostained somata becomes adult-like by around 1 postnatal month of age, but laminar distributions and densities of PV-immunolabeled chandelier cell axon terminations continue to exhibit refinements over postnatal development (Condé et al. 1996).

The development of PV-immunoreactivity follows a mostly similar sequence in other primate species, but there are some notable differences in comparison with that in macaques. For example, in the developing striate cortices of humans and marmoset monkeys (a species of New World monkey), a transient, overabundance of PV-immunoreactive somata – many of which possess morphological features of pyramidal cells – is evident in deep layers (Spatz et al. 1994; Cao et al. 1996). Such transient numbers of PV-immunoreactive somata decline to adult levels with increasing age, leaving an adult-like pattern by birth in humans (Cao et al. 1996), while changes in density and distribution continue into the postnatal period in marmoset monkeys.

In neonatal monkey area V1, PV-immunoreactive pericellular clusters typical of the terminations of large basket neurons are found in layers V and VI, but such immunoreactive clusters disappear in adolescent and adult monkey area V1 (Akil and Lewis 1992a). Whether such developmental patterns represent a transient population of PV-containing basket cells, or an enduring population of basket cells displaying transient PV immunoreactivity, is unknown.

Regionally, across the hierarchy of cortical areas devoted to visual processing, the emergence of adult-like patterns of PV-immunoreactive cells and processes follows a sequential progression in which primary visual cortex matures earliest, followed by secondary visual association areas and lastly by highest-order visual association areas of the parietal, temporal and prefrontal cortex (Condé et al. 1996).

6. MONOAMINERGIC AND CHOLINERGIC INNERVATION OF PRIMATE VISUAL CORTEX

6.1. INTRODUCTION

These systems are often referred to by their respective neurotransmitter content rather than the more conventional nomenclature referring to origin and termination site. Thus, there are distinct noradrenergic (NA), serotonergic (5-HT), dopaminergic (DA), and cholinergic (ACh) projections to neocortex. While these systems were originally described in rodent (see Moore and Bloom 1978, 1979 for review) reasonably detailed descriptions of the nuclei of origin (Felten and Sladek 1983) and the characteristics of each of these projections to the visual cortex of monkeys have been developed (see Foote and Morrison 1987).

Each monoamine projection originates from a separate and distinct nuclear complex at different levels of the neuraxis (Felten and Sladek 1983; Foote and Morrison 1987). The NA projection originates in the locus coeruleus (LC), a small compact nucleus in the dorsal pons. The 5-HT projection originates from the dorsal and median raphe, a nuclear complex that occupies the medial portion of the caudal mesencephalon, ventral to the aqueduct (Dahlström and Fuxe 1964; Kuhar et al. 1972; Steinbusch 1981; Lidov and Molliver 1982). The substantia nigra and ventral tegmental area (SN/VTA) in the rostral mesencephalon are the nuclei of origin of the DA projection (Berger et al. 1974; Fuxe et al. 1974; Hökfelt et al. 1974; Lindvall and Björklund 1978). The ACh projection originates from a nuclear complex in the basal telencephalon that occupies the medial septal nucleus through the diagonal band and further laterally into the nucleus basalis of Meynert (Shute and Lewis 1967; Fibiger 1982; Houser et al. 1983; Mesulam et al. 1984). The monoamine and ACh projections are highly divergent systems in that a given nucleus or individual neuron can innervate a vast expanse of neocortex (Foote and Morrison 1987).

The neocortical termination patterns of these systems differ fundamentally from the termination patterns of thalamocortical and corticocortical afferents that use glutamate as their neurotransmitter. Thalamocortical and corticocortical systems are organized to implement both serial and parallel processing of information through modules that are interconnected with precise topography, and they utilize excitatory amino acids as their neurotransmitter that exert powerful excitatory effects on discrete postsynaptic targets. The columnar organization of the visual cortex that is driven by the geniculo- and corticocortical afferents is in striking contrast to the tangentially organized extrathalamic afferents in which single axons may innervate not only different columns within a functional region but different functional areas (Morrison et al. 1979, 1982a,b). The available primate data suggest the following principles of organization for extrathalamic afferents to neocortex (Foote and Morrison 1987): A) Each extrathalamic afferent system exhibits pronounced regional and laminar specialization; B) These systems differ from each other in terms of which major cortical subdivisions, cytoarchitectonic areas, and cortical layers constitute their principal termination sites. They may also differ in their preferred cellular targets; C) In spite of these differences these afferents are likely to function in concert in the regulation of state in respect to cortical functions.

6.2. NORADRENERGIC INNERVATION OF VISUAL CORTEX

6.2.1. Anatomic organization

The NA innervation of neocortex arises solely from the locus coeruleus (Ungerstedt 1971; Morrison et al. 1981). In the rat it has been demonstrated that individual LC neurons not only innervate multiple cortical regions but often innervate widely separated brain regions, making these axons the most highly divergent axons in the central nervous system (Lindvall and Björklund 1974, 1978; Morrison et al. 1979, 1981; Loughlin et al. 1982). The NA fibers are primarily distributed in a tangential (i.e. parallel to the cortical surface) fashion, particularly in layer VI, where they are oriented predominantly in the anteroposterior plane, forming a continuous sheet of longitudinal fibers overlying the white matter (Morrison et al. 1978). In rat, these very fine caliber fibers branch to innervate all six layers of the neocortex, in a distinct laminar pattern that is consistent throughout the dorsolateral neocortex (Levitt et al. 1978; Morrison et al. 1978). Thus, the innervation pattern in rodent area 17 is very similar to that of other neocortical areas. Layer I possesses a lattice of rostrocaudal and mediolateral tangential fibers. The layer VI axons branch profusely into terminal axons that extend into layers V and layer IV, and to a lesser degree into layers II and III. The fibers in layers II and III are predominantly radial in orientation. NA synapses are present in all six layers, but predominate in layers IV and V.

The NA innervation of primate cortex does not possess the regional and laminar uniformity that is so prevalent in rat cortex (Morrison et al. 1982a,b; Levitt et al. 1984). However, even though greater regional and laminar specialization exists in the monkey cortex than in rat, the dominant tangential, intracortical trajectory is also a key feature of the NA innervation of the much more convoluted primate cortex (Morrison et al. 1982a,b, 1984). In primate cortex, the NA innervation is primarily directed at the primary motor and somatosensory cortices and their related association regions in the frontal and parietal lobes (Morrison et al. 1982a,b, 1984; Levitt et al. 1984; Lewis et al. 1989). These areas receive a much denser innervation than prefrontal cortex, or the visual areas of the temporal and occipital lobes. Second, when the brain regions subserving a particular modality such as vision are analyzed certain organizing principles in regard to cortical and subcortical NA innervation patterns emerge (Morrison and Foote 1986). For example, the innervation patterns in visual areas relate systematically to the dual stream parallel processing inherent to the visual system. In both New and Old World monkeys, primary visual cortex receives a fairly sparse NA innervation and has virtually no NA fibers in layer IV, the major thalamorecipient layer (Morrison et al. 1982a,b; Kosofsky et al. 1984; Fig. 38). Precisely at the border between areas V1 and V2, the laminar innervation patterns characteristic of NA fibers in area V1 shift such that layer IV of area V2 contains more fibers than layer IV of area V1, and the overall density of fibers in area V2 is higher (Morrison and Foote 1986). The visual regions of the inferotemporal cortex are fairly lightly innervated by NA fibers, while Brodmann area 7 of the parietal cortex is much more densely innervated. Visual thalamic nuclei also exhibit pronounced differences in NA innervation density in that the LGN is virtually devoid of NA innervation in New World and Old World monkeys, while the pulvinar-lateral posterior complex is densely innervated (Morrison and Foote 1986). This is in contrast to the rodent LGN, which does receive NA innervation (Lindvall et al. 1974; Parnavelas et al. 1989).

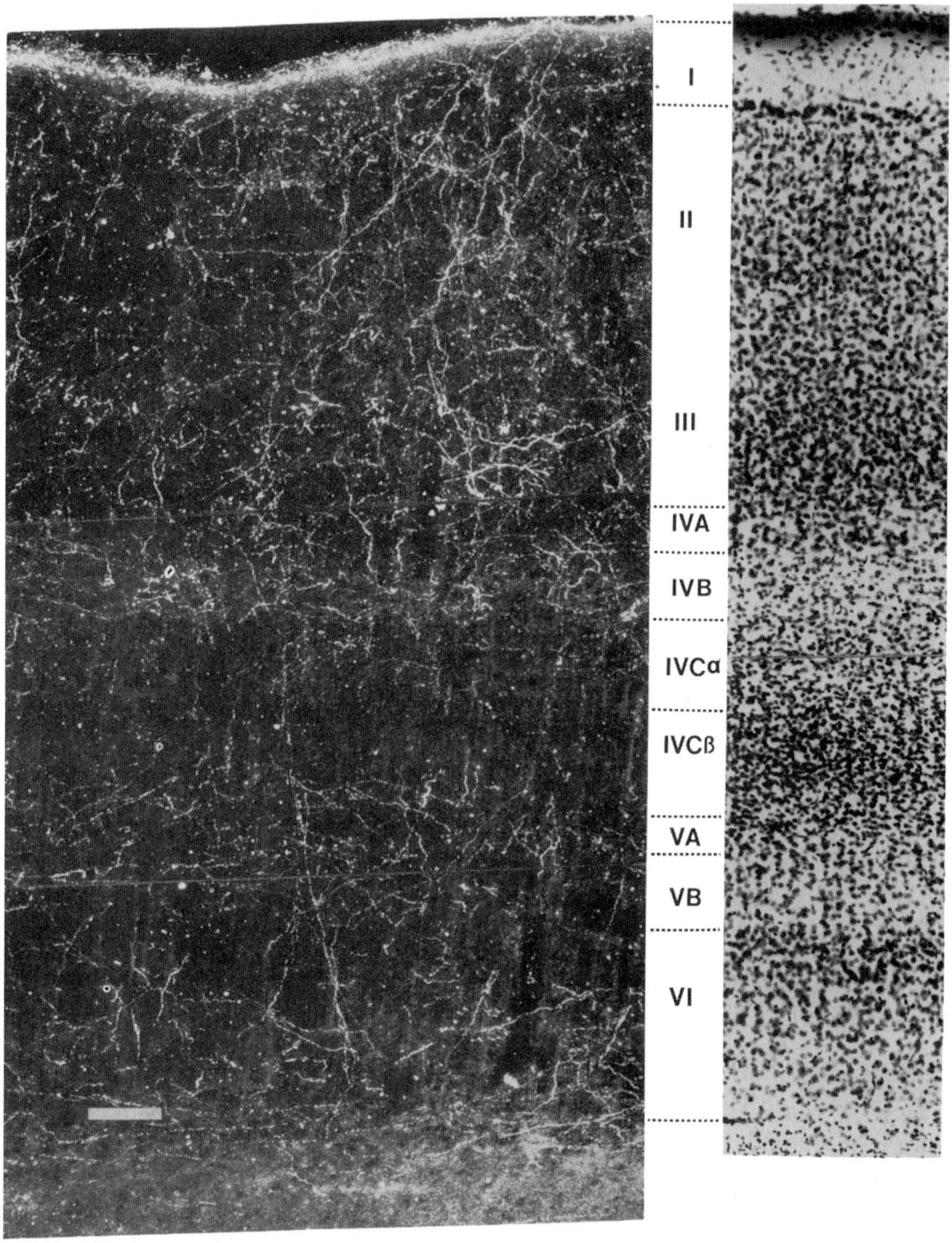

Fig. 38. Darkfield photomicrograph of immunocytochemical preparation of DBH-positive axons in area V1 of a macaque monkey. Note the laminar distribution of NA axons which form two broad bands of moderate density separated by a low density band in layer IVC. Cortical layers are indicated with reference to adjacent Nissl stain section on right. Scale bar = 100 μm. Reproduced from Kosofsky et al. (1984).

These NA patterns of innervation indicate that, functionally related visual regions share common and distinguishable densities of noradrenergic innervation. Specifically, tecto-pulvinar-extrastriate structures are more densely innervated than geniculostriate and inferotemporal structures (Morrison and Foote 1986). These relationships suggest that, within the visual system, NA fibers preferentially innervate regions involved in spatial analysis and visuomotor response rather than those involved in feature extraction and pattern analysis. This segregation within the visual system is consistent with the preferential NA innervation of regions involved in somatosensory-motor integration and the proposed involvement of the LC-NA system in attentional mechanisms (see Foote and Morrison 1987, Foote 1997).

6.2.2. Cortical target cells

The light microscopic innervation patterns, such as those evident in monkey primary visual cortex suggest that the preferred target of NA innervation is pyramidal cell dendrites of layers III, V and VI, as opposed to non-pyramidal cells within layer IV, or the distal dendrites of layer I. Electron microscopic data are scant in primate visual cortex, but the rodent data suggest that NA varicosities form synapses (Papadopoulos et al.. 1987, 989; Olschowka et al. 1981), and the synapses tend to be on a variety of targets: in decreasing frequency on spines, dendritic shafts, pyramidal, and non-pyramidal soma. This set of ultrastructural data taken together with light microscopic patterns suggest that although layer III and V pyramidal cells are likely to be the major target of the NA system in primary visual cortex (Morrison et al. 1982a,b; Kosofsky et al. 1984), non-pyramidal cells are also likely to be targeted by NA fibers.

6.2.3. Development of noradrenergic innervation in visual cortex

The noradrenergic innervation of macaque primary visual cortex is present at birth, but at a much lower density than in the adult (Foote and Morrison 1984). At birth, the fibers are evenly dispersed at a low density throughout layers II to VI, such that the lack of fibers in layer IV that is apparent at later ages is not pronounced at birth. By 18 days, the overall NA innervation density has increased in all layers except IV, where the density is even lower than at birth. By 60 days the lack of innervation of layer IV is even more apparent, and the overall density is nearly that present in the adult (Fig. 39).

6.2.4. Distribution of adrenergic receptors in the visual cortex

Autoradiographic analyses of α and β adrenergic receptors in the cerebral cortex demonstrate that overall prefrontal, motor and somatosensory regions have higher concentrations of binding sites than parietal, temporal and occipital areas. Using prazosine as a ligand for α_1 receptors, area V1 in macaque monkeys has a relatively modest density of these receptors, which peaks in layer I. Layers II-III, IVC, and V show intermediate levels of binding, while the lowest levels are found in layer IVA, IVB and VI (Rakic et al. 1988; Bigham and Lidow 1995). These patterns differ substantially from those observed using clonidine, an α_2 ligand that identifies high binding site densities in layers IVA, IVCβ, and VI, whereas layers I-III, IVB and V have intermediate levels and layer IVCα has the lowest densities of α_2 binding sites in macaque monkeys (Rakic et al. 1988; Bigham and Lidow 1995). Of note, the distribu-

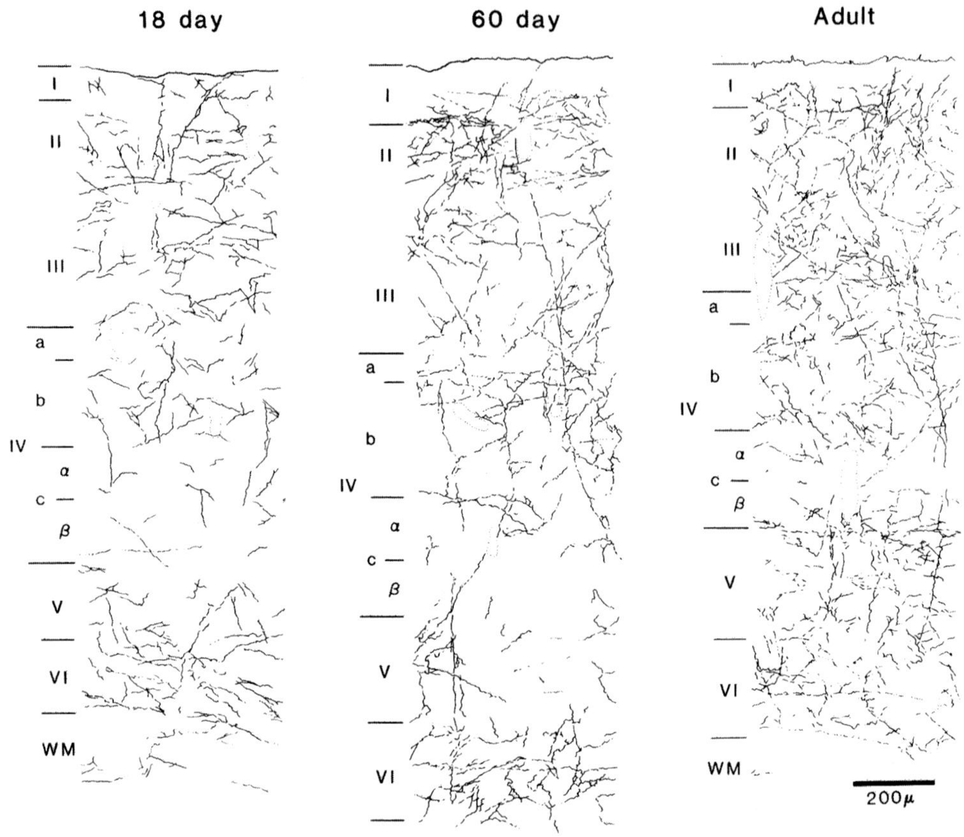

Fig. 39. Drawings based on photomontages of NA fibers as visualized with an anti-DBH antibody in area V1 of 18 day old, 60 day old, and adult macaque monkeys. Note the increased density and arborization in superficial and deep layers with increasing age, and the correlated appearance of a decreased density of fibers in layer IVC. WM, white matter. Scale bar = 200 μm. Reproduced from Foote and Morrison (1984).

tion patterns of α_1 receptor in the marmoset, a New World monkey, differ from those in Old World monkeys in that the highest level of binding are observed in layers I-III and IVC (Gebhard et al. 1993). In human area V1, α_2 receptors analyzed using the selective agonist bromoxidine are mostly concentrated in layers III and IVC (Pascual et al. 1992). In area V2 of the macaque monkey, α_2 receptors predominate in layer III, while the other layers show a relatively homogeneous and lower binding density. Alpha$_1$ display high densities in layer II and superficial III and lower concentrations in layers IV-VI (Rakic et al. 1988; Bigham and Lidow 1995).

The distribution of β receptors evidenced by iodopindolol is more homogeneous throughout the layers of area V1 of macaque monkeys with higher binding densities in layers II-III and V (Rakic et al. 1988; Bigham and Lidow 1995). In area V2, β receptors are mostly concentrated in layer II and superficial III, while lower densities are present in layers IV-VI (Rakic et al. 1988; Bigham and Lidow 1995). Interestingly, the presence of β receptors has been since reported on astrocytes in primate area V1 using immunohistochemistry, and the expression of these astrocytic β receptors is regulated by columnar activity (Aoki et al. 1994b).

It is worth noting that the distribution of adrenergic receptors partially matches that of noradrenergic fibers (see above), with high α_1 receptors in layer I of area V1 and a generally lower concentration of receptors in layer IV where noradrenergic fibers are sparse. However, the high densities of α_2 receptors in layer IVCβ and the low levels of adrenergic binding sites in layer VI which are relatively rich in noradrenergic fibers represent notable mismatches.

6.3. SEROTONERGIC INNERVATION OF VISUAL CORTEX

6.3.1. Anatomic organization

The 5-HT innervation of the primate neocortex arises primarily, if not exclusively, from the dorsal and median raphe nuclei, and exhibits a coarse topographic organization that is most pronounced in a rostrocaudal relationship between dorsal raphe and different cortical sites (Wilson and Molliver 1991a,b). In addition, the serotonergic neurons that innervate primary visual cortex are laterally situated and represent a separate set of neurons from those that innervate prefrontal, motor, or somatosensory areas (Wilson and Molliver 1991a,b). The 5-HT innervation of adult primate neocortex is generally more dense than is the NA innervation, and this is particularly evident in visual areas (Morrison et al. 1982a,b, 1986; Wilson and Molliver 1991a,b). It has been demonstrated most clearly in rat (Kosofsky and Molliver 1987) and cat (Mulligan and Törk 1988), that there are two classes of 5-HT fibers in neocortex; very fine, highly tortuous fibers with very small varicosities, and larger caliber fibers with larger, more beaded varicosities. In primate, two classes exist as well, and the fine fibers are more dense than the beaded fibers, but each has their own regional and laminar preferences (Wilson and Molliver 1991a,b). It appears that each class of fibers originates from distinct and different subnuclei within the raphe (Kosofsky and Molliver 1981; Wilson and Molliver 1991a,b). Molliver and coworkers (Molliver et al. 1989; Wilson et al. 1989; Mamounas et al. 1991) have shown that each class has differential vulnerability to certain amphetamine derivatives, in that these drugs cause a dramatic long-term degeneration of the fine fibers, yet do not appear to affect the beaded class of fibers. In addition, 5-HT 'baskets' have been described in cat neocortex that appear as a dense array of large varicosities that literally outline the soma and proximal dendrites of the postsynaptic neuron, most of which appear to be GABAergic interneurons (Mulligan and Törk 1988), but they are rare in primate neocortex, and absent from primate visual cortex (Wilson and Molliver 1991a,b; Morrison et al. 1982a).

The 5-HT innervation of primary visual cortex of the primate has been characterized in greater detail than that of any other region (Morrison et al. 1982a; Foote and Morrison 1984; Kosofsky et al. 1984; Morrison and Foote 1986; Berger et al. 1988). The 5-HT innervation of area V1 is very dense and distributed in a strictly laminated fashion. Serotonin fibers show a strong preference for layer IV in area V1, a tendency that is also evident in other cortical areas, whereas innervation of layer I is quite sparse (Fig. 40). Also, the morphology of 5-HT fibers in primate area V1 is quite heterogeneous, with a mixture of thick, nonvaricose fibers, large varicose fibers, and extremely fine, varicose fibers. Since the very thick fibers are most evident in white matter and the deep cortical layers, and are more evident in young animals, it is highly likely that they are fibers of passage. The thick fibers are usually tangential in orientation.

The first detailed characterization of the 5-HT innervation patterns in primate area

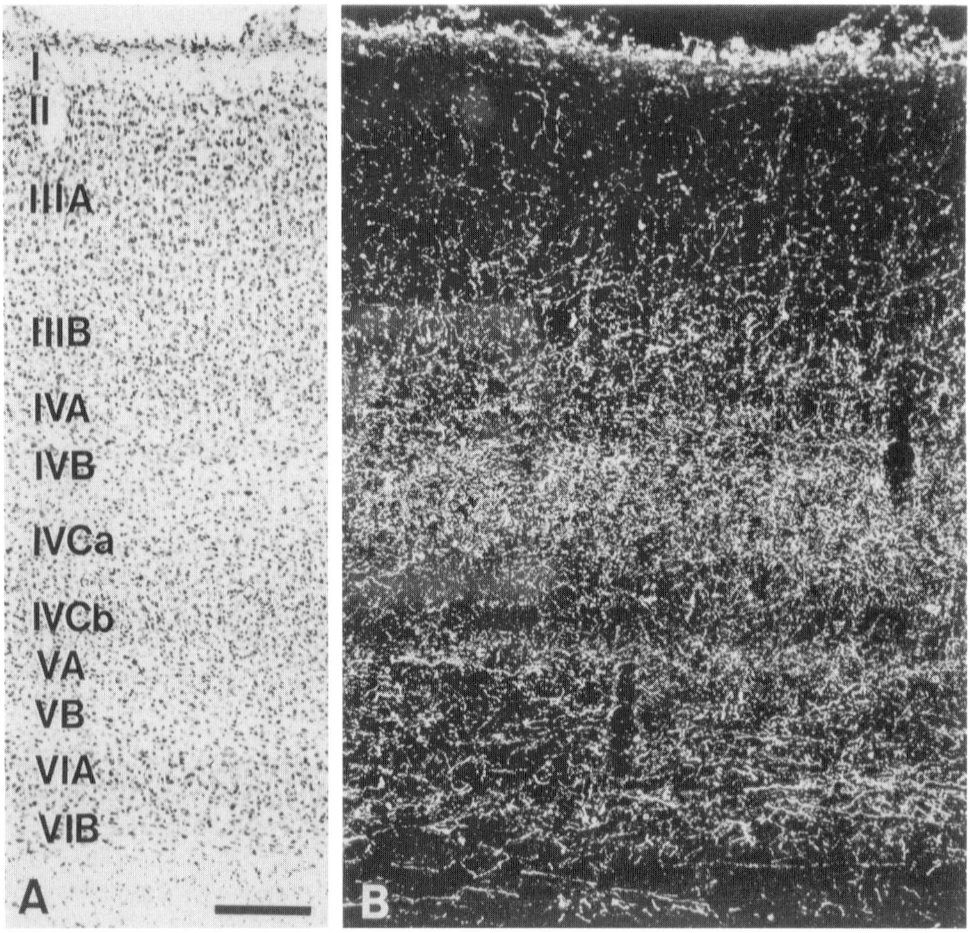

Fig. 40. Distribution of serotonin immunoreactive fibers in the primary visual cortex of the macaque monkey. *A*) Brightfield photomicrograph of the opercular surface of the primary visual cortex as seen in a Nissl stained section to delineate the layers. *B*) Darkfield photomicrograph of the corresponding field in an adjacent section immunostained for serotonin. Note the high density of innervation in IVA, IVB, and IVCα and the low density in IVCβ by comparison. Scale bar = 200 μm. Reproduced from de Lima et al. (1988).

V1 was done in a New World monkey, the Squirrel monkey (*Saimiri sciureus*; Morrison et al. 1982a). Area V1 of the squirrel monkey has a relatively low density of NA innervation, whereas the 5-HT innervation, particularly in layer IV, is among the densest of all neocortical regions. These two fiber systems exhibit a high degree of laminar specialization, and are, in fact, distributed in a complementary fashion in the squirrel monkey (Fig. 41): layers V and VI receive a moderately dense NA projection and a sparse 5-HT projection, whereas layers IVA and IVC receive a very dense 5-HT projection and are largely devoid of NA fibers (Morrison et al. 1982a). These patterns of innervation imply that the two transmitter systems may have different cellular targets and affect different stages of cortical information processing in squirrel monkey area V1: the raphe-cortical 5-HT projection is preferentially directed at the layer IVA and IVC (Fig. 41) which are most intimately linked to the primary sensory information relayed through the thalamus, whereas the NA projection is directed predominantly at

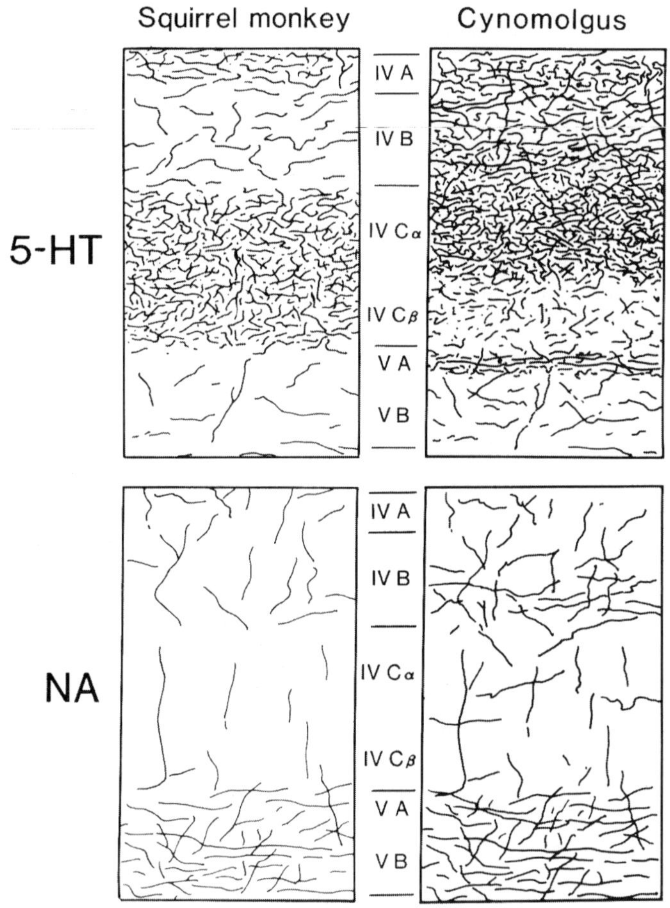

Fig. 41. Schematic illustrations of the laminar distribution of 5-HT and NA fibers in primary visual cortex of the squirrel monkey (a New World primate species) and the macaque monkey (an Old World primate species). The top two panels show 5-HT innervation in each species. Note that each sublamina of layers IVC and V in area V1 of the macaque monkey has a distinct terminal pattern and density, whereas in the squirrel monkey, the innervation of each of these laminae is homogeneous. The two bottom panels show the NA innervation in each species. Note the higher density in layer IVB of the macaque monkey compared to layer IVB of the squirrel monkey, and that in both species IVC has a low density. Reproduced from Foote and Morrison (1984).

the layers containing the majority of the cells which project out of area V1, the pyramidal cells of layers III and V (Morrison et al. 1982).

Based on cytoarchitectural characteristics, investigators have further subdivided layer IVc of area V1 into IVCα and IVCβ, and layer V has been further subdivided into VA and VB (Lund 1981). This extended laminar differentiation of layers IVC and V is very easily discernible in Nissl-stained sections of Old World monkey (i.e., macaque) area V1, whereas, in layers IVC and V of New World (squirrel monkey) area V1, these further laminar differentiations are very subtle, if they are present at all. The macaque monkey receives an even denser 5-HT innervation of area V1 than that found in squirrel monkey, whereas the overall density of the NA innervation is equivalent in the Old as compared to the New World monkeys (Fig. 41). In addition, variations of the 5-HT laminar patterns of innervation exist that are correlated with the further cortical specialization in the Old World species (Takeuchi and Sano 1983; Kosofsky et al. 1984; Morrison and Foote 1986). The 5-HT innervation of macaque monkey area V1 reflects the further laminar differentiation of layers IVC and V in that each of the sublaminae receives it own distinctive pattern and density of innervation. The very dense 5-HT innervation of IVC that is evident in the squirrel monkey is, in the Old World monkeys, largely directed at IVB and IVCα, which are far more densely innervated than IVCβ (Fig. 41). Furthermore, layer VA in Old World monkeys has a higher density than the superficial portion of layer V in the New World monkey (Takeuchi and Sano 1983; Kosofsky et al. 1984; Morrison and Foote 1986) (see Fig. 13 from Foote and Morrison). More specifically, a distinct band of tangential fibers is present at the IVCβ-VA border in the Old World monkeys which is not evident in the New World monkey. In addition, the overall density of 5-HT innervation in layers IVB, V, and VI of Old World monkeys is higher than that present in these layers of squirrel monkey primary visual cortex (Kosofsky et al. 1984).

The ultrastructural data on 5-HT innervation of macaque monkey area V1 support the innervation patterns apparent at the light microscopic level. In regard to species and region, the most precise data on synaptic sites and potential target cells has been obtained from Cynomolgus monkey area V1. In a combined light and electron microscopy study, it was demonstrated that each varicosity is likely to make one, and only one synapse, and thus the relative density and distribution of synapses can be surmised from varicosity counts (Fig. 42). Varicosity counts show that the highest densities are in layers IVB and IVCα, followed by VA, VIB and IIIA (de Lima et al. 1988). In addition, the ultrastructural study showed that relatively small diameter, distal dendrites are the primary target of the serotonin-immunoreactive synapses (Fig. 43). Thus, the primary consequence of the heterogeneous distribution of serotonin-immunoreactive fibers and synapses in area V1 is that cells with dendritic processes in layers IVB and IVCα may receive the densest 5-HT innervation. In addition, these data suggest that neurons possessing dendrites that extend radially across several layers of area V1 receive their serotonin-containing innervation in a preferred portion of their dendritic tree. Thus, small stellate cells of layer IVCβ, which have dendrites confined to this same layer, will be less frequently contacted than stellate cells in layer IVCα. On the other hand, layer VI pyramidal cells with basal dendrites in layer VIβ and apical dendritic arborization in layer IVCα are likely to be contacted on both the distal apical dendrites and the distal basal dendritic branches.

The high density of the serotonergic fibers in layer IVCα of area V1 is of interest because layer IVCα is the major target of geniculocortical fibers originating in the magnocellular layers of the LGN (see above). In the macaque monkey, the LGN is also heterogeneously innervated by serotonin. The magnocellular layers in this nucleus are slightly more densely innervated than the parvicellular layers (Morrison and Foote 1986). From layer IVCα the visual input is relayed to layer IVB and to VIB (Lund et al. 1994), which are also densely innervated by serotonin. Again, layer VIB is a recipient of geniculocortical afferents from the magnocellular layers (Lund et al. 1994). As described above, layer IVB is reciprocally related to a visual association area in the superior temporal sulcus referred to as area MT (Zeki 1974; Maunsell and Van Essen 1983). In the macaque and squirrel monkey a population of very large pyramidal neurons in layer VI also project to area MT, and layer VI, like IVB, receives a reciprocal projection from the same region. This indicates that within area V1 there is a stronger interaction between the serotonergic system and layers more dominantly

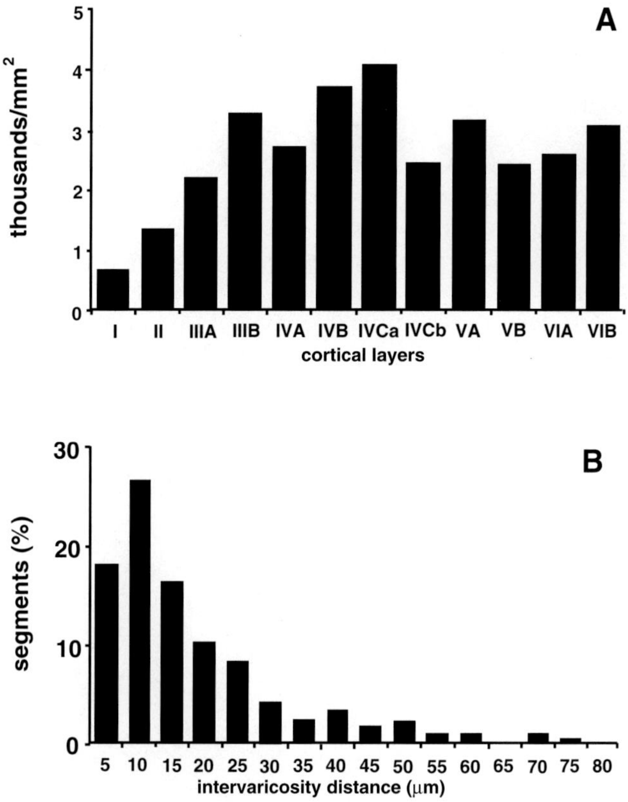

Fig 42. A) Histogram showing the laminar density of serotonin-immunoreactive varicosities within different cortical layers. These data are based on a light microscopic quantitative analysis. Each bar represents the mean value of four measurements. Note that the varicosity density peaks in IVB and IVCα matching the apparent distribution patterns in the light micrograph (Fig. 40B). *B*) Histogram showing that the intervaricosity distance of axonal segments is clustered between 5 and 20 μm, demonstrating the high preponderance of varicosities along most 5-HT axons. Reproduced from de Lima et al. (1988).

involved in the magnocellular stream than with layers involved in circuits related to the parvocellular stream (de Lima et al. 1988).

Thus, given the laminar patterns and propensity to synapse on distal dendrites of both pyramidal and non-pyramidal cells, the 5-HT system is well situated to modulate the activity of both interneurons and projection neurons that subserve visuospatial analysis (de Lima et al. 1988). Such targets are in keeping with the proposed role of serotonin in attentional mechanisms. This role for serotonin in primary visual cortex may be better developed in primate than in other species. In fact the cat area 17 possesses pericellular baskets (Mulligan and Törk 1988) that represent a striking morphologic specialization which is not present in macaque monkey area V1. Such baskets appear to be primarily on GABAergic interneurons, suggesting that in cat, 5-HT might have a profound influence on certain classes of inhibitory interneurons. Even though the monkey cortex lacks this light microscopic morphologic specialization, it appears that the distal dendrites of aspiny interneurons, as well as the distal dendritic shafts of pyramidal cells and spiny stellate cells, are all in a position to receive 5-HT innervation (de Lima et al. 1988).

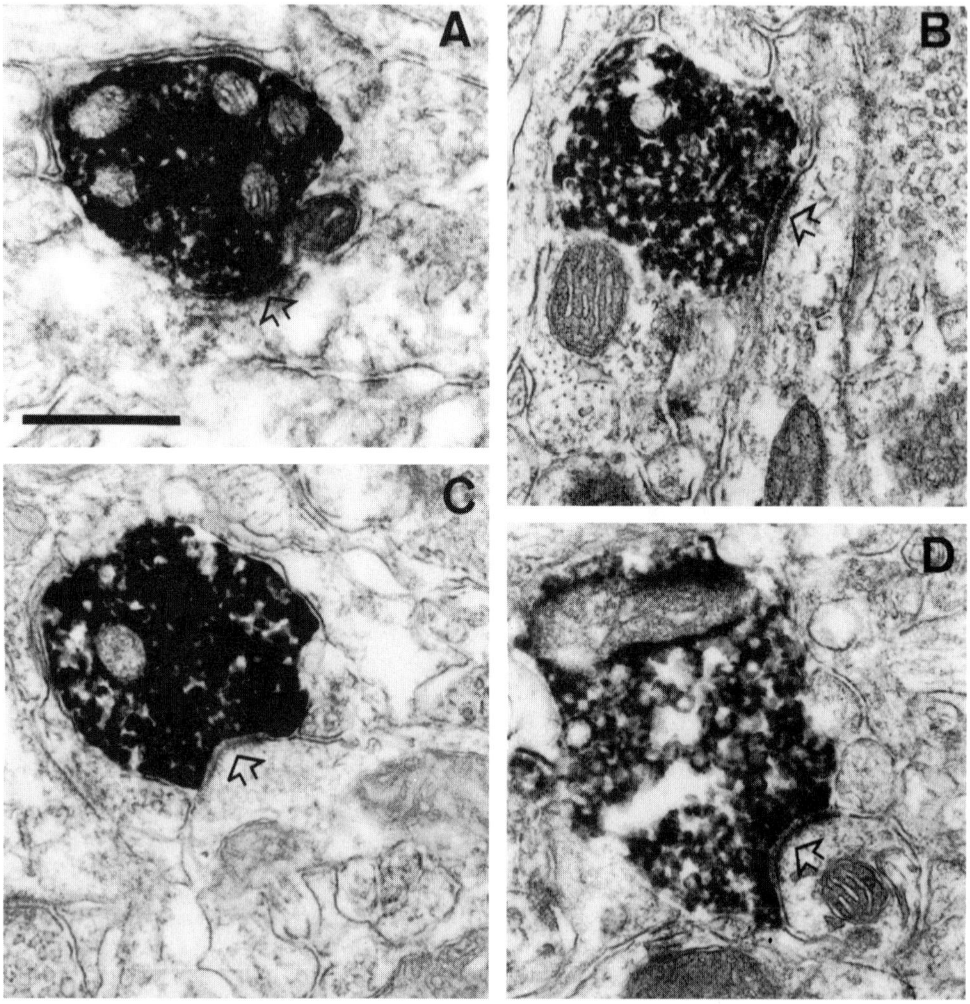

Fig. 43. Four typical serotonin-immunoreactive synapses as visualized with EM-immunocytochemical localization in the macaque. The presence of a postsynaptic density is designated by open arrows. Note that these synapses appear as type 1 synapses (i.e. asymmetrical). In all cases visualized here the post synaptic element is a small dendritic shaft, which was typical 5-HT target in area V1. *A and B*) layer IVB; *C*) layer IVCα; *D*) layer III. Magnification is the same for all four micrographs (scale bar = 0.5 μm). Reproduced from de Lima et al. (1988).

6.3.2. Development of 5-HT innervation of visual cortex

The preferential innervation of layer IV of area V1 is evident at birth, although the overall density of innervation is much lower than in the adult (Fig. 44). Similar to the NA innervation, the 5-HT innervation density of area V1 increases gradually over time, with no abrupt changes in density over the first two months of life. At all ages, the 5-HT innervation of monkey area V1 is much more dense than the NA innervation.

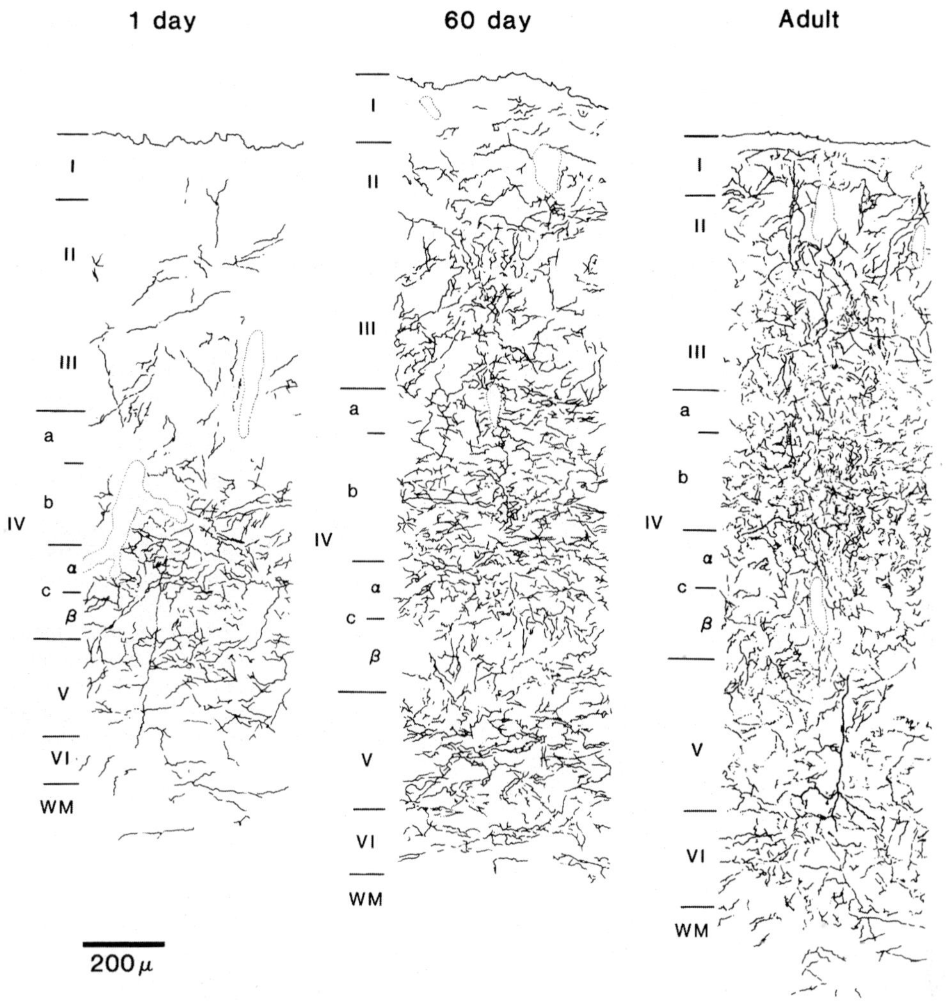

Fig. 44. Immunohistochemical distribution of serotonergic fibers in area V1 of 1 day old, 60 day old and adult macaque monkey. These illustrations are accurate tracings of serotonin-immunoreactive fibers in reconstructed photomontages. Cortical layers are labeled on the left of each panel. Note the high density in layers IVB and IVC, even at 1 day old. WM, subcortical white matter. Scale bar = 200 μm. Reproduced from Foote and Morrison (1984).

6.3.3. Distribution of serotonin receptors in the visual cortex

The distribution of $5HT_1$ receptors in the macaque monkey area V1 using serotonin as a ligand is characterized by a trilaminar pattern with peaks of binding density in layers IVA, IVCβ, and VI and relatively low levels of binding in layers I-III and V and intermediate levels in layers IVB-IVCα. A somewhat comparable pattern is observed with ketanserine binding to $5HT_2$ sites, except that low concentrations of receptors are seen in layers IVB and VI (Rakic et al. 1988). These patterns differ sharply from those seen in area V2, where $5HT_1$ receptors predominate in layers I, II and IV and $5HT_2$ receptors are more concentrated in the deep portion of layer III (Rakic et al. 1988). In

New World monkeys, the density of $5HT_1$ sites predominates in layers I-III and IVC (Gebhard et al. 1993). It is interesting to note that $5HT_2$ binding sites are the only receptors to decline with age in the macaque monkey visual cortex, whereas major differences between young and aged animals are observed for most receptor classes in other neocortical regions (Bigham and Lidow 1995). Finally, there is a rather weak correspondance between the distribution of serotonergic receptors and the presence of serotonergic fibers in the primate visual cortex (see above). Most notably, layer IVCβ that yields very low serotonergic synapse counts (de Lima et al. 1990), has a very high density of $5HT_1$ receptors, whereas layer IVCα which is very rich in 5-HT fibers, displays only moderate levels of binding sites.

6.4. DOPAMINERGIC INNERVATION OF VISUAL CORTEX

6.4.1. Anatomic organization

The DA innervation of rat cortex originates exclusively from the SN/VTA cell groups, and until recently was thought to be restricted to four discrete terminal fields: A) the medial surface rostral to the corpus callosum (anteromedial), B) a rostral region dorsal to the rhinal sulcus, C) a supragenual terminal field that is coincident with the anterior cingulate cortex, and D) a more caudally situated perirhinal terminal field (Lindvall and Björklund 1974, 1978). More recent data suggest that additional target regions in the visual and motor cortices exist in the rat, although they are very sparsely innervated (Berger et al. 1985).

As in the rat, the initial studies on DA innervation of primate neocortex were biochemical in nature. Biochemical studies demonstrated that dopamine levels as well as dopamine/noradrenaline ratios were elevated in various frontal and temporal regions of the monkey cortex, and suggested that occipital areas did not receive a significant DA innervation (Björklund et al. 1978; Brown et al. 1979). These studies offered the first evidence that the dopaminergic projection in the primate was likely to involve numerous discrete regions of association cortex within the frontal and temporal lobes. The biochemical studies were supported by fluorescence histochemical analyses of primate cortex (Levitt 1984), however, a detailed characterization of the regional and laminar patterns of DA innervation as distinct from NA innervation of primate cortex did not emerge until immunohistochemical (Lewis et al. 1987; Lewis and Morrison 1989), and radioautographic (Berger et al. 1988) analyses were carried out.

Immunohistochemical studies using an antibody to tyrosine hydroxylase (TH), a synthetic enzyme for catecholamines, suggest that the dopaminergic innervation extends throughout all four lobes of the primate neocortex, with dramatic regional variations in density that coincide with cytoarchitectonic and functional boundaries (Fig. 45). The DA projection is particularly dense in motor, cingulate, and certain prefrontal areas, whereas area V1 possessed the lowest density of labeled fibers of any neocortical region. The adjacent area V2 of the occipital lobe showed a small, but definite increase in fiber density which did not approach the density of the inferior temporal gyrus (Figs 45, 46). Within the temporal lobe, caudal portions of the superior temporal gyrus (including primary auditory cortex) contained the lowest density of DA fibers, whereas the rostral portion of the superior temporal gyrus (auditory association cortex) had the greatest density of DA innervation found within temporal neocortex. By comparison, all visual association regions of the inferior temporal gyrus

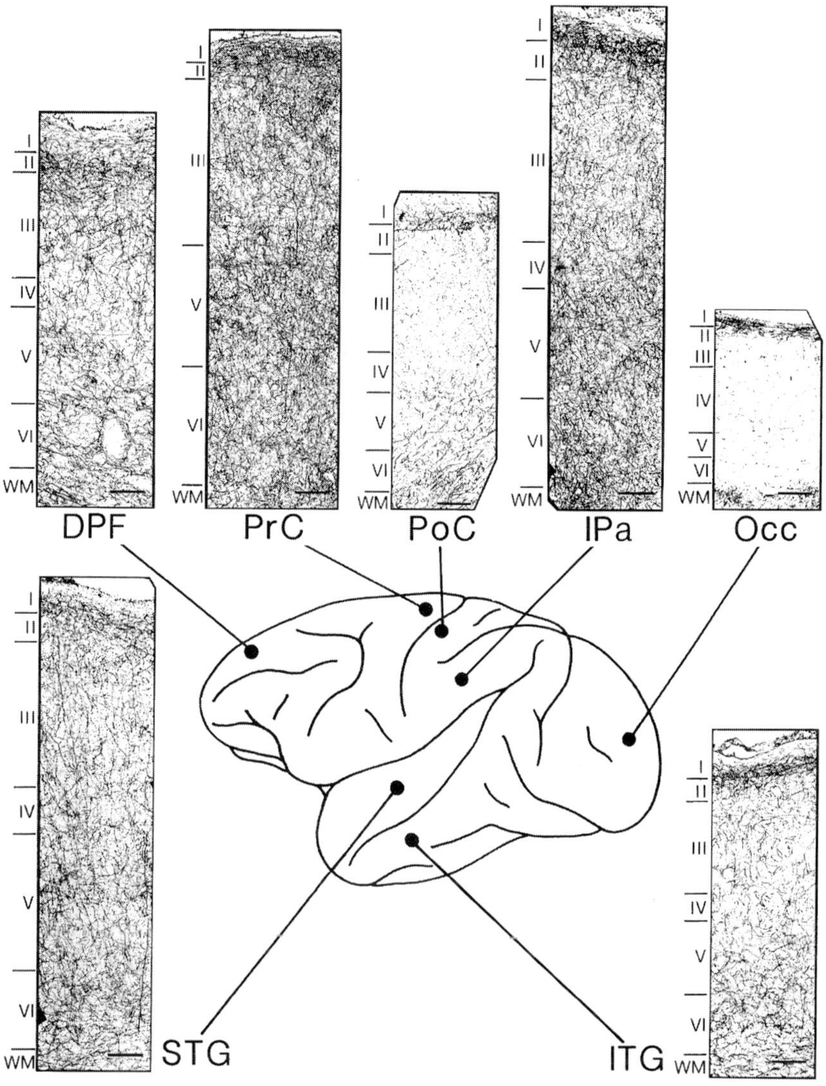

Fig. 45. This diagram shows accurate drawings based on photomontages of TH-immunoreactive fibers in 7 different cytoarchitectonic areas of macaque monkey cortex. Note that while the entire neocortex is innervated by DA fibers, there is extensive regional heterogeneity in the density and laminar distribution of labeled fibers. Each panel is a reversed image of dark field photomontages to increase the clarity of the fibers at this small size reproduction. DPF, dorsal medial prefrontal cortex (Brodmann area 9); PrC, precentral (primary motor cortex); PoC, postcentral (primary somatosensory cortex); IPa, inferior parietal cortex (area 7); Occ, occipital cortex (primary visual cortex); STG, rostral superior temporal gyrus (auditory association cortex); ITG, rostral inferior temporal gyrus (visual association cortex). Scale bar = 200 μm in all cases. Reproduced from Lewis et al. (1987).

had an intermediate density of labeled fibers (Fig. 45). These patterns are consistent with regional patterns in human cortex as well (Gaspar et al. 1989). The laminar pattern of fibers in a given region is correlated with fiber density. In very sparsely innervated regions such as area V1, DA fibers are limited to layer I, whereas area V2 has slightly more innervation in layers II (see Fig. 46), and in the temporal visual areas there are DA fibers in all layers, except IV (Lewis et al. 1987a).

6.4.2. Distribution of dopamine receptors in visual cortex

Several autoradiographic and immunohistochemical studies have revealed a rather widespread distribution of D_1 and D_2 dopamine receptors throughout the neocortex in rodents and primates. The visual cortex of the rat displays moderate levels of D_1 and D_2 receptors, with the D_1 receptors primarily located in the infragranular layers (Boyson et al. 1986; Dawson et al. 1986), whereas D_2 receptors are localized predominantly in layers I and V (Boyson et al. 1986); Bouthenet et al. 1987). In this species, D_1 receptors are always denser than D_2 in the visual cortex, particularly in the deep layers which contain most of the dopaminergic innervation (see above).

Comparably, in the macaque monkey neocortex, the densities of D_1 and D_2 receptors assessed by autoradiography using SCH23390 and raclopride binding, respec-

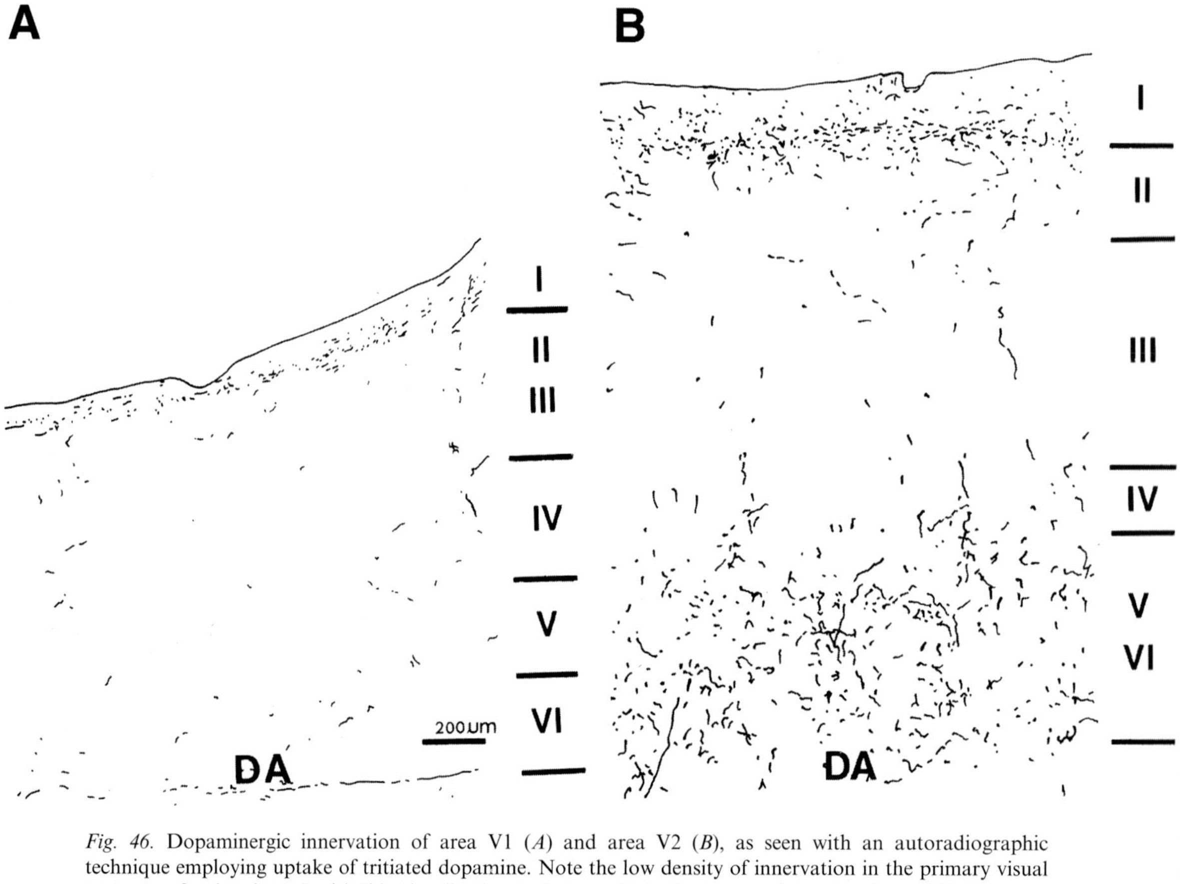

Fig. 46. Dopaminergic innervation of area V1 (*A*) and area V2 (*B*), as seen with an autoradiographic technique employing uptake of tritiated dopamine. Note the low density of innervation in the primary visual cortex confined to layer I with this visualization technique. Note the increased density of area V2, however, the fibers are still confined to layer I and the deep layers. Scale bar = 200 μm. These illustrations are accurate tracings of high magnification of the DA fibers as seen in photomontages, and then were secondarily reduced for reproduction. Cortical layers are indicated by roman numerals. Reproduced from Berger et al. (1988).

tively, display a rostro-caudal gradient with the lowest values in the occipital cortex. In area V1, D_1 binding sites show a very distinct trilaminar pattern with highest densities in layers I-II, IVA, and VB-VI (Lidow et al. 1991). This pattern stops abruptly at the V1/V2 border where area V2 is characterized by a bilaminar pattern seen in most neocortical regions with highest densities in layers I-IIIA and V-VI (Lidow et al. 1991). Analyses of mRNA distribution for D_1 receptors in the macaque monkey cerebral cortex including area V1 have shown that large numbers of neuronal perikarya in throughout layers II-VI contain D_1 receptor mRNA, with a peak in layer V (Brené et al. 1995). In contrast to the autoradiographic distribution of D_1 receptors, no region-specific distribution is observed for the D_2 binding sites that display consistently very high binding densities in layer V in the neocortex including area V1 (Lidow et al. 1991), although autoradiography using the highly selective antagonist epidepride reveals a dense band of D_2 sites in layer IVA as well (Lidow 1995). In addition, the adult pattern of receptor distribution is achieved as early as embryonic day E143 in area V1 for both D_1 and D_2 receptors (Lidow 1995).

Finally, a recent immunohistochemical study in the rat and macaque monkey brain using a specific antibody to the dopamine D_3 receptor, a member of the D_2 dopamine receptor family, demonstrated that in the monkey area V1, the neuropil is markedly devoid of stained dendrites, except in layers I, II, IIIA, and VI. In addition, pyramidal neuronal somata are lightly labeled, although Meynert cells are more intensely stained. This pattern contrasts with other neocortical areas that show prominent staining of apical dendrites in layer III and V as well as intensely labeled pyramidal cell perikarya (Nimchinsky et al. submitted). In the rat however, the primary visual cortex does not display a distribution pattern of D_3 receptors that is distinct from other neocortical regions.

6.5. CHOLINERGIC INNERVATION OF VISUAL CORTEX

6.5.1. Anatomic organization

In monkey, cortical ACh fibers appear to arise from the nucleus basalis of Meynert and, to a lesser extent, neurons within the diagonal band of Broca. In both monkey and human the general regional patterns for ACh innervation appear to be a high density of innervation in limbic cortex, less dense but very substantial innervation of area V1 and other primary sensory and motor areas, and a still less dense innervation of association cortex, including visual association areas (Mesulam et al. 1986, 1992). In area V1, layer I receives the most dense ACh innervation, followed by layers II and III. While less dense than area V1 overall, visual association areas do not exhibit the strong preference for superficial layers exhibited by area V1 (Mesulam et al. 1986, 1992; Fig. 47).

6.5.2. Distribution of cholinergic receptors in the visual cortex

Nicotinic and muscarinic receptors are present on the vast majority of cortical neurons as shown by autoradiographic and immunohistochemical investigations (Rakic et al. 1988; Schröder 1992; Bigham and Lidow 1995). In the primary visual cortex of macaque monkeys muscarinic receptor distribution evidenced by quinuclidinyl benzilate binding is rather homogeneous with high concentrations of binding sites in layers II-IVA, intermediate levels in layers I, IVCβ and V, and lowest levels in layers IVB,

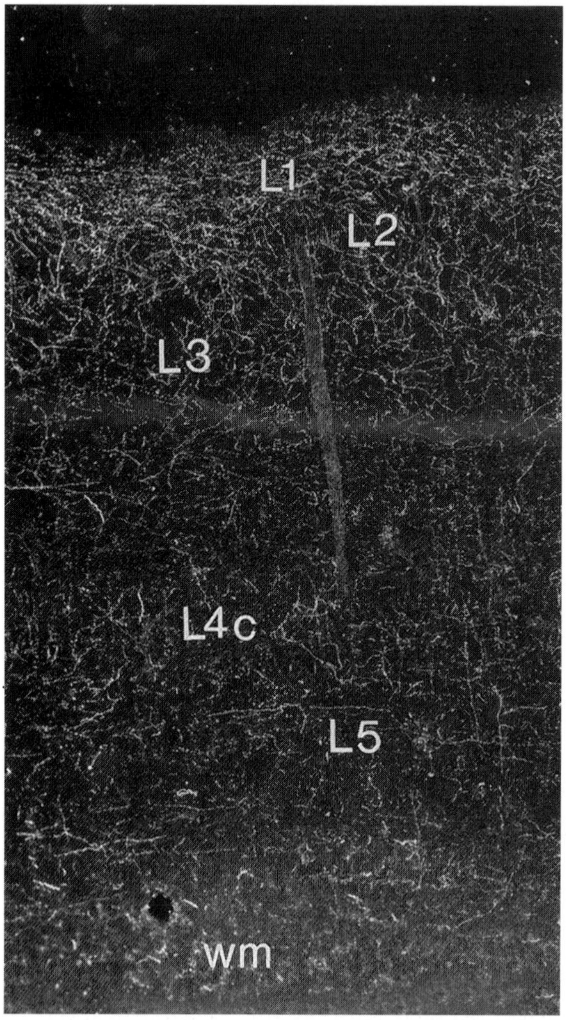

Fig. 47. Distribution of cholinergic fibers as visualized with an antibody to choline acetyltransferase in Brodmann area 17 of human cortex. Layers are designated by L1, L2, L3, L4c and L5; wm, white matter. Magnification = 266X.

IVCα and VI. In area V2, the highest densities are observed in layer II-III and the lowest in layer VI (Rakic et al. 1988). In area V1 there is a reasonably good match between the distribution of these receptors and that of cholinergic fibers which predominate in superficial layers in area V1 (see above). Immunohistochemical investigations have demonstrated that muscarinic receptors m1 and m2 are primarily associated with non-cholinergic asymmetric synapses as well as with symmtric cholinergic synapses. The m1 receptors occur postsynaptically in spines and dendrites at asymmetric synapses. In area V1, m1 receptor immunoreactivity is located predominantly in layer III and VI and labels principally pyramidal cells (Mrzljak et al. 1993). m2 receptor immunoreactivity is seen principally in presynaptic axons and also in subpopulations of non-pyramidal neurons. The distribution pattern for the m2 receptor-immunoreactivity is characterized by intense staining of layer IVA and IVCβ (Mrzljak et al. 1993). In addition, there is evidence that the m2 muscarinic receptor is selectively expressed in the parvocellular stream in the macaque monkey area V1, where it is enriched in the layer IV axons originating inthe parvocellular layers of the LGN and in the interblob domains in layer II and III, wher they are found in association with non-cholinergic synapses (Mrzljak et al. 1996). In addition, *in situ* hybridization studies have demonstrated that m1, m3 and m4 receptors have a highly heterogeneous laminar development pattern in the rat (Rossner et al. 1993), suggesting that these receptors have a differential involvement in the postnatal maturation of the visual function.

7. ACTIVITY-DEPENDENT PLASTICITY OF NEUROTRANSMITTERS, RECEPTORS AND OTHER NEUROCHEMICALS IN ADULT MONKEY AREA V1

7.1. INTRODUCTION

The idea that the structure and function of primate cerebral cortex can be altered by changing sensory inputs was shaped initially by studies of developing visual cortex. Monocular deprivation in young monkeys during their first 4–12 months of postnatal life results in various anatomical and physiological changes in the organization and connectivity of area V1, the most overt of which include: A) an expansion of the terminal arbor size of geniculocortical afferents representing the non-deprived eye, with a concomittant increase in the cortical territory from which neurons respond to non-deprived eye input; and B) a reduction in the terminal arbor size of the geniculocortical afferents representing the deprived eye, with a concomittant shrinkage of the cortical territory from which neurons respond to deprived eye input (Hubel et al. 1977; LeVay et al. 1980). After this time-period, monocular deprivation fails to produce the kinds of overt morphological changes to geniculocortical connectivity which can be readily induced earlier in postnatal life (LeVay et al. 1980), an observation which formed the traditional view that the capacity for plastic changes in area V1 was limited to a developmentally regulated critical period (Hubel and Wiesel 1977; Hubel et al. 1977; von Noorden and Crawford 1978). Nevertheless, LeVay, Wiesel and Hubel (1980) made the observation that subtle physiological changes in area V1 persisted even in monkeys subjected to monocular enucleation as adults. More recent studies support the idea that the mature visual cortex retains a remarkable capacity for considerable functional plasticity, including that of receptive field properties (Frégnec et al. 1988, 1992; Gilbert and Wiesel 1990; Pettet and Gilbert 1992; Shulz and Frégnec 1992) and receptive field organization (Kaas et al. 1990; Heinen and Skavenski 1991; Chino et al. 1992; Gilbert and Wiesel 1992; Darian-Smith and Gilbert 1995). In the section which follows, recent studies of adult monkey area V1 are reviewed which most likely pinpoint some of the anatomical and mechanistic correlates of the physiological effects of perturbing visual inputs in the mature brain. Such studies demonstrate that in adult macaque monkeys, manipulations of an eye which affect normal patterns of visual input which reach area V1 have profound activity-dependent effects on the neurochemical properties of intrinsic cortical neurons. These data indicate that visual experience plays a continous and fundamental role in shaping and modifying visual cortical function which extends throughout life, and that the neurochemical phenotype of neurons in visual cortex may be more plastic than previously throught.

7.2. CYTOCHROME OXIDASE

Histochemical staining for activity levels of the mitochondrial enzyme, cytochrome oxidase, is a cornerstone for anatomical studies of activity-dependent plasticity of neurochemically identified cells and pathways in primate area V1. There are two principal attributes of visualizing this metabolic marker for anatomical studies of neurochemical plasticity in area V1: A) the relative intensity of staining reflects the relative level of neuronal activity (Wong-Riley 1989); therefore increases or decreases in staining intensity are indicative of corresponding changes in functionally-driven activity; and B) monocular deprivation induces an altered, but characteristic pattern of staining which can be easily related to the unique laminar and compartmental segregation of afferents representing each eye (Horton and Hubel 1981; Horton 1984; Wong-Riley and Carroll 1984; Hendrickson and Tigges 1985; Hendry and Jones 1986; Trusk et al. 1990). Thus, by comparing immunocytochemically processed sections with adjacent ones stained for cytochrome oxidase activity, the eye dominance column in which specific neurochemical changes are evident can be readily identified.

Within 4–5 days of monocular deprivation by enucleation or intravitreal injections of the voltage-gated sodium channel blocker tetrodotoxin (TTX), dramatic changes in the pattern of cytochrome oxidase staining occur in adult primate visual cortex which are most evident in layer IVC, one of the principal thalamorecipient layers (Fig. 48; Hubel and Wiesel 1972; Hendrickson et al. 1978; Blasdel and Lund 1983). The staining pattern in layer IVC changes from a normal one of intense and homogeneous staining, to one in which the two sets of ocular dominance columns become easily distinguished (Horton and Hubel 1981; Hess and Horton 1984; Wong-Riley and Carroll 1984; Hendrickson and Tigges 1985; Hendry and Jones 1986; Edwards 1987; Trusk et al. 1990). In sections cut tangential to the flattened pial surface, ocular dominance columns driven by the deprived eye appear as lightly stained stripes 0.4–0.5 mm wide, which alternate with more-or-less equally wide stripes of normally intense staining which correspond to the ocular dominance columns driven by the intact eye (Fig. 48B, D). Such deprivation effects on deprived-eye columns extend into layer IVA, and can be discerned in supra- and infragranular layers as well.

In layers II and III, monocular deprivation affects the cytochrome oxidase staining pattern in both sets of ocular dominance columns. Normally in these layers, regularly-spaced, circular patches (also called blobs or puffs) of intense cytochrome oxidase staining form rows which lie along the centers of ocular dominance columns (Horton and Hubel 1981; Livingstone and Hubel 1982; Humphrey and Hendrickson 1983; Wong-Riley and Carroll 1984). Such patches are evident in most primates, including humans (Horton and Hedley-Whyte 1984; Wong-Riley et al. 1993). Surrounding the intensely-stained patches is a much paler-staining, inter-patch region. The physiology and connectivity of the neurons lying within the patches differ from those lying in interpatch regions: neurons within the patches receive direct geniculocortical input, are mostly if not exclusively monocular, and exhibit unique functional properties (Hendrickson et al. 1978; Weber et al. 1983; Livingstone and Hubel 1984; Tootell et al. 1988; Ts'o and Gilbert 1988; Silverman et al. 1989; Lachica and Casagrande 1992). Following periods of monocular deprivation by enucleation, TTX treatment, or by removing the cystalline lens, which produces a chronic blurring, rows of shrunken, pale cytochrome oxidase-stained patches are evident within the centers of deprived eye columns; these alternate with adjacent rows of patches with a normal intensity of staining, but in contrast to the normal pattern, such patches are often connected

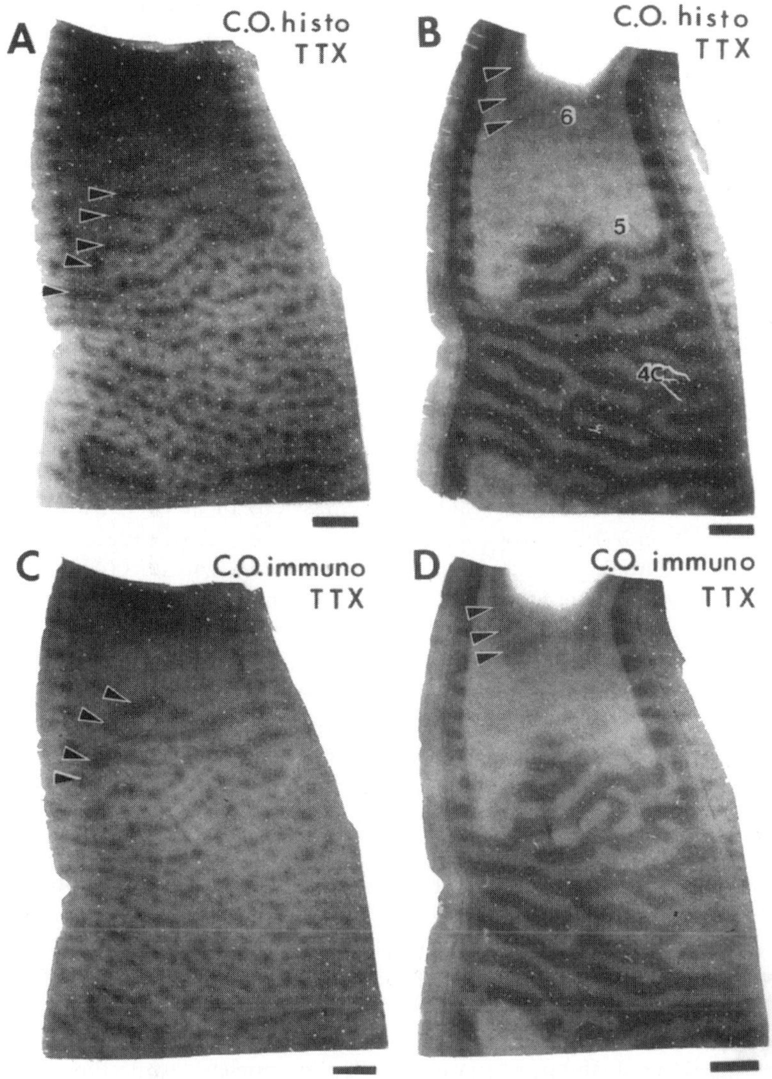

Fig. 48. Effects of monocular deprivation on laminar and compartmental patterns of staining for cytochrome oxidase activity in adult macaque monkey area V1. Cytochrome oxidase was visualized histochemically (*A, B*) or immunocytochemically (*C, D*) following 3 weeks of monocular treatment with tetrodotoxin. In superficial layers (*A, C*), rows of small, pale puffs corresponding to the deprived eye inputs alternate with rows of larger, darkly reactive puffs corresponding to the intact eye inputs. The alternating rows of lightly and darkly stained ocular dominance bands are visible in layers IVA (*arrowheads, A, C*) and VI (*arrowheads, B, C*), but are most distinct in layer IVC (*B, D*). Scale bar = 1 mm. Reproduced from Wong-Riley (1994).

together by thin bridges of intense staining which traverse the normally pale interpatch regions (Fig. 48A, C; Horton and Hubel 1981; Horton 1984; Wong-Riley and Carroll 1984; Hendrickson and Tigges 1985; Hendry et al. 1988a; Trusk et al. 1990). In layers V and VI, deprivation effects are less overt, but patches of less intense staining are present in deprived eye columns (Fig. 48B, D).

The pattern of cytochrome oxidase staining can also be disrupted by suturing together the eyelids of one eye (Horton 1984; Trusk et al. 1990). This manipulation reduces the light entering the eye by 1–4 log units (Crawford and Marc 1976) and eliminates all but the lowest spatial frequencies, thus producing a much milder form of deprivation in comparison with that produced by enucleation or TTX treatment. Under these conditions, the monocular deprivation-induced staining pattern in area V1 is largely similar to that described for the other manipulations, but longer time-periods of deprivation are required to produce overt effects.

7.3. GABA

Within 4 to 5 days following the elimination of retinal ganglion cell activity from one eye either by enucleation or TTX injection, immunocytochemical staining for GABA or its synthesizing enzyme GAD reveals an aberrant pattern of labeling in area V1 which is apparent in several cell-dense layers. The most overt effects on GABA and GAD staining are evident in layers IVC and IVA (Fig. 49), which are the principal thalamorecipient layers (Hubel and Wiesel 1972; Hendrickson et al. 1978; Blasdel and Lund 1983), and which also contain the greatest concentrations of GABA neurons in comparison with other layers (Fitzpatrick et al. 1987; Hendry et al. 1987). When viewed in sections cut tangential to the flattened pial surface, GABA and GAD staining in layers IVC and IVA is characterized by alternating lightly and darkly stained stripes, each approximately 0.5 mm wide, which correspond to deprived eye and non-deprived eye dominance columns, respectively (Fig. 49C, D; Hendry and Jones 1986). The impression of a lighter band of staining associated with the deprived eye columns is accounted for by two major effects at the cellular level: A) there is approximately a 50% reduction in the numbers of labeled GABA or GAD neurons and terminals (puncta) within deprived eye columns in comparison with intact eye columns (Fig. 49A, B), in which the numbers of GABAergic neurons are within the normal range; and B) the intensity of labeling of the remaining neurons is much lower in comparison with that of GABA- or GAD-containing neurons occupying the intact eye columns (Hendry et al. 1987; Hendry and Jones 1988a). The factors that determine which GABAergic cells lose their immunoreactivity entirely and which retain, albeit at less intense levels, their immunoreactivity may involve differences between these two GABAergic cell populations in their synaptic density and the origin of their synaptic inputs. Those that lose their immunoreactivity are sparsely innervated, but their synaptic inputs derive largely from the dorsal lateral geniculate nucleus, while those that retain their immunoreactivity are more densely innervated overall, but the proportion of contacts which derive from the thalamus is low (Hendry and Jones 1988b).

Fig. 49. Effects of monocular deprivation by tetrodotoxin treatment on GABA or GAD staining in adult monkey area V1. *A, B*) Matched sections, one stained immunocytochemically for GABA (*A*) taken through layer IV, showing one deprived and two flanking non-deprived ocular dominance columns, and the other (*B*) taken through a non-deprived column stained histochemically for cytochrome oxidase activity. GABA staining reveals the lower numbers of immunoreactive somata and puncta in deprived columns, particularly in layer IVC, in comparison with undeprived columns. *C, D*) Matched sections cut tangentially from a flattened block and stained immunocytochemically for GAD (*C*) or histochemically for cytochrome oxidase activity (*D*). In layer IVC, rows of intense GAD-immunostaining (intact eye dominance columns) alternate with rows of light GAD immunostaining (deprived columns). Deprivation effects on GAD immunoreactivity can also be seen in the superficial layers. *Circles* indicate identical blood vessels in the two photomicrographs. Scale bar = 1 mm. Reproduced from Hendry and Jones (1988).

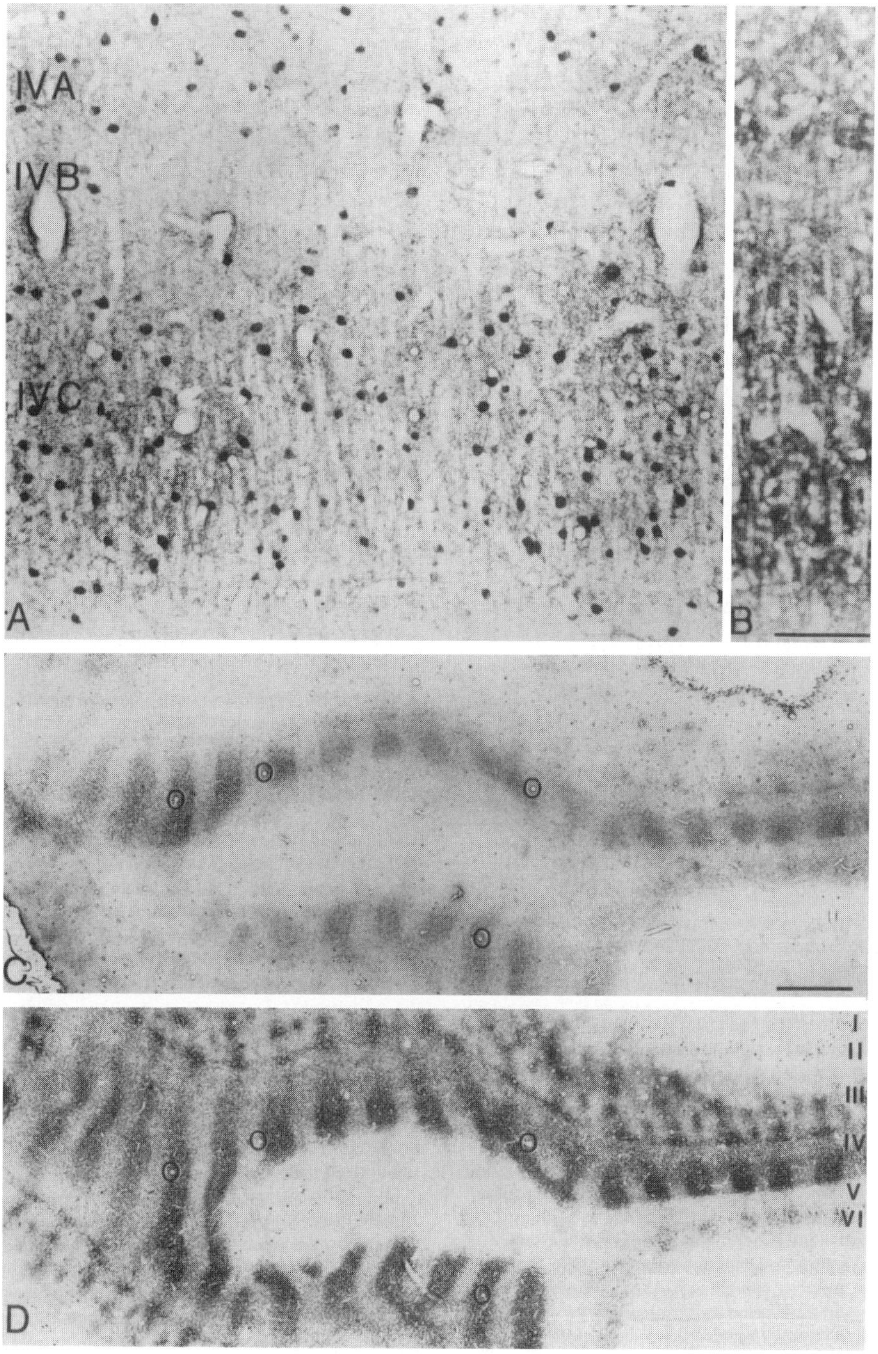
IVA
IVB
IVC
A
B
C
I
II
III
IV
V
VI
D

The pattern of GABA and GAD immunostaining in layers II and III is also affected by eliminating retinal activity of one eye by TTX (Hendry and Jones 1988a). However, unlike layer IV, where reduced immunostaining is restricted to deprived eye columns, changes in GABA and GAD staining in layers II and III affect parts of both deprived and intact eye columns. Within the cytochrome oxidase-enriched patches, the numbers of GABA-immunoreactive somata are quantitatively similar to those found in normal monkeys, regardless of whether the patches lay within the centers of deprived or non-deprived eye columns (Hendry and Jones 1988a). Within deprived eye dominance columns, however, reduced numbers of GABAergic somata are apparent in tissue surrounding the patches, and extend into the interpatch regions which span both deprived eye, and the adjacent, non-deprived eye columns (Hendry and Jones 1988a). There are no detectable changes in the pattern of GABA and GAD immunolabeling in layers V and VI (Hendry and Jones 1988a).

GABA and GAD immunostaining in area V1 is also affected when normal pattern vision is disrupted by suturing together the eyelids of one eye. Although this manipulation produces a less drastic effect on visual input, the changes in the numbers and intensity of labeled neurons are quantitatively similar to those described following enucleation or TTX treatment. However, much longer periods of lid suture-based deprivation (several weeks to months) are required to produce equally robust effects of immunocytochemical staining (Hendry and Jones 1986, 1988a).

The reductions in the numbers of GABA or GAD neurons following monocular deprivation are not due to transneuronal degeneration and cell death. Following lid suture or TTX treatment, cell counts of thionin-stained neurons within deprived eye columns of layer IVC reveal similar total numbers of neurons in comparison with those obtained from non-deprived eye dominance columns, and such neuron counts are within the normal range (Hendry and Jones 1988a). That such affected neurons remain viable is further suggested by reversal experiments, in which small biopsy specimens removed from the primary visual cortex of animals subjected to monocular lid suture or TTX treatment show the characteristic deprivation effects on GABA and GAD staining, while after two or more months of binocular vision restored by reopening the eyelids or ceasing TTX treatment, the patterns of GABA and GAD immunostaining return to normal (Hendry and Jones 1986; Hendry and Jones 1988a). These data suggest that levels of GABA or GAD within those neurons vulnerable to the effects of visual deprivation are dynamically modulated in response to changes in activity. Such modulation, however, appears to be dependent more on the relative imbalance of activity which arises between the normal and deprived eye inputs, rather than on a mere reduction or cessation of activity. This is supported by two principal lines of evidence. First, in the monocular segment of area V1 contralateral to a TTX-treated eye, GABA staining appears both quantitatively and qualitatively normal (Hendry and Carder 1992; Jones et al. 1994), which would not be expected if a simple reduction in activity was responsible for reducing GABA levels. Second, the patterns of GABA immunostaining are qualitatively and quantitatively normal in the primary visual cortex of monkeys in which the activity of geniculocortical afferents representing both eyes are eliminated equally by lesioning the LGN (Hendry and Carder 1992; Jones et al. 1994).

Changes in immunocytochemically detectable levels of GAD protein following monocular deprivation may have as their basis corresponding changes in gene expression for the mRNA encoding GAD. *In situ* hybridization histochemistry using transcript-specific cRNA probes reveals that following longer-term periods of TTX treatment of at least 15 days, corresponding decreases in GAD mRNA levels are evident (Benson et al. 1994). However, with short-term TTX treatment of 5–6 days, when immunocytochemical changes in GABA and GAD levels are maximal, there are no differences in the levels or patterns of GAD probe hybridization between normal and deprived eye columns (Benson et al. 1991a). These data suggest that activity-dependent effects on GAD and GABA levels may arise through a combination of transcriptional and posttranscriptional regulatory mechanisms, depending on the time-course of deprivation (Jones 1993).

7.4. GABA RECEPTORS

$GABA_A$ receptors are also robustly affected in an activity-dependent manner by visual deprivation. Radioligand binding studies in adult monkeys indicate that both ^{3}H-muscimol and ^{3}H-flunitrazepam binding is reduced by approximately 25% in deprived eye columns following monocular deprivation by intravitreal TTX injections, with a time-course for deprivation effects similar to that for the effects on GABA immunostaining (Hendry et al. 1990). This reduction appears to be due to decreased numbers of binding sites as opposed to decreased binding affinity (Hendry and Carder 1992), which suggests that levels of both GABA-binding subunits and benzodiazepine-binding subunits of the multimeric receptor-channel complex are affected by activity.

Immunocytochemical and *in situ* hybridization studies using subunit- or subunit class-specific antibodies and probes confirm and extend the autoradiographic findings by demonstrating that individual subunits which comprise the $GABA_A$ receptor complex are differentially regulated in a complex fashion in response to visual deprivation. Following 5–30 days of TTX-induced retinal impulse blockade, immunocytochemical staining in layer IVC for $\alpha1$, $\beta2/3$ and $\gamma2$ subunits changes from the normal pattern of high intensity and uniformity across columns, to an altered pattern in which lightly immunostained bands corresponding to the deprived eye columns alternate with intensely stained bands corresponding to the normal eye columns (Fig. 50; Hendry et al. 1990, 1994). The mechanism underlying the reduced protein levels may involve corresponding decreases in gene transcription of relevant subunit mRNAs. In comparison with intact eye columns, in which levels of probe hybridization for all subunits examined to date appear normal, deprived eye columns of layer IVC exhibit levels of probe hybridization which are reduced by 20–30% for a1 subunit mRNAs, and 9–17% for $\beta2$ and $\gamma2$ subunit mRNAs (Huntsman et al. 1994). The deprivation effects upon these subunit mRNAs extend outside of layer IVC, but are only evident for $\alpha1$ and $\gamma2$ subunit mRNAs, which exhibit reduced levels within corresponding deprived eye columns of layers II and III. There is little or no change observed in superficial layers for $\beta2$ subunit mRNAs, nor is there any change in levels observed, in any layer, for $\alpha2$, $\alpha4$ and $\beta1$ subunit mRNAs (Huntsman et al. 1994). Alpha5 subunit mRNAs, in contrast, exhibit normal levels and patterns of probe hybridization in all layers except layer VI, where lighter bands of hybridization corresponding to deprived eye columns alternate with more intense bands corresponding to intact eye columns (Huntsman et al. 1994). Thus, these data demonstrate that activity-dependent regulation of $GABA_A$ receptors occurs in a layer- and subunit-specific manner, indicating an enormous complexity in the cascade of effects upon GABAergic neurotransmission induced by alterations in visually driven input.

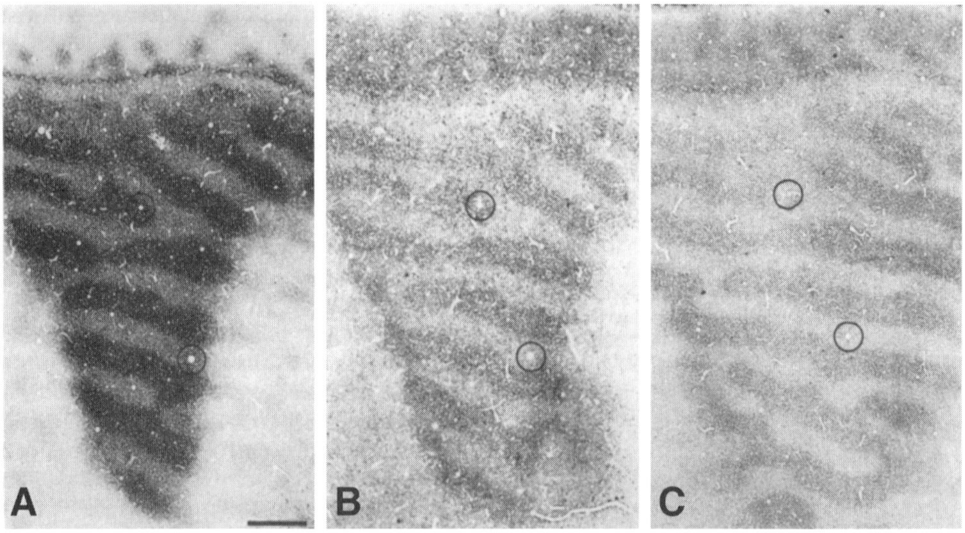

Fig. 50. Effects of monocular deprivation on subunits of the $GABA_A$ receptor. Photomicrographs of adjacent sections from adult macaque monkey area V1 showing reduced histochemical staining for cytochrome oxidase activity in deprived eye dominance columns of layer IVC (*A*), and a corresponding decrease in immunoreactivity for $\alpha 1$ (*B*) or $\gamma 2$ (*C*) subunits. *Circles* mark identical blood vessels. Scale bar = 1 mm. Reproduced from Jones (1993).

7.5. CALCIUM-BINDING PROTEINS

The three calcium-binding proteins most extensively studied in visual cortex – calbindin (CB), parvalbumin (PV) and calretinin (CR) – are differentially affected in response to visual deprivation in adult macaque monkey area V1 (Blümcke et al. 1994; Gutierrez and Cusick 1994; Carder et al. 1996).

7.5.1. Calbindin

Changes in CB immunoreactivity in response to monocular deprivation occur most rapidly in comparison with the other calcium-binding proteins, and are the least restricted, in that changes are evident across both laminar and eye-dominance column boundaries. Within 3–5 days of visual deprivation by intravitreal TTX injection, a dramatic reduction in levels of CB immunoreactivity is evident throughout the full width of deprived eye dominance columns in all sublayers of layer IV, including IVA, IVB and IVC, and extends for 100–150 μm into adjacent, normal eye dominance columns (Fig. 51). Thus, the only normal-appearing staining pattern in layer IV is found in a thin, darkly stained stripe situated at the centers of intact eye dominance

→

Fig. 51. Effects of monocular deprivation on calcium-binding protein immunoreactivity in adult macaque monkey area V1. Photomicrographs of adjacent sections, processed immunocytochemically for calbindin (*A*) or parvalbumin (*C*), or histochemically for cytochrome oxidase activity (*B*). *Arrows* in (*B*) demarcate intensely cytochrome oxidase-stained, intact eye dominance columns, which, by comparison with corresponding *arrows* in (*A*), reveal that thin stripes of intense calbindin immunoreactivity are present along the centers of such intact eye dominance columns. Scale bar = 2 mm. Reproduced from Carder et al. (1996).

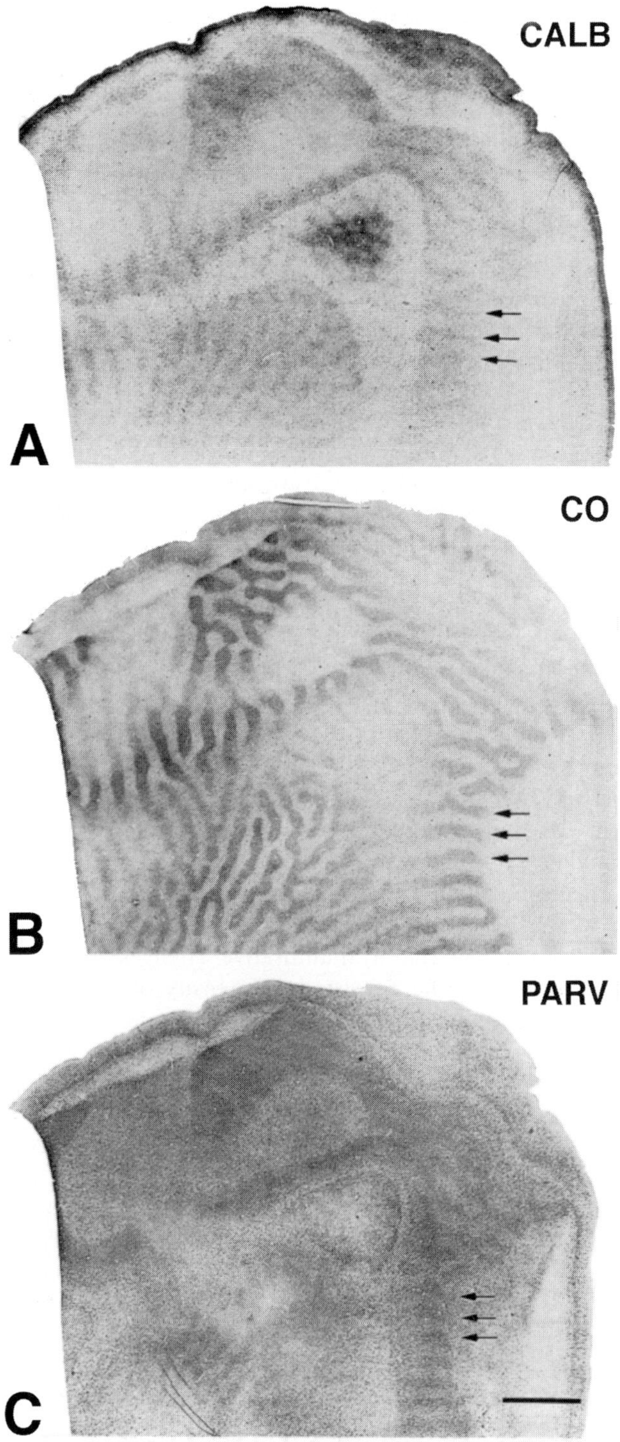
CALB
A
CO
B
PARV
C

columns (arrows, Fig. 51A, B). The CB-immunolabeled cells in layer IVC displaying reduced immunoreactivity are the ones which are normally only lightly stained, and which do not also contain GABA immunoreactivity (Hendry and Carder 1993). Thus, these CB-containing cells are most likely excitatory spiny stellate cells, which form the remaining, majority population of neurons in layer IVC of primate area V1 (Lund 1973; Tigges et al. 1977). Although dendrites of spiny stellate cells are mostly restricted to the ocular dominance column in which their parent soma is found, the axons of such cells traverse column boundaries (Katz et al. 1989), which likely plays a role in mediating deprivation effects upon flanking neurons in intact eye columns. In layers V and VI, a deprivation-induced pattern is evident, and similar to that in layer IV: pale, wide stripes encompassing both deprived eye columns and flanking border regions of normal eye columns interdigitate with thin, darkly-stained stripes situated at the centers of intact eye columns (Carder et al. 1996).

In supragranular layers, CB-immunoreactive neurons affected by monocular deprivation are strictly those which are intensely immunostained, which correspond to those which are also GABA-containing (Hendry and Carder 1993). In layer II, lightly stained bands 350–450 μm wide corresponding to deprived eye columns alternate with equally-wide bands of more intense immunoreactivity corresponding to the intact eye columns. Thus, unlike layer IVC, deprivation effects in layer II are restricted to the deprived eye columns, and are made up principally of a reduced density of stained processes and puncta.

Deprivation-induced changes to CB immunoreactivity in layer III are the most complex, and exhibit time-dependent changes which affect cells in all neuronal compartments (Carder et al., 1996). With only 3 days of deprivation, the staining pattern in layer III remains the normal one, which, when viewed in flattened tangential sections, is characterized by rows of circular, sparsely stained regions (corresponding to the cytochrome oxidase-enriched patches) containing few, lightly-labeled neurons and processes, surrounded by a matrix (the interpatch regions) composed of numerous, intensely labeled neuronal somata and processes (see section 5.5.2.; Celio et al. 1986; van Brederode et al. 1990; Hendry and Carder 1993; Blümcke et al. 1994). With 5 days of deprivation, however, two major quantitative changes occur which affect each compartment simultaneously: the density and intensity of labeled somata within the matrix (interpatch) region, which incorporates both eye dominance columns, are dramatically reduced, while within the circular regions (patches) of both deprived and intact eye dominance columns, the density and intensity of immunoreactive neuronal somata increase. With extended periods of deprivation of 10 days or longer, secondary changes occur in which the numbers and intensity of the novel CB immunoreactive somata are gradually reduced along the edges of the deprived eye patches, leaving by 30 days of deprivation staining in only a small, central portion of each deprived eye patch, while the CB immunoreactivity within the intact eye patches appears to fuse by a gradual expansion into the region bridging the patches lying along the intact eye columns. The overall impression of long-term deprivation effects on CB staining in layer III, then, is one characterized by three sets of parallel stripes: a narrow stripe lying along the centers of deprived eye columns consisting of small, interrupted spots of more intense staining; wider, pale bands of reduced staining incorporating the interpatch regions of both deprived and intact eye columns; and a thin, continous band of intense immunostaining lying along the centers of intact eye dominance columns (Carder et al. 1996).

7.5.2. Parvalbumin

Deprivation effects upon PV immunoreactivity are more subtle and occur with a slower time-course than those on CB staining (Blümcke et al. 1994; Gutierrez and Cusick 1994; Carder et al. 1996). With 5 days of deprivation, the numbers and intensity of PV-immunostained somata remain unchanged in deprived and intact eye dominance columns across all layers, and are within the normal range. However, a significant reduction in the density of immunostained puncta is evident in deprived eye dominance columns of layer IVC, with no effect observed for those lying in the adjacent, intact columns. It is likely that at least some of the immunostained puncta in layer IVC correspond to the terminals of geniculocortical afferents, which also display PV-immunoreactivity (Jones and Hendry 1989; Blümcke et al. 1990, 1994; Gutierrez and Cusick 1994; Johnson and Casagrande 1995). Similar changes are observed in layer III, where the density of PV-immunoreactive puncta at the edges of the patches within deprived eye columns is reduced, while within the centers of deprived eye patches, intensely stained puncta of a normal density remain. Such changes in immunoreactive puncta occur without any changes in the numbers or intensity of PV-immunostained somata in either eye column within layer III. The identity of the immunoreactive puncta susceptible to deprivation effects in layer III patches is less clear; while the patches receive a direct projection from the intercalated layers of the dorsal lateral geniculate nucleus (Hendrickson et al. 1978; Livingstone and Hubel 1982; Weber et al. 1983; Lachica and Casagrande 1992), the relay neurons within such layers are CB immunoreactive (Jones and Hendry 1989; Mize et al. 1992; Blümcke et al. 1994; Gutierrez and Cusick 1994; Johnson and Casagrande 1995). With longer-term periods of deprivation of 20–60 days, an effect in addition to decreased densities of puncta is observed in which the numbers of immunostained somata in layer IVC decrease by a slight, but quantitatively significant amount in deprived eye columns in comparison with those in intact columns or normal (non-deprived) area V1 (Fig. 51B, C). Similar reductions in PV immunostained somata within deprived eye columns can also be traced into layers IVA, IVB, V and VI with longer periods of deprivation, but are not apparently found in layer III.

7.5.3. Calretinin

No change in CR immunostaining can be detected in any layer over the period of 5–60 days of monocular deprivation (Carder et al. 1996).

It is clear that the regulation of calcium-binding proteins in area V1 by monocular deprivation is exceedingly complex, making it all the more difficult to determine the precise functional consequences of such regulation for the neurons in which these proteins are contained. One obvious possibility, however, is that changes in immunoreactivity correspond to changing levels of calcium binding proteins, which in turn could affect their role in buffering free intracellular calcium (Heizmann 1984; Mattson et al. 1991). Given the link between the magnitude of postsynaptic levels of free calcium ions and forms of synaptic plasticity such as long-term potentiation and depression (Mulkey and Malenka 1992; Perkel et al. 1993), it is not unreasonable to speculate that changes in levels of calcium binding proteins would have a major impact on the efficacy of synaptic transmission, particularly in the context of plasticity.

7.6. NEUROPEPTIDES

As detailed in Section 5.5.1., many neuroactive peptides are colocalized with subpopulations of GABA neurons throughout the monkey visual cortex (see Figs. 24, 26). Of these, only the tachykinin family of proteins displays overt changes in response to monocular deprivation (Hendry et al. 1988a). With 5 days of monocular deprivation by intravitreal TTX injection, approximately 80% of the GABA/tachykinin-immunoreactive cells in layers IVC and IVA lose their peptide immunoreactivity almost completely within deprived eye dominance columns. Likewise, there is approximately a 50% reduction in peptide-immunoreactive somata and puncta within the interpatch regions along the deprived eye rows within layer III, which also extends throughout the interpatch tissue which extends across rows, thereby affecting both sets of ocular dominance columns (Hendry et al. 1988a). Within the actual patches of either deprived or intact eye columns, the numerical density of immunostained neurons is within the normal range. It is likely that the basis for decreased peptide immunoreactivity is a corresponding decrease in the expression of β-preprotachykinin mRNA (Benson et al. 1994).

A different pattern of activity-dependent effects is observed when animals are subjected to monocular aphakia resulting from the removal of the crystalline lens. Under these conditions, tachykinin immunoreactivity in layer IVC remains normal following a period of 3–6 months of postoperative recovery (Hendry et al. 1988a). By contrast, both sets of ocular dominance columns are dramatically affected within layers II and III. In these layers, flattened sections cut tangential to the pial surface reveal two sets of parallel, alternating stripes: a wider one, corresponding to the intact eye columns, which contains expanded, intensely immunoreactive patches separated by small, lightly stained interpatch regions; and a thinner one, corresponding to the deprived eye columns, which contains small patches of intense immunostaining, separated by long interpatch gaps of very light staining. In comparison with normal (non-deprived) area V1, the numerical density of immunostained neurons in the wider rows is approximately 30% greater, while that of immunostained neurons within the thinner, shrunken rows is approximately 60% less (Hendry et al. 1988a).

7.7. GLUTAMATE

With a time-course similar to one producing deprivation-effects on staining for the major inhibitory neurotransmitter GABA, visually driven activity also regulates immunoreactivity for the major excitatory neurotransmitter glutamate (Carder and Hendry 1994). Five days of intravitreal injections of TTX into one eye produce alternating lightly- and darkly-stained stripes of immunoreactivity in layer IVC. The lighter stripes, which correspond to deprived eye columns, are composed of lower-than-normal numbers of glutamate-positive neurons through the full width of the deprived column, and a pale neuropil localized to the central three-quarters of the column. The intensity and composition of the neuropil composing the flanks of deprived eye columns appears similar to the pattern within intact-eye columns. It is likely that such neuropilar staining represents the axon terminals of both thalamocortical relay neurons, as well as terminals of intrinsic excitatory neurons. Such lightly-stained stripes extend into superficial layers as well, where regions corresponding to the rows of cytochrome-oxidase patches of deprived eye columns display reduced levels of immunostained somata and neuropil. Although glutamate also exists in a metabolic pool for protein synthesis as well as serves as a precursor for GABA synthesis, it is probably the neurotransmitter pool which is mostly recognized by the anti-glutamate antibody, since patterns of glutamate immunostaining in monkey area V1 do not always correspond to patterns of metabolic activity, nor do neurons which are glutamate immunoreactive also display immunoreactivity for GABA (Carder and Hendry 1994).

7.8. NITRIC OXIDE SYNTHASE

Nitric oxide synthase (NOS) is the synthesizing enzyme for nitric oxide (NO), a labile gas which in the brain functions as a diffusible, potent neurotransmitter strongly implicated in synaptic plasticity (e.g. Schuman and Madison 1991; Bredt and Snyder 1992; Haley et al. 1992; Izumi et al. 1992; Dawson and Snyder 1994). The majority of NOS-containing cells in the cerebral cortex corresponds to a subpopulation of GABAergic interneurons (Dawson and Snyder 1994). In adult monkeys subjected to monocular deprivation, immunocytochemically localized NOS is reduced within deprived eye columns of layer IVC (Sandell 1986; Aoki et al. 1993); this effect is due primarily to decreased staining of axon terminals of unknown origin (Aoki et al. 1993). There are no changes in the density, distribution or staining intensity of NOS-labeled somata in any layer (Aoki et al. 1993). The precise effects of reduced NOS staining in the context of monocular deprivation and whether this also represents reduced levels or functioning of NO are uncertain.

7.9. CHOLINESTERASES

Histochemical staining for acetylcholinesterase (AchE) and butrylcholinesterase (BuChE) activity reveals differential changes in staining intensity across layers following monocular deprivation. In layer IV, AchE staining is elevated within deprived eye columns of layer IVA and IVC, while BuChE staining is also elevated within deprived eye columns, but the effect is most overt in layer IVCα, the sublayer in which staining is normally the highest (Graybiel and Ragsdale 1982). In contrast, staining for these enzymes within the cytochrome oxidase patches of layers II/III is reduced in comparison with intact eye columns (Horton 1984). It remains undetermined whether increased staining represents greater quantites or activity of these enzymes within existing fibers, or the induction of new fibers expressing these enzymes.

7.10. CALCIUM-CALMODULIN DEPENDENT PROTEIN KINASE TYPE II (CAMII KINASE)

Perturbations of visual input can affect levels of intracellular proteins other than neurotransmitters/modulators or their receptors, and one example of such a protein is the multifunctional, neuron-specific protein kinase CaMII kinase. Following brief periods of monocular deprivation in adult monkeys, immunoreactivity for the α-subunit of CaMII kinase reveals alternating darkly and lightly immunostained stripes in layer IVCβ (Hendry and Kennedy 1986). A comparison of immunostained sections with those processed for cytochrome oxidase or from normal (non-deprived) monkeys reveals that the dark bands correspond to deprived eye columns, and arise from an increased intensity of immunoreactivity within individual neurons of deprived eye columns, while numbers of labeled cells in deprived eye columns remain unchanged in comparison with those in intact or normal eye columns. One interpretation of these data is that actual levels of the kinase are increasing within neurons, a possibility which is strengthened by corresponding *in situ* hybridization histochemistry which shows elevated mRNA levels in deprived eye columns (Fig. 52; Benson et al. 1991a). Upregulation of CaMII kinase α-subunit mRNA is most evident in layer IVCβ, but also extends above and below layer IV within deprived eye columns. Thus, in contrast to the deprivation-induced pattern described for GABA and related molecules, levels of CaM II kinase *are enhanced* in deprived eye columns.

7.11. CYTOSKELETON-RELATED PROTEINS

Microtubule associated protein 2 (MAP 2) is one of a family of proteins which functions to stabilize neuronal shape by promoting microtubule synthesis and cross-linking with other components of the cytoskeleton (Murphy et al. 1977; Matus 1988). Following five days of monocular deprivation in adult monkeys, immunoreactivity for MAP 2, which is localized to the soma and dendrites of most neurons in area V1 (Peters and Sethares 1991; Hendry and Bhandari 1992), as well as throughout the CNS (DeCamilli et al. 1984), is decreased in deprived eye columns in layer IVC as well as in regions corresponding to rows of cytochrome oxidase patches of layers II/III (Hendry and Bhandari 1992). In contrast, immunocytochemical staining in regions corresponding to the intact eye columns, as well as in layer IVA of both deprived and intact eye columns, appears normal (Hendry and Bhandari 1992). It remains to be determined whether reduced immunostaining for MAP 2 represents activity-dependent morphological changes to the structure or configuration of dendrites or spines (Fifkova and Morales 1989).

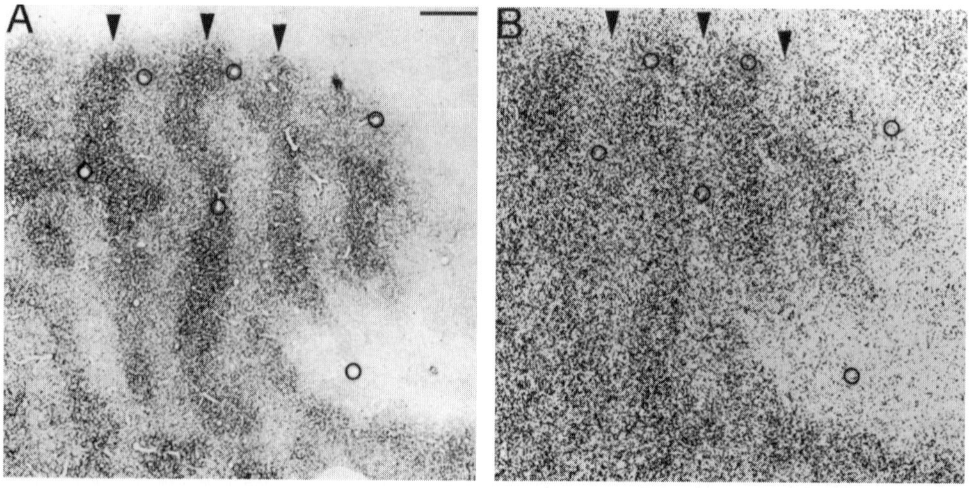

Fig. 52. Effects of monocular deprivation on levels of probe hybridization to CaMII kinase mRNA in adult macaque monkey area V1. Photomicrographs of adjacent sections through layer IVC processed for cytochrome oxidase activity (*A*) or by *in situ* hybridization for CamII kinase α-subunit mRNA (*B*). Comparisons of the positions of the *arrowheads*, which demarcate intact eye dominance columns, and the *circles*, which demarcate identical blood vessels lying within deprived eye dominance columns, show that the stripes of intense CaMII kinase probe hybridization correspond to the deprived eye dominance columns. Scale bar = 500 μm. Reproduced from Benson et al. (1991).

7.12. TRANSCRIPTION FACTORS

The level and pattern of immediate-early gene expression are very responsive to visual stimulation, which has been studied most thoroughly in rat and cat visual cortex (for review, see Kaczmarek and Chaudhuri, 1997). For example, brief exposure to visual stimution following early periods of dark-rearing induces upregulation in the expression of Fos (Beaver et al., 1993; Kaplan et al., 1996; Montero and Jian, 1995; Mower, 1994; Rosen et al., 1992), the protein product of the immediate-early gene c-Fos. In primates (adult vervet monkeys), as little as five hours of monocular deprivation leads to changes in labeling for the neuronal transcription factor Zif268 from a normal pattern in which intense immunoreactivity is homogeneously distributed throughout all cell-dense layers of area V1, to a pattern in which pia-to-white matter columns of reduced staining corresponding to the deprived eye dominance columns alternate with darker columns of staining representing the intact eye (Chaudhuri and Cynader 1993; Chaudhuri et al. 1997). Although speculative, it is likely that such changes in Zif268 immunoreactivity follow the rapid regulation of gene expression for NGFI-A, the immediate-early gene which encodes Zif268 (Milbrandt 1987; Lemarie et al. 1988). The rapid response of Zif268 and other, related proteins suggests that activity-coupled regulation of immediate-early gene expression may be a key component of a cascade of cellular changes induced by altered visual input.

8. CHEMOARCHITECTONICS: CHEMICAL MARKERS OF REGIONAL HETEROGENEITY

8.1. INTRODUCTION

The most informative chemoarchitectonic analyses of subcortical and cortical parcellation in the visual system have been based on several non-neurotransmitter markers. Immunohistochemical studies using an antibody to a chondroitin sulfate proteoglycan demonstrated that specific functional compartments and some of the visual areas in the occipitoparietal pathway are identifiable by high levels of labeling for this component of the neuronal surface and extracellular matrix (DeYoe et al. 1990; Hockfield et al. 1990), in spite of the fact that earlier studies have reported that it may not be strictly restricted to broad band neurons in the LGN and visual cortex in various mammalian species (Hendry et al. 1984, 1988). Comparably, immunohistochemical studies localizing the calcium-binding proteins PV, CB, and CR in the monkey and human visual areas have shown that these calcium-binding proteins are present within morphologically distinct groups of neurons that display specific regional and laminar distribution patterns. A striking degree of neurochemical specialization occurs in the LGN, where the magnocellular and parvocellular neurons contain PV, whereas the third (koniocellular) visual channel that emanates from the intercalated neurons, contains CB as well as the α subunit of type II calmodulin-dependent protein kinase (Hendry 1993; Hendry and Carder 1993; Casagrande 1994; Hendry and Yoshioka 1994). Analysis of the distribution of calcium-binding protein-immunoreactive neurons in several divisions of the occipitoparietal and occipitotemporal visual pathways in the macaque monkey, as well as chimpanzee and human (Kondo et al. 1994; Hof et al. 1994; Adams et al. 1996; Gattass et al. 1996) revealed that area V1 is characterized by very high numbers of PV and CB-immunoreactive cells, while CR-immunoreactive neurons are less numerous. In addition, PV immunoreactivity is concentrated in this area in layers IVA, IVCβ and VI, in a complementary fashion to CB and CR, which are concentrated in layers II, III, IVB, and IVCα (Blümcke et al. 1990; Van Brederode et al. 1990; Blümcke and Celio 1992; Hendry and Carder 1993; Johnson and Casagrande 1995; Carder et al. 1996). Also, in layer III, CB is present in a dense network of fibers and neurons in the interblob compartment, whereas PV is principally distributed within the blobs and CR shows no preferential localization to either compartment (Blümcke and Celio 1992; Blümcke et al. 1994; Johnson and Casagrande 1995; Carder et al. 1996). Areas V2, V3 and V3A display lower numbers than area V1 for the three neuronal populations. However, cell counts markedly increase in areas at higher levels in the cortical hierarchy. In particular, areas MT, VIP, V4 and TEO are characterized by very high densities of stained neurons. In addition, a subset of CB-immunoreactive pyramidal neurons located in layer III exhibit differential distribution among these cortical areas, with the highest densities in areas V4 and MT and slightly lower cell counts in areas TEO, TE, VIP and LIP. These results indicate that calcium-binding proteins define neuronal populations that are differentially represented among cortical regions of known function in the occipitoparietal and occipitotemporal pathways in the primate visual system. Thus, the detailed qualitative and quantitative analysis of the distribution of such non-neurotransmitter markers could be used as reliable tools to develop anatomic criteria for the characterization and precise localization of a given cortical field.

In this respect, the neurofilament protein triplet has been shown to be present in a subset of corticocortically-projecting pyramidal neurons in the macaque monkey neocortex, and to represent a reliable marker to differentiate neuronal populations (Campbell et al. 1991; Vickers and Costa 1992; Vickers et al. 1995). Neurofilament protein-containing neurons exhibit a high degree of regional and laminar selectivity in the primate neocortex in general (Campbell and Morrison 1989; Hof et al. 1990b; Hof and Nimchinsky 1992; Carmichael and Price 1994; Cusick et al. 1995; Hof et al. 1995b; Hof and Morrison 1995; Adams et al. 1996; Chaudhuri et al. 1996; Gattass et al. 1996; Nimchinsky et al. 1996), and previous analyses of neurofilament protein distribution in human areas V1 and V2 and in the inferior temporal cortex have demonstrated striking differences in the laminar and regional composition of these visual areas in particular (Campbell and Morrison 1989; Hof et al. 1990b; Hof and Morrison 1990). Although neurofilament protein function is not fully understood at the cellular level, it is a particularly interesting marker because its expression is neuron type-specific and is involved in determining the adult axonal and somatodendritic cytoskeleton (Riederer et al. 1995, 1996). It is implicated in the maintenance and stabilization of the axonal cytoskeleton, and its expression levels correlate with axonal size (Morris and Lasek 1982; Hoffman et al. 1987; Nixon et al. 1994; Pijak et al. 1996; Xu et al. 1996). The presence of high levels of nonphosphorylated neurofilament protein in the somatodendritic compartment of subsets of neocortical neurons may represent a neurochemical feature related to a distinctive role of these neurons in corticocortical systems (Campbell and Morrison 1989; de Lima et al. 1990; Campbell et al. 1991; Hof et al. 1995a, 1996; Hof and Morrison 1995; Nimchinsky et al. 1996). Of particular relevance is the fact that the presence of neurofilament protein in the neuron soma appears to be correlated to cell size and conduction velocity of nerve fibers (Lawson and Waddell 1991). Finally, neurofilament protein-enriched neurons have been shown to be exquisitely sensitive to the degenerative process in Alzheimer's disease (Hof et al. 1990b; Hof and Morrison 1990).

8.2. NEUROFILAMENT DISTRIBUTION: AN EXAMPLE OF QUANTITATIVE CHEMOARCHITECTONICS IN THE MACAQUE MONKEY VISUAL CORTEX

8.2.1. General overview

The antibody to neurofilament protein used in our parcellation studies (SMI-32), labels a population of pyramidal neurons with distinct regional and laminar distribution in the macaque monkey visual system (*Macaca fascicularis, Macaca mulatta*; Hof and Morrison 1995). Similar distribution patterns using the same antibody have been recently reported in the vervet monkey visual system (*Cercopithecus aethiops*; Chaudhuri et al. 1996), as well as in the chimpanzee (*Pan troglodytes*; Gattass et al. 1996), and human (Adams et al. 1996). The largest labeled neurons are located within the deep portion of layer III (IIIC) and the upper portion of layer V (VA) in most of the visual areas (Figs 53–55). These cells show extensive basal and apical dendritic arborizations. Layers II, IIIAB, VB, and VI contain somewhat smaller and lightly stained neurons, with a less exuberant dendritic arborization. Layer VI is characterized by a more heterogeneous population of neurofilament protein-immunoreactive neurons, with vertical and horizontal fusiform neurons, and very lightly labeled small pyramidal cells (Campbell and Morrison 1989; Hof and Morrison 1995).

Even though neurofilament protein immunoreactivity is a very sensitive index of regional specialization (see below), regional boundaries cannot be determined solely on labeling patterns, but rather in the context of data sets that include relevant published regional parcellations (Van Essen and Zeki 1978; Zeki 1978a,b; Maunsell and Van Essen 1983; Gattass et al. 1985, 1988; Van Essen 1985; Colby et al. 1988; Boussaoud et al. 1990; Morel and Bullier 1990; Van Essen 1990, 1992; Colby and Duhamel 1991; Felleman and Van Essen 1991; Distler et al. 1993; Nakamura et al. 1993; Galletti et al. 1996). In addition, while some areal boundaries are distinct and abrupt, others are difficult to discern but present, and still others occur as gradients rather than a precise boundary. Parallel series of Nissl and myelin stained sections are useful guides to establish regional features in spite of their lack of specificity at the cellular level. Immunohistochemical patterns are best presented as regional descriptors rather than in the context of boundaries alone to permit a more precise assessment of the neurofilament protein distribution, even where boundaries appear as gradients in cell distribution or density instead of clear borders. The particular morphologic features of labeled neurons also represent important guides in parcellation analysis and cell typology may be helpful to refine the quantitative analysis. Finally, the quantitative analysis itself relies on the use of stereologic methods that provide unbiased estimates of the regional and laminar cell numbers (Sterio 1984; Coggeshall and Lekan 1996). Based on qualitative and quantitative differences in neurofilament protein-immunoreactive neuron distribution about 50 possible boundaries among over 28 visual cortical areas as well as adjacent regions of cortex can be consistently distinguished with approximately 45% of the boundaries determined with high reliability, about 30% visible as transition zones between distinct regions of cortex, and 25% remaining as poorly defined based on these criteria (Hof and Morrison 1995). The immunohistochemical patterns are shown in Figures 53–55, and quantitative data are presented in Tables 2 and 3.

TABLE 2. *Optical disector analysis of the laminar densities of neurofilament protein-containing neurons in the macaque monkey visual cortex*

Layers	Areas		
	V1	V2	V3
III	53691.4 ± 3280.6	49864.0 ± 2861.3	72987.3 ± 4160.5
V-VI	20341.5 ± 1973.0	18873.9 ± 1655.5	30052.0 ± 2867.2
	V3A	V4	MT
III	53441.4 ± 2797.6	73467.4 ± 3767.6	76785.5 ± 4934.9
V-VI	20713.4 ± 3778.4	53875.5 ± 3086.6	70816.5 ± 4191.4
	MST	PO	DP
III	69370.9 ± 3440.8	70161.5 ± 3540.0	49712.9 ± 2273.3
V-VI	56067.2 ± 5685.0	28085.0 ± 1310.0	19387.7 ± 2166.4
	PIP	VIP	LIP
III	73457.7 ± 6491.0	67812.9 ± 4985.0	59104.0 ± 3329.8
V-VI	28624.2 ± 2662.4	24699.2 ± 2744.3	21868.5 ± 1974.5
	7a	MIP	MDP
III	48573.6 ± 3507.6	46432.7 ± 3461.7	52283.0 ± 6542.5
V-VI	17502.2 ± 2816.5	18773.1 ± 1778.4	20045.1 ± 1534.3
	FST	PGa	TPO
III	51800.1 ± 3380.7	55868.6 ± 4376.8	47705.2 ± 3128.2
V-VI	19130.1 ± 1290.0	21373.3 ± 2877.5	17901.1 ± 1973.8
	ITp	ITm	ITa
III	82585.6 ± 9370.0	99097.8 ± 10340.1	97856.1 ± 8566.7
V-VI	33034.0 ± 2975.5	37856.5 ± 4912.2	39142.4 ± 4014.2
	FEF	SEF	46
III	47426.8 ± 3895.0	73987.0 ± 3501.0	48282.2 ± 2140.6
V-VI	63612.4 ± 2356.0	71025.6 ± 1145.6	33991.7 ± 2774.8
	45	9	6
III	49026.4 ± 2762.0	49947.1 ± 4849.2	58746.0 ± 1294.0
V-VI	47696.9 ± 3121.7	35676.9 ± 1274.2	36997.8 ± 1156.2

Results represent neurofilament protein-immunoreactive neuron densities per mm^3 in each of layers III and V-VI, and are expressed as means ± SEM. Note the numerous trends in labeled neuron densities among the different cortical fields. The order in the Table does not reflect strictly the actual hierarchical pattern of connectivity between these areas. Compare the numerical data with the quantitative flat maps shown in Figures 56 and 57. The layers are numbered by Roman numerals. Nomenclature of visual regions is adapted from Felleman and Van Essen (1991; and references therein). It should be kept in mind that the precise anatomic boundaries and physiologic relevance of some of these areas have not yet been established with certainty. See text for details. Adapted from Hof and Morrison (1995).

8.2.2. Regional analysis

8.2.2.1. Primary and secondary visual cortex (areas V1 and V2)

The most prominent neurofilament protein-immunoreactive feature of area V1 is the presence of labeled layer IVB and Meynert cells (see Figs 2D, 53, 55A). The layer IVB (Meynert-Cajal) multipolar neurons form a dark staining stripe throughout area V1

TABLE 3. *Distribution of neurofilament protein-immunoreactive neurons in corticocortical projections between areas MT, V4, V1, and V2*

Projection/layer		NFP (%)	Projection/layer		NFP (%)
Feedforward to MT			*Feedforward to V4*		
V1 →	**MT**		**V1 →**	**V4**	
	IVB	100		III	25.1
	Meynert	100			
V2 →	**MT**		**V2 →**	**V4**	
	III	78.0		III	32.2
	V	–		V	–
V3 →	**MT**		**V3 →**	**V4**	
	III	74.8		III	35.9
	V	57.2		V	28.4
Feedback to V1			*Feedback to V2*		
MT →	**V1**		**MT →**	**V2**	
	II-III	60.0		II-III	61.8
	V-VI	79.1		V-VI	79.6
V4 →	**V1**		**V4 →**	**V2**	
	II-III	60.1		II-III	63.2
	V-VI	80.1		V-VI	75.9
V3 →	**V1**		**V3 →**	**V2**	
	II-III	64.8		II-III	67.2
	V-VI	72.8		V-VI	76.5

Data represent mean ± S.E.M. laminar percentages of double labeled neurons (i.e., retrogradely labeled and neurofilament protein-immunoreactive) in the feedforward projections from areas V1, V2, and V3 to the injections sites in areas V4 and MT (top part), and in the feedback projections from areas MT, V4, and V3 to the injections sites in areas V1 and V2 (bottom part). Note that layer IVB and Meynert cells in area V1 that project to area MT were all double labeled. Also there is a much higher percentage of double labeled neurons in the projection to area MT than in those to area V4. There is amore homogeneous distribution of double labeled neurons among the feedback connections compared to the feedforward projections. Adapted from Hof et al. (1996).

and are characterized by a very intense labeling of the perikarya and proximal portions of their dendrites. Meynert cells are observed at the interface between layers V and VI, either isolated or in small clusters of 2–4 cells. Meynert cells are easily recognized by their very large size, darkly stained thick apical dendrite climbing in some cases up to layer I, and extensive horizontal arborization of basal dendrites that may extend over 1 mm in layer VI. Layer IVB cells and Meynert cells are more densely packed within the calcarine portion of area V1 compared to the opercular cortex, indicating higher cellular densities in regions corresponding to the representaion of the periphery of the visual field (Hof and Morrison 1995). Layers II and IIIA-C show a population of small and lightly stained pyramidal neurons most numerous in the lower two-thirds of layer III. These cells display a homogeneous density throughout area V1. The dendrites of layer III neurofilament protein-immunoreactive neurons are very thin and their arborizations are restricted to this layer. Other than the Meynert-Cajal neurons of layer IVB, layer IV contains very few labeled cells. Layer V is characterized by a population of small, intensely labeled pyramidal cells with apical dendrites could be followed easily through layer IVC, and in sometimes up to layer IIIC. Layer VI has a heterogeneous population of small pyramidal and modified pyramidal neurons.

The transition zone between the primary and secondary visual areas contains giant pyramidal cells in continuity with layer IVB of area V1. These neurons were morphologically different from those in layer IVB, and are adjacent to a cluster of smaller, intensely labeled pyramidal neurons. The neurofilament protein-immunoreactive neuron counts in this cluster is less than the average neuron density in layer III of area V1. Neuron counts in layer III of area V2 compared to the V1/V2 border are higher, but not significantly different from those in area V1. These neurons have apical dendrites extending through layer III and branching within layer II (Figs 53, 55B). Neurofilament protein-immunoreactive pyramidal neurons in layer III of area V2 are smaller than in the V1/V2 border. Layer V in the V1/V2 transition and in area V2 is characterized by the presence of very large, evenly spaced neurofilament protein-immunoreactive pyramidal neurons, with a thick apical dendrite ascending to and dividing in layer I. Small, very lightly stained pyramidal and polymorphic neurons are also present in layers V and VI (Figs 53, 55B). The ventral portion of area V2 contains lower neurofilament protein-immunoreactive neuron densities particularly where it abuts prostriate area ProS and parahippocampal fields TF and TH. Also, the V1/V2 border is clearly distinguished by a population of large and densely labeled neurofilament protein-immunoreactive neurons in layer III. Callosal stripes of afferents lined up with the representation of the vertical meridian are known to characterize the boundaries between the visual areas within the prestriate cortex, representing important functional landmarks (Van Essen and Zeki 1978; Van Essen et al. 1982; Galletti et al. 1996). However only the V1/V2 border is visible based on neurofilament protein staining patterns. Comparable callosally-recipient zones between areas V4 and ITp, which may in part correspond to the area referred to as VOT (Van Essen et al. 1982), and in the occipitoparietal cortex [in the region referrred to as V6 by Galletti et al. (1996)] do not show a distinct staining pattern.

8.2.2.2. Areas V3, V4, V3A, and PO

A striking feature of area V3 compared to area V2 is the presence in layers II-IIIa of a small population of intermediate size, intensely labeled neurons with apical dendrites either reaching and dividing in layer I or confined to the upper portion of layer II. This pattern continues in areas V4 and in the subdivisions of area IT in the temporal cortex, with variation in the density of the labeled neurons (see below). The fact that neurofilament protein-immunoreactive neurons are rare in these layers renders their presence a very reliable marker for these regions. Neurofilament protein-immunoreactive neurons in layer IIIbc are smaller and more numerous than in area V2. Layers V and VI contain large and darkly labeled pyramidal neurons similar to those observed in area V2, but more numerous, particularly in layer VI (Figs 53, 55C).

Area V3A has fewer labeled neurons in layers II-IIIA and IIIBC than area V3 and slightly more neurofilament protein-immunoreactive neuron counts in layers V-VI, especially in the case of layer V where numerous large and intensely labeled pyramidal cells are found (Figs 53, 55D). These characteristics clearly differentiate area V3A from the dorsal parietal fields and the intraparietal sulcus neighboring regions, however, in some cases the boundary between areas V3A and V3d is difficult to establish with certainty, suggesting that the chemoarchitectural patterns in area V3A represent a transition zone between the lunate sulcus cortex and the parietal regions. The anatomic parcellation of these occipital and occipitoparietal areas based on the distribution of neurofilament protein-immunoreactive neurons generally agrees with existing data, although certain transition zones that have been established physiologically cannot be recognized with enough reliability using these morphologic criteria. In the occipital cortex, the staining patterns observed in areas V1, V2, V3, and V4 follows precisely the physiologic localization of these regions and changes in myelin patterns (Colby et al. 1988; Gattass et al. 1981, 1985, 1988).

The adjacent area PO in the parieto-occipital sulcus is distinguished by densely packed intermediate size neurofilament protein-immunoreactive pyramidal cells in layer IIIC with long and thin apical dendrites ascending to layer I. These cells are substantially more numerous than in layer III of area V3A. A few giant, darkly labeled pyramidal neurons are also present in layer IIIC, and layers II-IIIB are devoid of labeled neurons in area PO. Layers V-VI have a moderate numbers of labeled neurons, mostly located in layer V. These patterns clearly differentiate area PO from the dorsal part of area V3 on the medial aspect of the occipital lobe, which has lower neurofilament protein-immunoreactive neuron counts in both layers II-III and V-VI (Fig. 53). Area PO has been defined physiologically as receiving an important input from the representation of the periphery of the visual field from most of the occipital areas, the

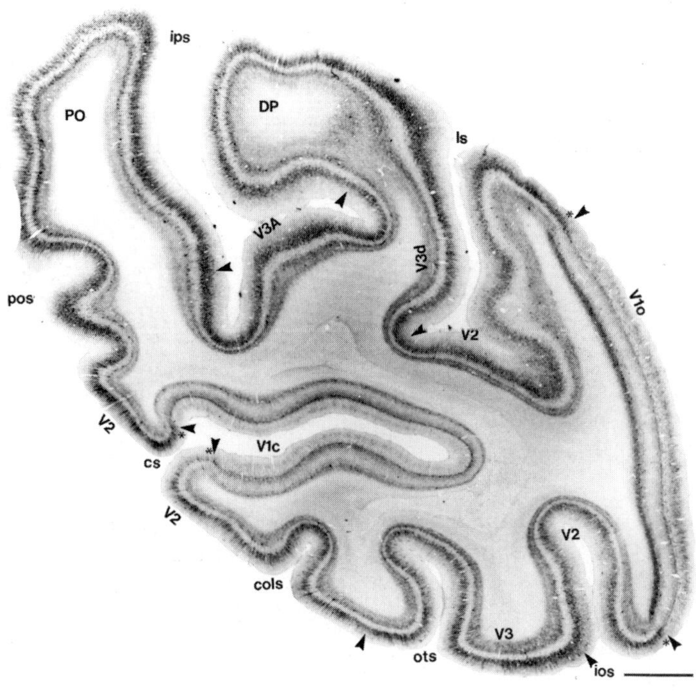

Fig. 53. Low power photomontage of a coronal section through the occipital cortex of a macaque monkey. Note the differences in regional and laminar distribution of neurofilament protein-containing neuron patterns among the cortical areas. Some of the borders (*arrowheads*) could not be defined on this section with certainty due to the angle of cut (V2/PO, V3d/DP). *Asterisks* mark the V1/V2 border. V1o, opercular portion of area V1; V1c, part of area V1 within the calcarine sulcus; cols, collateral sulcus; cs, calcarine sulcus; ios, inferior occipital sulcus; ips, intraparietal sulcus; ls lunate sulcus; ots, occipitotemporal sulcus; pos, posterior occipital sulcus. Cortical areas are defined in text. Scale bar = 1.5 mm. Adapted from Hof and Morrison (1995).

mediotemporal areas, and several parietal regions (Colby et al. 1988). The neurofilament protein staining pattern permits its clear delineation on the basis of labeling differences from area V2, the mediodorsal parietal cortex (areas MDP and 31), and medial intraparietal cortex (area MIP).

In area V4, the layer II-IIIA neurons are considerably more frequent than in area V3 and aggregate in small clusters. Area V4 exhibits much higher numbers of neurofilament protein-immunoreactive intermediate and small size pyramidal neurons in layers V-VI compared to area V3, and no large dark neurons in layer VA (Figs 54, 55E). These patterns permit an unambiguous recognition of area V4 border with V3. The transition area V4t represents a thin zone of cortex between areas V4 and MT on the ventral bank of the posterior portion of the superior temporal sulcus. This region lacks the layer II-IIIA cells typical of area V4, has lower labeled neuron counts in layers IIIbc and V-VI than area V4, and lacks the distinct band of neurofilament protein-immunoreactive neurons in layers V and VI that characterizes area MT (see below; Fig. 54).

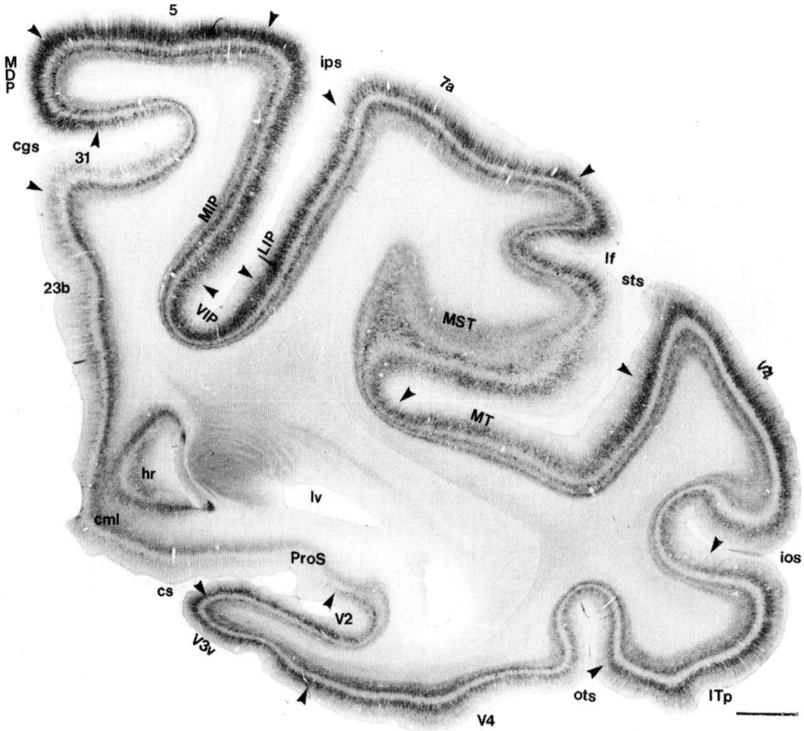

Fig. 54. Low power photomontage of a coronal section through the mediotemporal areas. Note the differences in regional and laminar distribution of neurofilament protein-containing neuron patterns among the cortical areas. The border between areas MST and 7a could not be defined with certainty on this section due to the angle of cut. ProS, prostriate area; cgs, cingulate sulcus; cml, caudomedial lobule; cs, calcarine sulcus; hr, hippocampal rudiment; ios, inferior occipital sulcus; ips, intraparietal sulcus; ots, occipitotemporal sulcus; lf, lateral fissure; lv, lateral ventricle; sts, superior temporal sulcus. Borders between regions are indicated by arrowheads. Cortical areas are defined in text. Scale bar = 1.5 mm. Adapted from Hof and Morrison (1995).

8.2.2.3. The mediotemporal areas MT and MST

These mediotemporal areas are characterized by a unique staining pattern in layers V and VI and by very high numbers of neurofilament protein-immunoreactive neurons in layer III. Area MT is readily distinguishable from the transition zone V4t and from area V4 itself by distribution of neurofilament protein in layer VI, characterized by a band of small and darkly labeled neurons with short dendrites, that is distinct from layer V that contains a sparse, very large neurofilament protein-immunoreactive neurons with long apical dendrites. This particular pattern in layer VI is seen only in areas MT and MST (Figs 54, 55F). The adjacent area MST in the depth and dorsal bank of the superior temporal sulcus has a similar staining pattern in the infragranular layers such that the MT/MST boundary appears as a gradual decrease in neurofilament protein-immunoreactive neuron counts. Layer III of areas MT and MST contains a dense population of large neurofilament protein-immunoreactive pyramidal neurons. Dorsally, area MST shows a progressive decline in the number of labeled neurons toward parietal area 7a. The mediotemporal regions are an example of optimal correlation between the physiologic and anatomic data, and neurofilament protein staining patterns, allowing an unequivocal determination of its borders with the parietal and temporal cortex. For instance, area MT has long been recognized as a small, densely myelinated field in the ventral bank and part of the fundus of the posterior portion of the superior temporal sulcus, that receives direct input from area V1, and it contains high numbers of neurons exhibiting preferential direction selectivities (Zeki 1974; Van Essen et al. 1981). This area is distinct from areas MST and FST, the latter located more anteriorly in the floor of the superior temporal sulcus (Newsome and Wurtz 1982; Desimone and Ungerleider 1986; Ungerleider and Desimone 1986), which can be differentiated on the basis of their connections with visual and polysensory regions in the occipital, parietal and temporal cortex (Boussaoud et al. 1990).

8.2.2.4. Areas of the superior temporal sulcus and inferior temporal cortex

In the superior temporal sulcus, areas FST, TPO, and PGa are all characterized by small neurofilament protein-immunoreactive neurons in layers III and V, and their borders are difficult to establish. However, they differ from adjacent regions by the number of immunoreactive neurons which is lower than in areas MST and MT, permitting a reliable definition of the border of fundal area FST and area MT. The dorsally located area TPO is distinguished from adjacent auditory association areas by their much higher neurofilament protein-immunoreactive neuron counts, but the border between areas TPO and FST cannot be determined precisely by the pattern of neurofilament protein immunoreactivity. The anteroventrally located area PGa differs from areas FST and TPO by its slightly higher neurofilament protein-immunoreactive neuron densities in layer V. Anteriorly, area PGa blends with the temporopolar region, and the neighbouring regions of the inferior temporal cortex are recognizable by their much higher overall density and characteristic laminar distribution of neurofilament protein-immunoreactive neurons, with a population of neurofilament protein-immunoreactive pyramidal neurons in layer IIIa comparable to that seen in areas V3 and V4.

In the inferior temporal cortex, the borders among the anterior, medial and posterior portions of area IT are reflected by gradual changes in labeled neuron counts (Fig. 54). In layer IIIa, the most posterior portion of area ITp has slightly higher neuron counts than area V4, the medial inferior temporal region ITm shows the highest numbers in layer II-IIIA, and the anterior region ITa exhibits a gradual decrease in cell counts toward the temporopolar areas. Layer IIIBC contains relatively smaller but much denser neurofilament protein-immunoreactive neurons compared to area V4 and superior temporal sulcus regions, with highest counts observed in areas ITm and ITa. Similarly, there is a progressive increase in the density of labeled neurons in layer V in the caudo-rostral direction, yet areas ITp, ITm, and ITa have lower neurofilament protein-immunoreactive neuron counts than area V4, and unlike V4, large neurofilament protein-immunoreactive neurons are present in this layer in the inferior temporal cortex regions. Ventrally, areas ITp and ITm show a declining number of labeled neurons in both supra- and infragranular layers in the direction of the parahippocampal regions.

This parcellation scheme can be compared to a hierarchical cluster analysis of the temporal cortex based on association between architectonics, response properties, and connections of these areas, reported by Baylis et al. (1987). These authors demonstrated robust associations between the inferior temporal cortex (areas TE1, TE2, TE3, TEm, and TEa), the superior temporal sulcus (areas PGa and IPa), and the the dorsal bank of the superior temporal sulcus (areas TPO, TS, and TAa). Interestingly, similar staining patterns are observed in areas ITp, ITm, and ITa which may closely correspond to TE1–3, our definition of area PGa may be comparable to PGa and IPa of Baylis et al. (1987), and area TPO appears clearly different from the cortical regions TS and TAa in the dorsal bank of the sulcus. Also, similar patterns of neurofilament protein immunoreactivity, that are tightly correlated to the distribution of corticocortical afferents have been reported in these areas (Cusick et al. 1995), and area ITp may correspond to area TEO of other authors (Baizer et al. 1991; Boussaoud et al. 1991; Distler et al. 1993), which joins area V4.

8.2.2.5. The parietal cortex (areas 7a, LIP, VIP, MIP, PIP, MDP, and DP)

All of the parietal cortex regions are characterized by the presence of intermediate-to-large size neurofilament protein-immunoreactive neurons in layers IIIbc and V and very light staining of other layers, and differences in labeled neuron numbers and morphology permits reliable definition among most of these regions. In addition, the immunohistochemical characteristics of the parietal areas are substantially different from the surrounding areas in the temporal and occipital cortex (Fig. 54).

Adjacent to area MST, area 7a in the lateral aspect of the parietal cortex ventral to the intraparietal sulcus is easily recognized by the abrupt discontinuation of the layer VI staining pattern typical of the mediotemporal areas. Most of the labeled neurons in area 7a are in layer V and the overall neuron counts in the infragranular layers of area 7a are lower than area MST. Also, layer IIIbc contains smaller pyramidal neurons than area MST, with lower densities in area 7a (Fig. 54).

The number of neurofilament protein-immunoreactive pyramidal neurons increases in area LIP and VIP on the lateral bank of the intraparietal sulcus and in the depth of this sulcus, compared to area 7a (Fig. 54). Area VIP is characterized by very large, intensely labeled neurofilament protein-immunoreactive pyramidal neurons with very long basal dendrites and conspicuous apical dendrites reaching layer I. Two areas, PIP and MIP, exist on the medial bank of the intraparietal sulcus, adjacent to area VIP. Posteriorly, area PIP contains the highest of neurofilament protein-immunoreactive neuron densities in the parietal cortex, whereas the medial area MIP exhibits much fewer labeled neurons in layers III and V-VI than areas PIP and VIP. There is a clear border between area MIP and the somatosensory area 5 on the dorsolateral aspect of the parietal cortex, which contains very high numbers of large neurofilament protein-immunoreactive neurons in both layers III and V. Also, area MIP has much lower neuron counts in layer III than area PO, which joins MIP posteriorly.

The mediodorsal area MDP is located on the medial aspect of the hemisphere and has staining patterns comparable to those observed in area 5, with large neurofilament protein-immunoreactive neurons in layers III and V-VI, with a density comparable to that found in areas LIP and 7a.

The extent of area VIP reported by Colby et al. (1988, 1993a,b) as a zone of heavy myelination in the depth of the intraparietal sulcus corresponding to the fundal projection zone from area MT (Maunsell and Van Essen 1983), corresponds quite precisely to the neurofilament protein staining patterns which allow for its differentiation from adjacent areas MIP, PIP, and LIP. However, area LIP displays a densely myelinated pattern, with some overlap with area VIP (Colby et al. 1988), whereas it is well defined by neurofilament protein labeling with a pattern distinct from that of area VIP. Furthermore, neurofilament protein staining also corresponds to the cytoarchitectonic analysis of areas 7a, LIP, and VIP reported by Blatt et al. (1990).

8.2.2.6. The prefrontal cortex (areas FEF, SEF, 45 and 46)

In neurofilament protein-stained sections, the frontal eye field (FEF) is a distinct cortical area in the anterior wall of the posterior portion of the prearcuate gyrus, at the level of the spur of the arcuate sulcus. The supplementary eye field (SEF) represents a transition zone between areas 6 and 9 on the superior aspect of the frontal cortex, dorsal to the ascending branch of the arcuate sulcus. The FEF and SEF show distinct neurofilament protein immunoreactivity patterns in both layer III and V, compared to premotor area 6 and prefrontal areas 9, 45 and 46. The FEF is characterized by a diffuse population of labeled pyramidal neurons in layers V and VI that outnumber those in layer III. Layer V contains some large isolated neurofilament protein-immunoreactive cells, however smaller and less numerous than those observed in area 6. The staining pattern of layer III is comparable to that in area 6. The border between the FEF and areas 46 and 45 is well visible due to much lower numbers of neurofilament protein-immunoreactive neurons in layers V-VI of the latter regions. Areas 45 and 46 display comparable populations of intermediate-to-large neurofilament protein-immunoreactive pyramidal neurons in both layers III and V-VI, with area 46 having higher neuron counts in layer III than area 45. The SEF is characterized by very high numbers of neurofilament protein-immunoreactive neurons in layer III, and cell counts in layers V-VI comparable to those in the FEF. The staining pattern in area 9 is comparable to that in the SEF, with intermediate size pyramidal neurons in supra- and infragranular layers. However, the SEF has much higher neurofilament protein-immunoreactive neuron counts than areas 9 and 6.

The FEF and the SEF are involved in smooth eye movements and target selection during saccadic eye movements, and are characterized by distinct cytoarchitecture (Huerta et al. 1987; Stanton et al. 1989; Gottlieb et al. 1993; Schall and Hanes 1993), and both fields have quite similar connectivity patterns with cortical areas and subcortical structures (Parthasarathy et al. 1992; Schlag and Schlag-Rey 1992; Schall et al. 1993a,b; Stanton et al. 1993). Interestingly, these two visuomotor regions exhibit a very similar staining pattern with neurofilament protein, with a much higher

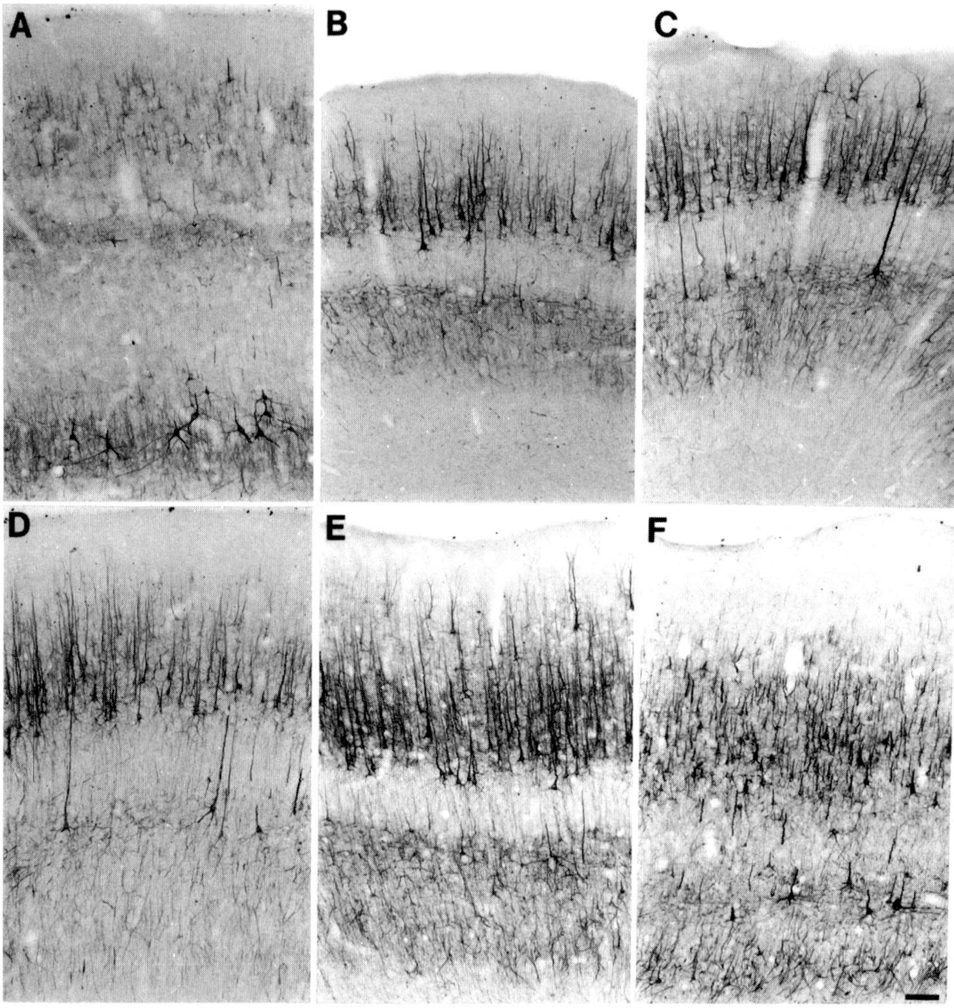

Fig. 55. Distribution of neurofilament protein in areas V1 (*A*), V2 (*B*), V3 (*C*), V3A (*D*), V4 (*E*), and MT (*F*) of the macaque monkey. There are clear laminar patterns of labeled neuron distribution among these areas. Note the prominent staining of large pyramidal neurons, including Meynert cells and layer IVB in area V1 (*A*), layer III and V neurons in areas V2, V3, and V3A (*B-D*), the denser and generally lighter pattern as well as layers II-superficial III neurons observed in area V4 (*E*). Some conspicuous layer II neurons are also labeled in area V3 (C). Area MT contains large layer III and V cells and a distinct band of immunoreactive neurons in layer VI (*F*). Scale bar (*on F*) = 100 μm. Reproduced from Hof et al. (1996).

density of labeled neurons in the infragranular layers than observed in the other visual areas. A comparably high density of neurofilament protein-immunoreactive neurons in the deep layers is also present in the ventrally located area 45. The inferior convexity of the frontal cortex part of area 45 may, in fact, include the lateralmost aspect of the FEF. It is interesting to note that the laminar density of neurofilament protein-containing cells is much lower in area 45 than in FEF, permitting the definition of the boundary between these fields. Areas 46 and 45 are strongly connected to the frontal eye fields (Huerta et al. 1987; Huerta and Kaas 1990; Stanton et al. 1993), and may represent the frontal extension of the occipitoparietal and occipitotemporal pathways (Bachevalier and Mishkin 1986; Funahashi et al. 1990, 1991, 1993; Wilson et al. 1993).

8.2.2.7. Regional and laminar variability in neurofilament protein-immunoreactive neuron counts

The quantitative analysis of the distribution of neurofilament protein-immunoreactive neurons in the visual cortex reveals a large degree of heterogeneity in the laminar localization of the labeled neurons. However, some cortical areas can be grouped according to common staining patterns or laminar neurofilament protein-immunoreactive neuron counts. Inter-regional variability can be assessed by the ratio of neurofilament protein-immunoreactive neuron counts in the infragranular layers and supragranular layers. In most visual areas, there are higher numbers of neurofilament protein-immunoreactive neurons in the supragranular layers than in the infragranular layers, resulting in low ratios (about 0.35; Figs 56, 57). In contrast, a few regions exhibit higher neurofilament protein-immunoreactive cell counts in the infragranular layers, and higher ratios, such as areas MT and MST, and area V4 (about 0.80), areas 45 and the SEF (about 1.00), and the FEF (1.35; Figs 56, 57). These differences in ratio are due to variations in the number of labeled neurons in the infragranular layers only, since the overall numbers of neurofilament protein-immunoreactive neurons in the supragranular layers are comparable for both the regions with low and high ratios. Thus, depending on the ratio of neurofilament protein-immunoreactive neuron counts, visual areas can be subdivided into four general hierarchical levels (Fig. 57), with areas V1, V2, V3, and V3A (low ratio) in the first level, areas MT, MST, and V4 (higher ratio) in the second level, the temporal and parietal regions (low ratio) in the third level, and the frontal regions in the fourth level (very high or inverted ratios). Interestingly, areas related to both the occipitoparietal and occipitotemporal pathways are represented in each of these levels, and that neurofilament protein-immunoreactive neurons are observed in comparable numbers in both pathways, suggesting that regional and laminar variation in labeled neuron counts alone may not account for the functional differences of these systems, whereas the specific distribution of morphological subtypes of neurofilament-containing neurons may be an important factor.

8.3. RELATIONSHIPS TO CONNECTIVITY PATTERNS

In both anatomic and functional terms, it is important to assess the relationship between the distribution of a neurochemical marker within distinct cortical areas and the connectivity of these areas. Using retrograde tract-tracing combined with immunohistochemistry, recent analyses of corticocortical projections have shown that overall, approximately 30% of all corticocortically projecting neurons contain neurofilament protein, although there is a large variability depending on the type of corticocortical connection considered (Campbell et al. 1991; Hof et al. 1995a, 1996; Nimchinsky et al. 1996). The regional and laminar distribution patterns of retrograde labeling in areas V1, V2, V3, and V3A indicate a clear pattern of neurochemical specialization among the neurons of origin of the feedforward projections to areas V4 and MT. In contrast, feedback projections from areas V3, V4, and MT are homogeneous with respect to their content of neurofilament protein (see Table 3; Hof et al. 1996).

There are large differences in percentages of double labeled (i.e. retrogradely traced

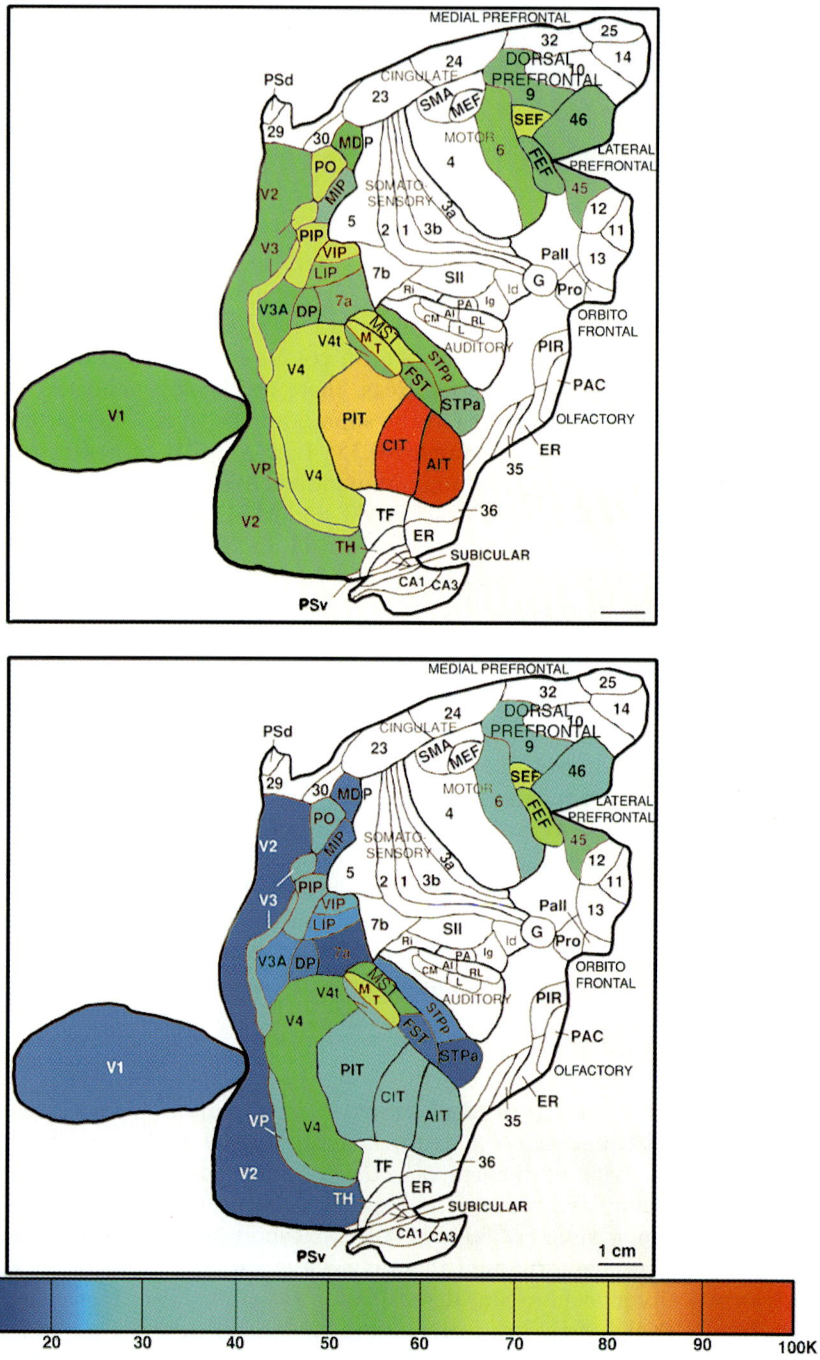
MEDIAL PREFRONTAL
32
25
14
24
DORSAL
10
PREFRONTAL
CINGULATE
PSd
23
SMA
MEF
9
SEF
46
29
30
MDP
MOTOR
6
LATERAL
PREFRONTAL
PO
4
FEF
MIP
SOMATO-
SENSORY
45
V2
3a
12
5
2
1
3b
11
V3
PIP
VIP
Pall
13
LIP
7b
SII
G
Pro
Ri
Id
Ig
7a
PA
V3A
DP
CM
AI
RL
L
ORBITO
FRONTAL
MST
V4t
M
T
AUDITORY
PIR
STPp
V4
FST
PAC
STPa
V1
PIT
OLFACTORY
CIT
ER
VP
AIT
35
V4
TF
36
V2
ER
TH
SUBICULAR
CA1
CA3
PSv
1 cm
0
10
20
30
40
50
60
70
80
90
100K

Fig. 56. Quantitative mapping of neurofilament protein-containing neurons in the visual cortical areas of the macaque monkey. Densities of immunoreactive neurons per mm^3 were obtained using a stereologic approach and mapped onto the flattened representation of the macaque monkey brain developed by Felleman and Van Essen (1991). Densities from layers II-III are shown in the *top panel* and those from layers V-VI are shown in the *bottom panel*. The color scale bar below each panel represents the average neuronal density (x 10^3). There are overall high densities in layer III, particularly in the inferior temporal cortex (*top panel*). Note the high density of neurofilament protein-containing neurons in layer V-VI of areas V4, MT, and MST, as well as in area 45 and the FEF-SEF, compared to the other regions. See text and Felleman and Van Essen (1991) for regional nomenclature. Scale bar = 1 cm.

neurons that contain neurofilament protein) projection neurons among the feedforward projections analyzed. Overall, 78% of the retrogradely labeled neurons projecting to area MT contain neurofilament protein, whereas only 32.1% do so in the projections to area V4. Among the projections to area MT, 100% of the retrogradely labeled cells in layer IVB and the Meynert cells in area V1 are neurofilament protein-immunoreactive. The projection neurons in layer III of areas V2 and V3 contain fewer double labeled neurons compared to the projection from area V1 (78 and 75%, respectively), and the projection from layers V-VI of area V3 to area MT contains fewer double labeled neurons than that from the superficial layers (57 vs 75%).

Conversely, the projections to area V4 are characterized by much lower numbers of neurofilament protein immunoreactivity than the projections to area MT, with percentages ranging from 20 to 40%. The neurofilament protein-immunoreactive neurons located within layers II and superficial III of area V3 are frequently retrogradely labeled following tracer injections in area V4, suggesting that these particular neurons, which represent a reliable a chemoarchitectonic marker for the occipitotemporal and inferior temporal areas (Hof and Morrison 1995), form short corticocortical connections.

Contrasting sharply with the heterogeneity of feedforward projections, the proportion of double labeled neurons is approximately 70% across all of the corresponding feedback projections to areas V1 and V2. Among the projections to areas V1 and V2, retrogradely labeled neurofilament protein-containing neurons are found in higher numbers in layers V-VI than in layers II-III in areas V4 and MT, but not in area V3 where double labeled neurons are present in both in supragranular and infragranular layers. Also, nearly 100% of the neurons participating in the interhemispheric projection from the V1/V2 border are double labeled (Hof et al. 1997b).

Thus, each of these visual projections is characterized by a specific neurofilament protein distribution profile. All of the neurons projecting from area V1 to area MT are double labeled, as are the interhemispheric neurons, whereas there is a progressive decrease in the percentages of immunoreactive neurons in the projections from areas V2 and V3. In contrast, the feedforward projections in the occipitotemporal pathway have a homogeneous, low proportion of double labeled neurons. The feedback projection, irrespective of their site of termination, all contain high levels of neurofilament protein. Interestingly, the high densities of neurofilament protein-immunoreactive neurons in the projections to area MT are comparable to those in long association pathways linking the temporal, parietal and prefrontal cortices. The proportion of double labeled neurons in the projections to area V4 however, is very similar to that seen in corticocortical connections between adjacent cortical domains (Campbell et al. 1991; Hof et al. 1995a). It is has been recently proposed that visual perception involves feedforward visual connections, whereas visual imagery may rely on sets of feedback

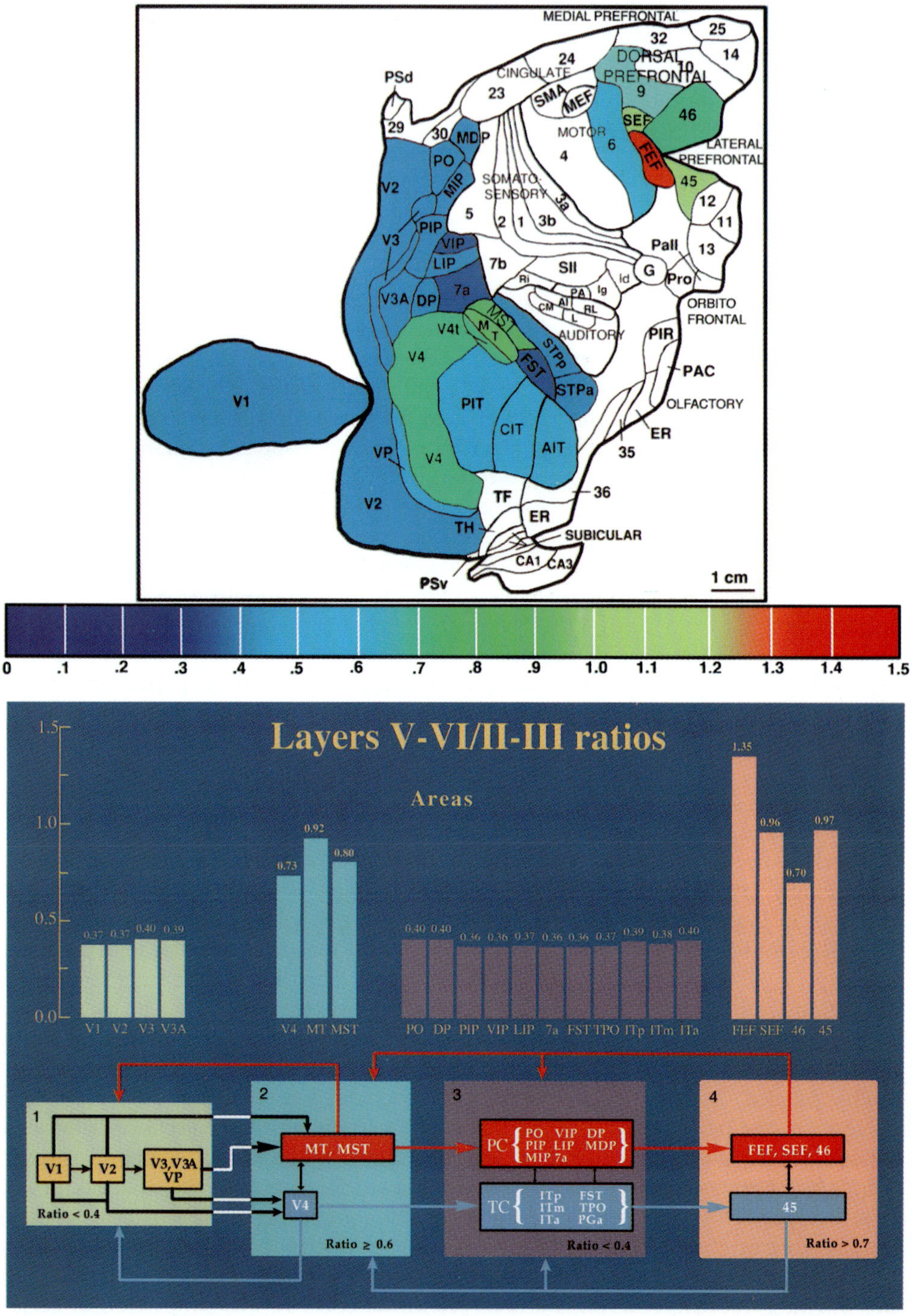
MEDIAL PREFRONTAL
DORSAL PREFRONTAL
CINGULATE
LATERAL PREFRONTAL
ORBITO FRONTAL
MOTOR
SOMATO-SENSORY
AUDITORY
OLFACTORY
SUBICULAR
PSd
PSv
V1
V2
V3
V3A
V4
V4t
VP
MDP
PO
MIP
PIP
VIP
LIP
DP
7a
7b
MST
MT
FST
STPp
STPa
PIT
CIT
AIT
TF
TH
ER
CA1
CA3
SMA
MEF
SEF
FEF
46
45
1 cm
0 .1 .2 .3 .4 .5 .6 .7 .8 .9 1.0 1.1 1.2 1.3 1.4 1.5
Layers V-VI/II-III ratios
Areas
0.37 0.37 0.40 0.39
V1 V2 V3 V3A
0.73 0.92 0.80
V4 MT MST
0.40 0.40 0.36 0.36 0.37 0.36 0.36 0.37 0.39 0.38 0.40
PO DP PIP VIP LIP 7a FST TPO ITp ITm ITa
1.35 0.96 0.70 0.97
FEF SEF 46 45
1
V1
V2
V3,V3A VP
Ratio < 0.4
2
MT, MST
V4
Ratio ≥ 0.6
3
PC { PO VIP DP PIP LIP MDP MIP 7a }
TC { ITp FST ITm TPO ITa PGa }
Ratio < 0.4
4
FEF, SEF, 46
45
Ratio > 0.7

Fig. 57. The *top panel* shows the mean ratios of neurofilament protein-containing neuron counts between the infragranular and supragranular layers in a series of visual cortical areas. Ratios were calculated from the data sets obtained using a stereologic approach and mapped onto the flattened representation of the macaque monkey brain developed by Felleman and Van Essen (1991). The color scale represent the value of the regional ratios. The *bottom panel* shows a schematic representation of hierarchical levels in the macaque monkey visual cortex according to the regional distribution and possible connections of neurofilament protein-containing neurons. The occipitoparietal stream appears in red and the occipitotemporal stream in blue. Areas in level 1 represents early stages of the cortical hierarchy and are not color-coded. Ratios of neurofilament protein-containing neuron counts between the infragranular and supragranular layers are indicated for each level. The differences in ratios could be related to higher density of feedback projections from levels 2 and 4 than from levels 1 and 3, as indicated by the color coded arrows. Compare with quantitative data in the corresponding color-coded histogram above and with the ratio flat map in the top panel. Note that areas V4, MT, MST, 46, 45 and the SEF have significantly higher ratios due to increased densities of labeled neurons in layers V and VI compared to all of the other cortical fields. An inverse distribution of neurofilament protein-containing cells between layers ratio is observed in the FEF. See text and Felleman and Van Essen (1991) for regional nomenclature. Scale bar = 1 cm. Adapted from Hof and Morrison (1995).

projections (Ishai and Sagi 1995; Miyashita 1995). In this context, neurofilament protein may be a useful marker to characterize certain populations of projection neurons that may have a distinctive role in visual processing.

9. RELATIONSHIPS BETWEEN NEUROCHEMICAL PHENOTYPE, CONNECTIVITY AND HUMAN NEUROPATHOLOGY: THE VISUAL CORTEX IN ALZHEIMER'S DISEASE

9.1. THE DISTRIBUTION OF PATHOLOGIC CHANGES IN THE VISUAL CORTEX IS HIGHLY RESTRICTED IN ALZHEIMER'S DISEASE

Selective neuronal vulnerability is a hallmark of all of the major neurodegenerative disorders. The selectivity is apparent at the level of affected brain regions, subregions or layers, and specific cell populations. In the case of Alzheimer's disease (AD), several analyses of the distribution of its characteristic alterations, senile plaques and neurofibrillary tangles, in neocortex have demonstrated striking regional and laminar correlations between the distribution of these lesions and the origins and terminations of long corticocortical and key hippocampal projections (Mutrux 1947; Pearson et al. 1985; Rogers and Morrison 1985; Hyman et al. 1986; Lewis et al. 1987; Hof et al. 1990b; Hof and Morrison 1990, 1994; Arnold et al. 1991). Along with senile plaque and neurofibrillary tangle formation, significant neuronal loss occurs in AD (Terry et al. 1981), which involves principally pyramidal cells in the cortical layers that furnish association and commissural projections, although all corticocortical systems are not equally vulnerable, with short projections from primary sensory to adjacent secondary sensory areas being less affected by the degenerative process than long association pathways (Pearson et al. 1985; Rogers and Morrison 1985; Lewis et al. 1987; Hof et al. 1990b; Hof and Morrison 1990, 1994; Vickers et al. 1996). Consequently, a global corticocortical disconnection may occur in AD and may constitute the primary anatomical substrate for dementia (Pearson et al. 1985; Rogers and Morrison 1985; Lewis et al. 1987; Hof et al. 1990b; Hof and Morrison 1990, 1994; De Lacoste and White 1993). In addition, high somatodendritic concentrations of neurofilament proteins represent a major risk factor for pyramidal cells in AD, in that such neurons are particularly vulnerable to neurofibrillary tangle formation (Vickers et al. 1992) and neurodegeneration (Hof et al. 1990b). It is instructive to put these relationships between neuronal vulnerability, connectivity (i.e., corticocortical circuits) and neurochemical phenotype (i.e., neurofilament protein) that have been demonstrated on a broad level in AD, in the context of our understanding of connectivity and neurochemical characteristics of the visual cortex in non-human primates.

Quantitative analyses of senile plaque and neurofibrillary tangle distribution in AD have shown that although visual association cortices contain a high density of senile plaques and neurofibrillary tangles, area V1 contains numerous senile plaques, but very few neurofibrillary tangles (Pearson et al. 1985; Rogers and Morrison 1985; Lewis et al. 1987; Braak et al. 1989; Hof et al. 1989, 1990b; Hof and Morrison 1990). In addition, in contrast to frontal, parietal, and temporal areas, the total neuronal counts in area V1 do not differ significantly between AD and control brains (Mountjoy et al. 1983). Also, neurofibrillary tangles are present in area V2, but unlike in temporal, parietal and frontal areas, they are predominantly located within layer III and are far less numerous (Lewis et al. 1987; Hof et al. 1989, 1990b; Hof and Morrison 1990). Thus, in respect to neurofibrillary tangles, areas V1 and V2 are relatively spared in AD as it typically presents, suggesting that corticocortical outflow from V1 to V2 is less affected than it is in temporal or parietal visual association areas.

The link between high neurofilament protein levels and neurodegeneration and/or neurofibrillary tangle formation is likely to figure prominently in the relative sparing of areas V1 and V2. Association areas such as inferior temporal and superior frontal cortices have very high densitites of neurofilament protein-containing pyramidal neurons. In these regions, there is extensive loss of these neurons in AD and the extent of the loss is correlated with neurofibrillary tangle density (Hof et al. 1990b). The non-human primate transport studies described above (see section 8.3.) have demonstrated that the majority of neurons participating in corticocortical connections between temporal, parietal and frontal association areas are furnished in large part by neurofilament protein-containing pyramidal neurons located in layers III and V. Thus, if one places the human neuropathological data on neurofibrillary tangle distribution and neurofilament protein-immunoreactive neuron loss into the context of the experimental analyses of the monkey cortex, these association areas are likely to suffer a severe corticocortical disconnection as the neurofilament protein-containing corticocortically-projecting neurons degenereate. In contrast, areas V1 and V2 have relatively low densities of neurofilament protein-containg pyramidal cells, and in a related fashion lower levels of neuron loss in AD (Mountjoy et al. 1983; Hof and Morrison 1990). Neurofilament protein-immunoreactive neuron loss is minimal in area V1 and localized primarily to layer III of area V2, since that neurofibrillary tangles are rare in area V1 and largely confined to layer III in area V2. It is known that layer IVB neurons and Meynert cells at the layer V-VI border give rise to the major long corticocortical projections emanating from area V1 (Freis et al. 1985; Van Essen 1985; Shipp and Zeki 1989). In addition, these layers contain the only heavily neurofilament protein-immunoreactive neurons present in area V1 (Campbell and Morrison 1989; Hof and Morrison 1990).

Therefore, if the long corticocortical projection neurons that are neurofilament protein-immunoreactive possess heightened vulnerability in AD, then the neurofilament protein-immunoreactive cell loss that does occur in area V1 should be most pronounced in layer IVB and in the Meynert cell class. This is, in fact, the case: in area V1, statistically significant differences in cell counts are found only in layer IVB and in layer VI in which the Meynert cell counts are lower in AD. As much as a 24.8%

neuronal loss exists in layer IVB, while this value reaches 26% for Meynert cells (Table 4; Fig. 58). In area V2, neurofilament protein-immunoreactive neuron loss is statistically significant only in the deep portion of layer III, with a 17.1% loss, and in layer VI, with a 19.4% loss (Hof and Morrison 1990; Table 4).These data demonstrate that cellular alterations in the visual system in dementia can be linked to specific circuits and to the particular biochemical phenotype of the neuronal populations at risk. In particular, an extensive loss of pyramidal cells likely to furnish corticocortical projections in layers III and V of inferior temporal cortex and in layer III of area V2, in absence of significant cell loss in layer III of area V1, and the marginal loss of neurofilament protein-enriched neurons in layer IVB and of the immunoreactive Meynert cells of layer VI in area V1 which provide the projection to area MT (Hof and Morrison 1990), suggest that much of the visual cortex and associated connections within the occipital lobe are spared initially in most AD cases, whereas the high order association visual areas of the inferior temporal cortex may be affected early in the course of the disease (Hof et al. 1990b; Hof and Morrison 1990, 1994). Such an increasing degree of neuropathologic lesions with ascension of the visual hierarchy is consistent with the deficits in high-order visual processing observed in AD, although not all modalities of visual function are affected (Cogan 1985; Hinton et al. 1986; Neary and Snowden 1987; Cronin-Golomb et al. 1993; Curcio and Drucker 1993; Kurylo et al. 1994, 1996; Parasuraman et al. 1995).

TABLE 4. *Neurofilament protein-immunoreactive neuron counts in area V1 and area V2 of control cases and Alzheimer's disease cases*

	NEURON COUNTS	
	Control cases	AD cases
Area V1		
LAYERS I	0.0	0.0
II	0.0	0.0
IIIs	2.72 ± 0.45	2.64 ± 0.45
IIId	6.92 ± 0.45	7.32 ± 0.80
IVA	0.24 ± 0.10	0.10 ± 0.10
IVB	8.24 ± 0.70	6.20 ± 0.40[b]
IVCα	1.92 ± 0.25	1.87 ± 0.25
IVCβ	0.24 ± 0.10	0.0
V	4.60 ± 0.17	4.54 ± 0.20
VI	7.36 ± 0.62	7.50 ± 0.80
M	3.92 ± 0.48	2.90 ± 0.20[b]
Area V2		
LAYERS I	0.0	0.0
II	0.10 ± 0.10	0.0
IIIs	1.24 ± 0.20	0.80 ± 0.20
IIId	23.04 ± 1.16	19.10 ± 0.85[a]
IV	1.08 ± 0.20	0.50 ± 0.10
V	3.92 ± 0.50	3.30 ± 0.30
VI	6.08 ± 0.40	4.90 ± 0.40

Data represent neurofilament protein-immunoreactive neuron counts per mm of cortical traverse and are means ± SEM from 5 control and 5 AD brains. Not the statistically significant differences in AD cases compared to controls for layer IVB and the Meynert cells in area V1 and in the deep portion of layer III in area V2: a, $p = 0.025$; b, $p < 0.05$, two-sided Mann-Whitney U test. M: Meynert cells; s and d refer to the superficial and deep portions of layer III, respectively. Adapted from Hof and Morrison (1990).

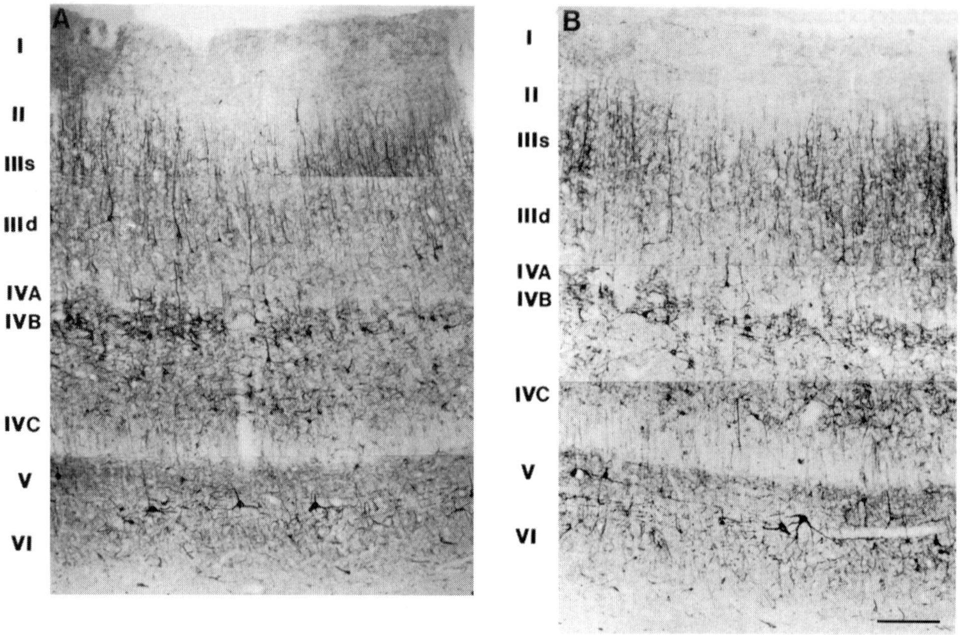

Fig. 58. Photomontages of neurofilament protein-immunoreactive neurons in area V1 of a control case (*A*) and an AD case (*B*). Note the presence of Meynert cells at the border of layers V and VI. There is only a marginal neurofilament protein-immunoreactive cell loss in the AD case compared to the control case. Scale bar = 200 μm. Reproduced from Hof and Morrison (1990).

9.2. DISRUPTION OF VISUAL PATHWAYS AND CLINICOPATHOLOGICAL CORRELATIONS: INSIGHTS THROUGH ATYPICAL VISUAL VARIANTS OF AD

9.2.1. Overview of posterior cortical atrophy

If in AD, senile plaque and neurofibrillary tangle formation represent the disconnection of specific corticocortical pathways, then some AD cases may have identifiable neurological deficits that could be directly linked to the disconnection that has occurred, as reflected by the distribution of the lesions. In recent years several studies in human have shown that a significant population of patients suffering from AD present with prominent visual symptoms. As the clinical signs indicate in these cases, the visual impairment is not a diffuse one, but rather affects specific aspects of visual function (Cogan 1985; Benson et al. 1988; Fletcher 1994; Nielson et al. 1996). Thus, it is important to discuss the selective involvement of particular components of the visual circuitry in the neurodegenerative process that occurs in AD. In this context, the observation in 1909 by the Hungarian physician Reszö Bálint of a complex visual syndrome characterized by a cortical paralysis of visual fixation associated with an optic ataxia and a disturbance of visual attention in patients with large bilateral parieto-occipital lesions of vascular origin (Bálint 1909), is of particular interest since similar visual signs have been evidenced in a subset of AD patients (Fletcher 1994). The occurrence of a Bálint's-like syndrome in AD patients was in fact first documented by Grünthal (1928) and Morel (1945), and these early studies have been followed by a growing number of reports. Most of the AD cases with focal occipital alterations display a complex neurologic symptomatology and variable disturbances of visual function that include alexia, anomia, agraphia, transcortical sensory aphasia, as well as Bálint's and Gerstmann's syndromes, with a relatively good preservation of memory in the early stages of their illness, instead of the memory impairment and temporospatial disorientation that dominates the usual presentation of AD. The cortical atrophy and metabolic deficits, as demonstrated in some of these cases by functional imaging or at autopsy, usually predominates in the parieto-temporo-occipital junction and occipital cortex (Magnani et al. 1985; Nissen et al. 1985; Benson et al. 1988; Hof et al. 1989, 1990a, 1993, 1997a; Jacquet et al. 1990; Mendez et al. 1990a,b,c; Berthier et al. 1991; Hof and Bouras 1991; Graff-Radford et al. 1993; Levine et al. 1993; Victoroff et al. 1994; Levy et al. 1995; Pietrini et al. 1996), matching the regions damaged by the vascular pathology in the case reported by Bálint. Deficits in motion perception and target tracing considered to be a usual, although not necessarily consistent, sign once dementia is established in AD cases, are a prominent feature in these cases with atypical visual presentation (Neary and Snowden 1987; Sadun et al. 1987; Fletcher and Sharpe, 1988; Katz and Rimmer, 1989; Kiyosawa et al. 1989; Kuskowski et al. 1989; Cronin-Golomb et al. 1993; Kurylo et al. 1994, 1996; Silverman et al. 1994; Butter et al. 1996; Furey-Kurkjian et al. 1996; Mentis et al. 1996). Such demented cases with visual deficits are frequently referred to as posterior cortical atrophy (Benson et al. 1988). The clinical symptomatology of AD patients with posterior cortical atrophy suggests the possible involvement of certain cortical pathways linking the primary visual occipital regions to the posterior parietal and cingulate visual association cortex in the early stages of the dementia (Hof et al. 1989, 1990a, 1993, 1997a; Hof and Bouras 1991).

The functional organization of the human occipital cortex and occipito-temporo-parietal junction has been documented in several recent studies and appears generally to correspond to the hierarchical schemes observed in nonhuman primates (Clarke and Miklossy 1990; Clarke 1993, 1994, 1995; Watson et al. 1993; Miklossy 1993; McCarthy et al. 1995; Orban et al. 1995; Sereno et al. 1995; Tootell et al. 1995; Tootell and Taylor 1995; DeYoe et al. 1996; Engel et al. 1997). These studies have provided evidence for the existence in human of a comparable organizational scheme of visual areas as seen in nonhuman primates and have defined the possible localization of areas V1, V2, V3/VP, V4, MT and additional fields in the parieto-occipital cortex. Using these neuroanatomic and brain imaging data as a guide to the study of cases with posterior cortical atrophy, it is possible to identify several putative extrastriate visual areas, relying on Nissl-stained sections and on the distribution and density of lesions. In these cases, progressive differences in lesion densities allowed for the distinction of several visual regions. Areas V1 in the calcarine sulcus and V2 were easily identified by their characteristic cytoarchitecture. The ventral portion of area V3 (referred to as VP) shows smaller pyramidal neurons in layers III and V than the adjacent fields V2 and V4, and the dorsal portion of area V3, and is much less myelinated (Clarke and Miklossy 1990). These fields have comparable lesion densities in posterior cortical atrophy cases. Area MT appeared as a restricted region with severe pathologic alterations located in the lateral inferior occipital gyri. This location corresponds to the heavily myelinated zone reported by Flechsig (1920), Clarke and Miklossy (1990), Clarke (1993), and Tootell and Taylor (1995), and to the area activated by motion detection tasks in functional imaging studies (Watson et al. 1993; Sereno et al. 1995; Tootell et al. 1995). The posterior parietal cortex and the cortex

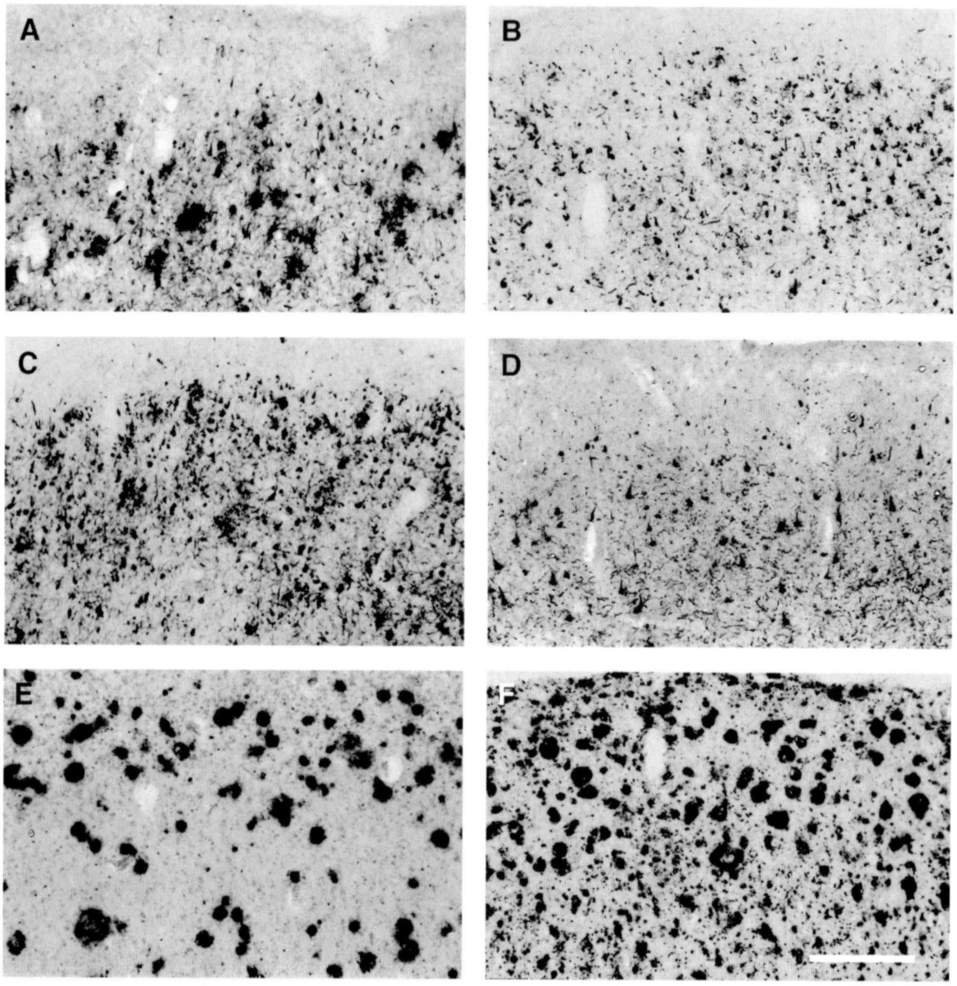

Fig. 59. Distribution of neurofibrillary tangles in areas V1 (*A*), V3 (*B*), 23 (*C*), and 9 (*D*), and senile plaques in areas V1 (*E*), and MT (*F*) in AD cases with posterior cortical atrophy. Note the very high numbers of lesions in these posterior cortical areas, and in particular in area V1, as well as the increase in lesion density in areas V3, MT, 7b and 23 compared to area V1. Also, prefrontal area 9 exhibits a lower density of neurofibrillary tangles in comparison to areas 23 and V3. Each panel shows layers I to III; the pial surface is to the top and the bottom of the panels is in layer IV. Tissues were stained with an antibody to the microtubule-associated protein tau (*A-D*), or with an antibody to the amyloid Aβ protein (*E-F*). Scale bar (*on F*) = 200 μm. Modified from Hof et al. (1997a).

within the precuneus (areas 7b/7m) is severely affected in posterior cortical atrophy cases and probably contains several visual association areas (Miklossy 1993; Tootell and Taylor 1995).

9.2.2. Neuropathologic analysis of posterior cortical atrophy cases

Quantitatively, the densities and distribution of senile plaques and neurofibrillary tangles in the posterior cortical atrophy cases that we had the opportunity to study (Hof et al. 1989, 1990b, 1993, 1997a; Hof and Bouras 1991), generally agrees with the finding of other investigators who also reported inordinately high densities of AD lesions in the occipital regions (Berthier et al. 1991; Jacquet et al. 1990; Mendez et al. 1990a,b,c; Graff-Radford et al. 1993; Levine et al. 1993; Victoroff et al. 1994). These observations are summarized in Figures 59 and 60. There is a gradient in neurofibrillary tangle densities from area V1 to the visual association regions in the parietal cortex, as well as in the posterior cingulate cortex in these cases who all suffered from visuomotor impairment resembling Bálint's syndrome, except in one who presented clinically with prosopagnosia and apperceptive visual agnosia (Hof and Bouras 1991). A comparable occipital shift in the lesion distribution is observed in this case, however, it differs from the others in that there is a preferential distribution of neurofibrillary tangles and senile plaques in the inferior temporal cortex (areas TEO and TE), whereas the inferior parietal and posterior cingulate cortex are less affected, demonstrating the specific involvement of certain components of the occipitotemporal visual pathway, consistent with the clinical symptomatology (Hof and Bouras 1991).

When compared to AD cases with the usual presentation of dementia, areas V1, V2 and V3 contain a much higher density of neurofibrillary tangles in the cases with posterior cortical atrophy (Figs 59, 60). Differences are most marked in the occipital areas and were predominant in areas V1 and V2 with increases over more conventional AD cases as high as 35-fold in posterior cortical atrophy cases. Smaller differences from conventional AD exist in areas V3/V4, MT, 7b/7m and 23 (Figs 59, 60). Neurofibrillary tangle counts do not differ significantly in posterior cortical atrophy cases in areas TE/TEO compared to AD cases, except in the case that presented with visual agnosia. In the prefrontal areas of posterior cortical atrophy cases, neurofibrillary tangle counts are always much lower than in conventional AD cases, clearly indicating the shift in lesion distribution onto cortical regions that are usually not severely involved in AD (Mutrux 1947; Pearson et al. 1985; Rogers and Morrison 1985; Hyman et al. 1986; Lewis et al. 1987; Hof et al. 1989, 1990a,b, 1993, 1997a; Hof and Morrison 1990, 1994; Arnold et al. 1991; Hof and Bouras 1991). Senile plaques are also much more numerous in the occipital cortex of posterior cortical atrophy cases than in conventional AD cases. Depending on the area, there is up to 5-fold increases in senile plaque density in the posterior cortical atrophy cases, with the largest differences in areas V1 and V2 followed by areas V3/V4, MT, 7b/7m and area 23. No difference between posterior cortical atrophy with visuomotor impairment and con-

Fig. 60. Quantitative maps of regional differences in density of neurofibrillary tangles (*A-C*) and senile plaques (*D-F*) between posterior cortical atrophy and AD cases. Average regional differences were calculated from large-scale density maps of series of sections throughout the cortical regions analyzed in the present study and are color-coded on the lateral (*A, D*), ventral (*B, E*), and medial (*C, F*) views of the brain from the patient studied by Morel 50 years ago (Morel 1945; Hof et al. 1993). The posterior atrophy is particularly visible on the lateral and medial views. The colors reflect an arbitrary scale since the differences in lesion densities from AD cases range from 10% to many-fold changes. Regions that have higher lesion densities in the AD cases appear in the blue-purple range, whereas regions with higher lesion densities in the posterior cortical atrophy cases appear in the red-yellow range. The *arrowhead* indicates the point where no differences are observed between posterior cortical atrophy and AD cases. Statistical significance levels are also indicated below each color scale. The hippocampal formation and entorhinal cortex show comparable lesion densities in AD and posterior cortical atrophy cases and are not reported on the maps. Note that the scales are not linear and that their range differs between the two panels (*neurofibrillary tangles, A-C; senile plaques, D-F*). The locations of the various cortical regions are indicated on each map as follows: 1, area V1; 2, area V2; 3, areas V3/VP/V4; 4, inferior temporal areas TE/TEO; 5, area MT; 6, posterior parietal area 7b/7m; 7, posterior cingulate area 23; 8, prefrontal areas 9/46/45; 9 (on *D*) represents the superior prefrontal area 9.

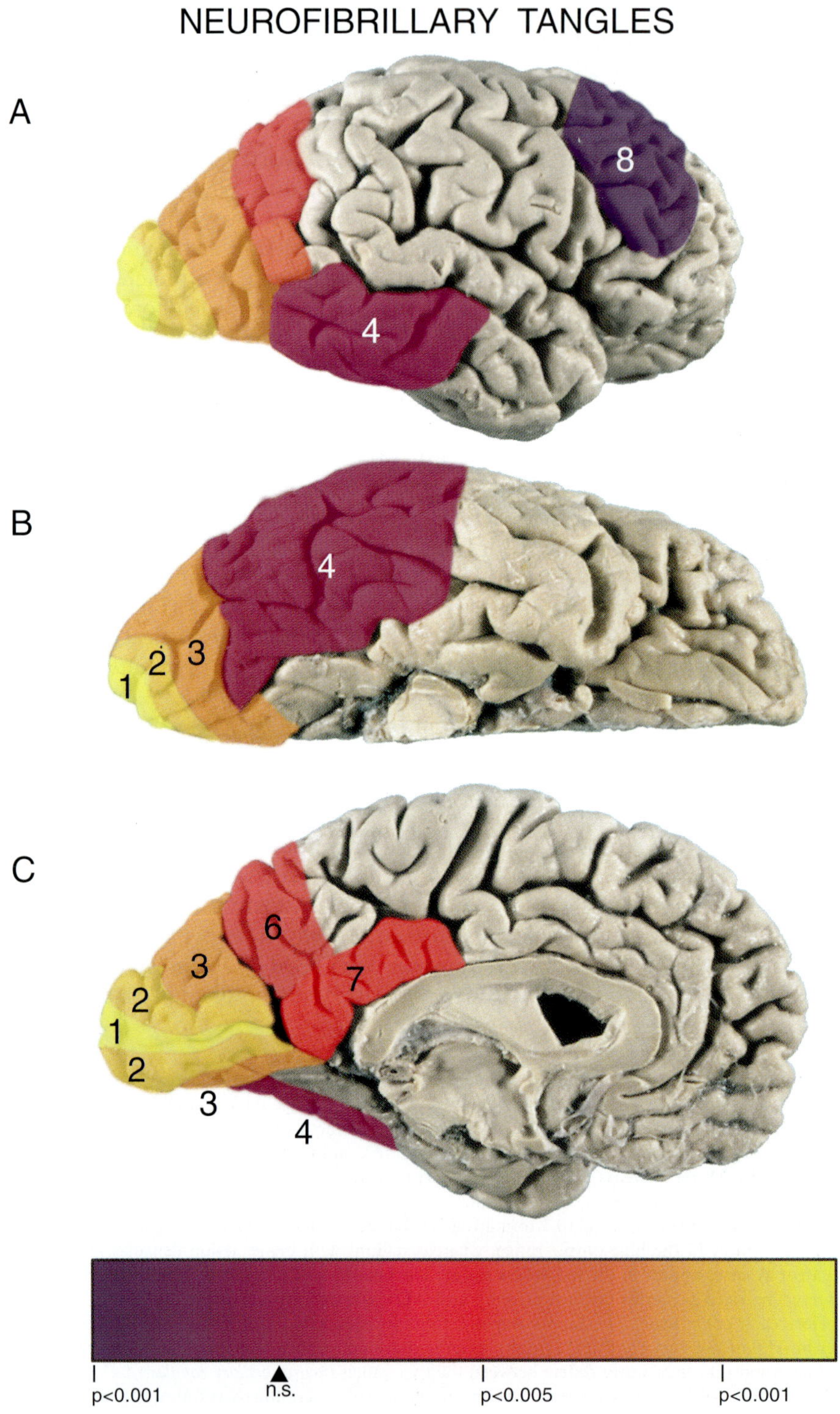
NEUROFIBRILLARY TANGLES
A
8
4
B
4
3
2
1
C
6
3
7
2
1
2
3
4
p<0.001
n.s.
p<0.005
p<0.001

Fig. 60.

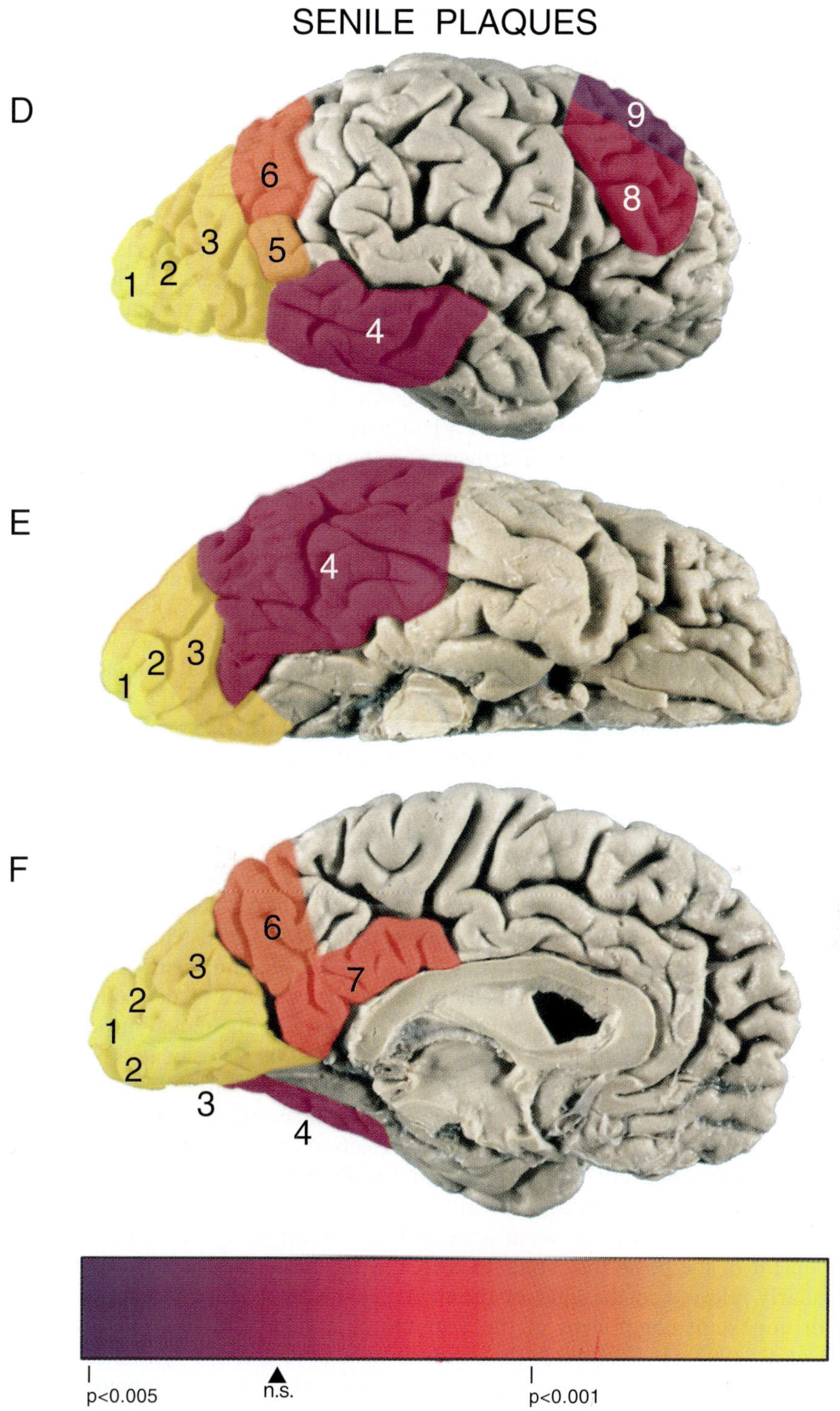

Fig. 60. (continued)

ventional AD cases is observed in areas TE/TEO. Area 9 has slightly higher senile plaque counts in posterior cortical atrophy, whereas no differences are observed for areas 45 and 46 (Figs 59, 60).

9.2.3. Relationship of neuropathologic patterns to cortical connectivity

Placed in the context of connectivity (Pearson et al. 1985; Rogers and Morrison 1985; Lewis et al. 1987), the distribution of neurofibrillary tangles and senile plaques in the posterior cortical atrophy cases strongly suggests that specific elements of visual corticocortical circuits are severely disrupted in this subset of cases. The high density of neurofibrillary tangles in layer III of area V1 in these cases suggests that the projection to area V2 is affected. The high density of neurofibrillary tangles in layer III of areas V2 and V3 similarly indicates that feedforward projections to areas V4, MT and 7b are affected in the posterior cortical atrophy cases. In addition, the high numbers of neurofibrillary tangles in layers V and VI of area V2 indicates that feedback projections to area V1 are also altered. Furthermore, the disruption of feedback connections, which terminates in both supra- and infragranular layers, is well correlated to the high senile plaque counts in all layers of area V1 in these cases, and increases in senile plaque density in area V2 suggest an involvement of feedback projections from area V3, as well as feedforward projections from area V1. The same logic can be applied to the very high senile plaque densities in areas MT, 7b/7m and 23, that may reflect the degeneration of feedforward connections from area V3 and V4, as well as to the prefrontal areas 9 and 46 onto which the occipitoparietal visual channel is likely to project (Wilson et al. 1993). Thus, the particular distribution of the neuropathologic alterations in the visual cortex of posterior cortical atrophy cases supports the notion that AD involves the breakdown of subsets of corticocortical projections (Lewis et al. 1987; Hof et al. 1990a,b, 1993, 1997a; Hof and Morrison 1990, 1994).

These patterns may correspond to organizational schemes that have been developed for the human occipital cortex (Clarke and Miklossy 1990; Miklossy 1993; Watson et al. 1993; Clarke 1993, 1994, 1995; McCarthy et al. 1995; Orban et al. 1995; Sereno et al. 1995; Tootell et al. 1995; Tootell and Taylor 1995; DeYoe et al. 1996; Engel et al. 1997). In particular, the region corresponding to area MT consistently contains higher neurofibrillary tangle densities than the cortex of the occipital fields. Furthermore, a recent study in human brains also demonstrated the presence of projections from the occipital cortex to the dorsal posterior parietal cortex and angular gyrus (areas 7b and 39) that may correspond to some of the visual association areas located in the parietal cortex of the monkey and specialized in visuomotor attention and visuospatial tasks (Miklossy 1993). Interestingly, Orban et al. (1995) have reported the existence in human of another center involved in motion detection that appears to be involved in detecting kinetic contours. It is likely that the infero-posterior and medial parietal areas 7b/7m that are severely affected in the posterior cortical atrophy cases encompass this region, which is distinct from area MT. In addition, the consistent and severe involvement of the posterior cingulate cortex in the posterior cortical atrophy cases is particularly relevant in the light of recent data indicating that this cortical area is in fact an important component of the visuomotor system (Olson and Musil 1992a,b; Olson et al. 1993). It is possible that clinical deficits in Bálint's syndrome are due in large part to the severe and early damage in area 23, since in nonhuman primates, there are direct projections from area V3 to area 23, as well as projections between parietal areas LIP and 7m and area 23 (Vogt et al. 1979; Baleydier and Mauguière

1987; Vogt and Pandya 1987). Area 23 responds to visual stimuli of large textured patterns and to changes in the orbital position of the eye, and lesions in posterior cingulate cortex impair spatial memory in monkeys (Murray et al. 1989; Olson and Musil 1992a,b; Olson et al. 1993). Altogether, these data support the involvement of cortical networks linking the parietal, cingulate and occipital regions in AD (Vogt et al. 1990, 1992; Minoshima et al. 1994; Reiman et al. 1996) and posterior cortical atrophy cases.

Altogether these data lend support to the hypothesis that long corticocortical projections are particularly vulnerable in AD, and in fact are much more vulnerable than a short projection such as that from area V1 to area V2. The fact that the lightly labeled neurofilament protein-immunoreactive cells in layer III of area V1 show no decrease in AD brains, along with the paucity of neurofibrillary tangles in this area (Lewis et al. 1987; Hof and Morrison 1990), suggests that the projection from area V1 to V2 remains fully intact in AD. In this respect, the degree to which corticocortically projecting cells are also neurofilament protein-immunoreactive neurons varies among corticocortical projections and both the length of a given corticocortical projection and the functional nature of the interconnected regions appear to represent determinant variables of the phenotype of these projection neurons (Campbell et al. 1991; Hof et al. 1995, 1996; Nimchinsky et al. 1996).

Thus, these atypical AD cases may serve as excellent examples of clinicopathological correlations between senile plaque and neurofibrillary tangle distribution and clinical symptoms that are likely to involve disruption of subsets of corticocortical connections and is consistent with the observation of Victoroff et al. (1994) suggesting that posterior cortical atrophy is clinically homogeneous but displays neuropathologic heterogeneity. The disturbances in motion perception and target tracing frequently observed in AD patients (Neary and Snowden 1987; Sadun et al. 1987; Fletcher and Sharpe 1988; Katz and Rimmer 1989; Kiyosawa et al. 1989; Kuskowski et al. 1989; Silverman et al. 1994; Butter et al. 1996; Furey-Kurkjian et al. 1996; Mentis et al. 1996), are more severe in posterior cortical atrophy and may be related to the very high densities of lesions in cortical regions such as areas MT, 7b/7m and 23. Although additional studies of such cases using functional imaging will provide the key information that is still missing between the neurologic symptoms and the cortical circuits and areas that are affected, these observations suggest that the characterization of relationships between functional role, connectivity patterns, and neurochemical characteristics of identified systems in the human cortex may lead to a better understanding of the anatomical substrate and pathogenetic mechanisms underlying neurodegenerative disorders such as AD.

10. ACKNOWLEDGEMENTS

We thank W.G.M. Janssen, E.A. Nimchinsky, and M.M. Adams for help during the preparation of this review, A.P. Leonard for help with computer graphics and photography, and S. Hernandez for secretarial assistance. This work was supported by NIH grants AG05138 (J.H.M., P.R.H.), AG06647 (J.H.M.), NS34659 (G.W.H.), and the Human Brain Project MHDA52154 (J.H.M.); the American Health Assistance Foundation (P.R.H.); and the Brookdale Foundation (P.R.H.).

11. REFERENCES

Adams MM, Ungerleider LG, Eden G, Bouras C, Bertini G, Morrison JH, Hof PR (1996): Neurofilament and calcium-binding proteins define visual cortical areas in human. *Soc. Neurosci. Abstr., 22*, 1059.

Akil M, Lewis DA (1992a): Differential distribution of parvalbumin-immunoreactive pericellular clusters of terminal boutons in developing and adult monkey neocortex. *Exp. Neurol., 115*, 239–249.

Akil M, Lewis DA (1992b): Postnatal development of parvalbumin immunoreactiviy in axon terminals of basket and chandelier neurons in monkey neocortex. *Prog. Neuro-Psychopharmacol. Biol. Psychiatry, 16*, 329–337.

Anderson SA, Classey JD, Condé F, Lund JS, Lewis DA (1995): Synchronous development of pyramidal neuron dendritic spines and parvalbumin-immunoreactive chandelier neuron axon terminals in layer III of monkey prefrontal cortex. *Neuroscience, 67*, 7–22.

Aoki C, Fenstemaker S, Lubin M, Go CG (1993): Nitric oxide synthase in the visual cortex of monocular monkeys as revealed by light and electron microscopic immunocytochemistry. *Brain Res., 620*, 97–113.

Aoki C, Venkatesan C, Go CG, Mong JA and Dawson TM, (1994a): Cellular and subcellular localization of NMDA-R1 subunit immunoreactivity in the visual cortex of adult and neonatal rats. *J. Neurosci., 14*, 202–5222.

Aoki C, Lubin M, Fenstemaker S (1994b): Columnar activity regulates astrocytic beta-adrenergic receptor-like immunoreactivity in V1 of adult monkeys. *Vis. Neurosci., 11*, 179–187.

Armstrong RA, Slaven A (1994): Does the neurodegeneration of Alzheimer's disease spread between visual cortical regions B17 and B18 via the feedforward or feedback short cortico-cortical projections? *Neurodegeneration, 3*, 191–196.

Arnold SE, Hyman BT, Flory J, Damasio AR, Van Hoesen GW (1991): The topographical and neuroanatomical distribution of neurofibrillary tangles and neuritic plaques in the cerebral cortex of patients with Alzheimer's disease. *Cereb. Cortex, 1*, 103–116.

Bachevalier J, Mishkin M (1986): Visual recognition impairment follows ventromedial but not dorsolateral prefrontal lesions in monkeys. *Behav. Brain Res., 20*, 249–261.

Baizer JS, Ungerleider LG, Desimone R (1991): Organization of visual inputs to the inferior temporal and posterior parietal cortex in macaques. *J. Neurosci., 11*, 168–190.

Baleydier C, Mauguière F (1987): Network organization of the connectivity between parietal area 7, posterior cingulate cortex and medial pulvinar nucleus: a double fluorescent tracer study in monkey. *Exp. Brain Res., 66*, 385–393.

Bálint R (1909): Seelenlähmung des 'Schauens', optische Ataxie, räumliche Störung der Aufmerksamkeit. *Mschr. Psychiat. Neurol., 25*, 51–81.

Barbaresi PB, Fabri M, Conti F, Manzoni T (1987): D-[^{3}H]Aspartate retrograde labelling of callosal and association neurones of somatosensory areas I and II of cats. *J. Comp. Neurol., 263*, 59–178.

Baskys A (1994): *Metabotropic Glutamate Receptors*. R.G. Landes, Austin, TX.

Baughman RW, Gilbert CD (1981): Aspartate and glutamate as possible neurotransmitters in the visual cortex. *J. Neurosci., 1*, 427–439.

Baylis GC, Rolls ET, Leonard CM (1987): Functional subdivisions of the temporal lobe neocortex. *J. Neurosci., 7*, 330–342.

Beaulieu C, Kisvárday Z, Somogyi P, Cynader M, Cowey A (1992): Quantitative distribution of GABA-immunopositive and -immunonegative neurons and synapses in the monkey striate cortex (area 17). *Cereb. Cortex, 2*, 295–309.

Beaulieu C, Somogyi P (1990): Targets and quantitative distribution of GABAergic synapses in the visual cortex of the cat. *Eur. J. Neurosci., 2*, 296–303.

Beaver,CJ, Mitchell DE, Robertson HA (1993): Immunohistochemical study of the pattern of rapid expression of c-Fos protein in the visual cortex of dark-reared kittens following initial exposure to light. *J. Comp. Neurol., 333*, 469–484.

Benevento LA, Rezak M (1976): Extrageniculate projections to layers VI and I of striate cortex (area 17) in the rhesus monkey (*Macaca mulatta*). *Brain Res., 96*, 51–52.

Benevento LA, Standage GP (1982): Demonstration of lack of dorsal lateral geniculate nucleus input to extrastriate areas MT and visual 2 in the macaque monkey. *Brain Res., 252*, 161–166.

Benson DF, Davis RJ, Snyder DB (1988): Posterior cortical atrophy. *Arch. Neurol., 45*, 789–793.

Benson DL, Huntsman MM, Jones EG (1994): Activity dependent changes in GAD and preprotachykinin mRNAs in visual cortex of adult monkeys. *Cereb. Cortex, 4*, 40–51.

Benson DL, Isackson PJ, Gall CM, Jones EG (1991a): Differential effects of monocular deprivation on glutamic acid decarboxylase and type II calcium-calmodulin-dependent protein kinase gene expression in the adult monkey visual cortex. *J. Neurosci., 11*, 31–47.

Benson DL, Isackson PJ, Jones EG (1991b): In situ hybridization reveals VIP precursor mRNA-containing neurons in monkey and rat neocortex. *Mol. Brain Res., 9,* 169–174.
Berger B, Gaspar P, Verney C (1991): Dopaminergic innervation of the cerebral cortex: unexpected differences between rodents and primates. *Trends Neurosci., 14,* 21–27.
Berger B, Tassin JP, Blanc G, Moyne MA, Thierry M (1974): Histochemical confirmation for dopaminergic innervation of the rat cerebral cortex after destruction of the noradrenergic ascending pathways. *Brain Res., 81,* 332–337.
Berger B, Trottier S, Verney C, Gaspar P, Alvarez C (1988): Regional and laminar distribution of the dopamine and serotonin innervation in the macaque cerebral cortex: a radioautographic study. *J. Comp. Neurol., 273,* 99–119.
Berger B, Verney C, Alvarez C, Vigny A, Helle KB (1985): New dopaminergic terminal fields in the motor, visual (area 18b) and retrosplenial cortex in the young and adult rat. Immunocytochemical and catecholamine histochemical analyses. *Neuroscience, 15,* 983–998.
Berthier ML, Leiguarda R, Starkstein SE, Sevlever G, Taratuto AL (1991): Alzheimer's disease in a patient with posterior cortical atrophy. *J. Neurol. Neurosurg. Psychiatry, 54,* 1110–1111.
Bigham MH, Lidow MS (1995): Adrenergic and serotonergic receptors in aged monkey neocortex. *Neurobiol. Aging, 16*, 91–104.
Björklund A, Divac I, Lindvall O (1978): Regional distribution of catecholamines in monkey cerebral cortex, evidence for a dopaminergic innervation of the primate prefrontal cortex. *Neurosci. Lett., 7,* 115–119.
Blakely RD, Coyle JT (1988): The neurobiology of *N*-acetylaspartylglutamate. *Int. Rev. Neurobiol., 30,* 39–100.
Blasdel GG, Lund JS (1983): Termination of afferent axons in macaque striate cortex. *J. Neurosci., 3,* 1389–1413.
Blatt GJ, Andersen RA, Stoner GR (1990): Visual receptive field organization and cortico-cortical connections of the lateral intraparietal area (area LIP) in the macaque. *J. Comp. Neurol., 299,* 421–455.
Blümcke I, Celio MR (1992): Parvalbumin and calbindin D-28k immunoreactivities coexist within cytochrome oxidase-rich compartments of squirrel monkey area 18. *Exp. Brain Res., 92*, 39–45.
Blümcke I, Hof PR, Morrison JH, Celio MR (1990): Distribution of parvalbumin immunoreactivity in the visual cortex of Old World monkeys and humans. *J. Comp. Neurol., 301*, 417–432.
Blümcke I, Weruaga E, Kasas S, Hendrickson AE, Celio MR (1994): Discrete reduction patterns of parvalbumin and calbindin D-28k immunoreactivity in the dorsal lateral geniculate nucleus and the striate cortex of adult macaque monkeys after monocular enucleation. *Vis. Neurosci., 11*, 1–11.
Boussaoud D, Desimone R, Ungerleider LG (1991): Visual topography of area TEO in the macaque. *J. Comp. Neurol., 306,* 554–575.
Boussaoud D, Ungerleider LG, Desimone R (1990): Pathways for motion analysis: cortical connections of the medial superior temporal and fundus of the superior temporal visual areas in the macaque. *J. Comp. Neurol., 296,* 462–495.
Bouthenet ML, Martres MP, Sales N, Schwartz JC (1987): A detailed mapping of dopamine D-2 receptors in rat central nervous system by autoradiography with [^{125}I]iodosulphide. *Neuroscience, 20,* 117–155.
Boyson SJ, McGonigle P, Molinoff PB (1986): Quantitative autoradiographic localization of the D_1 and D_2 subtypes of dopamine receptors in rat brain. *J. Neurosci., 6*, 3177–3188.
Bredt DS, Snyder SH (1992): Nitric oxide, a novel neuronal messenger. *Neuron, 8,* 3–11.
Brené S, Hall H, Lindefors N, Karlsson P, Halldin C, Sedvall G (1995): Distribution of messenger RNAs for D_1 dopamine receptors and DARPP-32 in striatum and cerebral cortex of the cynomolgus monkey: relationships to D_1 dopamine receptors. *Neuroscience, 67,* 37–48.
Brodmann K (1905): Beiträge zur histologischen Lokalisation der Grosshirnrinde. III. Dritte Mitteilung: die Rindenfeldern der niederen Affen. *J. Psychol. Neurol., 4,* 177–226.
Brodmann K (1909) *Vergleichende Lokalisationslehre der Grosshirnrinde.* Barth, Leipzig.
Brown RM, Crane AM, Goldman PS (1979): Regional distribution of monoamines in the cerebral cortex and subcortical structures of the rhesus monkey: concentrations and in vivo synthesis rates. *Brain Res., 168,* 133–150.
Bullier J, Kennedy H (1983): Projection of the lateral geniculate nucleus onto cortical area V2 in the macaque monkey. *Exp. Brain Res., 53,* 168–172.
Butter CM, Trobe JD, Foster NL, Berent S (1996): Visual-spatial deficits explain visual symptoms in Alzheimer's disease. *Am. J. Ophthalmol., 122*, 97–105.
Cajal SR (1899a): Apuntes para el estudio estructural de la corteza visual del cerebro humano. *Rev. Ibero-Americana Cienc. Med., 1,* 1–14. [Translated in: DeFelipe J, Jones EG (1988): *Cajal on the Cerebral Cortex.* Oxford University Press, London.]
Cajal SR (1899b): Estudios sobre la corteza cerebral humana I: Corteza visual. *Rev. Trim. Micrograf., 4,*

1–63. [Translated in: DeFelipe J, Jones EG (1988): *Cajal on the Cerebral Cortex*. Oxford University Press, London.]

Campbell MJ, Hof PR, Morrison JH (1991): A subpopulation of primate corticocortical neurons is distinguished by somatodendritic distribution of neurofilament protein. *Brain Res., 539,* 133–136.

Campbell MJ, Lewis DA, Benoit R, Morrison JH (1987): Regional heterogeneity in the distribution of somatostatin-28- and somatostatin-28_{1-12}-immunoreactive profiles in the monkey neocortex. *J. Neurosci., 7,* 1133–1144.

Campbell MJ, Morrison JH (1989): Monoclonal antibody to neurofilament protein (SMI-32) labels a subpopulation of pyramidal neurons in the human and monkey neocortex. *J. Comp. Neurol., 282,* 191–205.

Cao QL, Yan XX, Luo XG, Garey LJ (1996): Prenatal development of parvalbumin immunoreactivity in the human striate cortex. *Cereb. Cortex, 6,* 620–630.

Carder RK (1997): Immunocytochemical characterization of AMPA-selective glutamate receptor subunits: laminar and compartmental distribution in macaque striate cortex. *J. Neurosci., 17,* 3352–3363.

Carder RK, Hendry SHC (1994): Neuronal characterization, compartmental distribution, and activity-dependent regulation of glutamate immunoreactivity in adult monkey striate cortex. *J. Neurosci., 14,* 242–262.

Carder RK, Leclerc SS, Hendry SHC (1996): Regulation of calcium-binding protein immunoreactivity in GABA neurons of macaque primary visual cortex. *Cereb. Cortex*, 6, 271–287.

Carmichael ST, Price JL (1994): Architectonic subdivision of the orbital and medial prefrontal cortex in the macaque monkey. *J. Comp. Neurol., 346,* 366–402.

Casagrande VA (1994): A third parallel visual pathway to primate area V1. *Trends Neurosci., 17,* 305–310.

Casagrande VA, Kaas JH (1994): The afferent, intrinsic, and efferent connections of primary visual cortex in primates. In: A Peters, KS Rockland (Eds.): *Cerebral Cortex, Vol. 10, Primary Visual Cortex in Primates*, pp. 201–259. Plenum Press, New York.

Celio MR, Scharer L, Morrison JH, Norman AW, Bloom FE (1986): Calbindin immunoreactivity alternates with cytochrome c-oxidase-rich zones in some layers of the primate visual cortex. *Nature, 323,* 715–717.

Chaudhuri A, Cynader MS (1993): Activity-dependent expression of the transcription factor Zif268 reveals ocular dominace columns in monkey visual cortex. *Brain Res., 605,* 349–353.

Chaudhuri A, Zangenehpour S, Matsubara JA, Cynader MS (1996): Differential expression of neurofilament protein in the visual system of the vervet monkey. *Brain Res., 709,* 17–26.

Chaudhuri A, Nissanov J, Larocque S, Rioux L (1997): Dual activity maps in primate visual cortex produced by different temporal patterns of *zif268* mRNA and protein expression. *Proc. Natl. Acad. Sci USA, 94,* 2671–2675.

Chino YM, Kaas JH, Smith III EL, Langston AL, Cheng H (1992): Rapid reorganization of cortical maps in adult cats following restricted deafferentation in retina. *Vision Res., 32,* 789–796.

Clarke S (1993): Callosal connections and functional subdivision of the human occipital cortex. In: B Gulyás, D Ottoson, PE Roland (Eds.): *Functional Organisation of the Human Visual Cortex*, pp. 137–150. Pergamon Press, Oxford.

Clarke S (1994): Modular organization of human extrastriate visual cortex: evidence from cytochrome oxidase pattern in normal and macular degeneration cases. *Eur. J. Neurosci., 6,* 725–736.

Clarke S (1995): Association and intrinsic connections of human extrastriate visual cortex. *Proc. R. Soc. Lond. Ser. B [Biol. Sci.], 257,* 87–92.

Clarke S, Miklossy J (1990): Occipital cortex in man: organization of callosal connections, related myelo- and cytoarchitecture, and putative boundaries of functional visual areas. *J. Comp. Neurol., 298,* 188–214.

Cogan DG (1985): Visual disturbances with focal progressive dementing disease. *Am. J. Ophthalmol., 100,* 68–72.

Coggeshall RE, Lekan HA (1996): Methods for determining numbers of cells and synapses: a case for more uniform standards of review. *J. Comp. Neurol., 364,* 6–15.

Colby CL, Duhamel JR, Goldberg ME (1993a): The analysis of visual space by the lateral intraparietal area of the monkey: the role of extraretinal signals. *Prog. Brain Res., 95,* 307–316.

Colby CL, Duhamel JR, Goldberg ME (1993b): Ventral intraparietal area of the macaque: anatomic location and visual response properties. *J. Neurophysiol., 69,* 902–914.

Colby CL, Gattass R, Olson CR, Gross CG (1988): Topographical organization of cortical afferents to extrastriate visual area PO in the macaque: a dual tracer study. *J. Comp. Neurol., 269,* 392–413.

Condé F., Lund JS, Lewis DA (1996): The hierarchical development of monkey visual cortical regions as revealed by the maturation of parvalbumin-immunoreactive neurons. *Dev. Brain Res., 96,* 261–276.

Conti F, Rustioni A, Petrusz P, Towle AC (1987): Glutamate-positive neurons in the somatic sensory cortex of rats and monkeys. *J. Neurosci., 7,* 1887–1901.

Conti F, Fabri M, Manzoni T (1988): Glutamate-positive corticocortical neurons in the somatic sensory areas I and II of cats. *J. Neurosci., 8,* 2948–2960.

Conti F, DeFelipe J, Fariñas I, Manzoni T (1989): Glutamate-positive neurons and axon terminals in cat sensory cortex: a correlative light and electron microscopic study. *J. Comp. Neurol., 290,* 141–153.
Crawford MLJ, Marc RE (1976): Light transmission of cat and monkey eyelids. *Vision Res., 16,* 323–324.
Creutzfeldt OD, Ito M (1968): Functional synaptic organisation of primary visual cortex neurones in the cat. *Exp. Brain Res., 6,* 324–352.
Cronin-Golomb A, Sugiura R, Corkin S, Growdon JH (1993): Incomplete achromatopsia in Alzheimer's disease. *Neurobiol. Aging, 14,* 471–477.
Curcio CA, Drucker DN (1993): Retinal ganglion cells in Alzheimer's disease and aging. *Ann. Neurol., 33,* 248–257.
Cusick GC, Seltzer B, Cola M, Griggs E (1995): Chemoarchitectonics and corticocortical terminations within the superior temporal sulcus of the rhesus monkey: evidence for subdivisions of superior temporal polysensory cortex. *J. Comp. Neurol., 360,* 515–535.
Dahlström A, Fuxe K (1964): Evidence for the existence of monoamine-containing neurons in the central nervous system. I. Demonstration of monoamines in the cell bodies of brain stem neurons. *Acta Physiol. Scand., 62, (Suppl. 232),* 1–55.
Darian-Smith C, Gilbert CD (1995): Topographic reorganization in the striate cortex of the adult cat and monkey is cortically mediated. *J. Neurosci., 15,* 1631–1647.
Das A, Gilbert CD (1995): Long-range horizontal connections and their role in cortical reorganization revealed by optical recording of cat primary visual cortex. *Nature, 375,* 780–784.
Dawson TH, Gehlert DR, McCabe RT, Barnett A, Wamsley JK (1986): D-1 dopamine receptors in the rat brain: a quantitative autoradiographic analysis. *J. Neurosci., 6,* 2352–2365.
Dawson TM, Snyder SH (1994): Gases as biological messengers: nitric oxide and carbon monoxide in the brain. *J. Neurosci., 14,* 5147–5159.
De Lacoste MC, White III CL (1993): The role of cortical connectivity in Alzheimer's disease pathogenesis: a review and model system. *Neurobiol. Aging, 14,* 1–16.
de Lima AD, Bloom FE, Morrison JH (1988): Synaptic organization of serotonin-immunoreactive fibers in primary visual cortex of macaque monkey. *J. Comp. Neurol., 274,* 280–294.
de Lima AD, Morrison JH (1989): Ultrastructural analysis of somatostatin-immunoreactive neurons and synapses in the temporal and occipital cortex of the macaque monkey. *J. Comp. Neurol., 283,* 212–227.
de Lima AD, Voigt T, Morrison JH (1990): Morphology of the cells within the inferior temporal gyrus that project to the prefrontal cortex in the macaque monkey. *J. Comp. Neurol., 296,* 159–272.
DeCamilli P, Miller PE, Navone F, Theurkauf WE, Vallée RB (1984): Distribution of microtubule-associated protein 2 in the nervous system of the rat studied by immunofluorescence. *Neuroscience, 11,* 817–846.
DeFelipe, J. (1992): Neocortical neuronal diversity: chemical heterogeneity revealed by colocalization studies of classic neurotransmitters, neuropeptides, calcium-binding proteins, and cell surface molecules. *Cereb. Cortex 3,* 273–289.
DeFelipe J, Conti F, Van Eyck SL, Manzoni T (1988): Demonstration of glutamate-positive axon terminals forming asymmetric synapses in cat neocortex. *Brain Res., 455,* 162–165.
DeFelipe, J. and Fariñas, I. (1992) The pyramidal neuron of the cerebral cortex: morphological and chemical characterisitics of the synaptic inputs. *Prog. Neurobiol. 39,* 563–607.
DeFelipe J, Hendry SHC, Hashikawa T, Molinari M, Jones EG (1990): A microcolumnar structure of monkey cerebral cortex revealed by immunocytochemical studies of double bouquet cell axons. *Neuroscience, 37,* 655–673.
DeFelipe J, Hendry SHC, Jones EG (1986): A correlative electron microscopic study of basket cells and large GABAergic neurons in the monkey sensory-motor cortex. *Neuroscience, 17,* 991–1009.
DeFelipe J, Hendry SHC, Jones EG (1989a): Synapses of double bouquet cells in monkey cerebral cortex visualized by calbindin immunoreactivity. *Brain Res., 503,* 49–54.
DeFelipe J, Hendry SHC, Jones EG (1989b): Visualization of chandelier cell axons by parvalbumin immunoreactivity in monkey cerebral cortex. *Proc. Natl. Acad. Sci. USA, 86,* 2093–2097.
DeFelipe J, Hendry SHC, Jones EG, Schmechel D (1985): Variability in the terminations of GABAergic chandelier cell axons on initial segments of pyramidal cell axons in the monkey sensory-motor cortex. *J. Comp. Neurol., 231,* 364 384.
DeFelipe J, Jones EG (1985): Vertical organization of γ-aminobutyric acid-accumulating intrinsic neuronal systems in monkey cerebral cortex. *J. Neurosci., 5,* 3246–3260.
DeFelipe J, Jones EG (1992): High-resolution light and electron microscopic immunocytochemistry of colocalized GABA and calbindin D-28k in somata and double bouquet cell axons of monkey somatosensory cortex. *Eur. J. Neurosci., 4,* 46–60.
DeFelipe J, Sola RG, Marco P, del Río MR, Pulido P, Ramón y Cajal S (1993): Selective changes in the

microorganization of the human epileptogenic neocortex revealed by parvalbumin immunoreactivity. *Cereb. Cortex, 3,* 39–48.
Demeulemeester H, Vandesande F, Orban GA, Brandon C, Vanderhaeghen JJ (1988): Heterogeneity of GABAergic cells in cat visual cortex. *J. Neurosci., 8,* 988–1000.
Desimone R, Ungerleider LG (1986): Multiple visual areas in the caudal superior temporal sulcus of the macaque. *J. Comp. Neurol., 248,* 164–189.
Desimone R, Ungerleider LG (1989): Neural mechanisms of visual processin in monkeys. In: F Boller, J Grafman (Eds.): *Handbook of Neuropsychology, Vol. 2*, pp. 267–299. Elsevier, Amsterdam.
DeYoe EA, Carman GJ, Bandettini P, Glickman S, Wieser J, Cox R, Miller D, Neitz J (1996): Mapping striate and extrastriate visual areas in human cerebral cortex. *Proc. Natl. Acad. Sci. USA, 93*, 2382–2386.
DeYoe EA, Felleman DJ, Van Essen DC, McClendon E (1994): Multiple processing streams in occipitotemporal visual cortex. *Nature, 371,* 151–154.
DeYoe EA, Hockfield S, Garren H, Van Essen DC (1990): Antibody labeling of functional subdivisions in visual cortex: Cat-301 immunoreactivity in striate and extrastriate cortex of the macaque monkey. *Vis. Neurosci., 5,* 67–81.
DeYoe EA, Van Essen DC (1988): Concurrent processing streams in monkey visual cortex. *Trends Neurosci., 11,* 219–226.
Dinopoulos A, Dori I, Parnavelas JG (1991): Immunohistochemical localization of aspartatein corticofugal pathways. *Neurosci. Lett., 121,* 25–28.
Distler C, Boussaoud D, Desimone R, Ungerleider LG (1993): Cortical connections of inferior temporal area TEO in macaque monkeys. *J. Comp. Neurol., 334,* 125–150.
Donoghue JP, Wenthold RJ, Altschuler RA (1985): Localization of glutaminase-like and aspartate aminotransferase-like immunoreactivity in neurons of cerebral neocortex. *J. Neurosci., 10*, 2597–2608.
Dori I, Dinopoulos A, Cavanagh ME, Parnavelas JG (1992): Proportion of glutamate- and aspartate-immunoreactive neurons in the efferent pathways of the rat visual cortex varies according to the target. *J. Comp. Neurol., 319,* 191–204.
Dori I, Petrou M, Parnavelas JG (1989): Excitatory transmitter amino acid-containing neurons in the rat visual cortex: a light and electron microscopic immunocytochemical study. *J. Comp. Neurol., 290,* 169–184.
Engel SA, Glover GH, Wandell BA (1997): Retinotopic organization in human visual cortex and the spatial precision of functional MRI. *Cereb. Cortex, 7,* 181–192.
Fairén A, Valverde F (1980): A specialized type of neuron in the visual cortex of the cat: a Golgi and electron microscope study of chandelier cells. *J. Comp. Neurol. 194,* 761–779.
Fariñas I, DeFelipe J (1991a): Patterns of synaptic input on corticocortical and corticothalamic cells in the cat visual cortex. I. The cell body. *J. Comp. Neurol., 304,* 53–69.
Fariñas I, DeFelipe J (1991b): Patterns of synaptic input on corticocortical and corticothalamic cells in the cat visual cortex. II. The axon initial segment. *J. Comp. Neurol., 304,* 70–77.
Feldman ML (1984): Morphology of the neocortical pyramidal neuron. In: EG Jones, A Peters (Eds.): *The Cerebral Cortex, Vol. 5,* Sensory-motor Areas and Aspects of Cortical Connectivity, pp. 123–200. Plenum, New York.
Felleman DJ, Van Essen DC (1991): Distributed hierarchical processing in the primate cerebral cortex. *Cereb. Cortex, 1*, 1–47.
Felten DL, Sladek Jr. JR (1983): Monoamine distribution in primate brain V. Monoamine nuclei: anatomy, pathways and local organization. *Brain Res. Bull.,* 171–284.
Ferster D (1988): Spatially opponent excitation and inhibition in simple cells of the cat visual cortex. *J. Neurosci., 8,* 1172–1180.
Fibiger HC (1982): The organization and some projections of cholinergic neurons of the mammalian forebrain. *Brain Res. Rev., 4,* 327–388.
Fifkova E, Morales M (1989): Calcium-regulated contractile and cytoskeletal proteins in dendritic spines may control synaptic plasticity. *Ann. NY Acad. Sci., 568,* 131–137.
Fitzpatrick D, Itoh K, Diamond IT (1983): The laminar organization of the lateral geniculate body and the striate cortex in the squirrel monkey (*Samiri sciureus*). *J. Neurosci., 3,* 673–702.
Fitzpatrick D, Lund JS, Blasdel GG (1985): Intrinsic connections of macaque striate cortex: afferent and efferent connections of lamina 4C. *J. Neurosci., 5,* 3329–3349.
Fitzpatrick D, Lund JS, Schmechel DE, Towles AC (1987): Distribution of GABAergic neurons and axon terminals in in the macaque striate cortex. *J. Comp. Neurol.,* 264, 73–91.
Flechsig P (1920): *Anatomie des menschlichen Gehirns und Rückenmarks.* Thieme, Leipzig.
Fletcher WA (1994): Ophthalmological aspects of Alzheimer's disease. *Curr. Opin. Ophthalmol., 5,* 38–44.
Fletcher WA, Sharpe JA (1988): Smooth pursuit dysfunction in Alzheimer's disease. *Neurology, 38*, 272–277.
Fonseca M, Soriano E, Ferrer I, Martinez A, Tuñon T (1993): Chandelier cell axons identified by parval-

bumin-immunoreactivity in the normal human temporal cortex and in Alzheimer's disease. *Neuroscience, 55,* 1107–1116.
Foote SL (1997): The primate locus coeruleus: the chemical neuroanatomy of the nucleus, its efferent projections, and its target receptors. In: FE Bloom, A Björklund, T Hökfelt (Eds.): *Handbook of Chemical Neuroanatomy, Vol. 13, The Primate Nervous System Part I,* pp. 185–213. Elsevier, Amsterdam.
Foote SL, Morrison JH (1984): Postnatal development of laminar innervation patterns by monoaminergic fibers in monkey (*Macaca fascicularis*) primary visual cortex. *J. Neurosci., 4,* 2667–2680.
Foote SL, Morrison JH (1987): Extrathalamic modulation of cortical function. *Annu. Rev. Neurosci., 10,* 67–95.
Fox K, Sato H, Daw N (1989): The location and function of NMDA receptors in cat and kitten visual cortex. *J. Neurosci., 9,* 2443–2454.
Frégnec Y, Shulz D, Thorpe S, Bienenstock E (1988): A cellular analogue of visual cortical plasticity. *Nature, 333,* 367–370.
Frégnec Y, Shulz D, Thorpe S, Bienenstock E (1992): Cellular analogs of visual cortical epigenesis. I. Plasticity of orientation selectivity. *J. Neurosci., 12,* 1280–1300.
Freund TF, Martin KA Somogyi P, Whitteridge D (1985): Innervation of cat visual areas 17 and 18 by physiologically identified X- and Y- type thalamic afferents. II. Identification of postsynaptic targets by GABA immunocytochemistry and Golgi impregnation. *J. Comp. Neurol., 242,* 275–291.
Freund TF, Martin KAC, Soltesz I, Somogyi P, Whitteridge D (1989): Arborisation pattern and postsynaptic targets of physiologically identified thalamocortical afferents in striate cortex of the Macaque monkey. *J. Comp. Neuol., 289,* 315–336.
Freund TF, Martin KAC, Smith AD, Somogyi P (1983): Glutamate decarboxylase-immunoreactive terminals of Golgi-impregnated axoaxonic cells and of presumed basket cells in synaptic contact with pyramidal neurons of the cat's visual cortex. *J. Comp. Neurol., 221,* 263–278.
Freund TG, Maglóczky ZS, Soltész I, Somogyi P (1986): Synaptic connections, axonal and dendritic patterns of neurons immunoreactive for cholecystokinin in the visual cortex of the cat. *Neuroscience, 19,* 1133–1159.
Funahashi S, Bruce CJ, Goldman-Rakic PS (1990): Visuospatial coding in primate prefrontal neurons revealed by oculomotor paradigm. *J. Neurophysiol., 63,* 814–831.
Funahashi S, Bruce CJ, Goldman-Rakic PS (1991): Neuronal activity related to saccadic eye movements in the monkey's dorsolateral prefrontal cortex. *J. Neurophysiol., 65,* 1464–1483.
Funahashi S, Bruce CJ, Goldman-Rakic PS (1993): Dorsolateral prefrontal lesions and oculomotor delayed-response performance: evidence for mnemonic 'scotomas'. *J. Neurosci., 13,* 1479–1497.
Furey-Kurkjian ML, Pietrini P, Graff-Radford NR, Alexander GE, Freo U, Szczepanik J, Schapiro MB (1996): Visual variants of Alzheimer disease: distinctive neuropsychological features. *Neuropsychology, 10,* 294–300.
Fuxe K (1965): Evidence for the existence of monoamine neurons in the central nervous system. IV. Distribution of monoamine nerve terminals in the central nervous system. *Acta Physiol. Scand., 64, (Suppl. 247),* 39–85.
Fuxe K, Hökfelt T, Johansson E, Jonsson G, Lidbrink P, Ljungdahl A (1974): The origin of the dopamine nerve terminals in limbic and frontal cortex. Evidence for meso-cortical dopamine neurons. *Brain Res., 82,* 349–355.
Galletti C, Fattori P, Battaglini PP, Shipp S, Zeki S (1996): Functional demarcation of a border between areas V6 and V6A in the superior parietal gyrus of the macaque monkey. *Eur. J. Neurosci., 8,* 30–52.
Gattass R, Adams MM, Hof PR, Ungerleider LG (1996): Parcellation of visual cortical areas in the chimpanzee using neurofilament and calcium-binding proteins. *Soc. Neurosci. Abstr., 22,* 1059.
Gattass R, Gross CG, Sandell JH (1981): Visual topography on V2 in the macaque. *J. Comp. Neurol., 201,* 519–539.
Gattass R, Sousa ABP, Gross CG (1985): Cortical visual areas of the macaque: possible substrates for pattern recognition mechanisms. In: C Chagas, R Gattass, CG Gross (Eds.): *Pattern Recognition Mechanisms,* pp. 1–20. Pontifical Academy of Sciences, Vatican City.
Gattass R, Sousa APB, Gross (1988): Visuotopic organization and extent of V3 and V4 of the macaque. *J. Neurosci., 8,* 1831–1845.
Gatter KC, Powell TPS (1977): The projection of the locus coeruleus upon the neocortex in the macaque monkey. *Neuroscience, 2,* 441–445.
Gebhard R, Zilles K, Schleicher A, Everitt BJ, Robbins TW, Divac I (1993): Distribution of seven major neurotransmitter receptors in the striate cortex of the New World monkey *Callithrix jacchus. Neuroscience, 56,* 877–885.
Geiger JRP, Melcher T, Koh DS, Sakmann B, Seeburg PH, Jonas P, Monyer H (1995): Relative abundance

of subunit mRNAs determines gating and Ca^{2+} permeability of AMPA receptors in principal neurons and interneurons in rat CNS. *Neuron ,15,* 193–204.

Gilbert CD (1992): Horizontal integration and cortical dynamics. *Neuron, 9,* 1–13.

Gilbert CD, Wiesel TN (1989): Columnar specificity of intrinsic horizontal and corticocortical connections in cat visual cortex. *J. Neurosci., 9,* 2432–2442.

Gilbert CD, Wiesel TN (1990): The influence of contextual stimuli on the orientation selectivity of cells in primary visual cortex of the cat. *Vision Res., 30,* 1689–1701.

Gilbert CD, Wiesel TN (1992): Receptive field dynamics in adult primary visual cortex. *Nature, 356,* 150–152.

Giuffrida R, Rustioni A (1989): Glutamate and aspartate immunoreactivity in corticospinal neurons of rat. *J. Comp. Neurol., 288,* 154–164.

Gottlieb JP, Bruce CJ, MacAvoy MG (1993): Smooth eye movements elicited by microstimulation in the primate frontal eye field. *J. Neurophysiol., 69,* 786–799.

Graff-Radford NR, Bolling JP, Earnest IV F, Shuster EA, Caselli RJ, Brazis PW (1993): Simultanagnosia as the initial sign of degenerative dementia. *Mayo Clin. Proc., 68,* 955–964.

Gray EG (1969): Electron microscopy of excitatory and inhibitory synapses: a brief review. *Prog. Brain Res.,* 31, 141–155.

Graybiel AM, Ragsdale Jr. CJ (1982): Pseudocholinesterase staining in the primary visual pathway of the macaque monkey. *Nature, 299,* 439–442.

Grünthal E (1928): Zur Hirnpathologischen Analyse der Alzheimerschen Krankheit. *Psychiat.-Neurol. Wschr., 36,* 401–407.

Gutierrez C, Cusick CG (1994): Effects of chronic monocular enucleation on calcium binding proteins calbindin-D_{28k} and parvalbumin in the lateral geniculate nucleus of adult rhesus monkeys. *Brain Res., 651,* 300–310.

Haley JE, Wilcox GL, Chapman PF (1992): The role of nitric oxide in hippocampal long-term potentiation. *Neuron, 8,* 211–216.

Hayes TL, Lewis DA (1992): Nonphosphorylated neurofilament protein and calbindin immunoreactivity in layer III pyramidal neurons of human neocortex. *Cereb. Cortex, 2,* 56–67.

Heinen SJ, Skavenski AA (1991): Recovery of visual responses in foveal V1 neurons following bilateral foveal lesions in adult monkey. *Exp. Brain Res., 83,* 670–674.

Heizmann CW (1984): Parvalbumin, an intracellular calcium-binding protein: distribution, properties an possible roles in mammalian cells. *Experientia, 40,* 910–920.

Hendrickson AE, Hunt SP, Wu JY (1981): Immunocytochemical localization of glutamic acid decarboxylase in monkey striate cortex. *Nature, 292,* 605–607.

Hendrickson AE, Tigges M (1985): Enucleation demonstrates ocular dominance columns in Old World macaque but not in New World squirrel monkey visual cortex. *Brain Res., 333,* 340–344.

Hendrickson AE, van Brederode JFM, Mulligan KA, Celio MR (1991): Development of the calcium-binding proteins parvalbumin and calbindin in monkey striate cortex. *J. Comp. Neurol., 307,* 626–646.

Hendrickson AE, Wilson JR, Ogren MP (1978): The neuroanatomical organization of pathways between dorsal lateral geniculate nucleus and visual cortex in old and new world primates. *J. Comp. Neurol., 182,* 123–136.

Hendry SHC (1993): A neurochemically distinct third channel in the macaque lateral geniculate nucleus. In: D Minchiacchi, M Molinari, G Macchi, EG Jones (Eds.): *Thalamic Networks for Relay and Modulation,* pp. 251–265. Pergamon Press, Oxford.

Hendry SHC, Carder RK (1992): Organization and plasticity of GABA neurons and receptors in monkey visual cortex. *Prog. Brain Res., 90,* 477–502.

Hendry SHC, Bhandari MA (1992): Neuronal organization and plasticity in adult monkey visual cortex: immunoreactivity for microtubule-associated protein 2. *Vis. Neurosci., 9,* 445–459.

Hendry SHC, Carder RK (1993): Neurochemical compartmentation of monkey and human visual cortex: similarities and variations in calbindin immunoreactivity across species. *Vis. Neurosci., 10,* 1109–1120.

Hendry SHC, Fuchs J, de Blas AL, Jones EG (1990): Distribution and plasticity of immunocytochemically localized $GABA_A$ receptors in adult monkey visual cortex. *J. Neurosci., 10,* 2438–2450.

Hendry SHC, Hockfield S, Jones EG, McKay R (1984c): Monoclonal antibody that identifies subsets of neurones in the central visual system of monkey and cat. *Nature, 307,* 267–269.

Hendry SHC, Houser CR, Jones EG, Vaughn JE (1983a): Synaptic organization of immunocytochemically identified GABA neurons in the monkey sensory-motor cortex. *J. Neurocytol., 12,* 639–660.

Hendry SHC, Huntsman MM, Viñuela A, Möhler H, de Blas AL, Jones EG (1994): GABA-A receptor subunit immunoreactivity in primate visual cortex: distribution in macaques and humans and regulation by visual input in adulthood. *J. Neurosci., 14,* 2383–2401.

Hendry SHC, Jones EG (1981): Sizes and distribution of intrinsic neurons incorporating tritiated GABA in monkey sensory-motor cortex. *J. Neurosci., 1,* 390–405.
Hendry SHC, Jones EG (1986): Reduction in number of GABA immunostained neurons in deprived-eye dominance columns of monkey area 17. *Nature, 320,* 750–753.
Hendry SHC, Jones EG (1988a): Activity dependent regulation of GABA expression in the visual cortex of adult monkeys. *Neuron, 1,* 701–712.
Hendry SHC, Jones EG (1988b): Synaptic organization of GABA and GABA/tachykinin immunoreactive neurons in layer IVCβ of monkey area 17. *Soc. Neurosci. Abstr., 14,* 1123.
Hendry SHC, Jones EG, Beinfeld MC (1983b): Cholecystokinin-immunoreactive neurons in rat and monkey cerebral cortex make symmetric synapses and have intimate associations with blood vessels. *Proc. Natl. Acad. Sci. USA, 80,* 2400–2404.
Hendry SHC, Jones EG, Burstein N (1988a): Activity-dependent regulation of tachykinin-like immunoreactivity in neurons of the monkey primary visual cortex. *J. Neurosci., 8,* 1225–1238.
Hendry SHC, Jones EG, DeFelipe J, Schmechel D, Brandon C, Emson PC (1984b): Neuropeptide-containing neurons of the cerebral cortex are also GABAergic. *Proc. Natl. Acad. Sci. USA, 81,* 6526–6530.
Hendry SHC, Jones EG, Emson PC (1984a): Morphology, distribution, and synaptic relations of somatostatin- and neuropeptide Y-immunoreactive neurons in rat and monkey neocortex. *J. Neurosci., 4,* 2497–2517.
Hendry SHC, Jones EG, Emson PC, Lawson DEM, Heizmann CW, Streit P (1989): Two classes of cortical GABA neurons defined by differential calcium binding protein immunoreactivities. *Exp. Brain Res., 76,* 467–472.
Hendry SHC, Jones EG, Hockfield S, McKay RDG (1988b): Neuronal populations stained with the monoclonal antibody Cat-301 in the mammalian cerebral cortex and thalamus. *J. Neurosci., 8,* 518–542.
Hendry SHC, Kennedy MB (1986): Immunoreactivity for a calmodulin-dependent protein kinase is selectively increased in macaque striate cortex after monocular deprivation. *Proc. Natl. Acad. Sci. USA, 83,* 1536–1540.
Hendry SHC, Schwark HD, Jones EG, Yan J (1987): Numbers and proportions of GABA-immunoreactive neurons in different areas of monkey cerebral cortex. *J. Neurosci., 7,* 1503–1519.
Hendry SHC, Yoshioka T (1994): A neurochemically distinct third channel in the macaque dorsal lateral geniculate nucleus. *Science, 264,* 575–577.
Hess DT, Edwards MA (1987): Anatomical demonstration of ocular segregation in the retinogeniculocortical pathway of the New World capuchin monkey (*Cebus apella*). *J. Comp. Neurol., 264,* 409–420.
Hinton DR, Sadun AA, Blanks JC, Miller CA (1986): Optic-nerve degeneration in Alzheimer's disease. *New Engl. J. Med., 315,* 485–487.
Hockfield S, McKay RD, Hendry SHC, Jones EG (1984): A surface antigen that identifies ocular dominance columns in the visual cortex and laminar features of the lateral geniculate nucleus. *Cold Spring Harbor Symp. Quant. Biol., 35,* 877–889.
Hockfield S, Tootell RBH, Zaremba S (1990): Molecular differences among neurons reveal an organization of human visual cortex. *Proc. Natl. Acad. Sci. USA, 87,* 3027–3031.
Hof PR, Archin N, Osmand AP, Dougherty JH, Wells C, Bouras C, Morrison JH (1993): Posterior cortical atrophy in Alzheimer's disease: analysis of a new case and re-evaluation of an historical report. *Acta Neuropathol., 86,* 215–223.
Hof PR, Bouras C (1991): Object recognition deficit in Alzheimer's disease: possible disconnection of the occipito-temporal component of the visual system. *Neurosci. Lett., 122,* 53–56.
Hof PR, Bouras C, Constantinidis J, Morrison JH (1989): Balint's syndrome in Alzheimer's disease: specific disruption of the occipito-parietal visual pathway. *Brain Res., 493,* 368–375.
Hof PR, Bouras C, Constantinidis J, Morrison JH (1990a): Selective disconnection of specific association pathways in cases of Alzheimer's disease presenting with Balint's syndrome. *J. Neuropathol. Exp. Neurol., 49,* 168–184.
Hof PR, Cox K, Morrison JH (1990b): Quantitative analysis of a vulnerable subset of pyramidal neurons in Alzheimer's disease: I. Superior and inferior temporal cortex. *J. Comp. Neurol., 301,* 44–54.
Hof PR, Morrison JH (1990): Quantitative analysis of a vulnerable subset of pyramidal neurons in Alzheimer's disease: II. Primary and secondary visual cortex. *J. Comp. Neurol., 301,* 55–64.
Hof PR, Morrison JH (1994): The cellular basis of cortical disconnection in Alzheimer disease and related dementing conditions. In: RD Terry, R Katzman, KL Bick (Eds.), *Alzheimer Disease,* pp. 197–229. Raven Press, New York.
Hof PR, Morrison JH (1995): Neurofilament protein defines regional patterns of cortical organization in the macaque monkey visual system: a quantitative immunohistochemical analysis. *J. Comp. Neurol., 352,* 161–186.

Hof PR, Mufson EJ, Morrison JH (1995b): The human orbitofrontal cortex: cytoarchitecture and quantitative immunohistochemical parcellation. *J. Comp. Neurol., 359,* 48–68.
Hof PR, Nimchinsky EA (1992): Regional distribution of neurofilament and calcium-binding proteins in the cingulate cortex of the macaque monkey. *Cereb. Cortex, 2,* 456–467.
Hof PR, Nimchinsky EA, Morrison JH (1995a): Neurochemical phenotype of corticocortical connections in the macaque monkey: quantitative analysis of a subset of neurofilament protein-immunoreactive projection neurons in frontal, parietal, temporal, and cingulate cortices. *J. Comp. Neurol., 362,* 109–133.
Hof PR, Sailstad CA, Archin N, Edwards AM, Janssen WGM, Morrison JH (1994): Distribution of the calcium-binding proteins parvalbumin, calbindin and calretinin in the macaque monkey visual cortex. *Soc. Neurosci. Abstr., 20,* 1740.
Hof PR, Ungerleider LG, Webster MJ, Gattass R, Adams MM, Sailstad CA, Morrison JH (1996): Neurofilament protein is differentially distributed in subpopulations of corticocortical projection neurons in the macaque monkey visual pathways. *J. Comp. Neurol., 376,* 112–127.
Hof PR, Vogt BA, Bouras C, Morrison JH (1997a): Atypical form of Alzheimer's disease with prominent posterior cortical atrophy: a review of lesion distribution and circuit disconnection in cortical visual pathways. *Vision Res.*, 37, 3609–3625.
Hof PR, Ungerleider LG, Adams MM, Webster MJ, Gattass R, Blumberg DM, Morrison JH (1997b): Callosally-projecting neurons in the macaque monkey V1/V2 border are enriched in nonphosphorylated neurofilament protein. *Vis. Neurosci.*, 14, 981–987.
Hoffman PN, Cleveland DW, Griffin JW, Landes PW, Cowan NJ, Price DL (1987): Neurofilament gene expression: a major determinant of axonal caliber. *Proc. Natl. Acad. Sci. USA, 84,* 3472–3476.
Hökfelt T, Lungdahl A, Fuxe K, Johansson O (1974): Dopamine nerve terminals in the rat limbic cortex: aspects of the dopamine hypothesis of schizophrenia. *Science, 184,* 177–179.
Hollmann M, Hartley M, Heinemann SF (1991): Calcium permeability of KA-AMPA-gated glutamate receptor channels depends on subunit composition. *Science, 25,* 851–853.
Hollmann M, Heinemann S (1994): Cloned glutamate receptors. *Annu. Rev. Neurosci., 17,* 31–108.
Hornung JP, Fritschy JM (1996): Developmental profile of $GABA_A$-receptors in the marmoset monkey: expression of distinct subtypes in pre- and postnatal brain. *J. Comp. Neurol., 367,* 413–430.
Horton JC (1984): Cytochrome oxidase patches: a new cytoarchitectonic feature of monkey visual cortex. *Phil. Trans. R. Soc. Lond. Ser. B [Biol. Sci.], 304,* 199–253.
Horton JC, Hedley-Whyte ET (1984): Mapping of cytochrome oxidase patches and ocular dominance columns in human visual cortex. *Phil. Trans. R. Soc. Lond. Ser. B [Biol. Sci.], 304,* 255–272.
Horton JC, Hubel DH (1981): Regular patchy distributions of cytochrome oxidase staining in primary visual cortex of macaque monkey. *Nature, 292,* 762–763.
Houser CR, Crawford GD, Salvaterra PM, Vaughn JE (1983a): Organization and morphological characteristics of cholinergic neurons: an immunocytochemical study with a monoclonal antibody to choline acetyltransferase. *Brain Res., 266,* 97–119.
Houser CR, Hendry SHC, Jones EG, Vaughn JE (1983b): Morphological diversity of GABA neurons demonstrated immunocytochemically in monkey sensory-motor cortex. *J. Neurocytol., 12,* 617–638.
Houser CR, Vaughn JE, Hendry SHC, Jones EG and Peters A (1984): GABA neurons in the cerebral cortex. In: EG Jones, A Peters (Eds.): *Cerebral Cortex, Vol. 2, Functional Properties of Cortical Cells,* pp. 63–89. Plenum Press, New York.
Hubel DH, Wiesel TN (1972): Laminar and columnar distribution of geniculo-cortical fibers in the macaque monkey. *J. Comp. Neurol., 146,* 421–450.
Hubel DH, Wiesel TN (1977): Functional architecture of macaque monkey visual cortex. *Proc. R. Soc. Lond. B [Biol. Ser.], 198,* 1–59.
Hubel DH, Wiesel TN, LeVay S (1977): Plasticity of ocular dominance columns in monkey striate cortex. *Phil. Trans. R. Soc. Lond. Ser. B [Biol. Sci.], 278,* 377–409.
Huerta MF, Kaas JH (1990): Supplementary eye field as defined by intracortical microstimulation: connections in macaques. *J. Comp. Neurol., 293,* 299–330.
Huerta MF, Krubitzer LA, Kaas JH (1987): Frontal eye field as defined by intracortical microstimulation in squirrel monkeys, owl monkeys, and macaque monkeys. II. Cortical connections. *J. Comp. Neurol., 265,* 332–361.
Humphrey AL, Hendrickson AE (1983): Background and stimulus-induced patterns of high metabolic activity in the visual cortex (area 17) of the squirrel and macaque monkey. *J. Neurosci., 3,* 345–358.
Huntley GW, Hendry SHC, Killackey HP, Chalupa LM, Jones EG (1988): Temporal sequence of neurotransmitter expression by developing neurons of fetal monkey visual cortex. *Dev. Brain Res., 43,* 69–96.
Huntley GW, Jones EG (1990): Cajal-Retzius neurons in developing monkey neocortex show immunoreactivity for calcium binding proteins. *J. Neurocytol., 19,* 200–212.
Huntley GW, Rogers SW, Moran T, Janssen W, Archin N, Vickers JC, Cauley K, Heinemann SF, Morrison

JH (1993): Selective distribution of kainate receptor subunit immunoreactivity in monkey neocortex revealed by a monoclonal antibody that recognizes glutamate receptor subunits GluR5/6/7. *J. Neurosci., 13,* 2965–2981.

Huntley GW, Vickers JC, Morrison JH (1994b): Cellular and synaptic localization of NMDA and non-NMDA receptor subunits in cerebral cortex: organizational features related to cortical circuitry, function and disease. *Trends Neurosci., 17,* 536–543.

Huntley GW, Vickers JC, Janssen W, Brose N, Heinemann SF, Morrison JH (1994a): Distribution and synaptic localization of immunocytochemically identified NMDA receptor subunit proteins in the sensory-motor cortex and visual cortices of monkey and human. *J. Neurosci., 14,* 3603–3619.

Huntsman MM, Isackson PJ, Jones EG (1994): Lamina-specific expression and activity-dependent regulation of seven GABA-A receptor subunit mRNAs in monkey visual cortex. *J. Neurosci., 14,* 2236–2259.

Hyman BT, Van Hoesen GW, Kromer LJ, Damasio AR (1986): Perforant pathway changes and the memory impairment of Alzheimer's disease. *Ann. Neurol., 20*, 472–481.

Innocenti GM, Fiore I (1974): Postsynaptic inhibitory components of the response to moving stimuli in area 17. *Brain Res., 80,* 229–289.

Ishai A, Sagi D (1995): Common mechanisms of visual imagery and perception. *Science, 268,* 1772–1774.

Itaya SK, Itaya PW, Van Hoesen GW (1984): Intracortical termination of the retino-geniculo-striate pathway studied with transsynaptic tracer (wheat germ agglutinin-horseradish peroxidase) and cytochrome oxidase staining in the macaque monkey. *Brain Res., 304,* 303–310.

Izumi Y, Clifford DB, Zorumski CF (1992): Inhibition of long-term potentiation by NMDA-mediated nitric oxide release. *Science 257,* 1273–1276.

Jacquet MF, Boucquey D, Theaux R, Smidts M, Callewaerts A, Brucher JM, Kevers L (1990): L'atrophie corticale postérieure: variante anatomo-clinique de la maladie d'Alzheimer. *Acta Neurol. Belg., 90,* 265–273.

Johnson JK, Casagrande VA (1995): Distribution of calcium-binding proetins within the parallel visual pathways of a primate (*Galago crassicaudatus*). *J. Comp. Neurol., 356*, 238–260.

Johnson RR, Burkhalter A (1992): Evidence for excitatory amino acid neurotransmitters in the geniculocortical pathway and local projections within rat primary visual cortex. *Exp. Brain Res., 89,* 20–30.

Jonas P, Racca C, Sakmann B, Seeburg PH, Monyer H (1994): Differences in calcium permeability of AMPA type glutamate receptor channels in neocortical neurons caused by differential GluR B subunit expression. *Neuron 12,* 1281–1289.

Jones EG (1975): Varieties and distribution of non-pyramidal cells in the somatic sensory cortex of the squirrel monkey. *J. Comp. Neurol., 160,* 205–268.

Jones EG (1981): Anatomy of cerebral cortex: columnar input-output organization. In: FO Schmitt, FG Worden, G Adelman, SG Dennis (Eds.): *The Organization of the Cerebral Cortex*, pp. 199–235. MIT Press, Cambridge, MA.

Jones EG (1984): Laminar distribution of cortical efferent cells. In: A Peters, EG Jones (Eds.): *Cerebral Cortex, Vol. 1, Cellular Components of the Cerebral Cortex*, pp. 521–553. Plenum Press, New York.

Jones EG (1990) Determinants of the cytoarchitecture of the cerebral cortex. In: GM Edelman, WE Gall, and WM Cowan (Eds.): *Signal and Sense*, pp. 3–49. Wiley-Liss, New York, NY.

Jones EG (1993): GABAergic neurons and their role in cortical plasticity in primates. *Cerebral Cortex, 3*, 361–372.

Jones EG, DeFelipe J, Hendry SHC, Maggio JE (1988): A study of tachykinin-immunoreactive neurons in monkey cerebral cortex. *J. Neurosci., 8,* 1206–1224.

Jones EG, Hendry SHC (1986): Co-localization of GABA and neuropeptides in neocortical neurons. *Trends Neurosci., 9,* 71–76.

Jones EG, Hendry SHC (1989): Differential calcium binding protein immunoreactivity distinguishes classes of relay neurons in monkey thalamic nuclei. *Eur. J. Neurosci., 1,* 222–246.

Jones EG, Hendry SHC, DeFelipe J, Benson DL (1994): GABA neurons and their role in activity-dependent plasticity of adult primate visual cortex. In: A Peters, KS Rockland (Eds.): *Cerebral Cortex, Vol. 10, Primary Visual Cortex in Primates*, pp. 61–140. Plenum Press, New York.

Jones EG, Powell TPS (1969): Synapses on the axon hillocks and initial segments of pyramidal cell axons in the cerebral cortex. *J. Cell Sci., 5,* 495–507.

Kaas, J.H., Krubitzer LA, Chino YM, Langston AL, Polley EH, Blair N (1990): Reorganization of retinotopic cortical maps in adult mammals after lesions of the retina. *Science, 248,* 229–231.

Kaczmarek I, Chaudhuri A (1997): Sensory regulation of immediate-early gene expression in mammalian visual cortex: Implications for functional mapping and neural plasticity. *Brain Res. Rev., 23*, 237–256.

Kalb RG (1995): Current excitement about the glutamate receptor family. *Neuroscientist, 1*, 60–63.

Kaplan IV, Guo Y, Mower GD (1996): Immediate early gene expression in cat visual cortex during and after the critical period: Differences between EGR-1 and Fos proteins. *Mol. Brain Res., 36*, 12–22.

Kato N (1994): Long-term depression requiring tACPD-receptor activation and NMDA-receptor blockade. *Brain Res., 665,* 158–160.

Katz B, Rimmer S (1989): Ophthalmologic manifestations of Alzheimer's disease. *Surv. Ophthalmol., 34,* 31–34.

Katz LC, Gilbert CD, Wiesel TN (1989): Local circuits and ocular dominance in monkey striate cortex. *J. Neurosci., 9,* 1389–1399.

Kharazia VN, Weinberg RJ (1994): Glutamate in thalamic fibers terminating in layer IV of primary sensory cortex. *J. Neurosci., 14,* 6021–6032.

Kisvárday ZF, Cowey A, Somogyi P (1986): Synaptic relationships of a type of GABA-immunoreactive neuron (clutch cell), spiny stellate cells and lateral geniculate nucleus afferents in layer IVC of the monkey striate cortex. *Neuroscience, 19,* 741–761.

Kisvárday ZF, Martin KAC, Friedlander MJ, Somogyi P (1987): Evidence for interlaminar inhibitory circuits in the striate cortex of the cat. *J. Comp. Neurol., 260,* 1–19.

Kisvárday ZF, Cowey A, Smith AD, Somogyi P (1989): Interlaminar and lateral excitatory amino acid connections in the striate cortex of monkey. *J. Neurosci., 9,* 667–682.

Kisvárday ZF, Gulyas A, Beroukas D, North JB, Chubb IW, Somogyi P (1990): Synapses, axonal and dendritic patterns of GABA-immunoreactive neurons in human cerebral cortex. *Brain, 113,* 793–812.

Kiyosawa M, Bosley TM, Chawluk J, Jamieson D, Schatz NJ, Savino PJ, Sergott RC, Reivich M, Alavi A (1989): Alzheimer's disease with prominent visual symptoms. *Ophthalmology, 96,* 1077–1086.

Kondo H, Hashikawa T, Tanaka K, Jones EG (1994): Neurochemical gradient along the monkey occipito-temporal cortical pathway. *NeuroReport, 5,* 613–616.

Kosofsky BE, Molliver ME (1987): The serotonergic innervation of cerebral cortex: different classes of axon terminals arise from dorsal and median raphe nuclei. *Synapse, 1,* 153–168.

Kosofsky BE, Molliver ME, Morrison JH, Foote SL (1984): The serotonin and norepinephrine innervation of primary visual cortex in the Cynomolgus monkey (*Macaca fascicularis*). *J. Comp. Neurol., 230,* 168–178.

Kritzer MF, Cowey A, Somogyi P (1992): Patterns of inter- and intralaminar GABAergic connections distinguish striate (V1) and extrastriate (V2, V4) visual cortices and their functionally specialized subdivisions in the rhesus monkey. *J. Neurosci., 12,* 4545–4564.

Krnjevic K (1974): Chemical nature of synaptic transmission in vertebrates. *Physiol. Rev., 54,* 418–540.

Krnjevic K (1987): GABAergic inhibition in the neocortex. *J. Mind Behav., 8,* 537–547.

Krnjevic K, Phillis JW (1963): Iontophoretic studies of neurons in the mammalian cerebral cortex. *J. Physiol. (Lond) 165,* 274–304.

Krubitzer L, Kaas J (1990a): Convergence of processing channels in the extrastriate cortex of monkeys. *Vis. Neurosci., 5,* 609–613.

Krubitzer LA, Kaas JH (1990b): Cortical connections of MT in four species of primates: areal, modular, and retinotopic patterns. *Vis. Neurosci., 5,* 165–204.

Kubota Y, Jones EG (1992): Three populations of cortical GABAergic intrinsic neurons in the rat. *Soc. Neurosci. Abstr., 18,* 1149.

Kuhar MJ, Aghajanian GK, Roth RH (1972): Tryptophan hydroxylase activity and synaptosomal uptake of serotonin in discrete brain regions after midbrain raphe lesions: correlation with serotonin levels and histochemical fluorescence. *Brain Res., 44,* 165–176.

Kuljis RO (1994): Lesions in the pulvinar in patients with Alzheimer's disease. *J. Neuropathol. Exp. Neurol., 53,* 202–211.

Kuljis RO, Rakic P (1989a): Distribution of neuropeptide Y-containing perikarya and axons in various neocortical areas in the macaque monkey. *J. Comp. Neurol., 280,* 383–292.

Kuljis RO, Rakic P (1989b): Neuropeptide Y-containing neurons are situated outside cytochrome oxidase puffs in macaque visual cortex. *Vis. Neurosci., 2,* 57–62.

Kurylo DD, Corkin S, Dolan RP, Rizzo III JF, Parker SW, Growdon JH (1994): Broad-band visual capacities are not selectively impaired in Alzheimer's disease. *Neurobiol. Aging, 15,* 305–311.

Kurylo DD, Corkin S, Rizzo III JF, Growdon JH (1996): Greater relative impairment of object recognition than visuospatial abilities in Alzheimer's disease. *Neuropsychology, 10,* 74–81.

Kuskowski MA, Malone SM, Mortimer JA, Dysken MW (1989): Smooth pursuit eye movements in dementia of the Alzheimer type. *Alzheimer Dis. Assoc. Disord., 3,* 157–171.

Lachica EA, Casagrande VA (1992): Direct W-like geniculate projections to the cytochrome-oxidase (CO) blobs in primate visual cortex: axon morphology. *J. Comp. Neurol., 319,* 1–18.

Lawson SN, Waddell JP (1991): Soma neurofilament immunoreactivity is related to cell size and fibre conduction velocity in rat primary sensory neurons. *J. Physiol. (Lond) 435,* 41–63.

Lemarie P, Relevant O, Bravo R, Charnay P (1988): Two mouse genes encoding potential trancription

factors with identical DNA-binding domains are activated by growth factors in cultured cells. *Proc. Natl. Acad. Sci. USA, 85,* 4691–4695.
Lennie P (1980): Parallel visual pathways: a review. *Vision Res., 20,* 561–594.
Leranth C, Szeidemann Z, Hsu M, Buszaki G (1996): AMPA receptors in the rat and primate hippocampus: a possible absence of Glur2/3 subunits in most interneurons. *Neuroscience, 70,* 631–652.
Leuba G, Kraftsik R (1994): Visual cortex in Alzheimer's disease: occurrence of neuronal death and glial proliferation, and correlation with pathological changes. *Neurobiol. Aging, 15,* 29–43.
Leuba G, Saini K (1995): Pathology of subcortical visual centres in relation to cortical degeneration in Alzheimer's disease. *Neuropathol. Appl. Neurobiol., 21,* 410–422.
LeVay S, Wiesel TN, Hubel DH (1980): The development of ocular dominance columns in normal and visually deprived monkeys. *J. Comp. Neurol., 191,* 1–51.
Levine DN, Lee JM, Fisher CM (1993): The visual variant of Alzheimer's disease: a clinicopathologic case study. *Neurology, 43,* 305–313.
Levitt P, Moore RY (1978a): Noradrenaline neuron innervation of the neocortex in the rat, in the cat. *Anat. Embryol., 154,* 241–251.
Levitt P, Rakic P, Goldman-Rakic P (1984): Region-specific distribution of catecholamine afferents in primate cerebral cortex: a fluorescence histochemical analysis. *J. Comp. Neurol., 227,* 23–36.
Levy MH, Hart Jr WM, Sonstein FM, Ballinger Jr WE (1995): The incredible *shrinking* brain [Comments by A Sadun]. *Surv. Ophthalmol., 39,* 315–322.
Lewis DA, Campbell MJ, Foote SL, Goldstein M, Morrison JH (1987a): The distribution of tyrosine hydroxylase-immunoreactive fibers in primate neocortex is widespread but regionally selective. *J. Neurosci., 7,* 279–290.
Lewis DA, Campbell MJ, Foote SL, Morrison JH (1986): The monoaminergic innervation of primate neocortex. *Hum. Neurobiol., 5,* 181–188.
Lewis DA, Campbell MJ, Terry RD, Morrison JH (1987b): Laminar and regional distribution of neurofibrillary tangles and neuritic plaques in Alzheimer's disease: a quantitative study of visual and auditory cortices. *J. Neurosci., 7,* 1799–1808.
Lewis DA, Morrison JH (1989): The noradrenergic innervation of monkey prefrontal cortex: A dopamine-β-hydroxylase immunohistochemical study. *J. Comp. Neurol., 282,* 317–330.
Lewis DA, Lund JS (1990): Heterogeneity of chandelier neurons in monkey neocortex: corticotropin-releasing factor- and parvalbumin-immunoreactive populations. *J. Comp. Neurol., 293,* 599–615.
Lidov HGW, Molliver ME (1982): Immunohistochemical study of the development of serotonergic neurons in the rat CNS. *Brain Res. Bull., 9,* 559–604.
Lidow MS (1995): D_1- and D_2 dopaminergic receptors in the developing cerebral cortex of macaque monkey: a film autoradiographic study. *Neuroscience, 65,* 439–452.
Lidow MS, Goldman-Rakic PS, Gallager DW, Rakic P (1991): Distribution of dopaminergic receptors in the primate cerebral cortex: quantitative autoradiographic analysis using [^{3}H]raclopride, [^{3}H]spiperone and [^{3}H]SCH23390. *Neuroscience, 40,* 657–671.
Lindvall O, Björklund A (1974): The organization of the ascending catecholamine neuron sysems in the rat brain as revealed by the glyoxylic acid fluorescence method. *Acta Physiol. Scan. Suppl., 412,* 1–48.
Lindvall O, Björklund A (1978): Organization of catecholamine neurons in the rat central nervous system. In: LL Iversen, SD Iversen, SH Snyder (Eds.): *Handbook of Psychopharmacology, Vol. 9,* pp. 139–225. Plenum Press, New York.
Livingstone M, Hubel D (1988): Segregation of form, color, movement, and depth: anatomy, physiology, and perception. *Science, 240,* 740–749.
Livingstone MS, Hubel DH (1982): Thalamic inputs to cytochrome oxidase-rich regions in monkey visual cortex. *Proc. Natl. Acad. Sci. USA 79,* 6098–6101.
Livingstone MS, Hubel DH (1984): Anatomy and physiology of a color system in the primate visual cortex. *J. Neurosci., 4,* 309–356.
Loughlin SE, Foote SL, Fallon JH (1982): Locus coeruleus projections to cortex: topography, morphology and collateralization. *Brain Res. Bull., 9,* 287–294.
Lund JS (1973): Organization of neurons in the visual cortex, area 17, of the monkey (*Macaca mulatta*). *J. Comp. Neurol., 147,* 455–496.
Lund JS (1981): Intrinsic organization of the primate visual cortex, area 17, as seen in Golgi preparations. In: FO Schmitt, FG Worden, G Adelman, SG Dennis (Eds.): *The Organization of the Cerebral Cortex,* pp. 105–124. MIT Press, Cambridge, MA.
Lund JS (1984): Spiny stellate neurons. In: A Peters and EG Jones (Eds.): *Cerebral Cortex, Vol. 1, Cellular Components of the Cerebral Cortex,* pp. 255–308. Plenum Press, New York.
Lund JS (1987): Local circuit neurons of macaque monkey striate cortex: I. Neurons of laminae 4C and 5A. *J. Comp. Neurol., 257,* 60–92.

Lund JS, Boothe RG (1975): Interlaminar connections and pyramidal euron organisation in the visual cortex, Area 17, of the macaque monkey. *J. Comp. Neurol., 159,* 305–334.
Lund JS, Hendry GH, McQueen CL, Harvey AR (1979): Anatomical organization of the visual cortex of the cat: a comparison with area 17 of the macaque monkey. *J. Comp. Neurol., 184,* 559–618.
Lund JS, Hawken MJ, Parker AJ (1988): Local circuit neurons of macaque monkey striate cortex: II. Neurons of laminae 5B and 6. *J. Comp. Neurol. 276,* 1–29.
Lund JS, Yoshioka T, Levitt JB (1994): Substrates for interlaminar connections in area V1 of macaque monkey cerebral cortex. In: A Peters, KS Rockland (Eds.): *Cerebral Cortex, Vol. 10, Primary Visual Cortex in Primates,* pp. 37–60. Plenum Press, New York.
Magnani G, Bettoni L, Mazzuchi A (1982): Lesioni atrofiche bioccipitali di incerta natura all'origine di una agnosia visiva. *Riv. Neurol., 52,* 137–148.
Mamounas LA, Mullen CA, O'Hearn E, Molliver ME (1991): Dual serotoninergic projections to forebrain in the rat: morphologically distinct 5-HT axon terminals exhibit differential vulnerability to neurotoxic amphetamine derivatives. *J. Comp. Neurol., 314,* 558–586.
Marin-Padilla M (1969): Origin of the pericellular baskets of the pyramidal cells of the human motor cortex: a Golgi study. *Brain Res., 14,* 633–646.
Mates SL, Lund JS (1983): Neuronal composition and development in lamina 4C of monkey striate cortex. *J. Comp. Neurol., 221,* 60–90.
Matsubara J, Cynader M, Swindale NV, Stryker MP (1985): Intrinsic projections within visual cortex: evidence for orientation-specific local connections. *Proc. Natl. Acad. Sci. USA, 82,* 935–939.
Mattson MP, Rychlik B, Chu C, Christakos S (1991): Evidence for calcium-reducing and excitoprotective roles for the calcium-binding protein calbindin-D28K in cultured hippocampal neurons. *Neuron, 6,* 41–51.
Matus A (1988): Microtubule-associated proteins: their potential role in determining neuronal morphology. *Annu. Rev. Neurosci., 11,* 29–44.
Maunsell JHR, Newsome WT (1987): Visual processing in monkey extrastriate cortex. *Annu. Rev. Neurosci., 10,* 363–401.
Maunsell JHR, Van Essen DC (1983): The connections of the middle temporal visual area (MT) and their relationship to a cortical hierarchy in the macaque monkey. *J. Neurosci., 3,* 2563–2586.
McCarthy G, Spicer M, Adrignolo A, Luby M, Gore J, Allison T (1995): Brain activation associated with visual motion studied by functional magnetic resonance imaging in humans. *Hum. Brain Mapp., 2,* 234–243.
McGuire BA, Gilbert CD, Rivlin PK, Wiesel TN (1991): Targets of horizontal connections in macaque primary visual cortex. *J. Comp. Neurol. 305,* 370–392.
Meinecke DL, Rakic P (1992): Expression of GABA and $GABA_A$ receptor by neurons of the subplate zone in developing primate occipital cortex: evidence for transient local circuits. *J. Comp. Neurol., 317,* 91–101.
Mendez MF, Mendez MA, Martin R, Smyth KA, Whitehouse PJ (1990a): Complex visual disturbances in Alzheimer's disease. *Neurology, 40,* 439–443.
Mendez MF, Tomsak RL, Remler B (1990b): Disorders of the visual system in Alzheimer's disease. *J. Clin. Neuro-ophthalmol., 10,* 62–69.
Mendez MF, Turner J, Gilmore GC, Remler B, Tomsak RL (1990c): Bálint's syndrome in Alzheimer's disease: visuospatial functions. *Int. J. Neurosci., 54,* 339–346.
Mentis MJ, Horwitz B, Grady CL, Alexander GE, VanMeter JW, Maisog JM, Pietrini P, Schapiro MB, Rapoport SI (1996): Visual cortical dysfunction in Alzheimer's disease evaluated with a temporaly graded 'stress test' during PET. *Am. J. Psychiatry, 153,* 32–40.
Merigan WH, Maunsell JHR (1993): How parallel are the primate visual pathways? *Annu. Rev. Neurosci., 16,* 369–402.
Mesulam MM, Hersh LB, Mash DC, Geula C (1992): Differential cholinergic innervation within functional subdivisions of the human cerebral cortex: a choline acetyltransferase study. *J. Comp. Neurol., 318,* 316–328.
Mesulam MM, Mufson EJ, Levey AI, Wainer BH (1984): Atlas of cholinergic neurons in the forebrain and upper brainstem of the macaque based on monoclonal choline acetyltransferase immunohistochemistry and acetylcholinesterase histochemistry. *Neuroscience, 12,* 669–686.
Mesulam MM, Volicer L, Marquis JK, Mufson EJ, Green RC (1986): Systematic regional differences in the cholinergic innervation of the primate cerebral cortex: distribution of enzyme activities and some behavioral implications. *Ann. Neurol., 19,* 144–155.
Miklossy J (1993): The geniculocalcarine pathway in man, and some putative rostral visual areas involved in visuo-spatial attention. In: B Gulyás, D Ottoson, PE Roland (Eds.): *Functional Organisation of the Human Visual Cortex,* pp. 123–136. Pergamon Press, Oxford.
Milbrandt J (1987): A nerve growth factor-induced gene encodes a possible transcriptional regulatory factor. *Science, 238,* 797–799.

Minoshima S, Foster NL, Kuhl DE (1994): Posterior cingulate cortex in Alzheimer's disease. *Lancet, 344,* 895.
Mishkin M, Ungerleider LG, Macko KA (1983): Object vision and spatial vision: two cortical pathways. *Trends Neurosci., 6,* 414–417.
Miyashita Y (1995): How the brain creates imagery: projection to primary visual cortex. *Science, 268,* 1719–1720.
Mize RR, Luo Q, Tigges M (1992): Monocular enucleation reduces immunoreactivity to the calcium-binding protein calbindin 28kD in the rhesus monkey lateral geniculate nucleus. *Vis. Neurosci., 9,* 471–482.
Molliver ME, Mamounas LA, Wilson MA (1989): Effects of neurotoxic amphetamines on serotonergic neurons: immunocytochemical studies. *NIDA Res. Monogr., 1989,* 270–305.
Montero VM (1990): Quantitative immunogold analysis reveals high glutamate levels in synaptic terminals of retino-geniculate, cortico-geniculate, and geniculo-cortical axons in the cat. *Vis. Neurosci., 4,* 437–443.
Montero VM, Jian S (1995): Induction of c-fos protein by patterned visual stimulation in central visual pathways of the rat. *Brain Res., 690,* 189–199.
Monyer H, Burnashev N, Laurie DJ, Sakmann B, Seeburg PH (1994): Development and regional expression in the rat brain and functional properties of four NMDA receptors. *Neuron, 12,* 529–540.
Moore RY, Bloom FE (1978): Central catecholamine neuron systems: anatomy and physiology of the dopamine systems. *Annu. Rev. Neurosci., 1,* 129–169.
Moore RY, Bloom FE (1979): Central catecholamine neuron systems: anatomy and physiology of the norepinephrine and epinephrine systems. *Annu. Rev. Neurosci., 2,* 113–168.
Morel A, Bullier J (1990): Anatomical segregation of two cortical visual pathways in the macaque monkey. *Vis. Neurosci., 4,* 555–578.
Morel F (1945): Les aires striée, parastriée et péristriée dans les troubles de la fonction visuelle au cours de la maladie d'Alzheimer. *Confin. Neurol., 6,* 238–242.
Morris JR, Lasek RJ (1982): Stable polymers of the axonal cytoskeleton: the axoplasmic ghost. *J. Cell Biol., 92,* 192–198.
Morrison JH, Grzanna R, Molliver ME, Coyle JT (1978): The distribution and orientation of noradrenergic fibers in neocortex of the rat: an immunofluorescence study. *J. Comp. Neurol., 181,* 17–40.
Morrison JH, Molliver ME, Grzanna R (1979): Noradrenergic innervation of cerebral cortex: widespread effects of local cortical lesions. *Science, 205,* 313–316.
Morrison JH, Molliver ME, Grzanna R, Coyle JT (1981): The intra-cortical trajectory of the coeruleo-cortical projection in the rat: a tangentially organized cortical afferent. *Neuroscience, 6,* 139–158.
Morrison JH, Foote SL, Molliver ME, Bloom FE, Lidov HG (1982a): Noradrenergic and serotonergic fibers innervate complementary layers in monkey primary visual cortex: an immunohistochemical study. *Proc. Natl. Acad. Sci. USA, 79,* 2401–2405.
Morrison JH, Foote SL, O'Connor D, Bloom FE (1982b): Laminar, tangential and regional organization of the noradrenergic innervation of monkey cortex: dopamine-β-hydroxylase immunohistochemistry. *Brain Res. Bull., 9,* 309–319.
Morrison JH, Foote SL, Bloom FE (1984): Laminar, regional developmental and functional specificity of monoaminergic innervation patterns in monkey cortex. In: L Descarries, TA Reader, HH Jasper (Eds.): *Monoamine Innervation of Cerebral Cortex*, pp 61–75. Alan R. Liss, New York.
Morrison JH, Foote SL (1986): Noradrenergic and serotonergic innervation of cortical, thalamic, and tectal visual structures in old and new world monkeys. *J. Comp. Neurol., 243,* 117–138.
Morrison BM, Gordon JW, Ripps ME, Morrison JH (1996): Quantitative immunocytochemical analysis of the spinal cord in G86R superoxide dismutase transgenic mice: neurochemical correlates of selective vulnerability. *J. Comp. Neurol., 373,* 619–631.
Morrison JH, Siegel SJ, Gazzaley AH, Huntley GW (1996): Glutamate receptors: emerging links between subunit proteins and specific excitatory circuits in primate hippocampus and neocortex. *Neuroscientist, 2,* 272–283.
Mower GD (1994): Differences in the induction of Fos protein in cat visual cortex during and after the critical period. *Mol. Brain Res., 21,* 47–54.
Mrzljak L, Levey AI, Goldman-Rakic PS (1992): Association of m1 and m2 muscarinic receptor proteins with asymmetric synapses in the primate cerebral cortex: morphological evidence for cholinergic modulation of excitatory neurotransmission. *Proc. Natl. Acad. Sci. USA, 90,* 5194–5198.
Mrzljak L, Levey AI, Goldman-Rakic PS (1996): Selective expression of m2 muscarinic receptor in the parvocellular channel of the primate visual cortex. *Proc. Natl. Acad. Sci. USA, 93,* 7337–7340.
Mulkey RM, Malenka RC (1992): Mechanisms underlying induction of homosynaptic long-term depression in area CA1 of the hippocampus. *Neuron, 9,* 967–975.
Mulligan KA, Törk (1988): Serotonergic innervation of the cat cerebral cortex. *J. Comp. Neurol., 270,* 86–110.

Mulligan KA, van Brederode JF, Hendrickson AE (1989): The lectin *Vicia villosa* labels a distinct subset of GABAergic cells in macaque visual cortex. *Vis. Neurosci., 2,* 63–72.
Murphy DB, Johnson JA, Borisy GG (1977): Role of tubulin-associated proteins in microtubule nucleation and elongation. *J. Mol. Biol., 117,* 33–52.
Murray EA, Davidson M, Gaffan D, Olton DS, Suomi S (1989): Effects of fornix transection and cingulate cortical ablation on spatial memory in rhesus monkeys. *Exp. Brain Res., 74,* 173–186.
Mutrux S (1947): Diagnostic différentiel histologique de la maladie d'Alzheimer et de la démence sénile – Pathophobie de la zone de projection corticale. *Mschr. Psychiat. Neurol., 113,* 100–117.
Nakamura H, Gattass R, Desimone R, Ungerleider LG (1993): The modular organization of projections from areas V1 and V2 to areas V4 and TEO in macaques. *J. Neurosci., 13,* 3681–3691.
Nakanishi S (1992): Molecular diversity of glutamate receptors and implications for brain function. *Science, 258,* 597–603.
Neary D, Snowden JS (1987): Perceptuospatial disorder in Alzheimer's disease. *Semin. Ophthalmol., 2,* 151–158.
Newsome WT, Wurtz RH (1982): Identification of architectonic zones containing visual tracking cells in the superior temporal sulcus of the macaque monkey. *Invest. Ophthalmol. Vis. Sci., 22 (Suppl.),* 238.
Nielson KA, Cummings BJ, Cotman CW (1996): Constructional apraxia in Alzheimer's disease correlates with neuritic neuropathology in occipital cortex. *Brain Res., 741,* 284–293.
Nimchinsky EA, Hof PR, Janssen WGM, Moran T, Schmauss C, Morrison JH (submitted): Regional and cellular distribution of the dopamine D_3 receptor in the rat and macaque monkey brain: an immunocytochemical analysis. *Neuroscience.*
Nimchinsky EA, Hof PR, Young WG, Morrison JH (1996): Neurochemical, morphologic and laminar characterization of cortical projection neurons in the cingulate motor areas of the macaque monkey. *J. Comp. Neurol., 374,* 136–160.
Nissen MJ, Corkin S, Buonanno FS, Growdon JH, Wray SH, Bauer J (1985): Spatial vision in Alzheimer's disease – General findings and a case report. *Arch. Neurol., 42,* 667–671.
Nixon RA, Paskevich PA, Sihag RK, Thayer CY (1994): Phosphorylation on carboxyl terminus domains of neurofilament proteins in retinal ganglion cell neurons in vivo: influences on regional neurofilament accumulation, interneuronal spacing, and axonal caliber. *J. Cell Biol., 126,* 1031–1046.
O'Kusky J, Colonnier M (1982): A laminar analysis of the number of neurons, glia and synapses in the visual cortex (area 17) of adult macaque monkeys. *J. Comp. Neurol., 210,* 278–290.
Ogren MP, Hendrickson AE (1977): The distribution of pulvinar terminals in visual areas 17 and 18 of the monkey. *Brain Res., 137,* 343–350.
Olschowka JA, Molliver ME, Grzanna R, Rice FL, Coyle JT (1981): Ultrastructural demonstration of noradrenergic synapses in the rat central nervous system by dopamine-β-hydroxylase immunocytochemistry. *J. Histochem. Cytochem., 29,* 271–280.
Olson CR, Musil SY (1992a): Posterior cingulate cortex: sensory and oculomotor properties of single neurons in behaving cat. *Cereb. Cortex, 2,* 485–502.
Olson CR, Musil SY (1992b): Topographic organization of cortical and subcortical projections to posterior cingulate cortex in the cat: evidence for somatic, ocular and complex subregions. *J. Comp. Neurol., 323,* 1–24.
Olson CR, Musil SY, Goldberg ME (1993): Neurophysiology of posterior cingulate cortex in the behaving monkey. In: BA Vogt, M Gabriel (Eds.): *The Neurobiology of Cingulate Cortex and Limbic Thalamus,* pp. 366–380. Birkhäuser, Boston.
Orban GA, Dupont P, De Bruyn B, Vogels R, Vandenberghe R, Mortelmans L (1995): A motion area in human visual cortex. *Proc. Natl. Acad. Sci. USA, 92,* 993–997.
Ottersen OP, Fischer B, Storm-Mathisen J (1983): Retrograde transport of D-[^{3}H]aspartate in thalamocortica neurones. *Neurosci. Lett., 42,* 19–24.
Ottersen OP, Storm-Mathisen J (1987): Localization of amino acid neurotransmitters by immunocytochemistry. *Trends Neurosci., 10,* 250–255.
Parnavelas JG, Dinopoulos A, Davies SW (1989): The central visual pathways. In: A Björklund, T Hökfelt, LW Swanson (Eds.): *Handbook of Chemical Neuroanatomy, Vol. 7, Integrated systems of the CNS, Part II,* pp. 1–139. Elsevier Press, Amsterdam.
Papadopoulos GC, Parnavelas JG, Buijs RM (1987): Monoaminergic fibers form conventional synapses in the cerebral cortex. *Neurosci. Lett., 76,* 275–279.
Papadopoulos GC, Parnavelas JG, Buijs RM (1989): Light and electron microscopic immunocytochemical analysis of the noradrenaline innervation of the rat visual cortex. *J. Neurocytol., 18,* 1–10.
Parasuraman R, Greenwood PM, Alexander GE (1995): Selective impairment of spatial attention during visual search in Alzheimer's disease. *NeuroReport, 6,* 1861–1864.
Parthasarathy HB, Schall JD, Graybiel AM (1992): Distributed but convergent ordering of corticostriatal

projections: analysis of the frontal eye field and the supplementary eye field in the macaque monkey. *J. Neurosci., 12,* 4468–4488.

Pascual J, del Arco C, Gonzalez AM, Pazos A (1992): Quantitative light miscroscopic autoradiographic localization of alpha 2-adrenoceptors in the human brain. *Brain Res., 585,* 116–127.

Pearson RCA, Esiri MM, Hiorns R, Wilcock GK, Powell TPS (1985): Anatomical correlates of the distribution of the pathological changes in the neocortex in Alzheimer disease. *Proc. Natl. Acad. Sci. USA, 82,* 4531–4534.

Perkel DJ, Petrozzino JJ, Nicoll RA, Connor JA (1993): The role of Ca^{2+} entry via synaptically activated NMDA receptors in the induction of long-term potentiation. *Neuron, 11,* 817–823.

Peters A, Harriman KM (1990): Different kinds of axon terminals forming symmetric synapses with the cell bodies and initial axon segments of layer II/III pyramidal cells. I. Morphometric analysis. *J. Neurocytol., 19,* 154–174.

Peters A, Proskauer CC, Ribak CE (1982): Chandelier cells in rat visual cortex. *J. Comp. Neurol., 206,* 397–416.

Peters A, Sethares C (1991): Organization of pyramidal neurons in area 17 of monkey visual cortex. *J. Comp. Neurol., 306,* 1–23.

Petralia RS, Yokotani N, Wenthold RJ (1994): Light and electron microscope distribution of the NMDA receptor subunit NMDAR1 in the rat nervous system using a selective anti-peptide antibody. *J. Neurosci., 14,* 667–696.

Pettet MW, Gilbert CD (1992): Dynamic changes in receptive-field size in cat primary visual cortex. *Proc. Natl. Acad. Sci. USA, 89,* 8366–8370.

Pietrini P, Furey ML, Graff-Radford N, Freo U, Alexander GE, Grady CL, Dani A, Mentis MJ, Schapiro MB (1996): Preferential meatbolic involvement of visual cortical areas in a subtype of Alzheimer's disease: clinical implications. *Am. J. Psychiatry, 153,* 1261–1268.

Pijak DS, Hall GF, Tenicki PJ, Boulos AS, Lurie DI, Selzer MS (1996): Neurofilament spacing, phosphorylation, and axon diameter in regenerating and uninjured lamprey axons. *J. Comp. Neurol., 368,* 569–581.

Poggio GF, Fischer B (1977): Binocular interaction and depth sensitivity in striate and prestriate cortex of behaving rhesus monkey. *J. Neurophysiol., 40,* 1392–1405.

Rakic P (1982): Early developmental events: cell lineages, acquisition of neuronal positions, and areal and laminar development. *Neurosci. Res. Prog. Bull., 20,* 439–451.

Rakic P, Goldman-Rakic PS, Gallagher D (1988): Quantitative autoradiography of major neurotransmitter receptors in the monkey striate and extrastriate cortex. *J. Neurosci., 8,* 3670–3690.

Reid SNM, Daw NW (1997): Activity of metabotropic glutamate receptors has different effects in different layers of cat visual cortex. *Vis. Neurosci., 14,* 83–88.

Reid SNM, Romano C, Hughes T, Daw NW (1995): Immunohistochemical study of two phosphoinositide-linked metabotropic glutamate receptors (mGluR1α and mGluR5) in the cat visual cortex before, during and after the peak of the critical period for eye-specific connections. *J. Comp. Neurol., 355,* 470–478.

Reiman EM, Caselli RJ, Yun LS, Chen K, Bandy D, Minoshima S, Thibodeau SN, Osborne D (1996): Preclinical evidence of Alzheimer's disease in persons homozygous for the ε4 allele for apolipoprotein E. *New Engl. J. Med., 334,* 752–758.

Ribak CE (1978): Aspinous and sparsely-spinous stellate neurons in the visual cortex of rats contain glutamic acid decarboxylase. *J. Neurocytol., 7,* 461–478.

Riederer BM, Draberova E, Viklicky V, Draber P (1995): Changes of MAP2 phosphorylation during brain development. *J. Histochem. Cytochem., 43,* 1269–1284.

Riederer BM, Porchet R, Marugg RA (1996): Differential expression and modification of neurofilament triplet proteins during cat cerebellar development. *J. Comp. Neurol., 364,* 704–717.

Robinson DL, Petersen SE (1992): The pulvinar and visual salience. *Trends Neurosci., 15,* 127–132

Rockel AJ, Hiorns RW, Powell TPS (1980): The basic uniformity in structure of the neocortex. *Brain, 103,* 221–244.

Rockland KS, Pandya DN (1979): Laminar origins and terminations of cortical connections of the occipital lobe in the rhesus monkey. *Brain Res., 179,* 3–20.

Rockland K, Lund JS (1982): Intrinsic laminar lattice connections in primate visual cortex. *J. Comp. Neurol., 216,* 303–318.

Rogers J, Morrison JH (1985): Quantitative morphology and regional and laminar distributions of senile plaques in Alzheimer's disease. *J. Neurosci., 5,* 2801–2808.

Rosen KM, McCormack MA, Villa-Komaroff L, Mower GD (1992): Brief visual experience induces immediate early gene expression in the cat visual cortex. *Proc. Natl. Acad. Sci. USA, 89,* 5437–5441.

Rossner S, Kues W, Witzemann V, Schliebs R (1993): Laminar expression of m1-, m3- and m4-muscarinic cholinergic receptor genes in the developing rat visual cortex using in situ hybridization histochemistry. Effect of monocular visual deprivation. *Int. J. Dev. Neurosci., 11,* 369–378.

Sadun AA, Borchert M, DeVita E, Hinton DR, Bassi CJ (1987): Assessment of visual impairment in patients with Alzheimer's disease. *Am. J. Ophthalmol., 104,* 113–120.

Saint Marie RL, Peters A (1985): The morphology and synaptic connections of spiny stellate neurons in monkey visual cortex (area 17): a Golgi-electron microscopic study. *J. Comp. Neurol., 226,* 213–235.

Sandell JH (1986): NADPH diaphorase histochemistry in the macaque striate cortex. *J. Comp. Neurol., 251,* 388–397.

Sawatari A, Callaway EM (1996): Convergence of magno- and parvocellular pathways in layer 4B of macaque primary visual cortex. *Nature, 380,* 442–446.

Schall JD, Hanes DP (1993): Neural basis of saccade target selection in frontal eye field during visual search. *Nature, 366,* 467–469.

Schall JD, Morel A, Kaas JH (1993): Topography of supplementary eye field afferents to frontal eye feld in macaque: implications for mapping between saccade coordinate systems. *Vis. Neurosci., 10,* 385–393.

Schlag J, Schlag-Rey M (1992): Neurophysiology of eye movements. *Adv. Neurol., 57,* 135–147.

Schröder H (1992): Immunohistochemistry of cholinergic receptors. *Anat. Embryol., 186,* 407–429.

Schuman EM, Madison DV (1991): A requirement for the intercellular messenger nitric oxide in long-term potentiation. *Science, 254,* 1503–1506.

Schwartz ML, Meinecke DL (1992): Early expression of GABA-containing neurons in the prefrontal and visual cortices of rhesus monkeys. *Cereb. Cortex, 2,* 16–37.

Sereno MI, Dale AM, Reppas JB, Kwong KK, Belliveau JW, Brady TJ, Rosen BR, Tootell RBH (1995): Borders of multiple visual areas in humans revealed by functional magnetic resonance imaging. *Science, 268,* 889–893.

Shaw C, Cameron L, March D, Cynader M, Zielinski B, Hendrickson AE (1991): Pre- and postnatal development of GABA receptors in *Macaca* monkey visual cortex. *J. Neurosci., 11,* 3943–3959.

Shaw C, Cynader M (1986): Laminar distribution of receptors in monkey (*Macaca fascicularis*) geniculostriate system. *J. Comp. Neurol., 248,* 301–312.

Shulz D, Frégnec Y (1992): Cellular analogs of visual cortical epigenesis. II. Plasticity of binocular integration. *J. Neurosci., 12,* 1301–1318.

Shute CCD, Lewis PR (1967): The ascending cholinergic reticular system: neocortical, olfactory and subcortical projections. *Brain, 90,* 497–522.

Siegel SJ, Janssen WG, Tullai JW, Rogers SW, Moran T, Heinemann SF, Morrison JH (1995): Distribution of the excitatory amino acid receptor subunits GluR2(4) in monkey hippocampus and colocalization with subunits GluR5–7 and NMDAR1. *J. Neurosci., 15,* 2707–2719.

Sillito AM (1975): The contribution of inhibitory mechanisms to the receptive field properties of neurones in the striate cortex of the cat. *J. Physiol. (Lond.), 250,* 305–329.

Sillito AM (1977): Inhibitory processes underlying the directional specificity of simple, complex and hypercomplex cells in the cat's visual cortex. *J. Physiol. (Lond.), 271,* 699–720.

Sillito AM (1979): Inhibitory mechanisms influencing complex cell orientation selectivity and their modification at high resting discharge levels. *J. Physiol. (Lond.), 289,* 33–53.

Silverman MS, Grosof DH, De Valois RL, Elfar SD (1989): Spatial-frequency organization in primate striate cortex. *Proc. Natl. Acad. Sci. USA, 86,* 711–715.

Silverman SE, Tran DB, Zimmermann KM, Feldon SE (1994): Dissociation between the detection and perception of motion in Alzheimer's disease. *Neurology, 44,* 1814–1818.

Sledeczek F, Momiyama A, Takahashi T (1993): Presynaptic inhibitory action of a metabotropic glutamate receptor agonist on excitatory transmission in visual neurons. *Proc. R. Soc. Lond. Ser. B [Biol. Ser], 253,* 297–303.

Sloper JJ, Powell TPS (1979): A study of the axon initial segment and proximal axon of neurons in the primate motor cortex and somatic sensory cortices. *Phil. Trans. R. Soc. Lond. Ser. B [Biol. Sci.], 285,* 173–197.

Sommer B, Kohler M, Sprengel R, Seeburg PH (1991) RNA editing in brain controls a determinant of ion flow in glutamate-gated channels. *Cell, 67,* 11–19.

Sommer B, Seeburg PH (1992): Glutamate receptor channels: novel properties and new clones. *Trends Pharmacol. Sci., 13,* 291–296.

Somogyi P (1977): A specific 'axo-axonal' interneuron in the visual cortex of the rat. *Brain Res., 136,* 345–350.

Somogyi P (1979): An interneuron making synapses specifically on the axon initial segment (AIS) of pyramidal cells in the cerebral cortex of the cat. *J. Physiol. (Lond.), 296,* 18–19.

Somogyi P, Cowey A (1981): Combined Golgi and electron microscopic study on the synapses formed by double bouquet cells in the visual cortex of the cat and monkey. *J. Comp. Neurol., 195,* 547–566.

Somogyi P, Cowey A (1984): Double-bouquet cells. In: In: A Peters, EG Jones (Eds.): *Cerebral Cortex, Vol. 1, Cellular Components of the Cerebral Cortex,* pp. 337–360. Plenum Press, New York.
Somogyi P, Cowey A, Halasz N, Freund TF (1981): Vertical organization of neurones accumulating ^{3}H-GABA in visual cortex of rhesus monkey. *Nature, 294,* 761–763.
Somogyi P, Cowey A, Kisvárday ZF, Freund TF, Szentágothai J (1983a): Retrograde transport of γ-amino[^{3}H]butyric acid reveals specific interlaminar connections in the striate cortex of monkey. *Proc. Natl. Acad. Sci. USA, 80,* 2385–2389.
Somogyi P, Freund TF, Cowey A (1982): The axo-axonic interneuron in the cerebral cortex of the rat, cat and monkey. *Neuroscience, 7,* 2577–2607.
Somogyi P, Freund TF, Hodgson AJ, Somogyi J, Beroukas D, Chubb IW (1985): Identified axon-axonic cells are immunoeactive for GABA in the hippocampus and visual cortex of cats. *Brain Res., 332,* 143–149.
Somogyi P, Kisvárday ZF, Martin KAC, Whitteridge D (1983b): Synaptic connections of morphologically identified and physiologically characterized large basket cells in the striate cortex of cat. *Neuroscience, 9,* 261–294
Somogyi P, Soltész I (1986): Immunogold demonstration of GABA in synaptic terminals of intracellularly recorded, horseradish peroxidase-filled basket cells and clutch cells in the cat's visual cortex. *Neuroscience, 19,* 1051–1065.
Spatz WB, Illing RB, Vogt-Weisenhorn DM (1994): Distribution of cytochrome oxidase and parvalbumin in the primary visual cortex of the adult and neonate monkey, *Callithrix jacchus. J. Comp. Neurol., 339,* 519–534.
Stanton GB, Bruce CJ, Goldberg ME (1993): Topography of projections to the frontal lobe from the macaque frontal eye fields. *J. Comp. Neurol., 330,* 286–301.
Stanton GB, Deng SY, Golberg ME, McMullen NT (1989): Cytoarchitectural characteristic of the frontal eye fields in the macaque monkeys. *J. Comp. Neurol., 282,* 415–427.
Stefani FH, Mehraein P (1972): Histologische Veränderungen des Corpus geniculatum laterale in Alter, bei präseniler (Alzheimer) und seniler Demenz. *Graefes Arch. Klin. Exp. Ophthalmol., 184,* 314–321.
Steinbusch HWM (1981): Distribution of serotonin-immunoreactivity in the central nervous system of the rat – cell bodies and terminals. *Neuroscience, 6,* 557–618.
Sterio DC (1984): The unbiased estimation of the number and sizes of arbitrary particles using the disector. *J. Microsc., 134,* 127–136.
Stone J, Dreher B, Leventhal A (1979): Hierarchical and parallel mechanisms in the organization of visual cortex. *Brain Res. Rev., 1,* 345–394.
Streit P (1984): Glutamate and aspartate as transmitter candidates for systems of the cerebral cortex. In: EG Jones, A Peters (Eds.): *Cerebral cortex, Vol. 2, Functional Properties of Cortical Cells,* pp. 119–172. Plenum Press, New York.
Swanson LW, Hartman BK (1975): The central adrenergic system. An immunofluorescence study of the location of cell bodies and their efferent connections in the rat utilizaing dopamine-β-hydroxylase as a marker. *J. Comp. Neurol. 163,* 467–506.
Szentágothai J, Arbib MA (1974): Conceptual models of of neural organization. *Neurosci. Res. Prog. Bull., 12,* 306–510.
Takeuchi Y, Sano Y (1983): Immunohistochemical demonstration of serotonin nerve fibers in the neocortex of the monkey (*Macaca fuscata*). *Anat. Embryol., 166,* 155–168.
Terry RD, Peck A, DeTeresa R, Schechter R, Horoupian DS (1981): Some morphometric aspects of the brain in senile dementia of the Alzheimer type. *Ann. Neurol., 10,* 184–192.
Tieman SB, Neale JH, Teiman DG (1991): *N*-acetylaspartylglutamate immunoreactivity in neurons of the monkey's visual system. *J. Comp. Neurol., 313,* 45–64.
Tieman SB, Cangro CB, Neale JH (1987): *N*-acetylaspartylglutamate immunoreactivity in neurons of the cat's visual system. *Brain Res., 420,* 188–193.
Tigges M, Bos J, Tigges J, Bridges E (1977): Ultrastructural characteristics of layer IV neuropil in area 17 of monkeys. *Cell Tissue Res., 182,* 39–59.
Tigges J, Tigges M, Anschel S, Cross NA, Letbetter WE, McBride RL (1981) Areal and laminar distributions of neurons interconnecting the central visual cortical areas 17, 18, 19 and MT in the squirrel monkey (Saimiri). *J. Comp. Neurol., 202,* 539–560.
Tigges M, Tigges J, McDonald JK, Slattery M, Fernandez A (1989): Postnatal development of neuropeptide Y-like immunoreactivity in area 17 of normal and visually depreived rhesus monkeys. *Vis. Neurosci., 2,* 315–328.
Tootell RBH, Reppas JB, Kwong KK, Malach R, Born RT, Brady TJ, Rosen BR, Belliveau JW (1995): Functional analysis of human area MT and related visual cortical areas using magnetic resonance imaging. *J. Neurosci., 15,* 3215–3230.

Tootell RBH, Silverman MS, Hamilton SL, De Valois RL, Switzes E (1988): Functional anatomy of macaque striate cortex. III. Color. *J. Neurosci., 8,* 1569–1593.

Tootell RBH, Taylor JB (1995): Anatomical evidence for MT and additional cortical visual areas in humans. *Cereb. Cortex, 5,* 39–55.

Trick GL, Trick LR, Morris P, Wolf M (1995): Visual field loss in senile dementia of the Alzheimer's type. *Neurology, 45,* 68–74.

Trusk TC, Kaboord WS, Wong-Riley MTT (1990): Effects of monocular enucleation, tetrodotoxin, and lid suture on cytochrome oxidase reactivity in supragranular puffs of adult macaque striate cortex. *Vis. Neurosci., 4,* 185–204.

Ts'o DY, Gilbert CD (1988): The organization of chromatic and spatial interactions in the primate striate cortex. *J. Neurosci., 8,* 1712–1727.

Tsai CS, Ritch R, Schwartz B, Lee SS, Miller NR, Chi T, Hsieh FY (1991): Optic nerve head and nerve fiber layer in Alzheimer's disease. *Arch. Ophthalmol., 109,* 199–204.

Tsumoto T, Eckart W, Creutzfeldt OD (1979): Modification of orientation sensitivity of cat visual cortex neurones by removal of GABA mediated inhibition. *Exp. Brain Res., 34,* 351–363.

Ungerleider LG, Desimone R (1986): Cortical connection sof the visual area MT in the macaque. *J. Comp. Neurol., 248,* 190–222.

Ungerleider LG (1995): Functional brain imaging studies of cortical mechanisms for memory. *Science, 270,* 769–775.

Ungerstedt U (1971): Stereotaxic mapping of the monoamine pathways in the rat brain. *Acta Physiol. Scand., 367 Suppl.,* 1–48.

Valverde F (1971): Short axon neuronal subsystems in the visual cortex of the monkey. *Int. J. Neurosci., 1,* 181–197.

Valverde F (1985): The organizaing principles of the primary visual cortex in the monkey. In: A Peters, EG Jones (Eds.): *Cerebral Cortex, Vol. 3, Visual Cortex,* pp. 207–257. Plenum Press, New York.

Van Brederode JFM, Mulligan KA, Hendrickson AE (1990): Calcium-binding proteins as markers for subpopulations of GABAergic neurons in monkey striate cortex. *J. Comp. Neurol.,* 298, 1–22.

Van Essen DC (1985): Functional organization of primate visual cortex. In: A Peters, EG Jones (Eds.): *Cerebral Cortex, Vol. 3, Visual Cortex,* pp. 259–329. Plenum Press, New York.

Van Essen DC, Anderson CH, Felleman DJ (1992): Information processing in the primate visual system: and integrated systems perspective. *Science, 255,* 419–423.

Van Essen DC, Felleman DJ, DeYoe EA, Olavarria J, Knierim JJ (1990): Modular and hierarchical organization of extrastriate visual cortex in the macaque monkey. *Cold Spring Harbor Symp. Quant. Biol., 55,* 679–696.

Van Essen DC, Gallant JL (1994): Neural mechanisms of form and motion processing in the primate visual system. *Neuron, 13,* 1–10.

Van Essen DC, Maunsell JHR, Bixby JL (1981): The middle temporal visual area in the macaque: myeloarchitecture, connections, functional properties and topographic connections. *J. Comp. Neurol., 199,* 293–326.

Van Essen DC, Zeki SM (1978): The topographic organization of the rhesus monkey prestriate cortex. *J. Physiol. (Lond.), 277,* 193–226.

Varney MA, Jachec C, Deal C, Hess SD, Daggett LP, Skvoretz R, Urcan M, Morrison, JH, Moran T, Johnson EC, Velicelebi G (1996): Stable expression and characterization of recombinant human heteromeric *N*-Methyl-D-Aspartate receptor subtypes NMDAR1A/2A and NMDAR1A/2B in mammalian cells. *J. Pharmacol. Exp. Ther., 279,* 367–378.

Vickers JC, Chin D, Edwards AM, Sampson V, Harper C, Morrison JH (1996): Dystrophic neurite formation associated with age-related β amyloid deposition in the neocortex: clues to the genesis of neurofibrillary pathology. *Exp. Neurol., 141,* 1–11.

Vickers JC, Costa M (1992): The neurofilament triplet is present in distinct subpopulations of neurons in the central nervous system of the guinea pig. *Neuroscience, 49,* 73–100.

Vickers JC, Huntley GW, Edwards AM, Moran T, Rogers SW, Heinemann SF, Morrison JH (1993): Quantitative localization of AMPA/kainate and kainate glutamate receptor subunits in neurochemically identified subpopulations of neurons in the prefrontal cortex of the macaque monkey. *J. Neurosci., 13,* 2982–2992.

Vickers JC, Schumer RA, Podos SM, Wang RF, Riederer BM, Morrison JH (1995): Differential vulnerability of neurochemically identified subpopulations of retinal neurons in a monkey model of glaucoma. *Brain Res., 680,* 23–35.

Victoroff J, Ross W, Benson DF, Verity MA., Vinters HV (1994): Posterior cortical atrophy – Neuropathologic correlations. *Arch. Neurol., 51,* 269–274.

Vissavajjhala P, Janssen, WGM, Hu Y, Gazzaley AH, Moran T, Hof PR, Morrison JH (1996): Synaptic distribution of AMPA-GluR2 subunit and its colocalization with calcium-binding proteins in rat cerebral cortex: an immunohistochemical study using a GluR2-specific monoclonal antibody. *Exp. Neurol., 142,* 296–312.

Vogt BA, Crino PB, Vogt LJ (1992): Reorganization of cingulate cortex in Alzheimer's disease: neuron loss, neuritic plaques, and muscarinic receptor binding. *Cereb. Cortex, 2,* 526–535.

Vogt BA, Pandya DN (1987): Cingulate cortex of rhesus monkey: II. Cortical afferents. *J. Comp. Neurol., 262,* 271–289.

Vogt BA, Rosene DL, Pandya DN (1979): Thalamic and cortical afferents differentiate anterior from posterior cingulate cortex in the monkey. *Science, 204,* 205–207.

Vogt BA, Van Hoesen GW, Vogt LJ (1990): Laminar distribution of neuron degeneration in posterior cingulate cortex in Alzheimer's disease. *Acta Neuropathol., 80,* 581–589.

Von Noorden GK, Crawford MLJ (1978): Morphological and physiological changes in the monkey visual system after short-term lid suture. *Invest. Ophthalmol. Vis. Sci., 17,* 762–768.

Wang Z, McCormick DA (1993): Control of firing mode of corticotectal and corticopontine layer V burst-generating neurons by norepinephrine, acetylcholine, and 1S,3R-ACPD. *J. Neurosci., 13,* 2199–2216.

Watson JDG, Myers R, Frackowiak RSJ, Hajnal JV, Woods RP, Mazziotta JC, Shipp S, Zeki S (1993): Area V5 of the human brain: evidence from a combined study using positron emission tomography and magnetic resonance imaging. *Cereb. Cortex, 3,* 79–94.

Weber JT, Huerta MF, Kaas JH, Harting JK (1983): The projections of the lateral geniculate nucleus of the squirrel monkey: studies of the interlaminar zones and the S layers. *J. Comp. Neurol., 213,* 135–145.

Werner L, Winkelmann E, Koglin A, Neser J, Rodewohl H (1989): A Golgi deimpregnation study of neurons in the rhesus monkey visual cortex (areas 17 and 18). *Anat. Embryol., 180,* 583–597.

Williams SM, Goldman-Rakic PS, Leranth C (1992): The synaptology of parvalbumin-immunoreactive neurons in the primate prefrontal cortex. *J. Comp. Neurol. 320,* 353–369.

Wilson FAW, Ó Scalaidhe SP, Goldman-Rakic PS (1993): Dissociation of object and spatial processing domains in primate prefrontal cortex. *Science, 260,* 1955–1958.

Wilson MA, Molliver ME (1991a): The organization of serotonergic projections to cerebral cortex in primates: regional distribution of axon terminals. *Neuroscience, 44,* 537–553.

Wilson MA, Molliver ME (1991b): The organization of serotonergic projections to cerebral cortex in primates: retrograde transport studies. *Neuroscience, 44,* 555–570.

Wilson MA, Ricaurte GA, Molliver ME (1989): Distinct morphologic classes of serotonergic axons inprimates exhibit differential vulnerability to the psychotropic drug 3,4-methylenedioxymethamphetamine. *Neuroscience, 28,* 121–137.

Wong-Riley MTT (1989): Cytochrome oxidase: an endogeneous metabolic marker for neuronal activity. *Trends Neurosci., 12,* 94–101.

Wong-Riley MTT (1994): Primaty visual cortex – Dynamic metabolic organization and plasticity revealed by cytochrome oxidase. In: A Peters, KS Rockland (Eds.): *Cerebral Cortex, Vol. 10, Primary Visual Cortex in Primates,* pp. 141–200. Plenum Press, New York.

Wong-Riley MTT, Carroll EW (1984): Effect of impulse blockade on cytochrome oxidase activity in monkey visual cortex. *Nature, 307,*262–264.

Wong-Riley MTT, Hevner RF, Cutlan R, Earnest M, Egan R, Frost J, Nguyen T (1993): Cytochrome oxidase in the human visual cortex: distribution in the developing and adult brain. *Vis. Neurosci., 10,* 41–58.

Wörgötter F, Eysel UT (1991): Topographical aspects of intracortical excitation and inhibition contribution to orientation specificity in area 17 of the cat visual cortex. *Eur. J. Neurosci., 3,* 1232–1244.

Xing LS, Tieman SB (1993): Relay cells, not interneurons, of cat's lateral geniculate nucleus contain N-acetylaspartylgluatamate. *J. Comp. Neurol., 330,* 272–285.

Xu Z, Marszalek JR, Lee MK, Wong PC, Folmer J, Crawford TO, Hsieh ST, Griffin JW, Cleveland DW (1996): Subunit composition of neurofilaments specifies axonal diameter. *J. Cell Biol., 133,* 1061–1069.

Yamashita A, Hayashi M, Shimizu K, Oshima K (1989): Ontogeny of somatostatin in cerebral cortex of macaque monkey: an immunohistochemical study. *Dev. Brain Res., 45,* 103–111.

Yan XX, Cao QL, Luo XG, Garey LJ (1997): Prenatal development of calbindin D-28K in human visual cortex. *Cereb. Cortex, 7,* 57–62.

Yan XX, Zheng DS, Garey LJ (1992): Prenatal development of GABA-immunoreactive neurons in the human striate cortex. *Dev. Brain Res., 65,* 191–204.

Yan YH, Van Brederode JFM, Hendrickson AE (1995): Transient co-localization of calretinin, parvalbumin, and calbindin-D28K in developing visual cortex of monkey. *J. Neurocytol., 24,* 825–837.

Yoshioka T, Hendry SHC (1995): Compartmental organization of layer IVA in human primary visual cortex. *J. Comp. Neurol., 359,* 213–220.

Zeki S, Shipp S (1988): The functional logic of cortical connections. *Nature, 335,* 311–317.

Zeki SM (1974): Functional organization of a visual area in the posterior bank of the superior temporal sulcus of the rhesus monkey. *J. Physiol. (Lond.), 236,* 549–573.
Zeki SM (1978a): The cortical projections of foveal striate cortex in the rhesus monkey. *J. Physiol. (Lond.), 277,* 227–244.
Zeki SM (1978b): Uniformity and diversity of structure and function in rhesus monkey prestriate visual cortex. *J. Physiol. (Lond.), 277,* 273–290.
Zezula J, Cortes R, Probst A, Palacios JM (1988): Benzodiazepine receptor sites in the human brain: autoradiographic mapping. *Neuroscience, 25,* 771–796.

12. ABBREVIATIONS

Acronym	Full name (see Hof and Morrison 1995; and Felleman and Van Essen 1991)
V1	Visual area 1
V2	Visual area 2
V3	Visual area 3 (V3/VP)
V3A	Visual area 3 (annectent gyrus)
V4	Visual area 4
V4t	Transitional cortex between V4 and MT
MT	Mediotemporal area (includes MTp)
MST	Medial superior temporal area (includes MSTc and MSTp)
PP	Posterior parietal sulcal zone
DP	Dorsal prelunate area
PO	Parieto-occitpital area
PIP	Posterior intraparietal area
MIP	Medial intraparietal area
VIP	Ventral intraparietal area
LIP	Lateral intraparietal area
7a	Visual area 7a (inferior parietal cortex)
MDP	Medial dorsal parietal area
PG	Inferior parietal area (includes VIP, LIP, 7a, PP, and MST)
VOT	Ventral occipitotemporal area (VTF)
FST	Fundal superior temporal area
TPO	Superior temporal polysensory area (STP), dorsal part
PGa	Superior temporal polysensory area (STP), ventral part
ITp	Posterior inferior temporal cortex (PIT, TEO)
ITm	Medial inferior temporal cortex (CIT, TE, TEm, includes IPa)
ITa	Anterior inferior temporal cortex (AIT, TE, TEa)
ProS	Prostriate area (PSv/PSd)
TE, TF, TG, TH	Temporal cortex fields TE, TF, TG, and TH
FEF	Frontal eye field
SEF	Supplementary eye field
9, 6, 46, 45	Frontal cortex areas
35, 36	Perirhinal cortex

Subject index